Encyclopedia of American Wine

William I. Kaufman

Encyclopedia of American Wine

including Mexico and Canada

Jeremy P. Tarcher, Inc.
Los Angeles
Wine Appreciation Guild
San Francisco
Distributed by Houghton Mifflin Company
Boston

Library of Congress Cataloging in Publication Data

Kaufman, William Irving, 1922–
Encyclopedia of American wines, including Mexico and Canada.

1. Wine and wine making—United States—Dictionaries. 2. Wine and wine making—Canada—Dictionaries. 3. Wine and wine making—Mexico—Dictionaries. I. Title.
TP546.K38 1984 641.2'22'097 84-8447
ISBN 0–87477–323–7 (Jeremy P. Tarcher, Inc.)
ISBN 0–932664–39–3 (Wine Appreciation Guild)

Jeremy P. Tarcher, Inc.
9110 Sunset Blvd.
Los Angeles, CA 90069

Art Direction: Jeffrey Caldewey
Designer and Illustrator: Ronna Nelson

Eastern Maps: Courtesy of Association of American Vintners
and THE EASTERN GRAPE GROWER
Western Maps: Courtesy of Vintage Image

Manufactured in the United States of America
A 10 9 8 7 6 5 4 3 2 1

First Edition

Contents

Board of Advisors

I am indebted to the Board of Advisors for having taken the time to review, edit, correct, and comment on my original 900-page manuscript. Their expertise on the subject of vine and wine and their desire to help have made this volume what it is.

—William I. Kaufman

Jim Ahern • David A. Bailly • Greg Boeger • Jim Bundschu
Prof. Garth Cahoon • Richard Carey, Ph.D. • James Concannon
Charles Crawford • Albert Cribari • Bill Fuller • Norman Gates
Millie Howie • Lazare S. Kaufman • Louis P. Martini • Ed Mirassou
Peter Mondavi • R.G. "Dick" Peterson, Ph.D. • Bern Ramey
Sandra Rapazzini • Walter Schug • Dan Solomon • David Stare
Rodney Strong • Brother Timothy, F.S.C.
R. de Treville Lawrence, Sr. • Peter Watson-Graff • Fred E. Weibel
Carolyn Wente • Al Wiederkehr • John Wright

INTRODUCTION

The ENCYCLOPEDIA OF AMERICAN WINE had a "role model" (Frank Schoonmaker's Encyclopedia of Wine). To be without his book was a serious error for all persons interested in the wines of Europe. It was the easiest and most informative quick reference book on the subject of European wines. It had one serious omission in today's wine world and that was the minimal attention paid to American wines. This is also true of some of the other outstanding wine books that deal with today's world of wine. These omissions, or minimal attention to American wines, led me to the creation of this book which is ONLY devoted to the wines of America and includes those of Canada and Mexico. American wines have long since come of age and, with each passing year, they hold their own with any of the wines of the world. Witness the results of tastings that are constantly taking place where ever wine lovers get together. With major plantings of vineyards going on in Texas, North Carolina, Idaho, Georgia, Washington and Oregon, to mention a few, the fame of our American wines grows with each passing year.

Speaking of tastings, you will note that there are no subjective tasting notes in this volume. This is for two simple reasons: one, it would be impossible for anyone to be able to comment on the wines produced by the almost 1000 wineries in this encyclopedia (these wineries produce over 10,000 wines under various types and labels); and two, 90 percent of the wineries produce under 10,000 cases per year and most of the wine is sold either at the winery or within a small regional area. If you live in Southern California, you may be fortunate to taste John Culbertson's outstanding Champagne. In Ohio and the surrounding states a limited amount of Mon Ami's Gewurztraminer is available, Ingleside Plantations' Cabernet Sauvignon from Virginia or Duplin's Sparkling Scuppernong from North Carolina are but a few cases in point. Thus, my reason for omitting tasting notes is that they would be of no real advantage to wine lovers who are spread across America.

However, in order to help you with a tasting reference, I have included the results of what I consider to be the major wine competitions held across the United States from 1979-1984. I am certain that you will find this both interesting and helpful.

The purpose of this encyclopedia is to give you the kind of basic information that will help you expand your wine knowledge and serve as a reference to the wineries. The number of cases has been included to give you an identification of the winery's size. You will find a brief history, if there is one to tell; whether or not they estate bottle; where the vineyards are; the types of grapes grown; and the wines regularly produced. Designation of vineyard is a practice that is being included more and more on wine labels. Where the grapes are grown will be of interest to many. You the reader should be able to hear or see a winery name and be able to get a quick reference by simply looking for the winery in this encyclopedia.

The diversity of winery owner and winemaker's backgrounds has always fascinated me and, for that reason, I have selected a broad range of personalities, either because of expertise or background, for inclusion in my book. You will find doctors, lawyers, scientists, computer experts, engineers and college professors, along with trained oenologists and families with long history of winemaking. If I have omitted one of your favorites, you will have to forgive me.

Capsule histories, events and historical facts have been chosen for interest and, at the same time, limited because I would need another 1000 pages to cover the broad story of the American wine history.

Drink the wines of America . . .

Enjoy the wines of America . . .

Use the ENCYCLOPEDIA OF AMERICAN WINE to expand your wine knowledge.

WILLIAM I. KAUFMAN

ABBEY

The Trappist Abbey of Gethsemani, Nelson County, Kentucky; established by monks from France who planted vines and made wine. Starting in 1848, the vineyards' success encouraged others in Logan County to start vineyards and produce wine—in less than forty years 29 counties in Kentucky were growing grapes and making wine.

THE ABBEY

Founded 1973, Cuba, Missouri.

Under the direction of Brother Dressel. The winery produces wines from the Missouri Ozark Highlands.

Wines regularly bottled: Missouri Riesling, Concord, Catawba, Chablis, Burgundy, Champagne, Labrusca and Abbey White.

ABRAHAM, THE PLAINS OF—Quebec, Canada.

A high plain, above and adjoining the upper part of the city of Quebec. Quebec, in Canada, is the capital of the Province. (Montreal is the largest city.) Situated at the junction, the influence of the St. Lawrence and the St. Charles rivers. The population is predominately French-speaking. The Jesuit missionaries in 1636, at Quebec, were making Sacramental wine from the wild grapes that were in profusion along the St. Lawrence river. In 1917, all but one province adopted Prohibition laws ... Quebec; and, when Quebec finally adopted the laws, it only banned the sale of liquor. Wine and beer continued to be available. THE FIRST WINE MADE IN CANADA ... by Jesuit missionaries.

ACACIA WINERY

Founded 1979, Napa, Napa County, California.

Storage: Oak. Cases per year: 20,000.

Estate vineyard: Marina Vineyard.

Label indicating non-estate vineyard: St. Clair Vineyard, Madonna Vineyard, Winery Lake Vineyard, Lee Vineyard and Iund Vineyard.

Wines regularly bottled: Estate bottled, vintage-dated Chardonnay and vintage-dated Chardonnay and Pinot Noir.

Second label: Caviste. Occasionally bottles Zinfandel, Riesling, Sauvignon Blanc, Merlot and Cabernet.

acidity

Tasting term.

In wine, acidity is the word normally used to indicate the quality of sourness or tartness in the taste; i.e., the presence of agreeable fruit acids. Not to be confused with bitterness, dryness or astringency. An important and favorable element in wine quality.

acids

Natural acidity in grapes (and new wines) is mostly tartaric and malic acids; after malolactic "secondary" fermentation, the malic acid has changed into softer tasting lactic acid, but the tartaric acid remains unchanged through it all. Tartaric, in fact, remains unchanged through long-term bottle aging as well. (Tannins soften with age, but acids don't change unless the wine is chilled enough to cause cream of tartar to crystallize out of solution.) When that happens, you can see crystals on the cork or in the bottle but, even then, the taste of the wine is not usually changed significantly. Acidity helps to protect the wine from spoilage during fermentation.

acidulous
Tasting term.
Young, fresh, notable vivacity due to acid content.

ACKERMAN WINERY, INC.
Founded 1956, South Amana, Iowa County, Iowa.
Storage: Oak. Cases per year: 10,000.
History: Located in the Amana Colonies, the winery was founded by the grandfather of the current owner.
Wines regularly bottled: Apple, Cherry, Grape, Mulberry, Strawberry, Rhubarb, Dandelion, Plum, Elderberry, Blueberry, Cranberry, Blackberry and Peach.

acute
Tasting term.
Pungent, sharply insinuating.

AD MAJOREM DEI GLORIAM
Latin motto of the Jesuits—"To the Greater Glory of God . . ." Order of the Jesuits—winemakers, producers.

ADAMS COUNTY WINERY
Founded 1975, Ortanna, Adams County, Pennsylvania.
Storage: Oak and stainless steel. Cases per year: 2,500.
History: The winery building was built in 1875 by Levi Schwartz, a local farmer of German origin, and was a fruit farm for over 100 years. The winery is near Cashtown, an area of skirmishes during the Civil War. Gettysburg is 8 miles away. There is more fruit produced in Adams County than any other county in the United States.
Estate vineyard: Adams County Winery Vineyard.

Wines regularly bottled: Seyval Blanc, Vidal Blanc, Chelois Red, Chelois Rosé, Aurora. Also produced: Strawberry, Peach, Sour Cherry, Apple. Occasionally bottle: Pinot Chardonnay, Vidal Blanc, and Gewurztraminer (estate bottled, vintage-dated).

Adams County Winery Estate Vineyard
Adams County, Pennsylvania.
Pinot Chardonnay, Gewurztraminer, Seyval Blanc, Vidal Blanc and Marechal Foch vineyard.

ADAMS, JOHN (1735-1826)
Second President of the United States. Wine enthusiast, wine host in the White House. Strong interest in the wine grape due to his close friend, Thomas Jefferson. John Quincy Adams, the son of John Adams, was the sixth President of the United States. He, too, believed strongly that the wine grape was a more convivial offering at State functions than hard spirits.

ADAMS, LEON D.
Considered to be the Dean of American Journalists on the subject of wine and spirits, Leon Adams is internationally recognized as an authority in his field. He was founder of the Wine Institute in California and served as its secretary for twenty years. He also founded the Wine Advisory Board. He is the author of *The Commonsense Book of Wine, The Commonsense Book of Drinking,* and *The Wine Study Course.* His outstanding and fascinating story "The Wines of America" about *North American Wines and Winemakers* from the 16th century to the present is a book which no wine library should be without.

ADELAIDA CELLARS
Founded in 1983, San Luis Obispo County, California.
Storage: Oak and stainless steel. Cases per year: 4,000.
Label indicating non-estate vineyard: Estrella River Vineyard.
Wines regularly bottled: Vintage-dated Cabernet Sauvignon, Chardonnay; occasionally bottle Gamay Beaujolais.

Adelsheim Estate Vineyard
Yamhill County, Oregon.
Pinot Noir, Chardonnay, White Ries-

ling, Sauvignon Blanc, Pinot Gris and Gamay. Willamette Valley vineyard.

ADELSHEIM VINEYARDS

Founded 1971, Newberg, Yamhill County, Oregon.

Storage: Oak and stainless steel. Number of cases produced per year: 7,000.

History: The oldest winery on Chehalem Mountain, overlooking Chehalem Valley. The first area of the west to be farmed by Europeans in the 1830's. The area traditionally grew cherries, prunes, walnuts and peaches; now filberts and vinifera grapes.

Estate vineyard: Adelsheim Vineyard.

Label indicating non-estate vineyard: Sagemoor Farms Vineyard.

Wines regularly bottled: Estate bottled, vintage-dated Pinot Noir, Chardonnay, White Riesling. Also bottle vintage-dated Pinot Noir, Chardonnay, Semillon, Merlot. Occasionally bottle Sauvignon Blanc, Gamay Noir and Pinot Gris.

Second label: Adams Ranch.

ADLER FELS WINERY

Founded 1980, Santa Rosa, Sonoma County, California.

Number of cases produced per year: 3,000-6,000.

History: Adler Fels is the German translation of "Eagle Rock", a landmark next to the winery. One of the owners, David Coleman, is a graphic designer; and the co-owner/winemaker is Patrick Heck from the family that owns Korbel Champagne.

Label indicating non-estate vineyard: Salzgeber/Chan Vineyard, Nelson Vineyard, James Miller Vineyard, Bacigalupi Vineyard, Melaner a Deux Vineyard, and Lovin Vineyard.

Wines regularly bottled: Vintage-dated Cabernet Sauvignon, Chardonnay, Gewurztraminer, Johannisberg Riesling, Sauvignon Blanc, Pinot Noir and Champagne. Occasionally bottles Lovin Vineyard vintage-dated Johannisberg Riesling.

ADLUM, JOHN (1759-1836)

American horticulturist. Birthplace—York, Pennsylvania. He created and produced the CATAWBA grape from native varieties. His complete dedication to wine grapes started in 1814. Adlum was one of the first to recognize there was a future for a wine grape industry, if the native could be improved upon. He set up one of the first, if not the first, experimental areas to cross two vines; and from those attempts came the Catawba. In 1823 he published "Cultivation of the Vine."

aftertaste

Tasting term.

The feeling or lingering impression the wine leaves in the mouth after swallowing.

AGAWAM

A red labrusca grape grown mainly in Ontario.

age

The age of a wine may be approximated by careful tasting. Terms used to describe age are as follows:

Young: Fresh, without yeastiness or pronounced bouquet. Sometimes a young wine will have an overabundance of fresh fruit nose.

Mature: Possessing balanced bouquet and ready for bottling.

Aged: Wines possessing balance and bottle bouquet.

Wine aging cellar

aging

Wine develops smoothness, mellowness and character in aging. Everything that happens to wine during aging is not yet fully understood by scientists. However, many things are known: Some grape solids are deposited, the wine clarifies itself, some oxidation occurs, tannin and oak flavor are extracted from the cask and become part of the wine, and the many complex natural elements of the wine slowly interact, or "marry", for smoothness. Other complex natural changes also occur. These changes are the most mysterious. They create, in the wine, elements of flavor and bouquet, substances called aromatic esters and other compounds that are not found in grapes, grape juice or new wine.

Age is not a positive guide to quality. Most of the world's wines complete their aging quite early, even losing quality with further storage. Excessive aging results in the "passing out" of the wine. Some wines may age for decades and stay at their peak/plateau for many years. Some wine varieties prefer wood, others tanks lined with glass or stainless steel tanks. As wines mature, many producers complete the aging in smaller wood containers. Vintners choice: oak is favored by some, but in California redwood is also much in use. Casks of 1,000 gallons, oval shaped to make the lees deposit in a small space at the bottom, are preferred by many. Some finish the aging in even smaller casks or barrels. The smaller the container of wood, the greater the ratio of surface area through which the wine can "breathe" and take on the flavor characteristics of the container.

aggressive

Tasting term.

Very unbalanced, often caused by high alcohol acid or tannin content.

Agricola Ferrino

Coahuila, Mexico.

Carignane, Lenoir vineyard.

AHERN, JIM

Jim Ahern first became interested in wine when he traveled in Europe on a tour of duty with the United States Navy. When he settled in Los Angeles in 1968, he began his enological career with a home winemaking kit given to him by his wife. He founded the "Cellarmaster's" Home Winemaking Club in 1974, and served as Charter President, conducting semi-annual competitions. In 1978, Ahern decided to turn "pro", and he established the Ahern Winery in the San Fernando Valley. He specializes in Cabernet Sauvignon, and Chardonnay from Central Coast grapes grown in the Edna and Santa Maria Valleys.

AHERN WINERY

Founded 1978, San Fernando, Los Angeles County, California.

Storage: Oak. Cases per year: 6,000.

History: Jim Ahern is one of California's fine winemakers.

Labels indicating non-estate vineyards: Mac Gregor Vineyard, Paragon Vineyard and Bien Nacido Vineyard.

Wines regularly bottled: Vintage-dated Chardonnay, Zinfandel, Cabernet Sauvignon and Sauvignon Blanc.

AHLGREN VINEYARD

Founded 1976, Boulder Creek, Santa

Cruz County, California.

Storage: Oak. Cases per year: 1,400.

History: Very much a small family winery and vineyard. Dexter and Valerie, with their two daughters, operate the winery and care for the vineyards.

Estate vineyard: Ahlgren Vineyard.

Labels indicating non-estate vineyards: Ventana Vineyard, Ruby Hill Vineyard, York Creek Vineyard, Bonnydoon Vineyard, Bates Ranch Vineyard and Novitiate Vineyard.

Wines regularly bottled: Estate bottled, vintage-dated Chardonnay and Semillon. Also vintage-dated Chardonnay, Semillon, Zinfandel and Cabernet Sauvignon.

Ahlgren Estate Vineyard

Santa Cruz County, California.

Chardonnay and Semillon, Santa Cruz Mountain vineyard.

Ahollinger "Las Amigas" Vineyard

Napa County, California.

Chardonnay, Carneros vineyard.

Alaqua Estate Vineyard

Walton County, Florida.

Welder, Noble and Carlos vineyard.

ALAQUA VINEYARDS WINERY

Founded 1977, Freeport, Walton County, Florida.

Storage: Stainless steel. Cases per year: 3,000.

History: The first winery in Florida to use the Welder grape; and only Florida grapes are used to produce their wines. "Alaqua" is an Indian name for the sweet gum tree.

Estate vineyard: Alaqua Vineyards Winery.

Wines regularly bottled: Estate bottled, vintage-dated Welder, Noble and Carlos. September Welder, a late harvest, is produced in limited quantities.

ALATERA VINEYARDS

Founded 1977, Napa, Napa County, California.

Storage: Oak and stainless steel.

Label indicating estate vineyard: Vineyard Hill Farm.

Wines regularly bottled: Estate bottled, vintage-dated Cabernet Sauvignon, Gewurztraminer, Late Harvest Bunch Selected Johannisberg Riesling and Pinot Noir.

ALBA VINEYARD

Founded 1983, Milford, Warren County, New Jersey.

Storage: Oak and stainless steel. Cases per year: 3,000.

Wines to be regularly bottled: Vidal Blanc, Chardonnay, Gewurztraminer, Cayuga White, Foch, DeChaunac, Cabernet Sauvignon and Landell.

alcohol content

Average Percent Alcohol Content of Wine: (Red, White, Rosé) and Champagne: 12%-14%. Dessert: 14%-24%. Soft wines: 7%-10%. Rated on 100% alcohol, not "proof" of 200%. For example, a still wine of 12% alcohol is 24% proof. Under federal regulations, the alcoholic content must be stated on the label. Any variance of H degree on either side of the stated alcohol content is permissible in wines under 14%; 1 degree is permissible in wines over 14%. In lieu of stating the alcoholic content, the phrase "table wine" may be used on wines under 14%.

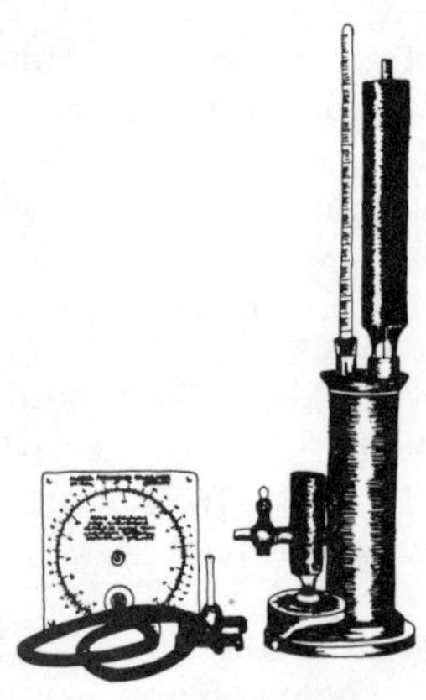

An "ebulliometer" determines the alcohol

alcohols

The major alcohol in wine is ethanol. Table wines with low ethanol content have a thin character and those with too high a concentration often have a "hot" taste. The other alcohols, never present in large concentrations, can be considered flavor components, especially in contributing something to the "nose" of the wine.

ALDERBROOK VINEYARDS

Founded 1981, Healdsburg, Sonoma County, California.

Storage: Oak and stainless steel. Cases per year: 7,000.

Estate vineyard: Rafanelli Vineyard.

Wines regularly bottled: Chardonnay, Sauvignon Blanc.

ALDERWOOD CELLARS (See Haviland Vintners.)

Burgundy, Chablis, Rhine and Claret.

ALEATICO

Wine grape.

A grape with a muscat character. Occasionally used for sweet table or dessert wine. Vino Santo, the natural sweet wine in Italy uses the Aleatico.

ALEXANDER VALLEY

Vineyard area in Sonoma County. Russian River flows through the valley towards the sea.

Alexander Valley Estate Vineyards

Sonoma County, California.

Chardonnay, Chenin Blanc, Gewurztraminer, Johannisberg Riesling, Cabernet Sauvignon, Pinot Noir and Zinfandel. Alexander Valley vineyard.

ALEXANDER VALLEY VINEYARDS

Founded in 1975, Healdsburg, Sonoma County, California.

Storage: Oak and stainless steel. Cases per year: 18,000.

History: The winery is on the original homesite of Cyrus Alexander, for whom the valley is named. The restored buildings and the graveyard where he is buried, are on the property. Alexander was deeded the property in 1842. The Wetzel family purchased the land in 1963.

Estate vineyard: Alexander Valley Vineyard.

Wines regularly bottled: Estate bottled, vintage-dated Chardonnay, Dry Chenin Blanc, Gewurztraminer, Johannisberg Riesling, Cabernet Sauvignon, Pinot Noir and Zinfandel.

Second label: Sin Zin.

Alexander's Crown Vineyard

Sonoma County, California.

Sonoma/Windsor estate Cabernet Sauvignon. Alexander Valley vineyard.

ALICANTE BOUSCHET

Wine grape.

A prolific grape, rarely used as a red table wine. Predominantly used in Burgundy blends for color. Originally came from Spain. Produced as a varietal by very few American wineries.

ALIGOTE

Wine grape.

A vinifera grape coming from Burgundy, France, that produces white fruity wine. It has never been successful in California.

Allegheny Mountains Vineyards
Prince Edward County, Virginia.
Seyval vineyard.

Allegro Estate Vineyard
New York County, Pennsylvania.
Cabernet Sauvignon, Merlot, Cabernet Franc, Chardonnay, Seyval Blanc, Vidal, Chambourcin, Chelois, Pinot Noir, Riesling, Petite Sirah, Gamay Beaujolais and Ravat vineyard.

ALLEGRO VINEYARDS
Founded 1973, Brogue, York County, Pennsylvania.

Storage: Oak and stainless steel. Cases per year: 1,500.

History: Winery is only interested in producing dry table wines made from Vinifera and French hybrids.

Estate vineyard: Allegro Vineyard.

Wines regularly bottled: Estate bottled, vintage-dated Cabernet Sauvignon, Chardonnay, Seyval Reserve, Seyval Blanc. Bottles proprietary wine called Opus I. Also produces Vin Blanc, Vin Rosé, Vin Rouge. Occasionally bottles Late Harvest Seyval Blanc.

Alma Vineyard
Los Gatos, California.
Semillon vineyard.

ALMADEN
Founded 1852, San Jose, Santa Clara and San Benito Counties, California.

Storage: Oak and stainless steel. Cases per year: 12.1 million.

History: The foundation for today's world-famed Almaden was Charles Lefranc. Relinquishing the security of his native France and plunging into the uncertainty of California in the 1800s, Lefranc joined his friend and older countryman, Etienne Thée, in the verdant valley of Santa Clara County. Thée, a farmer from Bordeaux, was the first to settle on the land now known as Almaden's home winery in San Jose.

The energetic Lefranc loved his new home. In only a few years, his brilliance would make it well known, and years after that, world renowned. But the area then was more famous for a nearby quicksilver mine called "Almaden", a Moorish word meaning "the mine". Lefranc would eventually christen the vineyards and winery New Almaden; but, first, he would see that the vineyards bore the fruit of Vitis vinifera, European grapevines yielding the luscious wines long established in France and Germany. Devoting his great energy to this task, Lefranc soon obtained fine varietal grape cuttings from his native France. He removed the common Spanish Mission grapevines at his new home and, gambled that transplanted cuttings would take hold and thrive in the soil of a new continent.

The planting was historical, marking one of the first successful commercial plantings of fine European wine grapes in northern California. Ironically, the original harvests were not pressed into wine, but sold as table grapes to meet an existing high demand for fruit.

Lefranc and Thée carved a wine cellar in the soil of Almaden and lined the thick adobe walls with oaken casks which, like the imported vines before them, rounded the Horn from France to California. Lefranc's dream for wines from the grapes of Almaden was to become a reality, and Almaden would take its place in modern history as the oldest producing winery in California.

By 1880, Lefranc, now married to Thée's daughter, Adele, had inherited Almaden Vineyards and was cultivating 130 acres in Santa Clara County. His vines from the districts of Champagne, Bordeaux, Burgundy, and the Rhone Valley were flourishing. By the end of the decade, the winery would produce 100,000 gallons a year.

Estate vineyards: Paicines, Cienega, King City, San Lucas and San Jose.

Wines regularly bottled: Under Charles Lefranc Founder's label are Chardonnay, Fumé Blanc, Late Harvest Johannisberg Riesling and Gewurztraminer, Maison Blanc, Cabernet Sauvignon, Pinot St. George, Zinfandel Royale, Maison Rouge and Founders Port. Under Almaden label are Chardonnay, Johannisberg Riesling, Sauvignon Blanc, Golden Chablis, Gewurztraminer, Chenin Blanc, Gray Riesling, French Colombard, Cabernet Sauvignon, Pinot Noir, Petite Sirah, Gamay Beaujolais, Zinfandel and Gamay Rosé. Generic wines produced are Monterey Burgundy, Monterey Chablis, Light Chablis, Light Rhine, Light Rosé, Carafe Chablis, Carafe Rhine, Carafe Burgundy and Carafe Rosé. Almaden "Mountain" wines are Mountain White Chablis, Mountain White Sauterne, Mountain Rhine, Mountain Red Burgundy, Mountain Red Claret, Mountain Red Chianti, Mountain Grenache Rosé and Mountain Nectar Vin Rosé. Sparkling wines are Almaden Blanc de Blanc, Chardonnay Nature, Eye of the Partridge (all three cuvée-dated), Brut and Extra Dry. Almaden also produces Centennial Brandy, Pale Triple Dry and Sweet Vermouths; Solera Cocktail, Golden, Cream and Flor Fino Sherries; Solera, Tinta, Tawny and Ruby Ports. Almaden's Le Domaine line includes Crown Chablis, Rhine, Burgundy and Rosé (bag-in-box wines); Champagnes: Brut, Extra Dry, Pink, Sparkling Burgundy and Cold Duck.

Almarla Estate Vineyards
Wayne County, Mississippi. Muscadines and French hybrids vineyards.

ALMARLA VINEYARDS
Founded 1979, Matherville, Wayne County, Mississippi.

Storage: Stainless steel. Cases per year: 12,000.

History: Part of a new movement to revive a wine industry that existed about 100 years ago. The winemaker was chief chemist for the BATF. Development will depend on whether vinifera can grow in the region.

Estate vineyard: Almarla Vineyards.

Wines regularly bottled: Cream Scuppernong, Cream Muscadine, Pink Muscadine, Noble. Semi-generic wines are: Sangria, Chablis, Chablis Plane Dry, Rosé, Vin Rosé, Sauterne, American Burgundy. Also bottle Almondetta.

ALMISSION
Wine grape.

A black grape that is a cross between a Mission and Alicante. Hybridized by L.O. Bonnet as a Port variety.

ALPHA
Wine grape.

A red wine grape that is grown in the Northern United States because of its ability to tolerate bitter cold.

Alpine Estate Vineyard
Benton County, Oregon.
Johannisberg Riesling, Pinot Noir, Cabernet Sauvignon, Chardonnay and Gewurztraminer Willamette Valley vineyard.

ALPINE VINEYARDS
Founded 1980, Monroe, Benton County, Oregon.
Storage: Oak and stainless steel. Number of cases produced per year: 3,0005,000.
History: Alpine is located in a panoramic setting in the foothills of the Coast Range Mountains.
Estate vineyard: Alpine Vineyards.
Wines regularly bottled: Estate bottled, vintage-dated White Riesling, Pinot Noir, Cabernet Sauvignon, Chardonnay, Gewurztraminer. Also estate bottled, vintage-dated Blanc de Blancs.

Alta Estate Vineyard
Napa County, California.
Chardonnay Napa Valley vineyard.

ALTA VINEYARD CELLAR
Founded 1878, Calistoga, Napa Valley, California.
Storage: Limousin oak. Cases per year: 2,000.
History: The original Alta Vineyard Cellar was established in 1878 by Captain C.T. McEachran, a Scottish sea captain who purchased the 40 acre tract of land for $480 in gold coin and began clearing the land and planting vines. In 1878, he built a stone winery having a capacity of 8,000 gallons.
Robert Louis Stevenson visited the winery in 1880, and his later description of it is in his "Silverado Squatters", the book about his two month honeymoon in the Napa Valley.
Stevenson and his wife rode by carriage from Calistoga to the road that led up the hill to Alta Cellars and there met Captain McEachran. The vintage wines Stevenson tasted were the very first produced. Stevenson wrote, "We went into his cellar and tasted his wines, red wines, respectively, one and two years old; the younger seemed to me more promising; but the quality of each was very pleasant, with a little gout du terrain that reminds one of a Burgundy."
Today, the wine from Alta Vineyard Cellar is not red, but white. In 1978, 100 years after McEachran's first wine, Alta produced a Chardonnay, the first vintage produced from this winery in the 20th century, from grapes grown on the original vineyard.
Estate vineyard: Alta Vineyard.
Label indicating non-estate vineyard: Monticello Vineyards, Cofran-Johnson Vineyards.
Wines regularly bottled: Estate bottled, vintage-dated Chardonnay.

ALTUS VITICULTURAL AREA
The viticultural area lies near the town of Altus and extends approximately five miles along a plateau situated between the Arkansas River bottomlands and the Boston Mountains. There are about 800 acres of grapes in the area.

AMADOR COUNTY
Sierra Foothills. Shenandoah Valley, Fiddletown.

AMADOR FOOTHILL WINERY
Founded 1980, Plymouth, Amador County, California.
Number of cases produced per year: 4,000.
Label indicating non-estate vineyard: Eschen Vineyard, Esola Vineyard.
Wines regularly bottled: White Zinfandel, Chenin Blanc, Sauvignon Blanc, Zinfandel.

AMADOR WINERY
Amador City, Amador County, California.
Storage: Oak.
Wines regularly bottled: Sutter's Gold, Spiced Mountain Jubilee, Sauterne, Chablis, Mountain Rhine, Burgundy and Madame Pink Chablis.

AMBASSADOR (See Lamont Winery.)

amber
Wine color.
The color of brass, a dark tone found only in dessert wines or in maderized wines.

AMERICA, THE FIRST WINES OF
The first wine production in America was accomplished by a small colony of French Huguenots in Florida, the year—1565. The Scuppernong was the source of the wine—the native grape of the South.

It is recorded by early French clerics and explorers that the North American Indians were making wine or fermenting the juices long years before the Huguenots of Florida.

From the 1870s to the late 1890s a wine rush, like the Gold Rush, hit the United States, a prosperity never before known. The wine states of the Union doubled and tripled their acreage and worked the wineries beyond capacity to try to meet the demand. This burgeoning period was caused by the devastation of the French vineyards due to the scourge of the dreaded phylloxera.

The vineyards of France were rotting, destroyed; the wineries shut down, a national disaster. (The wine industry of France did not recover until the man from Missouri, George Husmann, shipped tens of thousands of pest-resistant grapevines and rootstocks to help the French make a new start. In appreciation, Husmann was later awarded France's highest honor . . .).

While France was mourning its dying vineyards, California doubled its wine-grape acreage; the growers of Florida not only doubled their acreage but began to experiment with new grape varieties and blends of wine. One grower and winemaker after another felt the time had come to show the French what the Floridians can do . . . and they did. Vineyardist and winemaker from Tallahassee, E. Dubois, owner of the San Luis Vineyards was an experimenter, an innovator, a master winemaker and from his writings, detailed his experiments with the Norton and Cynthia grapes. In agricultural journals his name, his vineyards and his wines became known throughout the American wine industry. Dubois boasted that his Burgundies made from the Cynthia grapes were as fine as any imported from California which were produced from European graftings. E. Dubois believed in Florida viniculture and viticulture and he never wavered. There are many wine authorities who have compared Dubois' contributions to American wine development with the famed ones—Haraszthy of California, Longworth of Ohio, and Husmann of Missouri and later, of California. During this period

Florida is noted for having one of the very few woman vineyardists in the country, Mrs. M. Martin of Hermitage. Her acreage was second in size to Dubois' San Luis vineyards. Once again winegrape fever has taken over Florida; there are new vineyards and new wineries and hundreds and hundreds of home winemakers . . . and plantings of new varieties such as Lake Emerald, Stover and Black Spanish developed by Dr. Loren Stover and Dr. John Mortensen of the Grape Investigation Laboratory (Agricultural Research Center) at the University of Florida at Lessburg.

But let us not forget the Scuppernong . . . the native grape that made the first wine of Florida. The Scuppernong continues to be used to make a delightful, embracing white wine and who knows, with the new demands for the light white wine, the Scuppernong might make a repeat performance . . . a comeback. What could be more fitting than the revival of the Scuppernong, America's original wine.

AMERICAN

Wines simply stated as "American" are usually wines that have been made from grapes grown in more than one state, or grown in one state and bottled in another.

AMERICAN HYBRIDS

Grape varieties developed in America by cross-breeding. Historically, the cross-breeding has been between a native American labrusca (or one of its descendants) and traditional European varieties. The uniformity of this practice has been to give American hybrids a style or flavor common among them, which is generally described as foxy. Examples are Catawba, Delaware and Ives.

Full grape boxes awaiting transport

AMERICANS FOR WINE

A nationwide organization bringing together wine consumers interested in the well being of wine in the United States.

Americans for Wine, a voluntary grass-roots network, will provide a forum for consumers, growers, restaurants, retailers, wholesalers and wineries as a cohesive, independent group to express a collective view on public policy issues that affect wine. Members receive periodic newsletters and bulletins.

For information, contact the Wine Institute, San Francisco, California 94108.

AMERICAN WINE SOCIETY

A wine society that is open to all wine enthusiasts both professionals and consumers. Chapter meetings and tastings are held regularly. For membership information:

American Wine Society
RD. 3, Box 112, Rt. 89
Trumansburg, New York 14886

AMERICAN SOCIETY OF ENOLOGISTS

Started in 1950, the organization is

composed of enologists and viticulturists throughout the world. It publishes a quarterly scientific journal, holds an annual scientific conference.

AMERINE, DR. MAYNARD A.
(1911)

Professor of Enology Emeritus at the University of California at Davis, world-famous authority on winemaking. He is also a connoisseur and collector of fine wines and a prolific writer on wine and winemaking.

Dr. Amerine bases his endorsement of California wines on the evidence of his expert palate, not on the fact that he is a native Californian. He was born in San Jose and grew up on farms in the Modesto area. He attended Modesto High School and Junior College before entering the University of California.

After graduating from Davis in 1932, he joined the Davis staff as an enologist in 1935. He obtained his Ph.D. at Berkeley in 1936. Dr. Amerine became Junior Enologist in 1937 and Instructor of Enology in 1938. He was named Professor of Enology and Enologist in 1952. From 1957 to 1962 he was Chairman of the Department of Viticulture and Enology at Davis.

In addition to being a plant scientist and biochemist, Dr. Amerine is an expert on the sensory appreciation of food and wine. Dr. Amerine's published writings number over 330 books and articles, ranging from works on grape and wine technology to consumer magazine articles on the proper wines to serve. He has received many awards, in the United States and abroad.

Amity Estate Vineyard

Yamhill County, Oregon.

Pinot Noir, White Riesling, Chardonnay, Gewurztraminer and Muscat. Willamette Valley vineyard.

AMITY VINEYARDS

Founded 1974, Amity, Yamhill County, Oregon.

Storage: Oak and stainless steel. Cases per year: 8,000.

History: A small winery that produces many fine wines among which is Pinot Noir Nouveau. A light, fruity red wine styled like a French Beaujolais nouveau, but using Pinot Noir instead of Gamay grapes.

Estate vineyard: Amity Vineyard.

Label indicating non-estate vineyard: Wirtz Vineyard, Champoeg Vineyard, Feltz Vineyard, Wahle Vineyard, Sunnyside Vineyard, Red Hills Vineyard.

Wines regularly bottled: Vintage-dated Pinot Noir, Nouveau Pinot Noir, White Riesling, Gewurztraminer. Also proprietary vintage-dated Solstice Blanc and non-vintage Winemakers Reserve. Occasionally bottle Red Table Wine.

Second label: Redford Cellars.

ample

Tasting term.

Sensation of fullness, roundness; persistent and harmonious.

Amwell Valley Estate Vineyard

Hunterdon County, New Jersey.

Marechal Foch, Landot, Seyval, Villard Blanc, Aurora, Rayon d'Or and Ravat vineyard.

AMWELL VALLEY VINEYARDS
Founded 1982, Ringoes, Hunterdon County, New Jersey.
Storage: Oak and stainless steel. Cases per year: 1,000.
Estate vineyard: Amwell Valley Vineyard.
Wines regularly bottled: Estate bottled, vintage-dated Foch, Seyval, Aurora, Villard Blanc.

Anchor Acres Vineyard
Yates County, New York.
Glenora Wine Cellars estate Chardonnay Finger Lakes vineyard.

Anderson Valley Vineyard
Mendocino County, California.
Husch estate Pinot Noir, Chardonnay and Gewurztraminer vineyard.

ANDERSON VALLEY VITICULTURAL AREA
Anderson Valley is located in the southwestern part of Mendocino County and generally lies along the watershed of the Navarro River, stretching from its headwaters in the coastal range and extending northwest toward the Pacific Ocean. The total area is 57,600 acres with 600 acres of vineyards widely dispersed within its boundaries.
This area is located within the boundaries of another viticultural area to be called "Mendocino".

S. Anderson Estate Vineyard
Napa County, California.
Chardonnay, Pinot Noir and Pinot Blanc Napa Valley vineyard.

S. ANDERSON VINEYARD
Founded 1979, Napa, Napa County, California.
Storage: French cooperage, stainless steel. Cases per year: 4,000.
Estate vineyard: S. Anderson Vineyard.
Label indicating non-estate vineyard: Hoffman Ranch Vineyard.
Wines regularly bottled: Sparkling wines are Blanc de Noir and Brut (bottles are vintage-dated). Estate bottled, vintage-dated Chardonnay.
Second label: Heritage Hill Winery.

STANLEY F. ANDERSON
Founder of Wine-Art Sales Ltd., at Vancouver, British Columbia, Canada in 1959 to educate home winemakers on how to produce higher quality wines at home. Pioneered the use of varietal grape concentrates and adapted professional wine making technology for home winemakers. Founded Wine-Art of America in San Francisco, California in 1959. Numerous wine enthusiasts started making wine from Wine-Art classes and kits and moved on to start commercial wineries throughout the U.S. Author of THE ART OF MAKING WINE, the most popular handbook on home winemaking. Not to be confused with Stanley Anderson of S. Anderson Vineyard, Napa, California.

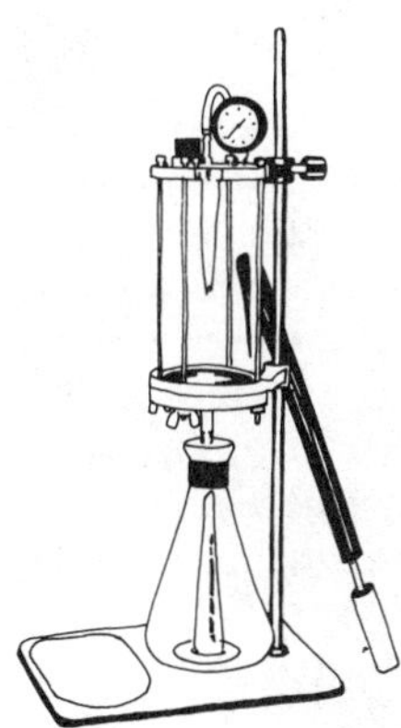

Laboratory filter for test bottling wine

ANDERSON WINE CELLARS
Founded 1980, Exeter, Tulare County, California.
Storage: Oak and stainless steel. Cases per year: 2,000.
Estate vineyard: S & K Vineyards.
Wines regularly bottled: Vintage-dated Chenin Blanc and Ruby Cabernet. Also estate bottled, vintage-dated French Colombard.

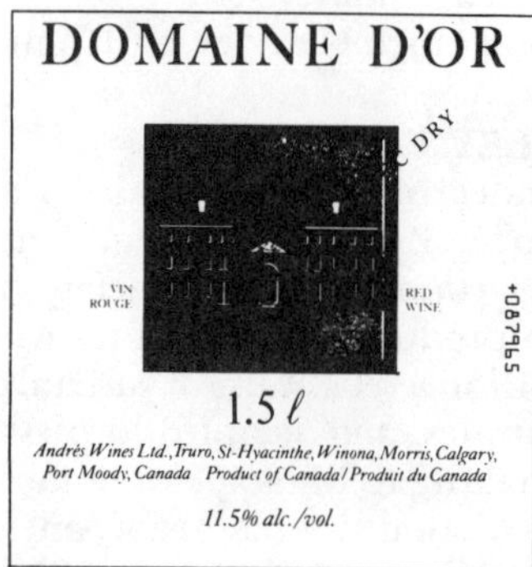

ANDRES

Winona, Ontario, Canada. Wineries: Port Moody, British Columbia.

Storage: Stainless steel. Cases per year: 900,000.

History: Andres Wines Ltd. was founded in British Columbia in 1961 by Mr. Andrew Peller. Mr. Peller recognized the future potential of the Canadian wine industry and accordingly set about building a truly national wine company. In 1964, winery operations were established in Calgary, Alberta and in Truro, Nova Scotia. In 1970, Andres entered the Ontario market with the purchase of Beau Chatel Wines in Winona. In 1974, Andres moved into Quebec with the founding of Les Vins Andres du Quebec at St. Hyacinthe. In 1975, Andres purchased the Valley Rouge winery located in Morris, Manitoba.

Label indicating non-estate vineyard: Reimer Mannhardt Vineyards, Inkameep Vineyard, Gehringer Brothers Vineyard.

Wines regularly bottled: Proprietary Sparkling wines: Baby Duck, Chante Rosé and Blanc, Richelieu, Sangria; and proprietary Portage du Fort, Moulin Blanc, Hochtaler, Sans Falcon, Cuvée du Marché.

Andrews Vineyard

Temecula, California.
Sauvignon Blanc vineyard.

ANGELICA

A white dessert wine, traditionally one of the sweetest wine types. It is either straw or amber colored and mild and fruity. Angelica originated in California and is produced from a number of grape varieties, including Grenache and Mission. Good with, or following, dessert with between-meals refreshments. Serve chilled or at room temperature. Most common Sacramental wine.

Annaberg Vineyard

Chalk Hill, California.
Balverne estate Scheurebe vineyard.

ANONYMOUS

"How is Champagne made?
By sheer genius, sir, sheer genius!"
—Conversation at White's Club, London.

And how's this for a description of the perfect wine?

"It's like the perfect wife—it looks nice and is nice, natural, wholesome, yet not assertive; gracious and dependable, but never monotonous."

"Wine improves with age—I like it more the older I get."

"When wine enlivens the heart
May friendship surround the table."

"There is no sounder purchase for a tired and depressed man than a bottle of good Burgundy."

"Sunbeams condensed from Nature's Holy Shrine
Are gently housed in every drop of wine."

"Nothing equals the joy of the drinker except the joy of the wine in being drunk."

Description of a corkscrew: "... the winelover's best friend and the rarest of tools."

In Vino Veritas

"In wine lies Truth, in Water nought
But Melancholy, dull and sour,
The Apple with old evil's fraught,
The Vine Brings Truth and Friendship's hour!

In Wine lies Truth. Its bright rays pass
Open and free from wiles and arts,
The glow of kindness lights the glass,
Men speak the thing that's in their hearts!"

"Good wine carrieth a man to heaven."

"Drink wine, and you will sleep well. Sleep well and you will not sin. Avoid sin, and you will be saved. Ergo, drink wine and be saved."

"Medieval Latin ...
Old men's milk ..."

"Old wine and old friends are enough provision."

Old Drinking Songs

"By wine we are generous made,
It furnishes fancy with wings;
Without it we ne'er should have had
Philosophers, poets or kings.

In wine, mighty wine, many comforts I spy;
If you doubt what I say, take a bumper and try!

For of all labors, none transcend
The works that on the brain depend;
Nor could we finish great designs
Without the pow'r of gen'rous wines."

Proverbs

Spanish ...
"With wine and hope, anything is possible."
French ...
"In water one's own face, but in wine one beholds the heart of another."
German ...
"There are more old wine drinkers than old doctors."
Irish ...
"Take the drink for the thirst that is yet to come."
Arabian ...
"Good wine praises itself."
Italian ...
"One barrel of wine can work more miracles than a church full of saints."
Russian ...
"Drink a glass of wine after your soup, and you steal a ruble from the doctor."

antique golden yellow
Wine color.
An intense yellow color with amber highlights.

Antuzzi's Estate Vineyard
Burlington County, New Jersey.
Baco Noir, Seyval Blanc, Cayuga, Ravat, Delaware, Concord, Catawba and Niagara vineyards.

ANTUZZI'S WINERY

Founded 1973, Delran, Burlington County, New Jersey.

Storage: Oak and stainless steel. Cases per year: 15,000.

History: A family tradition, started by his grandfather in Italy as a hobby, started Matthew Antuzzi when he was left all the winemaking equipment. A small but growing winery.

Estate vineyard: Antuzzi's Vineyard.

Wines regularly bottled: Baco Noir, Seyval Blanc, Cayuga, Ravat 51, Delaware, Concord, Pink Catawba, Niagara, Catawba White, May Wine. Also generics bottled are Rosé, White, Dry Red. And Proprietary Centennial White, Apple D'Apple. Also fruit wines: Blueberry, Cherry Royale, Strawberry, Raspberry, Blackberry. Also Sweet Claret.

Anzilotti Vineyard

Mendocino County, California.

Zinfandel vineyard.

APERITIF WINE

French word from the medieval Latin "Aperire," meaning "to open," and refers to wine and other drinks taken before meals to stimulate the appetite. In the strictest sense, it applies to vermouths and other wines flavored with herbs and other aromatic substances, but, in general usage, any wine, when served before a meal, may be referred to as an "aperitif."

Apalachee Vineyards

Oconee County, Georgia.

B & B Rosser estate Cabernet Sauvignon, Cabernet Franc, Merlot, Zinfandel, Carmine, White Riesling, Emerald Riesling, Colombard, Sauvignon Blanc and Chardonnay vineyard.

APPALACHIAN HARVEST (See MJC Vineyards.)

Estate bottled, vintage-dated Virginia Dutchess, Virginia Delaware.

appearance

Sensory evaluation term.

The appearance of a wine is judged by whether, or not, the wine seems clear, or contains sediment or suspended material of a colloidal or larger particle size.

appellation

Term which signifies in the United States the geographical origin of the grape used in a wine. When the appellation of origin appears on the label, 85 percent of the wine must come from grapes grown in that region. (See Viticultural Area.)

appellation controllee

The French law indication which states that not only site of grape origin is controlled but also cultural methods, production limits, etc. U.S. law only controls site of grape origin.

APPETIZER (APERITIF) WINES

Wines favored for before meals, or cocktail use. Sherry, Vermouth, White Wine and Champagne are most favored. They range from extra-dry to sweet, the drier types being more suitable when food is to follow. Alcohol stimulates the stomach's preparation for receiving food and is an aid to digestion.

appleness

Tasting term.

A frequent characteristic of fine white wine, resembling the flavor and aroma of fresh apple.

APTOS VINEYARDS (See Thomas Kruse Winery.)

ARAMON

Wine grape, seldom grown in the United States, but familiar to viticulturists because a cross between this vinifera grape and a native American grape is widely used for rootstock on which other viniferas are grafted.

Arata Vineyard
Santa Cruz County, California.
Cabernet Sauvignon Santa Cruz Mountain vineyard.

ARBOR CREST/WASHINGTON CELLARS
Founded 1982, Spokane, Spokane County, Washington.
Storage: Oak and stainless steel. Cases per year: 7,000-20,000.
Estate vineyard: Arbor Crest Vineyard.
Label indicating non-estate vineyard: Dionysus Vineyard, Bacchus Vineyard, Stewart's Sunnyside Vineyard, Stewart's Wahluke Vineyard, Sagemoor.
Wines regularly bottled: Vintage-dated White Riesling, Johannisberg Riesling, Late Harvest Johannisberg Riesling, Sauvignon Blanc, Chardonnay, Merlot, Cabernet Sauvignon, Late Harvest Gewurztraminer and occasionally bottles Gewurztraminer.

argols
The tartrate deposited by wines during aging, especially when chilled. Cream of tartar, used in baking, is made from it.

ARGONAUT WINERY
Founded 1976, Ione, Amador County, California.
Storage: Oak and stainless steel.
Estate vineyard: Willow Creek Vineyard.
Label indicating non-estate vineyard: John A. Ferrero Vineyard.
Wines regularly bottled: Estate bottled, vintage-dated Barbera. Also vintage-dated Zinfandel.

ARIZONA TERRITORY (See R.W. Webb Winery.)

Arensberg Vineyard
Sonoma County, California.
Zinfandel Alexander Valley vineyard.

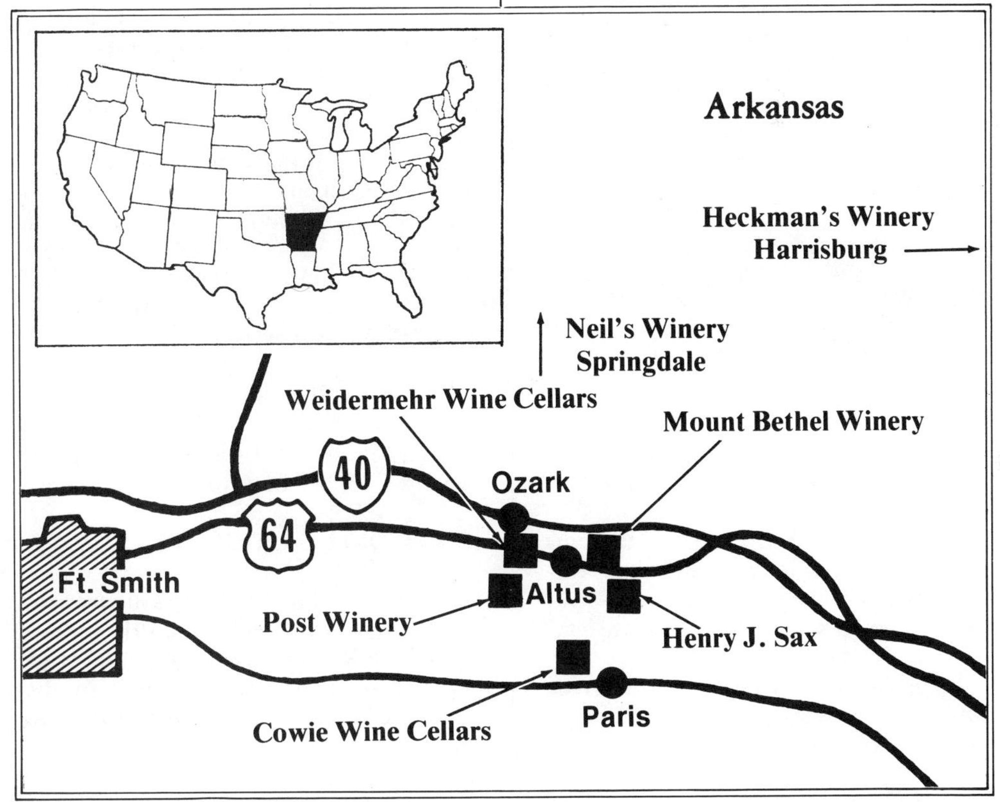

aroma

Tasting term.

Varietal: Certain of the grape varieties, when grown under optimum conditions, have aromas which are characteristic of that particular variety. While some of these characteristic varietal odors may be described as smelling "like some other fruit", or, in general terms, it is felt that the varietal aromas are basic to sensory examination and that they should stand as such. The wine enthusiast should have a built-in impression of the basic aromas of those varieties in which the aroma is easily detectable, so that when confronted with an unknown sample, he, or she, is capable of recognizing and identifying the aroma. Some of the varieties considered to have easily detectable characteristic aromas are:

Chardonnay, Muscat, Pinot Blanc, Sauvignon Blanc, Semillon, Gewurztraminer, White Riesling, Grey Riesling, Cabernet Sauvignon, Gamay, Pinot Noir, Pinot St. George, Zinfandel and some of the Labrusca varieties.

Distinct: An aroma sufficiently individual in character so as to permit differentiation from other wines—but not intense enough to permit varietal identification.

Vinous: This term is used to describe the smell of wine when no varietal or distinct aroma is detectable.

Arrendell Vineyard

Sonoma County/Green Valley, California.

Chardonnay and Pinot Noir vineyard.

Arrowhead Estate Vineyards

Atoka County, Oklahoma.

Cascade, Aurora vineyard.

ARROWHEAD VINEYARDS, INC.

Founded 1979, Caney, Atoka County, Oklahoma.

Estate vineyard: Arrowhead Vineyards.

Wines regularly bottled: Cascade, Aurora.

ARROWOOD, RICHARD L.

A native Santa Rosan, Richard Arrowood worked as a chemist and assistant manager with Korbel Champagne Cellars while attending Santa Rosa Junior College and California State University, Sacramento, where he received his B.A. in organic chemistry in 1968. His graduate work in fermentation science was done at California State University, Fresno.

After a year as production chemist and production control supervisor with the Italian Swiss Colony Winery, he was engaged as enologist by Sonoma Vineyards, in 1970, and appointed vice-president of production 1973. In 1974, he became the first employee of Chateau St. Jean as winemaster and vice president.

ARROYO (See Pendleton Winery.)

Arroyo Seco Vineyards

Monterey County, California.

Chardonnay, Cabernet Sauvignon and Muscat Canelli vineyard.

ARROYO SECO VITICULTURAL AREA

The 18,240 acres in Monterey County form a triangular-shaped area adjacent to the Arroyo Seco Creek which flows into the Salinas River near Soledad, California. Wine grapes were first planted in the Arroyo Seco vicinity at the nearby Mission Soledad during the 1830's.

ARTERBERRY, LTD.

Founded 1979, McMinnville, Yamhill County, Oregon.

Cases per year: 2,000.

Label indicating non-estate vineyard: The Red Hills Vineyard.

Wines regularly bottled: Vintage-dated Chardonnay, Rosé of Pinot Noir, Pinot Noir, Sparkling Chardonnay.

Second label: Red Hills, Arterberry Cellars, Arterberry Ciderworks. Under Red Hills Vineyards, produces vintage-dated Sparkling Wine.

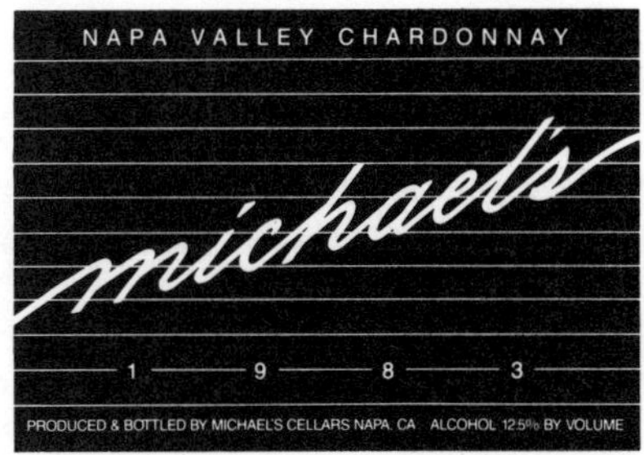

ARTISAN WINES

Founded in 1984, Napa County, California.

Number of cases per year: 10,000.

History: Founded by winemaker, Michael Fallow & graphic artist and marketing consultant, Jeffrey Michael Caldewey. Noted for their innovative approach to packaging and marketing and unswerving commitment to quality, Artisan encompasses three brands: Michael's, Ultravino, and Cru Artisan. Specializing in the Pacific Basin Markets of California, Hawaii, Hong Kong and Japan.

Wines regularly bottled: Vintage-dated Cabernet, Chardonnay and Sparkling Wine from Napa Valley, Bordeaux and Burgundy.

assemblage

The preliminary blending of wines from different vineyards after the first racking.

Ashby Vineyards

Elvira County, Missouri.

Concord vineyard.

Ashton Vineyards

Sonoma County, California.

Gewurztraminer, Pinot Noir vineyard.

astringency

The taste sensation of a mouth-drying, puckeriness or bitterness. A wine impresses the palate as being smooth, rough, puckery, or bitter, depending upon its tannin content and the types of tannins present. Alcohol, body, acidity, and particularly sugar have modifying effects upon the impression made by the tannin. As wine ages the degree of astringency decreases because of the oxidation and precipitation of the tannin substances.

Smooth or Soft: This term is used to describe wines of low astringency.

Slightly Rough to Very Rough: Terms used to describe increasing degrees of astringency.

astringent

Tasting term.

Rough, bitter, too rich in tannin. Not to be confused with sourness which is caused only by acids, not tannin.

atmosphere

The unit of measure for pressure of wine in a bottle of sparking wine or Champagne. 1 Atmosphere is roughly equal to 15 lbs per square inch. Air pressure at sea level is equal to 1 Atmosphere. At 50 degress F, sparkling wines have 3.5-6 Atmospheres. It increases to 6-10 Atmospheres at 70 degress F. This is why the bottles have metal cork retaining wires. Always point the bottle away from another individual when opening.

AUGUSTA VITICULTURAL AREA

The area is located in St. Charles County, Missouri. The first viticultural area to be approved by BATF.

AU NATUREL CHAMPAGNE

The dryest of all champagne styles. No "dosage" (sweetener) is added to this wine. The concept for the name was originated by Mirassou Sales Company in the early 1960's. They wanted to call the style "Nude" in reference to its being nude of dosage, which, like clothing, is sometimes used to cover up faults. When the BATF refused national label approval for "Nude", the Mirassous applied for and received approval for "Au Naturel" which means nude in French. Au Naturel is now a major part of the Mirassou line of fine Méthode Champenoise sparkling wines and is also produced by several other wineries who have adopted the name.

AURORE (AURORA)

A French hybrid grape that was developed by the French Viticulturist Albert Seibel in the last century. Aurore produces medium body, dry to fruity wine. Also blended for Champagne.

auslese

The German word for "selection". Under German law, a wine labeled Auslese refers to selecting only ripe grape bunches and discarding unripe berries.

austere

Tasting term.

Extreme dryness sometimes coupled with high acidity. Stronger than assertive. Strong, imposing, harsh, unbalanced.

AUSTIN CELLARS

Founded 1983, Solvang, Santa Barbara County, California.

Storage: Oak and stainless steel. Cases per year: 15,000.

Label indicating non-estate vineyards: Sierra Madre, Bien Nacido and Santa Maria Hills Vineyards.

Wines regularly bottled: Vintage-dated Sauvignon Blanc, Blanc Botrytis, Pinot Noir, Gewurztraminer, White Riesling and Chardonnay.

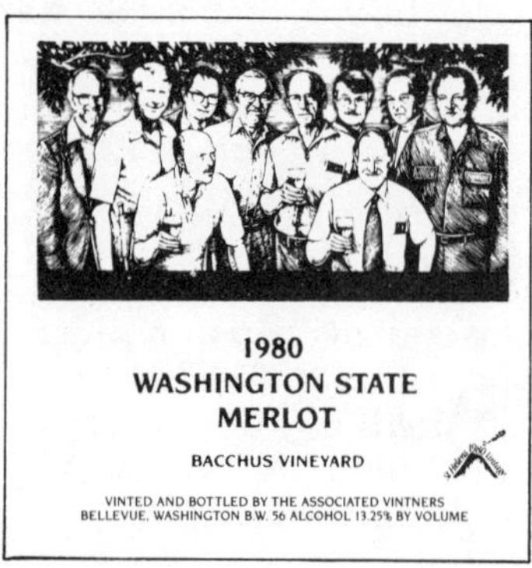

AV WINERY (COLUMBIA WINERY – 1984) p. 32

Founded 1962, Bellevue, King County, Washington.

Storage: Oak and stainless steel. Cases per year: 38,000.

History: In the late 1950's, Dr. Lloyd Woodburne, Dean, College of Arts and Sciences, (retired) University of Washington, was determined to make classic varietals. He was joined by five fellow professors from the university plus three businessmen and an engineer. Dr. Philip Church, meteorologist, University of Washington, compared the geographical and climatic conditions of Eastern Washington with the classic regions of Northern Europe and, based on the similarities between the mid-Yakima Valley and that of Alsace and Burgundy regions of France, the decision was made to plant the first vineyard in the heart of the valley. The first crush was in 1967 and the rest is wine history.

Estate vineyard: Yakima Valley.

Label indicating non-estate vineyard: Dionysus Vineyard, Bacchus Vineyard, Red Willow Vineyard, Jolona Vineyard.

Wines regularly bottled: Vintage-dated Dry Gewurztraminer, Dry Semillon, Dry White Riesling, Medium Dry Johannisberg Riesling, Chardonnay, Pinot Noir, Merlot, Cabernet Sauvignon, Gamay Beaujolais, Cascade Red. Also vintage-dated Proprietary Valley White.

B & W Vineyards

Wilcox, Arizona.

R.W. Welsh estate Cabernet Sauvignon, Chardonnay, Johannisberg Riesling, Zinfandel, Petite Sirah, French Colombard vineyard.

WILLIAM BACCALA WINERY

Founded 1981, Ukiah, Mendocino County, California.

Storage: Oak and stainless steel. Cases per year: 20,000.

Estate vineyard: Baccala Vineyards.

Labels indicating non-estate vineyard: Hillside Vineyard, B.J.L. Vineyards, C.S. Mendocino Vineyards Company.

Wines regularly bottled: Vintage-dated wines are Colombard, Chardonnay, Petite Sirah, Zinfandel and Cabernet Sauvignon.

BACCHANAL VINEYARDS

Founded 1978, Afton, Virginia.

Storage: Stainless steel.

Wines regularly bottled: Starting in 1984, varietal, estate bottled Zinfandel, Chardonnay, Cabernet Sauvignon, Chenin Blanc, Gewurztraminer, Semillon, Pinot Noir and Riesling.

BACCHUS

Roman god of wine. Dionysus (before Rome) was Greek god of wine.

Bacchus Vineyard

Franklin County, Washington.

Sauvignon Blanc, Merlot, Chenin Blanc, Cabernet Sauvignon and Chardonnay Columbia Valley Vineyard.

BACHMAN, PETER

Peter Bachman oversees all winemaking at Chateau Ste. Michelle. He came to the winery in 1981 as director of winery operations. He had a decade of winemaking experience earned at two of California's important wineries. In addition, Bachman tutored under wine industry notables, Dr. Richard G. Peterson and Andre Tchelistcheff. He supervises the efforts of three winemakers, each responsible for the winemaking at one of Chateau Ste. Michelle's wineries.

Bacigalupi Vineyards

Dry Creek, California.

Johannisberg Riesling vineyard.

Backus Vineyard

Napa County, California.

Joseph Phelps estate Cabernet Sauvignon vineyard.

BACO NOIR

Wine grape.

A French hybrid well grown in the Eastern United States. Produces dry, full bodied red wines of the Burgundy type.

bacterial

Spoiled wines in which it is possible to identify odd flavors, such as mousiness, butyric acid, ethyl acetate, or acetic acid may be generally described as bacterial.

STEPHEN BAHN WINERY

Founded 1978, Brogue, York County, Pennsylvania.

Storage: Oak and stainless steel. Cases per year: 800.

History: The Bahn family emigrated from Rheinpfalz, in 1731, to York County and were some of the initial settlers.

Labels indicating non-estate vineyard: Esdraelon Vineyard, Seven Valley Vineyard, Chanceford Vineyard, Marthur's.

Wines regularly bottled: Estate bottled, vintage-dated Pinot Noir, Chardonnay, Gewurztraminer, Johannisberg Riesling. Also vintage-dated Vidal Blanc, Seyval Blanc, Marechal Foch, DeChaunac Vin Gris, Niagara. Occasionally bottles fruit wines: Peach, Pear, Strawberry, Plum and Cherry.

Alexis Bailly Estate Vineyard
Dakota County, Minnesota.
Foch, Millot and Seyval Blanc vineyard.

ALEXIS BAILLY VINEYARD, INC.
Founded 1976, Hastings, Dakota County, Minnesota.
Storage: Oak. Cases per year: 2,000.
History: The winery is the first and only building ever constructed in the State of Minnesota for making of wine. Also the only winery that has ever made wine only from Minnesota grapes.
Estate vineyard: Alexis Bailly Vineyard.
Wines regularly bottled: Vintage-dated Marechal Foch, Millot, Seyval Blanc. Also Country Red, Country White. Proprietary vintage-dated wine Leon Millot, Foch.

Bainbridge Island Estate Vineyard
Kitsap County, Washington.
Muller Thurgau vineyard.

BAINBRIDGE ISLAND WINERY
Founded 1979, Bainbridge Island, Kitsap County, Washington.
Storage: Stainless steel. Cases per year: 1,250.
History: Produced the first commercial Madeleine Sylvaner in the United States.
Estate vineyard: Bainbridge Island Vineyard.
Labels indicating non-estate vineyard: Sagemoor Vineyards, Bartel Vineyards.
Wines regularly bottled: Estate bottled, vintage-dated Muller Thurgau. Also vintage-dated Madeleine Sylvaner, Gewurztraminer, White Riesling, Late Harvest Chardonnay. Also bottles proprietary Ferryboat White.

baked
A method of producing Sherry with a "flor flavor" by "baking" the wine through exposure to high temperature without employing the post-fermentation of "flor" yeast.

balance
Tasting term.
The harmonious relationship between sugar content and acidity. May also include the many odor and taste elements which are in such proportions as to produce a pleasant taste sensation.

Balcom & Moe Vineyard
Franklin County, Washington.
Sauvignon Blanc, Cabernet Sauvignon, Pinot Noir, Chardonnay and White Riesling vineyard.

Baldinelli Estate Vineyards
Amador County, California.
Zinfandel, Cabernet Sauvignon, Sauvignon Blanc vineyard.

BALDINELLI SHENANDOAH VALLEY VINEYARDS
Founded 1979, Plymouth, Amador County, California.
Storage: Oak and stainless steel. Cases per year: 5,000.

Estate vineyard: Baldinelli Vineyards.

Wines regularly bottled: Estate bottled, vintage-dated Zinfandel, White Zinfandel, Cabernet Sauvignon, Sauvignon Blanc. Also produce Red Table Wine.

Baldwin Estate Vineyard

Ulster County, New York.

Seyval, Ravat 51, Landot Noir vineyard.

BALDWIN VINEYARDS

Founded 1982, Pine Bush, Ulster County, New York.

Storage: Oak and stainless steel. Cases per year: 1,500.

History: Located on the Shawangunk Kill River, the winery is in a 200 year old stone house.

Estate vineyard: Baldwin Vineyard.

Wines regularly bottled: Estate bottled, vintage-dated Seyval, Ravat 51, Landot Noir. Also vintage-dated Chardonnay, Vidal Blanc, Riesling and vintage-dated generic Chablis.

BALLARD CANYON WINERY

Founded 1978, Solvang, Santa Barbara County, California.

Storage: Oak and stainless steel. Cases per year: 10,000.

Estate vineyard: Vintage Vineyards.

Label indicating non-estate vineyard: Tepusquet Vineyard.

Wines regularly bottled: Estate bottled, vintage-dated Johannisberg Riesling, Cabernet Sauvignon, Cabernet Sauvignon Blanc, "Rosalie" Cabernet Sauvignon Blanc, Johannisberg Riesling Reserve. Also produced are vintage-dated Chardonnay, Zinfandel, Fume Blanc, Muscat Canelli.

balling

The system for measuring soluble solids in grape juice, which are mostly sugars. Balling degrees indicate sugar content. Similar to Brix.

BALTIMORE, LORD

1622—instigated the growing of grapes for wine. The planting of European grape vines did not succeed.

Balto Vineyard

Sonoma County, California.

Cabernet Sauvignon vineyard.

Balverne Estate Vineyard

Sonoma County, California.

Chardonnay, Sauvignon Blanc, Gewurztraminer, Scheurebe, Cabernet Sauvignon, Zinfandel, Johannisberg Riesling vineyard.

BALVERNE WINERY AND VINEYARDS

Founded 1973, Windsor, Sonoma County, California.

Storage: Oak and stainless steel. Cases per year: 25,000.

History: The property of Balverne was part of a Spanish Land Grant made to the Mariano Vallejo Family. The stone waterways and planting beds surrounding the winery were built in 1880.

Estate vineyard: Balverne Estate Vineyard, Deerfield Vineyard, Stonecrest Vineyard, Pepperwood Vineyard, Oak Creek Vineyard, Annaberg Vineyard, Laurel Vineyard, Quartz Ridge Vineyard.

Wines regularly bottled: Estate bottled, vintage-dated Chardonnay, Sauvignon Blanc, Dry Gewurztraminer, Scheurebe, Healdsburger (a blend of Gewurztraminer, Johannisberg Riesling and Scheurebe), Cabernet Sauvignon, Zinfandel.

Bandiera Estate Vineyard
Sonoma County, California.

BANDIERA WINERY/CALIFORNIA WINE COMPANY

Founded 1937, Cloverdale, Sonoma County, California.

Storage: Oak and stainless steel. Cases per year: 75,000.

History: Emil and Ludina founded the winery and carried on traditional winemaking techniques until the founder's grandson changed the focus to varietals.

Estate vineyard: Bandiera Vineyards (Chiles Valley, Potter Valley, Dry Creek, Los Carneros), Sage Creek Vineyard.

Wines regularly bottled: Vintage-dated Cabernet Sauvignon, Pinot Noir, Zinfandel, Chardonnay, Sauvignon Blanc, Johannisberg Riesling. Under John Merritt label are vintage-dated Chardonnay, Sauvignon Blanc, Cabernet Sauvignon.

Second label: John B. Merritt.

BARBERA

Wine grape.

A grape that produces a red, deep-colored, full bodied, dry and tannic wine that ages well and softens with age. It is believed that Barbera was brought to California by Italian immigrants in the late 19th Century from the Piedmont district of Northern Italy.

Barboursville Estate Vineyard
Orange County, Virginia.
Cabernet Sauvignon, Chardonnay, Riesling, Merlot, Gewurztraminer vineyard.

BARBOURSVILLE WINERY, INC.

Founded 1976, Barboursville, Orange County, Virginia.

Storage: Oak and stainless steel.

History: The first Italian owned (Zonin) and operated commercial premium vineyard and winery in U.S. Winery is located on the former Barboursville Plantation, originally owned by statesman James Barbour, who served as Governor from 1812 to 1814. The site is a registered Virginia Historic Landmark; it includes the picturesque ruins of Barbours Mansion which was designed by Thomas Jefferson and burned Christmas Day, 1884.

Estate vineyard: Barboursville Vineyard.

Wines regularly bottled: Estate bottled Cabernet Sauvignon, Chardonnay, Riesling, Merlot, Gewurztraminer, Pinot Noir. Also proprietary estate bottled Rosé Barboursville.

BARCELONA (See Bardenheier's Wine Cellars.)

Port, Sherry, Muscatel, White Port.

Bardenheier Vineyards
Missouri.
French Hybrid vineyard.

BARDENHEIER'S WINE CELLARS
Founded 1873, St. Louis, St. Louis County, Missouri.

Storage: Stainless steel. Cases per year: 250,000.

History: Established in 1873 by John E. Bardenheier, who came to the United States from Germany in 1865. The winery has been in continuous operation, with the exception of the "Prohibition years", since its founding. One of the oldest family wineries in America still operated by the founding family.

Estate vineyard: Bardenheier's Vineyards.

Wines regularly bottled: Concord, Sweet Catawba, Pink Catawba. Also generics are Chablis Blanc, Pink Chablis, Sparkling Burgundy, Cold Duck, Burgundy, Chianti, Rhine, Sauterne, Vin Rosé, Pink Champagne. Proprietary wines bottled are: Vino Rosso, Vino Bianco, Strassenfest, Surrey Tawny Port, Surrey Cream Sherry, Surrey Dry Cocktail Sherry. Also bottle Spumante, Old Fashion Grape. Fruit wines are Cherry, Blackberry, Loganberry, Red Currant, 19% Apple, Hard Cider. Also bottle Sweet Vermouth, Dry Vermouth, Sangria. Dessert wines include Port, Sherry, Muscatel, White Port, Tokay, Cream Sherry, Almond Bavarian Cream.

Second label: Chateau Thayer, Grey Summit, Del Rio, Barcelona, Caladina, Frisco, 905, St. Johns, Brown Derby, N214, Delta Queen, Grapeland.

BARENGO/LOST HILLS VINEYARDS, dba Verdugo Vineyards
Founded 1934, Acampo, San Joaquin County, California.

Storage: Oak and stainless steel. Cases per year: 350,000.

History: In 1948, the first plantings of Ruby Cabernet were planted at Lost Hills. The grape was introduced by Professor Amerine, of U.C. Davis, and Dino Barengo.

Estate vineyard: Barengo Vineyard.

Labels indicating non-estate vineyard: Lost Hills Vineyard.

Wines regularly bottled: Under Los Hills label are French Colombard, Chenin Blanc, Grey Riesling, Cabernet Sauvignon, Zinfandel, White Zinfandel, Barbera Blanc. Under Lost Hills, generics are Chablis, Burgundy, Vin Rosé. Also produces vintage-dated Light Chenin Blanc, Light Zinfandel, Creme Marsala. Occasionally bottled wines are Dudenhoefer May and Spiced Wine, Barengo Holiday and Harvest Wine, 20 Year Old Anoushe Sherry.

Second label: Barengo Vintners Reserve, Dudenhoefer, Kossof.

BARENGO VINTNERS (See Barengo/ Lost Hills Vineyards.)

BARGETTO WINERY
Founded 1933, Soquel, Santa Cruz County, California.

Storage: Oak, stainless steel. Cases per year: 50,000.

History: Santa Cruz County's oldest operating winery. This family-owned winery recently celebrated its 50th Anniversary.

Labels indicating non-estate vineyards: Tepusquet Vineyard, Farview Farms Vineyard, St. Regis Vineyard.

Wines regularly bottled: Vintage-dated Dry and Medium Dry Johannisberg Riesling, Late Harvest Riesling, Chenin Blanc, Sauvignon Blanc, Chianti, Chardonnay, Zinfandel. 25% of production are fruit wines: Raspberry, Olallieberry, Apricot, Pomegranate, Brambleberry.

BARNES WINES, LTD.

Founded 1873, St. Catherines, Ontario, Canada.

Storage: Oak, stainless steel.

History: Barnes Wines, Limited is Canada's oldest winery, founded by George Barnes in 1873. The winery was originally known as the Ontario Grape Growers and Wine Manufacturing Company and was owned by the Barnes family until 1974. Barnes Wines began operations on the banks of the Old Welland Canal, its present day location

Wines regularly bottled: Proprietary Bon Appetit, Weinfest, Beauvois, Grand Celebration White Champagne, Grand Celebration Pink Champagne. Generics include Ontario Country White, Ontario Country Red, Sangria, Spumante Bianco.

Second label: Springwood, Heritage Estates.

barrel fermented

Wine fermented in barrel instead of in stainless steel or large wood tanks.

BARTHOLOMEW, FRANK

In 1942, Frank Bartholomew and his wife, Antonia, bought the abandoned Buena Vista Winery in Sonoma. They also purchased the adjacent vineyards of varietal grapes, originally planted by the Hungarian, Agoston Haraszthy in 1862. Both the vineyards and winery buildings required extensive restoration, but by 1949 the vineyards were in full production. By 1968, the winery had become so successful that Bartholomew, who worked full-time as the president of United Press International, was forced to sell the winery. He retained the vineyards. Then, in 1973, Bartholomew established Hacienda Wine Cellars and in 1977, sold controlling interest in the enterprise. Today, Bartholomew is retired and maintains an active interest in the Buena Vista vineyards.

BARTLETT MAINE ESTATE

Founded 1983, Gouldsboro, Hancock County, Maine.

Storage: Oak and stainless steel. Cases per year: 300.

History: Wines are made from domestic and wild fruit grown in the State of Maine.

Wines regularly bottled: Fruit and berry wines: Dry Blueberry, Dry Apple, Sparkling Apple, Strawberry, Mead (Sweet & Dry).

Bates Ranch

Santa Cruz Mountains, California.

Cabernet Sauvignon vineyard.

John B. Bates Vineyard

Santa Clara County, California.

Thomas Kruse estate Cabernet Sauvignon.

Barrel fermentation and aging cellar

BATF

U.S. Government Bureau of Alcohol, Tobacco and Firearms. A bureau of the Treasury Department. Regulates federal laws effecting wine labeling, taxes and winemaking.

Batto Ranch

Sonoma County, California.

Cabernet Sauvignon Sonoma Valley vineyard.

baume

The measure of the sugar content of the grape. One Baume is equal to approximately 1.75% of sugar content.

BAXTER, PHILLIP L.

Phillip L. Baxter began his winemaking career, following graduation from California State University, Fresno, in 1969, with a degree in enology. He worked first at the Charles Krug Winery, then with Lee Stewart at Souverain Cellars. He was winemaker at Souverain of Rutherford before becoming winemaker and partner of Rutherford Hill Winery at the same facility in 1976. Baxter produces traditionally styled wines, yet is interested in experimental winemaking techniques, such as rotating fermentation tanks. He is active in management of the winery and has developed a team approach to winemaking to assist in the production of quality wines. Baxter is involved with many wine industry associations, including the American Society of Enologists, the Napa Valley Wine Technological Group, and the Wine Institute.

BAY CELLARS

Founded 1982, Emeryville, Alameda County, California.

Storage: Oak and stainless steel. Cases per year: 1,500.

Labels indicating non-estate vineyards: Buena Vista Vineyards, Hazen Vineyards, Tepusquet Vineyards.

Wines regularly bottled: Vintage-dated Pinot Noir, Chardonnay, Merlot, Cabernet.

Bay View Vineyards, Carneros

Napa County, California.

Laird estate Chardonnay and Pinot Noir vineyard.

B & B ROSSER WINERY

Founded 1979, Athens, Oconee County, Georgia.

Storage: Oak and stainless steel.

Estate vineyard: Apalachee Vineyard.

Wines regularly bottled: Cabernet Franc, Carmine, White Riesling, Chardonnay.

Second label: Caledonia.

B & B Vineyard

Albuquerque, New Mexico.

La Chiripada estate French Hybrids and White Riesling vineyard.

bead

A bubble forming in or on a beverage; used to mean CO_2 bubbles in general or sometimes the ring of bubbles around the edge of the liquid.

Beatty Ranch

Napa Valley, California.

Zinfandel vineyard.

BEAULIEU VINEYARD

Founded 1900, Rutherford, Napa County, California.

Storage: Oak and stainless steel. Cases per year: 300,000-350,000

History: Founded by Georges de Latour, a young Frenchman who, after buying the estate, went back to France to buy the cuttings for his new vineyard. Latour survived Prohibition by selling altar wines. His wife, Fernande, named the winery using the French words "beautiful place". In 1937, his journey to France to find a new winemaker resulted in finding a young man, Andre Tchelistcheff, who not only changed the destiny of Beaulieu but also had a great impact on winemaking in the United States. Georges de Latour died in 1940, but the traditions established by Latour are still carried on and Andre Tchelistcheff is the winemaking consultant.

Wines regularly bottled: Estate bottled Cabernet Sauvignon, Pinot Noir, Gamay Beaujolais, Pinot Chardonnay, Johannisberg Riesling, Muscat Blanc, Sauvignon Blanc, Muscat de Frontignan; Generic Wines are Chablis, Burgundy, Brut Champagne, Champagne de Chardonnay.

BEAU VAL WINES

Founded 1979, Plymouth, Amador County, California.

Storage: Oak. Cases per year: 2,000.

Estate vineyard: Francis Vineyard.

Wines regularly bottled: Vintage-dated Zinfandel, Zinfandel Blanc.

Beauregard Ranch

Santa Cruz Mountains, California.

Cabernet Sauvignon, Chardonnay vineyard.

Beckstoffer "Los Amigos" Vineyards

Napa County, California.

Chardonnay, Cabernet Sauvignon, Chenin Blanc, Muscat de Frontignan, Pinot Noir, Sauvignon Blanc and White Riesling Carneros vineyard.

Beckwith Ranch Vineyard

Paso Robles, California.

Zinfandel, Grenache and Carignane vineyard.

beerenauslese

The German word for "berry selection". A special Auslese made from specially selected grapes that are picked berry by berry that have been affected by Botrytis. A rare and rich wine when produced.

BEL ARBRES

Founded 1975, Hopland, Mendocino County, California.

Storage: Oak and stainless steel. Cases per year: 15,000-20,000.

History: Bel Arbres, originally a second label for Fetzer Vineyards of Redwood Valley, is now a separate winery in Hopland.

Wines regularly bottled: Vintage-dated Chardonnay, Chenin Blanc, Sauvignon Blanc, White Zinfandel, Cabernet Sauvignon, Merlot, Zinfandel. Also produce Vin Rouge, Blanc de Blanc.

BEL CANTO (See Robert Stemmler Winery.)

BELL CANYON (See Burgess Cellars.)

Bell Hill Vineyard

Lake County, California.

Cabernet Sauvignon vineyard.

Bell Vineyards/Mira Monte Vineyards

Temecula, California.

Chenin Blanc, Sauvignon Blanc and Cabernet Sauvignon vineyard.

BELLA NAPOLI WINERY

Founded 1934, Manteca, San Joaquin County, California.

Storage: Oak and stainless steel.

Estate vineyard: Hat Family Vineyards.

Wines regularly bottled: Estate bottled Zinfandel, Chenin Blanc, French Colombard, Grenache, Pinot Chardonnay.

Bella Oaks Vineyards

Napa Valley, California.

Cabernet Sauvignon vineyard.

Belle Terre Vineyards

Alexander Valley, California.

Chardonnay, Johannisberg Riesling and Gewurztraminer vineyards.

BELLEROSE VINEYARD

Founded 1979, Healdsburg, Sonoma County, California.

Cases per year: 5,000.

History: Grapes have been grown and wine made on the property since 1887, when Captain Everett Wise had stones hauled by wagon from nearby Mill Creek to build the original winery. That building burned in the late thirties; the same stone walls form the present cellars, owned by Charles and Nancy Richard.

Estate vineyard: Bellerose Vineyard.

Wines regularly bottled: Vintage-dated Cabernet Sauvignon, Rosé du Val, Rouge du Val, Cabernet Sauvignon "Cuvee" Bellerose.

Beltane Ranch

Sonoma Valley, California.

Chardonnay vineyard.

Belvedere Estate Vineyard

Sonoma County, California.

Chardonnay vineyard.

BELVEDERE WINE COMPANY

Founded 1979, Healdsburg, Sonoma County, California.

Storage: Oak and stainless steel. Cases per year: 20,000.

History: Peter Friedman, one of the partners of Belvedere, was a co-founder of Sonoma Vineyards. Instead of a "Belvedere" brand with a vineyard designation, a single vineyard brand label has been created for each vineyard's wine. Their "Wine Discovery" label is Belvedere's negociant label.

Estate vineyard: Belvedere Vineyard, Wine Creek Vineyard.

Wines regularly bottled: Vintage-dated Bacigalupi Vineyards Chardonnay and Pinot Noir, Robert Young Vineyards Cabernet Sauvignon and Merlot, Winery Lake Vineyards Chardonnay and Pinot Noir, York Creek Vineyards Cabernet Sauvignon.

Second label: Wine Discovery.

BENMARL WINE COMPANY, LIMITED

Founded 1972, Marlboro, Ulster County, New York.

Storage: Oak and stainless steel. Cases per year: 10,000.

History: In the town of Marlboro the records indicate that a vineyard and winery existed on the property that noted illustrator Mark Miller purchased and named Benmarl. Ben means hill in Gaelic and Marl for its slaty soil. In 1867, the Dutchess grape was developed at the original vineyard.

Estate vineyards: Benmarl Estate Vineyard, Hampton Estate Vineyard.

Labels indicating non-estate vineyards: Mt. Zion Estate Vineyard.

Wines regularly bottled: Proprietary vintage-dated generics are Hudson Region White, Hudson Region Red, Marlboro Village Red, Marlboro Village White, Estate Reserve White, Estate Reserve Red, Cuvee Du Vigneron Red, Cuvee Du Vigneron White. Also produce Benmarl Estate vintage-dated Seyval Blanc and Benmarl Hudson Region Chancellor.

Second label: Cuvee Du Vigneron.

bentonite

Pure clay used to clarify wine.

Beresini Vineyard

St. Helena, California.

Chenin Blanc vineyard.

Bergstrom Vineyards

Paso Robles, California.

Ranchita Oaks estate Zinfandel, Cabernet Sauvignon, Petite Sirah, Chardonnay and Zinfandel vineyard.

BERINGER VINEYARDS

Founded 1876, St. Helena, Napa, Sonoma County, California.

Storage: Oak and stainless steel.

History: Jacob Beringer came to the Napa Valley, in 1869, with a determination to grow grapes and make fine wine. He arrived from New York via the newly completed transcontinental railway. Because of his extensive winemaking background in both Germany and France, Jacob found immediate employment in St. Helena as a cellarmaster for Charles Krug, at his pioneer winery.

In 1876, Beringer acquired 97 acres just north of the city limits of St. Helena. The winery, built in 1877 of local stone, abutting the carved-out hillside, is designed to operate by gravity flow. A road leading across the hillside to the third floor brought the grapes to be emptied into crushers and presses.

Fermenting tanks occupied the second floor, and aging casks the ground level space into which the two main tunnels opened.

While providing money and counsel, Jacob's brother, Frederick, did not leave New York until 1833, when he sold his business and moved to St. Helena, and built the famous Rhinehouse, a home that was very similar to the family home in Mainz.

The Beringer family continued to live on the winery property in the white farm house and Rhinehouse until the late 1960's.

Estate vineyards: Yountville Vineyard, Salvador Vineyard, Big Ranch Road Vineyard, Knights Valley Vineyard, Gasser Vineyard, Lemmon Ranch Vineyard, St. Helena Home Vineyard, Marolf Vineyard, De Carle Vineyard, Gamble Ranch Vineyard.

Wines regularly bottled: Vintage-dated Cabernet Sauvignon (Estate and Private Reserve), Pinot Noir, Zinfandel, Chardonnay (Estate and Private Reserve), Johannisberg Riesling, Fume Blanc, Chenin Blanc, Gamay Rosé, Gamay Beaujolais, Gewurztraminer, Malvasia-Amabile, Dry French Colombard. Dessert wines are Malvasia Bianca and Cabernet Sauvignon Port. Semi-generics are Chablis and Burgundy.

Second label: Los Hermanos.

BERKELEY WINE CELLARS

Founded 1975, Berkeley, Alameda County, California.

Storage: Oak. Cases per year: 4,200.

History: Winery is a division of a very successful wine and beer supply firm.

Labels indicating non-estate vineyard: Kelley Creek Vineyard.

Wines regularly bottled: Vintage-dated Zinfandel, Zinfandel Port, Chardonnay.

Second labels: Wine and the People, Berkeley Wine Company.

BERKELEY WINE COMPANY (See Berkeley Wine Cellars.)

BERNARDO WINERY, INC.

Founded 1889, San Diego, San Diego County, California.

Wines regularly bottled: Table, Fruit, Dessert Wines.

Bernstein Vineyard
Napa County, California.
Mt. Veeder estate Cabernet Sauvignon vineyard.

Berrien
Leelanau County, Michigan.
Seyval Blanc vineyard.

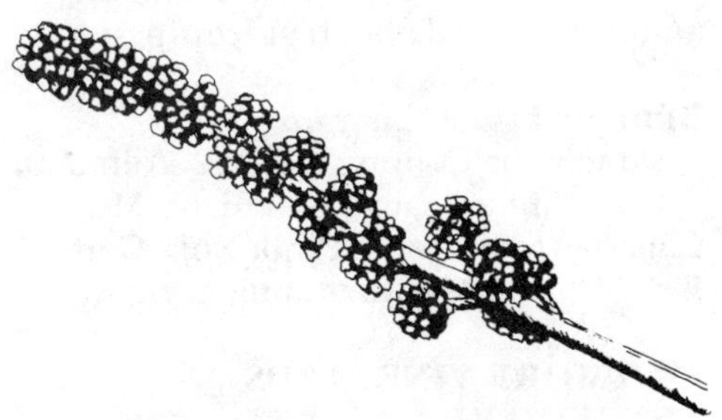

A cluster of young Chardonnay berries

berry
A fruit having a pulpy pericarp (skin or wall of the fruit) in which seeds are embedded as in the grape, the gooseberry, currant, the tomato.

BERRYWINE (See Berrywine Plantations Wine Cellars.)
Rosé, White, Mountain Red, Mountain Pink, Mountain White, Blackberry, Blackcherry, Damson Plum, Dandelion, Elderberry, Honey, Nectarine, Peach, Pear, Raspberry, Red Plum, Strawberry.

BERRYWINE PLANTATIONS WINE CELLARS
Founded 1976, Mt. Airy, Frederick County, Maryland.
Storage: Oak.
History: The plantation is operated by the Aellen family whose heritage goes back to the Rhine Valley on one side of the family and traditional Italian viticulture on the other.

Labels indicating non-estate vineyard: Linganore Vineyard.
Wines regularly bottled: Dry and Semi-sweet Grape and Fruit Wines.
Second label: Berrywine, Plantation, Linganore.

Besson Family Vineyard
Santa Clara County, California.
Thomas Kruse estate Zinfandel and Chardonnay vineyards.

beta
Wine grape.
An American hybrid that produces medium-bodied, tart, tannic, fruity red wine. Winter hardiest of all commercial grape varieties.

Bethel Heights Vineyard
Salem, Oregon.
Riesling, Pinot Noir, Gewurztraminer, Chenin Blanc and Chardonnay vineyards.

BIANCHI WINERY
Founded 1974, Kerman, Fresno County, California.
Cases per year: 243,113.
Estate vineyard: Bianchi Vineyards.
Wines regularly bottled: Chenin Blanc, French Colombard, Cabernet Sauvignon, Zinfandel, Grenache Rosé, generics are Chablis, Vin Rosé, Burgundy. Also produce Cosa Nostra, Casa Bianca.
Second label: Villa Sorrento.

Bias Estate Vineyards
Franklin County, Missouri.
DeChaunac, Foch, Seyval, Catawba vineyard.

BIAS VINEYARDS AND WINERY, INC.
Founded 1968, Berger, Franklin County, Missouri.
Storage: Stainless steel. Cases per year: 500.
Estate vineyard: Bias Vineyards.
Wines regularly bottled: Estate bottled DeChaunac, Foch, Seyval, Pink Catawba, Rosé Catawba. Also bottles mead.

Bible, Wine

"To eat, to drink and be merry."
Ecclesiastes 8:15

"Eat thy bread with joy and drink thy wine with a merry heart."
Ecclesiastes 9:7

"A feast is made for laughter and wine maketh merry."
Ecclesiastes 10:19

"Forsake not an old friend, for the new is not comparable to him. A new friend is as new wine: when it is old, thou shalt drink it with pleasure."
Ecclesiastes 9:10

"Wine was created from the beginning to make men joyful, and not to make men drunk. Wine drunk with moderation is the joy of the soul and the heart."
Ecclesiastes 31:35-36

"Wine which cheereth God and man."
Judges 9:13

"Give ... wine unto those that be of heavy heart."
Proverbs 31:6

"Wine maketh glad the heart of man."
Psalms 104:15

"The best wine ... that goeth down sweetly causing the lips of those that are asleep to speak."
Song of Solomon 7:9

"Drink no longer water but use a little wine for thy stomach's sake."
I Timothy 5:23

Bien Nacido Vineyards

Santa Barbara County, California.

Pinot Noir, Gewurztraminer, Merlot, Gamay Beaujolais, Cabernet Sauvignon, Chardonnay and Johannisberg Riesling vineyard.

Big Ranch Vineyard

Napa County, California.

Beringer estate Chardonnay, Pinot Noir and Grey Riesling vineyard.

Big Ranch Vineyard

Napa County, California.

Monticello estate Chardonnay, Sauvignon Blanc, Gewurztraminer and Semillon vineyard.

big wine

Tasting term to express ample body and fullness and apparent fruitiness.

Biltmore Estate Vineyard

Buncombe County, North Carolina.

Chardonnay, Sauvignon Blanc, Merlot, Cabernet Sauvignon, Pinot Noir, Gamay Beaujolais and Gewurztraminer vineyard.

BILTMORE VINEYARDS AND WINERY

Founded 1970, Ashville, Buncombe County, North Carolina.

Storage: Oak and stainless steel. Cases per year: 5,000.

History: Located on the grounds of Biltmore House and Gardens, a privately owned National Historic Landmark in Ashville, North Carolina. Biltmore is the former residence of George W. Vanderbilt and was completed in 1895. Biltmore House is open to the public. Biltmore

vineyards was begun in 1970, and the winery was bonded in 1977.

Estate vineyard: Biltmore Vineyard.

Wines regularly bottled: Estate bottled, vintage-dated proprietary wines are Biltmore Red, Biltmore White, Biltmore Rosé, Biltmore Estate Red, Biltmore Estate White, Biltmore Estate Rosé, Biltmore Estate Champagne.

Champagne bottles stacked in bins for bottle aging

binning

Bottle-aging of newly bottled wines, usually in bins, before release for sale. Newly disgorged Champagne is often stored in bins prior to labelling.

BIRKETT

Alexander Valley, California.

Fieldstone estate Johannisberg Riesling, Gewurztraminer and Sauvignon Blanc vineyard.

Birkmyer Vineyards

Napa Valley, California.

White Riesling vineyard.

BISCEGLIA BROTHERS WINE COMPANY

Founded 1880, Madera, Madera County, California.

Wines regularly bottled: Chenin Blanc, Cabernet Sauvignon. Semi-generics are Chablis, Rhine, Vin Rosé, Burgundy.

B & J Ranch

Napa County, California.

Shafer estate Cabernet Sauvignon, Chardonnay and Zinfandel vineyards.

BJELLAND FARMS (See Bjelland Vineyards.)

BJELLAND VINEYARDS

Founded 1969, Roseburg, Douglas County, Oregon.

Storage: Oak and stainless steel. Cases per year: 2,500.

History: Founder of the Oregon Wine Growers Association and the Annual Oregon Wine Festival, Paul Bjelland is one of the Oregon wine industry's pioneers. His was the second vinifera winery in the state.

Estate vineyard: Bjelland Vineyards.

Wines regularly bottled: Vintage-dated Johannisberg Riesling, Semillon, Sauvignon Blanc, Pinot Noir, Cabernet Sauvignon, Chardonnay, Gewurztraminer, Zinfandel. Occasionally bottle Brambleberry.

Second label: Bjelland Farms.

BJL Vineyard

Mendocino, California.

Kalin estate Botrytis Riesling and Semillon vineyard.

BLACK MALVOISIE

Wine grape.

A black grape used in California for blending in dessert wines.

Black Mountain Vineyard

Alexander Valley, California. Napa Cellars.

Chardonnay, Sauvignon Blanc, Zinfandel and Cabernet Sauvignon vineyard.

BLACK MUSCAT

A dessert wine made from one of the Muscat grapes. Sweet and high in alcohol content (12%-20%). A Ruby Port type taste, medium to deep red, rich fruity and full-bodied.

BLACK PEARL

Wine grape.

Casper Schraidt, a carpenter and cabinet maker by trade, left his homeland in Germany and settled at Put-in-Bay Island in Lake Erie in the early 1850's. He continued his trade during the Civil War years, but also became engaged in the pioneer efforts of the grape and wine industry to develop new and better grape varieties. Casper Schraidt developed some new kinds from seed, one of which had a very vigorous vine and bore small clusters of jet black berries, which, when vinified, made a wine of intense red color. He named the grape Black Pearl.

The Black Pearl is a truly American grape. Its botanical characteristics indicate parentage from the American wild grape Vitis Riparia. Probably from seedling of Clinton or Taylor.

The wines of Black Pearl were highly regarded, but it was not generally grown outside the Sandusky-Lake Erie Islands district. During the Prohibition Era its culture waned until but a few vines were in existence. The present planting at the Steuk Winery, in Sandusky, Ohio, has been slowly established from a single vine on Put-in-Bay Island in Lake Erie.

BLANC DE BLANCS

White wine made from white grapes. Term often used for Champagne description.

BLANC DE NOIR

White wine made from black grapes, by fermenting 'must' without the presence of skins. Usually imparts a slight pink color. Blanc de Noir wine types have increased in popularity. The better Blanc de Noirs use the noble black grapes (Cabernet Sauvignon, Zinfandel and Pinot Noir) or a blend of these grapes. Though the juices are separated differently at various wineries, the common denominator for all is speed.

BLANC VINEYARDS

Founded 1983, Redwood Valley, Mendocino County, California.

Storage: Oak and stainless steel. Cases per year: 1,500-2,000.

Estate vineyard: Blanc's Redwood Valley Vineyard.

Wine regularly bottled: Cabernet Sauvignon.

blending

Each wine may be deficient in one or more desirable qualities but due to the complex interaction of each wine with the other a synergistic response occurs. The skill of combining two or more wines to achieve a wine of high standards and quality. Called "cuvee" in French.

Blenheim Estate Vineyards

Albemarle County, Virginia.

Chardonnay vineyard.

BLENHEIM WINE CELLARS, LTD.

Founded 1979, Charlottesville, Albemarle County, Virginia.

Storage: Oak and stainless steel. Cases per year: 1,000.

Estate vineyard: Blenheim Vineyards.

Wines regularly bottled: Estate bottled, vintage-dated Chardonnay, Late Harvest Chardonnay.

BLOSSER, WILLIAM

William Blosser has been intimately involved in the development of the Oregon wine industry. He and his wife, Susan Sokol Blosser, started Sokol Blosser Winery, in 1971, while he was working as a planning department manager in Portland, Oregon. He served as winery manager until 1977, when he became the President. Today, he works full time at the winery. In 1971, Blosser helped found the organization that has become the Oregon Winegrowers Association. He served separate terms as President, Vice-President, Secretary and Treasurer of the Association between 1971 and 1976, and remains an active member. He is, also, a member of United Oregon Horticulture, and served as its Chairman in 1982-83.

Blue Mountain Vineyard

Monterey County, California.

Adjacent to Chalone Vineyard in the Pinnacles area, 1,700 feet above Salinas Valley. Vineyard of Chardonnay, Sauvignon Blanc, Pinot Noir and Cabernet Sauvignon.

BLUE RIDGE WINERY

Founded 1982, Carlisle, Cumberland County, Pennsylvania.

Storage: Oak. Cases per year: 1,000.

History: Located at the foot of the Blue Ridge Mountain Range. Winery is in 150 year old converted barn.

Estate vineyard: Verdekal Vineyard, Blue Ridge Winery Vineyard.

Wines regularly bottled: Estate bottled Aurora White, Aurora White Semi-Dry, Foch Red, Foch Rosé, Baco Red, Baco Rosé, DeChaunac Red, DeChaunac Rosé, Chelois Red, Chelois Rosé.

BLUFF POINT WINERY

Founded 1980, Penn Yan, Yates County, New York.

Wines regularly bottled: Delaware, Niagara, Baco, Cascade, Aurora, White Catawba.

BODEGAS FERRINO S.A.

Founded 1860, Coahuila, Mexico.

Cases per year: 35,000.

Labels indicating non-estate vineyards: Agricola Ferrino, Rancho Ferrino, La Laguna.

Wines regularly bottled: Vintage-dated proprietary Vino Tinto "Sangre de Cristo", also proprietary Vino Tinto Seco "Ferrino", Vino Blanco "Ferrino", Vino Generoso de Uva Moscatel "Ferrino".

BODEGAS DE SANTO TOMÁS, S.A.

Founded 1880, Mexico.

Cases per year: 500,000.

History: In the early 1880's, Francisco Andonegui, an Italian goldminer, and Miguel Omar, a Spaniard who had settled in the Santo Tomás Valley, became partners and began to reclaim the abandoned mission vineyards and winery. Virtually all of the winemaking equipment had disappeared and had to be replaced. The stone aqueducts and elaborate system of cisterns and dams painstakingly constructed by the mission Fathers, had crumbled and new methods to bring the spring water to the fields had to be devised. On a plateau in the north end of the fertile valley, the owners dug new cellars for aging and constructed grape-crushing works and sturdy wooden vats for fermentation. In 1888, Bodegas de Santo Tomás was established as a commercial winery.

Wine was hauled by horse wagon over the trail to the port city of Ensenada for the market and for ships in Todos Santos Bay. The trip to Ensenada, only 34 miles away, often took several days. Wine production from Bodegas de Santo Tomás quickly reached 800,000 litres per year. The winery continued with moderate success until the late 1920's. Andonegui by then the sole owner, was dying. He sold the assets of the winery to the Governor of Baja California, General Abelardo Lujan Rodriguez, who owned it for forty years. Rodriguez, the revolutionary general, who later became president of Mexico, was wealthy in his own right.

General Rodriguez moved the winery from the remote Santo Tomás Valley to the city of Ensenada. He also planted new vineyards in other fertile valleys, both inland and south of Santo Tomás, introducing new varieties of grapes for winemaking.

General Rodriguez died in 1964 and Bodegas de Santo Tomás was sold to the famous Elias Pando wine-importing firm of Mexico City.

Estate vineyard: Rancho Los Dolores, Rancho Santa Isabel.

Wines regularly bottled: Estate bottled, vintage-dated Chenin Blanc, Cabernet Sauvignon, Barbera, Blanco Pinot. Proprietary wines are Vina San Emilion, Blanco Espumoso Calvine. Also bottle Vino Tinto, Vino Blanco, Seco, Vino Rosado. Dessert wines are Port, Sherry, Moscatel.

body

Tasting term.

The non-sugar solids of a wine are referred to as body. Body is not primarily detected by taste but rather by the receptors sensitive to viscosity and possibly salt concentration. Since both alcohol and glycerol influence the viscosity, these substances complicate the estimation of body. It is the "feel" of the wine as it is swished about the mouth. Dry table wines are described as follows:

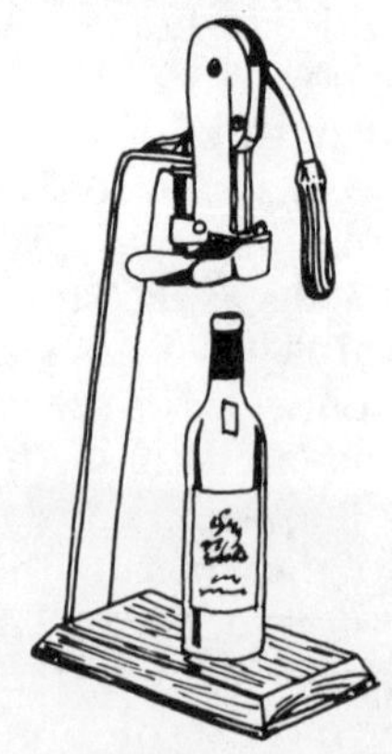

Commercial cork extractor

Light Body, Thin: In general, wines with less than 2.0 grams per 100 ml. extract are described as thin. Unbalanced or unharmonic are related terms.

Medium Body: This term refers to wines with extract content in the range of 2.1 to 3.0 grams per 100 ml.

Heavy Body, Rich, Robust, Full, Round: Terms such as these are used to describe wines containing more than 3 grams per 100 ml.

BOEGER, GREG

Greg Boeger's grandfather was Swiss-Italian winemaker Anton Nichelini, founder of the Napa Valley winery of the same name (which is still in operation today). As a child, Boeger worked at the family winery and vineyards, as well as at the farms and ranches of other relatives. This early experience influenced his choice of a winemaking career and, in 1968, he was graduated from the University of California, Davis, with a master's degree in agricultural economics, with a minor in viticulture. He was employed by the State of California as an agricultural statistician for two years as he searched for a location to begin his vineyard and winery. Boeger purchased land in El Dorado County. He planted and cared for his vineyards alone, with the exception of hired help during the peak seasons. The Boeger Winery in Placerville has grown since those early days, and Boeger now concentrates his efforts on winemaking. His wife, Susan, works in administration of the winery.

Boeger Winery Estate Vineyard

El Dorado County, California.

Cabernet Sauvignon and Merlot vineyard.

BOEGER WINERY, INC.

Founded 1972, Placerville, El Dorado County, California.

Storage: Oak and stainless steel. Cases per year: 10,000.

History: The Boeger Winery was the site of a winery and distillery founded in 1860. The stone wine cellar still exists and is used as a tasting room. The old winery and family home is on the National Register of Historic Buildings.

Estate vineyard: Boeger Winery Vineyard, Peck Vineyard.

Labels indicating non-estate vineyard: Walker Vineyard, Mirande Vineyard, Granite Hill Vineyard.

Wines regularly bottled: Estate bottled, vintage-dated Cabernet Sauvignon, Merlot. Also vintage-dated Zinfandel, Chenin Blanc, Sauvignon Blanc, White Zinfandel, Johannisberg Riesling, Chardonnay. Also vintage-dated Sierra Blanc, Hangtown Gold and non-vintage Hangtown Red. Occasionally bottles White Cabernet and Zinfandel Rosé.

Boepple Vineyards

Lubbock County, Texas.

Cabernet Sauvignon vineyard.

Bogle Estate Vineyards

Yolo County, California.

Chenin Blanc and Petite Sirah vineyard.

BOGLE VINEYARDS WINERY

Founded 1979, Clarksburg, Yolo County, California.

Storage: Oak and stainless steel. Cases per year: 4,000-6,000.

History: Grapes were first planted in the Clarksburg area in 1969 by Warren Bogle.

Estate vineyard: Bogle Vineyard.

Wines regularly bottled: Estate bottled, vintage-dated Chenin Blanc, Sarah's Blush (Rosé of Petite Sirah), and non-vintage Petite Sirah. Occasionally bottle Dry Chenin Blanc.

Bohan Vineyards

Sonoma County, California.

White Riesling, Pinot Noir, Gewurztraminer and Zinfandel vineyard.

BOISE

The capital of the state of Idaho, in the southwest part of the state. (Built on the site of Fort Boise, a post on the Oregon Trail.) Idaho had vineyards and wineries before the turn of the century. In 1898, at the Chicago World's Fair, a prize was given to Robery Schleiser for his wine. His vineyard was in the Clearwater River Valley near Lewiston.

Bonita Vineyard

Madera County, California.

Papagni estate Chardonnay vineyard.

Bonny Doon Vineyard

Santa Cruz County, California.

Cabernet Sauvignon Santa Cruz Mountain vineyard.

BONNY DOON VINEYARD

Founded 1983, Santa Cruz County, California.

Storage: Oak and stainless steel. Cases per year: 4,000.

Estate vineyard: Grahm Ranch.

Label indicating non-estate vineyard: Arrendell Vineyard, Bethel Heights Vineyard.

Wines regularly bottled: Estate bottled, vintage-dated Pinot Noir; vintage-dated Chardonnay, Claret, Syrah, Pinot Noir; also Vin Gris de Pinot Noir. Occasionally Vin Rouge.

Bonny's Vineyard

Napa Valley, California.

Cabernet Sauvignon vineyard.

BOORDY VINEYARDS

Founded 1945, Hydes, Baltimore County, Maryland.

Storage: Oak and stainless steel. Cases per year: 6,000.

History: Boordy is Maryland's largest winery. Philip and Jocelyn Wagner, Boordy's founders, were instrumental in the introduction of the French-American hybrid grapes into Eastern viticulture. Founded in Riderwood, Maryland, the

winery has developed a reputation far out of proportion to its modest size. In 1980, the winery passed into the hands of the R.B. Deford family.

Estate vineyard: Boordy Vineyard.

Label indicating non-estate vineyard: Cedar Point Vineyard.

Wines regularly bottled: Seyval Blanc. Also proprietary Cedar Point Red. Semi-generics are Noveau Red, Maryland Dry White, Maryland Rosé, Maryland Red.

Borel Vineyard

Temecula, California.

Cabernet Sauvignon vineyard.

Borelli Vineyard

San Joaquin County, California.

California estate Zinfandel, Chenin Blanc and Sauvignon Blanc vineyard.

CIRIACO BORELLI WINERY

Founded 1979, Stockton, San Joaquin County, California.

Storage: Oak. Number of cases produced per year: 36,500.

Estate vineyard: Borelli Vineyard.

Wines regularly bottled: Zinfandel, Rosé, Burgundy, Chablis.

BORRA'S CELLAR

Founded 1975, Lodi, San Joaquin County, California.

Storage: Oak. Cases per year: 800.

Estate vineyard: Borra's Cellar Vineyard.

Wines regularly bottled: Estate bottled, vintage-dated Barbera, White Zinfandel. Occasionally bottle Carignane, Mission (Angelica).

Borra's Cellar Estate Vineyard

San Joaquin County, California.

Barbera and Zinfandel vineyard.

John Bosche Vineyard

Napa Valley, California.

Cabernet Sauvignon and Merlot vineyard, historically used in Beaulieu Private Reserve, but recently in Freemark Abbey.

Boskydel Estate Vineyard

Leelanau County, Michigan.

Vignoles, Seyval Blanc, DeChaunac and Johannisberg Riesling vineyard.

BOSKYDEL VINEYARD

Founded 1976, Lake Leelanau, Leelanau County, Michigan.

Storage: Stainless steel. Cases per year: 3,000.

History: The first winery on the Leelanau Peninsula and the pioneer of grape growing in northwestern Michigan. Lake Leelanau affords a special micro-climate for grape culture.

Estate vineyard: Boskydel Vineyard.

Wines regularly bottled: Estate bottled, vintage-dated Vignoles, Seyval Blanc, DeChaunac, Johannisberg Riesling. Also semi-generic Red, White, Rosé.

BOTRYTIS (Noble Mold)

Botrytis Cinerea is the mold that grows on the surface of grapes under certain vineyard conditions. A state of overripeness caused by an organism which concentrates the sugar content of grapes. For some sweet wines, especially Sauternes and Rieslings, this is considered highly desirable. During growth, the mold shrivels the grapes, concentrating both sugar and flavor. Botrytis adds a very rich and lucious quality to Johannisberg Riesling, Sauvignon Blanc and Semillon and the results are worth the difficult slow process. Cinerea and other grape molds are often very deleterious and undesirable in grapes.

BOTRYTIZED

Wines made from grapes attacked by the mold, Botrytis cinerea, have a strong and distinctive odor. These wines are invariably very sweet.

bottled in bond

A term used to indicate bottled under Government supervision on bonded premises before alcohol tax has been remitted. Unflavored California brandies can be labeled "bottled in bond" and must be 100 proof and barrel-aged for four years or more.

Boucherie Mountain Vineyard

Okanagan, British Columbia.

Chasselas and Chenin Blanc vineyards.

BOUNTIFUL HARVEST WINERY

Founded 1983, Highland, Columbia County, Wisconsin.

Storage: Stainless steel.

Wines regularly bottled: Strawberry, Cherry, Cranberry-Apple, Honey. Also Red, White and Rosé.

bouquet

The odors that develop in wine, after the finish of the fermentation, are designated bouquet. These may be divided into two general groups tank aging bouquet and bottle bouquet.

Test-tasting wine from wooden tank

Tank and Barrel or Cask Aging Bouquet: The desirable odors produced in wine during storage in tanks or casks are known as aging bouquet. These may consist of the odoriferous substances extracted from oak, and the compounds formed from the aroma materials by their slow oxidation by air diffusing through the walls of the cask.

Bottle Bouquet: When the bottled wine is relatively plentiful in compounds in higher oxidation states, the slow oxidation-reduction interchanges that occur in the bottle give rise to new substances whose odor is designated bottle bouquet. It is recognized that bottle bouquet development is dependent upon the presence of basic and fermentation aroma substances, as well as bulk aging bouquet compounds, for its satisfactory development.

Bowman Vineyard

Amador County, California.

Zinfandel Shenandoah Valley vineyard.

Brae Burn Vineyards

Ontario, Canada.

Inniskillin estate Vidal, Gewurztraminer, Seyval Blanc, Chelois vineyard.

The Brander Estate Vineyard
Santa Barbara County, California.
Sauvignon Blanc vineyard.

THE BRANDER VINEYARD
Founded 1979, Los Olivos, Santa Barbara County, California.
Storage: Oak and stainless steel. Cases per year: 6,000.
Estate vineyard: Brander Vineyard.
Wines regularly bottled: Estate bottled Sauvignon Blanc.
Second label: St. Carl.

BRAREN & PAULI WINERY
Founded 1979, Potter Valley, Mendocino County, California.
Storage: Oak and stainless steel. Cases per year: 2,000.
Estate vineyard: Hawn Creek Vineyard, The Hill Vineyard.
Wines regularly bottled: Estate bottled, vintage-dated Chardonnay, Zinfandel, Sauvignon Blanc.

Brazo's Valley Vineyard
Hood County, Texas.
French/American Hybrid vineyard.

Allowing a wine to "breathe" before drinking

breathing
Opening and exposing a bottle of red wine to the air. A practice in dispute as to "how much" is necessary, if any.

breed
Wine Term.
Harmony and elegance with a distinctive personality.

BREITENBACH WINE CELLARS
Founded 1980, Dover, Tuscarawas-County, Ohio.
Storage: Oak and stainless steel.
History: Winery is located in the largest Amish settlement in the United States and the center of Ohio's swiss cheese industry.
Wines regularly bottled: Seyval Blanc, Vidal Blanc, Chancellor Noir, Baco Noir, Concord, Niagara, Catawba. Also produces Apple Wine.

BRENNER CELLARS
Founded 1979, Sonoma County, California.
Storage: Oak and stainless steel. Cases per year: 2,000.
Wines regularly bottled: Vintage-dated Zinfandel, Chardonnay, Cabernet Sauvignon.

BREWSTER, WILLIAM (1560-1644)
English colonist; a leader of the Pilgrims at Plymouth. He encouraged winemaking from the native grapes.

Briarcrest Vineyard
Alexander Valley, California.
Clos du Bois estate Cabernet Sauvignon vineyard.

brick red

Wine Color.

Typical of very old red wines; a color suggesting natural brick.

THE BRIDGEHAMPTON WINERY

Founded 1982, Bridgehampton, Suffolk County, New York.

Storage: Oak and stainless steel. Cases per year: 2,500.

History: Bridgehampton is the first winery on the south fork of Long Island known as "The Hamptons". This over 300 year old area was originally settled by the Dutch.

Estate vineyard: Greenfield Vineyard.

Labels indicating non-estate vineyard: Vineyards of Ken Conrad.

Wines regularly bottled: Vintage-dated Chardonnay, Riesling.

BRIDGES CREEK

An estate in Eastern Virginia on the Potomac. The birthplace of George Washington. In 1932, restored as a national monument. (Now called Wakefield.) Wine grapes grew at Bridges Creek.

T.G. BRIGHT & COMPANY, LTD.

Founded 1874, Niagara Falls, Ontario, Canada.

Storage: Oak. Cases per year: 2,000,000.

History: Founded by Thomas Bright, the winery is the second oldest winery in Canada. The oldest was founded in 1873. The winery was purchased, in 1933, by Harry Hatch whose son Douglas is president of the winery.

Wines regularly bottled: Vintage-dated Baco Noir, Gewurztraminer, Riesling, Pinot Chardonnay; Proprietary President Canadian Champagne, Entre-Lacs, Warnerhof, Lentre Cote, "74" and President (Sherry and Port), Liebes Heim, Mon Village and House Wine.

brilliant

Wines free of any visible solids and having a sparkling clarity.

BRIMSTONE HILL VINEYARD

Founded 1979, Pine Bush, Ulster County, New York.

Storage: Oak and stainless steel. Cases per year: 300-1,500.

Estate vineyard: Brimstone Hill Vineyard.

Wines regularly bottled: Vintage-dated Aurora, Seyval. Estate bottled, vintage-dated semi-generics Red, White, Rosé.

BRISTLE RIDGE VINEYARD

Founded 1979, Montserrat, Johnson County, Missouri.

Storage: Oak and stainless steel. Number of cases produced per year: 1,000.

Wines regularly bottled: Concord. Also semi-generics Rhine, Rosé, Burgundy.

BRISBANE, ARTHUR (1864-1936)

American editor and writer.

> "Wine is the most noble and beneficial of alcoholic drinks. Wine is for the sedentary whose work is thinking. Natural wines have been used without drunkenness by millions of human beings for ages. They supply the body with iron, tannin and vitamins."

brix

The system used for measuring the soluble solids in grape juice. A measure of sugar. Brix degrees range from zero degrees to about 50 degrees. Similar to Balling. Fresh grape juice measuring 20 degrees Brix will give a dry wine of about 11 percent alcohol.

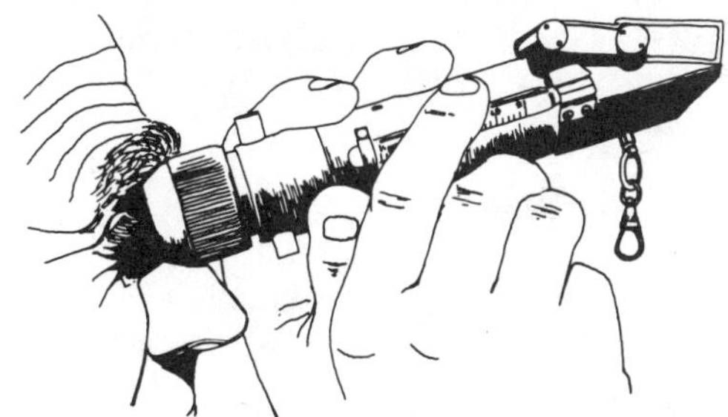

Checking grape sugar with refractometer

Bro-Cap Vineyard
Napa County, California.
Merlot, Napa Gamay, Sauvignon Blanc, Zinfandel vineyard.

J.F.J. BRONCO WINERY

Founded 1973, Ceres, Stanislaus County, California.

Storage: Stainless steel. Cases per year: 2,000,000.

Wines regularly bottled: Chablis, Ruby Rosé, Rich Burgundy, Rhinewine, Pink Chablis, Chablis Blanc, Sangria.

Second label: CC Vineyard Wines, Three Mountain Wines.

BRONTE CHAMPAGNE AND WINES COMPANY

Founded 1933, Hartford, Van Buren County, Michigan.

Storage: Oak and stainless steel. Cases per year: 50,000.

History: Bronte was founded the year of repeal of Prohibition, by Dr. T.W. Wozniak. The original winery was located in the old Columbia Brewery in Detroit and in 1951 was moved to Keeler, Michigan. The first French-American hybrid grapes planted in Michigan in 1953 were at the Bronte Vineyards.

Estate vineyard: Bronte Vineyard.

Wines regularly bottled: Baco Noir, Marechal Foch, Pink Delaware, Pink Catawba, Aurora Blanc, Vidal Blanc. Also Burgundy, Rosé, Pink Chablis, Scarlet Rosé, Rhine, Champagne (Extra Dry, Brut), Cold Duck, Sherry, Sparkling Burgundy, Champagne Cocktail. Dessert wines are Cream Sherry, Hartford Cream, Port, Hartford Port, Sweet Vermouth. Proprietary wines are all Beau Rouge, Rhinehaus.

Bronte Estate Vineyard
Van Buren County, Michigan.
Baco Noir, Marechal Foch, Delaware, Catawba, Aurora and Vidal vineyards.

BROTHERHOOD WINERY

Founded 1820's, Washingtonville, Orange County, New York.

Storage: Oak and stainless steel. Cases per year: 70,000.

History: Brotherhood, located in one of the oldest wine growing districts in the country, was created less than 50 years after the adoption of the Constitution of the United States and is America's oldest winery. Jean Jaques, a French emigre, selected the site near the Hudson River in Washingtonville, in 1816; and a vintage was documented in 1839.

Jesse Emerson, a thriving New York City wine merchant, bought the winery in 1880. During Prohibition, he continued to produce and sell altar wine, a practice in which Brotherhood still engages. Shortly before Repeal, Brotherhood changed hands for only the second time since its founding. It was acquired by L.L. Farrell, whose descendants produce the Brotherhood wines of today.

Estate vineyard: Penn Yan.

Wines regularly bottled: Cayuga, Dutchess, Chardonnay, Catawba. Proprietary wines produced are Holiday, May Wine, Rosario. Semi-generics include Chablis, Rhine, Sauterne, Rosé, Burgundy. Dessert wines are Golden Sherry, Cream Sherry, Ruby Port. Also produce Vintage Port, Celebration (Tawny) Port and Fino Sherry.

BROWN DERBY (See Bardenheier's Wine Cellars.)
Port, Sherry, Muscatel, White Port.

George Brown Vineyards
Sonoma County, California.
Gamay Beaujolais vineyard.

BRUCE, DAVID
David Bruce developed an interest in wine while he was a medical student at Stanford University. After obtaining training in dermatology in Oregon, Bruce returned to the Bay Area to purchase land appropriate for a vineyard. In 1961, he bought forty acres in the Santa Cruz Mountains and planted the land in Chardonnay and Pinot Noir. The winery was bonded in 1964. Bruce has carefully experimented with many different wine styles and methods in the past twenty years, and now emphasizes wine made from the grapes best suited to the best viticultural areas, including Pinot Noir, Chardonnay, Cabernet Sauvignon and Zinfandel. In 1964, he was the first to produce a white Zinfandel and was one of the first to make a 100% Petite Sirah and a dry Gewurztraminer. Bruce's first love, however, remains Pinot Noir. He is founder of the Pinot Noir Club, which is devoted to the promotion of the varietal wine, and he hopes someday to produce the greatest Pinot Noir ever made.

DAVID BRUCE WINERY
Founded 1968, Saratoga, Santa Cruz County, California.
Storage: Oak and stainless steel. Cases per year: 17,000.
History: One of the oldest wineries in the Santa Cruz Mountains.
Estate vineyard: David Bruce Vineyard.
Labels indicating non-estate vineyard: Wasson Vineyard, Monitz Vineyard, Scharffenberger Vineyard.
Wines regularly bottled: Estate bottled, vintage-dated Chardonnay, Pinot Noir. Also vintage-dated Chardonnay, Zinfandel and vintage-dated proprietary "Vintners Select" Cabernet Sauvignon. Occasionally bottles vintage-dated Petite Sirah, Gewurztraminer.

David Bruce Winery Estate Vineyard
Santa Cruz County, California.
Chardonnay and Pinot Noir vineyard.

BRUSHCREEK VINEYARDS
Founded 1977, Peebles, Highland County, Ohio.
Storage: Oak.
Wines regularly bottled: Catawba, Niagara, Concord, Chancellor. Also proprietary Dry Red, Dry White and Highland Fling.

A champagne flute displays "fine perlage"

brut
Usually applied to Champagne, which means that little, or no, "dosage" has been added to the wine. Usually 0 to 0.5% residual sugar. Legal definition in Champagne, France is "less than 15 grams per liter" which is 1.5%. Dry but "Natural" is drier.

Buccia Estate Vineyard
Ashtabula County, Ohio.
Baco, Seyval, Aurora, Chelois vineyard.

BUCCIA VINEYARD
Founded 1978, Conneaut, Ashtabula County, Ohio.
Storage: Oak, stainless steel. Cases per year: 1,100.

Estate vineyard: Buccia Vineyard.

Wines regularly bottled: Baco, Aurora, Seyval, Chelois. Also proprietary Terrace Red, Terrace White, Maiden's Blush.

Buckingham Valley Estate Vineyards

Bucks County, Pennsylvania.

Vidal Blanc, Seyval Blanc, Cayuga, Baco Noir, Chelois, Foch, DeChaunac and Niagara vineyard.

BUCKINGHAM VALLEY VINEYARDS

Founded 1966, Buckingham, Bucks County, Pennsylvania.

Storage: Oak and stainless steel. Cases per year: 7,000.

History: One of Pennsylvania's first vineyards and wineries.

Estate vineyard: Buckingham Valley Vineyards.

Wines regularly bottled: Estate bottled, vintage-dated Vidal Blanc, Seyval Blanc, Cayuga, Baco Noir, Chelois, Foch, DeChaunac, Niagara. Also estate bottled vintage-dated Concordia, Rosette.

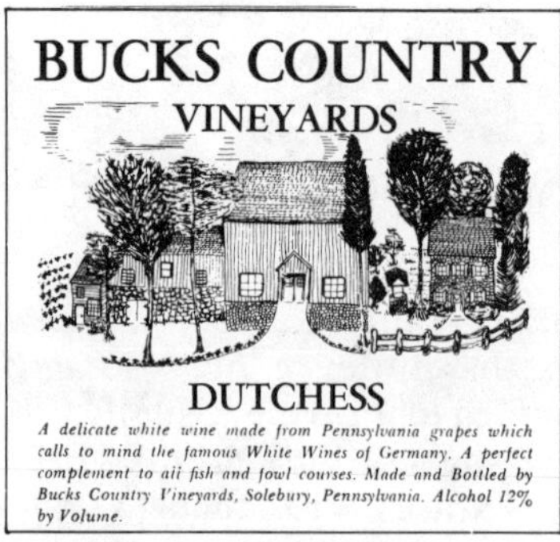

BUCKS COUNTRY VINEYARDS

Founded 1973, New Hope, Bucks County, Pennsylvania.

Storage: Stainless steel. Cases per year: 25,000.

History: Winery founded on property which was original land grant in 1716 from William Penn to Jacob Holcombe. Winery contains a unique Wine and Fashion Museum; wine artifacts, wine glass collection from New York Metropolitan Museum and 75 original costumes of stage and screen stars.

Wines regularly bottled: Pink Niagara, Concord, Pink Catawba, Blanc de Vidal, Aurora, Dutchess, Seyval Blanc, Dry Catawba, Chelois. Bottles proprietary Eye of the Pheasant, Pennsylvania Nouveau. Also bottles Dutch Apple Wine, Sangria, Country Red, Country Pink, Labrusca.

Grape vine bud break

bud

A small axillary or protuberance seen on the grape vine. The beginning of the shoot and the grape.

Buehler Estate Vineyards

Napa County, California.

Zinfandel, Pinot Blanc and Cabernet Sauvignon vineyard.

BUEHLER VINEYARDS

Founded 1972, St. Helena, Napa County, California.

Storage: Oak and stainless steel. Cases per year: 7,000-10,000.

History: Winemaker is Heide Peterson, daughter of Dr. Richard Peterson of The Monterey Vineyard.

Estate vineyard: Buehler Vineyard.

Wines regularly bottled: Estate bottled, vintage-dated Zinfandel, Pinot Blanc, Cabernet Sauvignon, White Zinfandel.

Buena Vista Estate Vineyards
Sonoma County, California.
Cabernet Sauvignon, Johannisberg Riesling, Gamay Beaujolais, Pinot Noir and Chardonnay vineyard.

Buena Vista Vineyards
Sonoma County, California.
Hacienda estate Cabernet Sauvignon, Pinot Noir, Chardonnay vineyard.

BUENA VISTA WINERY
Founded 1857, Sonoma, Sonoma County, California.
Cases per year: 90,000.
History: The Buena Vista vineyards were planted in the Sonoma Valley—the old "Valley of the Moon" of Jack London, in 1832, to produce wine for the Mission San Francisco Solano de Sonoma. Twenty years later the vineyards were acquired by Count Agoston Haraszthy, a titled vintner from the court of the Emperor of Hungary. The Count had fled to America after his participation in the Hapsburg Revolution, and he concluded his long flight in the Sonoma Valley, where he introduced European varietals and laid the foundation for the California wine industry. This extraordinary winemaker prospered in the New World. At Buena Vista his two sons married the daughters of General Mariano Vallejo, in a double wedding, at the old Mission. The winery and the vineyards, after falling into disuse, were restored by Frank Bartholomew, then the head of United Press International, in 1943. Today Buena Vista is owned by the Racke family of West Germany. The winery is a state historic landmark.
Estate vineyard: Buena Vista Estate Vineyard.
Wines regularly bottled: Estate bottled, vintage-dated Cabernet Sauvignon, Johannisberg Riesling, Gamay Beaujolais, Pinot Noir Rosé, Late Harvest Gamay Beaujolais, Pinot Noir, Chardonnay.
Also bottle vintage-dated Cabernet Sauvignon, Green Hungarian, Pinot Noir Rosé, Chardonnay, Spiceling, Zinfandel, Johannisberg Riesling. Also vintage-dated semi-generics produced are Chablis, Burgundy.

BUFFALO VALLEY WINERY
Founded 1979, Lewisburg, Union County, Pennsylvania.
Storage: Oak, stainless steel.
Wines regularly bottled: Grape and fruit wines.

bulk wines
Wines which are stored, shipped or packaged in containers usually having a capacity of five gallons, or more.

BULL, EPHRAIM WALES
In 1853, at the Massachusetts Horitcultural Society, he first presented the Concord grape, named for the town where he lived from 1836 to his death there in 1895. He worked on the development of grapes for many years. The Concord sprouted from a wild seed he planted.

Bully Hill Estate Vineyard
Steuben County, New York.
Baco Noir, Chelois Noir, Chancellor Noir, Cascade Noir, Marechal Foch, Colobel Noir, Seyval Blanc, Verdelet Blanc, Aurora Blanc, Vidal Blanc, Cayuga Blanc Finger Lakes vineyard.

BULLY HILL WINE AND CHAMPAGNE COMPANY

Founded 1970, Hammondsport, Steuben County, New York.

Storage: Oak and stainless steel.

History: Bully Hill is the site of the original Taylor Wine Company established in 1880. The original vineyard was 70 acres. In 1970 Greyton H. Taylor (son of the founder) and Walter S. Taylor (the grandson) established the Bully Hill Wine Company on the wine property owned by Walter Taylor from 1880 to 1926. Walter S. Taylor is the sole owner of the winery. The Greyton H. Taylor Wine Museum established near the winery is the first wine museum in America.

Estate vineyard: Bully Hill Vineyard.

Wines regularly bottled: Baco Noir, Chancellor Noir, Chelois Noir, Aurora Naturel, Aurora Blanc, Seyval Blanc, Marechal Foch, Vidal Blanc, Cayuga Blanc, Verdelet Blanc. Proprietary wines are Walter's Red, Walter's White, Bully Hill Red, Bully Hill White, Space Shuttle Rosé, Workers Wine, Pink Lady Rosé, Founders Red, Founders White, Old Barnyard Red.

bung

Barrel stopper. Bungs are usually made of wood, silicone or glass.

bunghole

The opening, or hole, of the cask through which it is filled, then closed tight with the bung.

BURGAW

American muscadine grape that produces sweet white wine.

BURGER

A vinifera grape that produces light bodied white wines. Mainly used for blending.

BURGESS CELLARS

Founded 1889, St. Helena, Napa County, California.

History: The winery site is part of the original Rossini homestead. The building was constructed in 1875. In 1943, the property was purchased by J. Leland Stewart who operated the winery for 25 years and then sold it to the Burgess family

Wines regularly bottled: Estate bottled, vintage-dated Chardonnay, Cabernet Sauvignon, Zinfandel.

Second label: Bell Canyon.

BURGUNDY

Burgundy is a district in France where red wines are made from Pinot Noir and Gamay. The famous Burgundy Whites are made from Chardonnay. In America, the name used to describe generous, full-bodied, dry red dinner wines, with a pronounced flavor, body, and bouquet and a deep red color. In California, Burgundy may be made from a number of different grape varieties, including Gamay, Petite Sirah, Pinot Noir, Carignane and Zinfandel.

Burkittsville Vineyard
Catoctin County, Maryland.
Catoctin estate Chardonnay, Cabernet Sauvignon, Riesling, Sauvignon Blanc, Seyval, Vidal vineyard.

burnt
Wine Term.
This term is used to describe the burned sugar or cooked grape juice taste and odor. When burned concentrate or overheated wines are used to sweeten other wines, the unpleasant flavors frequently become very objectionable in the wine.

butt
A wine cask with the capacity of 100 to 140 gallons.

BUZZARD LAGOON VINEYARD (See Cook-Ellis Winery.)
Chablis, Mountain White Table Wine.

DAVIS BYNUM WINERY
Founded 1965, Healdsburg, Sonoma County, California.
Storage: Oak and stainless steel. Cases per year: 18,000.
Estate vineyard: Davis Bynum Vineyard.
Wines regularly bottled: Vintage-dated wines produced are Zinfandel, Pinot Noir, Cabernet Sauvignon, Fumé Blanc, Chardonnay and Gewurztraminer. Occasionally bottle Merlot and Carignane.
Second label: River Bend.

Byrd Estate Vineyards
Frederick County, Maryland.
Chardonnay, Sauvignon Blanc, Cabernet Sauvignon, White Riesling, Gewurztraminer, Seyval Blanc and Vidal Blanc vineyard.

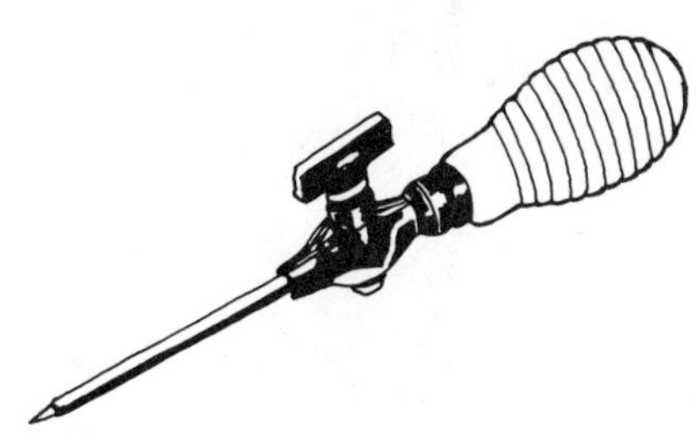

Air pressure cork opener

BYRD VINEYARDS
Founded 1976, Myersville, Frederick County, Maryland.
Storage: Stainless steel. Cases per year: 3,000.
History: The first winery in Maryland to produce a commercial Sauvignon Blanc and Gewurztraminer.
Estate vineyard: Byrd Vineyard.
Wines regularly bottled: Estate bottled, vintage-dated Chardonnay, Sauvignon Blanc, Cabernet Sauvignon, White Riesling, Gewurztraminer, Seyval Blanc, Vidal Blanc. Occasionally bottle Vidal Blanc, Merlot, Chambourcin.
Second Label: Church Hill.

CABERNET FRANC

One of several Cabernet wine grape varieties. Franc is a leading variety of the St. Emilion District of Bordeaux. Used mainly in blending with Cabernet Sauvignon and Merlot.

CABERNET PFEFFER

Wine grape named for vinegrower and breeder William Pfeffer who came to California in the 1860's. Highly popular until the turn of the century when it was destroyed by disease and drought. In 1908 California wine pioneer, Dr. Harold Ohrwall planted Cabernet Pfeffer in the Cienega Valley. The Charles Lefranc Cellars bottle a small amount from this rare stand of grapes from the San Benito County vineyard. Cabernet Pfeffer is a great curiosity that has been saved by the Charles Lefranc Cellars.

CABERNET SAUVIGNON

Wine grape capable of producing prestigious wines. This wine is dry, medium to full-bodied and capable of great complexity if allowed to age. When young, Cabernet Sauvignon has a dominant tannic characteristic. Originally from the Bordeaux region of France, where it is one of the principal grapes for the great chateau Clarets. The Roman poet, Pliny the Elder, mentioned Cabernet Sauvignon in his First Century writings, as a Bordeaux region wine.

CACHE CELLARS

Founded 1978, Davis, Solano County, California.

Cases per year: 4,000.

Labels indicating non-estate vineyard: Baldinelli Vineyard, Ventana Vineyard, Wetzel Vineyard, Rancho Tierra Rejada Vineyard, La Reina Vineyard, San Saba Vineyard.

Wines regularly bottled: Vintage-dated Zinfandel, Sauvignon Blanc, Chardonnay, Pinot Noir, Carnelian "Nouveau".

CADENASSO WINERY

Founded 1906, Fairfield, Solano County, California.

Storage: Oak and stainless steel.

Wines regularly bottled: Cabernet Sauvignon, Grey Riesling, Pinot Noir, Grignolino, Zinfandel, Chenin Blanc. Also proprietary Passionatta. Semi-generics produced are Burgundy, Chablis, Rosé, Haut Sauterne.

Cagnasso Estate Vineyards

Ulster County, New York.

Aurora, Seyval Blanc, Leon Millot, Marechal Foch, Chelois, DeChaunac, Hudson River Region vineyard.

CAGNASSO WINERY

Founded 1977, Marlboro, Ulster County, New York.

Storage: Oak. Cases per year: 3,000.

Estate vineyard: Cagnasso Vineyards.

Wines regularly bottled: Estate bottled Aurora, Seyval Blanc, Leon Millot, DeChaunac, Marechal Foch, Chelois; generics, Moselle, Chianti; also Labrusca, Vino Rosé, Rosso Amabile, Bianco Classico, White Favorite.

CAIN CELLARS

Founded 1981, St. Helena, Napa County, California.

Storage: Oak and stainless steel. Cases per year: 5,000.

History: Property was the 542 acre

landmark McCormick Ranch overlooking St. Helena, which has been in same ownership since 1840.

Estate vineyard: Cain Mountain Vineyards.

Wines regularly bottled: Vintage-dated Sauvignon Blanc, Cabernet Sauvignon.

CAKEBREAD CELLARS

Founded 1973, Rutherford, Napa County, California.

Storage: Oak and stainless steel. Cases per year: 28,000.

History: In 1982, the winery received the Award of Honor for Design Excellence from The American Institute of Architects.

Estate vineyard: Cakebread Vineyard.

Label indicating non-estate vineyard: Beatty Ranch.

Wines regularly bottled: Vintage-dated Cabernet Sauvignon, Zinfandel, Sauvignon Blanc, Chardonnay.

CAKEBREAD, JACK

Jack and Dolores Cakebread were born and raised in the California Bay Area, and attended the University of California, Berkeley. After their marriage in 1950, Jack joined the Strategic Air Command and they moved to Europe, where he practiced his photography skills. Upon their return to California, Jack worked in the family garage business in Oakland, and as a free-lance photographer. One photography assignment, for Nathan Chroman's book *The Treasury of American Wines,* took the Cakebreads to most of the Napa Valley wineries. While there, he spotted 22 acres of centrally located vineyards near Rutherford, and decided to expand his agricultural and gardening skills to viticulture. He purchased the land in 1973, and he and Dolores spent evenings replanting ten of the acres in Sauvignon Blanc. Today, 13 acres of Cabernet Sauvignon vines have been added to their original holding. Cakebread Cellars is run by Jack and Dolores, their son Bruce, and his wife Rosemary, who works as enologist.

CALADIA (See Bardenheier's Cellars.)

Port, Sherry, Muscatel, White Port.

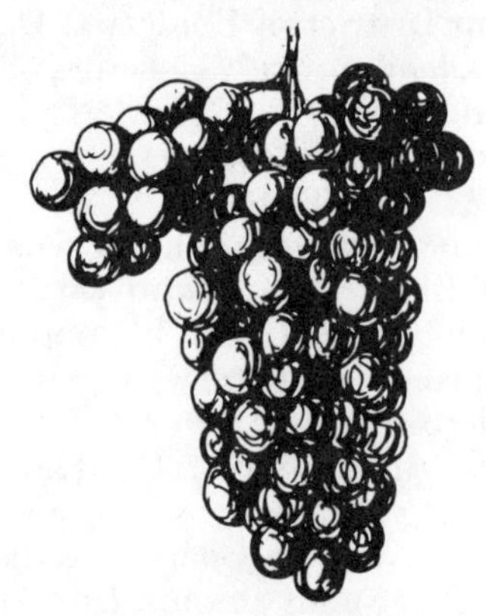

Cabernet Sauvignon

CALAFIA CELLARS

Founded 1979, St. Helena, Napa County, California.

Storage: Oak and stainless steel. Cases per year: 2,500.

History: The name is derived from a legend concerning Queen Calafia and her matriarchal society of Amazons on an island named Matinino in the West Indies. This island was chronicled in a daily log by a companion of Columbus on his first voyage to the New World. The legend was popularized by Garcia Ordoñez de Montalvo, a Spanish author in his novel, Los Sergas de Esplandian, published about 1503. It told of the stately Queen Calafia and her highly organized and energetic society. Her great wealth was guarded by griffins, symbol of protection in Europe. The Griffin logo protects the wealth that the vines and the land bear. Calafia and her fierce warriors went to other lands with their griffins and engaged in battle. The most famous of which was their aiding of the Turks against the Christians in the battle of Constantinople in 1453.

The book of Montalvo was popular with the Spanish explorers, including Cortez, and was probably aboard his ship as he moved up the west coast of Mexico.

Upon landing at the top of Baja, with its rugged mountains and barren terrain, the men were reminded of the California island description in the book and, believing they were on an island, called it Calafia's Land or California. Most historical scholars today agree that this is how California was named. The winery adopted its name from this piece of California history.

Estate vineyard: Johnson Vineyard.

Label indicating non-estate vineyard: Kitty Hawk Vineyard, Pickle Canyon Vineyard, Foureeminette Vineyard, HNW Vineyard.

Wines regularly bottled: Vintage-dated Chardonnay, Sauvignon Blanc, Merlot, Zinfandel, Cabernet Sauvignon.

Second label: Redwood Canyon Cellars.

Calcaire Vineyard

Alexander Valley, California.

Clos de Bois estate Chardonnay vineyard.

Wooden barrel spigot

Caldwell Vineyard

Santa Ynez, California.

Chardonnay and Cabernet Sauvignon vineyard.

CALERA WINE COMPANY

Founded 1975, Hollister, San Benito County, California.

Storage: Oak. Cases per year: 8,000.

History: The name of the winery comes from the fact that it was built on the site of an old lime kiln (calera in Spanish). Winery is completely operated on gravity-flow.

Estate vineyard: Jensen Vineyard, Selleck Vineyard, Reed Vineyard.

Labels indicating non-estate vineyard: Los Alamos Vineyard, Cienega Vineyard, Templeton Vineyard, Doe Mill Vineyard.

Wines regularly bottled: Estate bottled, vintage-dated Pinot Noir. Vintage-dated Zinfandel, Pinot Noir, Chardonnay. Occasionally bottle Sweet Late Harvest Zinfandel, Essence Zinfandel.

Califania Vineyard

Santa Maria/Paso Robles, California.

Chardonnay vineyard.

CALIFORNIA CELLER MASTERS

Founded 1974, Lodi, San Joaquin County, California.

Storage: Oak and stainless steel. Cases per year: 5,000.

Wines regularly bottled: Grape and fruit wines under three labels: Coloma Cellars, Mother Lode, Gold Mine.

CALIFORNIA CELLARS, TAYLOR

Founded 1978, Gonzales, Monterey County, California.

Storage: Stainless steel. Cases per year: over 6,000,000.

Wines regularly bottled: Burgundy, Chablis, Rhine, Rosé, Chardonnay, Sauvignon Blanc, Chenin Blanc, French Colombard, Johannisberg Riesling, Cabernet Sauvignon, Zinfandel, Light Rosé, Light Chablis, Light Rhine.

Second label: Vivante generic wines.

Taylor California Cellars began as a second label of The Monterey Vineyard. It quickly outgrew its smaller "mother winery" and was split off as a separate entity in 1981. It was sold in 1983 by Coca Cola to Seagrams.

CALIFORNIA FRUIT AND BERRY WINE COMPANY (See San Benito

Chablis, Vin Rosé, Rhine, Champagne.

Second label: L. LeBlanc Vineyards.

CALIFORNIA GROWERS WINERY
Founded 1936, Cutler, Tulare County, California.

Storage: Oak and stainless steel. Cases per year: 300,000.

Wines regularly bottled: Burgundy, Chablis, Vin Rosé, Rhine, Champagne.

Second label: L. LeBlanc Vineyards.

CALIFORNIA MEADERY/WINERY OF THE ROSES
Founded 1979, Napa, Napa County, California.

Storage: Oak and stainless steel. Cases per year: 1,000.

History: The Meadery keeps each nectar flow (flower type) separate and vinifies accordingly.

Label indicating non-estate vineyard: Littlefield Apiaries.

Wines regularly bottled: Mead (honey wine). Occasionally bottle Chardonnay, Cabernet Sauvignon and Champagne.

Second label: Winery of the Roses.

CALIFORNIA VILLAGES (See Gibson Wine Company.)
Light Chablis, Light Rhine, Light Rosé.

CALIFORNIA WINE TRADE BEGINNING
Commercial wine-grape growing was started about 1824 by Joseph Chapman, one of the first settlers in California. At the Pueblo of Los Angeles, he set out about 4,000 vines. In 1831, Jean Louis Vignes, a Frenchman from the Bordeaux wine district, started a commercial vineyard approximately where the Los Angeles Union Station now stands, importing cuttings of different varieties of grapes directly from Europe. Other plantings soon followed, and within a generation, wine-grape growing became the principal agricultural industry of the Los Angeles district.

By 1840, Vignes was chartering ships which he loaded at San Pedro with wines and brandies for Santa Barbara, Monterey and San Francisco. As early as 1860, California wine firms had established agencies in New York and shipped wines around Cape Horn to the eastern states. The first transcontinental railroad, in 1869, opened the remainder of the country to the wine firms. Soon California wine growers were shipping wines to Europe and Latin America and Australia. Records show that Germany, Denmark, England and Canada were buying California Port in 1867.

CALISTOGA
The town is in the northern vineyard area of Napa County.

Calistoga Ranch Vineyard
Napa River, California.

Cabernet Sauvignon, Napa Gamay, Muscat Canelli vineyard.

CALISTOGA VINEYARDS
(See Cuvaison.)

Callaway Estate Vineyard
Riverside County, California.

Chardonnay, Sauvignon Blanc, Chenin Blanc vineyard.

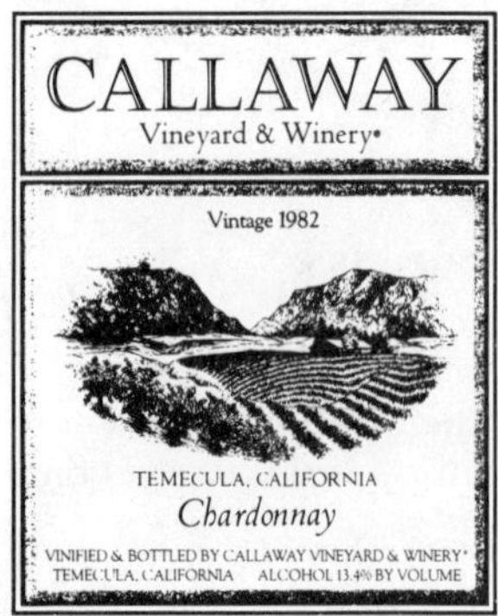

CALLAWAY VINEYARD AND WINERY
Founded 1974, Temecula, Riverside County, California.

Storage: Oak and stainless steel. Cases per year: 100,000.

History: One of the first quality table wine producers in Southern California from vinifera grapes grown on vinifera root.

Estate vineyard: Callaway Vineyard.

Wines regularly bottled: Estate bottled, vintage-dated Chardonnay, Sauvignon Blanc, Fume Blanc, Chenin Blanc. Also vintage-dated White Riesling. Occasionally bottle Sweet Nancy (a botrytised Chenin Blanc Dessert Wine), Santana (a botrytised White Riesling Dessert Wine), Port (a dessert wine from Petite Sirah, Cabernet Sauvignon, Zinfandel).

CALONA WINES LIMITED

Founded 1932, Kelowna, British Columbia, Canada.

Storage: Oak and stainless steel. Cases per year: 2,000,000.

History: The first winery founded in the Okanagan Valley, by W.A.C. Bennett (former Premier of British Columbia) and Cap Capozzi to use a surplus of apples in the depression years of the 1930's.

Wines regularly bottled: Vintage-dated Johannisberg Riesling, Fume Blanc, Chardonnay, Chenin Blanc, Gewurztraminer; proprietary Schloss Laderheim, Haut Villages, Sommet Rouge and Blanc, Royal red and white, Tiffany, La Scala Spumante, Fontano Bianco. Occasionally bottle Chancellor, Rougeon, Champagne, Sweet and Dry Vermouth.

Second label: OK Cellars, Frazer Valley Distributors, Winemaster's selection.

CALVARESI WINERY

Founded 1982, Reading, Berks County, Pennsylvania.

Storage: Stainless steel. Cases per year: 1,000.

Labels indicating non-estate vineyards: Cloverhill Farm Vineyards.

Wines regularly bottled: Vintage-dated Concord, Niagara, Catawba, DeChaunac, Aurora, Cayuga, Chelois, Chambourcin Rosé, Vidal, Chancellor, Delaware, Apple, Strawberry.

Cambiaso Estate Vineyard

Sonoma County, California.

Cabernet Sauvignon and Chardonnay vineyard.

CAMBIASO VINEYARDS

Founded 1934, Healdsburg, Sonoma County, California.

Storage: Oak and stainless steel. Cases per year: 70,000.

Estate vineyard: Cambiaso Vineyard.

Label indicating non-estate vineyards: Warren Bogle Vineyard, Jack Sorocco Vineyard, Oakhill Vineyards, R & J Cook Vineyards, Tom Johnson Vineyard.

Wines regularly bottled: Vintage-dated Sauvignon Blanc, Chardonnay, Chenin Blanc, Zinfandel, Cabernet Sauvignon, Petite Sirah, Barbera. Also semi-generic Chablis, Burgundy, Vin Rosé.

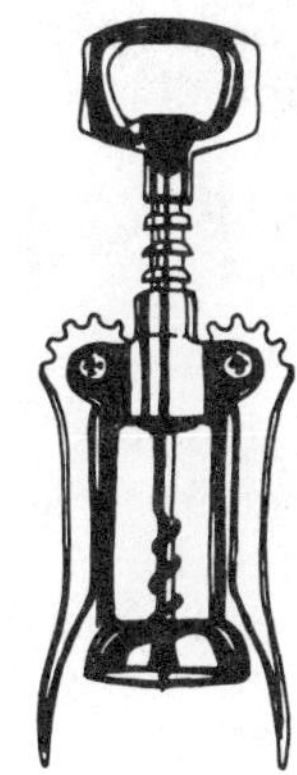

Butterfly-lever rack and pinion corkscrew

CAMPBELLS EARLY

A native American grape that produces wine similar to Concord. Used primarily for sweet wines and blending.

CANADA

Winemaking in Canada began with the early settlement of the French. The first known record is contained in Le Jeune's Relation of the Jesuits for the year 1636.

"In some places there are many wild vines loaded with grapes; some have made wine of them through curiosity. I tasted it and it seems to be very good."

No doubt the early settlers, French and English, continued to make wine but the next known record occurs nearly two hundred years after Le Jeune's relation.

In 1811, it is reported that John Schiller, an ex-corporal of the German army, settled in Cooksville, Ontario, where he planted a vineyard and made wine for sale to his neighbors. After this first report of commercial planting and winemaking in Canada, there is no record of any other until about forty years later.

Commencing about 1850 and for a number of years thereafter, articles on winemaking appeared in both government papers and agricultural publications.

The earliest commercial vineyards in the Okanagan Valley of British Columbia were planted near Kelowna in the late 1920's but the history of grape production goes back to the initial meeting called in 1895 to form the British Columbia Fruit Grower's Association.

In 1982, Canada had 39 producing wineries; nine in British Columbia, four in Alberta, one in Manitoba, twelve in Ontario, ten in Quebec, one in New Burnswick and two in Nova Scotia.

Situated between the 43rd and 50th parallels, like the chief wine growing regions of the world, are the Niagara Peninsula of Ontario and the Okanagan Valley of British Columbia. In 1880, there were only 2,000 acres of grapes in all of Canada. Today there are more than 28,000 acres. More important is the fact that today's growth in cultivation is directed at the improvement and increase in varieties grown for their superior winemaking qualities.

CANADA MUSCAT

A white grape with the characteristic flavor of Muscat, used for sweet dessert and sparkling wines. Rarely varietally labeled.

CANANDAIGUA WINE COMPANY, INC.

Founded 1945, Canandaigua, Ontario County, New York.

Storage: Oak and stainless steel. Cases per year: 5,000,000.

Wines regularly bottled: Proprietary Richards Wild Irish Rose, Richards Wild Irish Rose White Label, Roget Sparkling Wines (Dry, Spumante, Pink, Cold Duck, Sparkling Burgundy, Alicante), Sonnenberg White.

Second label: Chateau Martin Vermouths.

Candelstick Estate

Ulster County, New York.

Benmarl estate Hybrids Hudson River Region vineyard.

CAPARONE WINERY

Founded 1980, Paso Robles, San Luis Obispo, California.

Cases per year: 3,000-3,500.

Estate vineyard: Caparone Vineyard.

Label indicating non-estate vineyard: Tepusquet Vineyard.

Wines regularly bottled: Vintage-dated Cabernet Sauvignon, Merlot.

Cape Sandy Vineyards

Marion County, Indiana.

Easley estate Marechal Foch, DeChaunac, Chelois, Baco Noir, Aurora Blanc, Siebel 13053, Seyval Blanc and Dutchess

CAPISTRO (See Casabello, Canada.)

carbon dioxide, CO_2

The gas produced during fermentation. If it is retained a sparkling wine or Champagne will result. This gas is sometimes used to shield grapes, must and wine from oxygen.

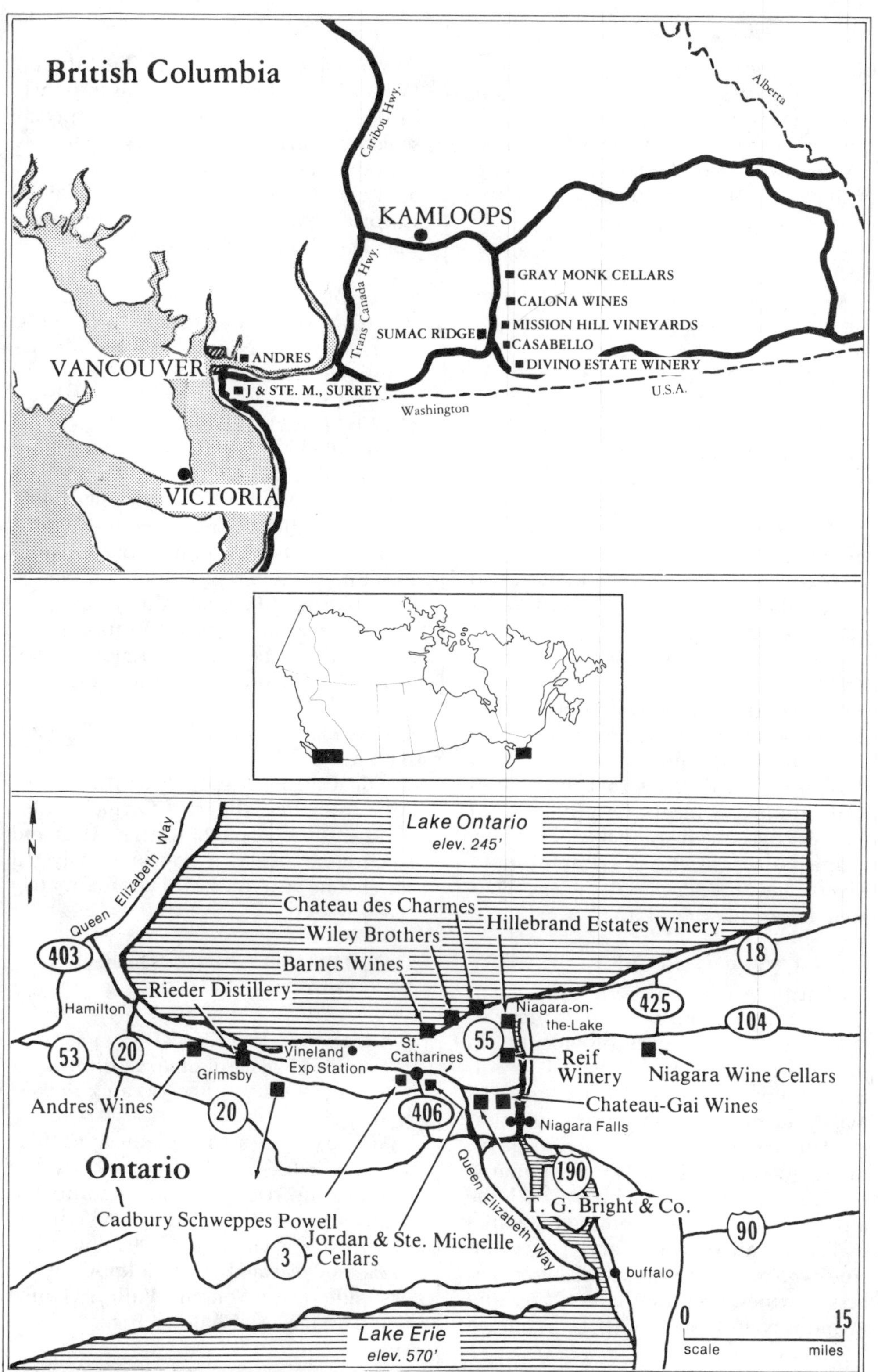
British Columbia
Caribou Hwy.
Alberta
KAMLOOPS
Trans Canada Hwy.
GRAY MONK CELLARS
CALONA WINES
SUMAC RIDGE
MISSION HILL VINEYARDS
CASABELLO
ANDRES
VANCOUVER
DIVINO ESTATE WINERY
J & STE. M., SURREY
U.S.A.
Washington
VICTORIA
Lake Ontario
elev. 245'
Queen Elizabeth Way
Chateau des Charmes
Wiley Brothers
Hillebrand Estates Winery
Barnes Wines
Rieder Distillery
Hamilton
Niagara-on-the-Lake
Vineland Exp Station
St. Catharines
Grimsby
Reif Winery
Niagara Wine Cellars
Andres Wines
Chateau-Gai Wines
Niagara Falls
Ontario
T. G. Bright & Co.
Cadbury Schweppes Powell
Jordan & Ste. Michellle Cellars
buffalo
Lake Erie
elev. 570'
0 15
scale miles
403
18
425
104
53
20
55
406
190
90
3

carbonic maceration

Whole grapes are placed in a closed vat in which air is replaced by carbon dioxide; this stops oxidation. An intensely colored wine with a light berry-like odor and fresh young flavor results. This is the process for "nouveau" wines.

carboy

A large glass bottle, especially one protected by basketwork.

J. CAREY CELLARS

Founded 1978, Solvang, Santa Barbara County, California.

Storage: Oak and stainless steel. Cases per year: 6,500.

Estate vineyard: J. Carey Vineyards.

Wines regularly bottled: Estate bottled, vintage-dated Cabernet Sauvignon, Merlot, Chardonnay, Cabernet Blanc. Also vintage-dated Sauvignon Blanc. Proprietary labeled wine: Mariage Blanc.

CARIGNANE

A wine grape that is predominantly used for blending in common wines. Full of tannin, good color, clean flavor and heavy body best described as robust. Goes well with hearty meals. Originally from the Mediterranean region of Europe. Spanish in origin, it is also important in Algeria and Southern France where it has been grown since the 12th Century.

CARISETTI, DOMENIC A.

Domenic A. Carisetti, born in Brooklyn, was graduated from the City College of New York. Following employment with the Browne Vintners division of Joseph E. Seagram & Sons, Carisetti obtained his masters degree in food science/enology at the University of California, Davis. He then joined the Taylor Wine Company, in 1975, as an assistant winemaker. Nine months later, he was promoted to winemaker for all Taylor brand wines, including New York State sparkling wines, dessert wines, premium table wines and vermouths. In 1980, Carisetti was named senior winemaker for all Taylor and Great Western brand wines. His responsibilities include vintage-dated and special selection varietal table wines. He has introduced several new wine blends, such as Lake County Chablis and Taylor Empire Cream Sherry, and one new product line, Lake County Soft Wines.

CARLOS

White Muscadine grape producing fruity wines in North Carolina, Mississippi and Florida.

CARMEL BAY WINERY

Founded 1977, Carmel, Monterey County, California.

Storage: Oak. Cases per year: 500-1,000.

Labels indicating non-estate vineyards: Shandon Valley Vineyards, D & M Junction Vineyards, Sleepy Hollow Vineyards, J & L Farms, El Dorado Vineyards.

Wines regularly bottled: Vintage-dated Zinfandel, Cabernet Sauvignon, Pinot Noir, Dry Chenin Blanc, Pinot Blanc.

CARMEL VALLEY VITICULTURAL AREA

The Carmel Valley viticultural area runs from the village of Carmel Valley southeasterly along the Carmel River and Cachugua Creek for approximately ten miles. The area consists of approximately 9,200 acres.

CARMEL VALLEY (See Durney Vineyard.)

CARMENET VINEYARDS

Founded 1983, Sonoma, California.

Storage: Oak. Number of cases: 15,000/year.

History: Vines first planted in 1972. First called Glen Ellen Vineyards but name changed in 1983 to avoid confusion with pre-existing Glen Ellen Winery. New winery facility has 15,000 square feet of caves excavated out of a knoll on the east side of the Sonoma Valley. Owned and managed by Chalone, Inc.

Labels indicating non-estate vineyards: Edna Valley.

Wines regularly bottled: Estate bottled Cabernet Sauvignon, vintage-dated. Also vintage-dated Sauvignon Blanc.

Carmenet Vineyards

Sonoma County, California.

Carmenet estate Cabernet Sauvignon Sonoma mountain vineyard.

CARMINE

A new California grape that is a hybrid of Ruby Cabernet and Merlot, developed at U.C. Davis.

CARNELIAN

A new grape variety developed by University of California at Davis. A cross between Cabernet Sauvignon, Grenache and Carignane. Yields a zesty, robust, dry red wine with complex aroma.

CARNEROS

This appellation region is in the southern part of both Napa and Sonoma Counties, alongside San Francisco Bay. Some of California's finest wineries have vineyards in the region.

CARNEROS CREEK WINERY

Founded 1972, Napa, Napa County, California.

Storage: Oak and stainless steel. Cases per year: 21,000.

Estate vineyard: Carneros Creek Winery.

Label indicating non-estate vineyard: Nathan Fay Vineyards, Truehard Vineyards, Hyde Vineyards, Giles Vineyards, Turnbull Vineyard.

Wines regularly bottled: Estate bottled, vintage-dated Pinot Noir. Also vintage-dated Cabernet Sauvignon, Sauvignon Blanc, Fume Blanc. Also Chardonnay. Red Table Wine and White Table Wine. Occasionally bottles Zinfandel.

CAROLINA WINERY (See Duplin Winery.)

Rosé, Red, Country Scuppernong.

Carpenter Vineyards

Jackson County, Oregon.

Cabernet Sauvignon and Chardonnay vineyard.

Carpy Conolly Vineyard

Napa Valley, California.

Freemark Abbey estate Chardonnay, Sauvignon Blanc, Johannisberg Riesling and Merlot vineyard.

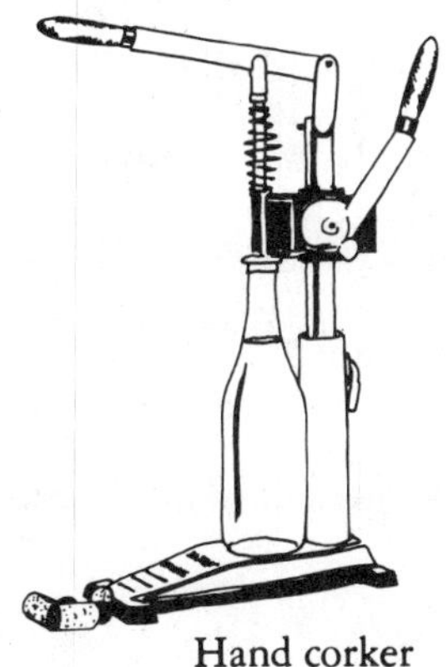

Hand corker

CARROUSEL WINERY

Founded 1981, Santa Clara County, California.

Cases per year: 750.

Estate vineyard: De Santis Vineyard.

Label indicating non-estate vineyard: Rest & Be Thankful Vineyard, Chamisal Vineyard.

Wines regularly bottled: Vintaged-dated Merlot, Chardonnay; occasionally Zinfandel.

CARVER WINE CELLARS

Founded 1977, Rollo, Phelps County, Missouri.

Storage: Oak and stainless steel. Cases per year: 2,000.

Estate vineyard: Carver Vineyards.

Wines regularly bottled: Estate bottled, vintage-dated White Riesling, Chardonnay, Gewurztraminer, Chancellor, Vidal Blanc, Seyval Blanc, Cayuga, White Vignoles. Occasionally bottles Catawba, Cabernet Sauvignon.

Second label: Chateau Ste. Genevieve.

Carver Wine Cellars Estate Vineyards

Phelps County, Missouri.

White Riesling, Chardonnay, Gewurztraminer, Chancellor, Vidal Blanc, Seyval Blanc, Cayuga, Vignoles vineyard.

CASABELLO WINES

Founded 1966, Penticton, British Columbia, Canada.

Cases per year: 425,000. One of several Ridout wineries, owned by John Labatt Ltd.

Wines regularly bottled: Estate selection, vintage-dated Chenin Blanc, Johannisberg Riesling, Gewurztraminer, Chardonnay. Also Burgundy, Pinot Noir, Fleur de Blanc, Marechal Foch, Burgonay, Burgeon Rouge, Summerland Riesling, Vino Rosso, Chablis Blanc, Rinegarten, Alpenweiss.

Second label: Casabello, Gala, Capistro, San Gabriel.

CASA DE FRUTA

Founded 1908, Hollister, San Benito County, California.

Storage: Oak and stainless steel. Cases per year: 8,700.

Estate vineyard: Casa de Fruta Vineyards.

Wines regularly bottled: Estate bottled, vintage-dated Gewurztraminer, Chenin Blanc, Chenin Blanc Soft, Zinfandel Rosé, Black Hamburg, Black Hamburg Rosé. Dessert wines bottled are Black Muscat, Apricot, Blackberry, Strawberry, Pomegranate.

Casa de Fruta Estate Vineyard

San Benito County, California.

Gewurztraminer, Chenin Blanc, Zinfandel, Black Hamburg vineyard.

CASA GRANDE (See Casa Madero, Mexico.)

Casa Larga Estate Vineyards

Monroe County, New York.

Pinot Chardonnay, Johannisberg Riesling, Pinot Noir, Cabernet Sauvignon, Gewurztraminer, Aurora, DeChaunac, Merlot, Chelois, Rougeon, Catawba, Niagara and Delaware Finger Lakes vineyard.

CASA LARGA VINEYARDS, INC.

Founded 1978, Fairmont, Monroe County, New York,

Storage: Oak and stainless steel. Cases per year: 4,000.

History: Casa Larga, meaning "Large House", is Andrew Colarmotolo's fulfillment of a lifetime ambition since working the family vineyards as a young man.

Estate vineyard: Casa Larga Vineyard.

Wines regularly bottled: Estate bottled, vintage-dated Chardonnay, Johannisberg Riesling, Cabernet Sauvignon, Gewurztraminer, DeChaunac, Aurora, Delaware; also Rosé, Red and White Table Wine.

CASA MADERO, S.A.

Founded 1597, Coahuila, Mexico.

Storage: Oak and stainless steel.

History: Established in the Valley of Parras (or "Valley of the Vines") site where wine was first produced in the New World from native vines of the area around 1560. The winery was founded by Don Lorenzo Garcia who arrived in Parras in the early 1580's to establish a vineyard and winery. On August 15, 1597 he received, from Philip II, King of Spain, a land grant for the formal establishment of his vineyards and winery. It is the oldest winery in the Americas in continuous operation.

Estate vineyards: San Lorenzo, San Judas, Santa Barbara.

Wines regularly bottled: Estate bottled, vintage-dated Zinfandel, Cabernet Sauvignon, Malbec, White Riesling, Colombard, Chenin Blanc, Petite Sirah.

Second label: Madero, San Lorenzo, Reserva de la Casa, Sagargnac, Evaristo, Casa Grande, Parras.

CASA NUESTRA

Founded 1980, St. Helena, Napa County, California.

Storage: Oak and stainless steel. Cases per year: 1,000.

Estate vineyard: Casa Nuestra Vineyard.

Wines regularly bottled: Estate bottled, vintage-dated Dry Chenin Blanc. Also estate bottled, vintage-dated proprietary wine: Tinto.

Casa Nuestra Estate Vineyard

Napa County, California.

Chenin Blanc vineyard.

CASA PINSON HERMANOS, S.A.

Founded 1979, Mexico, D.F.

Storage: Oak and stainless steel. Cases per year: 55,000.

Estate vineyard: Valle de Guadalupe.

Wines regularly bottled: Proprietary Don Eugenio Ruby Cabernet, Pinot Noir, Gamay, Grenache Rosé, Ernestino Chiatto Dry Red and White Lambrusco, Foylenmilch White Riesling, Spumante Sparkling White.

Second label: Alcalde, Pinson.

CASABELLO (See Casa Bello, Canada.)

CASCADE

A French hybrid wine grape producing fruity, young, crisp Red and Rosé wines.

CASCADE MOUNTAIN VINEYARDS

Founded 1977, Amenia, Dutchess County, New York.

Storage: Oak and stainless steel. Cases per year: 7,000.

History: The Wetmore name has been associated with American wine for over 100 years. William Wetmore is a relative of Charles Wetmore who planted the first Cresta Blanca vineyards in California.

Estate vineyard: Cascade Mountains Vineyard.

Wines regularly bottled: Estate bottled, vintage-dated Summertide, Reserve Red, New Harvest, Cascade Rosé; also Little White and Cascade Red.

Second label: Esprit.

"Casks"

cask

Any round, bulging wooden container for wine. Includes barrels, puncheons, pipes, butts, tuns, hogsheads, all of which signify various measures of capacity in different countries. A container used for fermenting wine usually is called a tank or vat. The two terms, in the U.S., usually are applied to containers which stand upright and are straight sided rather than barrel shaped. A vat is an open container, while a tank is closed. In U.S. wineries the capacity is stencilled on the side.

CASK (See Warner Vineyards.)

Cold Turkey, Imperial Cranberry, Champagnes, Cold Duck, King Solomon Kosher Wine, Sweet Red, Sherry, Port, Muscatel.

CASQUIERO, RICK

Rick Casquiero graduated from California State University, Hayward, in 1973, with a bachelor's degree in chemistry. He joined Weibel Vineyards in 1978 and became assistant winemaker in 1980. Working at Weibel's Mission San Jose Winery, Casquiero is responisble for the production of all premium table wines, champagnes and the 800 private labels bottled by the winery.

CASSAYRE-FORNI CELLARS

Founded 1976, Rutherford, Napa County, California.

Storage: Oak and stainless steel. Cases per year: 7,000.

History: The owners are third and fourth generation Napa Valley families.

Wines regularly bottled: Vintage-dated Cabernet Sauvignon, Chardonnay, Chenin Blanc, Zinfandel.

CASWELL VINEYARDS

Founded 1981, Sonoma County, California.

Storage: Oak and stainless steel. Cases per year: 1,000-10,000.

Estate vineyard: Gavilan Vineyard, Gold Ridge Vineyard, Winter Creek Vineyard.

Label indicating non-estate vineyard: Morelli Ranch, Hillcrest Ranch, Sonoma Vineyards.

Wines regularly bottled: Estate bottled, vintage-dated Zinfandel, Rosé of Zinfandel, vintage-dated Johannisberg Riesling, Petite Sirah, Rosé of Zinfandal, also Claret and Cyder.

Second label: Winter Creek.

CATAWBA

The American hybrid grape that produces white and pink medium to sweet fruity wines. Also used for sparkling wines.

CATOCTIN VINEYARDS

Founded 1983, Brookeville, Montgomery County, Maryland.

Storage: Oak and stainless steel. Cases per year: 4,000.

History: Formed by two grower families (the Milnes and the Wolfs) and a winemaker (Robert Lyon).

Estate vineyard: Burkettsville Vineyard, Wolf Hill Vineyard.

Wines regularly bottled: Estate bottled, vintage-dated Chardonnay, Cabernet Sauvignon, Riesling, Sauvignon Blanc, Seyval Blanc, Seyval, Vidal; also estate bottled, vintage-dated Rosé, Red and White Table Wines.

CATOCTIN VITICULTURAL AREA

Catoctin is in parts of Maryland's Frederick and Washington Counties. The viticultural area lies west of the town of Frederick, in Western Maryland, and encompasses 265 square miles (170,000 acres). It consists of a large intermountain valley and upland area bounded on the east by the Catoctin Mountains, on the west by South Mountain, on the north by the Maryland-Pennsylvania state line and on the south by the Potomac River.

CAVAS DE SAN JUAN, S.A.

Founded 1959, Queretaro, Mexico.

Cases per year: 155,000-200,000.

History: Winery is in the oldest wine producing region of the American Continent.

Estate vineyard: Viñedos Hidalgo, Viñedos Queretanos, Viñedos San Isidro.

Wines regularly bottled: Estate bottled, vintage-dated proprietary Hidalgo Cabernet Sauvignon, Pinot Noir, estate-bottled proprietary Hidalgo, Blanco Amabile, Blanco Seco, Rosado Seco, Rosé de Cabernet, Tinto San Isidro; also Sparkling Brut and Sparkling Blanc de Blancs.

Second label: La Casona.

CAVISTE (See Acacia.)

Occasionally bottled Zinfandel, Riesling, Sauvignon Blanc, Merlot and Cabernet.

Caymus Estate Vineyards

Napa County, California.

Cabernet Sauvignon, Chardonnay and Pinot Noir vineyard.

CAYMUS VINEYARDS

Founded 1972, Rutherford, Napa County, California.

Storage: Oak. Cases per year: 30,000.

History: In 1906, Charles Wagner's father purchased the land to produce grapes and prunes. In 1972, Charles and his son Chuck founded Caymus, named for a tribe of Indians who lived in the area during California's first land grants.

Estate vineyard: Caymus.
Wines regularly bottled: Estate bottled, vintage-dated Pinot Noir, Pinot Noir Blanc, Chardonnay, Cabernet Sauvignon. Also vintage-dated Fume Blanc, Zinfandel. Occasionally bottle Late Harvest Johannisberg Riesling.
Second label: Liberty School.

Cayote Creek Vineyard
Santa Clara County, Califorina.
Zinfandel vineyard.

CAYUGA WHITE
Wine grape named after one of the Finger Lakes. A New York hybrid developed at the New York Experimental Station. Produces a delicately flavored light to medium bodied white wine.

CBC (See Cordtz Brothers Cellars.)

CC VINEYARDS (See JFJ Bronco.)
Chablis, Pink Chablis, Rhinewine, Vin Rosé, Burgundy.

CEDAR HILL WINE COMPANY
Founded 1974, Cleveland Heights, Cuyahoga County, Ohio.
Storage: Oak. Cases per year: 2,000.
Wines regularly bottled: Vintage-dated Seyval Blanc, Vidal Blanc, Steuben, Dutchess, Marechal Foch, Chancellor Noir, Chambourcin, Leon Millot, Chardonnay and Sparkling Wine (Seyval/Vidal).

Second label: "Critter Wines" Terminal Red, Orchestra Red.

Cedar Lane Vineyard
Shenandoah County, Virginia.
Shenandoah Vineyard estate Vidal, Seyval, Riesling, Chardonnay, Chambourcin, Cabernet Sauvignon, Pinot Noir Shenandoah Valley vineyards.

Cedar Point Vineyard
Talbot County, Maryland.
Foch, Chancellor, Chelois, Seyval Blanc and Vidal Blanc vineyards.

Cellarmaster Vineyard
Mendocino County, Califorina.
Parducci estate Chardonnay, Cabernet Sauvignon, Merlot, Petite Sirah, Pinot Noir and Zinfandel vineyard.

CENTRAL COAST
Lying between Santa Barbara on the south and San Francisco on the north, the Central Coast encompasses Alameda, Santa Barbara, San Luis Obispo, Monterey, San Benito, Santa Clara, Santa Cruz and San Mateo counties.

CENTRAL VALLEY
One could almost call the valley the "California Grape Bowl"; covering the San Joaquin Valley in the south to the Sacramento Valley in the north. Few outstanding vineyards, but a great contributor to the "everyday" wines of America.

CENTURION
Wine grape developed at U.C. Davis. A cross pollination of Carignane, Cabernet Sauvignon and Grenache. Produces a more full bodied and darker wine than the Carnelian grape. Grown in the Central Valley of California.

Cerro Vista Vineyard
Napa County, California.
Chardonnay vineyard.

CENTRAL COAST MAP

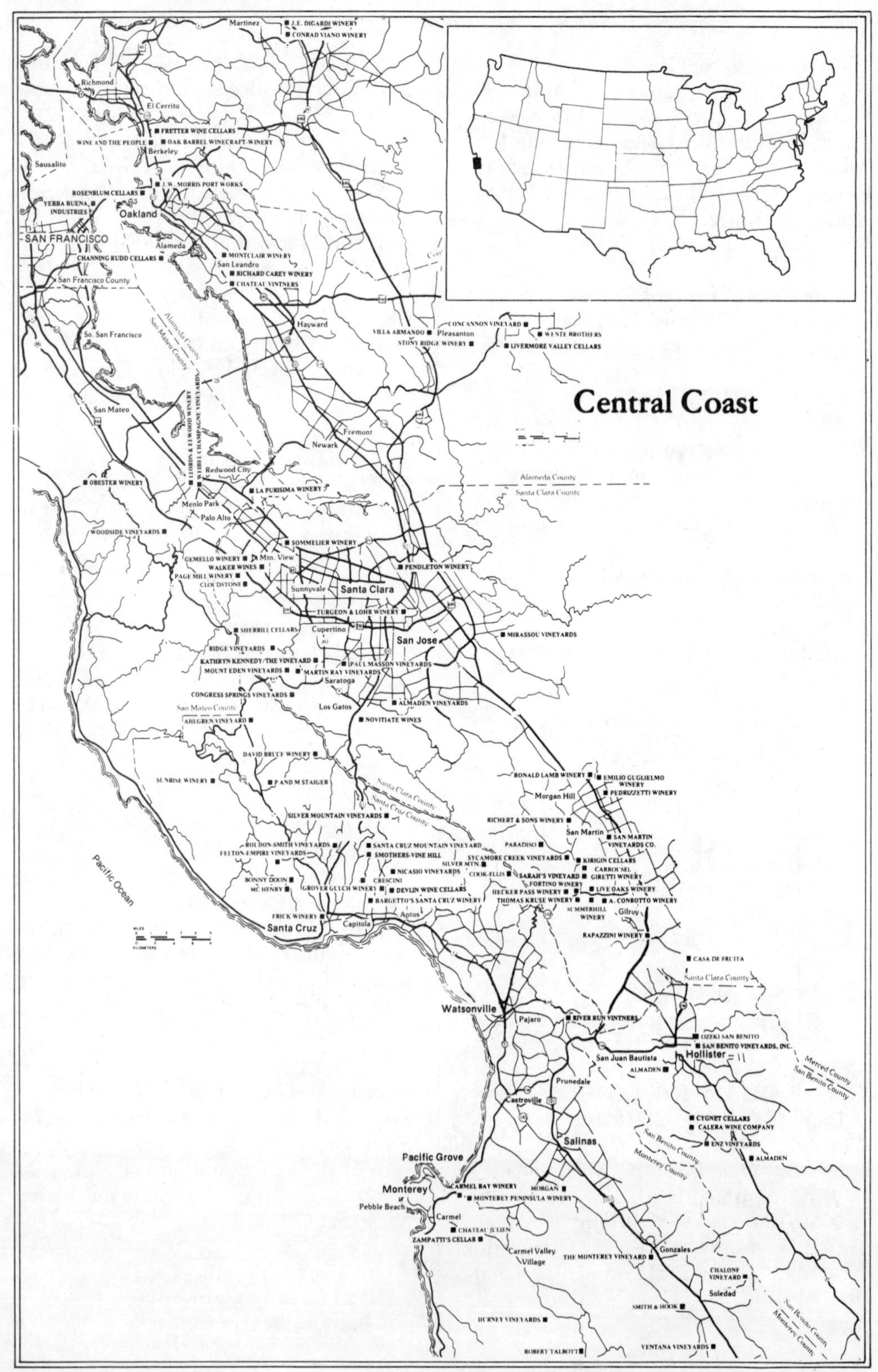

CHABLIS
A village in Burgundy where by French government regulations the wine must be 100% Chardonnay. A dry, white dinner or table wine, California Chablis has a fruity flavor, but is less tart than Rhine wine and can be made from any variety. It is delicate, light to medium straw in color and light to medium bodied.

CHADDS FORD WINERY
Founded 1982, Chadds Ford, Chester County, Pennsylvania.
Storage: Oak and stainless steel. Cases per year: 3,000.
History: Eric Miller was previously a partner (with his parents) in Benmarl Vineyards in Hudson Valley. In partnership with Hudson Cattell, he edited and published *The Pennsylvania Grape Letter and Wine News* and *Wine East Magazine*. He also served as a winemaker before starting his own winery with his wife Lee.

Wines regularly bottled: Vintage-dated Chardonnay, Chambourcin, Niagara, Steuben Rosé, Apple; and vintage-dated House Red and White, Spring Wine. Occasionally Blanc de Blancs (sparkling).
Second label: Longwood Vineyards.

Chalais Vintners
York County, Pennsylvania.
French American and Vinifera vineyard.

CHALET DEBONNÉ VINEYARDS, INC.
Founded 1971, Madison, Lake County, Ohio.
Storage: Oak and stainless steel. Cases per year: 17,000.
Estate vineyard: Chalet Debonné Vineyards, Debevc Vineyards.
Wines regularly bottled: Niagara and proprietary River Rouge; also Delaware, Pink Catawba; Red and White Wines. Occasionally Chardonnay, Foch and Cabernet Sauvignon.
Second label: Debevc Vineyards.

Flat Vineyard
Monterey County, California.
Robert Talbott estate Sauvignon Blanc Carmel Valley vineyard.

Chalk Hill Vineyard
Sonoma County, California.
Donna Maria estate Pinot Noir, Gewurztraminer, Chardonnay, Sauvignon Blanc and Cabernet Sauvignon vineyard.

Chalk Hill Vineyard
Chalk Hill, California.
Sonoma/Windsor estate Chardonnay vineyard.

CHALK HILL VITICULTURAL AREA
Chalk Hill comprises approximately 33 square miles and is located eight miles north of Santa Rosa. There are approximately 1,600 acres of producing vineyards in the viticultural area. The Russian River Valley viticultural area encompasses all of the Chalk Hill viticultural area except for a portion which overlaps into the Alexander Valley viticultural area.

CHALK HILL WINERY (See Donna Maria Vineyards.)

Chalone Estate Vineyard
Monterey County, California.
Pinot Noir, Chardonnay, Pinot Blanc and Chenin Blanc vineyard.

CHALONE VINEYARD
Founded 1920, Soledad, Monterey County, California.
Storage: Oak and stainless steel.
History: Vineyard is near the Pinnacles National Monument at 2,000 feet elevation. "Chalone" is a BATF appellation.
Estate vineyard: Chalone Vineyard.
Wines regularly bottled: Estate bottled, vintage-dated Pinot Noir, Pinot Blanc, Chardonnay, Chenin Blanc, Pinot Noir Reserve, Pinot Blanc Reserve and Chardonnay Reserve.

CHALONE VITICULTURE AREA
This area consists of 8,640 acres located on a geological bench in the Gabilan (or Gavilan) Mountain Range of Central California.

CHAMBOURCIN
French Hybrid wine grape that produces light bodied red wine. Few wineries produce as a varietal.

Chamisal Estate Vineyards
San Luis Obispo, California.
Chardonnay, Cabernet Sauvignon, Zinfandel and White Riesling Edna Valley vineyard.

CHAMISAL VINEYARD
Founded 1979, San Luis Obispo, San Luis Obispo County, California.
Storage: Oak and stainless steel. Cases per year: 2,000.
History: Norman Goss is a former concert cellist and member of the Los Angeles Philharmonic. He also founded the Stuft Shirt restaurants. The inspiration for his vineyard occurred when, as a young, man he met the famous pianist Ignace Paderewski and tasted the first bottling from Paderewski's San Luis Obispo vineyard.
Estate vineyard: Chamisal Vineyard.
Wines regularly bottled: Estate bottled, vintage-dated Chardonnay, Cabernet Sauvignon. Also occasionally bottles Sauvignon Blanc, Pinot Noir.

CHAMPAGNE BOTTLE SIZES
(Sometimes used for other wines as well.)

Name	*Servings*	*Fl. Oz.*	*Metric*
Split	1	6.5	200 ml (F bottle) (187 ml)
Tenth	2	13	400 ml (H bottle) (375 ml)
Fifth	4-5	26	800 ml (one bottle) (750 ml)
Magnum		52	1.6 liter (2X bottle) (1.5)
Jeroboam		104	3.2 liter (4X bottle) (3.0)
Rehoboam		156	4.8 liter (6X bottle)
Methuselah		208	6.4 liter (8X bottle)
Salmanzar		312	9.6 liter (12X bottle)
Balthazer		416	12.8 liter (16X bottle)
Nebuchadnezzar		520	16.0 liter (20X bottle)

CHAMPAGNE, "CHARMAT" METHOD

Charmat involves fermenting wine in large tanks and filtering and bottling in the same manner as the transfer process. This product must carry the designation "Charmat" or "Bulk Process" on the label.

Champagne Production "Naturally Fermented in This Bottle"; French: Methode Champenoise

The individual bottle-fermented method in which every stage of production takes place in the individual bottle, and the resulting product reaches the consumer in its original container. This process is referred to as the "traditional method." By law, only champagne produced by this method may bear the inscription "NATURALLY FERMENTED IN THIS BOTTLE".

The fermentation in the bottle is the art of the champagne cellarmaster or winemaker. Before bottling the blended still wines that are destined for champagne, it is necessary to add the exact amount of sugar in order to produce—supported by addition of a very active, pure cultured yeast—6 atmospheres of carbon dioxide at a temperature of 50 degrees. The carbon dioxide makes the wine sparkle as soon as the secondary fermentation in the bottle has finished.

The bottles are stacked in cool cellars of steady temperature, where they await ripening and bottle-aging.

After storage and aging of several years, the Champagne is placed on racks with holes for the necks of the bottles, that are always directly downward. Each bottle is shaken and turned, alternately, in the right and left direction, at regular intervals during a period of 8-10 weeks, until the sediment has settled in the neck of the downwardly-directed bottle, and the wine is absolutely transparent. For most Champagne procedures, this process is now fully automated, from fermentation to disgorging and corking.

Only 3-6 months before shipment takes place, the Champagne is freed from the sediment. For this purpose, the bottles with their necks placed downwards, are brought into a freezing solution and cooled below 12 degrees F. A few minutes of this immersion are sufficient to freeze the sediment and a small amount of the wine in the neck. It is now possible to turn the bottle upright and get the crown cap off the bottle. The gas pressure forces the sediment out. This is called "Disgorgement."

Before the finishing cork is put into the bottle, a small additional dosage is added. Cane sugar is dissolved in well-balanced wine and sometimes aged grape brandy, making this dosage. This creation of the dosage and the amount is the art of the winemaker. The "style" reflects the individual winemaker. The quantity of the dosage depends on the taste to suit the consumer. For "Natural" there is no addition of sugar.

Special heavy bottles made to withstand pressure of 628 atmospheres and only the finest corks are used. After being disgorged, the champagne is stored for several months before being made available for shipment.

Champagne, "Transfer Method"

The "Transfer Method" is a variation of the traditional method, which involves fermenting the champagne in the bottle, thus qualifying for a "FERMENTED IN THE BOTTLE" label designation. After fermenting in the bottle, however, the contents are removed under counter-pressure and placed in large pressurized tanks, mechanically filtered into another tank, and then refilled into empty bottles. Method used for medium priced Champagnes.

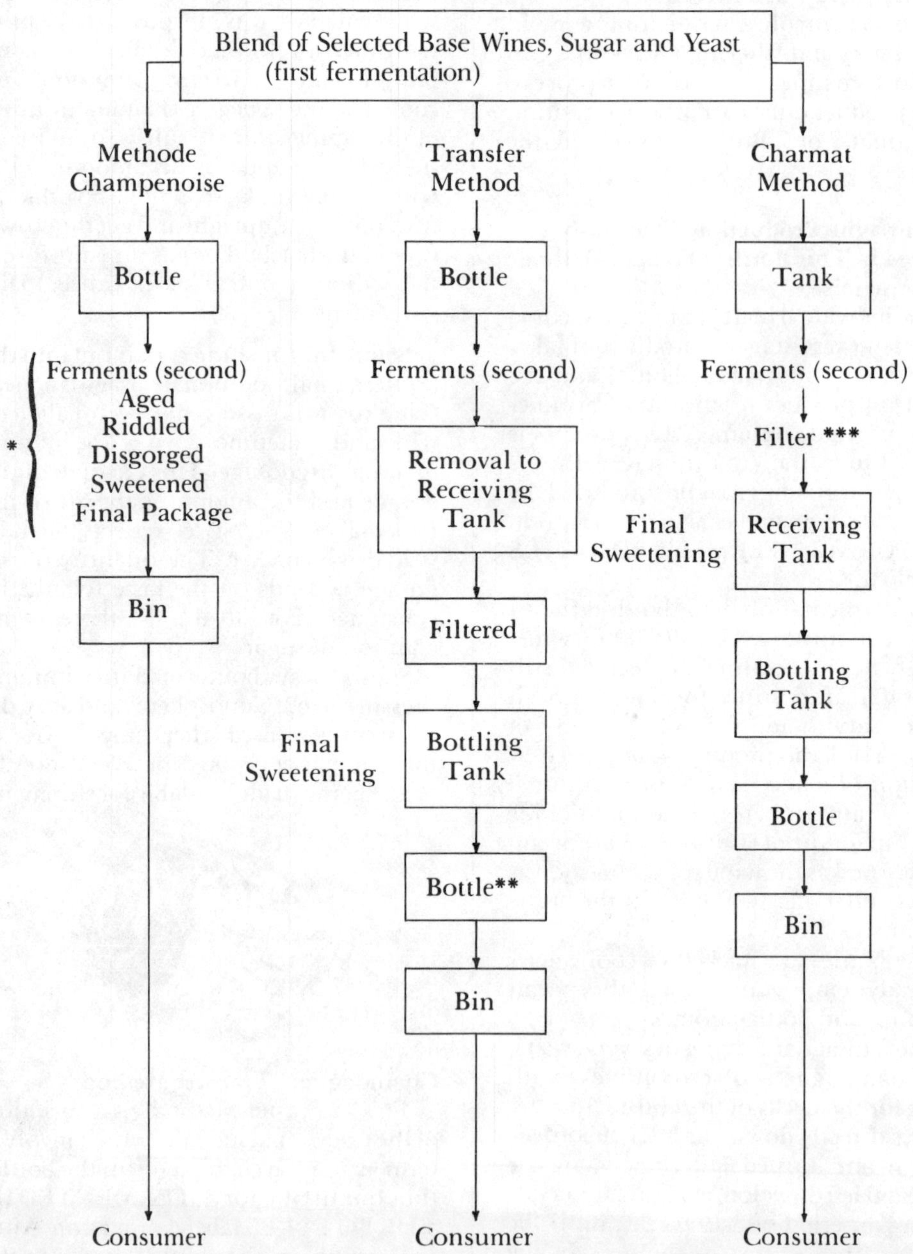

* Most of these processes now totally automated and no longer done by hand.
** Double bottling of transfer process may lead to aeration.
*** Filtering done isobarometrically in an atmosphere excluding air.

CHAMPAGNE, LEVELS OF SWEETNESS

(Either French or English term is used depending on the producer.)

French Term	*Term Used in English*	*% Sugar* Ranges Found in the Literature*	*Taste Description*
Nature, Nautural, Au Natural	Same	Sometimes used to refer to a level of sweetness lower than brut; a confusing term to use in America since in France, in general, "vin nature" refers to natural, unsweetened wine; while in the champagne district, it refers to still wine. (Wines not to be used in champagne production are called "vins nature de la Champagne.")	Very dry tasting
Brut	Brut	Up to 1.5	Dry tasting to a hint of sweetness
Extra-sec	Extra Dry ("Off-dry")	1 to 2 (sometimes up to 3)	Just barely to noticeably sweet
Sec	Dry	2 to 4	Noticeably sweet
Demi-sec	Semi-dry	4 to 6	Sweet
Doux	Sweet	8 to 10	Very sweet

*Some sparkling wines exceed the above ranges in given categories while still using the same term.

Champoeg Vineyard
Willamette Valley, Oregon.
Chardonnay vineyard.

Champs de Brionne Estate Vineyard
Grant County, Washington.
Cabernet Sauvignon, Pinot Noir, Gewurztraminer, Chenin Blanc, Chardonnay and Riesling vineyard.

CHAMPS DE BRIONNE WINERY
Founded 1983, Quincy, Grant County, Washington.
Storage: Oak and stainless steel. Cases per year: 8,000-20,000.
Estate vineyard: Champs de Brionne Vineyard.
Wines regularly bottled: Estate bottled, vintage-dated Cabernet Sauvignon, Pinot Noir, Gewurztraminer, Chenin Blanc, Chardonnay, White Riesling. Occasionally estate bottled Semillon, Merlot.

CHANCELLOR
Wine grape developed in France during the last century. The French-American grape produces a dry, medium to full bodied, fruity red wine. Grown throughout Eastern America.

CHANTER
Founded 1982, Napa Valley, California.
Storage: Oak and stainless steel. Cases per year: 3,000.
History: This is a new winery and estate vineyard being developed by Douglas and Virginia Johnson in a new lower Napa Valley vineyard area.
Wines regularly bottled: Chardonnay, Merlot.

Chanter Estate Vineyard
Napa Valley, California.
Chardonnay, Merlot Lower Napa Valley vineyard.

CHAPMAN, JOSEPH
Joseph Chapman was the first American viticulturist in California. Working at many professions in the Los Angeles area between 1820 and 1830, Chapman became acquainted with the local mission friars and their wines. He planted four thousand vines from 1824 to 1826. It is assumed that the first California vintage, by an American was produced in 1827. Chapman is known to have moved from the area by 1836.

Chappellet Estate Vineyard
Napa County, California.
Cabernet Sauvignon, Merlot, Chardonnay, Chenin Blanc and Johannisberg Riesling vineyard.

CHAPPELLET VINEYARD
Founded 1968, St. Helena, Napa County, California.
Storage: Oak and stainless steel. Cases per year: 25,000.
Estate vineyard: Chappellet Vineyard.
Wines regularly bottled: Vintage-dated Cabernet Sauvignon, Chardonnay, Chenin Blanc, Johannisberg Riesling.

chaptalizing
A sugaring of the must before, or during, fermentation to provide adequate alcohol for the type of wine being produced. Illegal in California, but is done in France and Germany for grapes which did not reach maturity. It is legally practiced in the U.S. outside California.

character
The wine's "personality" as revealed by the senses of taste and smell. The combination of vinosity, balance, style and varietal identity.

characteristic
Typical of a particular wine or type.

CHARAL WINERY AND VINEYARDS, INC.
Founded 1975, Blenheim, Kent County, Ontario, Canada.
Storage: Stainless steel. Cases per year: 30,000.

History: Charal marks the rebirth of winemaking in Kent County bordered on the south by Lake Erie.

Wines regularly bottled: Estate bottled Seyval Blanc, Marechal Foch, Dutchess, Baco Noir, Chardonnay, Riesling; also Leon Millot, estate bottled proprietary Sziegfried Rebe; and proprietary Chandelle Blanc, Rosé, Rouge.

CHARBONO

Wine grape.

A grape that Dr. Albert Winkler of U.C. Davis determined was not Barbera, but a separate varietal. A full-bodied, distinct tannic, robust, earthy wine. Very few acres planted in California. Originally from Italy. Fuller bodied than Barbera. Ages very well (best at 4 to 6 years).

charcoal

The BATF limits the use of charcoal. Instead, "activated carbon", which is clean, neutral and standardized, is chosen.

Chardonnay

CHARDONNAY

Wine grape.

The great white Burgundy vinifera grape that produces rich, dry, medium to full-bodied white wine. Even though it is believed to be older, the Chardonnay grape has been dated to 1200 A.D.

charmat

One of three methods of producing sparkling wine. Wine is fermented in stainless steel or glass lined tanks, the natural CO_2 gas generated is captured in the wine, then the wine is filtered and bottled.

CHASSELAS, WHITE OR GOLDEN

A vinifera grape that produces a light white wine.

CHATEAU BACHER (See Fretter Wine Cellars.)

CHATEAU BENOIT

Founded 1972, Carlton, Yamhill County, Oregon.

Storage: Oak and stainless steel. Cases per year: 10,000.

Estate vineyard: Chinquopin Vineyard, Lafayette Vineyard.

Wines regularly bottled: Vintage-dated Sauvignon Blanc, White Riesling, Pinot Noir and Benoit Blanc. Occasionally Chardonnay and Blanc de Blanc (Sparkling).

Second label: Crystal Creek.

CHATEAU BOSWELL

Founded 1979, St. Helena, Napa County, California.

Cases per year: 2,000.

Estate vineyard: Chateau Boswell.

Wines regularly bottled: Vintage-dated Cabernet Sauvignon.

CHATEAU BOUCHAINE

Founded 1980, Napa, Napa County, California.

Storage: Oak and stainless steel. Cases per year: 25,000.

History: The pre-Prohibition Garetto Winery was purchased and has been completely renovated.

Estate vineyard: Chateau Bouchaine.

Label indicating non-estate vineyard: Carneros Napa Valley Vineyard.

Wines regularly bottled: Vintage-dated Chardonnay, Pinot Noir, Sauvignon Blanc.

Second label: Poplar Vineyards.

Chateau Bouchaine Estate Vineyard

Napa County, California.

Chardonnay vineyard.

Chateau Camelia

Alameda County, California.

Fretter estate Pinot Noir vineyard.

Chateau Chevalier Estate Vineyard

Napa County, California.

Cabernet Sauvignon, Merlot, Pinot Noir, Cabernet Franc Napa Valley vineyard.

CHATEAU CHEVALIER WINERY

Founded 1884, St. Helena, Napa County, California.

Storage: Oak and stainless steel. Cases per year: 10,000.

Estate vineyard: Chateau Chevalier.

Wines regularly bottled: Estate bottled, vintage-dated Pinot Noir, Cabernet Sauvignon, Chardonnay.

CHATEAU CHEVRE

Founded 1979, Yountville, Napa County, California.

Storage: Oak and stainless steel. Cases per year: 4,000.

Estate vineyard: Hazen's Vineyard, Mueller Vineyard.

Wines regularly bottled: Estate bottled, vintage-dated Merlot. Soon will bottle Fumé Blanc.

CHATEAU DE LEU WINERY

Founded 1981, Suisun, Solano County, California.

Storage: Oak and stainless steel. Cases per year: 10,000-25,000.

History: The present ranch was originally planted in vineyards by the Capell family early in the 1880's. In 1954, it was acquired by the Volkhardts. In 1981, the decision was made to create their own wines and the Chateau was built on the property.

Estate vineyards: Volkhardt Vineyards.

Wines regularly bottled: Estate bottled, vintage-dated Chardonnay, Gamay, French Colombard, Fume Blanc. Also vintage-dated Chardonnay. And a proprietary vintage-dated De Leu Blanc.

CHATEAU DIANA

Founded 1981, Sonoma County, California.

Storage: Stainless steel.

History: Bulk wine is bought and bottled.

Wines regularly bottled: Vintage-dated Chardonnay, Cabernet, Petite Sirah, Chenin Blanc. Also N.V. Gewurztraminer. Some generics produced are Chablis, Rosé, Burgundy.

CHATEAU DU LAC

Lakeport, Lake County, California.

Wines regularly bottled: Vintage-dated Chardonnay, Johannisberg Riesling, Sauvignon Blanc, Cabernet Sauvignon.

Chateau Elan Estate Vineyards
Jackson County, Georgia.
White Riesling, Chardonnay and Sauvignon Blanc vineyard.

CHATEAU ELAN LIMITED
Founded 1983, Braselton, Jackson County, Georgia.
History: Ed Friedrich, who is President of Chateau Elan, was one of California's most prestigious winemakers. Chateau Elan's first crush will be in 1984.
Estate vineyard: Chateau Elan Vineyard.

Chateau Esperanza Estate Vineyard
Yates County, New York.
Chancellor Noir, Seyval Blanc and Riesling Finger Lakes vineyard.

CHATEAU ESPERANZA WINERY, LTD.
Founded 1979, Bluff Point, Yates County, New York.
Storage: Oak and stainless steel. Cases per year: 20,000.
History: Only winery in the east to be owned and operated by women. Winery is housed in a 150 year old Greek Revival stone mansion built in 1838.
Estate vineyard: Chateau Esperanza Vineyard.
Label indicating non-estate vineyard: Grow Vineyards, John Henry Vineyard, Hosmer Vineyard, Smith Vineyard, Plane's Vineyard.
Wines regularly bottled: Estate bottled, vintage-dated Johannisberg Riesling, Seyval Blanc and Chancellor Noir. Vintage-dated Chardonnay, Ravat, Aurora Blanc, Cayuga White, Foch, Cabernet Sauvignon and Gewurztraminer.

CHATEAU FILIPPI (See J. Filippi Vintage Company.)

CHATEAU-GAI WINES
Founded 1966, Scoudouc, Westmorland County, New Brunswick, Canada.
Storage: Oak and stainless steel.
Wines regularly bottled: Proprietary Alpenweiss, Capistra, San Gabriel, Cavallo, Chianno, Imperial Champagne, Spumante Classico.

CHATEAU GRAND TRAVERS
Founded 1975, Grand Traverse County, Michigan.
Storage: Oak and stainless steel. Cases per year: 20,000.
Estate vineyard: Grand Traverse Vineyard, Old Mission Vineyard.
Wines regularly bottled: Estate bottled, vintage-dated Chardonnay, Johannisberg Riesling, Merlot, Petite Sirah, Scheurebe, Pinot Blanc, Gamay Beaujolais, Chenin Blanc; vintage-dated Pinot Noir, Johannisberg Riesling; Proprietary estate bottled vintage-dated Bunch Select Ice Wine. Occasionally bottles bunch selected Riesling, Chardonnay and Chardonnay Ice Wine.

CHATEAU JULIEN/GREAT AMERICAN WINERIES, INC.
Founded 1982, Carmel, Monterey County, California.
Storage: Oak and stainless steel. Cases per year: 25,000.
History: "Julien" is derived from the Julien District of Bordeaux, the role model for the winery.
Wines regularly bottled: Vintage-dated Merlot, Sauvignon Blanc, Private Reserve Chardonnay, Cabernet Sauvignon. Also proprietary Julien Dry Sherry and Carmel Cream Sherry.

Chateau La Caia Estate Vineyard
Madison County, Alabama.
Chambourcin and Cabernet Sauvignon vineyard.

CHATEAU LA CAIA VINEYARD & WINERY
Founded 1982, Madison County, Alabama.
Cases per year: 1,500.
Estate vineyard: Chateau La Caia Vineyard.

Wines regularly bottled: Estate bottled, vintage-dated Chambourcin, Cabernet Sauvignon.

CHATEAU LAFAYETTE (See Weibel Vineyards.)

CHATEAU MONTELENA WINERY

Founded 1882, Calistoga, Napa County, California.

Storage: Oak. Cases per year: 25,000.

History: Founded by Alfred L. Tubbs, whaling tycoon and California State Senator, the winery continued operations until Prohibition. Reopened on a much more limited basis afterwards, it remained active until the late 40's. Chateau Montelena was reinstituted as a modern facility in 1972 and observed its centennial celebration in 1982.

Estate vineyard: Chateau Montelena.

Label indicating non-estate vineyard: (Grapes do not appear on label) Hanna Vineyards, Curtiss Ranches, Hafner Vineyards, Gauer Ranch.

Wines regularly bottled: Estate bottled, vintage-dated Zinfandel, Cabernet Sauvignon. Also vintage-dated Chardonnay, Johannisberg Riesling.

Second label: Silverado Cellars.

CHATEAU MORRISETTE/ WOOLWINE WINERY

Founded 1983, Woolwine, Floyd County, Virginia.

Storage: Oak and stainless steel.

Estate vineyard: Weathervane Vineyard.

Wines regularly bottled: Estate bottled, vintage-dated Seyval Blanc, Marechal Foch, Baco Noir, Vidal Blanc, Chelois, Niagara, Chardonnay, Merlot, Pinot Noir, Gamay Beaujolis.

CHATEAU NAPOLEON (See Weibel Vineyards.)

CHATEAU NOUVEAU

Founded 1980, St. Helena, Napa County, California.

Storage: Stainless steel. Cases per year: 1,000.

History: The wine is made in the "Nouveau" style using the carbonic maceration technique in which whole, uncrushed grapes are fermented in closed stainless steel tanks. This procedure reduces tannin and accentuates the fruitiness of the grapes.

Wines regularly bottled: Vintage-dated Gamay Beaujolais.

CHATEAU REIEM (See Meier's Wine Cellars.)

Champagne, Pink Champagne,

CHATEAU ST. JEAN

Founded 1973, Kenwood, Sonoma County, California.

Storage: Oak and stainless steel. Cases per year: 100,000.

Estate vineyard: St. Jean Vineyard.

Labels indicating non-estate vineyard: Belle Terre Vineyards, Robert Young Vineyards, Murphy Ranch, Jimtown Ranch, Hunter Ranch, Forrest Crimmins Ranch, Frank Johnson Vineyards, McCrea Vineyards, La Petite Etoile.

Wines regularly bottled: Estate bottled, vintage-dated Pinot Blanc, Fume Blanc, Chardonnay, Muscat Canelli. Also vintage-dated Pinot Blanc, Fume Blanc, Chardonnay, Gewurztraminer, Johannisberg Riesling, Late Harvest Johannisberg Riesling, Blanc de Blanc and Brut Champagne. Also bottle Vin Blanc.

Chateau St. Jean Estate Vineyard
Sonoma County, California.
Pinot Blanc, Sauvignon Blanc, Chardonnay and Muscat Canelli vineyard.

CHATEAU STE. MICHELLE (See also River-Ridge Winery.)
Founded 1967, Woodinville, County, Washington.
Storage: Oak and stainless steel. Cases per year: 350,000.
History: The headquarters winery is styled after a French country chateau. The winery estate was once part of the Hollywood Farm estate of Fred Simson, built in 1912 and carefully preserved. The winery pioneered the planting of vinifera grapes in the Yakima Valley. The 3,200 acre vineyards are in the Yakima Valley and the Columbia River Basin.
Estate vineyard: Cold Creek Vineyard, Hahn Hill Vineyard, Paterson Vineyard.
Wines regularly bottled: Vintage-dated Cabernet Sauvignon, Merlot, Johannisberg Riesling, Chardonnay, Chenin Blanc, Fume Blanc, Gewurztraiminer, Grenache Rosé, Semillon Blanc, Muscat Canelli; vintage Sparkling Blanc de Noir, and vintage Sparkling Brut.
Second label: Farron Ridge.

CHATEAU THAYER
(See Bardenheier's Wine Cellars.)
Vidal, Marechal Foch, Baco Noir, Chelois; proprietary Mont Rosé, Chablis, Light Chablis and Light Rosé.

Chauvet Vineyard
Sonoma Valley, California.
Zinfandel vineyard.

CHEHALEM MOUNTAIN WINERY/MULHAUSEN VINEYARDS
Founded 1973, Newberg, Washington County, Oregon.
Storage: Oak and stainless steel.
Estate vineyard: Mulhausen Vineyards.
Labels indicating non-estate vineyard: Maresh Vineyards.
Wines regularly bottled: Chardonnay, Pinot Noir, White Riesling.

CHELOIS
An American grown French hybrid grape producing dry and aromatic red wines in some districts east of the Rockies. Ages well.

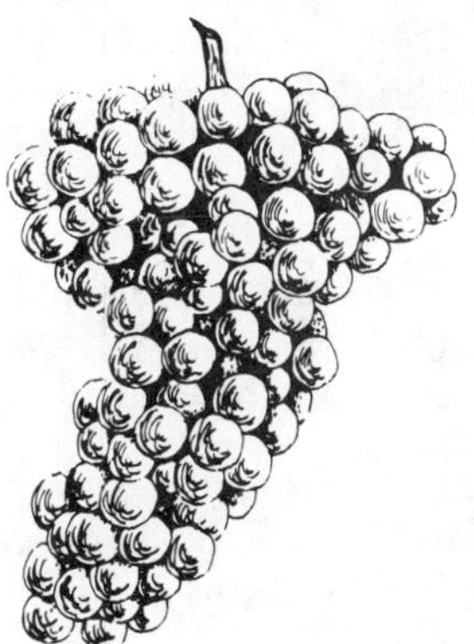
Chenin Blanc

CHENIN BLANC
The classic vinifera grape of the Loire region. Produces fresh fruity, dry to sweet white wine. The name is believed to come from Mount-Chenin in the Loire. It was known to be growing along the Loire River near Anjou around 845 A.D. Grown widely in California. Occasionally grown in the East.

CHERMONT WINERY, INC.
Founded 1978, Esmont, Albemarle County, Virginia.

Storage: Oak and stainless steel. Cases per year: 1,200-2,000.

Estate vineyard: Chermont Vineyard.

Wines regularly bottled: Estate bottled, vintage-dated Chardonnay, Cabernet Sauvignon, Riesling.

Chermont Winery Estate Vineyard

Albemarle County, Virginia.

Chardonnay, Riesling and Cabernet Sauvignon vineyard.

cherry

Wine color and taste description.

The color and/or taste of ripe red cherries.

CHERRYWOOD FARMS

(See Arbor Crest.)

Burgundy, Chablis, Cherry, proprietary Dragon Flower.

CHEVRIER (See Stony Ridge.)

Estate bottled, vintage-dated Dry Semillon; vintage-dated Dry Semillon.

CHIANTI WINE

A full bodied, ruby red wine, originally from Tuscany, strongly flavored, fruity with a medium tartness. Traditionally made from Sangiovese grapes, but other grapes are often used. Especially good with red meats and pastas. Serve at room temperature or slightly chilled.

Chicama Estate Vineyards

Dukes County, Massachusetts.

Chardonnay, Pinot Noir, Cabernet Sauvignon, Gewurztraminer, White Riesling and Merlot "Martha's Vineyard."

CHICAMA VINEYARDS

Founded 1971, West Tisbury, Dukes County, Massachusetts.

Storage: Oak and stainless steel. Cases per year: 5,000.

History: Five miles off the coast of Massachusetts lies the Island of Martha's Vineyard. The Mathieson family founded the first bonded winery in the state. The vineyards are the first commercial planting of European wine grapes in Massachusetts since Colonial days.

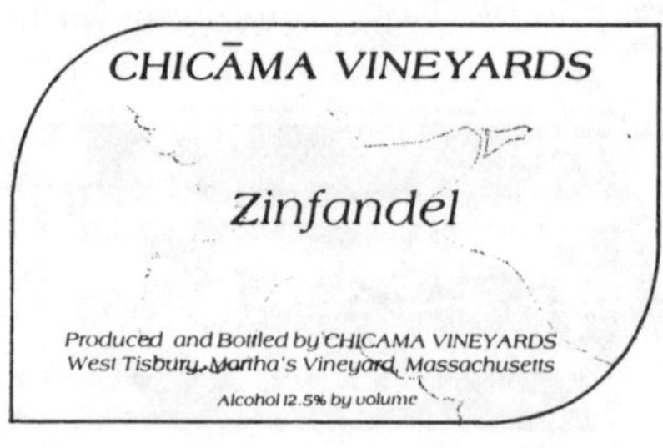

Estate vineyard: Chicama Vineyards.

Wines regularly bottled: Estate bottled, vintage-dated Chardonnay, Merlot, Gewurztraminer, Pinot Noir, Cabernet Sauvignon, Riesling; vintage-dated Ruby Cabernet, Zinfandel, Chenin Blanc; vintage-dated proprietary Summer Island Rosé, Sea Mist Sparkling Wine (Brut).

chilling wine

Many wines taste better chilled. Rosés and light-bodied reds, such as Gamay, taste fruitier and more refreshing when chilled. The crispness and character of dry and off-dry white wines is enhanced with chilling. Dry and off-dry white wines and Rosés should be served at between 50 and 55 degrees F. Light-bodied red should be served at between 55 and 60 degrees F.

The length of time it takes to properly chill wine in a refrigerator where the temperature is 42 degrees F is:

Time	*Wine Temp.*
0	65
H hr.	59
1 hr.	55
1H hr.	52
2 hr.	49

Chinqudpin Vineyard
Lane County, Oregon.
Chateau Benoit estate Pinot Noir, Chardonnay and White Riesling vineyard.

CHISPA CELLARS

Founded 1976, Murphys, Calaveras County, California.

Storage: Oak and stainless steel. Cases per year: 300-500.

Label indicating non-estate vineyard: Clockspring Vineyard.

Wine regularly bottled: Zinfandel.

THE CHRISTIAN BROTHERS

Founded 1882, Napa, Napa County, California.

Storage: Oak and stainless steel. Cases per year: 1,500,000.

History: The Brothers of the Christian Schools, popularly The Christian Brothers, is a lay religious teaching order of the Roman Catholic Church, founded in Rheims, France, in 1680, by a French nobleman and priest, St. Jean Baptiste de La Salle. The Christian Brothers are dedicated to the education of students, primarily on the secondary and collegiate level. They are not priests or monks, but laymen who dedicate themselves to a religious life, taking vows of poverty, chastity and obedience and sharing a communal life.

The Brothers established their first institution in the United States at Baltimore, Maryland, in 1848. Just twenty years later, in 1868, the Brothers moved west to California to found their first community there, and opened a novitiate, or training school, at Martinez in 1879.

Already planted on the property at Martinez were 12 acres of vineyards. The Brothers crushed the grapes and in 1882 began producing wine for their own table, and later for Sacramental purposes. Gradually, some table wine began to be sold in the neighborhood and the word spread as to its excellent quality. On a very small scale, as the demand increased, the Brothers at Martinez found themselves, by 1887, in the wine business, in addition to their primary work in education and religion.

With the growth of the Order's work on the Pacific Coast, the facilities at Martinez proved inadequate. In 1932, the Brothers moved into beautiful, new, mission-style quarters which they constructed on a large estate in the Mayacamas foothills of the Napa Valley. At Mont La Salle, besides training young men for the Brotherhood and conducting a residence school for elementary students, they have continued their winemaking.

In 1957, Mont La Salle Vineyards was incorporated as separate and distinct from but wholly owned by De La Salle Institute, the Brothers' educational and religious corporation. Like any privately owned business firm, Mont La Salle Vineyards pays federal and state income, excise and property taxes, as well as all other applicable levies. After-tax net profit of the winery is used by De La Salle Institute for the educational and religious work of the Brothers; twelve institutions of learning in California and Oregon.

Estate vineyard: The Christian Brothers.

Wines regularly bottled: Estate bottled, vintage-dated Johannisberg Riesling, Pinot St. George. Also vintage-dated Cabernet Sauvignon, Chardonnay. More varietals bottled are Johannisberg Riesling, Grey Riesling, Chenin Blanc, Chardonnay, Napa Fumé Blanc, Cabernet Sauvignon, Zinfandel, Pinot Noir. A few proprietary wines are La Salle Rosé, Chateau La Salle. The generics include Burgundy, Claret, Chablis, Rhine Wine, Sauterne, Vin Rosé, Napa Rosé, Brut Champagne, Extra Dry Champagne, Extra Cold Duck, Champagne Rosé. Dessert wines are Golden Muscatel, Treasure Port, Tawny Port, Ruby Port, Cream Sherry, Golden Sherry, Dry Sherry, Cocktail Sherry and Vintage Port. Also bottles Sweet and Extra Dry Vermouth.

The Christian Brothers Estate Vineyard
Napa County, California.

Johannisberg Riesling, Pinot St. George and Chenin Blanc vineyard.

CHRISTINA, QUEEN—Of Sweden (1632-1689)

Daughter of Gustavus Adolphus.

Her famous orders to her governor John Printz in New Sweden in the New World in 1643, "Plant grapevines in the New Sweden. I order you, John Printz . . ." New Sweden (1638-55) included parts of what are now Pennsylvania, New Jersey and Delaware.

CHRISTINA WINE CELLARS

Founded 1979, La Crosse, LaCrosse County, Wisconsin.

Storage: Oak.

Wines regularly bottled: Catawba, Niagara, Concord Grape, Montmorency Cherry, Natural Apple; generics, Chablis, Burgundy, proprietary Octoberfest Wine.

CHRISTINA WINE CELLARS

Founded 1975, McGregor, Clayton County, Iowa.

Storage: Oak and stainless steel. Cases per year: 1,000.

History: Winery is located in 100 year old former office of Diamond Jo Reynolds Steam Boat Lines along the Mississippi River. Building is on the National Historic Register. Christine Lawler is one of the "new wave" young women with a degree in enology.

Labels indicating non-estate vineyard: Terra Vineyard.

Wines regularly bottled: Niagara, Concord, Marechal Foch, Catawba; a proprietary Octoberfest; also Apple, Cranberry Apple; a semi-generic, Chablis. Occasionally other fruit wines.

CHURCH HILL (See Byrd Vineyards.)

Church Hill Rosé.

Ciapusci Vineyard

Mendocino County, California.

Zinfandel vineyard.

CIDER, HARD

Hard Cider is a fresh, fruity alcoholic beverage fermented from apple juice.

Ciel du Cheval Vineyard

Benton County, Washington.

Johannisberg Riesling, Gewurztraminer, Merlot and Chardonnay vineyard.

CIENEGA VALLEY VITICULTURAL AREA

Cienega Valley, California, located in San Benito County, is approximately five miles south of Hollister and west of the Paicines viticultural area.

Cienega Vineyard

San Benito County, California.

Almaden estate Palomino, Sauvignon Vert, Sylvaner, Johannisberg Riesling, Cabernet Sauvignon, Gamay Beaujolais, Pinot Noir, Pinot St. George, Zinfandel and Cabernet Pfeffer vineyard.

Cienega Vineyards

San Benito County, California.

Zinfandel vineyard.

CILURZO VINEYARD AND WINERY

Founded 1978, Temecula, Riverside County, California.

Storage: Oak and stainless steel. Cases per year: 8,000.

History: Vincenzo and Audrey planted their vineyard in Temecula in 1968.

Estate vineyard: Cilurzo Vineyards.

Label indicating non-estate vineyard: La Cresta Vineyard, Long Valley Vineyards, Mira Monte Vineyards, Kaarup Vineyards.

Wines regularly bottled: Estate bottled, vintage-dated Petite Sirah, vintage-dated Gamay Beaujolais, Cabernet Sauvignon, Chardonnay, Fumé Blanc, Chenin Blanc. Occasionally bottles Chenette (Petite Sirah and Chenin Blanc).

CINCINNATI, OHIO

A city in southwest Ohio, on the Ohio River.

Nicholas Longworth came to Cincinnati in 1804; rich from real estate investments and an avid horticulturist, he planted vineyards, in Cincinnati, of Alexander grapes. He later planted Catawba cuttings secured from John Adlum. By 1859, 2,000 acres of Catawbas were planted in the Cincinnati area. The city became a viticultural center. Longworth made the first champagne in America.

CLAREMONT ESTATE WINERY AND VINEYARDS

Founded 1979, Peachland, British Columbia, Canada.

Storage: Oak. Cases per year: 12,000.

Estate vineyard: Claremont Vineyards.

Wines regularly bottled: Estate bottled, vintage-dated Riesling, Muscat Riesling, Rougeon, occasionally Pinot Blanc, Gewurztraminer, Sauvignon Blanc, Johannisberg Riesling, Marechal Foch.

claret

Claret applies to any dry, pleasantly-tart, light and medium-bodied dinner wine of ruby-red color. Originated by the British to describe acceptable Bordeaux wines.

CLARKSBURG VITICULTURAL AREA

The Clarksburg viticultural area is located 12 miles south of Sacramento, California and 2½ miles from the center of quaint Clarksburg, encompassing 64,640 acres. The Sacramento River and the Delta waterways have a thermal effect on the area, and the microclimate is Region II and III. The Merritt Island viticultural area is located entirely within the Clarksburg viticultural area.

clean

A well made wine, with no alien tastes, well stored.

clear

Tasting term.

Wines free of any visible solids, but lacking the sparkling clarity of brilliant wines. As opposed to brilliance, the difference may be absence of a final filtration (which some winemakers are opposed to, believing it robs flavor).

climates and wine

European vineyardists have known for centuries varieties like Riesling, Gewurztraminer and Pinot Noir do best when planted in naturally cool climates, like Germany, Alsace and Burgundy. These varieties are never planted in warmer places like Bordeaux. Neither do the Burgundians or Germans plant Cabernet Sauvignon in their cool areas. Each region has a multitude of "microclimates", small, local areas which are warmer, cooler, wetter or drier than the general areas surrounding them.

The University of California at Davis, developed the "heat summation" method, separating the grape growing "regions" of California to determine where each variety might best be planted. The five regions relate to the number of "degree days" of each. Degree days are computed by totaling the difference between the mean temperature of each day and 50 degrees F. during the growing season (April through October 31) and then multiplying that number by the actual number of days. For example: If the average temperature for a five day period were 80 degrees F., the number of degree days would be (80-50) × 5 = 150 degree days.

The base of 50 degrees F. is used because that is the temperature at which grapes begin growing.

The general climatic character of a region can thus be determined through computing the number of degree days during the total growing season. California's five growing regions are broken into numbers of degree days. The "regions" are not geographical.

Region #1, coolest region, less than 2500 degree days.

Region #2, moderately cool region, less than 3000 degree days.

Region #3, warm region, 3000-3500 degree days.

Region #4, moderately hot, less than 4000 degree days.

Region #5, hot region, more than 4000 degree days.

CLINE CELLARS

Founded 1983, Oakley, Contra Costa County, California.

Storage: Oak. Cases per year: 4,000.

Labels indicating non-estate vineyard: Doridon Vineyard.

Wines regularly bottled: Semillon, Zinfandel.

CLINTON

An American hybrid grape that dates back to Colonial times. Produces a sweet, spicy moderately colored red wine.

Clinton Estate Vineyards

Dutchess County, New York.

Estate Seyval Blanc Hudson River Region vineyard.

CLINTON VINEYARDS, INC.

Founded 1977, Schultzville, Dutchess County, New York.

Storage: Stainless steel. Cases per year: 5,000.

Estate vineyard: Clinton Vineyard.

Wines regularly bottled: Seyval Blanc, Seyval Natural (Sparkling).

Clockspring Vineyards

Amador County, California.

Sauvignon Blanc and Zinfandel Shenandoah Valley vineyard.

clones

A clone is a group of plants which have identical genetic composition. It is seen in a single example of a variety, which is propagated for their improvement of the original variety. Clones are not separately named but are numbered with a variety name. Sometimes clones are identified on a winery's back label.

CLOS DU BOIS

Founded 1974, Healdsburg, Sonoma County, California.

Storage: Oak and stainless steel. Cases per year: 50,000.

Estate vineyard: Flintwood Vineyard, Calcaire Vineyard, Woodleaf Vineyard, Briarcrest Vineyard, Marlstone Vineyard, Cherry Hill Vineyard.

Wines regularly bottled: Estate bottled, vintage-dated Cabernet Sauvignon, Cabernet Sauvignon/Merlot, Merlot, Pinot Noir, Early Harvest Gewurztraminer, Early Harvest Johannisberg Riesling, Late Harvest Johannisberg Riesling, Chardonnay, Chardonnay proprietor's reserve. Also vintage-dated Sauvignon Blanc. Occasionally bottle Vin Rouge, Vin Blanc.

Second label: River Oaks Vineyard.

Clos du Bois Estate Vineyard

Sonoma County, California.

Cabernet Sauvignon, Merlot, Pinot Noir, Gewurztraminer, Johannisberg Riesling, Chardonnay vineyard.

CLOS DU VAL

Founded 1972, Napa, Napa County, California.

Storage: Oak. Cases per year: 28,000.

History: In 1970, Bernard Portet, a Bordelais, was asked to make a study of the best grape growing areas of the world. He stopped in Napa for a few weeks and never left.

Estate vineyard: Clos du Val.

Wines regularly bottled: Estate bottled, vintage-dated Merlot, Cabernet Sauvignon, Zinfandel. Also vintage-dated Chardonnay.

Second label: Gran Val.

Clos du Val Estate Vineyard

Napa County, California.

Merlot, Cabernet Sauvignon and Zinfandel vineyard.

CLOUDSTONE VINEYARDS

Founded 1981, Los Altos Hills, Santa Clara County, California.

Storage: Oak. Cases per year: 500.

Estate vineyard: Cloudstone Vineyards.

Wines regularly bottled: Estate bottled, vintage-dated Zinfandel. Also vintage-dated Chardonnay, Cabernet Sauvignon.

cloudy

Tasting term.

A wine containing suspended solids is cloudy.

Cloverhill Farm Vineyard

Breirysville, Pennsylvania.

French Hybrids vineyard.

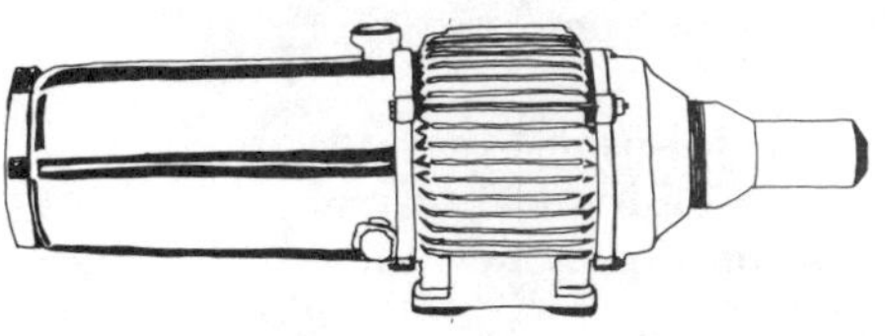

Table top hand loading foil spinner

Clovis Ranch Vineyard

Madera County, California.

Papagni estate Alicante Bouschet vineyard.

coarse and harsh

Tasting terms.

The term, coarse, is used to describe the odor and taste in wines of poor balance in which astringency, or acidity, is excessive. A harsh wine is usually similar to a coarse wine, but more unpleasant.

COAST RANGE NEGOCIANTS

Founded 1979, San Leandro, Alameda County, California.

Cases per year: 10,000.

Wines regularly bottled: Cabernet Sauvignon, Chardonnay, Fumé Blanc, Sauvignon Blanc, Zinfandel. Also California Premium Red, California Premium White. Occasionally bottle Merlot, Zinfandel.

COASTAL MIST (See Knudsen-Erath.)

Cobblestone Vineyard
Monterey County, California.
Chardonnay, Pinot Blanc vineyard.

COCETTI WINERY, LTD.
Founded 1982, St. Helena, Napa County, California.
Storage: Oak. Cases per year: 3,500.
Labels indicating non-estate vineyard: Rustridge Farms, Phil Morisoli Vineyard, Haynes Vineyards.
Wines regularly bottled: Vintage-dated Chardonnay. Also Sauvignon Blanc. Occasionally estate bottled Chardonnay, Sauvignon Blanc.

Cofran-Johnson Vineyards
Napa Valley, California.
Chardonnay vineyard.

Cohn Vineyard
Napa, California.
Cabernet Sauvignon vineyard.

Cold Creek Vineyard
Benton County, Washington.
Chateau Ste. Michelle estate Cabernet Sauvignon, and Chardonnay Columbia Valley vineyard.

COLD DUCK
A blend of White Champagne and Sparkling Burgundy and a little Concord grape wine. Cold Duck is a semi-sweet, ruby-red, light and festive wine.

cold fermentation
(See controlled fermentation.)

Cole Ranch Vineyard
Mendocino County, California.
Cabernet Sauvignon and Johannisberg Riesling vineyard.

COLE RANCH VITICULTURAL AREA
Cole Ranch in Mendocino County is located in a small, narrow mountain valley approximately one mile long and a half mile wide. There are 61 acres planted with Cabernet Sauvignon, Johannisberg Riesling and Chardonnay grapes.

Collins Vineyard
Napa Valley, California.
Conn Creek estate Cabernet Sauvignon, Zinfandel, Merlot and Cabernet Franc vineyard.

COLOMA CELLARS
Founded 1860, Escalon, San Joaquin County, California.
Storage: Stainless steel. Cases per year: 5,000.
History: The winery's first owner, Martin Allhoff joined the Gold Rush in 1849. Mining did not work out and he started to plant vines. By 1867 the wines were in great demand in Virginia City, Nevada. Owned today by California Cellar Masters.
Wines regularly bottled: Zinfandel, Gamay Beaujolais. Generics are C.M. Blanc, C.M. Rhine, Brut Champagne. Also Cream Sherry, Tinta Madera.

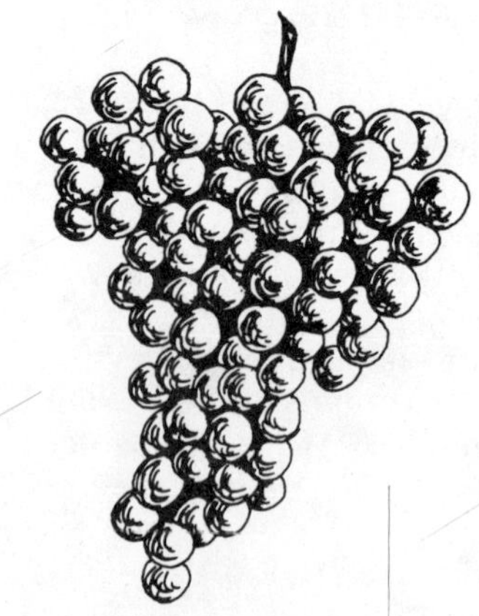

French Columbard

COLOMBARD (See French Colombard.)

Colonial Estate Vineyards
Warren County, Ohio.
DeChaunac, Baco Noir, Foch, Villard Blanc, Seyval Blanc and Aurora vineyard.

COLONIAL VINEYARDS

Founded 1974, Lebanon, Warren County, Ohio.

Storage: Oak and stainless steel. Cases per year: 1,250.

History: Winery in a barn that was built in 1857.

Estate vineyard: Colonial Vineyards.

Wines regularly bottled: Estate bottled DeChaunac, Baco Noir, Foch, Villard Blanc, Seyval Blanc, Aurora.

COLONY (ITALIAN SWISS)

Founded 1881, Asti, Sonoma County, California.

History: Originally founded as a cooperative in 1881 under the name Italian Swiss Colony later absorbed by California Wine Association and after Repeal bought by Louis Petri and sold to Allied Grape Growers (a co-op) in 1949; became a part of the second largest winemaking company in America, United Vintners. In the late 1960's the winery was purchased by Heublein. Today Colony belongs to the North Coast Growers group, another co-op.

Wines regularly bottled: French Colombard, Chenin Blanc, Rhine, Riesling, Cabernet Sauvignon, Zinfandel, Moselle, Grenache, Vin Rosé, Burgundy, Chianti, Sherry and Port.

COLONY VILLAGE WINERY

Founded 1976, Williamsburg, Iowa County, Iowa.

Storage: Oak. Cases per year: 2,000.

Wines regularly bottled: Fruit and berry Wines, Apple, Grape, Blackberry, Rhubarb, Dandelion, Red Clover, Blueberry, Cranberry, Cherry, Elderberry, Apricot.

color

The juice of most grapes has little or no color. Wines obtain their color from the presence of pigments that are present in the skins and flesh of the grapes and are released by alcohol and heat during fermentation.

White Wines. Contain several flavonoids which have been definitely identified. The interaction of oxygen with these substances, especially in the presence of the trace amounts of metallic ions, produces materials having amber or brown colors. Terms used in describing the colors of white wines follow:

1. Almost colorless or very light straw; 2. Light yellow or light straw with or without greenish tint; 3. Medium yellow; 4. Light gold; 5. Medium gold.

Amber Wines. Wines containing varying amounts of brown color, modifying the yellow, are amber colored. Amber colored white wines usually result from the action of oxygen on the wine and consequently an amber color in white table wines indicates overaging, overaeration, heating or the use of overripe grapes. Amber is naturally preferred, however, in such dessert wines as Sherry. Terms used in describing the colors of amber wines follow:

1. Light amber; 2. Medium amber; 3. Dark amber.

Red Wines. The color of red wines results from the presence of anthocyanins. In the *vinifera* grape the principal anthocyanin is malvidin monoglucoside. Differences in the reddish hue, of course, result from the color of the anthocyanins in varying amounts and from the fact that the color of the anthocyans varies with the hydrogen ion concentration of the wine. Oxygen, as in the case of white wines, also affects the color of red wines by producing brown colored oxidation products. Red wine colors are described by the following terms:

1. Pink or Rosé: Wines containing only a small amount of the anthocyans have a pink color. Most genuine Rosé wines fall in this group. An orange tint, modifying the pink, results from overaging or overoxidation.

2. Light Red: The depth of this color is above that accepted for Rosé wines, but lighter than most of the standard types of red wines.

3. Medium Red: The depth of this color applies to most Standard Red Wines.

4. Dark Red: This color is characteristic of red wines which have value for blending. They frequently have a blue or purple tint when young.

5. Tawny Color: Red wines when aged for a long time, when heated or when overoxidized, acquire a color that is a mixture of brown with red of the original. It is characteristic of Tawny Ports.

Colorado Mountain Estate Vineyards

Mesa and Fremont Counties, Colorado.

Chardonnay, White Riesling, Gewurztraminer, Pinot Noir, Sauvignon Blanc, Semillon, Zinfandel, Merlot and Cabernet Sauvignon vineyards.

COLORADO MOUNTAIN VINEYARDS

Founded 1978, Palisade, Mesa County, Colorado.

Storage: Oak and stainless steel. Cases per year: 4,000.

History: Colorado's first winery using Colorado grown grapes since Prohibition. Winery is built into the hillside and overlooks the Grand Valley with a view of Mount Farfield.

Estate vineyard: Colorado Mountain Vineyards.

Wines regularly bottled: Estate bottled, vintage-dated White Riesling, Pinot Noir Blanc, Chardonnay, Cabernet Sauvignon.

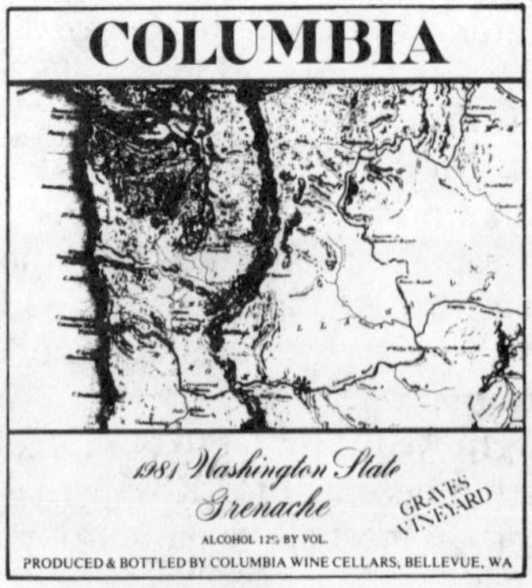

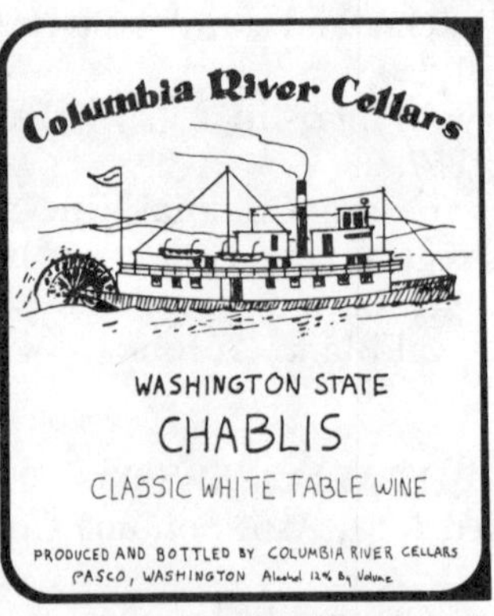

Commonwealth Estate Vineyards

Plymouth County, Massachusetts.

Aurora, Cayuga, Seyval Blanc, Foch, DeChaunac, Riesling and Chardonnay vineyard.

COMMONWEALTH WINERY

Founded 1978, Plymouth, Plymouth County, Massachusetts.

Storage: Oak and stainless steel. Cases per year: 15,000.

History: Winery is located in a colonial revival library, that was designed in 1899, and overlooks Plymouth Bay.

Estate vineyard: Commonwealth Vineyards.

Wines regularly bottled: Estate bottled, vintage-dated Chardonnay; also Cayuga White, Aurora, Seyval Blanc, Vidal Blanc, Riesling, DeChaunac, Rosé de Chaunac; also proprietary Harvest Red and White, Plymouth Red and Cranberry Apple.

CONCANNON FAMILY

Concannon Vineyard was established in 1883 when James Concannon purchased 47 acres of land in the Livermore Valley. After arriving in the United States from Ireland, Concannon held several different jobs in Boston and San Francisco. On the advice of Archbishop Alemany, the first Archbishop of San Francisco, he entered the wine business in order to supply Sacramental wines to the Catholic Church. He imported grape cuttings and oak casks from Bordeaux, and began both the production of wine and exportation of cuttings to Mexico. After the Phylloxera period, new bud cuttings on resistant rootstock from Chateau d'Yquem and Chateau Lafite were purchased from Montpellier Nursery. In 1911, James passed away but his son Joseph carried on. During Prohibition Sacramental and medicinal wines continued to be produced. This makes Concannon one of the oldest continuous wineries in America. On the death of Joseph in 1961, his sons James and Joseph continued in the family tradition. They were the first to bottle Petite Sirah as a varietal wine in America. In 1983, the family sold the winery to Distillers Company Limited, an English company. James remains as president of the winery.

CONCANNON VINEYARDS

Founded 1883, Livermore, Alameda County, California.

Storage: Oak and stainless steel.

History: James J. Concannon established the winery, in 1883, in Livermore. (See Concannon Family History.)

Estate vineyard: Concannon Vineyards.

Label indicating non-estate vineyard: Greenfield Vineyard, Arroyo Seco Vineyard, Wilson Vineyard, Tepusquet Vineyard, Rancho Tierra Rejada, Paragon Vineyard, Noble Vineyard.

Wines regularly bottled: Vintage-dated Sauvignon Blanc, Petite Sirah, Cabernet Sauvignon, Chardonnay, Chenin Blanc, Zinfandel Rosé, Livermore Riesling. Also two vintage-dated semi-generics, Chablis and Burgundy.

CONCORD

An American native labrusca grape that produces full bodied, grapey, and often, sweet wine, having what is described as a "foxy" taste. (See Ephraim Bull.)

Concord

condition

A wine's clarity or soundness.

CONESTOGA VINEYARDS, INC.

Founded 1963, Lancaster, Lancaster County, Pennsylvania.

Storage: Stainless steel. Cases per year: 2,000.

History: Oldest winery in Pennsylvania. First Pennsylvania winery to use French hybrids, which it planted in the late 1950's.

Estate vineyard: Landey Vineyards.

Wines regularly bottled: Estate bottled, vintage-dated Vidal Blanc, Chambourcin; vintage-dated Seyval Blanc, Foch, Millot; also Concord, Catawba Rosé, Niagara, Dutchess, Apple; and Red, White, Rosé, Red Rosé and Pink Rosé.

Second label: Distelfink, Landey Vineyards.

CONGRESS SPRINGS VINEYARDS

Founded 1976, Saratoga, Santa Clara County, California.

Storage: Oak and stainless steel. Cases per year: 5,000.

History: The vineyards were established in 1892. The main winery was constructed in 1912; the present winery was constructed in 1923. The current owner's first crush was in 1976.

Estate vineyard: Monmartre Vineyards.

Label indicating non-estate vineyard: St. Charles Vineyard.

Wines regularly bottled: Estate bottled, vintage-dated Chardonnay, Zinfandel. Vintage-dated Chardonnay, Pinot Blanc, Cabernet Sauvignon, Pinot Noir. Occasionally bottle Chenin Blanc, Sauvignon Blanc, Gewurztraminer and Johannisberg Riesling. Proprietary wines are Mont Blanc, Mont Rouge.

CONN CREEK WINERY

Founded 1974, St. Helena, Napa County, California.

Storage: Oak and stainless steel. Cases per year: 20,000-25,000.

History: Among the winery partners are the proprietors of La Tour Haut-Brion and La Ville Haut-Brion.

Estate vineyard: Collins Vineyard, Los Niños Vineyard.

Wines regularly bottled: Estate bottled, vintage-dated Zinfandel. Also vintage-dated Cabernet Sauvignon, Chardonnay and proprietary Chateau Maja Chardonnay.

Conn Ranch Vineyard

Napa County, California.

Cabernet Sauvignon vineyard.

CONNEAUT CELLARS WINERY

Founded 1982, Conneaut Lake, Crawford County, Pennsylvania.

Storage: Stainless steel. Cases per year: 4,000.

Wines regularly bottled: Chardonnay, Cabernet Sauvignon, Seyval Blanc, Vidal Blanc, DeChaunac, Foch; Proprietary, Princess Snowater Catawba, Wolf Island Delaware, Huidekopen Niagara, Midway Rosé, Rougeon.

CONQUISTADOR

A Florida grown grape producing deep, fruity red table wine. Developed by Dr. John Mortensen.

Ken Conrad Vineyard

Long Island, New York.

Riesling and Chardonnay vineyard.

A. CONROTTO WINERY

Founded 1926, Gilroy, Santa Clara County, California.

Storage: Oak. Cases per year: 5,000.

History: In 1906 Anselmo Conrotto migrated from the Piedmont area in Italy to Gilroy, California. He returned to Italy, but the lure of his own winery and vineyard brought him back to the Santa Clara Valley to establish his winery in 1926. His son took over the winery and vineyard in 1957.

Estate vineyard: Conrotto Vineyards.

Wines regularly bottled: Vintage-dated Barbera, Zinfandel, Petite Sirah. Also semi-generic Burgundy, Chablis and Vin Rosé.

Continente Vineyard

Contra Costa County, California.

Zinfandel vineyard.

controlled fermentation

The aim is to speed up or slow down the process as needed, usually with chilling, to prevent oxidation or damage to the delicate flavors of the wine.

Conway Vineyard

New Mexico.

Ruby Cabernet, French Colombard, Zinfandel and Carignane vineyard.

R & J COOK
Founded 1978, Clarksburg, Yolo County, California.
Storage: Oak and stainless steel. Cases per year: 58,000.
History: The first major winery in the delta growing district of Clarksburg.
Estate vineyard: Monticello Vineyard, Miller Association.
Wines regularly bottled: Estate bottled, vintage-dated Chenin Blanc (Very Dry, Semi Dry, Extra Dry), Sauvignon Blanc, Merlot Blanc, Cabernet Sauvignon, Petite Sirah, Merlot. Also Varietal White and estate bottled Varietal Red. Occasionally bottles Orange Muscat.

COOK, ROGER W.
Roger W. Cook is a fourth generation farmer in the Sacramento Delta region. A graduate of California State Polytechnic University, San Luis Obispo, Cook farmed tomatoes, sugar beets, grain, alfalfa and safflower with his father until 1971. Beginning in 1968, he also planted varietal grapes on what was to become his home ranch. In 1978, Cook met his wife Joanne. Although she had her own career in the land title business, he "lied" to her about how great the farming life was, and she agreed to marry him and help him establish a winery. The R. & J. Cook winery produced 1,000 gallons of Cabernet Sauvignon that first year. They have added several other varietals, and sales for 1983 were projected at 60,000 cases in 27 states and 2 foreign countries.

COOK-ELLIS WINERY, INC.
Founded 1981, Corralitos, Santa Cruz County, California.
Storage: Oak. Cases per year: 400.
Label indicating non-estate vineyards: Sleepy Hollow Vineyard, Vinco Vineyard.
Wines regularly bottled: Vintage-dated Chardonnay, Pinot Noir, Chardonnay, Fumé Blanc.
Second label: Buzzard Lagoon.

Cook Estate Vineyard
Yakima Valley, Washington.
The Hogue Cellars estate Riesling, Chardonnay, Gewurztraminer, Chenin, Cabernet, Merlot, Sauvignon Blanc and Muscat Canelli vineyard.

Cook's Delta Vineyard
Clarksburg, California.
Chenin Blanc and Cabernet Sauvignon vineyard.

COOK'S IMPERIAL CHAMPAGNE
Founded in 1956 by Isaac Cook, the same person Chicago's Cook County is named for. The winery was located in St. Louis until after Prohibition. Resurrected by Guild Wineries in 1970.

Cooley Ranch Vineyard
Cloverdale, California.
Zinfandel vineyard.

Cross arm trellis type vineyard expands vine canopy

cooperage

The general term used to designate containers in which wines are stored and aged. It includes casks and wooden or stainless steel aging tanks. The term is derived from the occupation of cooper—one who makes or repairs wooden containers. The cooper's art has recently been revived; several small shops assemble, repair and shave fine oak barrels from Europe. The actual manufacture of small American Oak barrels is mainly in Missouri, Arkansas, Kentucky and California.

A "cooper" forms a barrel

Cope Vineyards

Napa Valley, California.

Cabernet Sauvignon vineyard.

COPENHAGEN CELLARS—VIKING FOUR

Founded 1965, Solvang, Santa Barbara County, California.

Storage: Stainless steel. Cases per year: 7,000.

Labels indicating non-estate vineyard: La Presa Vineyard, Viña de Santa Ynez Vineyard.

Wines regularly bottled: Vintage-dated Chenin Blanc, Cabernet Blanc, Johannisberg Riesling. Occasionally bottle Chardonnay.

copper

Wine color.

A Rosé with reddish copper tone.

Coppola Vineyard

Napa County, California.

Niebaum-Coppola estate Cabernet Sauvignon, Cabernet Franc, Merlot vineyard.

CORBETT CANYON VINEYARDS

Founded 1978, San Luis Obispo, San Luis Obispo County, California.

Storage: Oak and stainless steel. Cases per year: 100,000.

History: Located on the original Pedro de Coralles land grant, deeded by the King of Spain, long before California entertained thoughts of statehood.

Label indicating non-estate vineyard: Bien Nacido Vineyards, French Camp Vineyards.

Wines regularly bottled: Vintage-dated Fumé Blanc, Chardonnay, Chenin Blanc, Gewurztraminer, Gewurztraminer Rosé, Pinot Noir, Merlot, Cabernet Sauvignon; also semi-generic Rhine, Chablis; and Red, White and Rosé Table Wine.

CORDTZ BROTHERS CELLARS

Founded 1979, Cloverdale, Sonoma County, California.

Storage: Oak. Cases per year: 10,000-15,000.

History: The land and the winery are located on the Old Musalacon Land Grant. Cordtz has been allowed to operate under the original bonded winery number 328 which was established in 1906.

Wines regularly bottled: Vintage-dated Zinfandel, Cabernet Sauvignon, Sauvignon Blanc, Chardonnay. Occasionally bottles vintage-dated Gewurztraminer.

Second label: CBC.

corky

Wine term.

A musty, unpleasant smell and taste that are present when a wine has a faulty and/or mouldy cork.

CORRAL DE PIEDRA (See Chamisal Vineyard.)

CORTEZ
Governor of Mexico who ordered the planting of grape vines in 1525. King of Spain outlawed new plantings, or replacement in 1595.

CORTI BROTHERS
Sacramento, Sacramento County, California.
History: Highly respected retailers that select and have wines finished to their specifications.
Wines regularly bottled: Vintage-dated proprietary Stony Hill Semillon de Soleil, Edmeades Cabernet Sauvignon. Also vintage-dated Amador County Zinfandel.

CORTI, DARRELL
Darrell Corti is a discriminating wine merchant. He is one of seven wine buyers for the four respected gourmet supermarkets owned by the Corti Brothers in the Sacramento, California area. After receiving a Bachelor's degree in Spanish in 1964 from St. Mary's College, Corti joined the family firm. Fluent in Spanish, French, Italian and Portuguese, Corti participates in the selection and merchandising of wines from around the world. He considers good everyday table wine as important as the rare, fine wines that are stocked. The 2,000 labels are selected for quality and interest. Corti shares his knowledge and love of wine and food in frequent speaking engagements.

COSENTINO SELECT
(See Crystal Valley Cellars.)
Bottled as Chardonnay "the Sculptor", Cabernet Sauvignon "the Creator", Merlot "the Poet".

Costello Estate Vineyard
Napa County, California.
Gewurztraminer, Chardonnay Napa Valley vineyard.

COSTELLO VINEYARDS WINERY
Founded 1982, Napa County, California.
Storage: Oak and stainless steel. Number of cases produced per year: 8,000.
Estate vineyard: Costello Vineyards.
Wines regularly bottled: Dry Gewurztraminer, Chardonnay.

Cote des Colombes Estate Vineyard
Banks, Washington County, Oregon.
Pinot Noir, Cabernet Sauvignon, Gewurztraminer Willamette Valley vineyard.

A decanting funnel

Cote des Colombes Estate Vineyard
Banks, Washington County, Oregon.
Pinot Noir, Cabernet Sauvignon, Gewurztraminer Willamette Valley vineyard.

COTE DES COLOMBES VINEYARD
Founded 1977, Washington County, Oregon.
Storage: Oak and stainless steel. Cases per year: 2,500.
Estate vineyard: Cote des Colombes Vineyard.
Label indicating non-estate vineyard: Springdale Vineyard, Graves Vineyard.
Wines regularly bottled: Estate bottled, vintage-dated Pinot Noir, Cabernet Sauvignon; vintage-dated Chardonnay, White Riesling, Gewurztraminer, Chenin Blanc. Occasionally bottle Semillon, Zinfandel, Pinot Blanc.
Second label: Les Colombes.

Cottage Estate Vineyards
Ulster County, New York.
Seyval, Foch and Chelois Hudson River Region vineyard.

COTTAGE VINEYARDS

Founded 1981, Marlboro-on-the-Hudson, Ulster County, New York.

Storage: Stainless steel. Cases per year: 250-2,500.

Estate vineyard: Cottage Vineyards.

Wines regularly bottled: Estate bottled, vintage-dated Seyval and non-vintage Red Wine.

H. COTURRI AND SON, LTD.

Founded 1979, Glen Ellen, Sonoma County, California.

Storage: Oak. Cases per year: 2,000.

Estate vineyard: Coturri Vineyard.

Label indicating non-estate vineyard: Horne Vineyards, Freiberg Vineyard, Los Vignerons Vineyard, Sobre Vista Vineyards, Cooke Vineyard, Rancho Alta Vineyard, Glen Ellen Vineyard, Quail Hill Vineyard.

Wines regularly bottled: Estate bottled, vintage-dated Zinfandel. Also vintage-dated Cabernet Sauvignon, Zinfandel, Semillon, Chardonnay, Gewurztraminer, Pinot Noir. Occasionally bottles vintage-dated Johannisberg Riesling, Chenin Blanc, Sylvaner, Apple wine.

Second label: King Wine Company, Enterprise Cellars, The Kings.

Cowan Vineyard

Amador County, California.

Zinfandel Shenandoah Valley vineyard.

COWIE WINE CELLARS

Founded 1967, Paris, Logan County, Arkansas.

Storage: Oak and stainless steel. Cases per year: 2,500.

History: Many immigrants contributed to the grape wine tradition of Arkansas; among the most notable was Joseph Bachman, the great-great-uncle of the Cowie family, who journeyed from Switzerland to Altus, Arkansas, where he devoted a lifetime to propagating new varieties of grapes. Any serious student of viticulture in Arkansas knows of his many creations, one of which won a

silver medal and diploma at the Louisiana Purchase Exposition at St. Louis in 1904.

Wines regularly bottled: Vintage-dated Cynthiana; also Dry Niagara, Noble Muscadine, Concord, Chancellor, Strawberry, Lavacaberry; a semi-generic Burgundy; a proprietary Southern Rosé (Muscadine Rosé).

Second label: River Valley Winery.

Cox Family Vineyard

Lubbock County, Texas.

Pheasant Ridge estate Sauvignon Blanc, Chenin Blanc, Chardonnay, Cabernet Sauvignon, Ruby Cabernet Sauvignon, Carignane and French Colombard vineyard.

CRAWFORD, CHARLES M.

The University of California class of 1940 included many notable winemakers. Charles M. Crawford was amongst them, graduating with a Bachelor of Science degree in Food Science. While obtaining his Master's degree in Microbiology at Cornell, Crawford began his winemaking career by working as a consultant for eight wineries through the New York State Agricultural Experiment Station. He also worked as a winemaker for Urbana Winery. He joined E. & J. Gallo Winery in 1942. He has remained there since, working in winemaking, production management, product research and development, and quality assurance. He is currently vice-president and secretary, as well as the Gallo spokesman on matters relating to wine and the technical aspects of the winery. Crawford is a member of the California Wine Institute, is past president and charter member of the American Society of Enologists, recipient of that society's Merit Award in 1966, and is past chairman of the State Department of Agriculture's Wine Grape Inspection Committee.

Ed Crawford Vineyard
Yakima Valley, Washington.
Lemberger vineyard.

cremant
French term for Champagne that is about one-half the standard effervescence. Very lightly sparkling.

CRESCENT GOLD (See Stony Ridge.)

CRESCINI WINES
Founded 1980, Soquel, Santa Cruz County, California.
Storage: Oak and stainless steel. Cases per year: 650.
Label indicating non-estate vineyard: Frank Woods Vineyard, Curtis Ranch Vineyards, Ventana Vineyards.
Wines regularly bottled: Vintage-dated Cabernet Sauvignon, Merlot, Petite Sirah, Chenin Blanc.

CRESTA BLANCA
Ukiah, Mendocino County, California.
Storage: Oak and stainless steel.
History: Established in 1882 by Charles Wetmore, a colleague of Samuel Clemens, in Livermore, California, and relocated to Mendocino County in 1972 when it was purchased by Guild.
Estate vineyard: Cresta Blanca Vineyards.
Wines regularly bottled: Estate bottled, vintage-dated Chenin Blanc, Zinfandel, Pinot Noir, Cabernet Sauvignon, Chardonnay. Also bottles vintage-dated Chardonnay, Pinot Noir, Gewurztraminer, Johannisberg Riesling, Petite Sirah, Gamay Beaujolais, French Colombard, Chenin Blanc, Gamay Rosé and vintage-dated Chablis, Blanc de Blanc Champagne (Extra Dry and Brut), Chardonnay Champagne. Dessert wines are Dry Watch Sherry, Triple Cream Sherry, Triple Dry Sherry, Tinta Port and Brandy (vintage-dated).
Second label: Mendocino Vineyards.

Cresta Blanca Estate Vineyards
Mendocino County, California.
Chenin Blanc, Chardonnay, Cabernet Sauvignon, Pinot Noir and Zinfandel vineyards.

Creston Manor Estate Vineyards
San Luis Obispo County, California.
Sauvignon Blanc, Chardonnay and Cabernet Sauvignon vineyard.

CRESTON MANOR VINEYARDS AND WINERY
Founded 1981, Creston, San Luis Obispo County, California.
Storage: Oak and stainless steel. Cases per year: 2,000.
Estate vineyard: Creston Manor Vineyards.
Wines regularly bottled: Estate bottled, vintage-dated Sauvignon Blanc, Chardonnay, Cabernet Sauvignon.

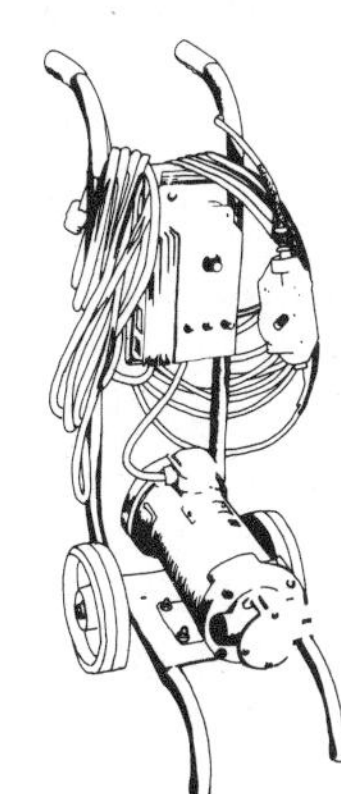

Portable wine transfer pump

CRIBARI, ALBERT B.

Albert B. Cribari was born in 1920 and lived on, or near, the family dairy farm in San Bruno, California, until 1922, when he moved to New York where his father had gone to sell grapes, which had become a big business. An interesting side note is that his Grandfather, Beniamino, had established a wine business in Morgan Hill in 1904, but had set up a dairy farm as a back-up business during Prohibition. His dairy was on the site of what is today's San Francisco International Airport.

In 1936, Albert started working at the Cellar-in-the-Sky, the Cribari bottling and Champagne operation. At the same time, he attended Fordham Prep and Fordham College. In 1941, he moved back to California and attended University of Santa Clara, graduating with a degree in chemistry. After returning from the Army in 1946, he moved to Fresno and was the Cribari winemaker until 1954. Then he became national sales manager when Cribari merged with Guild. After several years in sales, Albert returned to the winery to become winemaster in 1968.

CRIBARI AND SONS WINERY

Founded 1904, Fresno, Fresno County, California.

Storage: Oak and stainless steel.

History: Founded in 1904 by Beniamino Cribari, who was born and raised in Aprillano, Italy. He established a vineyard 5 miles west of Morgan Hill, California, in the nearby foothills of Paradise Valley. In 1924, a winery was purchsed in Fresno to produce grape concentrate. In 1946, it became the headquarters of the company.

Wines regularly bottled: Vintage-dated Napa Gamay Rosé, French Colombard, Chenin Blanc, Cabernet Sauvignon, Pinot Chardonnay, Zinfandel. Semi-generics include: Light Chablis, Chablis, Mountain Chablis, Mountain Rhine, Mountain Vin Rosé, Mountain Burgundy, Mendocino Burgundy, Mellow Burgundy, Chianti. Also Champagne, Pink Champagne, Cold Duck, Spumante, Vino Bianco, Vino Rosso, Marsala, Madeira, Sherry, Port, Vermouth (Sweet and Dry), Brandy. Vino Fiamma is a proprietary wine.

Forrest Crimmins Ranch

Sonoma County, California.

Sauvignon Blanc vineyard.

CRONIN VINEYARDS

Founded 1980, Woodside, San Mateo County, California.

Storage: French oak and stainless steel. Cases per year: 500.

Estate vineyard: Cronin Vineyards.

Label indicating non-estate vineyard: Ventana Vineyards.

Wines regularly bottled: Estate bottled, vintage-dated Cabernet Sauvignon, Chardonnay. Also vintage-dated Chardonnay, Cabernet Sauvignon, Pinot Noir.

CROSS CANYON VINEYARDS (See Ranchita Oaks Winery.)

Crossroads Vineyards

Napa County, California.

Flora Springs estate Chardonnay and Sauvignon Blanc Napa Valley vineyards.

Crosswoods Estate Vineyards

New London County, Connecticut.

Chardonnay, Vidal Blanc, Gamay Beaujolais, Gewurztraminer, Pinot Noir, Johannisberg Riesling vineyard.

CROSSWOODS VINEYARDS

Founded 1981, New London County, Connecticut.

Storage: Oak and stainless steel. Cases per year: 8,000.

History: The winery and vineyards are on a hill overlooking the Atlantic Ocean near the Mystic Seaport.

Estate vineyard: Crosswoods Estate Vineyards.

Wines regularly bottled: Estate bottled, vintage-dated Chardonnay, Johannisberg Riesling, Gamay Noir, Gewurztraminer

CROWN REGAL WINE CELLARS
Founded 1981, Brooklyn, Kings County, New York.
Storage: Oak and stainless steel. Cases per year: 10,000.
Wines regularly bottled: Concord; also Crown Classic Red and Red Wine.
Second label: Kesser.

cru
A vineyard or growth. French term.

Dumping grapes into stemmer-crusher

crush
The process of stemming and crushing grapes for wine at harvest time. The purpose is to break the skins and release the juice. Not to be confused with pressing, which comes later.

crust
Deposit of sediment, by wine, while aging in the bottle; the deposit adheres to the inside of the bottle as a crust. Crusted wines are old, bottle-aged.

CRYSTAL CREEK (See Chateau Benoit.)

Crystal Hill/Cedar Lane/Willow Run Vineyard
Shenandoah Valley, Virginia.
Shenandoah estate Vidal, Seyval, Riesling, Chardonnay, Chambourcin, Cabernet Sauvignon and Pinot Noir vineyard.

CRYSTAL SPRINGS
(See Pedrizzetti Winery.)

CRYSTAL VALLEY CELLARS/ COSENTINO WINE COMPANY
Founded 1980, Modesto, Stanislaus County, California.
Storage: Oak and stainless steel. Cases per year: 12,000.
History: Spumante d'Francesca is named for the Cosentino's 92-year-old Grandmother who made wicker baskets for Tipo Chianti in 1917.
Label indicating non-estate vineyard: Deer Creek Vineyards, Robert Young Vineyards.
Wines regularly bottled: Vintage-dated Limited Reserve Merlot, Chardonnay, Cabernet Sauvignon. Proprietary wines are Crystal Fumé (vintage-dated) and Robins Glow Blanc de Noir (non-vintage) and Spumante d'Francesca. Also bottle Champagne Rosé, Extra Dry Champagne.
Second label: Cosentino Wine Company.

CUCAMONGA
Indian meaning: Land of Many Waters. Once a key California wine region southeast of Los Angeles.

CUCAMONGA VINEYARDS
Cucamonga, San Bernardino County, California.
Storage: Oak and stainless steel.
Wines regularly bottled: Cabernet Sauvignon, Chenin Blanc, Petite Sirah, Johannisberg Riesling, Moscato de Primo. Semi-generic wines are Burgundy, Rosé, Chablis. Sparkling wines produced are Limited Edition Brut and Extra Dry Champagnes, Moscato Spumante, Cuvée d'Or Extra Dry Champagne, Pink Champagne, Sparkling Burgundy, Cold Duck.

JOHN CULBERTSON WINERY
Founded 1981, Fallbrook, San Diego County, California.
Storage: Stainless steel. Cases per year: 5,000.

Estate vineyard: Rancho Regalo del Mar.

Wines regularly bottled: Champagne (Brut, Natural, Cuvée Tranquille) all vintage-dated.

Cullinan Vineyard

Sonoma County, California.

Zinfandel vineyard.

Curtis Ranch

Napa Valley, California.

Chardonnay, Cabernet Sauvignon, Cabernet Franc, Pinot Noir, Sauvignon Blanc, Gewurztraminer and Merlot vineyard.

Sonoma-Cutrer Vineyards

Founded 1981, Sonoma County, California.

Sonoma-Cutrer Vineyards estate Chardonnay Russian River Valley vineyard.

cutting

In viticulture a segment of the cane, or branch, of a grapevine cut during the dormant season and used for asexual propagation of new vines identical to the parent (clonal propagation). Most winegrape experimental stations work with cuttings. Placed in earth or other suitable medium, cuttings produce roots and grow into mature vines.

CUVAISON

Founded 1970, Calistoga, Napa County, California.

Storage: Oak and stainless steel. Cases per year: 20,000.

Estate vineyard: Cuvaison Carneros Vineyard.

Wines regularly bottled: Vintage-dated Chardonnay, Cabernet Sauvignon, Zinfandel.

cuvee

Literally, the contents of a cask of wine—usually refers to an especially prepared blend of wines such as a blend of still wines before secondary fermentation, to produce champagne.

CUVÉE DU VIGNERON (See Benmarl Wine Company, Ltd.)

CYGNET CELLARS

Founded 1977, Hollister, San Benito County, California.

Storage: Oak and stainless steel. Cases per year: 2,000.

Wines regularly bottled: Vintage-dated Zinfandel, Chardonnay, Cabernet Sauvignon, Carignane, Petite Sirah, Pinot Noir, Pinot St. George.

CYNTHIANA

Wine grape.

An American hybrid grape, grown mostly in Missouri and Arkansas, producing good red wine that has a non-foxy aroma. Ages well.

Cypress Valley Estate Vineyard

Blanco County, Texas.

French Colombard and Riesling vineyard.

CYPRESS VALLEY WINERY

Founded 1978, Cypress Mill, Blanco County, Texas.

Storage: Oak and stainless steel. Cases per year: 15,000.

Estate vineyard: Cypress Valley Vineyard.

Wines regularly bottled: Estate bottled, vintage-dated French Colombard, Riesling; vintage-dated Red and White Table Wine.

D'Agostini Vineyard
Amador County, California.
Zinfandel vineyard.

Dal Porto Vineyards
Amador County, California.
Zinfandel vineyard.

Dante Dusi Vineyard
San Luis Obispo County, California.
Zinfandel vineyard.

DAQUILA WINES

Founded 1981, Seattle, King County, Washington.

Storage: Oak and stainless steel. Cases per year: 2,500.

Label indicating non-estate vineyard: Sagemoor Vineyards.

Wines regularly bottled: Vintage-dated Sauvignon Blanc, Gewurztraminer, Semillon, Merlot, Muscat Blanc, Rosé.

DARE, VIRGINIA

Daughter of Ananias and Elenor Dare, members of the Roanoke Island colony, North Carolina. Virginia Dare was the first child born of English parents in North America. What fate befell Virginia Dare and her family and the remaining colonists is only speculation. Virginia Dare is the oldest wine label (brand) in the United States. It is owned by Canandaigua Wine Company.

DATES, WINE HISTORY

1524 Hernando Cortez, governor of New Spain (Mexico) ordered every Spaniard (with a land grant and Indian labor) to plant 1,000 grape vines per 100 Indians.

1524 Verranzo, the Italian navigator, notes the Scuppernong grape while exploring Cape Fear River Valley of North Carolina.

1562 The Scuppernong grapes are used by the French Huguenots in Florida to make the first American wine.

1584 Sir Walter Raleigh found the Scuppernong on Roanoke Island in Virginia.

1593 Mexico's first commercial winery is established in Coahuila.

1593 Vinicola del Marques de Aguayo, the oldest winery in Mexico (in Parras de la Fuente) begins. It is rebuilt in 1965.

1605 -1614 Mission grape vineyards are planted in Albuquerque, New Mexico.

1619 Lord Delaware brings vinifera grape vines, and French vignerons to tend them, to Virginia.

1622 Lord Baltimore instigates the growing of grapes for wine but the plantings of European grape vines do not succeed.

1623 A law is passed by the

Virginia Assembly compelling every homeowner to plant ten vinifera vines.

1626 Bodegas de San Lorenzo of Casa Madero, Mexico's second winery, is built. And rebuilt in 1962.

1632 George Winthrop begins paying for the island Martha's Vineyard with wine (a hogshead of wine each year).

1632-1773 Attempts to grow the European wines in Massachusetts, Maryland, Pennsylvania, New Sweden, South Carolina, Rhode Island, New York are all unsuccessful.

1636 Jesuit missionaries of Quebec are producing wine for Sacramental purposes, using wild-grapes.

1683 William Penn tries, but fails, to start a vineyard of imported French and Spanish grapes.

1700 Wine making begins in Chile, Argentina, Peru.

1750 200,000 vines are planted in the middle Rio Grande Valley.

1769 Franciscan Fathers establish Mission San Diego; plant first California vineyard there.

1771 Indians press grapes in California's oldest winery located behind Mission San Gabriel.

1777 Spanish padres cultivate Santa Clara County's first grapes at the Santa Clara Mission.

1797 Spanish padres plant Alameda County's first cultivated vines, brought from Spain via Mexico and Mission San Diego.

1801 There are now 12 missions in California making wine.

1811 The Scuppernong grape gets its name from a newspaper article about James Blount's census of the town of Scuppernong and the grapes grown there.

1816 William Prince, of New York, introduces the Blue Isabella of South Carolina.

1818 John Eichelberger begins growing the hardy red Alexander (discovered by John Alexander) for wine, making Pennsylvania the first commercial wine growing state in North America.

1819 *The America Farmer*, October 1819, stated:

"Many farmers near Fayetteville in North Carolina have for years past drunk excellent wine of their own making from the native grape . . . Wine is made along the Cape Fear River from Fayetteville to the Sea, a distance of nearly seventy miles, and the farmers use it freely as cider is used in New England."

1823 Major John Adlum, of Washington DC, introduces the Pink Catawba.

1824 Nicholas Longworth creates America's first Champagne (from the Catawba grape). Sparkling Catawba inspires Longfellow's "Ode to Catawba Wine".

1824 Joseph Chapman, early settler, sets out 4,000 vines at Pueblo of Los Angeles.

1824 Padre Jose Altimira, Spanish founder of Sonoma Mission, sets out more than 1,000 vines. (Sonoma County's first cultivated vines were planted by Russian colonists at Fort Ross in 1812. They left 32 years later without having had any appreciable effect on California winegrowing.)

1826 Commercial winemaking begins in Los Angeles.

1831 Jean Louis Vignes, Frenchman from Bordeaux, starts commercial vineyard on present site of Los Angeles Union Railroad Station, becoming first person to make winegrowing a business in California.

1832 Brookside Vineyard Company is founded by Marius Biane.

1836 General Mariano Vallejo, last Mexican military commandant at Sonoma, takes over Sonoma Mission's vineyards; replants and revives vines unattended after Mission was appropriated by Mexican government in 1834.

1838 George C. Yount establishes the first homestead in Napa County and plants vines from Sonoma Mission.

1839 Tiburcio Tapia, on a Cucamonga land grant from the governer of Mexico, starts California's first winery, Thomas Vineyards.

1839 Brotherhood Winery, the first commercial winery in New York opens in the Hudson River Valley. Founded by a French shoemaker, Jean Jacques, it is the oldest still active winery in the United States.

1840-1883 Wine families Biane, Mirassou, Wente, Concannon enter the wine business and continue to date in California.

1841 Captain John Sutter, originator of Sacramento County's wine industry, begins his cultivation of native wild grapes at Sutter's Fort.

1843 Ephraim Bull plants a grapevine (that he names Concord) that is still growing next to his cottage in Concord, Massachusetts.

1846 Dr. John Marsh presses first vintage in Contra Costa County, California.

1850 Captain Charles M. Weber founds the city of Stockton, California, and lays out San Joaquin County's first vineyard.

1850 California wines are being shipped around the world.

1852 The Thompsons prove that irrigation isn't necessary to successfully grow grapes in Napa Valley.

1852 Charles LeFranc makes first commercial planting of European wine grapes in Santa Clara County, California.

1852 Etienne Thée establishes Almaden Winery in California.

1854 Pierre Pellier establishes vineyards and winery in Santa Clara County, California.

1854 George Krause plants first vines in Stanislaus County, California.

1854 Ephraim W. Bull, of Massachusetts, introduces the Purple Concord.

1854 Charles Kohler and John Frohling, two musicians from Germany, start the first winery in San Francisco.

1855 California wines are exhibited for the first time at California State Fair.

1856 Colonel Agoston Haraszthy, "father of California wine growing," purchases vineyard in Sonoma and transplants 13,000 vine cuttings, including six foreign varieties.

1856 Pierre Pellier returns from France with a wealth of grape varieties, some of them the same "stock" that are cultivated by the Mirassou family today.

1856 Joseph Osborn establishes Oak Knoll Farm which is (today owned by the Beringer Brothers) the oldest continuing vineyard and winery in Napa County, California.

1857 Missouri now produces sparkling wine, calling it Sparkling Catawba.

1857 Pierre Sansevaine creates the first sparkling wine from Califronia grapes.

1859 General Don Mariano Guadelupe Vallejo dominates the Northern California wine industry (first non-clerical winegrower in Sonoma Valley.)

1860 Charles Krug establishes his winery in Napa Valley.

1860 Stevenson P. Stockton. Stock Life Oak Vineyard, acquires his first land and within two years has 12 acres of wine grapes.

1860 Missouri surpasses even Ohio as leader in wine producing.

1861 Legislature of California enacts resolution for a commission to improve the grape and wine industry in California. This farsighted action results in appointment of Agoston Haraszthy to the commission and his subsequent voyage to Europe. He brings back 100,000 cuttings of 300 grape varieties to distribute throughout California.

1861 Famed Trinity Vine (named for the Holy Trinity because of its three main trunks) is planted at Mission San Gabriel near Los Angeles on site of early Franciscan Mission and Winery.

1862 Jacob Schram founds his winery south of Calistoga, California.

1863 Pleasant Valley Wine Company begins making New York's first Champagne, calling it Sparkling Catawba.

1868 The Niagara grape, a cross between the Concord and the Cassady, is created.

1869 Leland Stanford, railroad builder, governer, and founder of Stanford University, starts a long association with grapes and wine by purchasing land at Warm Springs, in Alameda County, California, where 350 acres of vines are planted and a winery is constructed. In 1881, Senator Stanford acquires Vina in Tehama County, which becomes the largest vineyard in the world, having, by 1888, a total of 3,575 acres of vines. (Since destroyed.)

1870 Georgia catches up and takes the lead in wine producing.

1870 Jacob Rommel, of Missouri, introduces the green Elvira.

1873 Barnes Winery, the oldest in Canada, is founded near the original Welland Canal.

1874 In Chautauqua, New York, plans are made for forming the Womens Christian Temperance Union in Cleveland (which evolved into the Prohibition Act of 1920).

1874 Sutter Home Winery at St. Helena, California, is built by John Thomann, a Swiss-German winemaker.

1877 Beringer Brothers founded their winery at St. Helena.

1878 Eastern wine producers accuse California producers of putting French and German labels on some California wines, and California labels on Eastern wines. To retaliate, they labeled the *worst* European wines with California labels.

1879 Original Alta Vineyard Cellar established by Captain C.T. McEachran, a Scottish sea captain.

1879 Gustav Niebaum, a Finnish sea captain begins Inglenook Winery in Napa County, California.

1880 Italian Swiss Colony established by Andrea Sbarbaro.

1880 Experimental grape growing station and courses in viticulture are established at the College of Agriculture of the University of California at Berkeley. Today, on the Davis campus of the university, there exists the most complete collection of grape varieties and species in the world. America's best library of foreign and domestic literature on grapes, wines and brandies, chemical and microbiological laboratories and fermentation and conditioning rooms, cellars, pilot plants and extensive vineyards. The teaching staff is composed of experts of worldwide reputation.

1880 United States starts a special department of wine research.

1881 Stanford acquires Vina and makes it the largest vineyard in the world.

1882 Chateau Montelena, in Napa County, California, built by whaling tycoon Alfred Tubbs, is designed to resemble a French chateau.

1882 Charles A. Wetmore founds Cresta Blanca Winery in Alameda County, California.

1882 Walter Taylor plants his first vines in Hammondsport, New York, and begins the empire of The Taylor Wine Company.

1883 The Wente family establishes its first winery in California.

1885 George Hearst (United States Senator, miner, founder of San Francisco Examiner) acquires Madrone Vineyards.

1886 Jose DaRosa begins the family tradition of winegrowing continued today by his son Edward DaRosa.

1886 Korbel begins producing champagnes, in Sonoma County, California.

1886 United States Senator Fair (of Fairmont Hotel) invests in a large vineyard and winery, on the Petaluma River, which includes the first continuous brandy distillery on the West Coast.

1889 James Concannon introduces better wine grapes in Mexico.

1889 Cresta Blanca wines win two gold medals at the International Exposition in Paris.

1889 Thirty-five gold, silver and bronze medals are

awarded to California wines at the Paris Exposition.

Late 1800's Virginia claret (the Norton grape produced by Dr. Norton) becomes world famous.

1890 There is a glut on the market of Niagara grapes and production all but stops until 1960.

1900 California wine wins gold medal at Paris Exposition.

1900 Secondo Guasti plants vines in the desert to start the Cucamonga desert wine industry in Southern California.

1900 Beaulieu Vineyard is started by Georges de Latour at Rutherford, California.

1904 Captain Paul Garrett wins grand prize for his sparkling wine at the Louisiana Purchase Exposition held in St. Louis. (Sparkling Scuppernong.)

1908 Ernest Wente is the second student ever to enroll at the University of California at Davis.

1910 The Mexican wine industry is booming.

1917 Anna Held takes her famous Champagne bath.

1919 Ohio votes for Prohibition one year before rest of nation.

1920 National Prohibition goes into effect.

1929 "Captain" Paul Garrett (of Virginia Dare) starts a combine of California and New York wineries, called Fruit Industries, Inc., to get around Prohibition by selling grape concentrate ("Vine-Glo") to the consumers who make wine from it.

1929 Chateau Gai in Canada is the first winery in North America to make Champagne by the Charmat process.

1929 On election day, Nazario Ortiz Garza begins his career as Governer and as the largest producer of wine and brandy in Mexico.

Depression Pasquale Capozzi, a grocer, and William A.C. Bennett, a hardware merchant, begin Calona Winery in Canada, first with unsaleable apples and then with California grapes.

1933 Prohibition is repealed.

1933 The first wineries begin to operate in Washington State.

1934 The wines of British Columbia are beginning to be made from California grapes.

1934 Wine Institute is founded, an independent, nonprofit, voluntary membership association of the California wine industry. It exists to enable wine growers to accomplish collectively what cannot be done alone.

1934 Vinicola del Vergel, the oldest winery at Gomez Palacio, is built. It has one of "the most spectacular Old World underground cellars in North America." (Leon Adams).

1934-1940 Wine consumption grows 200%.

1938 The Grape Protectorate Law, requiring growers to convert 45% of their grapes into brandy, helped save the California wine industry and launch its brandy industry.

1939 Frank Schoonmaker's Selections adds grape names never heard of before and starts a huge list of "varietal" wines.

1939 Society of Medical Friends of Wine is organized in San Francisco to foster research and interest in the medical uses of wine.

1941-1945 During World War II years, wine industry produces chemicals, backs bond drives, provides brandy for first-aid kits and turns over tank cars to government.

1941 Henry Sonneman buys vineyards on one of the Bass Islands of Ohio and with the help of research, revives the Ohio wine industry by promoting research and study on grape growing along the Ohio River.

1945 Daily service of California wines is a feature at the United Nations Conference on International Organizations held in San Francisco.

1946-52 Garza, as Secretary of Agriculture, establishes vineyards throughout a large section of Mexico by distributing grapevines to anyone willing to start a vineyard.

1947 Mogen-David has its start because its kosher wine was all that was left to sell when the Wine Company of America tanks of bad wine all had to be dumped.

1951 High Tor Vineyard is started by playwright Everett Crosby.

1956 James Zellerbach ages his Chardonnay in French oak barrels and changes the whole flavor of California Chardonnay.

1957 Louis Petri builds the first American wine tanker for seven million gallons.

1958 The Gallos build a glass factory, the first and only by a winery.

1960 European growers begin putting varietal labels on their wines.

1960 Ohio Agriculture Research and Development Center establishes research vineyards in Southern Ohio. The first major attempt to restore the grape industry again to Southern Ohio.

1962 Ficklin Vineyard produces the first fine table wines from Region V (where only dessert wines were produced).

1963 Bern Ramey and James Allen begin producing bottle fermented champagne in a wine cellar located beneath an Illinois Central Railroad Station near Chicago.

1963 Dr. Konstantin Frank establishes his Vinifera Wine Cellars Vineyard on Keuka Lake, Hammondsport, New York. He spearheaded vinifera growing east of California.

1964 Dr. Wozniak, of Bronte Cellars, claims to be the world's first producer of "Cold Duck"; a fact disputed by Germany's claim of selling "Kalte Ente" in the late 1940s.

1965 Chateau Gai becomes Canada's first exporter of wine to Great Britain.

1965 The only winery in New Hampshire, White Mountain Vineyards, is started by the Canepas.

1968 Vines planted at Highburg, the Plains, Virginia. Vinifera Wine Growers Association experiment.

1970 Real estate taxes on vineyards destroyed the huge Santa Clara wine industry (land values soared as a result of the great migration west after World War II.)

1970 Beginning of the modern wine grape explosion into Monterey County. Between 1970-1974, wine grape acreage in this new region increased 30 fold, from about 1,000 acres to over 30,000!

1972 Alaska's first winery is started by a Catholic priest, Emet R. Engel, producing milk wine.

1972 A group of nine poor families in Oklahoma are the innovators of a project that ended with 300 welfare families successfully planting vineyards.

1973 Research discovers that 9/10's of the kosher wine drinkers are not Jewish.

Barrel aging in caves

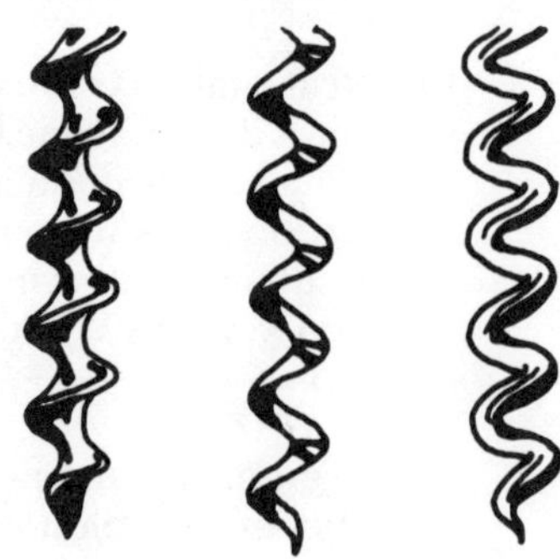

Three types of corkscrews: web-helix, wire helix and an auger

DAUGHTERS WINE CELLAR

Founded 1979, Madison, Lake County, Ohio.

Storage: Oak and stainless steel. Cases per year: 2,000.

Label indicating non-estate vineyard: Green Farm, Presque Isle; Walter Fruit Press; Grand River Vineyards; Wine and the People.

Wines regularly bottled: Agawam, Dutchess, Delaware, Catawba, Chancellor Rosé, Cabernet Sauvignon; also proprietary Rosanna Rosé.

THE DAUMÉ WINERY

Founded 1982, Ventura County, California.

Storage: Oak and stainless steel. Cases per year: 2,000.

Label indicating non-estate vineyard: Tepusquet Vineyards.

Wines regularly bottled: Vintage-dated Chardonnay, Pinot Noir. Also Vin Gris.

DAVIES, JACK L.

Jack L. Davies began his career in wine when he purchased the historic Schramsberg Vineyard's in 1965. He had previously held several marketing and managing positions, after graduating from Harvard University in 1950 with a Master's in Business Administration. Serving as General Manager at Schramsberg, he has restored the winery as a producer of premium California champagne. In addition to his speaking and writing on wine and management topics, Davies is involved in many industry activities. He is vice-Chairman and director of the California Wine Institute, past president of the Napa Valley Vintners, past president of Wine Service Cooperative Association, member of the San Francisco Wine and Food Society and the Society of Bacchus. He is a Supreme Knight in the Brotherhood of the Knights of the Vine.

DAVIS

The home of University of California, Davis (world famous for its Department of Viticulture and Enology).

DAVIS MOUNTAIN WINES, INC.

Founded 1982, Jeff Davis County, Texas.

Storage: Oak and stainless steel.

History: Gretchen Glasscock found a premium microclimate for her vineyard through research including thermal scans from a satellite.

Estate vineyard: Glasscock Vineyards.

Wines regularly bottled: Estate bottled, vintage-dated proprietary Chenin Blanc, Sauvignon Blanc, Cabernet, Chardonnay.

De Carle Vineyard

Napa County, California.

Beringer estate Cabernet Sauvignon vineyard.

DE CHAUNAC

Wine grape.

A French hybrid grape producing dry, medium to light bodied red wine. Similar to Gamay. Named after Aldemar DeChaunac, an Ontario wine pioneer.

De Loach Estate Vineyards

Sonoma County, California.

Zinfandel, Pinot Noir and Gewurztraminer vineyard.

DE LOACH VINEYARDS
Founded 1975, Santa Rosa, Sonoma County, California.
Storage: Oak and stainless steel. Cases per year: 22,000.
Estate vineyard: De Loach Vineyards.
Wines regularly bottled: Estate bottled, vintage-dated Zinfandel, Pinot Noir, Gewurztraminer, White Zinfandel. Also vintage-dated Chardonnay, Fumé Blanc.

DE LUCA, JOHN A.
John A. DeLuca was appointed president of the California Wine Institute in 1975. He has brought his broad experience in agricultural, governmental and academic affairs to the job. After receiving a master's degree in Soviet studies from Harvard University in 1958, DeLuca traveled throughout the U.S.S.R. as part of a cultural exchange program. In 1964-1965, he taught courses in American Foreign Policy and International Relations at California State University, San Francisco. The following year, he served as a White House Fellow, then worked in 1966-67 as a special assistant to Senator Frank Church of Idaho. He obtained a Ph.D. in political science from the University of California, Los Angeles, in 1968. For the next seven years, DeLuca served as Deputy Mayor of San Francisco and lectured at several Bay Area universities on issues in municipal government, urban and state politics and fiscal policy. His involvement in California State agriculture has included presidency of the California Agri-Council on International Trade, and Membership on the Agricultural Policy Advisory Committee.

De May Estate Vineyards
Steuben County, New York.
Concord, Delaware, Niagara, Baco Noir and Landot Keuka Lake vineyard.

DE MAY WINE CELLARS
Founded 1975, Hammondsport, Steuben County, New York.
Storage: Stainless steel. Cases per year: 3,000.
Estate vineyard: De May Vineyard.
Wines regularly bottled: Estate bottled, vintage-dated Delaware, Niagara, Landot, Baco Noir; Champagne; also estate bottled vintage-dated Vin Rosé and Chablis Blanc.

De Santis Vineyard
Santa Clara County, California.
Carrousel estate Cabernet Sauvignon vineyard.

DE TREVILLE LAWRENCE, R., SR.
After retiring from a position in the U.S. Department of State in 1972, R. De Treville Lawrence, Sr. was able to devote his time to serious experimentation in growing premium wine grapes on the East Coast. In 1973, he created the non-profit Vinifera Wine Grower's Association. Operating out of Virginia, the organization has members in nearly every state and in nine foreign countries. The "VWGA Journal" has been published quarterly for ten years. VWGA activities include annual conferences, winegrowing seminars, a free information service and the Monteith Trophy Award. Lawrence serves as both president and editor for the Association, at Highbury, experimental vineyard planted in 1968.

Deaver Ranch Vineyard
Amador County, California.
Zinfandel vineyard.

DEBEVC VINEYARDS
Founded 1971, Lake Erie, Ohio.
Storage: 75,000 gallons.
Chalet Debonne estate Seyval Blanc, Vidal Blanc and Villard Blanc vineyard.

decant
To pour wine gently from the bottle in which crust or sediment has been deposited, for the purpose of obtaining clear wine for serving. The container into which the wine is poured is called a decanter. Decant also means to pour wine from a large container into a small container for more convenient handling.

decanting

This operation consists of separating, in one continous movement, the maximum of limpid wine from its lees. The carafe is tilted so that the wine flows smoothly along the inner walls; pouring it roughly or vertically could over-aerate it. The first third of the bottle is poured out very carefully, indeed, to avoid turbulence; you will see how the air, afterwards, enters freely at the neck before the surface of the wine reaches the punt of the bottle.

Decanting wine with a candle, and funnel

The lees will thus stay all in one place. Continue pouring, watching the progression of the lees against the light of a candle which you will have placed behind the bottle. Leave the empty bottles nearby so that your guests may identify the labels, but if you are serving several wines you should identify your decanters. There are short cork chains for this, with a needle-point at each end for sticking into each end of the cork that are then placed around the neck of the carafe.

deep ruby

Wine color.

Red with a component of blue, resembling the gem.

deep straw

Wine color describing some white wines.

Intense yellow color.

Deer Creek Vineyards

Sacramento City, California.

Chardonnay vineyard.

DEER PARK WINERY

Founded 1979, Deer Park, Napa County, California.

Storage: Oak and stainless steel. Cases per year: 5,000.

Labels indicating non-estate vineyard: Morton Vineyards, La Jota Vineyards, Navoni Vineyard, Le Blanc Vineyard.

Wines regularly bottled: Vintage-dated Sauvignon, Zinfandel, Chardonnay, Petite Sirah.

Deer Run Vineyard

Connecticut.

Hopkins estate Marechal Foch and Seyval Blanc vineyard.

Deerfield Vineyard

Chalk Hill, California.

Balverne estate Chardonnay vineyard.

degorgement

The removal of the sediment which collected in the neck of the inverted champagne bottle during riddling.

degree days (See heat, heat summation.)

DEHLINGER WINERY

Founded 1976, Sebastopol, Sonoma County, California.

Storage: Oak and stainless steel. Cases per year: 7,500.

Estate vineyard: Dehlinger Winery Vineyard.

Label indicating non-estate vineyard: Hillside Vineyards.

Wines regularly bottled: Vintage-dated Chardonnay, Zinfandel, Pinot Noir, Cabernet Sauvignon. Occasionally bottles Petite Sirah.

Del Mar River Road Vineyard

Monterey County, California.

Chardonnay vineyard.

DEL RIO (See Bardenheier's Wine Cellars.)

Port, Sherry, Muscatel, White Port, Rhine, Chablis, Burgundy, Rosé, Pink Chablis.

Del Rio Vineyard/Quemado Vineyard

Texas.

Val Verde estate Lenoir and Herbemont vineyard.

Del Vista Estate Vineyards

Hunterdon County, New Jersey.

Seyval Blanc, Aurora, Vidal Blanc, Riesling, Chardonnay, Sylvaner, Chelois, DeChaunac, Chancellor, Delaware, Cabernet Sauvignon, Pinot Noir, Zinfandel and Merlot Delaware Valley vineyard.

DEL VISTA VINEYARD

Founded 1982, Frenchtown, Hunterdon County, New Jersey.

Storage: Stainless steel. Cases per year: 900-1,000.

Estate vineyard: Del Vista Vineyards.

Labels indicating non-estate vineyard: Great Hill Vineyard, Seabrook Vineyard, Kings Road Vineyard.

Wines regularly bottled: Estate bottled, vintage-dated Chardonnay, Sylvaner-Riesling, Cabernet Sauvignon, Claret; vintage-dated DeChaunac Nouveau, Aurora, Villard Blanc, Vidal Blanc, Seyval Blanc; and Red Table Wine.

Delaware

DELAWARE

Wine grape.

An American hybrid grape. First brought to attention around 1850 at the town of Delaware, Ohio where it gets its name. Produces medium bodied, aromatic white wine. Also used for champagne.

DELAWARE PROPHET

An 18th Century North American Indian Cult Leader. He was called Delaware Prophet and no Indian knew him by any other name. He was a fire and brimstone preacher among the Delaware Indians in the Muskingum Valley in Ohio; he denounced inter-tribal warfare, polygamy and use of magic; his strongest condemnation was against the crushing of wild grapes and the potent liquid that Indians brewed (really fermented wine) that caused drunkenness, debauchery and neglect to crops and family. He implored the Indians to heed his words and if they did, The Great Spirits would give them renewed strength (and skills) to fight off the white man. He constantly reminded the women of the tribes, wherever they saw grapes growing, to tear out the Evil spirit's medicine by the roots. He left, at each village, deerskins with symbols that told of his message and each day they were to teach others that good and strength would come to those who heeded his symbols. The Prophet's converts were many and his religious preachings were so intense and passionate that the tribal medicine men feared for their lives whenever he appeared in their villages and preached his message. The Prophet's words finally influenced the great Pontiac, chieftain of the Ottawas, to rally the other tribes, the Wyandots, Potawatomis and the Ojibwas, to attack the British forts and settlements, starting with a surprise attack on Detroit. The Prophet, in the meantime, was urging the Delawares, the Senecas and the Shawnees to make war on other British outposts. The siege of Detroit failed; the tribes, after costly losses, dispersed and some signed treaties of peace. Pontiac, with a few faithful, tried another rebellion in the west but it failed; he, too, signed a peace treaty in 1766. The DELAWARE PROPHET and his cult were shattered and he seemed to vanish and was never heard from again. The tribes went back to making wine from the grapes. But then came THE MUNSEE PROPHET and his message and, in turn succeeded by, the SHAWNEE PROPHET, all preached against the curse of the wild grape and the results therefrom. The American Redman preached Prohibition long before the white man ...

delicate
Tasting term.
Soft, refined, pleasing.

DELICATO VINEYARDS
Founded 1935, Manteca, San Joaquin County, California.
Storage: Oak and stainless steel. Cases per year: 1,000,000.
History: Founded in 1935 by Gaspare Indelicato and Sebastiano Lupino. Now owned by Gaspare's sons, Frank, Anthony and Vincent.
Estate vineyard: Clements Ranch, Delicato Estate Vineyard.
Wines regularly bottled: Vintage-dated Petite Sirah, Cabernet Sauvignon, Zinfandel, Chardonnay, Chenin Blanc, Green Hungarian, French Colombard, Sauvignon. Also bottle generics: Burgundy, Chablis Blanc, Rhine, Vin Rosé. And Light Wine.
Second label: Settlers Creek.

DELTA
The triangular area at the mouth of the San Joaquin and Sacramento rivers.

DELTA QUEEN (See Bardenheier's Wine Cellars.)
Champagne.

demi-sec
"Half-dry." A term to describe a fairly sweet Champagne with a residual sugar of over 2.5%.

Demostene
Sonoma, California.
Kalin estate Pinot Noir vineyard.

DER WEINKELLER
Founded 1973, Amana, Iowa County, Iowa.
Storage: Oak.
Wines regularly bottled: Fruit and berry wines.

dessert wines
Sweet, full-bodied wines served with desserts or as refreshments are called dessert wines. Their alcohol content is 14%-24%. They range from medium-sweet to sweet and from pale gold to red. The popular types are Angelica, Madera, Marsala, Muscatel, Port, Sherry and Tokay. Sweet wines of less than 14% alcohol are categorized as table wines although often used as dessert wines. Sweet Rieslings or other late-harvest type wines usually are 14.5 or 16% alcohol.

DEVLIN WINE CELLARS
Founded 1978, Soquel, Santa Cruz County, California.
Storage: Oak. Cases per year: 3,000.
Wines regularly bottled: Chardonnay, Cabernet, Merlot, Chenin Blanc.

DIABLO VISTA WINERY
Founded 1977, Benicia, Solano County, California.
Estate vineyard: Polson Vineyards.
Wines regularly bottled: Cabernet Sauvignon, Merlot, Zinfandel, Chardonnay, Chenin Blanc.

DIAL, ROGER
Roger Dial holds a Ph.D. from the University of California at Berkeley and teaches politics at Dalhousie University in Halifax. He became involved in wine in the mid-60's with the Davis Bynum Winery in California. Moving to Canada in 1969, he planted his first commercial vineyard, east of Niagara. He has pioneered the development of vinifera and amurenisis (Russian) varieties in Eastern Canada.

Roger is president of Grand Pré Wines and is currently the president of the Winegrowers Association of Nova Scotia.

DIAMOND

An American hybrid grape that produces light, crisp, fruity white wine. Also used for champagne.

DIAMOND CREEK VINEYARDS

Founded 1968, Calistoga, Napa County, California.

Storage: Oak. Cases per year: 2,000.

Estate vineyard: Volcanic Hill Vineyard, Red Rock Terrace Vineyard, Gravelly Meadow Vineyard.

Wines regularly bottled: Estate bottled, vintage-dated Cabernet Sauvignon.

Diamond Hill Estate Vineyard

Providence County, Rhode Island.

Pinot Noir and Chardonnay vineyard.

DIAMOND HILL VINEYARD

Founded 1976, Cumberland, Providence County, Rhode Island.

Storage: Oak.

Estate vineyard: Diamond Hill Vineyard.

Wines regularly bottled: Estate bottled proprietary Bernston Pinot Noir, Chardonnay; also produce Apple, Peach, Pear, Blueberry.

Diamond Mountain Vineyard

Napa County, California.

Roddis estate Cabernet Sauvignon vineyard.

DIAMOND OAKS VINEYARDS

Founded 1978, Cloverdale, Sonoma County, California.

Storage: Oak and stainless steel. Cases per year: 24,000.

Wines regularly bottled: Vintage-dated Chardonnay, Fumé Blanc, Cabernet Sauvignon.

Second label: Thomas Knight.

Diamond T Ranch

Monterey County, California.

Robert Talbott Vineyard and Winery estate Chardonnay, Pinot Noir and Sauvignon Blanc Carmel Valley vineyard.

Dickerson Vineyard

Napa County, California.

Zinfandel vineyard.

Dionysus Vineyard

Franklin County, Washington.

Chardonnay, Semillon, Riesling, Pinot Noir, Gewurztraminer, Cabernet Sauvignon Columbia Valley vineyard.

D & M Junction Vineyards

Monterey County, California.

Cabernet Sauvignon vineyard.

Doe Mill Vineyard

Butte County, California.

Zinfandel vineyard.

DOLAN VINEYARDS

Founded 1980, Redwood Valley, Mendocino County, California.

Storage: Oak. Cases per year: 2,000.

Estate vineyard: Hillside Vineyard.

Label indicating non-estate vineyard: Lolonis Vineyard.

Wines regularly bottled: Vintage-dated Chardonnay.

DOMAINE CHANDON

Founded 1973, Yountville, Napa County, California.

Storage: Stainless Steel. Cases per year: 220,000.

Estate vineyard: Domaine Chandon.

Wines regularly bottled: Champagne: Napa Valley Brut, Brut Special Reserve, Blanc de Noirs. Also bottle Panache (a proprietary label aperitif).

Domaine Chandon Estate Vineyards

Napa County, California.

Pinot Noir, Chardonnay and Pinot Blanc vineyards in Carneros, Mt. Veeder and Yountville.

DOMAINE DE GIGNOUX

Founded 1983, Charlottesville, Albemarle County, Virginia.

Storage: Oak. First crop in 1985.

Estate vineyard: Domaine de Gignoux.

Wines regularly bottled: Red Bordeaux Cabernet, Chardonnay (Macon style), Beaujolais.

DOMAINE DE LA VENNE (See Oasis Vineyard.)

Domaine Elucia

Napa Valley, California.

Shaw estate Chardonnay, Sauvignon Blanc and Napa Gamay vineyard.

DOMAINE LAURIER

Founded 1978, Forestville, Sonoma County, California.

Storage: Oak and stainless steel. Cases per year: 7,000.

Estate vineyard: Domaine Laurier Estate Vineyard.

Wines regularly bottled: Estate bottled, vintage-dated Pinot Noir, Cabernet Sauvignon. Also vintage-dated Chardonnay, Sauvignon Blanc.

Domaine Laurier Estate Vineyard

Sonoma County, California.

Pinot Noir and Cabernet Sauvignon vineyard.

DON EUGENIO (See Casa Pinson Hermanos.)

DONATONI WINERY

Founded 1979, Inglewood, Los Angeles County, California.

Storage: Oak and stainless steel. Cases per year: 1,200.

Label indicating non-estate vineyard: Nepenthe Vineyard.

Wines regularly bottled: Vintage-dated Cabernet Sauvignon, Chardonnay.

DONNA MARIA VINEYARDS

Founded 1980, Healdsburg, Sonoma County, California.

Storage: Oak and stainless steel. Cases per year: 12,000-35,000.

Estate vineyard: Donna Maria Vineyards.

Wines regularly bottled: Estate bottled, vintage-dated Pinot Noir, Gewurztraminer, Chardonnay, Sauvignon Blanc, Cabernet Sauvignon. Occasionally bottle Late Harvest Semillion (Chalk Hill Label).

Second label: Chalk Hill Winery.

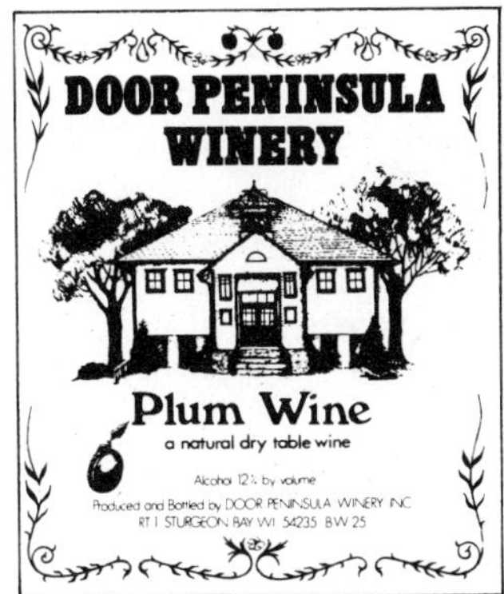

DOOR PENINSULA WINERY

Founded 1974, Sturgeon Bay, Door County, Wisconsin.

Storage: Stainless steel. Cases per year: 6,000.

Wines regularly bottled: Fruit wines; Apple, Cherry, Plum, Pear.

Doridon Vineyard

Oakley, California.

Semillon and Zinfandel vineyard.

dosage

The addition of sugared wine and brandy to another wine in order to make it conform to established standards of sweetness. Only used for champagne and sparkling wines.

DOVER VINEYARDS, INC.

Founded 1932, Westlake, Cuyahoga County, Ohio.

Storage: Oak. Cases per year: 50,000.

History: Originally founded as a grower cooperative, its present owner comes from an old Hungarian wine making tradition.

Wines regularly bottled: Dry Concord, Pink Catawba, Sweet Catawba, Cream Concord, Cream Niagara, Blackberry; generics: Rhine, Burgundy, Sauterne, Haut Sauterne; also Labrusca, Half & Half, Vin Rosé; Port, Sherry, Sangria; White and Pink Champagne, Sparkling Burgundy, Spumante, Cold Duck; and Ohio Rosé.

Downing/D'Agostini Brothers/Cowan Family Vineyards

Shenandoah Valley, California.

Zinfandel vineyard.

DOWNS, EDGAR B.

Edgar B. "Pete" Downs was graduated with a Bachelor's degree in Fermentation Science from the University of California, Davis, in 1973. He was hired by Italian Swiss Colony, where he worked for three years as chief production chemist and cellar operations supervisor. In 1976, Downs became enologist and research director at Korbel Champagne Cellars. Three years later, he assumed the position of winemaker in charge of sparkling wines at Chateau St. Jean Vineyards and Winery in Sonoma County, where he continues to work. Downs is a member of the American Society of Enologists.

Isabel Downs Vineyard

Lake County, California.

Konocti estate Sauvignon Blanc Clear Lake vineyard.

DREYFUS, BENJAMIN

There are few today who know the name, Benjamin Dreyfus, yet each year millions of people visit the city he helped to found and was its first mayor ... Anaheim, California ... home of Disneyland.

Dreyfus was once a dominant figure in the American wine industry and, yet, rarely is his name mentioned by wine authorities or wine writers when they extol the contributions of wine industry pioneers.

Born in Westheim, this German-Jewish immigrant came to America and settled in Baltimore, Maryland; he knew little of the language but soon learned and, in 1851, he became a naturalized American citizen, the same year he decided to make the trek to California and seek his future and fortune.

Dreyfus, instead of heading for the gold rush country, came to Los Angeles; after a few years, he was one of the first oil refiners in Southern California; he then organized a brewery, another successful venture; from the beer business he partnered with Felix Bachman and, later, August Landenberger, in 1857, and established the Los Angeles Vineyard Company, located on a tract of land on the Santa Ana River twenty-five miles south of Los Angeles city proper. 1,200 acres were purchased from the Ontiveros family and divided into 20 acre parcels to contain 10,000 grape vines. The project was noted in the press as—"the most important ever contemplated in the southern country . . ." The Mother Colony was then renamed Anaheim (German for "Our Home"). As the "dream of Dreyfus" began to be realized, the grape vines and olive trees planted, he and Landenberger opened the first general merchandising store, in 1860, and soon were designated as an accredited Wells Fargo Agency.

The Anaheim wine tycoon was producing more than just the original white wines; soon the Anaheim Winery was making Angelica, Port and Muscatel.

By the year 1880, in addition to those at Anaheim, Dreyfus had vineyards and wineries in Cucamonga, San Gabriel, and Napa Valley producing more than 800,000 gallons a year.

The American Grocer of New York in 1876 declared, "No wines placed in the Eastern market are held in higher esteem than those of the house of B. Dreyfus of Anaheim, California."

During that period, Benjamin Dreyfus was as well known as any man in the State of California. Today, his major role and rich contribution are all but forgotten . . .

Albert F. Kercheval wrote a poem describing the Mother Colony, Anaheim, but it really is about a German-Jewish immigrant, a naturalized American citizen, a planter of grape vines, the Wine Tycoon of Anaheim . . .

"And further still toward tropic clime
Looks down on Lovely Anaheim.
No fairer scene, by rainbow spanned,
Or Sweeter Grapes hath Fatherland.
Here plenty dwells; and mirth and wine
Are mingled with the songs of Rhine,
And silvery patriarchs recline
Beneath the olive and the vine . . ."

dry

The opposite of sweet; free of sugar. Fermentation converts the natural sugar of the grape into alcohol and carbon dioxide gas. A wine becomes dry when all the sugar has been consumed by fermentation. Dryness should not be confused with astringency, acidity, tartness or sourness; it simply means lacking in sweetness. The wines, which uninformed individuals are apt to call "sour" are dry or tart, are made with these flavor characteristics especially to blend with the flavors of main course foods. (A Champagne or Sherry labeled "dry" is actually semi-dry, and even an "extra-dry" Champagne is slightly sweet.) Really dry Champagne is labeled "Brut" or "Nature"; the driest sherries are labeled "Extra Dry".

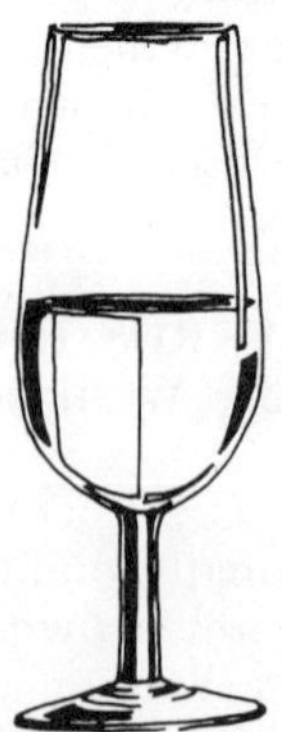

The traditional "Sherry" glass

Dry Creek Bench Land Vineyard
Sonoma County, California.
Zinfandel vineyard.

Dry Creek Ranch
Sonoma County, California.
Seghesio Wineries estate Petite Sirah,

Napa Gamay, Zinfandel, Carignane, Chenin Blanc, French Colombard, Dry Creek Valley vineyard.

DRY CREEK VALLEY
The valley is near Geyserville and Healdsburg in Sonoma County.

DRY CREEK VALLEY VITICULTURAL AREA
Dry Creek Valley in Sonoma County is located in north central Sonoma County, California, northeast of the town of Healdsburg, in an arm of the Russian River Valley. It extends from the Dry Creek/Russian River confluence south of Healdsburg to the Warm Springs Creek/Dry Creek confluence east of the Warm Springs Dam.
The viticultural area encompases approximately 80,000 acres with 5,000 acres planted with grapes.

Dry Creek Valley/Woodleaf
Dry Creek Valley, California.
Clos du Bois estate Cabernet Sauvignon vineyard.

Dry Creek Vineyard
Napa County, California.
Cabernet Sauvignon vineyard.

Dry Creek Vineyard #2
Napa County, California.
Pinot Noir vineyard.

DRY CREEK VINEYARD, INC.
Founded 1972, Healdsburg, Sonoma County, California.
Storage: Oak and stainless steel. Cases per year: 50,000.
Estate vineyard: Dry Creek Vineyard.
Label indicating non-estate vineyard: Robert Young Vineyard.
Wines regularly bottled: Estate bottled, vintage-dated Merlot, Fumé Blanc, Chardonnay. Also vintage-dated Dry Chenin Blanc, Fumé Blanc, Chardonnay, Cabernet Sauvignon, Zinfandel. Occasionally bottles Petite Sirah, Late Harvest Gewurztraminer.
Second label: Idlewood White.

Dry Creek Vineyard, Inc. Estate Vineyard
Napa County, California.
Merlot and Chardonnay vineyard.

GEORGES DUBOEUF & SON
Founded 1980, Healdsburg, Sonoma County, California.
Cases per year: 14,000.
Wines regularly bottled: Vintage-dated Gamay Beaujolais, French Colombard.

Du Pratt Vineyard
Mendocino County, California.
Zinfandel vineyard.

DUCKHORN VINEYARDS dba St. Helena Wine Company
Founded 1976, St. Helena, Napa County, California.
Storage: Stainless steel. Cases per year: 7,000.
Estate vineyard: Duckhorn Vineyards.
Label indicating non-estate vineyard: Three Palms Vineyard.
Wines regularly bottled: Vintage-dated Napa Valley Merlot, Napa Valley Cabernet Sauvignon, Napa Valley Sauvignon Blanc.

DUDENHOEFER (See Barengo/Lost Hills Vineyard.)

Duff Vineyard
Lake Erie, Ottawa County, Ohio.
Concord, Catawba, Niagara, Ives and Delaware vineyard.

dull
Tasting term.
Wines having an easily seen, distinctly colloidal, haze, but free of visible suspended material.

Dundee Hills Vineyard
Willamette Valley, Oregon.
Pinot Noir, Chardonnay, Riesling and Cabernet vineyard.

DUNN VINEYARDS
Founded 1982, Angwin, Napa County, California.
Storage: Oak. Cases per year: 2,000.
History: The vineyards in the Howell Mountain region where, in the 1890's, there were several hundred acres of vineyards. Dunn produces its Cabernet in old world style.
Estate vineyards: Dunn Vineyards, Mission Vineyards.
Wines regularly bottled: Vintage-dated Cabernet Sauvignon.

Duplin Estate Vineyards
Duplin County, North Carolina.
Magnolia, Carlos, Scuppernong and Noble vineyard.

DUPLIN WINE CELLARS
Founded 1973, Rose Hill, Duplin County, North Carolina.
Storage: Stainess steel. Cases per year: 300,000.
Wines regularly bottled: Estate bottled, vintage-dated Magnolia, Scuppernong, Noble, Carlos; vintage-dated American Port and Sherry; generic Chablis and Burgundy; Champagne, Sparkling Scuppernong.
Second label: Carolina Winery, Olde North Vineyard.

Durant Vineyards
Yamhill County, Oregon.
White Riesling, Pinot Noir and Chardonnay vineyard.

Durney Estate Vineyard
Monterey County, California.
Cabernet Sauvignon, Gamay Beaujolais, Chenin Blanc and Johannisberg Riesling vineyard.

DURNEY VINEYARD
Founded 1968, Carmel, Monterey County, California.
Storage: Oak, stainless steel. Cases per year: 15,000.
History: The first vineyard and winery in the Carmel Valley.
Estate vineyard: Durney Vineyard.
Wines regularly bottled: Estate bottled, vintage-dated Cabernet Sauvignon, Gamay Beaujolais, Chenin Blanc, Johannisberg Riesling.
Second label: Carmel Valley Winery.

Dusschee Vineyard
Dallas, Oregon.
Pinot Noir, Chardonnay, Gewurztraminer.

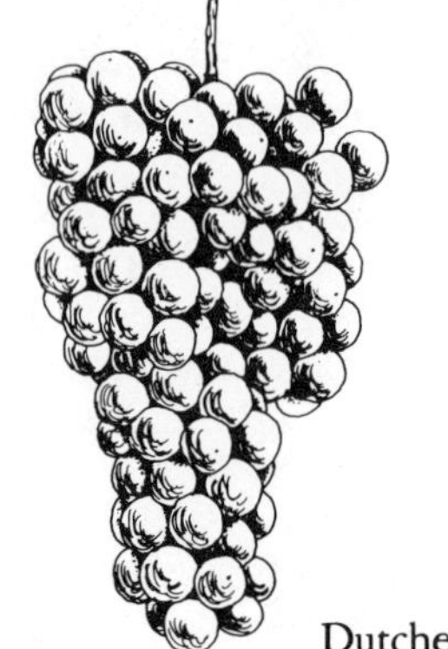
Dutchess

DUTCHESS
Wine grape.
An American hybrid grape developed by A.J. Caywood in the Hudson Valley last century. Produces fruity, slightly sweet white wine. Also used for Champagne.

Dutton Ranch
Sonoma County, California.
Pinot Noir, Chardonnay vineyard.

DYER, DAWNINE SAMPLE
Dawnine Sample Dyer obtained a Bachelor's degree in Biology from the University of California, Santa Cruz, in 1974. She became interested in wine through summer travel in Europe and part-time work at nearby Bargetto Winery. After graduation, she joined the enology team at the Robert Mondavi Winery, then worked at Inglenook Vineyards until 1976. She then joined Domaine Chandon, the California subsidary of Moet-Hennessy. She made several study trips to Moet & Chandon in Epernay, France, to expand her knowledge of methode champenoise wines.

Eagle Crest Estate Vineyards

Livingstone County, New York.

Niagara and Catawba Finger Lakes vineyard.

EAGLE CREST VINEYARDS, INC.

Founded 1872, Conesus-on-Hemlock Lake, Livingstone County, New York.

Storage: Oak and stainless steel. Cases per year: 35,000.

History: Founded by Bishop Bernard McQuaid, first Roman Catholic Bishop, Rochester Diocese, New York, for producing Sacramental wines for his clergy. Originally called O-Neh-Da Vineyards which is Seneca dialect for Hemlock Tree. O-Neh-Da Sacramental wines are still produced, as are table wines, under the Barry Label.

Estate vineyard: Eagle Crest Vineyard.

Wines regularly bottled: Vintage-dated Aurora, Riesling, White Delaware, Rosé of Iona; Niagara, Pink Delaware; vintage-dated Select Red and White Wines; two semi-generics: Chablis and Mellow Burgundy; also Cream Sherry.

EAGLE POINT (See Scharffenberger Cellars.)

Vintage-dated Chardonnay, Blanc de Noir.

Eagle Point Vineyard

Mendocino County, California.

Zinfandel vineyards.

early harvest

The equivalent of the "trocken" & "halbtrocken" wines of Germany. Early harvest wines are generally White Rieslings and are produced only in the coolest years. Wines produced from these grapes will usually contain low levels of alcohol and residual sugar, coupled with high total acidity. By law, may not have more than 19.9 degrees Brix at harvest.

EASLEY WINERY

Founded 1974, Indianapolis, Marion County, Indiana.

Storage: Oak and stainless steel.

History: One of two vineyards in America that are directly on the banks of the Ohio River.

Estate vineyard: Cape Sandy Vineyards.

Wines regularly bottled: Estate bottled Marechal Foch, DeChaunac, Chelois, Baco, Aurora Blanc, Siebel 13053, Seyval Blanc, Dutchess; semi-generics: Chablis, Burgundy, Pink Chablis; also Red, White, Rosé, Mulled White, May Wine, Altar Wine.

Eaton Estate Vineyard

Columbia County, New York.

Seyval Blanc vineyard.

EATON VINEYARDS

Founded 1980, Pine Plains, Columbia County, New York.

Cases per year: 800-1,000.

History: Jerome Eaton has been a professional horticulturist for thirty years. Former director of Duke Gardens and director of Old Westbury Gardens.

Estate vineyard: Eaton Vineyard.

Wines regularly bottled: Estate bottled, vintage-dated Seyval Blanc.

ECKERT'S SUNNY SLOPE WINERY AND VINEYARDS

Founded 1981, Washington, Franklin County, Missouri.

Storage: Oak and stainless steel. Cases per year: 1,000.

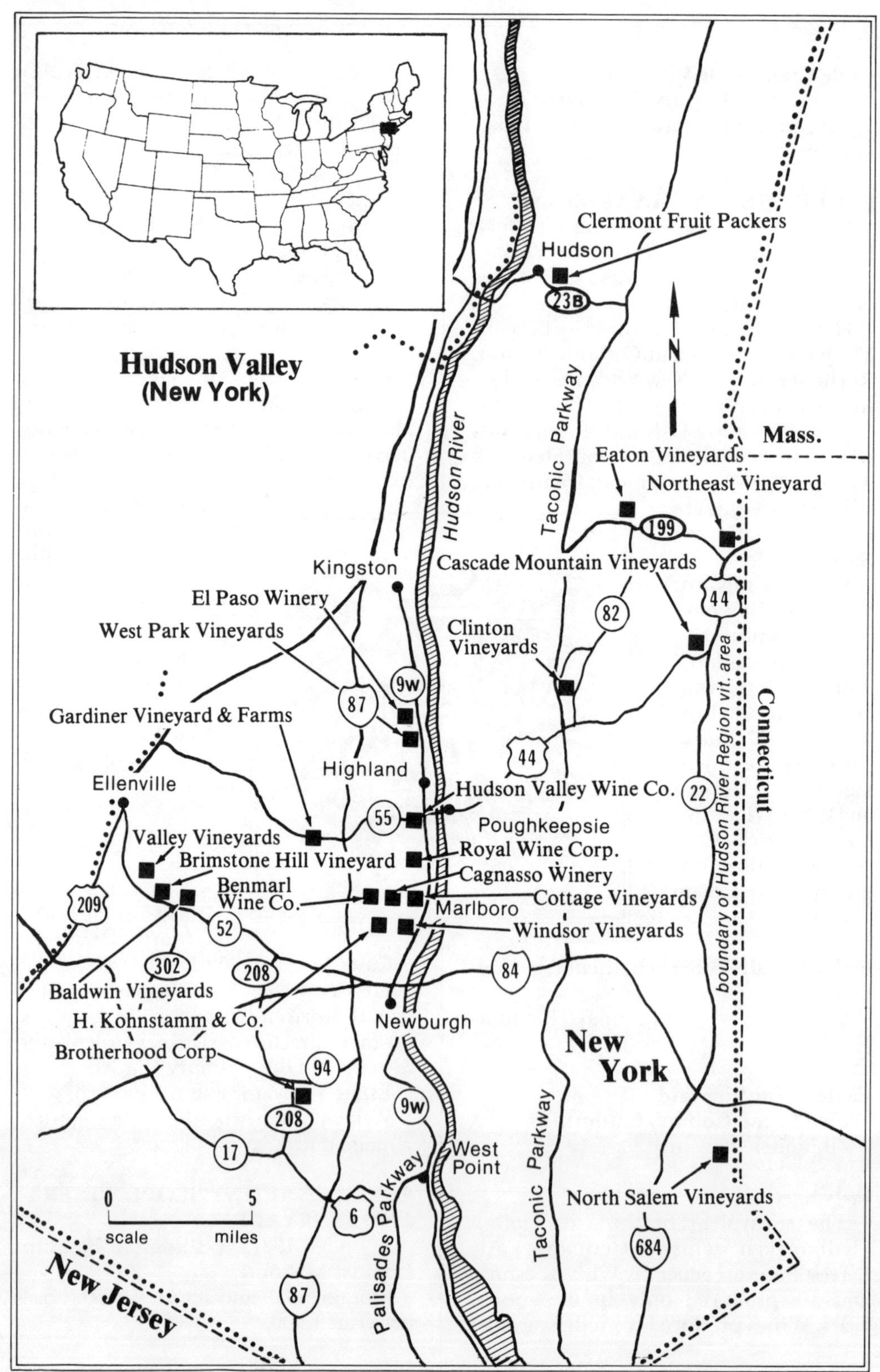
Hudson Valley
(New York)
Clermont Fruit Packers
Hudson
23B
N
Taconic Parkway
Hudson River
Mass.
Eaton Vineyards
Northeast Vineyard
199
Kingston
Cascade Mountain Vineyards
El Paso Winery
44
82
West Park Vineyards
Clinton
Vineyards
9W
87
Gardiner Vineyard & Farms
Connecticut
boundary of Hudson River Region vit. area
44
Highland
Ellenville
Hudson Valley Wine Co.
22
55
Poughkeepsie
Valley Vineyards
Royal Wine Corp.
Brimstone Hill Vineyard
Cagnasso Winery
Benmarl
Wine Co.
Cottage Vineyards
209
Marlboro
52
Windsor Vineyards
302
208
84
Baldwin Vineyards
H. Kohnstamm & Co.
Newburgh
New
York
Brotherhood Corp
94
9W
208
17
Palisades Parkway
West
Point
Taconic Parkway
0
5
scale
miles
6
North Salem Vineyards
684
New Jersey
87

History: Winery is located on a historic farm, of the early 1800's, in a two story log house. Also at the winery is a two story 1860's (Federal style) brick house.

Estate vineyard: Sunny Slope Vineyards.

Wines regularly bottled: Red, White, Rosé, Fruit Wines, Mead.

EDDINS, JAMES C.

James C. Eddins was graduated in 1957 from the United States Naval Academy and served for twenty-four years in the active and reserve Marine Corps. As a civilian, he has worked as an engineer and salesman for the Bendix Corporation and for the IBM Corporation and taught data processing and computer programming at Troy State and Faulkner State Junior College. In 1972, he and his wife, Marianne, established Perdido Vineyards and Winery in Perdido, Alabama. It was the first winery constructed in Alabama since Prohibition and holds the honor of being Alabama Bonded Winery Number 1. Eddins specializes in the production of Muscadine wines made from native Scuppernong grapes.

EDDY JR., THOMAS G.

Thomas G. Eddy, Jr. knew he wanted to be a winemaker from the age of 14. During high school, he would often cook and hold wine tastings for his parents' dinner parties. His professional education began when he enrolled in the enology program at the University of California, Davis. After graduation, Eddy held an internship during the 1974 harvest at Wente Bros. Winery. Following this experience, he became the table winemaker at LaMont Winery for two years. Next, he accepted the larger challenge of producing table wines for both Italian Swiss Colony and Inglenook under United Vintners. During a five year period, Eddy worked as enologist, cellar supervisor, champagnemaster and production master for the company. In 1981, he left to become general manager and director of winemaking at Souverain Cellars. In 1984 he joined The Christian Brothers as director of winemaking. Eddy is active in several wine industry organizations, including membership in the American Society of Enologists, and as board member of the California Wine Institute.

EDELWEIN

A sweet wine produced from Johannisberg Riesling by a process similar to German Beerenauslese or Trockenbeerenauslese.

EDELWEISS

A fruity, white American hybrid grape developed in Minnesota by Elmer Swensen.

EDELWEISS WINERY

Founded 1982, New Haven, Franklin County, Missouri.

Storage: Oak and stainless steel. Cases per year: 2,500.

History: The Heck family is originally from the Rhine, in Germany, and look back over a 200 year tradition of winemaking.

Estate vineyard: Stone Church Vineyards.

Wines regularly bottled: Estate bottled, vintage-dated Chancellor, Seyval Blanc, Vidal, Villard Blanc and proprietary Golden September.

EDELZWICKER

A white wine being produced in the Midwest from a blend of Gewurztraminer and White Riesling grapes. ("Edel"-noble, "Zwicker"-mixture or blend.)

EDMEADES VINEYARDS

Founded 1972, Philo, Mendocino County, California.

Storage: Oak and stainless steel. Cases per year: 24,000.

History: Although the winery was established in 1972, the first varietal wine grapes were planted by the Edmeades family in the Anderson Valley in 1964

Estate vineyard: Edmeades Vineyards.

Label indicating non-estate vineyard

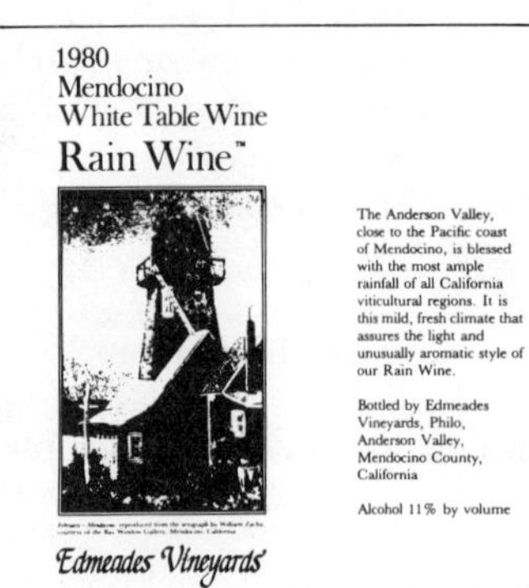

Du Pratt Vineyard, Ciapusci Vineyard, Pacini Vineyard, Anzilotti Vineyard, B.J. Carney Vineyard.

Wines regularly bottled: Vintage-dated Cabernet Sauvignon, Chardonnay, Zinfandel, Chardonnay Reserve, Pinot Noir, Gewurztraminer. Also proprietary Rain Wine, Whale Wine.

Second label: Mendocino Wine Guild, Anderson Valley Winery, Mendocino Wine Company.

EDNA VALLEY CELLARS (See Edna Valley Vineyards.)

Edna Valley Estate Vineyards

San Luis Obispo County, California.

Pinot Noir, Chardonnay, Sauvignon Blanc and Chenin Blanc vineyard.

EDNA VALLEY VINEYARDS

Founded 1980, San Luis Obispo, San Luis Obispo County, California.

Storage: Oak. Cases per year: 25,000.

Estate vineyard: Edna Valley Vineyard.

Labels indicating non-estate vineyard: Johnson Vineyard.

Wines regularly bottled: Estate bottled, vintage-dated Pinot Noir, Vin Gris of Ponit Noir, Chardonnay, Chardonnay Reserve, Chenin Blanc.

Second label: Edna Valley Cellars.

Edna Valley Estate Vineyards

San Luis Obispo County, California.

Cabernet Sauvignon Edna Valley vineyard.

EDNA VALLEY VITICULTURAL AREA

Edna Valley is an elongated valley of approximately 35 miles. The upper end merges into the Los Osos Valley just beyond the city of San Luis Obispo. It projects into the surrounding uplands to the Santa Lucia Mountains and to the San Luis Range.

Robert Egan Vineyard/Bur Mar Vineyard

Yountville, California.

Cabernet Sauvignon vineyard.

egg whites

Used to clarify wine.

EHRENFELSER

Wine grape.

A vinifera cross of Riesling and Sylvaner grapes. Produces mellow, rich golden white wine with a flowerly bouquet.

EHRLE BROTHERS, INC.

Homestead, Iowa County, Iowa.

Storage: Oak.

Wines regularly bottled: Proprietary Homestead fruit wines.

Eisele Vineyards

Napa Valley, California.

Cabernet Sauvignon, Zinfandel, Gamay and Chardonnay vineyard.

Eisert Vineyard

Goodhye County, Minnesota.

Millot, Seyval Blanc, DeChaunac, Sigfried Rebe vineyard.

E & K WINE COMPANY

Founded 1860, Sandusky, Erie County, Ohio.

Storage: Oak and stainless steel. Cases per year: 2,000.

Label indicating non-estate vineyard: Kelley's Island.

Wines regularly bottled: Concord, Catawba, Delaware, Niagara.

Second label: Kelley's Island Wine Company.

EL DORADO COUNTY
Sierra Foothills, California.

El Dorado Vineyard
Monterey County, California.
Chenin Blanc vineyard.

EL DORADO VINEYARDS
Founded 1975, Camino, El Dorado County, California.
Storage: Oak. Cases per year: 500.
Label indicating non-estate vineyard: Herbert Vineyards, Madrona Vineyards, Walker Vineyard, Stonebarn Vineyard.
Wines regularly bottled: Vintage-dated Chardonnay, Sauvignon Blanc, Zinfandel. Also bottle Chenin Blanc, Cabernet Sauvignon.

EL DORADO VITICULTURAL AREA
El Dorado viticultural area in California's El Dorado County is located on the western slope of the central Sierra Nevada Mountains. Overlying bedrock in many places are mantles of river gravel and volcanic debris, which are well-drained and particularly suitable for wine grape production. The higher average elevation of the viticultural area, as opposed to the lower foothills and the Central Valley, offers a cooler growing climate with a higher average rainfall. Appellation limits: elevation, lower 1,200 ft., upper 3,500 ft.

elegant
Tasting term.
Subtle; harmonious; noble.

El-Hi Vineyard
Pierce County, Washington.
Vierthaler Winery estate White Riesling vineyard.

EL PASO DE ROBLES WINERY AND VINEYARD
Founded 1981, Paso Robles, El Dorado County, California.

EL PASO WINERY
Founded 1977, Ulster County, New York.
Storage: Oak and stainless steel. Cases per year: 1,300.
Wines regularly bottled: Cream Niagara; proprietary Vino de Mesa, Rosado, Clarete, Blanco, Tino del Valle; also Chablis, Mellow Red and Hudson Valley Gold.

Elk Cove Estate Vineyards
Yamhill County, Oregon.
Pinot Noir, Riesling, Chardonnay and Gewurztraminer vineyard.

ELK COVE VINEYARDS
Founded 1974, Yamhill County, Oregon.
Storage: Oak and stainless steel. Cases per year: 12,000.
History: The vineyard is named after the Roosevelt elk which migrates to the area each Spring.
Estate vineyard: Elk Cove Vineyards.
Label indicating non-estate vineyard: Dundee Hills Vineyard, Wind Hill Vineyard.
Wines regularly bottled: Estate bottled, vintage-dated Pinot Noir, Riesling, Chardonnay, Gewurztraminer; also vintage-dated Cabernet Sauvignon and Riesling.

Ellendale Estate Vineyards
Polk County, Oregon.
Pinot Noir, Chardonnay, Riesling, Merlot and Cabernet Franc vineyard.

ELLENDALE VINEYARDS
Founded 1981, Polk County, Oregon.
Storage: Oak and stainless steel. Cases per year: 5,000.
Estate vineyard: Ellendale Vineyards.
Label indicating non-estate vineyard: Sagemore Farms.
Wines regularly bottled: Estate bottled, vintage-dated Pinot Noir, vintage-dated Merlot, Cabernet Sauvignon, Riesling, Chardonnay, Gewurztraminer; also Niagara, Concord, Mead and Fruit wines (Apple, Blackberry, Cherry, Loganberry, Plum, Strawberry, Rhubarb).

Ellis Ranch
Sonoma County, California.
Seghesio Wineries estate Napa Gamay, Carignane Alexander Valley vineyard.

Elm Valley Vineyard
Leelanau County, Michigan.
Mawby estate Seyval Blanc and Vignoles vineyard.

Elvira

ELVIRA
Wine grape.
An American hybrid grape that produces neutral white wine. Used exclusively in popular wine blends.

Elysian Farms
Alexander Valley, California.
Zellerbach estate Cabernet Sauvignon and Merlot vineyards.

Emerald Acres Vineyard
Yakima Valley, Washington.
Aligoté vineyard.

EMERALD RIESLING
Wine grape.
A California hybrid grape developed at the University of California, Davis. It is the child of White Riesling and Muscadelle—one of the Muscat family. A light, fruity white wine.

EMERSON, RALPH WALDO (1803-1882)
American essayist.

> "I think wealth has lost much of its value if it have not wine. I abstain from wine only on account of the expense. When I heard that Mr. Sturgis had given up wine, I had the same regret that I had lately in hearing that Mr. Bowditch had broken his hip . . ."

> "Wine, which Music is—
> Music and Wine are one."

> "A man will be eloquent, if you give him good wine."

EMILE'S (See Emilio Guglielmo Winery.)
Chablis, Blanc Sec, Vin Rosé, Mellow Burgundy, Burgundy, Sherry, Pale Dry Sherry, Cream Sherry, Port, Champagne, Sweet and Dry Vermouths, Fruit Wines, Brandy.

HENRY ENDRES WINERY
Founded 1935, Oregon City, Clackamas County, Oregon.
Storage: Oak.
History: Henry Sr. began the winery; Henry C. took over in 1951 and Henry F. will take over in 1985—same family for 49 years to date.
Wines regularly bottled: Apple, Mead, Raspberry, Blackberry, Currant, Elderberry, Plum, Rhubarb and Labrusca.
Second label: Henry's.

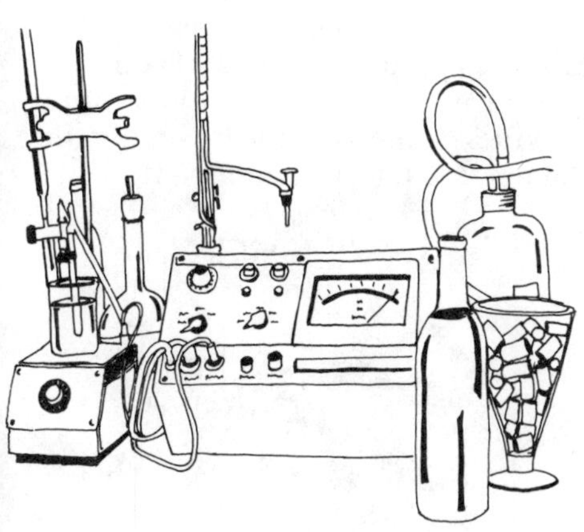

An enologist's worktable

ENOLOGY

The science of the study of winemaking; related to viticulture, which is the science of grape culture.

ENTERPRISE CELLARS (See H. Coturri & Sons.)

ENZ VINEYARDS

Founded 1973, Hollister, San Benito County, California.

Cases per year: 4,000.

Wines regularly bottled: Pinot St. George, Sauvignon Blanc, Zinfandel, Orange Muscat, French Colombard.

Eola Hills Vineyards

Yamhill County, Oregon.

Hidden Springs estate Riesling, Chardonnay, Pinot Noir, Cabernet Sauvignon and Sauvignon Blanc vineyards.

ERATH, DICK

Dick Erath was originally trained as an electronic engineer. However, his interest in wine goes back to his father and grandfather, who were both winemakers in Germany. His arrival in Oregon from Walnut Creek, California, was due to the prospects of an electronics job. The combination of a chance meeting with an Oregon enthusiast and a discussion of Oregon grapes led him to make the move to Oregon in the late sixties and start his own vineyard. He and his wife Kina founded their vineyard in 1968. In 1971 they joined Carl Knudsen and Knudsen Erath was founded. Located in the Red Hills of Dundee, in the Northern Willamette Valley. Knudsen Erath is Oregon's largest vineyard operation today.

ERIE-CHAUTAUQUA REGION

The moderating influence of Lake Erie, plus long hours of summer sunlight, well-drained gravel and shale soils have a most favorable influence on this region for grape growing. Westfield and Fredonia are the main wine towns.

ERNESTINO CHIALTO (See Casa Pinson Hermanos.)

Eschen Vineyard

Amador County, California.

Zinfandel and Black Muscat vineyards.

Escola Vineyard

Amador County, California.

Zinfandel vineyard.

Esdraelon Vineyard/Seven Valley Vineyard/Chanceford Vineyard/Marthur's Vineyard

York County, Pennsylvania.

Vidal, Seyval, DeChaunac, Foch and Niagara vineyard.

Henry Espinoza's Vineyard

Lake County, California.

Cabernet Sauvignon vineyard.

estate bottled

In order to be allowed to state on a wine label that the wine is "Estate Bottled", the winery must conform to the following specifications:

1. One hundred percent (100%) of the wine must originate from the listed appellation, and the smallest area and most specific designation should be used (i.e.,

use of an approved or pending viticultural area as opposed to a county line [political boundaries] which would encompass a larger area).

2. The winery must own or control 100% of the vineyards from which the wine was produced. If a vineyard is "controlled", the winery must perform all viticultural practices as if it owned the vineyard.

Of lesser importance since a very large co-op winery may call any of the vineyards of the members "estate bottled". Now replaced in many wineries by the more meaningful term "Grown, produced and bottled by".

esters

Aromatic substances brought about by the reactions of alcohols and acids in wines, which contribute to bouquet.

Estrella River Estate Vineyards

San Luis Obispo County, California.

Chardonnay, Sauvignon Blanc, Chenin Blanc, Johannisberg Riesling, Muscat Canelli, Cabernet Sauvignon, Zinfandel, Barbera and Syrah Paso Robles vineyard.

ESTRELLA RIVER WINERY

Founded 1977, Paso Robles, San Luis Obispo County, California.

Storage: Oak. Cases per year: 100,000.

Estate vineyard: Estrella River Winery.

Wines regularly bottled: Estate bottled, vintage-dated Fumé Blanc, Chardonnay, Chenin Blanc, Johannisberg Riesling, White Zinfandel, Zinfandel Rosé, Zinfandel, Cabernet Sauvignon, Barbera, Syrah, Late Harvest Muscat Canelli. Also bottles vintage-dated Johannisberg Riesling. Also special blended Selections of Cabernet Sauvignon, Chardonnay, Chenin Blanc, Zinfandel. Occasionally bottles Cabernet Sauvignon and Chardonnay Reserves.

EVENSON VINEYARDS

Founded 1979, Oakville, Napa County, California.

Storage: Stainless steel. Cases per year: 750.

History: The Evenson families are descendents of pioneer Sonoma and Napa families dating back to the 1800's.

Estate vineyard: Evenson Vineyards.

Wines regularly bottled: Estate bottled, vintage-dated Gewurztraminer.

extra dry

In Champagne, less dry than Brut. In Sherry, the driest.

THE EYRIE VINEYARDS

Founded 1966, Dundee, Yamhill County, Oregon.

Storage: Oak and stainless steel. Cases per year: 5,000.

History: The first producing winery in the county since before Prohibition. Produces the unusual Pinot Gris, Muscat Ottonel and Pinot Meunier.

Estate vineyard: The Eyrie Vineyards.

Wines regularly bottled: Pinot Noir, Pinot Gris, Chardonnay, Pinot Meunier, Muscat Ottonel.

FACELLI, LOUIS

Louis Facelli decided to become a winemaker in 1979 after thirteen years of a home winemaking hobby. He and his wife, Sandy, spent 18 months looking for appropriate vineyard property in the Treasure Valley of Idaho. They built a small winery specializing in fruit wines near Wilder. During the following years, they also experimented with locally grown varietal grapes, including production of the Gewurztraminer in Idaho. In 1982, the Facellis formed a partnership with Norman and Fred Batt, prominent agriculturists in the area. A larger facility was constructed, and the Louis Facelli Winery opened three months later. Production has increased to 40,000 gallons, emphasizing Riesling, Chardonnay, Gewurztraminer and Pinot Noir. Fruit wines are also made. Facelli's philosophy is that the winemaker must preserve nature's fruit with an uncompromised quality.

LOUIS FACELLI WINERY

Founded 1981, Caldwell, Canyon County, Idaho.

Storage: Oak and stainless steel.

Estate vineyard: Louis Facelli Vineyard.

Label indicating non-estate vineyard: Sagemoor Farms.

Wines regularly bottled: Estate bottled, vintage-dated Riesling, Gewurztraminer, Pinot Noir and Chardonnay. Vintage-dated Riesling, Chardonnay, Cabernet Blanc, Gewurztraminer; also Fruit wines: Apricot, Cherry, Pear, Plum and Blackberry.

FADIMAN, CLIFTON (1904-)

American editor and writer.

"The drinking of wine seems to me to have a moral edge over many pleasures and hobbies in that it promotes love of one's neighbor."

FAIR, JAMES GRAHM

James Grahm Fair was a 'forty-niner' who made his fortune in the Comstock Lode. He was also an early pioneer in the California wine industry. While a U.S. Senator, in the 1880's, Fair bought a large vineyard, winery and brand distillery on the Petaluma River in Sonoma County. His Fair ranch wine sold for between 20 cents and 50 cents a gallon, and he found the business more profitable than mining. At the time of his death, in 1894, the winery and distillery were producing 300,000 gallons of wine and brandy each year. As quoted by Irving McKee, Fair is reported to have said "There will always be a market for wine."

Fair Haven Estate Vineyard

Schuyler County, New York.

Concord, Niagara and Aurora Finger Lake vineyard.

FAIR HAVEN WINERY

Founded 1982, Valois, Schuyler County, New York.

Storage: Oak and stainless steel. Cases per year: 450.

Estate vineyard: Fairhaven Vineyard.

Wines regularly bottled: Vintage-dated Blanc de Noir, proprietary Seneca Vista Red and Lake Hill Red; also Red Table Wine.

Fairacre Nursery

Benton County, Washington.

Late Harvest Riesling vineyard.

FAIRMONT VINEYARDS

Founded 1978, Napa County, California.

Storage: Oak.

History: George Kolarovich, the owner and winemaker, spent 25 years in Australia where he was responsible for the development of Barossa Cooperative Winery and Kaiser Stahl Cellars. He and his wife, who is also involved in Fairmont, moved to California in 1977.

Wines regularly bottled: Vintage-dated Cabernet Sauvignon, Chardonnay, Johannisberg Riesling, Sauvignon Blanc and Pinot Noir.

FALCON CREST (See Spring Mountain Vineyards.)

Falken Hausen Farms

Chautauqua County, New York.

Chardonnay vineyard.

FALL CREEK VINEYARDS

Founded 1975, Tow, Llano County, Texas.

Storage: Oak and stainless steel. Cases per year: 3,500.

History: The winery and vineyards are located in the rugged Highland Lakes area of Central Texas, on the northwest shore of Lake Buchanan.

Estate vineyard: Fall Creek Vineyards.

Wines regularly bottled: Vintage-dated Emerald Riesling, Villard Blanc, Carnelian, Chenin Blanc, Ruby Cabernet, Zinfandel, Sauvignon Blanc.

FARFELU VINEYARD

Founded 1975, Flint Hill, Rappahannock County, Virginia.

Storage: Oak and stainless steel. Cases per year: 200.

Estate vineyard: Farfelu.

Wines regularly bottled: Seyval Blanc, Dry Red Table Wine.

FAR NIENTE WINERY

Founded 1885, Oakville, Napa County, California.

Storage: Oak and stainless steel. Cases per year: 15,000.

History: Construction started in 1882 and completed in 1885 by John Benson and architect Hamden McIntyre; winery operated until Prohibition; closed and remained vacant until 1979 when Gil Nickel purchased the property and began restoration. The building has been placed on the National Register of Historic Places.

Estate vineyard: Stelling Vineyard.

Wines regularly bottled: Estate bottled, vintage-dated Chardonnay, Cabernet Sauvignon. Also vintage-dated Chardonnay.

FARRELL, ELOISE D.

Mrs. Eloise D. Farrell, president, Brotherhood Winery, married Frank L. Farrell, and fell in love with the wine industry.

During World War II, Mrs. Farrell was in charge of radio broadcasting for Chester Bowles, then OPA czar. After the war, she continued working with him at the United Nations at Lake Success, New York.

Mrs. Farrell studied winemaking under Julius Fessler, the renowned oenologist and wine chemist in California. She returned to New York and was winemaker for Brotherhood for many years. She, also virtually went on the road for Brotherhood.

Mrs. Farrell was graduated from the University of Washington. She taught school for several years, then moved into radio programming as director of the American School of the Air for CBS radio.

Mrs. Farrell became president of Brotherhood upon the death of her husband in 1974. She and her only child, Anne, have set their sights on continued growth of the winery and the renaissance of the New York State industry.

FARRON RIDGE
(See Chateau Ste. Michelle and River Ridge Winery.)

FARVIEW FARM VINEYARD

Founded 1979, Templeton, San Luis Obispo County, California.

Cases per year: 7,000.

History: Farview Farm Vineyard was first planted in 1972 and currently includes 36 acres of Zinfandel and 14 acres of Merlot.

Estate vineyard: Farview Farm Vineyard.

Wines regularly bottled: Estate bottled, vintage-dated Merlot, Zinfandel, White Zinfandel; occasionally bottle vintage-dated Zinfandel Reserve.

Farview Farm Vineyard

San Luis Obispo County, California.

Zinfandel and Merlot vineyard.

Fay Vineyard

Napa Valley, California.

Cabernet Sauvignon and Merlot vineyard.

FBF WINERY/FITZPATRICK

Founded 1980, Somerset, El Dorado County, California.

Storage: Oak and stainless steel. Cases per year: 4,000.

Estate vineyard: Fitzpatrick Vineyard.

Label indicating non-estate vineyard: Clockspring Vineyard, Jehling Vineyard, Rock Hill Vineyard.

Wines regularly bottled: Vintage-dated Sauvignon Blanc, Zinfandel, Cabernet Sauvignon and Chardonnay.

FELTON-EMPIRE VINEYARDS

Founded 1976, Felton, Santa Cruz County, California.

Storage: Oak and stainless steel. Cases per year: 22,000.

Estate vineyard: Hallcrest Vineyard.

Label indicating non-estate vineyard: Bohan Vineyards, Vinifera Association, Ashton Vineyard, Bates Vineyard, Beauregard Ranch.

Wines regularly bottled: Estate bottled, vintage-dated White Riesling, Cabernet Sauvignon. Also vintage-dated White Riesling, Gewurztraminer, Chenin Blanc, Chardonnay, Pinot Noir, Cabernet Sauvignon. Occasionally bottles Pinot Noir and Cabernet Sauvignon N.V.

FELTZ SPRINGS (See Mill Creek Vineyards.)

Feltz Vineyard

Willamette Valley, Oregon.

Pinot Noir and Riesling vineyard.

FENESTRA WINERY

Founded 1976, Livermore, Alameda County, California.

Storage: Oak and stainless steel. Cases per year: 2,200.

History: The winery was originally built in 1889.

Label indicating non-estate vineyard: La Reina Vineyard, Smith & Hook Vineyard.

Wines regularly bottled: Vintage-dated Sauvignon Blanc, Zinfandel, Chardonnay.

Fenn Valley Estate Vineyards
Allegan County, Michigan.
Seyval Blanc, Vidal, Vignoles, Chancellor, Foch, Baco Noir, Aurora, Colobel, Riesling, Chardonnay and Gewurztraminer Fennville District vineyard.

FENN VALLEY VINEYARDS
Founded 1973, Fennville, Allegan County, Michigan.
Storage: Stainless steel. Cases per year: 12,000.
Estate vineyard: Fenn Valley Vineyards.
Wines regularly bottled: Estate bottled, vintage-dated Vidal Blanc Reserve, Seyval Blanc, Vignoles, Johannisberg Riesling, Chancellor; also Fruit wines: Peach, Blueberry, Apple; and Ruby Red, Vin Blanc, Regal, Red and White, Rosé. Occasionally bottles Foch, Gewurztraminer and Blanc de Blanc (Champagne). Also fruit wines.

Fennville
Michigan.
Seyval, Vignoles and Foch vineyard.

FENNVILLE VITICULTURAL AREA
The boundaries of the area are the Kalamazoo River in the north, the Black River in the south, Lake Michigan to the west and a line about a mile east of Fennville.

FENTON ACRES WINERY
Founded 1979, Healdsburg, Sonoma County, California.
Storage: Oak. Cases per year: 2,000.
Estate vineyard: Westside Vineyards.
Wines regularly bottled: Estate bottled, vintage-dated Chardonnay, Pinot Noir.

fermentation, wine
The changing of sugars into alcohol and carbon dioxide in the presence of yeast. Yeasts are introduced into tanks of must to start the process; it stops when the sugars are depleted or when the alcohol level reaches about 15-16 percent and stuns the yeast. Secondary fermentation takes place in sparkling wines to give them their distinctive carbonation. Not to be confused with malolactic fermentation.

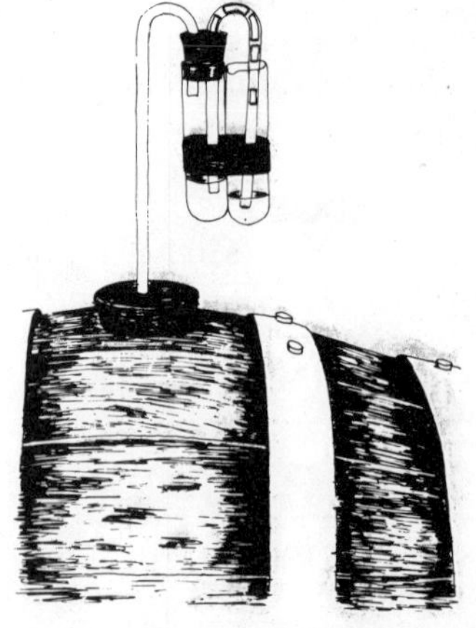

A fermentation bung allows the escape of carbon dioxide

FERRELL, TOM
Tom Ferrell attended the University of California at Davis where he earned his degree in enology. He was the winemaker for Inglenook Vineyards for eleven years, prior to his joining Franciscan Vineyards as vice-president of winemaking and production in 1982. Today he is president and winemaster of Franciscan.

John A. Ferrero Vineyards
Amador County, California.
Zinfandel vineyard.

Ferrigno Estate Vineyards
Phelps County, Missouri.
Vidal Blanc, Seyval Blanc, Chelois, DeChaunac, Baco Noir, Chancellor and Concord Ozark Plateau vineyard.

FERRIGNO VINEYARDS
Founded 1922, St. James, Phelps County, Missouri.
Storage: Oak and stainless steel.
Estate vineyard: Ferrigno Vineyards.
Wines regularly bottled: Estate bottled Chelois, Dry DeChaunac, Vidal Blanc, Seyval Blanc, Concord, Semi-Sweet Catawba; also Medium Dry Vin Rosé and proprietary Vino di Famiglia.

FETZER, JOHN EDWARD

John Edward Fetzer grew up on the family ranch in the Redwood Valley, California. He and his ten younger siblings were responsible for the care and expansion of the ranch vineyards. The Cabernet Sauvignon, Sauvignon Blanc and Pinot Noir grapes were sold to home winemakers throughout the country. When reports came back to the family of the distinctive, high quality wines that were being made from their Mendocino County grapes, the Fetzer family decided to build their own winery. Work began on the renovation and construction of the facilities when Fetzer graduated from high school in 1966. The winery was completed in time for the 1968 vintage. Fetzer became president of the winery after the death of his father in 1981. The winemaking approach of Fetzer Vineyards has been formed by the close relationship Fetzer has had with the growth and development of the vineyards and winery. He produces both everyday and premium table wines and has recently experimented with sparkling wines.

FETZER VINEYARDS

Founded 1968, Redwood Valley, Mendocino County, California.

Storage: Oak and stainless steel. Cases per year: 400,000.

History: The winery is located in Redwood Valley, one of the oldest vineyard sites in the County. The old stage road to the Mendocino Coast used to run through the property. The ranch was first settled by a gold miner, Anson Jedediah Seward, who came west after mining did not work out.

Label indicating non-estate vineyard: Cole Ranch, Lolonis Vineyard, Scharffenberger Vineyard, Ricetti Vineyard, Redwood Valley Vineyards, Brutaco Vineyard, McFarland Vineyard, Arrendell Vineyard.

Wines regularly bottled: Estate bottled, vintage-dated Zinfandel, Cabernet Sauvignon. Also vintage-dated Pinot Blanc, Fumé Blanc, Chardonnay, French Colombard, Johannisberg Riesling, Chenin Blanc, Gewurztraminer, Muscat Canelli, Gamay Beaujolais, Pinot Noir, Zinfandel, Cabernet Sauvignon, Petite Syrah. Also bottles Premium Rosé, Premium White and Premium Red.

FICKLIN, DAVID

David Ficklin was born in Fresno, California in 1918. Prior to World War II, he studied chemistry at the University of California, Los Angeles. In a family decision made after the war, Ficklin, his father and his brother established a winery devoted to the production of Port. While his brother, Walter Jr., planted the vineyards in Portuguese grape varieties, Ficklin studied viticulture and enology at the University of California, Davis. He returned to the Madera vineyards to build the winery, and the firsst Ficklin Vineyard's Port was produced in 1948. Ficklin served as president of the winery, his brother operated the vineyards and his father, Walter Sr., worked in sales. Upon his retirement in 1983, Ficklin's wife, Jean, assumed the presidency. Her son, Peter, is now winemaker and nephew Steve is in charge of the vineyards. Ficklin is a member of the American Society of Enologists, the San Joaquin Valley Winegrower's Association and is a member of the board of directors for the California Wine Institute.

FICKLIN VINEYARDS

Founded 1946, Madera, Madera County, California.

Storage: Oak and stainless steel. Cases per year: 2,000.

Estate vineyard: Ficklin Vineyards.

Wines regularly bottled: Estate bottled Port (Tinta Cao, Tinta Madeira, Touriga and Souzao).

FIDDLETOWN VITICULTURAL AREA

Fiddletown in Amador County, California, differs from the neighboring Shenandoah Valley of California viticultural area because of its higher elevations of 1,500 to 2,500 feet, colder nighttime temperatures and higher rainfall of 30 to 40 inchs per year. The area surrounding the northern and eastern boundaries is above 2,500 feet and, for the most part, too rugged and too cold for grape growing.

FIELD, EUGENE (1850-1895)

American poet.

> When I demanded of my friend what viands he preferred, he quoth: "A large cold bottle and a small hot bird."

> "There is a glorious candor in an honest quart of wine,
> A certain inspiration which I cannot well define."

FIELD STONE WINERY

Founded 1976, Healdsburg, Sonoma County, California.

Storage: Oak and stainless steel. Cases per year: 10,000.

History: Winery and vineyard are located on the original historic Cyrus Alexander farm tract for which the valley is named.

Estate vineyard: Turkey Hill Vineyard, Terra Rosa Vineyard, Birkitt Vineyard, Hoot Owl Creek Vineyard, Indian Ridge Vineyard, Home Ranch.

Wines regularly bottled: Estate bottled, vintage-dated Chenin Blanc, Johannisberg Riesling, Gewurztraminer, Petite Sirah, Cabernet Sauvignon, Rosé of Petite Sirah. Occasionally bottle Late Harvest Riesling, White Table Wine, Red Table Wine.

FIELDBROOK VALLEY WINERY

Founded 1976, Fieldbrook, Humboldt County, California.

Label indicating non-estate vineyard: Lowenthal Ranch, Gardner Ranch.

Wines regularly bottled: Dry Chenin Blanc, Zinfandel, Petite Sirah, Merlot. Also bottle Trinity River Blanc (blend of Sauvignon Blanc and Semillon).

J. FILIPPI VINTAGE COMPANY

Founded 1932, Mira Loma, San Bernardino County, California.

Storage: Oak and stainless steel. Cases per year: 200,000.

History: In 1922, Giovanni and his son Joseph came to the United States from Italy and settled in what was known as the Cucamonga Valley. In 1923, they planted their first 20 acre vineyard and by 1934 they had expanded to 300 acres and built their winery.

Wines regularly bottled: Gamay Beaujolais, Pinot Noir, Johannisberg Riesling, Riesling, Gewurztraminer, Chenin Blanc, Green Hungarian, Pinot Chardonnay, Cabernet Sauvignon. Also Sherries, Ports, Champagnes, Cold Duck, Moscato Spumante, Marsala, Mead, Vermouth, fruit wines and mellow table wines.

Second labels: Chateau Filippi, Old Rancho and Thomas Wines.

Filsinger Estate Vineyard

Riverside County, California.

Chardonnay, Sauvignon Blanc, Emerald Riesling, Zinfandel and Petite Sirah vineyard.

FILSINGER VINEYARDS AND WINERY

Founded 1979, Temecula, Riverside County, California.

Storage: Oak and stainless steel. Cases per year: 8,000.

Estate vineyard: Filsinger Vineyards.

Label indicating non-estate vineyard: Santa Gertrudis Vineyard.

Wines regularly bottled: Estate bottled, vintage-dated Chardonnay, Sauvignon Blanc, Emerald Riesling, Zinfandel, Petite Sirah. Also vintage-dated Cabernet Sauvignon and Johannisberg Riesling. Occasionally Beaujolais Nouveau.

fine

Fining, the process of clearing young wines by adding beaten egg white, bentonite, or heavy gelatin, which absorb impurities and settle to the bottom of the tank with them, to be removed by racking or filtration.

FINGER LAKES REGION

The vineyards in the hills surrounding the Finger Lakes owe a great deal of their good fortune to Mother Nature. The deep waters of the Finger Lakes moderate the climate. The excellent soil drainage is due to the shale beds. The slow warming lakes retard spring growth against the dangers of frost and keep the vines warmer on chilly fall nights. Hammondsport, Naples, Conesus and Canandaigua are the major wine towns.

FINGER LAKES VITICULTURAL AREA

"Finger Lakes" in New York consists of the area immediately around the 11 Finger Lakes, encompasses an area with a relative uniform growing-season length, and covers approximately 4,000 square miles.

FINGER LAKES WINE CELLARS

Founded 1981, Branchport, Yates County, New York.

Storage: Oak and stainless steel. Cases per year: 4,000.

History: In the early 1800's, Adam Hunt came to Yates County from eastern New York and purchased part of the current farm in the town of Jerusalem. In 1973, Arthur and Joyce Hunt moved to the farm, the 6th generation of Hunts to work the same land. In 1981 they began their winery.

Estate vineyard: Hunt Farms.

Labels indicating non-estate vineyard: Knapp Vineyards, Morehouse Vineyards.

Wines regularly bottled: Estate bottled, vintage-dated Niagara, Aurora Special Reserve, Aurora, Delaware; also vintage-dated Seyval, Cayuga, Seyval Special Selection Late Harvest, Golden Muscats, Marechal Foch, Dutchess Special Reserve, DeChaunac, and proprietary Vin D'Or and Classic Red.

The "Ah So" corkpuller

fining

A traditional process of clarification, The fining medium settles to the bottom, carrying with it the fine suspended particles.

finish

Tasting term.

Referring to the palate sensation, or aftertaste, remaining after the wine is swallowed.

FINK WINERY

Founded 1975, Dundee, Monroe County, Michigan.

Storage: Oak. Cases per year: 1,000.

Wines regularly bottled: Vintage-dated White Riesling, Chambourcin; also fruit wines and mead.

Second label: Crest; Park Garden.

fino

Denoting the driest, lightest Sherry.

FIRESTONE VINEYARD

Founded 1973, Los Olivos, Santa Barbara County, California.

Storage: Oak and stainless steel. Cases per year: 75,000.

Estate vineyard: Arroyo Perdido Vineyard, Ambassador's Vineyard.

Wines regularly bottled: Estate bottled, vintage-dated Cabernet Sauvignon, Merlot, Pinot Noir, Chardonnay, Sauvignon Blanc, Gewurztraminer, Rosé of Cabernet Sauvignon, Rosé of Merlot, Johannisberg Riesling. Occasionally bottles Selected Harvest Johannisberg Riesling, Vintage Reserve Cabernet Sauvignon, Vintage Reserve Pinot Noir.

FISHER RIDGE WINE COMPANY

Founded 1977, Liberty, Putnam County, West Virginia.

Storage: Oak and stainless steel. Cases per year: 1,200.

History: The first winery in West Virginia and the first commercial producer of wine in the state in over 100 years.

Estate vineyard: Fisher Ridge Vineyard.

Wines regularly bottled: Estate bottled, vintage-dated Blanc de Blanc, proprietary Choix du Cochon; proprietary Cochon Rouge, West Virginia Apple. Occasionally estate bottled Vidal, Chardonnay.

FISHER VINEYARDS

Founded 1979, Santa Rosa, Sonoma County, California.

Storage: Oak and stainless steel. Cases per year: 4,000.

History: The winery was constructed of Redwood and Douglas Fir, timbered and milled on the hillside above the site. The winery was selected in 1982 by the American Institute of Architects California Council as representing the best design work of California architects done throughout the world since 1975.

Estate vineyards: Whitney's Vineyard, The Wedding Vineyard.

Wines regularly bottled: Vintage-dated Chardonnay, Cabernet Sauvignon.

flat

Tasting term.

Lacking acidity; in sparkling wines, without effervescence.

flavors

For want of a more suitable term, flavor is used to describe those complex impressions on the palate created when the wine is worked over in the mouth.

Varietal: The impression of varietal aroma is frequently conformed and emphasized by a taste exemplified by the varietal grape.

Alcohol: The sensory impression of the alcohol content of a wine may best be determined by tasting.

Flintwood

Dry Creek Valley, California.

Clos du Bois estate Chardonnay vineyard.

flinty

Wine term.

Often used to describe wine that is dry, clean, sharp.

Originally applied to Chablis and some other white Burgundies. The smell of flintstone.

flor

A selected yeast culture which, under suitable conditions, grows on the surface of wine and produces the flavor characteristics in Sherries, so named, or is used in the Crowther submerged method.

flor process

There are two methods of producing "flor" sherries. One is the traditional Spanish flor way where the yeast grows naturally on the surface. The second method is the creation of Crowther who developed the technique of the submerged flor process. A method in which the flor (yeast) is circulated through the wine, thus making the yeast work and developing the flor taste in days and not years.

FLORIDA HERITAGE WINERY

Founded 1981, Anthony, Marion County, Florida.

Storage: Stainless steel. Cases per year: 7,000.

History: Winery's Muscadine vineyards are a touch of the Old South.

Estate vineyard: Florida Heritage Vineyard.

Wines regularly bottled: Estate bottled, vintage-dated Noble, Welder, Carlos; also estate bottled vintage-dated Private Cellars.

FOCH

Marechal Foch Wine Grape.

Blue-black Burgundy type French hybrid, which makes deep hearty red, medium to fully bodied, dry wine. Delightful as a "Beaujolais Nouveau" type.

THOMAS FOGARTY WINERY

Founded 1982, Portola Valley, San Mateo County, California.

Storage: Oak and stainless steel. Cases per year: 3,000.

Estate vineyard: Portola Valley Vineyards.

Wines regularly bottled: Vintage-dated Chardonnay, Pinot Noir.

FOLIE A DEUX WINERY

Founded 1981, St. Helena, Napa County, California.

Storage: Oak. Cases per year: 1,000-5,000.

Estate vineyard: Folie a Deux Vineyard.

Label indicating non-estate vineyard: John Pun Vineyard, Robert Egan Vineyard, Barressini Vineyard.

Wines regularly bottled: Vintage-dated Dry Chenin Blanc, Cabernet Sauvignon, Chardonnay.

FOLLE BLANCHE

Wine grape.

A grape that produces a dry to medium-dry, tart, fruity and fresh tasting white wine with a mild aroma of both apple and grape. Originally from France where it once was the Cognac grape.

E.B. FOOTE WINERY

Founded 1978, King County, Washington.

Storage: Oak and stainless steel.

Wines usually bottled: Chardonnay, Chenin Blanc, Cabernet Sauvignon, Pinot Noir, Riesling, Gewurztraminer.

FOPPIANO, LOUIS J.

Louis J. Foppiano, president of Foppiano Vineyards, was born into a winemaking family. His grandfather, John, came to California from Genoa during the Gold Rush in search of a fortune.

Instead, he settled in the Russian River Valley in Sonoma County, and made a living farming vegetables. In 1896, he and his son, Louis A. Foppiano, purchased the Smith Winery. They planted additional vineyards and ran the winery jointly until 1910, when John retired. Under the direction of Louis A. Foppiano, the vineyards survived Prohibition through the cultivation of prune trees planted between the rows of vines. After the Repeal, and the death of his father, Louis J. Foppiano became Winemaker. He produced hearty, high-alcohol red wines which were sold by the barrel or jug, filled for the customer at the spigot. The present winery was constructed in 1937, and the vineyards were doubled just after World War II, with the purchase of an adjoining ranch. Older grape varieties were replaced with premium vinifera grapes, although some vines, planted eighty years ago, are still in production. In the 1970's, Foppiano and his two sons, Louis M. Foppiano, and Rod, initiated major production improvements. The result was the release of the Louis J. Foppiano label in 1979, a premium line of table wines. Rod Foppiano died early in 1984.

Milk carton protects young vine from animals

FOPPIANO WINE COMPANY

Founded 1896, Healdsburg, Sonoma County, California.

Storage: Oak and stainless steel. Cases per year: 150,000.

History: One of California's oldest winemaking families. The family owns and operates the vineyards and winery.

Estate vineyard: Foppiano Ranch.

Wines regularly bottled: Vintage-dated Dry Chenin Blanc, Pinot Noir, Chardonnay, Zinfandel, Petite Sirah, Cabernet. Also vintage-dated Sonoma White Burgundy.

Second label: Riverside Farm.

Forestville Ranch

Russian River Valley, California.

River Road estate Chardonnay and Johannisberg Riesling vineyard.

Forgeron Estate Vineyard

Lane County, Oregon.

White Riesling, Pinot Noir, Chenin Blanc, Pinot Gris and Cabernet Sauvignon Willamette Valley vineyard.

FORGERON VINEYARD

Founded 1972, Elmira, Lane County, Oregon.

Storage: Oak and stainless steel. Cases per year: 6,500.

Estate vineyard: Forgeron Vineyards.

Wines regularly bottled: Estate bottled, vintage-dated Willamette Valley White Riesling, Rosé of Pinot Noir, Blanc de Pinot Noir, Cabernet Sauvignon; also vintage-dated White Riesling, Chenin Blanc.

fortified

The *term* "fortified" wine has not been used for over 30 years, but it originally referred to wines which are strengthened by the addition of wine spirits to raise the alcohol content from less than 14% to over 18% by volume. Sherry and Port are examples of todays fortified wines. These wines known today as "dessert" wines although generally consumed before or after mealtime.

Table wines (those wines meant to be consumed with a meal) are under 14% alcohol and have no wine spirits added.

FORTINO WINERY

Founded 1970, Gilroy, Santa Clara County, California.

Storage: Oak.

Wines regularly bottled: Zinfandel Blanc, Cabernet Sauvignon Blanc, Johannisberg Riesling, Carignane, Zinfandel, Ruby Cabernet, Petite Sirah, Charbono, Cabernet Sauvignon, Grenache Ruby, Rosé of Cabernet Sauvignon, White Grenache.

Foss Valley Ranch

Napa County, California.

William Hill estate Chardonnay and Cabernet Sauvignon Napa Valley vineyard.

FOUNTAIN GROVE (See Martini & Prati Wines.)

Four Chimneys Farm Estate Vineyard

Yates County, New York.

Delaware, Catawba, Chenin Blanc, Elvira, Concord, Riesling, Cascade, Cabernet Sauvignon, Chardonnay and Pinot Noir Finger Lakes vineyard.

FOUR CHIMNEYS FARM WINERY

Founded 1980, Himrod, Yates County, New York.

Storage: Oak and stainless steel. Cases per year: 2,000.

History: All grapes are grown under the certification program of the State Natural Foods Association. The chateau, which is Four Chimneys, and its barns were built in the early 1860's in the Italian Villa style. The Grape House was built to keep table grapes out of the sun while waiting pickup at the barge landing. Grapes were packed and shipped, from the lake through canals, to markets in New York and beyond.

Estate vineyard: Four Chimneys Farm Vineyard.

Wines regularly bottled: Estate bottled, vintage-dated Delaware (Late Harvest), Catawba, Seyval, Landot Noir, proprietary. Kingdom White (non-vintage) and Red, Golden Crown, Dayspring, Eye of the Bee, Eye of the Dove, Kingdom Rosé (non-vintage). Also vintage-dated Cabernet Sauvignon, Pinot Noir, Chardonnay, Johannisberg Riesling, Chenin Blanc and Gamay.

FOURNIER, CHARLES

Charles Fournier was born, in 1902, into a winemaking family in Reims, France. He spent summers in the vineyards and cellars of Veuve Cliquot Ponsardin Champagnes, then received a formal education in chemistry at the University of Paris, and in winemaking at schools in France and Switzerland. He joined Veuve Cliquot, in 1926, as winemaker, and was promoted to production manager four years later. In 1933, Fournier was asked to assist in improving the post-Prohibition position of Gold Seal Vineyards in the United States. He accepted the challenge of rebuilding the winery and stayed, despite the shock of East Coast weather and the taste of American wine grapes. He became convinced that American wines, especially champagnes, could be improved. In 1936, with the help of Maryland journalist Philip Wagner, he began experimentation with French-American vines. He discovered many methods of improving grape quality and yield, and within ten years, Gold Seal was producing well-respected sparkling wines. In 1951, Fournier was promoted to president of the winery. He subsequently developed Catawba Pink and the Henri Marchant line of Champagnes. In 1953, Fournier was the first to recognize the Finger Lakes region of New York state as having a microclimate that could support vinifera grapes. In the 1960's, he realized his dreams of producing vinifera wines on the East Coast, when Gold Seal released a Riesling and a Chardonnay. Amongst his many awards was the 1982 Leon D. Adams "Man of the Year" citation and the Merit award of the American Society of Enologists. In 1982, the New York State Assembly passed a resolution thanking and congratulating him on his achievements. Fournier passed away in 1983.

Fourreeminette Vineyards
Napa County, California.
Chardonnay, Cabernet Sauvignon and Pinot Noir vineyards.

foxiness
Tasting term.
The "grapey" smell and taste of labrusca varieties, notably Concord and Niagara.

Field grape crusher

FOXWOOD WINE CELLARS
Founded 1978, Woodruff, Spartanburg County, South Carolina.
Storage: Stainless steel.
Estate vineyard: Walnut Grove Vineyards.
Wines regularly bottled: Scuppernong; also Red, White and Rosé.
Second label: South Carolina Wine Company, Ramblin Rose Wine Company.

FOYLENMILCH (See Casa Pinson Hermanos.)

Francis Vineyard
Shenandoah Valley, California.
Beau Val estate Sauvignon Blanc, Barbera and Zinfandel vineyard.

FRANCISCAN VINEYARDS
Founded 1973, Rutherford, Napa County, California.
Storage: Stainless steel. Cases per year: 100,000.
Estate vineyard: Oakville Vineyard, Alexander Valley Vineyard.
Wines regularly bottled: Estate bottled, vintage-dated Merlot, Zinfandel, Brut Sparkling Wine, Cabernet Sauvignon, Johannisberg Riesling, Chardonnay, Fumé Blanc, vintage-dated Charbono. Also Cream Sherry. And semi-generic Cask 321 Burgundy and Chablis.

FRANK, DR. KONSTANTIN
Russian-born of German parents emigrated to the United States in 1951. A major force in the Eastern winegrape industry. Imported winegrape stocks from Europe that were successfully planted in New York State. First recipient of the vinifera Monteith Award. His Riesling and Chardonnay wines made him famous.

FRANKEN RIESLING
Wine grape.
Not a true Riesling (See Sylvaner).

FRANKLIN, BENJAMIN (1706-1790)
American statesman, scientist, inventor, and writer.

> "We hear of the conversion of water into wine at the marriage, in Çana, as a miracle. But this conversion is, through the goodness of God, made every day before our eyes. Behold the rain, which descends from Heaven upon our vineyards, and which enters into the vine-roots to be changed into wine; a constant proof that God loves us."

Franklin Hill Estate Vineyard

Northampton County, Pennsylvania.

Seyval, DeChaunac, Vidal, Chardonnay vineyard.

FRANKLIN HILL VINEYARD

Founded 1982, Northampton County, Pennsylvania.

Storage: Oak and stainless steel. Cases per year: 1,200.

Estate vineyard: Franklin Hill Vineyard.

Wines regularly bottled: Estate bottled, vintage-dated Seyval, Seyval Reserve, DeChaunac Rosé, Vidal, Chardonnay.

Franks Ranch Vineyard

Sonoma County, California.

Seghesio Wineries estate Pinot Noir, Zinfandel, Golden Chasselas, Carignane vineyard.

Franks Vineyard

Shenandoah Valley, Amador County, California.

Port varietals.

FRANZIA WINERY

Ripon, San Joaquin County, California. Part of The Wine Group of San Francisco, California.

Storage: Stainless steel. Cases per year: 10,000,000 (est.)

Wines regularly bottled: Zinfandel, Chenin Blanc, French Colombard, Cabernet Sauvignon, Grenache Rosé. Also semi-generics: Chablis, Rhine, Pink Chablis, Vin Rosé, Burgundy. Sparkling wines include Sparkling Rosé, Champagne, Pink Champagne.

FRASINETTI WINERY

Founded 1897, Sacramento, Sacramento County, California.

Cases per year: 55,000.

Wines regularly bottled: Chenin Blanc, French Colombard, Zinfandel, Johannisberg Riesling, Gamay Beaujolais, Cabernet Sauvignon, Petite Sirah. Also produces Chablis, Burgundy, Rosé. And Angelica, Marsala, Cream Sherry and Port.

CHRIS A. FREDSON WINERY

Founded 1911, Geyserville, Sonoma County, California.

Storage: Oak.

Wines regularly bottled: Zinfandel, Chablis and Burgundy. Only for distribution through other wineries.

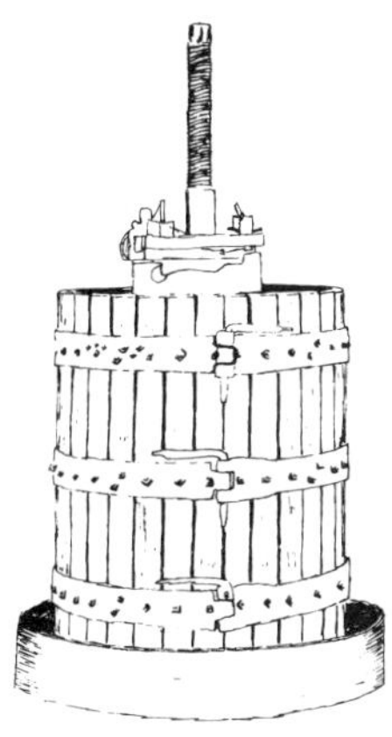

Wooden grape "basket press"

free run

The wine that runs from the tank or press before pressure is applied to the pomace. Free run usually makes up four-fifths of the total volume. When the free run is drawn off, the pomace is pressed to extract press wine, which contains more tannin. Both press and free run are often blended for balance. Fine wines are usually produced solely from free run.

FREEMARK ABBEY WINERY

Founded 1967, St. Helena, Napa County, California.

Storage: Oak and stainless steel. Cases per year: 27,000.

Estate vineyards: Red Barn Vineyard, Carey Conolly Vineyard, Curtis Ranch.

Label indicating non-estate vineyard: Bosche Vineyard.

Wines regularly bottled: Estate bottled, vintage-dated Cabernet Sauvignon, Chardonnay, Johannisberg Riesling, vintage-dated Cabernet Bosche (Cabernet Sauvignon), Edelwein (Late-harvest Johannisberg Riesling).

freezing wine

The age old problem of what to do with that half bottle of uncorked wine can best be solved by freezing. However, there are some ground rules; (Be aware that sometimes tartrates will be deposited and the taste of the wine will be flatter.)

1. Mature or well-aged wines, and delicate wines, suffer the most. For the most part, wine drinkers have very little left after the wines have been opened.

2. Full bodied, young, high tannic wines, are the best for freezing.

3. Do not keep wines frozen for more than several weeks.

4. A *must.* Allow adequate space in the bottle to allow for expansion. Bottle should be minus one glass full. Lay bottle on its side in the freezer so that cork is covered.

5. Allow to stand at room temperature for four to six hours before serving, or let stand in refrigerator during the day, and take out for thawing a few hours before serving.

6. Do not freeze sparkling wines. They lose their bubbles.

7. Try it.

Freiberg Vineyards/Sobre Vista Vineyard

Sonoma Valley, California.

Chardonnay vineyard.

Freites Ranch

Napa County, California.

Pinot Noir vineyard.

French Camp Vineyard

Paso Robles County, California.

Zinfandel, Cabernet Sauvignon and Sauvignon Blanc vineyard.

FRENCH COLOMBARD

Wine grape.

This grape was brought to California, in the 1870's, from France where it was used mainly for brandy. It has been used for blending and sparkling wines. In recent years the French Colombard has gained popularity as a varietal.

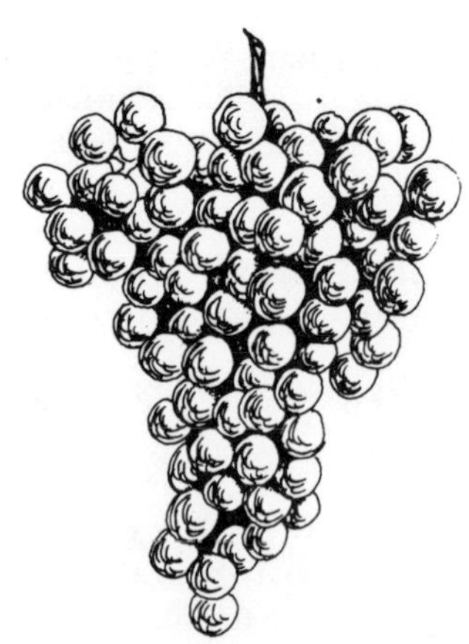

French Columbard

FRENCH HYBRIDS

Also referred to as French/American hybrids.

Grape varieties developed in France from cross-breeding native American varieties with the traditional European varieties. Technically, the only difference between a French hybrid and an American hybrid is: one is developed by a Frenchman and the other by an American (both using grapes from both sides of the ocean). However, this distinction is more than one of form as the two groups of grapes have vastly different styles. The French, disliking the "foxy" quality of American hybrids, tried to avoid the native American Labrusca in their breeding process. So as to develop wines more common to their taste, the French used the native American Rupestris and Riparia. The more successful of these crosses have been used in the colder parts of North America to produce wine more common in taste to the familiar European wines. Examples are: Aurora Blanc, Marechal Foch, Leon Millot, Seyval Blanc, Vignoles, Villard Blanc.

FRENCH VALLEY VINEYARDS (See Hart Winery.)

fresh

Tasting term.

The term fresh is used to describe a fruity, tart, young wine.

FRESNO COUNTY
Central San Joaquin Valley, California.

FRETTER WINE CELLARS, INC.
Founded 1977, Berkeley, Alameda County, California.
Storage: Oak. Cases per year: 1,000.
Estate vineyard: Chateau Camelia.
Label indicating non-estate vineyard: Leaky Lake Vineyard. Narsai David Vineyards, Bacher Vineyard.
Wines regularly bottled: Vintage-dated Cabernet Sauvignon, Chardonnay, Gamay, Gamay Rosé, Merlot. Occasionally bottles Bacher Blanc, vintage Pinot Noir.
Second label: Chateau Bacher, Chateau Camelia.

Frey Estate Vineyards
Mendocino County, California.
Cabernet Sauvignon, Sauvignon Blanc, Chardonnay and Grey Riesling vineyard.

FREY VINEYARDS
Founded 1980, Redwood Valley, Mendocino County, California.
Storage: Oak and stainless steel. Cases per year: 4,000.
Estate vineyard: Frey Vineyards.
Label indicating non-estate vineyard: Guntly Vineyards, Ponnequin Vineyards.
Wines regularly bottled: Estate bottled, vintage-dated Cabernet Sauvignon, Sauvignon Blanc, Grey Riesling. Also vintage-dated Chardonnay, Zinfandel, French Colombard.

FRICK WINERY
Founded 1976, Santa Cruz, Santa Cruz County, California.
Storage: Oak and stainless steel. Cases per year: 3,500.
Label indicating non-estate vineyard: Cayote Creek Vineyard.
Wines regularly bottled: Vintage-dated Pinot Noir, Petite Sirah, Zinfandel, Chardonnay.
Second label: Full City.

FRIEDRICH, ED
Born in the Mosel Valley of Germany, Friedrich studied at the famed Institute of Viticulture and Enology in Trier, Germany. After graduation in 1958, he came to the United States and joined Paul Masson Vineyards in California, ultimately managing the company's Soledad winemaking facility.
He joined the Wiederkehr Wine Cellars in Arkansas in 1973 as senior enologist and quality control manager. While there he was involved in producing vinifera wines in Arkansas.
Friedrich joined the San Martin Winery in 1974 as chief enologist and quality control manager, and rose to vice president/general manager and winemaster. He left San Martin in the spring of 1982 and is today president of Chateau Elan, a new winery and vineyard in Georgia.
Friedrich was a member of the board of directors of The Wine Institute in California. In 1977 he served as president of the Santa Clara Valley Wine Growers Association and a member of the board of directors of the Monterey Grape Growers Association.

FRISCO (See Bardenheier's Wine Cellars.)
Port, Sherry, Muscatel, White Port.

FRITZ CELLARS
Founded 1979, Cloverdale, Sonoma County, California.

Storage: Oak and stainless steel. Cases per year: 15,000.

Estate vineyard: Fritz Estate Vineyard.

Label indicating non-estate vineyard: Gauer Ranch.

Wines regularly bottled: Estate bottled, vintage-dated Chardonnay, Fumé Blanc, Zinfandel. Occasionally bottles Pinot Noir and Pinot Noir Blanc.

Fritz Estate Vineyard

Sonoma County, California.

Chardonnay, Sauvignon Blanc and Zinfandel vineyard.

FROG'S LEAP WINERY

Founded 1981, St. Helena, Napa County, California.

Storage: Oak and stainless steel. Cases per year: 3,000.

Wines regularly bottled: Vintage-dated Sauvignon Blanc, Chardonnay.

Frontenac Point Estate Vineyard

Seneca County, New York.

Chardonnay, Riesling, Pinot Noir, Chelois, Marechal Foch, Vidal, Seyval Blanc, Ravat, Gamay Beaujolais Finger Lake vineyard.

FRONTENAC POINT VINEYARD

Founded 1978, Trumansburg, Seneca County, New York.

Storage: Oak. Cases per year: 1,500-3,000.

History: Owner, Jim Doolittle provided an economic report and assisted in drafting legislative language which resulted in the law permitting the establishment of small farm wineries in New York State. Organized the first New York State Commercial Wine Competition.

Estate vineyard: Frontenac Point Vineyard.

Wines regularly bottled: Estate bottled, vintage-dated Chardonnay, Chelois Rosé, Blanc Sec, Nouveau Red, Reserve Red.

FRUIT OF THE WOODS WINE CELLAR, INC.

Founded 1971, Eagle River, Vilas County, Wisconsin.

Storage: Stainless steel. Cases per year: 5,000.

Wines regularly bottled: Fruit wines: Cranberry, Raspberry-Apple, Apple, Cranberry-Apple, Loganberry, Blackberry.

FRUIT WINES

Wines made from fruits other than grapes not to be confused with the wine term, fruity. Popular fruit wines include Apple, Cherry, Blueberry and Strawberry. Fruit wines require considerable skill and care in the making so they taste like the fruit from which they came. The success with fruit wines is noticeable at first sip!

FRUIT WINES OF FLORIDA

Founded 1972, Tampa, Hillsborough County, Florida.

Storage: Oak and stainless steel. Cases per year: 9,500.

History: The first winery in Florida to commercially produce grape wines. The sandy soil vineyard is where once only palmettos would grow, but the sandy soil is rich with a clay base, about six feet down, that holds the moisture.

Label indicating non-estate vineyard: Midulla Vineyards.

Wines regularly bottled: Lopez Blanc (Stover), Jean de Noir (Conquistador), Classic Florida (Semi-dry White), Orange, Tangerine, Grapefruit and Pineapple Wine.

Second label: Midulla Vineyards, Palmetto Country, Floriana, Mr. Dude 44.

fruity

Tasting term.

Describes the fresh, tart, generally pleasant fruit-like impression given by well made young wines.

full bodied

Tasting term.

Structured, rich in all components.

FULL CITY (See Frick Winery.)

FULLER, BILL

Bill Fuller began his career in the wine industry when he worked at Italian Swiss Colony in Cloverdale, California. He returned to school to obtain his master's degree in food technology, specializing in enology, from the University of California, Davis, in 1962. For the next ten years, Fuller was assistant winemaker at the Louis Martini Winery. He was responsible for supervising the crushing and the blending of the wines, amongst other duties. In 1973, he joined with Bill Malkmus to found Tualatin Vineyards in Forest Grove, Oregon. The winery capacity has grown to 60,000 gallons; the vineyards now comprise 83 acres. Fuller works as winemaker, and assists in marketing in Oregon, California and Washington. He has been an active member of the California Wine Institute, the Napa Valley Wine Technical Committee, the Table Wine Research Board and the Oregon Wine Advisory Board. Fuller has served three terms as president of the Oregon Winegrower's Association and on the board of the American Society of Enologists.

FUMÉ BLANC (See Sauvignon Blanc.)

This term has become a popular way to describe this increasingly consumed dry, crisp white wine. The name usually implies a dryer type of Sauvignon Blanc. Refers to somewhat smoky character.

Fuqua Vineyards

Yamhill County, Oregon.

White Riesling and Chardonnay vineyard.

FURNESS, ELIZABETH

Eighty-five year old Elizabeth Furness did not become involved in the wine industry until just eleven years ago. She started life as Elizabeth Merrill, raised in California and Britain, educated in London, Paris and Boston. As a young woman, Furness was admired for her beauty, athletic abilities, and singing voice. Family friends included Rodin and Edward Steichen. In 1942, she married Thomas Furness and moved to a 400 acre estate in Middleburg, Virginia. The land was used exclusively as a farm until Furness heard a lecture by Dr. Hamilton Mowbry in 1972. She sought his advice in starting a vineyard, and sent soil samples to the University of Mississippi to determine which vinifera grapes would best produce. Vine cuttings were ordered from California, and Piedmont Vineyards was founded. It was the first commercial vinifera vineyard and winery in Virginia, and now boasts 30 acres of grapes.

GAIGE, CROSBY (1882-1949)
American theatrical producer and writer.

"If, as and when we become a nation of wine drinkers, we will be a healthier and a happier people.

"The slowly unfolding story of wine and food contains the saga of human culture, good manners and well being.

"It is axiomatic that good wine and good food go hand in hand."

GALLEANO WINERY
Founded 1933, Mira Loma, Riverside County, California.

Storage: Oak. Cases per year: 63,000.

History: The home purchased by Domenico Galleano in 1927 from Mrs. Estaban Cantu, wife of Colonel Estaban Cantu, Governor of Baja California, Mexico (1910-1917), is now the residence of Donald D. Dominico's grandson. It is on the register of historical places.

Estate vineyards: Hofer Ranch, Shue Ranch, Merrill Ranch, Hunt Ranch.

Wines regularly bottled: Zinfandel, Mission, Mosacto, Tokay. Also some semi-generics: Burgundy, Chablis, Sauterne, Chianti, Vis Rosé Rhine. And also Port, Sherry, Muscatel, Marsala, Vermouth, Vinno Rosso.

Second label: Green Valley.

GALLICO (See Honeywood Winery.)
Pinot Noir, Chardonnay, Riesling, Gewurztraminer, Chenin Blanc, Chablis.

GALLO, ERNEST AND JULIO
The story of Ernest and Julio Gallo is proof of the "Great American Dream". Like Carnegie, Ford and other great American pioneers of industry, the Gallo brothers started with an idea, the willingness to work at their dream, and the desire to create a business that would not only succeed, but would have as a basic premise the principals of giving the consumer an excellent product at a fair price.

Born near Modesto, the brothers grew up working the small vineyard owned by their father, an immigrant from Italy's northern Piedmont Region. "We had a tractor in the barn, but we didn't have enough money to buy gas", recalls Ernest. "Instead, we used four mules and worked the vineyards seven days a week from daylight to dusk." With the first stirrings of Repeal, they dug up $5,900.23 in capital and set out to produce their own wines. They rented a railroad shed for $60 a month, bought a $2,000 grape crusher and redwood tanks on 90 to 180 day credit terms.

There was one nettlesome problem: though they had plenty of experience growing grapes, they did not know how to make wine. In the Modesto Public Library, Ernest found a pair of pre-Prohibition pamphlets, one on fermentation and the other on the care of wine. He and Julio then visited the author of these pamphlets, Professor William Vere Cruess, at the University of California to get answers to questions developed by reading the pamphlets. Thus enlightened, they made the rounds of local grape growers and soon had enough grapes to make all the wine that the tanks could hold. A few days before Prohibition ended, the brothers received a form letter from a would-be wine distributor in

Chicago. Ernest Gallo hastened to Chicago and sold the distributor 6,000 gallons at 50 cents each. Emboldened, he continued East and found enough customers to take his entire production.

From then on, the Gallo brothers winery grew steadily, first as small winemakers producing wine in bulk for local bottlers, then franchise bottling in Los Angeles, New York and New Orleans under their own label until 1943, when they first bottled at Modesto under the Gallo label.

Ernest and Julio were always interested in grape quality. They began research on varietal grapes in 1946. Eventually, more than 400 different varieties were planted and replanted in their experimental vineyards during the 1950's and 1960's. These varieties were tested for their ability to produce fine table wines in the different regions of California. Now, it was no small task to convince growers to convert from common grape varieties (remnants of Prohibition), to the delicate, thin-skinned varietals. It takes at least four years for a vine to begin bearing and, more years to develop typical, varietal characteristics. In 1968, therefore, Gallo offered long-term contracts to growers, guaranteeing top prices for their grapes every year, provided they met Gallo quality standards. With a guaranteed long-term "home" for his crop, a grower could borrow the needed capital to finance the costly replanting.

Julio Gallo established a grower relations staff of skilled viticulturists to aid contract growers. This staff still counsels growers on the latest viticultural techniques.

The E & J Gallo Winery remains a private family-owned winery. Ernest and Julio Gallo, and members of their immediate family, personally supervise every aspect of winemaking and distribution. Julio, with his son and son-in-law: enology, viticulture and winery operations. Ernest, with his two sons: marketing and sales.

E & J GALLO WINERY

Founded 1933, Modesto, Stanislaus County, California.

Storage: Oak and stainless steel. Cases per year: 50,000,000.

History: While Modesto remains winery headquarters, no wine is vinified there. Wines are created in, or near, each viticultural region in wineries owned by or completely dedicated to production for Gallo alone. For example, Napa Valley grapes are crushed under Gallo supervision at the Napa Valley Co-op winery. While grapes from Sonoma and Mendocino Counties are crushed at the Gallo Winery in the Dry Creek Valley, just outside the town of Healdsburg, Northern Sonoma County. Monterey grapes are crushed and vinified at Gallo's Livingston Winery.

From grape to bottle, each Gallo wine is carefully attended by a winemaker responsible for that wine, under the direction of Julio Gallo. Each winemaker, in effect, operates a small winery under the larger Gallo umbrella. Each wine receives the personal care required for quality, yet the winemaker can draw upon the larger corporate resources.

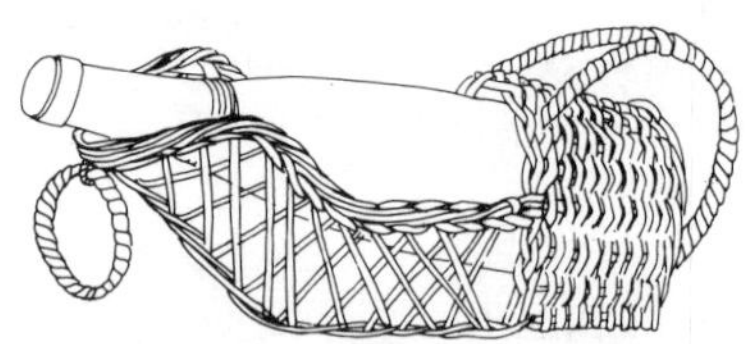

Wicker wine basket

Wines regularly bottled: Cabernet Sauvignon, Sauvignon Blanc, Chenin Blanc, French Colombard, Gewurztraminer, Chardonnay, Johannisberg Riesling. The generic wines bottled are Burgundy, Hearty Burgundy, Rhine, Chablis Blanc, Pink Chablis, Red Rosé, Vin Rosé and some dessert wines: Port, Cream Sherry, Tawny Port, Very Dry Sherry. Also Sweet and Dry Vermouths and Andre Champagnes and E & J Brandy.

GAMAY
Wine grape.
A grape that was planted as far back as the 14th Century in France. Produces light, fruity, fresh and slightly tart wine. Originally from the Burgundy region of France where it is the dominant grape in Beaujolais. Also referred to as Napa Gamay. Should not be confused with Gamay Beaujolais. Should be drunk when still young—one or two years old. Some enjoy this red wine slightly chilled.

Gamay

GAMAY BEAUJOLAIS
Wine grape.
A varietal grape that was introduced in California from the Beaujolais region of France and produces a light, fresh wine very much in the style of Gamay. The Gamay Beaujolais is a sub-variety of Pinot Noir. Not to be confused with Gamay or Napa Gamay.

GAMAY NOIR
A red California named wine made from the Gamay.

GAMAY ROSÉ
A varietal pink table wine that is fresh, fruity and slightly sweet, made from the Gamay.

Gamble Vineyard
Napa County, California.
Beringer estate Cabernet Sauvignon, Chardonnay and Merlot vineyard.

Garden Creek Ranch
Sonoma County, California.
Gewurztraminer and Cabernet Sauvignon Alexander Valley vineyard.

Gardner Ranch
Humboldt County, California.
Sauvignon Blanc and Semillon vineyard.

Gardiner Estate Vineyards
Ulster County, New York.
Vidal, Seyval, Chancellor, Leon Millot, Cascade, Riesling and Chardonnay Hudson River Region vineyard.

GARDINER VINEYARDS AND FARMS CORPORATION
Founded 1973, Gardiner, Ulster County, New York.
Storage: Stainless steel. Cases per year: 12,000.
Estate vineyard: Gardiner Vineyards.
Wines regularly bottled: Estate bottled, vintage-dated proprietary Chateau Georges Chancellor, Seyval, Vidal, Leon Millot and Northern Lights White. Occasionally bottles Villard.

Garfield Vineyard
Sonoma County, California.
Windsor Vineyards estate Chardonnay vineyard.

garnet
Wine color.
A red combined with yellow and ochre. Dark red like pomegranate.

GARRET, CAPTAIN PAUL
The man from North Carolina who made millions with his wine made from Scuppernong grapes, VIRGINIA DARE. He wanted to pay tribute to a favorite daughter of North Carolina, Virginia Dare, and he did, "Say it again with Virginia Dare..." His father, Dr. Farancis Garret, and an uncle, in 1865, decided they wanted to be in the wine business so they bought the Medoc Vineyard, the first commercial winery in North Carolina, started in 1835.

Garth
Amador County, California.
Karly estate Sauvignon Blanc, Shenandoah Valley vineyard.

Gasser Vineyard
Napa County, California.
Beringer estate Chenin Blanc vineyard.

gassy
Wines containing dissolved carbon dioxide are recognized by the slightly biting sensation on the palate. When warmed, such wines will liberate small bubbles of gas.

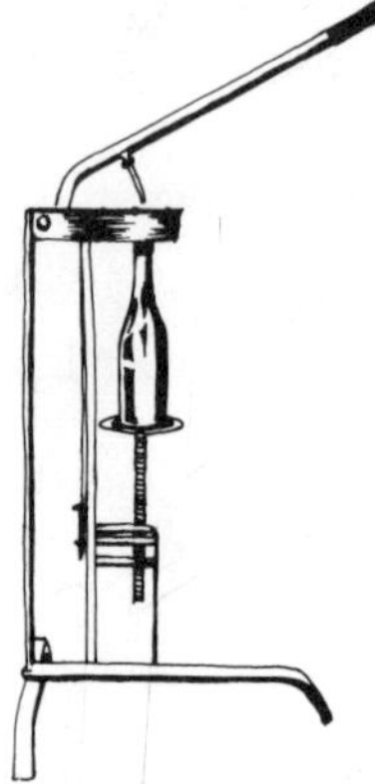

Hand corker

GATES, NORMAN E.
Norman Gates, Grand Commander and founder of The Knights of the Vine, the American wine brotherhood, was born and raised in Nevada City, California.

After serving in the Air Force during World War II, he returned to Morocco, where he spent 11½ years in a civilian capacity with the Air Force. While in Morocco, Norman met and married Nicole. From Morocco, back to the United States and then on to France, his interest in wine continued. While in France, Norman joined many wine brotherhoods and began a study of them. This study led him to the idea of a wine brotherhood devoted to the wines of America.

Being a man of great tenacity, no matter what the obstacles were, he stayed with it and, in May 1971 founded The Knights of the Vine. Both men and women may join. The women are called Gentle Ladies.

Today there are 15 chapters in the major cities of the United States devoted to the promotion and joys of drinking the wines of America. The Gates dream came true.

Gauer Ranch
Sonoma County, California.
Chardonnay Alexander Valley vineyard.

Gavilan
Sonoma County, California.
Caswell estate Zinfandel Green Valley vineyard.

GAVILAN CREEK
(See Summerhill Vineyards.)

Gaylord Vineyard
Litchfield County, Connecticut.
Chardonnay and Seyval vineyard.

Gehringer Brothers Vineyard
Okanagan Valley, British Columbia Verdelet vineyard.

GEHRS, DAN
Dan Gehrs attended Pacific Lutheran University and earned a bachelor's degree in political science in 1973. In 1974, he joined Paul Masson Vineyards Champagne Cellar in their Public Relations Department. While at Paul Masson, Gehrs made some experimental batches of wine from grapes grown in the Santa Cruz Mountains. He became convinced that the region was capable of producing first class grapes and wines and, in 1975, he and his wife Robin moved into an abandoned winery estate near Saratoga. They were responsible for the care of the old Zinfandel vineyards.

Toward the end of that year, Gehrs formed a partnership with the owner of the property, Vic Erickson, and reopened the winery, calling it Congress Springs Vineyards. In 1976, he left his job at Paul Masson and harvested the first Zinfandel from the estate. The following year, five acres of Chardonnay were planted, producing the first crop in 1981. Additional vineyards in the Santa Cruz Mountains are leased from the Novitiate Winery, and some grapes are purchased from nearby Santa Clara Valley vineyards. Gehrs is dedicated to revitalizing the Santa Cruz Mountain wine growing region and participated in securing the appellation of origin for the area.

gelatin
A protein used in wine to clarify the wine and remove excess tannin.

GEM CITY VINELAND COMPANY, INC.
Founded 1857, Nauvoo, Hancock County, Illinois.
Storage: Oak. Cases per year: 4,000.
Estate vineyard: Gem City Vineland Company.
Wines regularly bottled: Proprietary Old Nauvoo Concord White, Niagara, Rosé, Sauterne, Burgundy.

GEMELLO WINERY
Founded 1934, Mountain View, Santa Clara County, California.
Storage: Oak and stainless steel. Cases per year: 2,000.
History: John Gemello founded the winery and his son Mario became the winemaker in the 1940's. Today the winemaker is John's granddaughter, Sandy Obester. (See also: Obester Winery.)
Label indicating non-estate vineyard: Scott Knight Smith Vineyard, Redwood Ranch.

Wines regularly bottled: Vintage-dated Cabernet Sauvignon, Zinfandel, Merlot, Petite Sirah, Pinot Noir. Also bottle Barbera. Occasionally bottle Spring Fumé Blanc.
Second label: Mountain View.

Gena's Vineyard
San Luis Obispo County, California.
French Colombard and Chenin Blanc vineyard.

generic/semi-generic
Wine type names which stand for general type characteristics are called generic. Semi-generic names of geographic origin originally applied to the wines of specific Old World Viticultural Districts. Burgundy, Chablis, Sauterne, Rhine are some of the type names of geographic origin. Generic are names like Mountain Red Wine and Claret.

Tying cane to grape stake

J.H. GENTILI WINES
Founded 1981, Redwood City, San Mateo County, California.
Storage: Oak and stainless steel. Cases per year: 300.
Wines regularly bottled: Vintage-dated Chardonnay, Cabernet Sauvignon, Red and White Zinfandel. Occasionally bottles Johannisberg Riesling, Red and White Pinot Noir.

Gerber Vineyards
Josephine County, Oregon.
Pinot Noir and Gewurztraminer vineyard.

Germanton Estate Vineyard
Stokes County, North Carolina.
Seyval Blanc, Niagara and Fredonia vineyard.

GERMANTON VINEYARD AND WINERY, INC.
Founded 1981, Germanton, Stokes County, North Carolina.
Storage: Stainless steel.
Estate vineyard: Germanton Vineyards.
Wines regularly bottled: Estate bottled Seyval, Niagara, Fredonia; Sweet and Dry Red, Sweet and Dry White.

GERWER (See Stoney Creek Vineyards.)

Gewurztraminer

GEWURZTRAMINER
Wine grape.
A clone of the original Traminer, it is a native of the Pfalz and Alsace regions of Germany. This varietal grape produces an aromatic, medium-bodied, spicy white wine, often with the slightest touch of sweetness. In German, "gewurz" means spicy—and that actually describes the wine. It has a unique floral spiciness like that of carnation—rather than that of cinnamon or ginger. Mostly preferred before dinner, with appetizers. Late-harvest is also produced and is sweet, floral and spicy, used as a dessert wine.

Geyser Peak Estate Vineyards
Sonoma County, California.
Cabernet Sauvignon, Sauvignon Blanc, Gamay Beaujolais, Chardonnay, Pinot Noir, Johannisberg Riesling, Gewurztraminer, Zinfandel, Pinot Noir, Chenin Blanc Sonoma Valley and Russian River Valley vineyards.

GEYSER PEAK WINERY
Founded 1880, Geyserville, Sonoma County, California.
Storage: Oak and stainless steel. Cases per year: 2,000,000.
History: As a producer of fine wines, Geyser Peak leads the nation in innovative packaging techniques. Besides the traditional glass bottle, "wine in a box" and wine in aluminum cans.
Estate vineyards: Geyser Peak Vineyards.
Wines regularly bottled: Estate bottled, vintage-dated Cabernet Sauvignon, Fumé Blanc, Chardonnay, Pinot Noir, Soft Johannisberg Riesling, Gewurztraminer, Pinot Noir Blanc, Chenin Blanc, Soft Chenin Blanc, Rosé of Cabernet Sauvignon. Also Brut Champagne, Blanc de Noir Champagne. Occasionally bottles Petite Sirah, Gamay Beaujolais, Grand Rosé.
Second label: Summit, Nervo.

GEYSERVILLE
This town is in Sonoma County, California on the edge of the Alexander Valley.

GIBSON WINE COMPANY
Founded 1934, Sanger, Elk Grove, Fresno and Sacramento County, California.
Storage: Stainless steel.
Wines regularly bottled: Zinfandel, Tokay. Generics are Chablis, Rhine, Vin Rosé, Burgundy, Chianti, Chablis Blanc, Pink Chablis, Sauterne. Dessert wines: Sherry, Port, Muscatel, Marsala. Also, fruit, berry wines and hard cider.

Second label: Oreon, California Villages, Romano, Farley's Hard Cider and Silverstone Cellars Bag-in-Box table wines.

GIFFORD, JIM

After Jim Gifford was graduated from Indiana University, in 1971, with a bachelor's degree in English, he served four years in the United States Navy. In 1975, he returned to school, studying first at the University of California, San Diego, then attending the graduate program in food science/enology at California State University, Fresno. In 1978, Gifford worked for the Guild Wineries and Distilleries in their grower relations department. The following year, he accepted the position of winemaker in charge of white wines at Fetzer Vineyards in Mendocino County. In 1980, Gifford became winemaker at Gold Seal Vineyards in New York, and was promoted to head winemaker one year later. Gifford has won many awards for his wines at competitions and exhibitions. He is an active lecturer on wine-related topics and has presented many technical papers to industry organizations. He is a member of the American Society of Enologists, and has served as a sparkling wine consultant for Joseph E. Seagram and Sons at the Crillon Winery in Mendoza, Argentina.

Giles Vineyard

Sonoma County, California.

Chardonnay Carneros Region vineyard.

GILROY

Town, South of San Jose, in Santa Clara County, near the center of the Santa Clara Valley, base of Hecker Pass district.

Girard Estate Vineyard

Napa County, California.

Chardonnay, Cabernet Sauvignon, Chenin Blanc vineyard.

GIRARD WINERY

Founded 1980, Oakville, Napa County, California.

Storage: Oak. Cases per year: 14,000.

Estate vineyard: Girard Vineyard, Viridian Vineyard.

Wines regularly bottled: Estate bottled, vintage-dated Chardonnay, Cabernet Sauvignon, Chenin Blanc. Also vintage-dated Sauvignon Blanc. Occasionally Botrytised Semillon.

Second label: Stephens Winery.

GIUMARRA, GIUSEPPE

The Giumarra family had grown grapes for generations in their native Sicily, where wine was more plentiful than water. In the early 1900's, Giuseppe "Joe" Giumarra and his father came to North America. Giumarra's first job was selling fruit from a pushcart in Toronto; then, after a brief stay on the east coast, he moved west to southern California. He became familiar with the San Joaquin Valley as he hauled wholesale produce to the markets in the Los Angeles area. After the arrival of the rest of his family in 1922, Giumarra purchased farmland in Kern County, where he successfully cultivated both table and wine grapes. In 1946, he realized his dreams with the founding of the Giumarra Vineyard's winery near his Edison, California property. "Papa Joe", as he became known, initially planned to produce only bulk wines for sale to other wineries. His commitment to quality resulted in the table wines now bottled under the Giumarra name. Currently, three generations of Giumarras, including "Papa Joe", continue the winemaking tradition in the San Joaquin Valley.

Giumarra Estate Vineyards

Kern County, California.

Cabernet Sauvignon, Gamay Beaujolais, Zinfandel, Johannisberg Riesling, Chenin Blanc, French Colombard, Green Hungarian, Chardonnay, Pinot Noir, Carnelian, Grenache vineyard.

GIUMARRA VINEYARDS

Founded 1946, Edison, Kern County, California.

Storage: Oak and stainless steel. Cases per year: 600,000.

Estate vineyard: Giumarra Vineyard.

Label indicating non-estate vineyards: Rancho Sisquoc, Tepesquet Vineyard.

Wines regularly bottled: Estate bottled, vintage-dated Cabernet Sauvignon, Gamay Beaujolais, Nouveau Zinfandel, Johannisberg Riesling, Chenin Blanc, French Colombard, Green Hungarian, Chardonnay. Also non-vintage Gamay Beaujolais, Carnelian Rosé, Cabernet Sauvignon Rosé, Grenache Rosé, Petite Sirah, Pinot Noir, Carnelian. Also vintage-dated semi-generics: Mountain Burgundy, Vin Rosé, Chablis, Rhine.

GLASSCOCK VINEYARDS (See Davis Mountain Wines.)

Glasscock Vineyards, Inc.

Ft. Davis, Texas.

Davis Mountain estate Chenin Blanc, Sauvignon Blanc, Chardonnay, Cabernet Sauvignon and Merlot vineyard.

"Taster's" glass

glassware

For carafes, decanters and glasses—always look for simple shapes and avoid thick or cut-glass because this has a prism effect giving brilliant but also sombre effects to the wine and it is thus quite impossible to appreciate its true hue or robe.

The decanter should be rounded, without any angles or corners to create turbulence. The neck should be straight and of medium length, with a stopper. The decanter should have a capacity about 25% more than its contents.

A carafe should never be filled right up; there should always be sufficient surface area in contact with the oxygen of the air. For decanting magnums you may perhaps have some difficulty in finding a decanter with a capacity of 1.7 or 1.8 litres.

The glasses should have stems and a foot. The stem is an invention from the XV century. In those days, the fork being unknown, the fingers were used to pick up food and, when drinking, the glass was held in the hollow of the hand, with the stem passing between the ring-finger and the little-finger, so as to avoid making greasy marks. Today, the stem has the advantage of showing off the robe of the wine and the height it gives allows easy service. The rim of the glass should be slightly closed-in, tulip-style, so as to allow the wine to be twirled in the glass without risk and also to concentrate the bouquet.

Glen Creek Estate Vineyard

Polk County, Oregon.

Chardonnay, Gewurztraminer vineyard.

GLEN CREEK WINERY

Founded 1982, Salem, Polk County, Oregon.

Cases per year: 2,500.

Estate vineyard: Glen Creek Winery Vineyard.

Label indicating non-estate vineyard: Sagemoor Farms.

Wines regularly bottled: Vintage-dated Chardonnay, Gewurztraminer, Sauvignon Blanc; occasionally Pinot Noir.

Second label: Wood Hill Cellars.

Glen Ellen Estate Vineyards

Sonoma County, California.

Cabernet Sauvignon, Sauvignon Blanc, Merlot, Cabernet Franc and Semillon.

Glen Ellen Vineyards
Sonoma County, California.
Merlot, Zinfandel, Pinot Noir and Cabernet Sauvignon vineyard.

GLEN ELLEN WINERY
Founded vineyards 1968, winery 1980, Glen Ellen, Sonoma County, California.
Cases per year: 15,000.
Estate vineyard: Glen Ellen Vineyards.
Label indicating non-estate vineyard: Les Pierres Vineyard, Hanford Vineyards, Larson Vineyards.
Wines regularly bottled: Estate bottled Cabernet Sauvignon, Sauvignon Blanc. Also Chardonnay, Sauvignon Blanc, two proprietary wines: Proprietors Reserve Red, Proprietors Reserve White.

Glen Oak Vineyard
Napa County, California.
Louis Martini estate Zinfandel, Chenin Blanc, Cabernet Sauvignon, Cabernet Franc and Petite Sirah, Chiles Valley vineyard.

GLENOAK HILLS WINERY
Founded 1978, Temecula, Riverside County, California.
Storage: Oak and stainless steel. Cases per year: 600.
Label indicating non-estate vineyard: Knole Vineyard, Thomas Vineyard, St. Gertrudis Vineyard.
Wines regularly bottled: Vintage-dated Cabernet Sauvignon, Zinfandel, Chardonnay, Sauvignon Blanc, Gamay Beaujolais, White Riesling.

GLENORA WINE CELLARS, INC.
Founded 1977, Dundee, Yates County, New York.
Storage: Oak and stainless steel. Cases per year: 15,000.
History: The winery was conceived by four independent grape growers in 1976. Shortly after New York State passed the Farm Winery Bill.
Estate vineyard: Anchor Acres Vineyard, Spring Ledge Vineyard.

Wines regularly bottled: Estate bottled, vintage-dated Chardonnay, Johannisberg Riesling, Ravat Blanc, Seyval Blanc, Cayuga. Chardonnay methode champenoise occasionally produced.

Goat Hill Vineyard
Hunterdon County, New Jersey.
French Hybrid vineyard.

GOLD MINE (See California Cellar Masters.)
Produce Apricot, Plum, Pomegranate, Blackberry and Raspberry fruit wines.

Gold Ridge
Sonoma County, California.
Chardonnay Green Valley vineyard.

GOLD SEAL VINEYARDS, INC.
Founded 1865, Hammondsport, Steuben County, New York.
Storage: Oak and stainless steel. Cases per year: 200,000.
History: Founded as the Urbana Wine Company producing under the brand name "Gold Seal". In 1957 the company name was officially changed to Gold Seal Vineyards, Inc. Closed down in early 1984 with little hope of ever reopening.
Estate vineyard: Seneca Lake E Vineyard, Keuka Lake E Vineyard, Keuka Lake W Vineyard.

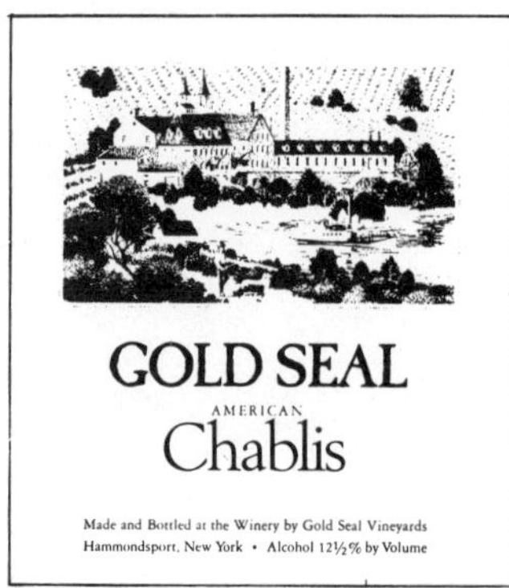

Wines regularly bottled: Estate bottled, vintage-dated Chardonnay, Johannisberg Riesling, Gewurztraminer; Champagnes (Blanc de Noir, Blanc de Blancs, Brut, Extra Dry); semi-generics: Chablis, Burgundy, Rhine, Dry Sauterne, Vin Rosé.

Second label: Henry Marchant, Pierre Corbeau, Charles Fournier, Mountain Lake.

GOLDEN CHASSELAS

Wine grape.

Produces a light wine, dry to medium sweet.

GOLDEN HILLS (See Pedrizzetti Winery.)

Golden Rain Tree Estate Vineyard

Posey County, Indiana.

Estate Seyval Blanc, Vidal Blanc, Chancellor Noir, Ravat, Baco Noir, Chelois, Marechal Foch, DeChaunac, Concord, Catawba and Niagara vineyard.

GOLDEN RAIN TREE WINERY, INC.

Founded 1975, Wadesville (at St. Wendel), Posey County, Indiana.

Storage: Oak and stainless steel. Cases per year: 10,000.

Estate vineyard: Golden Rain Tree Vineyard.

Wines regularly bottled: Estate bottled Seyval Blanc; also proprietary: Criterion White and Red; Director's Choice; Harmonite Blanc, Rosé, Burgundy and Chablis; Spirit of '76; St. Wendel Rosé and White; Shanti; Thunder on the Ohio; Soft Red; Dandi (Apple and Citrus).

GOLDMAN, MAX

Max Goldman began his fifty year career in the wine industry shortly after he graduated from Whittier College with a bachelor's degree in chemistry and physics in 1933. He started at the Roma Wine Company in Lodi, California, as chief chemist and winemaker. During his two years there, Goldman introduced the Lot Systems of recordkeeping to the industry. In 1936, he joined the Petri Wine Company in Escalon, California as chief chemist, winemaker and plant manager. He experimented with cold fermentation and malo-lactic fermentation techniques, and worked to derive industrial alcohol from molasses during World War II. Goldman left in 1944 to take the position of chief chemist, winemaker and general manager at Sanger Wine Company in Sanger, California, where he worked until 1950. He then rejoined Petri Wine Company, this time in Madera, California, again as chief chemist, winemaker and plant manager. He designed and supervised construction of their large laboratory during his two year stay. Goldman spent the next six years at the Waterford Winery, in Waterford, California, where he served as chief chemist, winemaker and general manager. It was there that he began working with sparkling wine production. He continued developing new sparkling wine techniques at Great Western Winery-Pleasant Valley Wine Company in Hammondsport, New York, where he worked as chief chemist, winemaker and vice-president in production, from 1958 until 1965. He was also responsible for setting up a Solera system for sherries and ports. Goldman returned to California to join Bohemian Distributing Company as chief chemist and vice-president in production. In 1972, Goldman purchased the 1882

York Mountain Winery in Templeton, California. He has restored both the winery and 40 acres of vineyards, and has plans to plant another 30 acres, principally in Chardonnay. York Mountain has recently received a viticultural area appellation from the BATF. Goldman has been active in many industry associations, including President of the American Society of Enologists, Chairman of the Wine Industry Panel at Cornell University. He is also a member of the Board of Directors of the California Wine Institute.

Good Harbor Estate Vineyard

Leelanau County, Michigan.

Ravat, Seyval Blanc, DeChaunac, Aurora, Riesling and Chardonnay Leelanau Peninsula vineyard.

GOOD HARBOR VINEYARDS

Founded 1980, Lake Leelanau, Leelanau County, Michigan.

Storage: Stainless steel. Cases per year: 5,000.

Estate vineyard: Good Harbor Vineyards.

Wines regularly bottled: Estate bottled, vintage-dated Seyval Blanc, Vignoles; vintage-dated DeChaunac Rosé, Foch, Riesling.

Gordon Ranch Vineyards

Napa County, California.

Chenin Blanc and Sauvignon Blanc vineyard.

GOTTESMAN, ROBERT G.

Robert G. Gottesman was born in the southwestern Pennsylvania town of Brownsville. After graduation from high school there, he moved to New York City and took a temporary job as a messenger with Schenley Distillers.

His "temporary" job lasted 21 years including a five year hitch in the United States Army in World War II. After leaving Schenley, he purchased Paramount Distillers of Cleveland, Ohio in 1957.

In 1976, Paramount expanded into the wine business with the acquisition of Meier's Wine Cellars in Silverton, Ohio. In 1979, Paramount acquired Lonz Winery on Middle Bass Island in Lake Erie, Ohio; and in 1980, two northern Ohio wineries, Mantey and Mon Ami, were added. Grapes for these wineries are grown on Isle St. George, also called North Bass Island.

Today Gottesman is 100% committed to solving the competitive problems of an Ohio winery—new plantings, new products, new packages. His outlook today is, "We will keep trying until we find what it takes to attract new customers and new markets."

GRAF, MARY ANN

Winemaker at Simi Winery, California from 1973-1979. Partner in the enological consulting firm Vinquiry.

The Graham Farm

Madison, Arkansas.

Muscadines: Magnolia, Carlos and Noble vineyard

Graham Ranch Vineyard

Santa Cruz County, California.

Bonny Doon estate Chardonnay, Pinot Noir, Cabernet Sauvignon, Merlot, Cabernet Franc Santa Cruz Mountains vineyard.

GRAND CRU VINEYARDS

Founded 1970, Glen Ellen, Sonoma County, California.

Storage: Oak and stainless steel. Cases per year: 40,000.

History: Founded in 1886 by Francis Lamoine; original vineyard established in 1896; parts of the original stone winery building in current use.

Label indicating non-estate vineyard: Garden Creek Ranch, Cook's Delta Vineyard.

Wines regularly bottled: Vintage-dated Gewurztraminer, Chenin Blanc, Sauvignon Blanc, Cabernet Sauvignon. Occasionally bottles Zinfandel and Late Harvest Gewurztraminer.

GRAND PRÉ WINES

Founded 1979, Grand Pré, Nova Scotia, Canada.

Storage: Oak. Cases per year: 10,000.

History: The village of Grand Pré dates back to the 1680's as the home of the Acadians who migrated north from Champlain's first colony in the southern Annapolis Valley. The owner-winemaker is Roger Dial, who is a professor of political science at the University in Halifax. Most interesting are two Russian varieties grown, Michuvnitz and Severnyi; both produce red wines of character.

Wines regularly bottled: Estate bottled, vintage-dated Marechal Foch, Seyval, Severnyi, Cuvée d'Amur; also proprietary L'Acadie Blanc and Rouge.

GRAND RIVER VALLEY VITICULTURAL AREA

Grand River Valley in northeastern Ohio is approximately 125,000 acres and consists of all the land within two miles, in any direction, of the Grand River, from its origin to the point at which it flows into Lake Erie. Approximately one-third of the viticultural area is located inside the Lake Erie viticultural area.

Grand River Vineyards

Lake Erie, Ohio.

Chancellor vineyard.

Grand Traverse Vineyard/Old Mission Vineyard

Grand Traverse, Michigan.

Chateau Grand Traverse estate Chardonnay, Riesling, Merlot, Petite Sirah, Pinot Noir, Scheurebe, Pinot Blanc, Gamay Beaujolais and Chenin Blanc vineyard.

Granite Hill Vineyard

El Dorado County, California.

Sauvignon Blanc and Semillon vineyard.

GRANITE SPRINGS WINERY

Founded 1981, Somerset, El Dorado County, California.

Storage: Oak and stainless steel. Cases per year: 3,500.

Estate vineyard: Granite Springs Vineyard.

Label indicating non-estate vineyard: Twin Rivers Vineyards, Madrona Vineyards.

Wines regularly bottled: Vintage-dated White Zinfandel, Chenin Blanc, Cabernet Sauvignon, Zinfandel, Sauvignon Blanc. Occasionally bottles vintage-dated Red Table Wine.

Second label: Dry Diggins, Harvest Cellars.

Granval

Napa Valley, California.

Clos du Val estate Chardonnay and Pinot Noir vineyard.

GRANVILLE VINEYARD

Founded 1979, Granville, Licking County, Ohio.

Storage: Oak and stainless steel.

Estate vineyard: Granville Vineyard.

Wines regularly bottled: Seyval Blanc.

grape shot

A cluster of small cast iron balls used as a charge for a cannon.

grape stone

The seed of the grape.

THE GRAPE VINE WINERY

Founded 1982, Amana, Iowa County, Iowa.

Wines regularly bottled: Fruit and Berry wines: Grape, Cranberry, Blackberry, Strawberry-Rhubarb, Plum, Elderberry, Cherry and proprietary Piestengel (Rhubarb).

GRAPERY

An 1800's building in which grapes were grown. A plantation of vines. A winery.

"GRAPES OF WRATH"

The lyric words from the "Battle Hymn of the Republic" by Julia Ward Howe. "He is trampling out the vintage where the grapes of wrath are stored."

GRAS, HERNAN J.

Hernan J. Gras was born and raised in Chile. He attended the School of Agronomy in the Catholic University from 1962 to 1965, and spent the following two years in the Enology and Viticulture Department, earning a master's degree in 1967. Gras spent the next year in France, continuing his studies at the Ecole Superieure d'Enologie, Université de Bordeaux. Upon his return to Chile, he worked for one year as a research scientist in the Enology Department of the Catholic University, then spent five years as a technical advisor in the Chilean Ministry of Agriculture. In 1974, Gras emigrated to Canada, and joined the T.G. Bright & Co., Ltd., in Ontario, producer of Brights Wines. He began in viticulture research, and then worked for one year in the laboratory. In 1976, Gras became quality control manager, and in 1979 he was promoted to winemaker.

Gravelly Meadow Vineyard

Napa County, California.

Diamond Creek estate Cabernet Sauvignon Napa Valley vineyard.

Graves Vineyard

Klickitat County, Washington.

Chenin Blanc and Cabernet Sauvignon vineyard.

GRAY MONK CELLARS, LTD. ESTATE WINERY

Founded 1982, Okanagan Centre, British Columbia.

Storage: Stainless steel. Cases per year: 4,800.

History: The most northerly estate winery in North America. The only winery producing Kerner.

Estate vineyard: G & T Heiss Vineyards.

Wines regularly bottled: Estate bottled, vintage-dated Johannisberg Riesling, Pinot Gris, Pinot Auxerrois, Bacchus, Gewurztraminer, Kerner, Marechal Foch. Occasionally Scheurebe, Rotburger, proprietary Schonburger, Lemburger.

Wind machine in vineyard for frost protection

GREAT WESTERN WINERY

Founded 1860, Hammondsport, New York.

History: Great Western Winery (Pleasant Valley Wine Company) is the oldest winery in the Finger Lakes wine region. The old stone vaults and solid oak archways are reminders that it was once one of the first American wineries to win an award in international competitions. Great Western Winery produced one of the first wines in the country from the French-American grapes and is among the few wineries with a Solera system for aging sherries.

Wines regularly bottled: Estate bottled, vintage-dated Aurora, Aurora Blanc, Cayuga White, Delaware, Diamond, Dutchess, Rosé of DeChaunac, Rosé of Isabella, Seyval Blanc, Verdelet, Vidal Blanc, Vidal (Ice Wine), Catawba (Ice Wine). Also Baco Noir, DeChaunac, Seyval Blanc, Pink Catawba; semi-generics: Chablis, Rhine, Burgundy; Champagnes (Natural, Brut, Extra Dry, Pink, Pink Cold Duck, Sparkling Burgundy); Sherries, Port, Tawny Port, Sweet and Dry Vermouth.

green

Tasting term.

Term applied to a young wine.

Green Acres

Sonoma Valley, California.

Sebastiani estate Gamay Beaujolais, Traminer, Sauvignon Blanc, Chardonnay, Green Hungarian and Barbera vineyard.

GREEN, ALISON

In 1969, when Alison Green was 14, her father purchased the Simi Winery in Healdsburg, California. During the summers, she worked as "cellar rat", until 1973, when she was promoted to lab technician. While in the laboratory, she met winery consultant Andre Tchelistcheff. He persuaded her to work the crush rather than return to her studies in humanities at the University of California, Davis. When she did return to school in 1974, Green changed her major to fermentation science. In 1975, she took another university leave of absence to spend five months at L'Institut National des Recherches Agronomiques (L'INRA) in Alsace, where she continued her studies and participated in the crush. Upon her return in 1976, Green went to work, first at Hoffman Mountain Ranch in Paso Robles, then at Firestone Vineyards in the Santa Ynez Valley. She began as an enologist at that Santa Barbara County winery, and completed her bachelor's degree that year by commuting to Davis once a week. Since 1981, Green has served as winemaker, and is responsible for production of all the Firestone wines.

Green Farm
Lake Erie, Ohio.
Agawam vineyard.

GREEN HUNGARIAN
Wine grape.
A grape that produces a light, neutral fresh white wine; drink young.

Green Pastures Vineyard
Sonoma County, California.
Chardonnay vineyard.

Green and Red Estate Vineyard
Napa County, California.
Zinfandel and Chardonnay vineyard.

GREEN AND RED VINEYARD
Founded 1972, St. Helena, Napa County, California.
Storage: Oak and stainless steel. Cases per year: 1,500.
Estate vineyard: Green and Red Vineyard.
Wines regularly bottled: Estate bottled, vintage-dated Zinfandel.

GREEN VALLEY (See Galleano Winery.)

Green Valley Estate Vineyards
Callaway County, Missouri.
Foch, Chancellor, Chambourcin, Aurora, Seyval Blanc, Baco Noir and Villard Noir vineyard.

GREEN VALLEY (SOLANO COUNTY) VITICULTURAL AREA
"Green Valley" qualified by the words "Solano County" will distinguish this viticultural area from the viticultural area in Sonoma County, California to be called "Green Valley" qualified by the words "Sonoma County." Both valleys have been known historically as Green Valley.
The words "Solano County" must appear in direct conjunction with the name "Green Valley".

GREEN VALLEY (SONOMA COUNTY) VITICULTURAL AREA
Sonoma County Green Valley Viticultural area is completely encompassed by the boundaries of each of three other viticultural areas: Russian River Valley, Northern Sonoma, and North Coast.

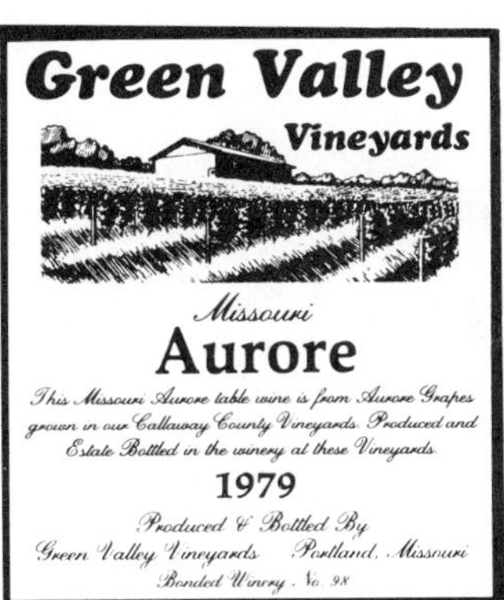

GREEN VALLEY VINEYARDS
Founded 1972, Portland, Callaway County, Missouri.
Storage: Oak and stainless steel.
Estate vineyard: Green Valley Vineyards.
Wines regularly bottled: Estate bottled, vintage-dated Foch, Chancellor, Chambourcin, Aurora, Seyval Blanc, Baco, Villard Noir; also semi-generics: Chablis, Rhine, Burgundy, Sauterne.

Green Valley Vineyards
Sonoma County, California.
Pommeraie estate Cabernet Sauvignon Green Valley vineyard.

GREENE, ALBERT GORTON
(1802-1868)
American lawyer and poet.

"Fill every beaker up, my men,
Pour forth the cheering wine;
There's life and strength in every drop—
Thanksgiving to the vine!"

Greenfield Vineyard
Suffolk County, New York.

Bridgehampton estate Chardonnay, Pinot Noir, Riesling and Sauvignon Blanc Long Island vineyard.

Greenfield Vineyards
Monterey County, California.
Turgeon & Lohr estate Chardonnay, Johannisberg Riesling, Pinot Blanc, Sauvignon Blanc and Monterey Gamay vineyard.

GREENSTONE WINERY AND VINEYARD
Founded 1980, Ione, Amador County, California.
Storage: Oak and stainless steel. Cases per year: 8,500.
Estate vineyard: Greenstone Winery and Vineyards.
Label indicating non-estate vineyard: Potter-Cowan Vineyards.
Wines regularly bottled: Estate bottled, vintage-dated California Colombard. Also vintage-dated Zinfandel and White Zinfandel. And non-vintage wines: Burgundy, Dry French Colombard, Amador Cream Sherry, Dry Chenin Blanc and Zinfandel Rosé.

Greenwood Ridge Estate Vineyards
Mendocino County, California.
White Riesling, Cabernet Sauvignon, Merlot vineyard.

GREENWOOD RIDGE VINEYARDS
Founded 1980, Philo, Mendocino County, California.
Storage: Oak and stainless steel. Cases per year: 2,000.
Estate vineyard: Greenwood Ridge Vineyards.
Wines regularly bottled: Estate bottled, vintage-dated White Riesling, Cabernet Sauvignon.

Gregory Vineyards
Sonoma County, California.
Cabernet Sonoma Valley vineyard.

GRENACHE
Wine grape.
A grape that is predominately used in California for Rosé, because of its pale color and strong flavor. A native of Spain, where it is called Garnacha, it is an important variety in France and Spain, where it is also used to produce medium red wines. It was first introduced in California in 1941.

GRENACHE ROSÉ
When the Grenache grape is used for Rosé, the grapes' strong character is much tempered. A fruity, light, tart wine that has a noticeable sweetness.

GREY RIESLING
Wine grape.
A California named grape that produces a light, fresh, slightly sweet, but delicate wine. Originally from France, under the name Chauché gris.

GREY SUMMIT (See Bardenheier's Wine Cellars.)
Rhine, Chablis, Burgundy, Rosé, Pink Chablis.

GREYSTONE
Early name of the historic old stone winery in St. Helena purchased in 1950 by The Christian Brothers; condemned in 1984 as an earthquake hazard, its future is uncertain. When it was built in 1889, it was the largest stone winery in the world.

GRGICH HILLS CELLAR
Founded 1977, Rutherford, Napa County, California.
Storage: Oak and stainless steel: Cases per year: 15,000-20,000.
History: Grgich is the winemaker who, while at another winery, created the Chardonnay that set Paris on its ear in 1976.
Wines regularly bottled: Vintage-dated Chardonnay, Fumé Blanc, Johannisberg Riesling, Late Harvest Riesling, Zinfandel, Cabernet Sauvignon.

GRGICH, MILJENKO
Miljenko "Mike" Grgich was born into a winemaking family in Croatia, Yugoslavia, in 1923. He received a degree in enology from the University of Zagreb before coming to the United States in 1958. After his arrival in the Napa Valley, Grgich continued his education under the direction of Lee Stewart at Souverain. He worked briefly at the Christian Brothers, then spent nine years at Beaulieu, with Andre Tchelistcheff. The following five years he worked at Robert Mondavi. Grgich then joined Chateau Montelena as Winemaker, producing the award-winning 1973 Chardonnay. Four years later, on July 4th, 1977, Grgich Hills Cellars was founded, in partnership with Austin E. Hills of the San Francisco coffee family. Grgich creates distinctive, individual wines, specializing in Chardonnay, which accounts for over half of the winery's production. He is a Master Knight of the Vine.

GRIGNOLINO
Wine grape.
A grape that produces a tart wine with an orange pigment character. Should be drunk two to five years old. Originally from Northern Italy. Excellent with rich meat dishes. A relatively rare variety, something of a curiosity.

Griffin Russian River Ranch
Sonoma County, California.
Hop Kiln estate Chardonnay, French Colombard, Gewurztraminer, Johannisberg Riesling, Napa Gamay, Zinfandel, Petite Sirah Russian River vineyard.

GRILLO
Wine grape.
White grape used for blending in dessert wines.

Gross' Highland Estate Vineyards
Atlantic County, New Jersey.
Villard Blanc, Vidal Blanc, Niagara, Catawaba, Dutchess, Delaware and Villard vineyards.

GROSS' HIGHLAND WINERY
dba Bernard D'Arcy Wines
Founded 1934, Absecon, Atlantic County, New Jersey.
Storage: Oak and stainless steel.
History: At the turn of the century, John Gross came to America, from Germany, to continue a family tradition of winemaking. He established one of the early wineries in Atlantic County around Vineland. After his death, his son-in-law Bernard D'Arcy took over the responsibilities.
Estate vineyard: Gross' Highland Vineyards.
Wines regularly bottled: Johannisberg Riesling, Gamay Beaujolais, Sauvignon Blanc, Niagara, Pink Catawba, Sparkling Pink Catawba, Champagnes, Spumante; also Ronay Blanc, Ronay Noir, Country Red, White and Gold, Sweet Claret, Vin

Rosé, Schillerwein, Rhine Wine, Cream Sherry, Cream Almonique.

Second label: Bernard D'Arcy.

GROTH VINEYARDS AND WINERY

Founded 1981, Oakville, Napa County, California.

Storage: Oak and stainless steel. Cases per year: 22,000-30,000.

Estate vineyards: Oakcross Vineyards, Hillview Vineyard.

Wines regularly bottled: Estate bottled, vintage-dated Sauvignon Blanc, Chardonnay, Cabernet Sauvignon.

GROVER GULCH WINERY

Founded 1979, Soquel, Santa Cruz County, California.

Storage: Oak. Cases per year: 1,000.

Wines regularly bottled: Cabernet Sauvignon, Petite Sirah, Carignane, Zinfandel.

Grow Vineyards

Yates County, New York.

Johannisberg Riesling, Chardonnay and Ravat Finger Lakes vineyard.

Label pasting machine

GSS Vineyard

Belen, New Mexico.

La Chiripada estate French Hybrids and White Riesling vineyard.

After seventeen years of toil and saving, Emilio was able to purchase 15 acres. He built his home, constructed a small winery in the deep basement and established a vineyard.

Success came quickly because Emilio made honest wines that were very well received by the French and Italian communities in San Francisco. Emilio hauled it to San Francisco by horse and wagon, later by truck; he had a son, George Guglielmo who helped distribute the wine.

Prohibition was a blow to the family enterprise, but, with Repeal, Emilio and George began construction of a larger winery. Following World War II, they further expanded and modernized the winery.

Estate vineyard: Emilio Guglielmo Vineyard.

Wines regularly bottled: Estate bottled, vintage-dated Gamay Beaujolais, Zinfandel, Barbera, Petite Sirah, Cabernet Sauvignon, Grignolino Rosé, Dry Semillon. Vintage-dated Chardonnay, Fumé Blanc, Johannisberg Riesling. Also estate bottled, vintage-dated Claret, Fumante and Burgundy.

Second label: Emile's.

GUADALUPE VALLEY WINERY

Founded 1975, New Braunfels, Comal County, Texas.

Storage: Oak and stainless steel. Cases per year: 3,000.

Wines regularly bottled: Peach, Strawberry, Red, White, Rosé. Occasionally Hill Country Rosé, River Valley White, Texas Red and White, Prince of New Braunfels.

Second label: Poteet Strawberry, Schlaraffenland.

GUASTI (See Lamont Winery.)

GUENOC VALLEY VITICULTURAL AREA

The Guenoc Valley, in Lake County, is located south of McCreary Lake and east of Desert Reservoir and has approximately 250 acres planted in grapes. There is one winery located within this viticultural area.

Geunoc Estate Vineyard
Lake County, California.
Cabernet Sauvignon, Merlot, Malbec, Cabernet Franc, Petite Sirah, Zinfandel, Chardonnay, Chenin Blanc, Sauvignon Blanc, Semillon Guenoc Valley Vineyard.

Guenoc Vineyard
Lake County, California.
Sauvignon Blanc and Semillon vineyard.

GUENÒC WINERY
Founded 1981, Middleton, Lake County, California.
Storage: Oak and stainless steel. Cases per year: 52,000-68,375.
History: Winery located in historic Guenoc Valley. The land was formerly owned by Lillie Langtry, famous actress, who hired a winemaker from Bordeaux to oversee her vineyards and produce wine.
Estate vineyard: Guenoc Valley Vineyard.
Wines regularly bottled: Vintage-dated Cabernet Sauvignon, Petite Sirah, Chardonnay, Sauvignon Blanc, Chenin Blanc, Zinfandel.

Emilio Guglielmo Estate Vineyards
Santa Clara County, California.
Semillon, Zinfandel, Petite Sirah, Grignolino, Gamay Beaujolais vineyard.

EMILIO GUGLIELMO WINERY, INC.
Founded 1925, Morgan Hill, Santa Clara County, California.
Storage: Oak and stainless steel. Cases per year: 50,000.
History: Guglielmos, for centuries, had been synonymous with winemaking. From Roman times the family had tilled the soil and tended vines in the hills of Piedmonte, northern Italy, the country's foremost winegrowing district.
In 1883, with the birth of Emilio Guglielmo who, raised in the vitrous environment of his ancestors, followed in their footsteps and learned well their art. Emilio was, however, drawn by the lure of America, the land of opportunity where it was said that, in California, grape vines provided bountiful crops and there was potential for fine wine.
In 1908, Emilio set out for the New World. Making his way to California, he reached the beautiful Santa Clara Valley.

GUILD WINERIES
Founded 1948, Lodi, San Joaquin County, California.
History: One of the nation's largest grape grower owned wine companies.
Wines regularly bottled: Vintage-dated Blanc de Blanc, Cabernet Sauvignon, Chablis, Chardonnay, Chenin Blanc, French Colombard, Gamay Rosé, Gamay Beaujolais, Gewurztraminer, Grey Riesling, Johannisberg Riesling, Petite Sirah, Pinot Noir and Zinfandel. Also Sherry, Port, Extra Dry, Brut and Chardonnay Champagne.
Labels: Cresta Blanca, Cribari, Ceremony, St. Mark, Roma, Cooks, Garretts, Vintners Choice and Mendocino.

GUNDLACH-BUNDSCHU WINERY
Founded 1858, Sonoma, Sonoma County, California.
Storage: Oak and stainless steel. Cases per year: 35,000.
History: Winery has been growing grapes and producing wines from the same vineyards longer than any other winery in California. Founded by Jacob Gundlach, its name was changed a few years later, to its current one, when the present day owner's great-grandfather became a partner.
Estate vineyard: Rhinefarm Vineyards.
Label indicating non-estate vineyard: Sangiacomo Vineyards, Batto Ranch, Gregory Vineyards.
Wine regularly bottled: Estate bottled, vintage-dated Cabernet Sauvignon, Merlot, Pinot Noir, Zinfandel, Gewurztraminer, Kleinberger. Also vintage-dated Chardonnay, Cabernet, Sonoma Riesling.

Habichshos Vineyards
Napa County, California.
Cabernet Sauvignon, Chenin Blanc, Sauvignon Blanc and Zinfandel Napa Valley vineyards.

Hacienda Vineyard
Monterey County, California.
Chardonnay-Late Harvest vineyard.

HACIENDA WINE CELLARS
Founded 1973, Sonoma, Sonoma County, California.
Storage: Oak and stainless steel. Cases per year: 23,000.
History: Winery is in the midst of the vineyards Agoston Haraszthy established in 1862. The Oat Valley vineyard has been in A. Crawford Cooley's family since the 1850's.
Estate vineyard: Oat Valley Vineyard, Buena Vista Vineyard.
Wines regularly bottled: Estate bottled, vintage-dated Cabernet Sauvignon, Chardonnay, Selected Reserve Pinot Noir. Also vintage-dated Chardonnay, Gewurztraminer, Dry Chenin Blanc, Zinfandel, Johannisberg Riesling.

HADLEYS PRESTIGE VINEYARDS
Founded 1965, San Diego County, California.
Number of cases produced per year: 12,000.
Wines are purchased from major producers and marketed through their own retail stores.
Wines regularly sold: Chenin Blanc, French Colombard, Cabernet Sauvignon, White Zinfandel, Chardonnay, Johannisberg Riesling, Green Hungarian, Zinfandel, Grenache Rosé. Also generics: Burgundy, Chablis, Extra Dry Champagne, Vin Rosé, Rhine, Pink Chablis. Dessert wines: Cream and Pale Dry Sherry, Ruby and Tawny Port, Marsala. Fruit wines sold are Apricot, Blackberry, Cherry, Concord, Loganberry, Raspberry, Red Currant, Strawberry, Plum, Pomegranate. Also Mead wine.

Hafle Estate Vineyards
Clark County, Ohio.
Baco Noir, DeChaunac and Concord vineyard.

HAFLE VINEYARDS
Founded 1974, Springfield, Clark County, Ohio.
Storage: Oak and stainless steel. Cases per year: 2,000.
Estate vineyard: Hafle Vineyards.
Wines regularly bottled: Estate bottled Baco Noir, DeChaunac, Concord; also Sangria, Vin Rosé; Haut Sauterne, Hafle Red, Hafle White.

HAFNER VINEYARD
Founded 1982, Healdsburg, Sonoma County, California.
Storage: Oak and stainless steel. Cases per year: 8,000.
Wines regularly bottled: Estate bottled, vintage-dated Cabernet Sauvignon and Chardonnay.

Hafner Vineyards
Sonoma County, California.
Chardonnay Alexander Valley vineyard.

HAGAFEN CELLARS
Founded 1980, Napa, Napa County, California.
Cases per year: 2,500.

History: Founded to produce premium Napa wines that are Kosher. Kashrut supervision by Orthodox Rabbinical Council of San Francisco.

Label indicating non-estate vineyard: Winery Lake Vineyard.

Wines regularly bottled: Vintage-dated Johannisberg Riesling, Chardonnay, Cabernet Sauvignon.

Hageus Vineyard
Lubbock County, Texas.
Chardonnay, Johannisberg Riesling and Gewurztraminer vineyard.

Hahn Hill Vineyard
Benton County, Washington.
Chateau Ste. Michelle estate Johannisberg Riesling, Chenin Blanc, Semillon, Cabernet Sauvignon, Pinot Noir Columbia Valley vineyard.

Hahn Hill Vineyard
Benton County, Washington.
River Ridge estate White Riesling, Cabernet Sauvignon, Semillon, Pinot Noir, Chenin Blanc and Merlot Columbia Valley vineyard.

HAIGHT JR., SHERMAN POST

Sherman Post Haight, Jr., was graduated from Trinity College in Connecticut in 1946 and began a long career in the textile printing business. He worked for companies throughout the United States and Canada before he retired to found his own winery. He is currently president of the Haight Vineyard in Litchfield, Connecticut. Haight is a member of the American Society of Enologists.

Haight Estate Vineyards
Litchfield County, Connecticut.
Johannisberg Riesling, Chardonnay, Seyval, Marechal Foch and Pinot Noir Litchfield Hills vineyard.

HAIGHT VINEYARDS

Founded 1975, Litchfield, Litchfield County, Connecticut.

Storage: Oak and stainless steel. Cases per year: 5,000.

History: Winery is located in one of New England's old historic villages.

Estate vineyard: Haight Vineyard, Ripley Vineyard, Gaylord Vineyard, Haverstick Vineyard.

Wines regularly bottled: Estate bottled, vintage-dated Riesling, Chardonnay; estate bottled Marechal Foch; Sparkling wine: proprietary Covertside Red, Covertside White.

Tom Hallam Vineyard
London City, Tennessee.
Concord and Niagara vineyard.

Hallcrest Vineyard
Santa Cruz County, California.
Felton-Empire estate White Riesling and Cabernet Sauvignon Santa Cruz Mountains vineyard.

Hamlet Hill Estate Vineyards
Windham County, Connecticut.
Muscadet, Pinot Meunier, Pinot Noir, Riesling, Chardonnay, Gewurztraminer, Seyval, Vérdelet, Aurora, Vignoles, Vidal, Chancellor, Baco Noir, Foch, Cascade, White Rouge vineyards.

HAMLET HILL VINEYARDS

Founded 1975, Pomfret, Windham County, Connecticut.

Storage: Oak and stainless steel. Cases per year: 5,000.

History: First winery in east to grow Muscadet and Pinot Meunier for sparkling wine base. The exclusive producer of private label wines for Brown University.

Estate vineyard: Hamlet Hill Vineyards.

Wines regularly bottled: Estate bottled, vintage-dated Riesling, Seyval Blanc; estate bottled, vintage-dated proprietary Charter Oak Red, Brunonian Reserve; vintage-dated proprietary Promfret White, Woodstock White, Drumlin Rosé, Redcastle, White Reel; also Creamery Brook Champagne.

Second label: Brunonian Reserve.

Hampton Estate

Ulster County, New York.

Benmarl estate Hybrids Hudson River Region vineyard.

HANDLEY CELLARS

Founded 1982, Philo, Mendocino County, California.

Storage: Oak and stainless steel. Cases per year: 250.

Label indicating non-estate vineyard: B.J. Carney Vineyard, Handley Vineyard.

Wines regularly bottled: Vintage-dated Chardonnay.

Handley Ranch

Sonoma County, California.

Sauvignon Blanc vineyard.

Hanford Vineyards

Sonoma County, California.

Sauvignon Blanc Sonoma Valley vineyard.

HANNA, PHIL TOWNSEND (1887-1958)

American editor and writer.

"Conversion of the endurable, unfermented juice of the grape into the long-lived, stimulating and nourishing beverage we know as wine, is a process in the marginal area that borders on the miraculous."

"Like the charms of a beautiful woman, wine disdains clinical appraisal, the while it provokes ecstatic adoration."

"Good wine is of the substance of life itself and its be-all and end-all is indeed in the precincts of the metaphysical."

Hanna Vineyards

Napa County, California.

Chardonnay and Johannisberg Riesling Napa Valley vineyard.

Hanzell Estate Vineyards

Sonoma County, California.

Pinot Noir, Chardonnay Sonoma Valley vineyard.

HANZELL VINEYARDS

Founded 1957, Sonoma, Sonoma County, California.

Storage: Oak and stainless steel. Cases per year: 2,000.

History: Winery was founded by the late Ambassador James T. Zellerbach and designed after the Clos de Vougeot in Burgundy.

Estate vineyard: Hanzell Vineyards.

Wines regularly bottled: Estate bottled, vintage-dated Chardonnay and Pinot Noir.

HARASZTHY, AGOSTON

"Count" Agoston Haraszthy fled from

Hungary to America after participating in the Hapsburg Revolution. He was an experienced winemaker, and, in 1857, came to the Sonoma Valley in California in search of vineyard property. Pleased with the climate and geography, he purchased the Buena Vista vineyards, planted by General Vallejo to provide grapes for Mission San Francisco Solano de Sonoma. Haraszthy expanded his vineyards to 400 acres, and produced many award-winning wines. He encouraged his fellow winemakers, in his many articles and speeches, to recognize the potential of the region for great wine, and to constantly improve their viticultural and enological techniques. In 1861, he travelled through Europe at the request of the State of California, to collect vinifera cuttings. He brought back 100,000 plants, including nearly 300 different varietals, which were distributed to vineyards throughout the area. He also wrote a book, titled "Grape Culture, Wines and Winemaking". His family was as successful socially as he was professionally, and his two sons married the daughters of General Mariano Vallejo, in a double wedding. His enthusiasm, dedication and influence on the development of California winemaking have led many to consider Haraszthy the 'father of the premium wine industry'.

HARBOR WINERY

Founded 1972, West Sacramento, Yolo County, California.

Storage: Oak and stainless steel. Cases per year: 1,500.

Label indicating non-estate vineyard: Spring Lane Vineyard.

Wines regularly bottled: Vintage-dated Cabernet Sauvignon, Chardonnay. Also vintage-dated proprietary wine Mission del Sol. Occasionally bottle Zinfandel.

hard

Tasting term.

Wine taster's term for a wine with excessive tannin. Not necessarily a fault in a young wine, where it may indicate a long maturity.

Hargrave Estate Vineyard

Suffolk County, New York.

Cabernet Sauvignon, Chardonnay, Pinot Noir, Sauvignon Blanc, Pinot Noir, Riesling, Gewurztraminer and Merlot North Fork of Long Island vineyard.

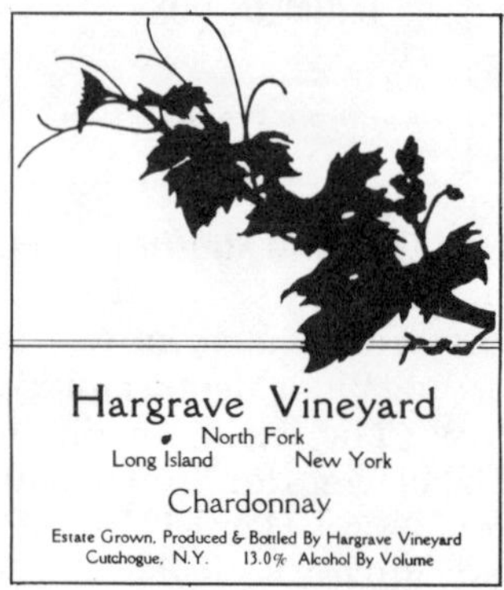

HARGRAVE VINEYARD

Founded 1973, Cutchogue, Suffolk County, New York.

Storage: Oak and stainless steel. Cases per year: 6,000.

History: The first commercial winery ever established on Long Island. Hargrave's growing season averages 210 days per year, the same as Bordeaux, and the North Fork gets 20%-30% more sunshine than upstate New York. It is the only wine growing region in the world surrounded on three sides by water. The background of a Harvard degree in Chinese studies and then publishing, before winemaking, makes Alex Hargrave another one of the unique American winemakers with "left-field" backgrounds.

Estate vineyard: Hargrave Vineyard.

Wines regularly bottled: Estate bottled Cabernet Sauvignon, Cabernet Sauvignon Reserve, Chardonnay, Chardonnay Collectors Series, Sauvignon Blanc, Fumé Blanc, Blanc de Noir, Pinot Noir, Riesling, Gewurztraminer, Cabernet/Merlot, Merlot.

harmonious
Tasting term.
Balanced, nicely acidic, soft and tannic in terms of balance.

Hart Estate Vineyard
Riverside County, California.
Cabernet Sauvignon vineyard.

HART WINERY
Founded 1980, Temecula, Riverside County, California.
Estate vineyard: Hart Vineyard.
Label indicating non-estate vineyard: Miramonte Vineyard, Andrews Vineyard.
Wines regularly bottled: Estate bottled, vintage-dated Cabernet Sauvignon. Also vintage-dated Chenin Blanc, Sauvignon Blanc. Occasionally bottle Petite Sirah, Zinfandel, Gamay Beaujolais.
Second label: French Valley Vineyards.

HARVEST CELLARS
(See Granite Springs Winery.)
Apple wine.

Harvey Ranch
Napa County, California.
Flora Springs estate Chardonnay, Sauvignon Blanc and Pinot Blanc Napa Valley vineyard.

Hat Family Vineyards
San Joaquin County, California.
Bella Napoli estate Zinfandel, Chenin Blanc, French Colombard, Grenache and Pinot Chardonnay vineyard.

HAVELOCK GORDON NEGOCIANTS
San Francisco, California.
Wines regularly sold: Chardonnay, Cabernet Sauvignon, Zinfandel.

Haverstick Vineyard
Litchfield County, Connecticut.
Ravat vineyard.

HAVILAND VINTNERS
Founded 1981, Lynnwood, Snohomish County, Washington.
Storage: Oak and stainless steel. Cases per year: 12,600.
Estate vineyard: Haviland Vineyard.
Label indicating non-estate vineyard: Dionysus Vineyard.
Wines regularly bottled: Dry White Riesling, Select and Late Harvest Riesling, Chardonnay, Sauvignon Blanc, Cabernet Sauvignon, Pinot Noir, Merlot.
Second label: Alderwood Cellars.

HAWK CREST (See Stag's Leap Wine Cellars.)
Gewurztraminer.

HAWKINS, SIR JOHN (1532-1595)
British Admiral, member of Parliament, kinsman to Sir Francis Drake; led his ship, The Victory, into battle against the Spanish Armada and helped inflict heavy losses on the "Invincible Armada". When a young Captain, on an expedition to the New World, he lifted the siege against the French Huguenots at their colony in Fort Caroline, Florida. His hogsheads were almost dry but he soon learned that the Huguenots had wine to help fill his ship's supply. The Huguenots offered Hawkins wine from their "Twenty hogheads" made entirely from "Native grapes". The wine grapes no doubt were Muscadines or the Florida Scuppernong. So, in 1565, Hawkins and his men drank and filled their hogsheads with "the first

American wine" . . . made by white men, that is. The Indians were wine makers of wild grapes long before the arrival of the Europeans.

The Hawks Vineyards
Alexander Valley, California.
Cabernet Sauvignon vineyard.

Hawn Creek Vineyard
Mendocino County, California.
Chardonnay vineyard.

HAYWOOD WINERY

Founded vineyard 1974, winery 1980, Sonoma, Sonoma County, California.

Storage: Oak and stainless steel. Cases per year: 8,000.

Estate vineyard: Chamisal Vineyards.

Wines regularly bottled: Estate bottled, vintage-dated Chardonnay, Cabernet Sauvignon, Zinfandel, White Riesling. Also an estate bottled, vintage-dated White Wine. And a vintage-dated Spaghetti Red.

Hazen Vineyards
Napa County, California.
Merlot Napa Valley vineyard.

HEALDSBURG

In Sonoma County, the town near the Alexander and Dry Creek Valleys.

HEARST, GEORGE

George Hearst, miner and founder of the San Francisco Examiner, was also an early supporter of the California wine industry. Around 1885, Hearst purchased the Madrone Vineyard in Glen Ellen, Sonoma County. Although it had been planted in table wine grapes by the previous owner, the vines were destroyed by phylloxera just a few years after Hearst took over. He replanted 350 acres on resistant root stock from Medoc and Gironde, and added two cellars and a distillery to the original small winery. He promoted his wines to his frequent guests, and believed in the temperate table use of wine. Hearst was also a strong advocate of improving the quality of California wine to offset low prices due to overproduction. He operated the Madrone winery until his death in 1891; the winery is still in production today.

HECK, ADOLF L.

Adolf L. Heck comes from a champagne making family. His grandfather produced sparkling wines in Alsace-Lorraine, and his father was owner of Cook's Imperial Champagne Winery in St. Louis, Missouri. After studying bacteriology at St. Louis University, Heck learned the art of champagne making at the Geisenheim Wine Institute in Germany. He returned to the United States to become champagne maker at his father's winery. Following World War II and a job with National Distillers in the Defense Program, Heck joined Italian Swiss Colony in California. He was promoted to president of the winery in 1951. In 1954, he returned to his first love, champagne making, with the purchase of Korbel Winery. Already well-known for its production of California champagne, Heck added his own cuvées to the Korbel line. He also developed modern production techniques, including the automatic riddling machine, which he designed and patented. Under Heck's direction, Korbel introduced several new champagnes, notably Korbel Natural, a Blanc de Noirs and a Blanc de Blancs. Today, Heck serves as chairman of the board for F. Korbel and Brothers. He continues to be an active member of the California Wine Institute, and is Chairman of the Academy of Master Wine Growers.

HECKER PASS
The pass through the Coast Range Mountains west of Gilroy.

Hecker Pass Estate Vineyard
Santa Clara County, California.
Grenache, Petite Sirah, Carignane, Zinfandel, Ruby Cabernet Santa Clara Valley vineyard.

HECKER PASS WINERY
Founded 1972, Gilroy, Santa Clara County, California.
Storage: Oak. Cases per year: 6,000.
Estate vineyard: Hecker Pass Vineyard.
Wines regularly bottled: Estate bottled, vintage-dated Grenache Rosé, Grenache, Petite Sirah Select, Zinfandel, Carignane, Petite Sirah. Also vintage-dated French Colombard; non-vintage Cream Sherry, Medium Dry Sherry, Ruby Port and one semi-generic: Chablis.

Heineman Estate Vineyard
Ottawa County, Ohio.
Concord, Catawba, Niagara, Ives and Delaware Lake Erie vineyard.

HEINEMAN WINERY
Founded 1888, Put-in-Bay, Ottawa County, Ohio.
Storage: Oak and stainless steel. Cases per year: 10,000.
History: Founded by Gustav Heineman who was an immigrant from the Baden region of Germany. The winery has been in the family since then. The grandson, Louis, now operates the winery and great-grandson, Ed, is the winemaker.
Estate vineyard: Heineman Vineyard.
Label indicating non-estate vineyard: Duff Vineyard.
Wines regularly bottled: Pink Catawba, Delaware, Dry Catawba, Seyval Blanc, DeChaunac, Vidal Blanc, Marechal Foch; semi-generics: Burgundy and Sauterne; proprietary SW Belle, SW Concord, SW Catawba. Also Rosé and Claret.

G & T Heiss Vineyards
Okanagan Centre, British Columbia.
Gray Monk estate Kerner, Pinot Gris, Pinot Auxerrois, Bacchus, Gewurztraminer, Rotburger, Siegerrebe, Marechal Foch vineyard.

HEITZ, JOSEPH
Joseph Heitz entered the wine industry while stationed in Fresno, California, during World War II. After a day at the base, Heitz worked a second job in the cellars of Italian Swiss Colony. Following the war, he completed a bachelor's degree in enology from the University of California, Davis, and then stayed on to do graduate research. Beginning in 1949, he worked successively at Gallo, the Winegrower's Guild in Lodi and at Mission Bell Winery. In 1951, Heitz accepted a position at Beaulieu Vineyards, where he remained for seven years. He returned to the academic life in 1958, and helped establish the Enology Department at California State University, Fresno. It was in 1961 that Heitz was finally able to purchase his own winery. The original Napa Valley winery building serves, today, as sales room for Heitz Cellar, and the wine is produced in an early stone winery purchased in 1964. Heitz believes that red wines are the strength of the wine industry, and he is best known for his Cabernet Sauvignon. Most of his grapes are purchased, allowing him to concentrate on winemaking, instead of grape growing. Heitz Cellar is a family operation: son David serves as vice president and winemaker and daughter Kathleen works in sales and marketing. His wife, Alice, has assisted him from the beginning and currently works in public relations and promotion. Heitz is renowned for his quality wines, named after the vineyard from which they are produced, priced at what they are worth in the world market.

Heitz Estate Vineyards
Napa County, California.
Chardonnay, Grignolino and Zinfandel vineyards.

HEITZ WINE CELLARS
Founded 1961, St. Helena, Napa County, California.
Storage: Oak and stainless steel. Cases per year: 37,000.
History: Heitz Cellars began, in 1961, in the tiny building that now serves as its sales room. In 1964, the family bought its present home ranch in Spring Valley. Spring Valley was developed as a vineyard and winery in the 1880's by Anton Rossi and his family. Early records indicate that the Rossis were making about 10,000 gallons of wine a year in the fine old stone winery that now holds Heitz white wines from fermentation until they are ready for bottling. The Heitzs added a second building in 1972.
Estate vineyard: Heitz Vineyards.
Label indicating non-estate vineyard: Martha's Vineyard, Bella Oaks Vineyard.
Wines regularly bottled: Estate bottled, vintage-dated Chardonnay and Grignolino. Vintage-dated Cabernet Sauvignon, Pinot Noir, Grignolino, Chardonnay, Johannisberg Riesling, Sweet Johannisberg Riesling, Gewurztraminer; a non-vintage Barbera. Also bottle Chablis, Burgundy, Champagne (Brut and Extra Dry) and "Cellar Treasure" Port and Sherry.

Counterscrew type corkpuller

Carl Helmholz Vineyard
Sonoma County, California.
Cabernet Sauvignon Alexander Valley vineyard.

HEMINGWAY, ERNEST (1898-1961)
American writer.

"Wine is one of the most civilized things in the world and one of the natural things of the world that has been brought to the greatest perfection, and it offers a greater range for enjoyment and appreciation than possibly any other purely sensory thing which may be purchased."

"This wine is too good for toast-drinking, my dear. You don't want to mix emotions up with a wine like that. You lose the taste." (From "The Sun Also Rises," Count Mippipopolous to Brett.)"

HEMPHILL, ALLAN J.
Allan J. Hemphill is currently president of Chateau St. Jean Vineyards and Winery in the Sonoma Valley. He began his professional career while he was a student in enology at California State University, Fresno, working part-time in production and in the laboratory at Korbel Champagne Cellars. After earning his bachelor's degree in 1962, Hemphill continued at Korbel. He was wine chemist and assistant champagne maker for three years, production manager from 1966 to 1972, and vice president of operations for the following four years. In 1977, Hemphill became president of Chateau St. Jean. He is an active member of the American Society of Enologists and is past president of the Sonoma County Wine Grower's Association. He has also served as a director of both the California Wine Grower's Foundation and the Council of California Growers.

HENRI MARCHANT (See Gold Seal.)
Champagne, Labrusca Red, Labrusca White.

John Henry
Yates County, New York.
Aurora Blanc, Foch, Cabernet Sauvignon, Gewurztraminer Finger Lakes vineyard.

Henry Estate Vineyard
Douglas County, Oregon.
Riesling, Chardonnay, Gewurztraminer, Pinot Noir Umpqua Valley vineyard.

HENRY WINERY
Founded 1978, Umpqua, Douglas County, Oregon.
Storage: Oak and stainless steel. Cases per year: 6,000.
Estate vineyard: Henry Vineyard.
Wines regularly bottled: Estate bottled, vintage-dated Pinot Noir, Chardonnay, Gewurztraminer: and Table Wine.

HENRY'S (See Henry Endres Winery.)

Herbert Estate Vineyard
El Dorado County, California.
Zinfandel, Sauvignon Blanc vineyard.

HERBERT VINEYARDS
Founded 1982, Somerset, El Dorado County, California.
Cases per year: 1,000.
Estate vineyard: Herbert Vineyards.
Wines regularly bottled: Estate bottled, vintage-dated Zinfandel, White Zinfandel.

Hercules Vineyard
St. Charles County, Missouri.
Winery of the Little Hills estate Seyval Blanc, Villard, Vidal Blanc, DeChaunac, Chelois, Chancellor, Concord vineyard.

HERITAGE (See Barnes Wines, Ltd.)
Very Pale Dry Sherry, Dry Sherry, Golden Sherry, Cream Sherry.

Heritage Estate Vineyards
Miami County, Ohio.
Seyval Blanc, Niagara, Aurora, DeChaunac, Baco Noir, Catawba, Concord and Vidal vineyard.

HERITAGE ESTATES
(See Barnes Wines, Ltd.)
Canada Chablis, Burgundy, Claret, Rosé, Rhine Wine.

HERITAGE HILL
(See S. Anderson Vineyard.)

HERITAGE VINEYARDS
Founded 1978, West Milton, Miami County, Ohio.
Storage: Oak and stainless steel.
Estate vineyard: Heritage Vineyards.
Wines regularly bottled: Estate bottled, vintage-dated Seyval Blanc, Niagara, Delaware, DeChaunac, Baco Noir, Foch, Chelois, Catawba, Concord; also Vin Blanc Heritage Rouge, Red and Rosé.

HERITAGE WINE CELLARS
Founded 1978, North East, Erie County, Pennsylvania.
Storage: Stainless steel.
History: D.C. Bostwick purchased the original 100 acre fruit farm back in 1833. His grandson, Kenneth, converted the farm to grape production. Today, the great-grandsons operate the winery.
Wines regularly bottled: Niagara, Delaware, Dutchess, Chablis, Seyval Blanc, Concord, Pink Catawba, Gladwin 113, Vidal Blanc, DeChaunac, DeChaunac Rosé, Marechal Foch; also Apple, Cherry, Elderberry; and semi-generics Burgundy, Chablis; carbonated Niagara, Catawba, Chablis; and Sweet Country White, Holiday Spice.

HERMANN VITICULTURAL AREA
Hermann (central Missouri) viticultural area is located in the northern portion of Gasconade and Franklin counties, south of the Missouri River. It consists of approximately 80 square miles with 102 acres of wine grapes. The history of grape growing and wine production in the area of Hermann goes back to 1843 when it was the largest wine-producing area in the State.

Hermannhof Estate Vineyards
Gasconade County, Missouri.
Seyval, Vidal, Delaware and Vidal Blanc Hermann area vineyard.

HERMANNHOF WINERY
Founded 1852, Hermann, Gasconade County, Missouri.
Storage: Oak and stainless steel. Cases per year: 10,000.
History: The winery's ten stone cellars and brick superstructures are among the early Hermann buildings placed on the National Register of Historic Site by the Federal Government.
Estate vineyard: Hermannhof Vineyards.
Label indicating non-estate vineyard: Bacchus Vineyard.
Wines regularly bottled: Vidal Blanc, Delaware, Villard, Concord, Seyval Blanc; proprietary White Lady, Settlers Pride, Founders Reserve, Virginia Seedling; and Rosé. Also Blackberry and Cherry Wines.

Heron Hill Estate Ingle Vineyard
Steuben County, New York.
Riesling Finger Lakes vineyard.

Heron Hill Estate Vineyard
Steuben County, New York.
Chardonnay, Riesling and Seyval Blanc Finger Lakes vineyard.

HERON HILL VINEYARDS
Founded 1977, Hammondsport, Steuben County, New York.
Storage: Oak. Cases per year: 8,000.
Estate vineyard: Heron Hill Vineyards, Heron Hill Ingle Vineyard.
Wines regularly bottled: Estate bottled, vintage-dated Riesling, Chardonnay, Seyval, Aurora; vintage-dated Ravat, Cayuga. Occasionally Apple Wine.
Second label: Otter Spring.

HEUBLEIN WINES (See Inglenook and Colony.)
San Francisco County, California.

HFH Ranch
Amador County, California.
Sauvignon Blanc Shenandoah Valley vineyard.

HIARING, PHILIP
The editor and publisher of "Wines and Vines" magazine, Philip Hiaring, has had a long and successful writing career. He was graduated, in 1937, with a bachelor's degree in journalism, from the University of Idaho. He subsequently held several positions as a journalist, starting with two years at the Salt Lake Tribune, followed by 16 years with Associated Press. During this period, Hiaring also saw active duty in the United States Air Force in World War II and the Korean War. From 1955 to 1969, Hiaring worked in the public relations office of Bank of America. He was then employed by the California Wine Institute for eight years, prior to a job at Calcan. His interest in wine began in 1959, when Hiaring founded the Order of Military Wine Tasters. He wrote a syndicated wine column from 1962 to 1974. In addition, Hiaring initiated the Wine Industry Technical Symposium, which has held annual events for ten years. He is also responsible for creating the Wines and Vines Award for Excellence in Wine Writing.

HIDALGO (See Cavas de San Juan.)

Hidalgo Vineyards.
San Juan, Mexico.
Cavas de San Juan estate Cabernet Sauvignon, Merlot, Pinot Noir, Malbec, Carignane, Grenache, Feher Szagos, Ugni Blanc, Chenin Blanc, Traminer vineyard.

HIDDEN CELLARS
Founded 1981, Talmage, Mendocino County, California.
Cases per year: 8,500.
Label indicating non-estate vineyard: Anderson Valley Vineyard, Mark Turla Vineyard, B.J. Lovin Vineyard, Stanford Road Ranch, Goodrich Vineyard, Potter Valley Vineyard, Talmage Vineyard.
Wines regularly bottled: Vintage-dated Gewurztraminer, Sauvignon Blanc, Zinfandel, Johannisberg Riesling, Late Harvest Riesling.

HIDDEN SPRINGS WINERY
Founded 1980, Salem, Yamhill County, Oregon.
Storage: Oak. Cases per year: 3,000.
Estate vineyard: Eola Hills Vineyard, Spring Valley Vineyard.
Wines regularly bottled: Vintage-dated Pinot Noir, Chardonnay, White Riesling, Pinot Noir Blanc; and vintage-dated Red Table Wine.

HIGGINS VINEYARD (See Sierra Vista Winery.)
Vintage-dated Zinfandel.

High Tor Estate Vineyard
Rockland County, New York.
Chancellor Noir, Aurora, Seyval Blanc and DeChaunac vineyard.

HIGH TOR VINEYARDS
Founded 1950, New City, Rockland County, New York.
Storage: Oak. Cases per year: 2,750.
History: The land of High Tor was granted to the Van Orden family by King George III. Wine grapes have grown on the property, with some vineyards dating back to the 18th century. There is also a sacred Indian burial ground on the property.
Estate vineyard: High Tor Vineyard.
Wines regularly bottled: Estate bottled Chancellor Noir, Aurora, Seyval Blanc, DeChaunac; proprietary Beacon White, Rockland Red and White; also Sherry.

HIGHLAND MANOR WINERY
Founded 1979, Jamestown, Fentress County, Tennessee.
Storage: Oak and stainless steel. Cases per year: 1,000.
History: The first Tennessee licensed winery in the 20th century. Located on the Cumberland Plateau area in the northeast of Tennessee.
Estate vineyard: Wheeler Vineyard.
Label indicating non-estate vineyard: Malinda's Vineyard.

Wines regularly bottled: Alwood, Catawba, Concord, Cabernet Sauvignon, Muscadine; also Red and White.

Hildebrand Vineyards
Lubbock County, Texas.
Chardonnay vineyard.

The Hill Vineyard
Mendocino, California.
Braren Pauli estate Zinfandel vineyard.

WILLIAM HILL WINERY

Founded 1976, Napa, Napa County, California.

Storage: Oak and stainless steel. Cases per year: 20,000.

Estate vineyard: Veeder Peak Vineyard, Silverado-Altus Ranch, Foss Valley Ranch.

Wines regularly bottled: Vintage-dated Chardonnay, Cabernet Sauvignon.

Hillcrest Ranch
Sonoma County, California.
Johannisberg Riesling, Green Valley vineyard.

Hillcrest Estate Vineyard
Douglas County, Oregon.
White Riesling, Cabernet Sauvignon, Pinot Noir, Gewurztraminer, Chardonnay and Zinfandel Umpqua Valley vineyard.

HILLCREST VINEYARD

Founded 1960, Roseburg, Douglas County, Oregon.

Storage: Oak and stainless steel. Cases per year: 10,000.

History: Richard Sommer, owner and operator, planted the vinifera varietal grape vineyard in Oregon in 1960-61. He produced some of the first varietal wine and sparkling wine in Oregon.

Estate vineyard: Hillcrest Vineyard.

Wines regularly bottled: Estate bottled, vintage-dated Cabernet Sauvignon, White Riesling, Gewurztraminer, Pinot Noir. Occasionally Merlot, Ice Wine, Sauvignon Blanc, Semillon.

HILLCREST WINERY, LTD.

Founded 1982, Greensburg, Westmoreland County, Pennsylvania.

Storage: Oak and stainless steel. Cases per year: 2,500.

Label indicating non-estate vineyard: Presque Isle Wine Cellars.

Wines regularly bottled: Riesling, Cabernet Sauvignon, Vidal Blanc, Seyval Blanc, Chardonnay.

Hildebrand Estate Vineyard
Ontario, Canada.
Seyval Blanc, Baco Noir, Riesling, Chardonnay vineyard.

HILLEBRAND ESTATES WINERY, LTD.

Founded 1980, Niagara-on-the-Lake, Ontario, Canada.

Storage: Oak and stainless steel. Cases per year: 50,000.

Estate vineyard: Hildebrand Vineyards.

Label indicating non-estate vineyard: Joseph E. Pohorly Vineyards.

Wines regularly bottled: Gewurztraminer, Chardonnay, Riesling, Seyval Blanc, Baco Noir; Proprietary Schloss Hillebrand Le Baron Rouge and Blanc, Chevalier Rouge, Comtesse Blanche, Elizabeth Rosé. Occasionally Canadian Sherry and Cream Sherry, proprietary Wellington, Lady Ann.

Hillside Vineyard
Napa County, California.
Sauvignon Blanc vineyard.

Hillside Vineyards
Sonoma County, California.
Chardonnay, Cabernet Sauvignon and Zinfandel vineyard.

Hillside Vineyards/Cobblestone Vineyard
Monterey County, California.
Chardonnay vineyard.

Hillview Vineyard
Napa County, California.
Groth estate Sauvignon Blanc, Merlot vineyard.

Hinman Estate Vineyards
Lane County, Oregon.
White Riesling, Gewurztraminer and Pinot Noir vineyard.

HINMAN VINEYARDS

Founded 1979, Eugene, Lane County, Oregon.

Storage: Oak and stainless steel. Cases per year: 6,000.

Estate vineyard: Hinman Vineyards.

Label indicating non-estate vineyard: Markham Vineyards, McKenzie River Vineyards.

Wines regularly bottled: Vintage-dated Oregon White, Pinot Noir, Gewurztraminer, Pinot Noir, Riesling, Cabernet Sauvignon; White Table Wine. Occasionally Chardonnay and Sauvignon Blanc.

Hinzerling Estate Vineyards
Benton County, Washington.
Gewurztraminer, Chardonnay, White Riesling, Merlot and Cabernet Sauvignon vineyard.

HINZERLING VINEYARDS

Founded 1972, Prosser, Benton County, Washington.

Cases per year: 6,000.

Estate vineyard: Hinzerling.

Label indicating non-estate vineyard: Mercer Ranch, Sagemoor Vineyard.

Wines regularly bottled: Estate bottled Gewurztraminer, White Riesling, Cabernet Sauvignon, Chardonnay, proprietary Ashfall White; Cabernet Sauvignon; occasionally Merlot, Cabernet/Merlot, Blanc de Noir, Blanc de Blanc and Die Sonne (individual select cluster Gewurztraminer).

HISTORY OF VINEYARDS OF EARLY CALIFORNIA

The Jesuit Fathers carried Spanish colonization and wine growing up the western coast into the Mexican Peninsula of Baja, California. Their successors, the Franciscans, advanced into what is now the State of California. As each new settlement or mission was established, vines were planted, as one of the first steps in transforming wilderness into civilization.

The Franciscans and their leader, Padre Junipero Serra, established Mission San Diego, in 1769, and planted wine grapes there. Thus, it was discovered that California was a land especially favored for wine-growing. In Northern Mexico and Baja California these pioneers of western wine-growing had suffered many hardships in cultivating the arid lands and in trying to supply themselves with wines needed for Sacramental and table use. At San Diego, the grapevines thrived and the wines were better.

As the Franciscans moved northward, establishing new missions, they found the same results. They had discovered a new wine-growing region. More than a century later, it was destined to become one of the premier wine regions of the world.

Eventually, Padre Serra's missionaries built a chain of 21 missions, from San Diego to Sonoma. Sonoma was the northernmost point of their El Camino Real or "Kings Highway." They planted vineyards and made wine at nearly all of the missions. San Gabriel Mission, near Los Angeles, was the site of their largest winery. There they had three wine presses.

HMR Estate Vineyard

San Luis Obispo County, California.

Pinot Noir, Sylvaner, Cabernet Sauvignon and Chardonnay Paso Robles vineyard.

HMR LIMITED

Founded 1965, Paso Robles, San Luis Obispo County, California.

Storage: Oak and stainless steel.

Estate vineyard: HMR Vineyard.

Wines regularly bottled: Estate bottled, vintage-dated Pinot Noir, Chardonnay, Riesling, Pinot Noir Blanc, Cabernet. Also vintage-dated Cabernet Sauvignon, Chardonnay, Chenin Blanc, Johannisberg Riesling, Sauvignon Blanc; and a non-vintage Zinfandel.

HNW CELLARS (Louis Honig, HNW Vineyards.)

Founded 1980, Rutherford, Napa County, California.

Storage: Oak and stainless steel. Cases per year: 6,000.

History: HNW Vineyards were part of the Caymus land grant originally held by George Yount, then passed to the Fealey family, the Wagner family and, finally, to the Honig family. In 1966 Louis Honig with his family founded HNW.

Estate vineyard: HNW Vineyard.

Wines regularly bottled: Estate bottled, vintage-dated Sauvignon Blanc.

HNW Estate Vineyards

Napa County, California.

Cabernet Sauvignon, Chardonnay, Sauvignon Blanc, Semillon, Pinot Noir and Zinfandel Napa Valley vineyard.

hock

A dry, white table wine usually make with Riesling grapes, in which, as with other white wines, the fermentation takes place after the skins have been separated. Hock was originally a term used by the English in reference to wine shipped from Hockheim, Germany. The name is also used to describe the tall wine bottle shape commonly used for Riesling wines.

Hofer Ranch/Shue Ranch/ Merrill Ranch/Hunt Ranch
Cucamonga, California.
Galleano estate Zinfandel, Matero, Muscato Carignane, Grenache, Burger and Mission vineyard.

Hogshead

hogshead
Container that holds 63 wine gallons. Large hogshead—holds from 63 to 140 wine gallons. A large cask for wine of varying capacities (actually old measurements). Early settlers in America spoke and wrote of hogsheads of wine.

THE HOGUE CELLARS
Founded 1982, Prosser, Benton County, Washington.
Storage: Oak and stainless steel. Cases per year: 17,000.
Estate vineyard: Schwartzman Vineyard, Roza Estates Vineyard, Cook Estate Vineyard.
Wines regularly bottled: Vintage-dated Chenin Blanc, White Riesling; proprietary Schwartzman Vineyards, White Riesling.

Hogue Ranches
Spokane County, Washington.
Latah Creek estate Johannisberg Riesling, Gewurztraminer, Chardonnay, Fumé Blanc, Chenin Blanc and Semillon Yakima Valley vineyard.

Holdenreid Vineyard
Lake County, California.
Cabernet Sauvignon vineyard.

HOLMES, OLIVER WENDELL
(1809-1894)
American man of letters.

"Among the great whom heaven has made to shine
How few have learned the art of arts—to dine."

"I give you one health in the juice of the vine,
The blood of the vineyard shall mingle with mine;
Thus let us drain the last dew drops of gold,
And empty our hearts of the blessings they hold."

"Then a smile and a glass and a toast and a cheer,
For all the good wine, and we've some of it here."

"'Tis nature's law that wine should flow
To wet the lips of friends.
Then once again, before we part
My empty glass shall ring."

"And let the loving cup go round,
The cup with blessed memories crowned."

Home Ranch
Mendocino County, California.
Parducci estate French Colombard, Petite Sirah and Cabernet Sauvignon vineyard.

Home Ranch
Sonoma County, California.
Fieldstone estate Cabernet Sauvignon, Johannisberg Riesling and Chenin Blanc vineyard.

Home Ranch
Sonoma County, California.
Seghesio estate Zinfandel, Carignane, Pinot Noir, Petite Sirah, Chenin Blanc, Chardonnay Alexander Valley vineyard.

Home Vineyard
Mendocino County, California.
Fetzer estate Cabernet Sauvignon, Sauvignon Blanc, Zinfandel, Pinot Noir and Semillon vineyard.

Home Vineyard
Monterey County, California.
Jekel estate Johannisberg Riesling, Chardonnay, Cabernet Sauvignon, Pinot Noir, Pinot Blanc and Muscat Canelli vineyard.

THE HONEY HOUSE WINERY
Founded 1976, Salem, Lane County, Oregon.
Storage: Oak. Cases per year: 416-833 cases Mead Wine.
Wines regularly bottled: Mead, Blackberry, Cherry, Plum, Rhubarb, Tomato, Squash.

HONEYWOOD WINERY
Founded 1934, Salem, Marion County, Oregon.
Storage: Stainless steel. Cases per year: 16,000.
History: One of Oregon's oldest producing wineries and one of its largest. Oregon's only producer of Kosher wine (Star of Israel), and the States leading fruit winery. The Salem Wine Festival is hosted by Honeywood.
Wines regularly bottled: Fruit and Berry Wines (Loganberry, Blackberry, Red Currant, Concord, Raspberry, Apricot, Gooseberry, Boysenberry, Apple, Hard Cider, Plum, Rhubarb, Mead.
Second label: Gallico; Star of Israel.

Hood River Estate Vineyard
Hood River County, Oregon.
White Riesling, Gewurztraminer vineyard.

HOOD RIVER VINEYARDS
Founded 1981, Hood River, Hood River County, Oregon.
Estate vineyard: Hood River Vineyard.
Wines regularly bottled: Estate bottled White Riesling, Gewurztraminer and Perry; also White Riesling, Pinot Noir, Cabernet Sauvignon, Raspberry and Blanchet Blanc.

HOODSPORT WINERY
Founded 1979, Hoodsport, Mason County, Washington.
Storage: Stainless steel. Cases per year: 12,000.
History: Only winery bottling Island Belle Grape. Wine is produced from the original 1800's vineyard.
Estate vineyard: Stretch Island Vineyards.
Label indicating non-estate vineyard: Person Orchards, Rocking Chair Vineyard.
Wines regularly bottled: Estate bottled Island Belle, also Gewurztraminer, Chenin Blanc, Riesling, Chardonnay, Semillon, Rhubarb, Gooseberry, Loganberry, Raspberry and occasionally Merlot.

HOP KILN WINERY
Founded 1975, Healdsburg, Sonoma County, California.
Cases per year: 8,000.
History: Hop Kiln is housed in a California historical landmark, a structure which served the important hop industry of California's North Coast region; once

the major hop growing area in the west. Built in 1905 by stone masons, it represents the finest example of its type.

Estate vineyard: Griffin Russian River Ranch.

Wines regularly bottled: Estate bottled, vintage-dated Johannisberg Riesling, Late Harvest Riesling, Napa Gamay, Petite Sirah. Also vintage-dated Chardonnay, Gewurztraminer, Zinfandel, Primitivo Zinfandel; proprietary A Thousand Flowers, Marty Griffins Big Red. Occasionally bottles Late Harvest Primitivo Zinfandel, Weihnachten Riesling, Verveux (dry Sparkling Johannisberg Riesling).

Second label: Sweetwater Springs.

Hopkins Estate Vineyard

Litchfield County, Connecticut.

Marechal Foch, Leon Millot, Aurora, Cayuga, Ravat and Seyval vineyards.

HOPKINS VINEYARD

Founded 1979, New Preston, Litchfield County, Connecticut.

Storage: Oak and stainless steel. Cases per year: 666.

History: The land has been farmed by the Hopkins family for almost 200 years (1787). The winery is located in 18th Century barn. Proclaimed a Connecticut Century farm in the mid 1970's.

Estate vineyard: Hopkins Vineyard, Deer Run Vineyard, Strawberry Ridge Vineyard.

Wines regularly bottled: Seyval Blanc, Ravat Blanc; proprietary wines: Barn Red (Marechal Foch and Leon Millot), Sachem's Picnic, Waramaug White (Aurora), Yankee Cider (Apple Wine).

Hopland Ranch Vineyard

Mendocino County, California.

Seghesio estate Pinot Noir, Flora vineyard.

HORIZON

A new grape variety introduced in 1983 and developed at Geneva Experiment Station. Produces a fruity white wine.

HORIZON WINERY

Founded 1977, Santa Rosa, Sonoma County, California.

Storage: Oak. Cases per year: 1,000.

Label indicating non-estate vineyard: Polson Vineyards.

Wines regularly bottled: Vintage-dated Zinfandel, Petite Sirah, Cabernet Sauvignon.

Horne Vineyards

Sonoma County, California.

Cabernet Sauvignon Sonoma Valley vineyard.

Hosmer Vineyards

Seneca County, New York.

Aurora, Chardonnay, Riesling, Cayuga White, Delaware, Seyval Blanc and Baco Noir Finger Lakes Region vineyard.

Filling a sommelier's taste vin

hot

Tasting term.

There are apparently two types. First, table wines of excessively high alcohol content produce a hot burning sensation. Second, dessert wines with a high amount of aldehyde and related compounds are said to be hot.

HOVEY, RICHARD (1864-1900)

American poet.

> "Comrades, pour the wine tonight
> For the parting is with dawn;
> Oh, the clink of cups together,
> With the daylight coming on!"

Howard T. Vineyards
Napa County, California.
Sauvignon Blanc Napa Valley vineyard.

Howell Mountain Estate Vineyard
Napa County, California.
Cabernet Sauvignon, Cabernet Franc and Merlot vineyard.

HOWELL MOUNTAIN VINEYARD
Founded 1983, St. Helena, Napa County, California.
Storage: Oak. Cases per year: 2,000.
Estate vineyard: Howell Mountain Vineyard, Adamson Vineyard.
Wines regularly bottled: Estate bottled Cabernet Sauvignon, Chardonnay.

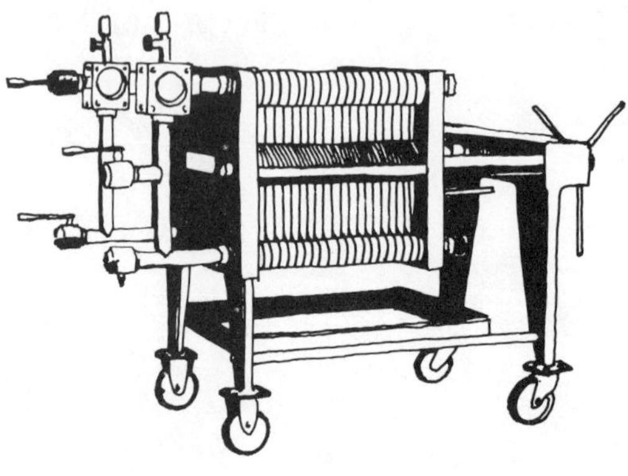

A plate filter is used to improve the clarity of wine

HOWELL MOUNTAIN VITICULTURAL AREA
Howell Mountain in Napa County, California is part of the Napa Valley viticultural area. It consists of approximately 14,080 acres.

HOWIE, MILDRED
After attending the University of California, Berkeley in Journalism, Mildred Howie began her writing career at ABC, San Francisco. Ten years later, she left to enter the public relations field and in 1971, opened her own firm, Howie Public Relations. Her clients were primarily in the food or entertainment business, including Geyser Peak Winery from 1974 to 1978. In 1979 she began a weekly wine column in the Healdsburg Tribune. A second wine column for the San Rafael Independent Journal was started in 1981. Howie has also written many articles for "Wine Country Magazine", "Vintage" and "California Living". She is co-author of "Pride of the Wineries", with Harvey Steiman and Bob Thompson. In 1982, Howie acquired the "Redwood Rancher", which she has renamed "Wine West". She is founder of the Russian River Wine Road, and creator of the California Wine Experience (with the Golden Gate Restaurant Association). Howie spearheaded the idea of the Sonoma County Wine Library, and currently serves as president of the SCWL association. She is a Gentle Lady in the Brotherhood of the Knights of the Vine, and is a member of Les Amis du Vin, W.I.N.O. and the Society of Wine Educators.

Hoy Vineyard
Templeton County, California.
Zinfandel vineyard.

Huber Estate Vineyard
Clark County, Indiana.
Vidal, Seyval Blanc, Aurora, Chancellor, Chelois, Concord, Niagara and Catawba vineyard.

HUBER ORCHARD WINERY
Founded 1978, Borden, Clark County, Indiana.
Storage: Oak and stainless steel. Cases per year: 6,300.
History: Winery is owned and operated by the sixth generation to live on this farm.
Estate vineyard: Huber Vineyard.
Wines regularly bottled: Aurora, Seyval Blanc, Vidal, DeChaunac, Chelois, Chancellor, Catawba, Niagara, Concord; Fruit wines; proprietary Starlite White, St. Johns Red.

HUDSON RIVER REGION

The Hudson River Valley is one of the nations oldest wine producing regions. When Henry Hudson sailed up the river in 1609, he found its banks covered with wild native grapes. Since that time, traditional wines have been made in the region from those native grapes. In the 1950's French hybrid vines were introduced. Highland, Marlboro and Washingtonville are the main wine towns.

HUDSON RIVER REGION VITICULTURAL AREA

The area includes all of Columbia, Dutchess and Putnam Counties, the eastern portions of Ulster and Sullivan Counties, nearly all of Orange County, and the northern portions of Rockland and Westchester Counties. There are nearly 1,000 acres of vineyards in the viticultural area.

Hudson Valley Estate Vineyard

Ulster County, New York.

Catawba, Delaware, Iona, Concord, Chelois and Baco Noir Hudson Valley vineyard.

HUDSON VALLEY WINE COMPANY, INC.

Founded 1907, Highland, Ulster County, New York.

Storage: Oak. Cases per year: 50,000.

History: The original vineyards and land were purchased by Aldo Bulognesi in the early 1900's. He built an estate with stables, which is now the winery, and established vineyards. When he lost all of his money in the Wall Street crash, he turned to his winery as a means of support. All through Prohibition he sold "Sacramental Wine". The present owners bought the winery in the 1970's.

Estate vineyard: Hudson Valley Vineyard.

Wines regularly bottled: Estate bottled Pink Catawba, Delaware, semi-generic Chablis, Burgundy, Haut Sauterne, White Burgundy, Rosé, Champagnes (Pink. Brut, Extra Dry, Cold Duck, Blanc de Blanc, Sparkling Burgundy). Occasionally bottles Chelois and proprietary Hot Rumour (wine flavored with Cinnamon and Clove).

The Hughes Vineyard

Napa County, California.

Chardonnay Napa Valley vineyard.

HULTGREN AND SAMPERTON

Founded 1978, Healdsburg, Sonoma County, California.

Storage: Oak. Cases per year: 10,000.

Wines regularly bottled: Vintage-dated Chardonnay, Cabernet Sauvignon, Gamay, Gamay Beaujolais, Petite Sirah, Pinot Noir.

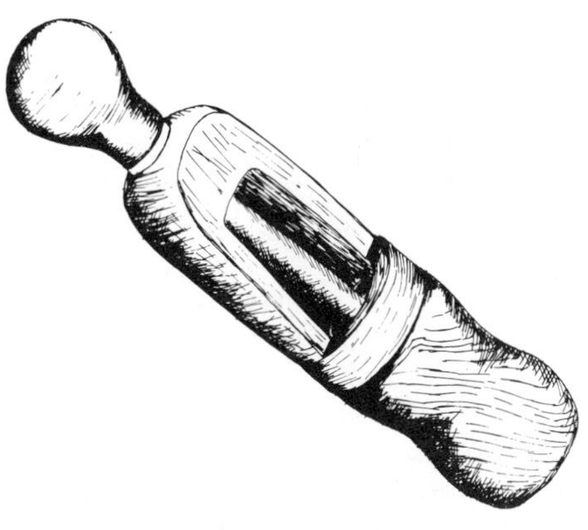

Wooden hand corker

Hunt Farms Vineyard

Yates County, New York,

Finger Lake Wine Cellar estate Niagara, Aurora and Delaware Finger Lakes vineyard.

Hunter Ranch

Sonoma County, California.

Chardonnay Sonoma Valley vineyard.

Hunter Vineyard

Mendocino County, California.

Pinot Noir vineyard.

HUSCH VINEYARDS

Founded 1971, Philo, Mendocino County, California.

Storage: Oak and stainless steel. Cases per year: 10,000.

Estate vineyard: Anderson Valley Vineyard, La Ribera Ranch.

Wines regularly bottled: Estate bottled, vintage-dated Gewurztraminer, Sauvignon Blanc, Chardonnay, Pinot Noir, Cabernet Sauvignon.

HUSMANN, GEORGE

George Husmann is an unrecognized hero from the early days of the wine industry. Born in Germany, he came to America at the age of nine, in 1836, and settled with his family in Missouri. Twelve years later, he established his own vineyard. Over the next two decades, he expanded the Bullfton Wine Company to include 1,700 vineyard acres, and thus developed the state of Missouri into a major wine producing region. In the 1870's, Husmann began shipping his resistant Dutchess Fitz-James root stock to help combat the phylloxera epidemic in France. He is credited by George Ordish in his book, *The Great Wine Blight,* as a major contributor to saving the French vineyards. In 1878, Husmann was appointed professor of pomology, forestry and viticulture at the University of Missouri, despite a formal education that was limited to grade school. He moved to California in 1881 and was hired by G. Simonton of Napa to operate his vineyard and winery in the Carneros region. He worked there for the next six years. Then, in 1889, Husmann created and organized the exhibit at the Paris World's Fair that established the Napa Valley as the premier wine region in the United States. In addition, Husmann was a prolific wine writer, publishing many articles and printing several definitive books of the time, including *American Grape Growing and Wine Making.* Despite the many important contributions he made in both France and the United States, Husmann died in 1902, condemned for his erroneous conviction that Riparia and Lenoir root stocks would solve the California phylloxera problem.

Hyatt Vineyards

Sonoma County, California.

Chardonnay vineyard.

HYBRIDS

Crosses, usually purposeful, between one species and another; intended to produce fruit or wine superior to that otherwise commonly available. Hybridization has been practiced in earnest all over the world since the 1800's, with the result that many early hybrids have been improved, considering the objectives originally sought. So-called French hybrids are the products of Frenchmen (Seibel, Baco, Seyve and Villard, Ravat among the better known) who developed many of the hybrid vines growing in America and France today. American breeders, Munson, Moore, Ephraim Bull (who developed Concord in the mid-19th century), Caywood and others were responsible for producing the early American hybrid varieties of today. The New York Experimental Station at Geneva, New York and the Horticultural Research Institute of Ontario at Vineland have been responsible for originating most of the new wine varieties of this half-century. Breeding is in progress at numerous university centers around the nation, with the likely result that more crosses will be forthcoming in the future. Hybridization is not the same as cloning. Seyval Blanc, Vidal Blanc, Ravat Blanc and Aurora are examples.

Hyde Vineyard

Napa County, California.

Chardonnay Carneros vineyard.

Hyland Vineyards

Yamhill County, Oregon,

Pinot Noir, White Riesling and Chardonnay vineyard.

ICE WINE
Wine that has been made from grapes that have been allowed to freeze and are harvested and pressed when frozen. A very sweet wine.

IDLEWILD WHITE
(See Dry Creek Vineyards, Inc.)

IMPERATOR (See Robin Fils &. Cie Ltd.)
Champagne, Sparkling Burgundy.

Indian Ridge
Sonoma County, California.
Fieldstone estate Chenin Blanc Alexander Valley vineyard.

INDIANA WINE COUNTRY
The Ohio River Valley proved itself favorable for viticulture (grape growing) as early as 1802 when Swiss immigrants led by John James Dufour, purchased land which is now Switzerland County and established the city of Vevay. Until the late 1820's Indiana was the foremost wine growing area in the United States. By 1845 the phylloxera had taken its toll and heralded the demise of the Indiana winegrowing industry.

Indiana is producing distinctive country wines from the French-American hybrid grapes. These vines possess the winter hardiness of eastern wild grapes coupled with the delicate and subtle wine making characteristics associated with the better California wine grapes.

From the sand dune hills and microclimate of the Lake Michigan area through the highlands of south central Indiana to the escarpment along the Ohio River, Hoosiers are developing both red and white wines.

INDUSTRIAS DE LA FERMENTACION S.A. DE C.V.
Aquascalientes, Mexico.
Estate vineyard: Vinedos Donhuis, Vinedos La Estancia, Vinedos La Escondida, Vinedos La Esperanza.
Wines regularly bottled: Estate bottled Cabernet Sauvignon, Chardonnay, Chenin Blanc, Sylvaner, Pinot Noir, Malbec, Merlot, Semillon.

Inglenook Estate Vineyards
Napa County, California.
Charbono, Cabernet Sauvignon and Muscat Blanc vineyard.

INGLENOOK VINEYARDS
Founded 1879, Rutherford, Napa County, California.
Storage: Oak and stainless steel. Cases per year: 334,600.
History: Inglenook means "a warm and cozy corner," and it was this ambiance that drew founder, Gustave Niebaum, to the Napa Valley. Niebaum, a Finnish sea captain, at the age of 37 quit the sea to become a vineyardist ... and to make wines. In typical Niebaum fashion, they were not to be merely "good" wines, they were to be "exquisite." He voyaged to European winelands, where he obtained vine cuttings and learned the winemaking techniques of the famed chateaux. He then began the task of building the "finest winery in America."

Captain Niebaum had ideas of his own, ideas that, a century later, were to become proven winemaking concepts. "To produce the finest wines, to equal and excel the most famous vintages of Europe," he said, "it is necessary to have the right kinds of vines, grown on suitable soils, to maintain the most perfect cleanliness in the winery, and to give the wines constant care and proper age."

The Captain's dream was first realized at the Paris Exposition of 1889, and by the time of his death in 1908, Inglenook wines had attained an international reputation.

With the passing of Gustave Niebaum, his great-nephew, John Daniel, Jr., assumed the helm of Inglenook, after Prohibition. During his 25 years of devotion to the winery, he raised its reputation to even higher levels of distinction. This was a new era for Inglenook, and among many innovations pioneered by John Daniel were the vintage-dating of wines to aid consumers in their selections, the use of Napa Valley as an appellation of origin, and the emphasis on varietal designations. He, also, was among the first to stress "estate bottlings."

John Daniel was once quoted as saying, "Wines are like children. No matter how much love and attention you give each of them, some always turn out better than others." It was this philosophy that eventually led to the designation of "Limited Cask" wines at Inglenook, in addition to the "Estate Bottling."

Approaching retirement age in 1964, Daniel sold the winery to a grower's cooperative, United Vintners. In 1969, the Heublein Corporation purchased United Vintners and assumed management of the winery.

Estate vineyard: Inglenook, Yountville Island, Yountville Ranch.

Wines regularly bottled: Estate bottled, vintage-dated Chardonnay, Fumé Blanc. Gewurztraminer, Chenin Blanc, Johannisberg Riesling, Muscat Blanc, Gamay Beaujolais, Pinot Noir, Zinfandel, Charbono, Cabernet Sauvignon, Petite Sirah, "Limited Cask" Cabernet Sauvignon. Occasionally bottles vintage-dated Cabernet Rosé, Burgundy, Chablis.

Ingleside Plantation Estate Vineyard
Westmoreland County, Virginia.
Chardonnay, Cabernet Sauvignon, Riesling, Seyval, Vidal, Chancellor and Chambourcin vineyard.

INGLESIDE PLANTATION WINERY

Founded 1980, Oakgrove, Westmoreland County, Virginia.

Storage: Oak and stainless steel. Cases per year: 10,000.

Estate vineyard: Ingleside Plantation Vineyard.

Wines regularly bottled: Estate bottled, vintage-dated Cabernet Sauvignon, Chancellor, Chardonnay, Seyval, Aurora, Riesling; vintage-dated proprietary Wirtland Rosé, Roxbury Red; also vintage Nouveaù Red, Fraulein; and Champagne.

Second label: Virginia Rose.

Inkameep Vineyards
Okanagan Valley, British Columbia.
White Riesling Scheurebe Ehrenfelser vineyard.

INNISKILLIN WINES, INC.

Founded 1974, Niagara-on-the-Lake, Ontario, Canada.

Storage: Oak and stainless steel. Cases per year: 100,000.

History: Colonel Cooper, a member of the Inniskillin regiment, served in North America during the War of 1812 and was granted crown land on the completion of his service. The Inniskillin farm was the site of the original winery and the historic name was derived from the Irish regiment—the Inniskilling Fusiliers. In 1978, the winery expanded and moved from the original site to the historic Brae Burn Estate, located adjacent to the Niagara River.

Estate vineyard: Brae-Burn Vineyards, Montegue-Inniskillin Vineyard.

Label indicating non-estate vineyard: Seeger Vineyard.

Wines regularly bottled: Estate bottled, vintage-dated Marechal Foch, Chelois, Millot-Chambourcin, Vidal, Seyval Blanc, Gamay Blanc, Riesling, Chardonnay, Gewurztraminer and proprietary Brae Blanc; also proprietary non-vintage Brae Rouge and vintage Rosé.

intense

Tasting term.

Lively, full-flavored.

Iron Horse Estate Vineyard

Sonoma County, California.

Cabernet Sauvignon, Zinfandel and Sauvignon Blanc.

IRON HORSE VINEYARDS

Founded 1979, Sebastopol, Sonoma County, California.

Storage: Oak and stainless steel. Cases per year: 24,000.

History: One of the original homestead wineries of Sonoma County, established in 1876.

Estate vineyard: Iron Horse Vineyards (Alexander Valley and Green Valley).

Wines regularly bottled: Estate bottled, vintage-dated Pinot Noir, "Blanc de" Pinot Noir, Chardonnay, Zinfandel, Sauvignon Blanc, Cabernet Sauvignon. Also estate bottled, vintage-dated sparkling wines (Blanc de Blancs, Brut, Blanc de Noir). Occasionally bottles proprietary wines Tin Pony White and Red Table Wine.

Isabella

ISABELLA

An American hybrid wine grape that produces a tangy Concord type wine. One of the earliest known native North American grapes. Used a great deal in New York State Champagnes.

ISC WINES OF CALIFORNIA, INC.

(See Colony.)

Subsidiary of Allied Grape Growers founded in early 1974. Headquartered in San Francisco with wineries at Asti, Escalon, Lodi and Reedly, California. Reported to be the second largest winery operation in America.

Includes: Colony, Italian Swiss Colony, Petri, Lejon, Jacques Bonet, G & D, and Annie Green Springs.

Isle St. George Vineyards
Isle St. George, Ohio.
Meier's estate Catawba, Concord, Johannisberg Riesling, Chardonnay, Delaware, Baco Noir, Seyval and Gewurztraminer vineyard.

ISLE ST. GEORGE VITICULTURAL AREA
The area is located in Lake Erie about 18 miles from Port Clinton, Ohio. Grape growing has been the principal concern of the islanders since the mid 1880's. Isle St. George has approximately 350 acres of grapes which are sent to the mainland by ferry for processing.

ITALIAN SWISS COLONY
(See ISC and Colony.)

Iund Vineyard
Napa County, California.
Pinot Noir Carneros vineyard.

IVES
An American hybrid wine grape that produces dry to medium sweet, fruity labrusca type wine. Discovered in 1840 by Henry Ives of Cincinnati, Ohio.

Jacobs Vineyards
Napa County, California.
Chardonnay Carneros District vineyard.

JAEGER, HERMANN
In 1867, grape breeder Jaeger of Neosho, Missouri took wild Ozark vine roots to France, seventeen carloads. Later awarded the Cross of the French Legion of Honor.

Jaeger Vineyards
Napa County, California. Rutherford Hill.
Chardonnay Napa Valley vineyard.

JANACA VINEYARDS
(See Thomas Kruse Winery.)

Jarvis Vineyard
San Mateo County, California.
Santa Cruz Mountains estate Pinot Noir and Chardonnay Santa Cruz Mountains vineyard.

Jasper Long Vineyard
Sonoma County, California.
Merlot Dry Creek Valley vineyard.

JEFFERSON ESTATE BRAND
(See Valley View Vineyard.)
Rouge Red, Rouge White, Rouge Rosé, American Cabernet Sauvignon.

JEFFERSON, THOMAS
The third President of the United States was a great wine and food enthusiast. While ambassador to France he brought the finest wines back with him. He also learned how to grow vines and make wine. For over 30 years he attempted to grow vinifera vines at his estate "Monticello". Although unsuccessful with vinifera, he never waned his belief that America was a land for grape growing and wine making.

He finally planted the red Alexander at "Monticello" and encouraged its plantings in Pennsylvania, Ohio and Indiana during the early 1800's. The red Alexander could sustain the cold and pests that the vinifera could not tolerate.

In 1808, Jefferson said, "We can produce in the United States as many varieties of wines as Europe does; not the same ones, but undoubtedly of the same quality." He was Virginia's original wine pioneer.

Jehling Vineyard
Amador County, California.
Zinfandel Shenandoah Valley vineyard.

JEKEL VINEYARD
Founded 1978, Greenfield, Monterey County, California.
Storage: Oak and stainless steel. Cases per year: 50,000.
Estate vineyard: Home Vineyard.
Wines regularly bottled: Estate bottled, vintage-dated Johannisberg Riesling, Late Harvest Johannisberg Riesling, Pinot Blanc, Chardonnay, Private Reserve Chardonnay, Pinot Noir, Cabernet Sauvignon, Private Reserve Cabernet Sauvignon, also vintage-dated Muscat Canelli.

Jensen Vineyard
San Benito County, California.
Calera estate Pinot Noir vineyard.

Jensen Vineyards
San Luis Obispo County, California.
Cabernet Sauvignon Paso Robles vineyard.

Jimtown Ranch
Sonoma County, California.
Chardonnay Alexander Valley vineyard.

Louis Jindra Estate Vineyard
Jackson County, Ohio.
Vidal Blanc, Seyval Blanc, Chelois, Baco Noir, Marechal Foch and Catawba Ohio Valley vineyard.

LOUIS JINDRA WINERY
Founded 1979, Jackson, Jackson County, Ohio.
Number of cases produced per year: 2,000.
Estate vineyard: Louis Jindra Vineyard.
Wines regularly bottled: Estate bottled, vintage-dated Vidal Blanc, Seyval Blanc, Chelois, Baco Noir, Marechal Foch; Catawba; proprietary Ohio Valley Red, White and Rosé.

J & L Farms
Monterey County, California.
Famous company which farms vineyards for several different owners in Monterey County.

JOANNES-SEYVE
A French-hybrid wine grape that produces white wine.

JOHANNISBERG RIESLING
A grape that produces a fruity-floral, tart, piquant wine, usually made semi-sweet or sweet. Johannisberg is a village in Germany. The grape is the White Riesling or Riesling. The name Johannisberg is mainly a California name for the same grape. The White Riesling is the premier grape of the great vineyards of Germany. The late harvest versions, made from over-ripe grapes, are thick, sweet and long lasting social wines comparable with German Ausleses and Beerenansleses.

JOHLIN CENTURY WINERY
Founded 1870, Oregon, Lucas County, Ohio.
Storage: Oak and stainless steel. Cases per year: 1,200.
Estate vineyard: Johlin Century.
Label indicating non-estate vineyard: King Vineyards.

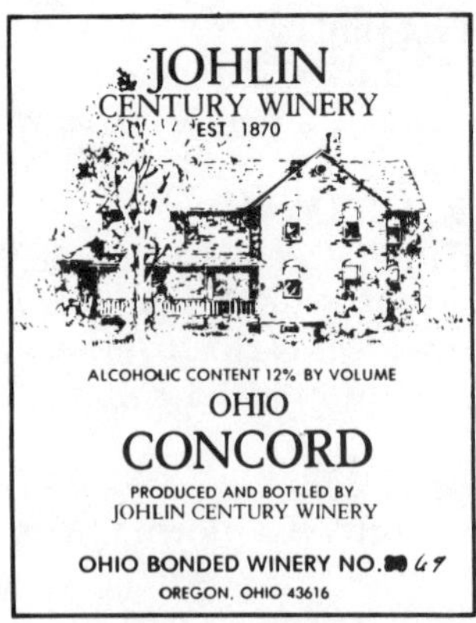

Wines regularly bottled: Concord, Catawba, Niagara; Haut Sauterne and Vin Rosé.

JOHNSON ESTATE (FREDERICK S. JOHNSON VINEYARDS)
Founded 1961, Westfield, Chatauqua County, New York.
Storage: Stainless steel. Cases per year: 10,000.
History: The estate has been in the Johnson family for 70 years. Not only the oldest estate winery in New York State but also the first in the area to produce wine from French-American grapes.
Estate vineyard: Johnson Estate Vineyard.
Wines regularly bottled: Estate bottled, vintage-dated Seyval Blanc, Delaware, Chatauqua Blanc, Cascade Rosé, Chancellor Rosé, Ives Noir, Chancellor Noir; also Blanc de Blanc Sec, Vin Rosé, Doux, Robust Red and Liebestropfchen.

Johnson Estate Vineyard
Chatauqua County, New York.
Aurora Blanc, Seyval Blanc, Delaware, Catawba, Vidal Ives, Chancellor, Cascade, Chelois and Concord Lake Erie vineyard.

JOHNSON, FREDERICK S.
Frederick S. Johnson was graduated from Cornell University, in 1946, with a degree in agriculture; he went on to become an expert in tropical agricultural development. He has worked for, and been a consultant to, such international organizations as W.R. Grace, Chase Manhattan, UNDP, USAID, and countless others. Today he spends his time between his vineyards and his consulting work. The inheritance of a Concord vineyard in Westfield, New York, where he was born, led him to change the vineyard over to French hybrids and start his winery in 1961. The Frederick S. Johnson Vineyard Winery was Chatauqua's first to produce estate-bottled wines.

Johnson Turnbull Estate Vineyards
Napa County, California.
Cabernet Sauvignon Napa Valley vineyard.

JOHNSON TURNBULL VINEYARDS
Founded 1977, Oakville, Napa County, California.
Storage: Oak. Cases per year: 2,000.
Estate vineyard: Johnson Turnbull Vineyard.
Wines regularly bottled: Estate bottled, vintage-dated Cabernet Sauvignon.

Johnson Vineyard
Napa County, California.
French Colombard, Chardonnay and Golden Chasselas Napa Valley vineyard.

Frank Johnson Vineyards
Sonoma County, California.
Chardonnay and Gewurztraminer vineyard.

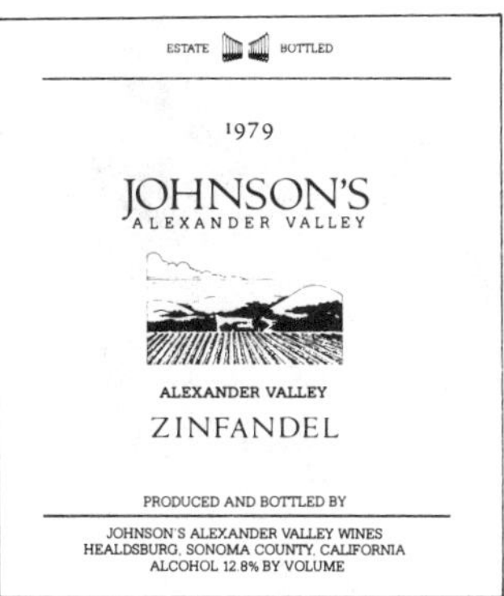

JOHNSON'S ALEXANDER VALLEY
Founded 1975, Healdsburg, Sonoma County, California.
Storage: Oak and stainless steel. Cases per year: 8,000-10,000.
Estate vineyard: Johnson's Alexander Valley Vineyard.
Wines regularly bottled: Estate bottled, vintage-dated Cabernet Sauvignon, Zinfandel, Pinot Noir, Chenin Blanc, Johannisberg Riesling, Gewurztraminer, Chardonnay.

Johnson's Alexander Valley Estate Vineyard
Sonoma County, California.
Cabernet Sauvignon, Zinfandel, Pinot Noir, Chenin Blanc, Johannisberg Riesling, Gewurztraminer and Chardonnay Alexander Valley vineyard.

Jolona Vineyard
Yakima Valley, Washington.
Chardonnay vineyard.

Jones Ranch
San Luis Obispo County, California.
Duriff vineyard.

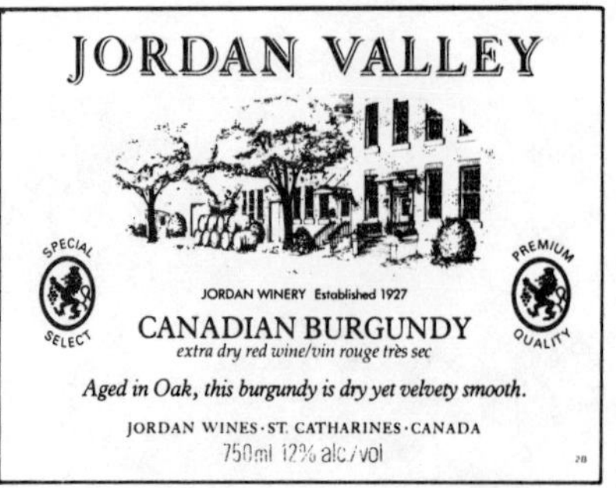

JORDAN OF CANADA
(See Ste. Michelle.)

Jordan Estate Vineyard
Sonoma County, California.
Cabernet Sauvignon and Chardonnay vineyard.

Jordan Vineyard
Amador County, California.
Karly estate Sauvignon Blanc Shenandoah Valley vineyard.

JORDAN VINEYARD AND WINERY
Founded 1976, Healdsburg, Sonoma County, California.
Storage: Oak and stainless steel. Cases per year: 45,000.
Estate vineyard: Jordan Vineyard.
Wines regularly bottled: Estate bottled, vintage-dated Cabernet Sauvignon, Chardonnay.

jug wines
The dictionary definition of jug is "a vessel in various forms for holding liquids commonly having a handle, often a lip or spout, sometimes with a narrow neck stopped by a cork." My definition is a fair to excellent table wine selling at a moderate price, in 1.5, 3 and 4 liter bottles. Homage must be paid to E. & J. Gallo for their outstanding contribution in making jug wines far superior to the general inferior European "vin ordinaire." Today we find not only generic but also varietals and often vintage-dated "jugs."

Julian Vineyards
Riverside County, California.
Menghini estate Chardonnay Temecula vineyard.

K (See Karly Wines.)

Kaarup Vineyards
Temecula, California.
Gamay Beaujolais vineyard.

KALIN CELLARS
Founded 1977, Novato, Marin County, California.
Storage: Oak. Cases per year: 6,000.
Label indicating non-estate vineyard: BJL Vineyard, Stony Ridge Vineyard, Dutton Ranch, Long Vineyard.
Wines regularly bottled: Vintage-dated Pinot Noir, Chardonnay, Cabernet Sauvignon, Merlot, Semillon, Zinfandel, Botrytis Riesling.

Karly Estate Vineyards
Amador County, California.
Sauvignon Blanc Shenandoah Valley vineyard.

KARLY WINES
Founded 1979, Plymouth, Amador County, California.
Storage: Oak and stainless steel. Cases per year: 5,000.
Estate vineyard: Karly Vineyard, Garth Vineyard, Jordan Vineyard.
Wines regularly bottled: Estate bottled, vintage-dated Fumé Blanc; also vintage-dated Zinfandel, Chardonnay, Fumé Blanc. Occasionally estate bottle Petite Sirah and Barbera.
Second label: K; Motherlode Vineyards.

KASTENBAUM, ROBERT
Ph.D., Health researcher.

"Wine has a tradition of bridging the generation gap."

"Early and wise introduction to wine, within the network of the family, may help the young to appreciate that they can obtain and share real pleasures within the scope of family life . . ."

KEDEM/ROYAL WINE CO.
Founded 1848 in Czechoslovakia, 1900's, Milton, New York.
Storage: Wood and stainless steel.
Wines regularly bottled: Only Kosher wines. Vintage-dated Zinfandel, Seyval Blanc and Chenin Blanc. Also Concord, Tokay, White Concord, Malaga, Sparkling Burgundy, Cold Duck, Pink Champagne, Champagne, Rhine, Cream Sherry and traditional Sweet Wines.

KEEHN, RICHARD & KAREN
Richard and Karen Keehn entered the wine industry when they purchased the McDowell Vineyards, in Mendocino County, in 1970. Richard had just completed 14 years of service in the U.S. Army, working as a test pilot, Aviation Safety Officer, Company Commander in Viet Nam and Director of ROTC Aviation for 22 Universities. Karen had been working in the family lumber business. In addition to operating the vineyards, the Keehns also developed Western Styles, Inc., a retail merchandizing firm and ADCO Redwood, Inc., a wood product exporting company. In 1978, they decided to dedicate their combined efforts to founding a winery. McDowell Cellars, Inc. was opened in 1982. Richard serves as president, Karen works as vice president and treasurer. Their winemaking philosophy is to produce only wines that can be enjoyed with food, and they plan to operate two series of cooking classes at

the winery. McDowell wines are currently exported to Germany, the Netherlands, New Zealand and Australia. The Keehns have been active in many wine industry organizations. Richard was the founding chairman of the California Association of Winegrape Growers, is on the board of directors of the American Vineyard Foundation, and is a member of the California Wine Institute and the California Council of Growers. In 1983, he was the U.S. delegate in Paris for the International Governmental Congress of Enology and Viticulture. Karen is a member of the California Wine Institute, the California Association of Winegrape Grower's and the Mendocino Vintners Association.

Keenan Estate Vineyards
Napa County, California.
Chardonnay, Merlot and Cabernet Sauvignon Spring Mountain vineyard.

ROBERT KEENAN WINERY
Founded 1977, St. Helena, Napa County, California.
Storage: Oak. Cases per year: 8,000.
Estate vineyard: Keenan Vineyards.
Wines regularly bottled: Estate bottled, vintage-dated Chardonnay, Merlot, Cabernet Sauvignon.

Keene Dimick
Napa County, California.
Chenin Blanc and Chardonnay vineyard.

Keig Vineyard
Napa County, California.
Cabernet Sauvignon vineyard, mostly planted on its own roots. Now used as a model for following the spread of phylloxera by aerial photography.

Keith Vineyard
Sonoma County, California.
Chardonnay Alexander Valley vineyard.

Kelley Creek Vineyard
Sonoma County, California.
Zinfandel Dry Creek Valley vineyard.

Kelley's Island
Erie County, Ohio.
Catawba vineyard.

Kathryn Kennedy Estate Vineyard
Santa Clara County, California.
Cabernet Sauvignon Santa Cruz Mountain vineyard.

KATHRYN KENNEDY WINERY
Founded 1979, Saratoga, Santa Clara County, California.
Storage: Oak. Cases per year: 1,000.
Estate vineyard: Kathryn Kennedy Vineyard.
Wines regularly bottled: Estate bottled, vintage-dated Cabernet Sauvignon. Vintage-dated Dry Whites (100% varietals).
Second label: Saratoga Cellars.

KENWOOD VINEYARDS
Founded 1970, Kenwood, Sonoma County, California.
Storage: Oak and stainless steel. Cases per year: 50,000.
History: The winery was built in 1906, as the Pagani Brothers Winery. The current owners purchased it in 1970.
Label indicating non-estate vineyard: Beltane Ranch, Jack London Ranch.
Wines regularly bottled: Vintage-dated Chenin Blanc, Sauvignon Blanc, Chardonnay, Zinfandel, Pinot Noir; vintage-dated Vintage Red and Vintage White. Occasionally bottles Pinot Noir Blanc and Johannisberg Riesling.

KENWORTHY VINEYARDS
Founded 1979, Plymouth, Amador County, California.
Wines regularly bottled: Zinfandel, Cabernet Sauvignon, Chardonnay.

Keswick Vineyard
Leelanau County, Michigan.
Mawby estate Seyval Blanc, DeChaunac and Foch vineyard.

Keuka East
Steuben County, New York.
Gold Seal estate White Riesling, Aurora, Seyval Blanc, Marechal Foch and Leon Millot Finger Lakes vineyard.

Keuka West
Steuben County, New York.
Gold Seal estate Catawba, Coloel, Vidal Blanc, Chardonnay, Concord and Ravat Finger Lakes vineyard.

Keyhole Ranch Vineyard
Sonoma County, California.
Seghesio estate Zinfandel, Pinot Noir, Cabernet Sauvignon, French Colombard Russian River Valley vineyard.

King City Vineyard
Monterey County, California.
Almaden estate Chardonnay, Chenin Blanc, Sauvignon Blanc, Semillon, Folle Blanche, Johannisberg Riesling, Cabernet Sauvignon, Napa Gamay, Merlot, Pinot St. George, Ruby Cabernet and Zinfandel vineyard.

King Vineyards
Lucas County, Ohio.
Concord, Catawba and Niagara Lake Erie vineyard.

KING WINE COMPANY (See H. Coturri & Sons.)

King's Road Vineyard
Hunterdon County, New Jersey.
French Hybrids vineyard.

Kiona Estate Vineyard
Benton County, Washington.
Chenin Blanc, Chardonnay, Cabernet Sauvignon, White Riesling and Merlot vineyard.

KIONA VINEYARDS
Founded 1975, West Richland, Benton County, Washington.
Storage: Oak and stainless steel. Cases per year: 5,000.
Estate vineyard: Kiona Vineyards.
Wines regularly bottled: Estate bottled, vintage-dated Chenin Blanc, White Riesling, Chardonnay, Merlot Rosé, Lemberger.

KIRIGIN CELLARS
Founded 1976, Gilroy, Santa Clara County, California.
Storage: Oak and stainless steel.
History: Winery stands on site of the historic Solis Ranch homestead. First part of the existing building was built in 1827 from redwood hauled and cut from nearby Mt. Madonna.
Estate vineyard: Uvas Valley Vineyard.
Wines regularly bottled: Estate bottled, vintage-dated Malvasia Bianca, Pinot Noir, Cabernet Sauvignon; vintage-dated White Riesling, Gewurztraminer, Sauvignon Blanc, Zinfandel; also Opol Rose and Vino DeMocca (proprietary), Champagne, Rhine, Burgundy.

KIRKPATRICK CELLAR WINERY
Founded 1982, Eureka, Humboldt County, California.
Cases per year: 150.
History: The Kirkpatrick fruit wines are unique in that a whole fruit, pear or apple, is in each bottle.
Wines regularly bottled: Pear and Apple Wines.

Kistler Estate Vineyard
Sonoma County, California.
Chardonnay and Cabernet Sauvignon Sonoma Valley vineyard.

KISTLER VINEYARDS
Founded 1979, Glen Ellen, Sonoma County, California.
Storage: Oak and stainless steel. Cases per year: 6,000.
Estate vineyard: Kistler Vineyards.
Label indicating non-estate vineyard: Winery Lake Vineyard, Veeder Hills Vineyard, Sonoma-Cutrer Vineyards, Dutton Ranch.
Wines regularly bottled: Vintage-dated Chardonnay, Cabernet Sauvignon, Pinot Noir.

Pruning back the vine canes

Kitty Hawk Vineyard
Napa County, California.
Pinot Noir, Cabernet and Merlot vineyard.

KLEINBERGER RIESLING
A grape which produces a light white and delicate wine.

Klingshirn Estate Vineyard
Lorain County, Ohio.
Concord and Niagara vineyard.

KLINGSHIRN WINERY, INC.
Founded 1935, Avon Lake, Lorain County, Ohio.
Storage: Oak and stainless steel. Cases per year: 5,000.
Estate vineyard: Klingshirn Vineyards.
Wines regularly bottled: Dry Catawba, Niagara, Dry Concord, Pink Catawba, Sweet Concord; also Vin Rosé, Dry and Haut Sauterne.

KNAPP, KENNETH
Kenneth Knapp has been involved in the wine industry for more than fifty years. Beginning in 1933, he worked in a variety of different positions, learning the business at several California Central Valley wineries. In 1937, he purchased a small vineyard in the Fresno area. Three years later, he joined Sunnyside Winery as production manager, where he worked until he was drafted into the United States Army. He served in Europe during World War II as an engineer. His experiences in France and Germany convinced him that California had the ideal growing climate for wine grapes, and he opened the Selma Winery, Inc., in 1947. In 1954, Knapp purchased the old Sebastiani Winery in Woodbridge, California, and renamed it the Rio Vista Winery. He operated two wineries until 1978, producing wines that "taste like the ripe grape," not of wood, other fruits or herbs. Knapp then joined the Turner Winery in Woodbridge, where he continues to serve as a consultant. In addition to promoting California as the finest wine grapegrowing region in the world, Knapp has been active in many industry organizations, including the San Joaquin Wine Grower's Association and the California Wine Institute.

Knapp Farms
Steuben County, New York.

Ravat, Seyval Blanc, Riesling and Chardonnay Finger Lakes vineyard.

THOMAS KNIGHT (See Diamond Oaks Vineyard.)
Red (Cabernet and Merlot), White (Sauvignon Blanc, Semillon, Gewurztraminer).

KNIGHTS OF THE VINE
A wine brotherhood (national organization) dedicated to the appreciation of American wines. Founded by Norman E. Gates (see biography).

Knights Valley Vineyard
Sonoma County, California.
Beringer estate Gamay, Sauvignon Blanc, Chenin Blanc and occasional Riesling with botrytis mold vineyard.

KNIGHTS VALLEY VITICULTURAL AREA
Knights Valley in northeastern Sonoma County, California, encompasses approximately 36,240 acres or 57 square miles with more than 1,000 acres of grapes. It is bounded on the north by Pine Mountain, on the south by the petrified forest area immediately to the north of Porter Creek, on the west by the boundaries for Alexander Valley and Chalk Hill viticultural areas, and on the east by the Sonoma County line bordering Lake County and Napa County.

Knole Vineyard
Riverside County, California.
Riesling vineyard.

KNOWLES, LEGH
Legh Knowles was hired for Beaulieu Vineyards in 1962, by Helen de Pins, daughter of founder Georges de Latour. He had previously been employed by the Wine Advisory Board, the Taylor Wine Company, E. & J. Gallo Winery, and as a trumpet player for Glen Miller during the Big Band era. When Heublein, Inc., acquired Beaulieu in 1969, Knowles was promoted to Vice President and General Manager. He became President in 1975, and Chairman in 1983. In 1982, Knowles was presented with the first John Wayne Commemorative Medal, the Gold Vine Award.

Knudsen Erath Estate Vineyard
Yamhill County, Oregon.
Pinot Noir, Chardonnay, White Riesling, Gewurztraminer and Merlot Willamette Valley vineyard.

KNUDSEN ERATH WINERY
Founded 1968, Dundee, Yamhill County, Oregon.
Storage: Oak and stainless steel.
Estate vineyard: Knudsen Erath Vineyard.
Wines regularly bottled: Estate bottled, vintage-dated Pinot Noir, Chardonnay, White Riesling, Gewurztraminer, Merlot; also non-vintage Pinot Noir; and Blanc de Blancs and proprietary Dundee Villages Red and Dundee Villages White.
Second label: Coastal Mist; Dundee Villages.

KOLLN VINEYARDS AND WINERY
Founded 1971, Bellefonte, Centre County, Pennsylvania.
Storage: Oak and stainless steel. Cases per year: 2,000.
Estate vineyard: Kolln Vineyards.
Wines regularly bottled: White, Red, Rosé Table Wines; Apple Wine.

Komes Ranch

Napa County, California.

Flora Springs estate Chardonnay, Cabernet Sauvignon, Cabernet Franc and Merlot Napa Valley vineyard.

KONOCTI WINERY

Founded 1974, Kelseyville, Lake County, California.

Storage: Oak and stainless steel. Cases per year: 40,000.

History: In 1974, twenty-six growers decided to form a cooperative winery venture. Previously, they had been selling their grapes to premium wineries in Sonoma, Napa and Mendocino Counties. They custom crushed their first wine, a Cabernet Sauvignon, at the Robert Mondavi Winery in 1974. In 1979, they built the present winery facility, Please. The vineyards are owned and operated by grower-members of the cooperative. Parducci is also part owner.

Estate vineyard: Isabel Downs Vineyard, Richard Tennis Vineyard plus 24 other vineyards.

Wines regularly bottled: Estate bottled, vintage-dated Cabernet Sauvignon, Fumé Blanc, Johannisberg Riesling, Zinfandel, Cabernet Sauvignon Blanc. Occasionally bottles White Zinfandel, Zinfandel Alegre.

F. KORBEL AND BROTHERS

Founded 1882, Guerneville, Sonoma County, California.

Storage: Oak and stainless steel. Cases per year: 700,000.

History: Founded by the Korbel Brothers, who were originally in the redwood logging business in the Russian River area. Once the trees were gone, they planted vines to begin their famous Champagne winery. (See Adolf L. Heck biography.)

Estate vineyard: Korbel Vineyards.

Wines regularly bottled: Estate bottled, vintage-dated Champagnes (Blanc de Blancs, Blanc de Noir, Natural, Brut, Extra Dry, Sec, Rosé); and Brandy. Occasionally bottles some table wines.

Smudge pot in winter vineyard

HANNS KORNELL CHAMPAGNE CELLARS

Founded 1952, St. Helena, Napa County, California.

Cases per year: 85,000.

History: Champagne Master Hanns Kornell is another wonderful American success story. Arriving from Germany in 1940, he started his cellar 12 years later. Hanns is a third generation Champagne maker.

Wines regularly bottled: Vintage-dated Champagnes (Sehr Trocken, Brut, Extra Dry, Demi-Sec, Rosé, Rouge, Muscat Alexandria.)

KOSSOF (See Barengo/ Lost Hills Vineyard.)

Charles Krug Estate Vineyard

Napa County, California.

Cabernet Sauvignon, Sauvignon Blanc, Chardonnay, Gamay Beaujolais, Pinot Noir, Gewurztraminer, Merlot, Johannisberg Riesling, Chenin Blanc, Grey Riesling and Muscat Canelli Napa Valley vineyard.

CHARLES KRUG WINERY

Founded 1861, St. Helena, Napa County, California.

History: The oldest operating winery in the Napa Valley. Charles Krug established his winery in 1861. The winery is a historic landmark. Since 1943, it has been owned by C. Mondavi and Sons.

Estate vineyard: Charles Krug Vineyard.

Wines regularly bottled: Estate bottled, vintage-dated Fumé Blanc, Chardonnay, Zinfandel, Gamay Beaujolais, Pinot Noir, Johannisberg Riesling, Grey Riesling, Gewurztraminer, Muscat Canelli. Also Charles Krug Vintage Selection Cabernet Sauvignon. Also Chablis and Burgundy.

Second label: C.K. Mondavi.

Kruger Estate Vineyard

Saline County, Missouri.

Concord, Catawba, Niagara and Riesling vineyard.

KRUGER'S WINERY AND VINEYARDS, LTD.

Founded 1977, Nelson, Saline County, Missouri.

History: The family had been making wine since coming from Germany in 1855.

Estate vineyard: Kruger Vineyards.

Wines regularly bottled: Estate bottled Concord, Catawba, Niagara, Riesling.

THOMAS KRUSE WINERY

Founded 1971, Gilroy, Santa Clara County, California.

Storage: Oak and stainless steel. Cases per year: 4,000.

Estate vineyard: Besson Family Vineyard, John Bates Vineyard.

Wines regularly bottled: Vintage-dated French Colombard, Chardonnay, Zinfandel, Zinfandel Rosé, Cabernet Sauvignon; and Gilroy White Table Wine, Gilroy Red Table Wine. Occasionally Grignolino and Carignane.

Second label: Aptos Vineyards, Janaca Vineyards.

LUCIE KUHLMAN

A French hybrid developed by Lucie Kuhlman's father, Eugene Kuhlman, in Colmar, France. A sister seedling to the Foch and Millot. Produces a light neutral red wine.

LA ABRA FARM AND WINERY, INC.
Founded 1973, Lovingston, Nelson County, Virginia.
Storage: Stainless steel. Cases per year: 2,000.
Estate vineyard: Mountain Cove Vineyards.
Wines regularly bottled: Estate bottled, vintage-dated proprietary Villard Blanc, Baco Noir, Skyline Red and White, Dry Red (All under Mountain Cove Vineyard Label.)

La Buena Vida Estate Vineyard
Parker County, Texas.
Rayon D'Or, Vidal, Chambourcin vineyard.

LA BUENA VIDA VINEYARDS
Founded 1978, Springtown, Parker County, Texas.
Storage: Stainless steel. Cases per year: 8,000-10,000.
History: The winery is in Springtown, a dry precinct; wine can be made there, but cannot be sold on the grounds. Retail and tasting room is in a Fort Worth suburb 15 miles south of the winery, in a wet precinct.
Estate vineyard: La Buena Vida Vineyards.
Label indicating non-estate vineyard: Brazos Valley Vineyards.
Wines regularly bottled: Estate bottled, vintage-dated Rayon D'Oro, Vidal Blanc, Chambourcin, Reserve Port, Vintage Port; proprietary Texas Gold Premium Red and White and Select Rosé, Vida del Sol; also Blanc de Blanc (vintage and non-vintage). Occasionally Vintage Texas Port, Vida del Sol Red Dessert Wine.

LA CASONA (See Cavas de San Juan.)
Blanco, Tinto.

La Chripada Estate Vineyard
Rio Arriba County, New Mexico.
Baco Noir, Leon Millot, Marechal Foch, White Riesling, Chancellor Noir vineyards.

LA CHIRIPADA WINERY
Founded 1981, Dixon, Rio Arriba County, New Mexico.
Storage: Oak and stainless steel. Cases per year: 1,000-5,000.
Estate vineyard: La Chiripada Vineyards.
Wines regularly bottled: Vintage-dated Baco Noir Rosé, Leon Millot, Marechal Foch, Chancellor Noir, Baco Noir, Riesling; also vintage Blanc de Blancs, Claret and Rio Arriba County Blanco.

LA CREMA VINERA
Founded 1979, Petaluma, Sonoma County, California.
Storage: Oak and stainless steel. Cases per year: 10,000.
Label indicating non-estate vineyard: Winery Lake Vineyards, Ventana Vineyards, Vineburg Vineyards, Dulton Vineyards, Arendell Vineyards.
Wines regularly bottled: Vintage-dated Chardonnay, Pinot Noir. Occasionally bottles Vin Gris de Pinot Noir, Gonzo Chardonnay.
Second label: Petaluma Cellars.

La Cresta and Long Valley/Mira Monte Vineyards
Riverside County, California.
Cabernet Sauvignon vineyard.

La Estancia Vineyard
Monterey County, California.
Chardonnay, Johannisberg Riesling and Pinot Noir vineyard near Gonzales.

La Gae Vineyard
Clackamas County, Oregon.
Pinot Noir and Chardonnay Willamette Valley vineyard.

La Jota Vineyard Company
Napa County, California.
Zinfandel Napa Valley vineyard.

La Laguna Vineyard
Coahiula, Mexico.
Rubirard, Carignane vineyard.

La Loma and Las Amigas Vineyards
Napa County, California.
Louis Martini estate Chardonnay, Pinot Noir and Gamay Beaujolais Carneros

LA MONTANA
(See Martin Ray Vineyards.)

La Petite Etoile Vineyard
Sonoma County, California.
Sauvignon Blanc vineyard.

La Presa Vineyard
Santa Barbara County, California.
Copenhagen Cellars estate Chenin Blanc, Cabernet Sauvignon and Chardonnay Santa Ynez Valley vineyard.

La Reine Vineyard
Monterey County, California.
Chardonnay vineyard.

La Ribera Ranch
Mendocino, California.
Cabernet Sauvignon, Sauvignon Blanc and Chardonney vineyard.

LA ROCCA WINE COMPANY
Founded 1968, Fayetteville, Cumberland County, North Carolina.
Storage: Stainless steel. Cases per year: 3,000.
Wine regularly bottled: Vintage-dated Muscadine.

La Vina Estate Vineyard
Chamberio County, New Mexico.
Cabernet Sauvignon, Ruby Cabernet, Zinfandel, Johannisberg Riesling, French Colombard, Carignane vineyard.

LA VINA WINERY
Founded 1977, Chamberino, Chamberino County, New Mexico.
Storage: Oak and stainless steel. Cases per year: 3,000-4,000.
History: Winery is in an area where vineyards once flourished, the fertile Mesilla Valley of the Rio Grande.
Estate vineyard: La Vina Estate.
Wines regularly bottled: Johannisberg Riesling, Zinfandel Rosé, Carignane, Ruby Cabernet, Zinfandel; also semi-generics: Chablis, Burgundy; also Blanco and Vin Rosé.

LABRUSCA
Vitis Labrusca is a grape species native to America. Concord, Catawba and Niagara are examples of American hybrids with strong Labrusca qualities.

Lafayette Vineyard
Yamhill County, Oregon.
Chateau Benoit estate White Riesling and Muller-Thurgau Willamette Valley vineyard.

Lagiss Vineyard
Livermore County, California
Livermore Valley Cellars estate Riesling, Colombard, Chardonnay, Pinot Blanc, Golden Chasselas and Servant vineyard.

LAIRD VINEYARD
Founded 1980, San Rafael, Napa County, California.
Storage: Oak. Cases per year: 10,000.
Estate vineyard: Bayview Vineyards.
Wines regularly bottled: Estate bottled, vintage-dated Chardonnay, Cabernet Sauvignon.

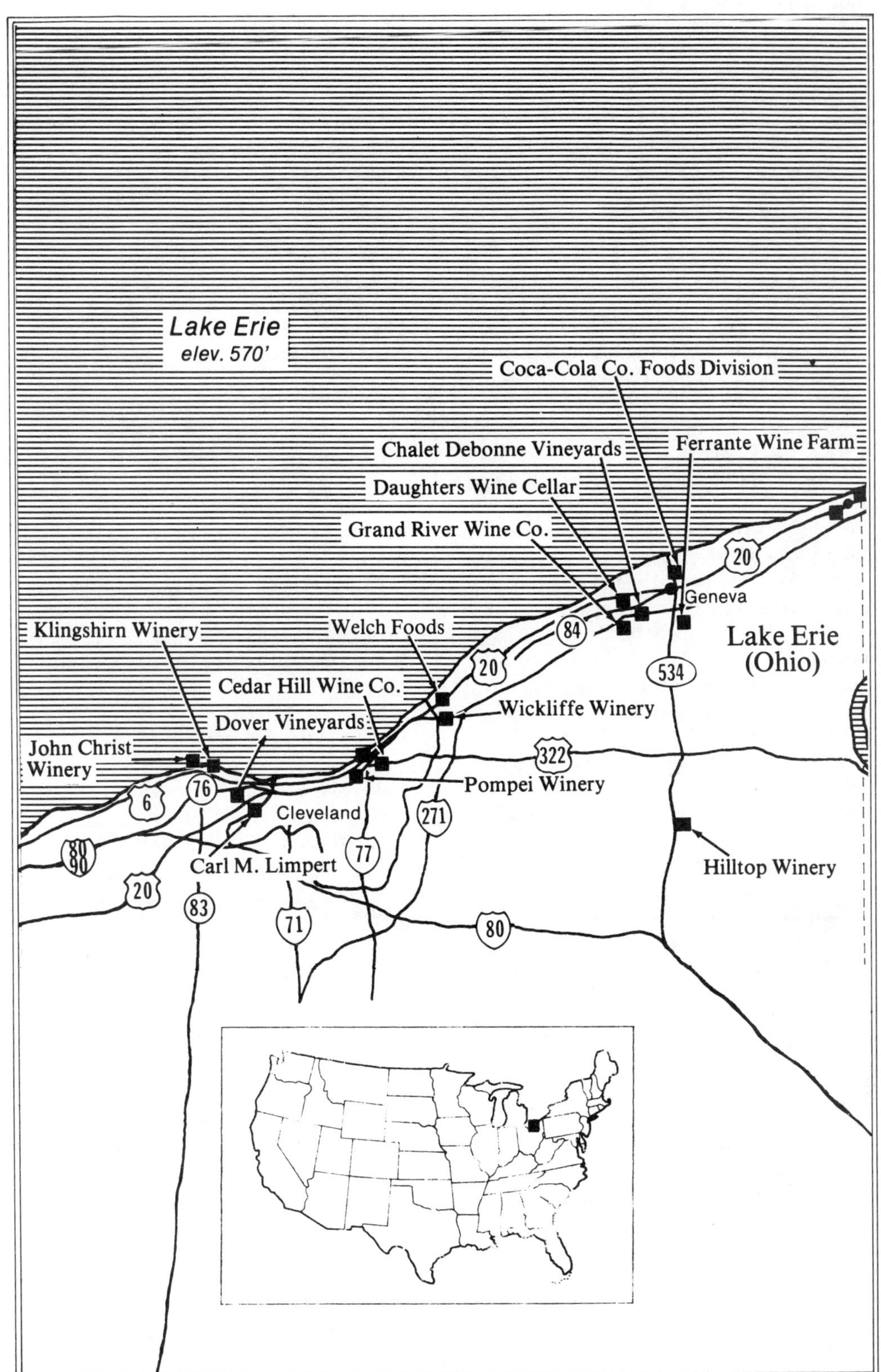
Lake Erie
elev. 570'
Coca-Cola Co. Foods Division
Chalet Debonne Vineyards
Ferrante Wine Farm
Daughters Wine Cellar
Grand River Wine Co.
20
Geneva
Klingshirn Winery
Welch Foods
84
Lake Erie
(Ohio)
20
534
Cedar Hill Wine Co.
Wickliffe Winery
Dover Vineyards
John Christ
Winery
322
76
6
Pompei Winery
Cleveland
271
80
90
77
Carl M. Limpert
Hilltop Winery
20
83
71
80

LAKE, DAVID
David Lake began his wine career in Great Britain, where he worked for ten years prior to coming to the United States. Following a year at the University of California, Davis, in the Department of Viticulture and Enology, Lake spent another year as a consultant to several Oregon wineries, including Eyrie Vineyards and Amity Vineyards. In 1979, he joined A.V. Winery as winemaker. Lake is the only member of the British Institute of Masters of Wine currently making wine in America.

LAKE ERIE VITICULTURAL AREA
The viticultural area in the states of New York, Pennsylvania and Ohio to be known as "Lake Erie" is located on the lake plain bordering the southern and eastern shores and on the island archipelago of Lake Erie. There are approximately 40,000 acres of commercial vineyards of one acre or more located in every county, except Sandusky, along the lakeshore from near Toledo, Ohio to south of Buffalo, New York.

The area encompasses the boundaries of Isle St. George viticultural area in Ohio and overlaps with the boundaries of the Grand River Valley viticultural area in Ohio.

LAKE MICHIGAN SHORE VITICULTURAL AREA
The 2,000 square-mile area encompasses the counties of Berrien and Van Buren and portions of Allegan, Kalamazoo and Cass Counties. The area includes 14,472 acres of grapes.

Lakeshore Estate Vineyard
Seneca County, New York.
Chardonnay, Riesling, Gewurztraminer and Cabernet Sauvignon Finger Lakes vineyard.

LAKESHORE WINERY
Founded 1978, Romulus, Seneca County, New York.
Cases per year: 1,200.
Estate vineyard: Lakeshore Vineyard.
Wines regularly bottled: Estate bottled Chardonnay, Riesling, Gewurztraminer; and Cabernet Sauvignon. Occasionally Baco Noir.

Lakeside Estate Vineyard
Berrien County, Michigan.
Concord, Niagara, Vidal Blanc, Seyval Blanc, Chardonnay and Johannisberg Riesling vineyard.

LAKESIDE VINEYARD, INC.
Founded 1934, Harbert, Berrien County, Michigan.
Storage: Oak.
History: One of Michigan's first wineries that was originally known as the Molly Pitcher Winery.
Estate vineyard: Lakeside Vineyard.
Wines regularly bottled: Estate bottled Concord, Niagara, Vidal Blanc, Seyval Blanc, Johannisberg Riesling, Chardonnay.
Second label: Leonardo Da Vino; Molly Pitcher; Olson Family Wine Cellars.

Lakespring Estate Vineyard
Napa County, California.
Chardonnay Napa Valley vineyard.

LAKESPRING WINERY
Founded 1980, Napa, Napa County, California.

Storage: Oak and stainless steel. Cases per year: 15,000.
Estate vineyard: Lakespring Vineyard.
Wines regularly bottled: Vintage-dated Chenin Blanc, Sauvignon Blanc, Chardonnay, Merlot, Cabernet Sauvignon.

Ronald Lamb Estate Vineyard
Santa Clara County, California.
Chardonnay vineyard.

RONALD LAMB WINERY
Founded 1976, Morgan Hill, Santa Clara County, California.
Storage: Oak and stainless steel. Cases per year: 750.
Estate vineyard: Ronald Lamb Vineyard.
Label indicating non-estate vineyard: Ventana Vineyards, George Brown Vineyards, Upton Ranch.
Wines regularly bottled: Vintage-dated Chenin Blanc, Gamay Beaujolais, Zinfandel, Johannisberg Riesling.

Lambert Bridge Estate Vineyard
Sonoma County, California.
Chardonnay, Cabernet Sauvignon, Merlot Dry Creek Valley vineyard.

LAMBERT BRIDGE
Founded 1975, Healdsburg, Sonoma County, California.
Storage: Oak and stainless steel. Cases per year: 15,000.
Estate vineyard: Lambert Vineyards.
Wines regularly bottled: Vintage-dated Chardonnay, Cabernet Sauvignon, Merlot.

LA MONT WINERY INC.
Founded 1946, Di Giorgio, Kern County, California.
Storage: Oak and stainless steel. Cases per year: 2,000,000.
Wines regularly bottled: Chenin Blanc, French Colombard, Ruby Cabernet, Cabernet Sauvignon; generic Chablis, Rhine, Burgundy, Rosé.
Second label: Ambassador, Guasti, Mountain Gold.

Lancaster County Estate Vineyard
Lancaster County, Pennsylvania.
Aurora, Seyval, Vidal, Rayon D'Or, Cascade, Chancellor, Foch, Chelois vineyard.

LANCASTER COUNTY WINERY, LTD.
Founded 1979, Willow Street, Lancaster County, Pennsylvania.
Storage: Stainless steel. Cases per year: 3,800.
History: The winery is established on a land grant dating back to 1735. Old stone farmhouse (1819) is on property.
Estate vineyard: Lancaster County Winery Vineyards.
Wines regularly bottled: Estate bottled Seyval Blanc, Vidal Blanc, Foch, Cascade, Chancellor; proprietary Colonial Concord, Colonial Spiced Apple; semi-generic White Chablis, Pink Chablis; and Dry Red and Dry White Wine Special Reserve.

LANCASTER VALLEY VITICULTURAL AREA
The area is located in southeastern Pennsylvania, primarily in Lancaster County, with a small portion extending into Chester County. Lancaster Valley is approximately 31 miles long, 12 miles wide, and contains approximately 225,000 acres.

LANDEY (See Conestoga Vineyards.)
Vidal Blanc, Catawba Rosé, Dutchess, Seyval Fleur.

Landey Vineyard
Lancaster County, Pennsylvania.
Conestoga estate Vidal Blanc, Chambourcin and Seyval Blanc Lancaster Valley vineyard.

Landmark Estate Vineyard
Sonoma County, California.
Cabernet vineyard in Alexander Valley; Chardonnay vineyards in Alexander Valley, Russian River Valley, Sonoma

LANDMARK VINEYARDS

Founded 1974, Windsor, Sonoma County, California.

Storage: Oak and stainless steel. Cases per year: 22,000.

Estate vineyard: Landmark Vineyards.

Label indicating non-estate vineyard: Gauer Ranch.

Wines regularly bottled: Estate bottled, vintage-dated Chardonnay, Cabernet Sauvignon. Also produce a White Table Wine. Occasionally bottles Zinfandel.

LANDOT

A French hybrid grape producing light, fruity red wine.

LANGBEHN, LARRY

After being graduated with a bachelor's degree in chemistry, from the University of California, Davis, Larry Langbehn became an environmental toxicology chemist. Five years later, Langbehn decided he needed a career change, and he returned to U.C. Davis for graduate training in viticulture and enology. In 1975, he worked in the Napa Valley under Jerry Luper and Brad Webb, at Freemark Abbey Winery, in an internship program. When Luper left the winery five months later, Langbehn was asked to become winemaker trainee. He was soon made winemaker, a position he has continued to hold at the winery. He is responsible, not only for operation of the winery, but, for the selection and picking schedule of grapes grown at the Freemark Abbey vineyards. Langbehn is most proud of his 1976 Edelwein Gold.

FRANZ WILHELM LANGGUTH WINERY

Founded 1982, Mattawa, Grant County, Washington.

Storage: Stainless steel.

History: In 1789 Franz Wilhelm Langguth started making wine near Trier in Germany. Seven generations have carried on this winemaking tradition. In 1982, after years of searching for an American vineyard, he decided on the Columbia River basin. The new winery is under the direction of Johann Wolfgang Langguth.

Estate vineyard: Weinbau Vineyards.

Wines regularly bottled: Estate bottled, vintage-dated Johannisberg Riesling, White Riesling, Gewurztraminer; vintage-dated Johannisberg Riesling, White Riesling, Gewurztraminer, Sauvignon Blanc, Cabernet Sauvignon.

MARIO LANZA WINES (See Wooden Valley Winery.)

Pinot Chardonnay, French Colombard, Johannisberg Riesling, Riesling, Green Hungarian, Gewurztraminer, Sauvignon Blanc, Pinot Noir Blanc, Chenin Blanc, Zinfandel Blanc, Malvasia Bianca, Cabernet Rosé, Gamay Beaujolais, Pinot Noir, Cabernet Sauvignon, Zinfandel.

LAPIC WINERY

Founded 1977, New Brighton, Beaver County, Pennsylvania.

Storage: Stainless steel.

Wines regularly bottled: Seyval Blanc, Baco Noir, Concord, Delaware White, Pink Catawba; and proprietary Valley Red and White, Daugherty Red and White.

Largo Vineyard

Mendocino County, California.

Parducci estate Chardonnay, Sauvignon Blanc and Chenin Blanc vineyard.

Larson Vineyards
Napa County, California.
Chardonnay Carneros vineyard.

Las Tables Estate Vineyard
San Luis Obispo County, California.
Zinfandel Paso Robles vineyard.

LAS TABLAS WINERY
Founded 1865, Templeton, San Luis Obispo County, California.
Storage: Stainless steel. Cases per year: 3,000.
Estate vineyard: Las Tablas.
Wines regularly bottled: Zinfandel. Also Rosé and White Table Wine.

LATAH CREEK WINE CELLARS LTD.
Founded 1983, Spokane, Spokane County, Washington.
Storage: Oak and stainless steel. Cases per year: 12,000.
Estate vineyard: Hogue Ranches.
Wines regularly bottled: Estate bottled, vintage-dated Johannisberg Riesling, Chenin Blanc, Chardonnay, Semillon, Fumé Blanc, Gewurztraminer, Pinot Noir Rosé, May Wine.

late harvest
Equivalent to the Spätlese wines of Germany. Johannisberg Riesling or Gewurztraminer grapes are picked in an over-ripe condition with some incidence of Botrytis. The German word Spätlese means "late picking". By law must have a minimum of 24 degrees Brix at harvest.

Laurel Glen Estate Vineyard
Sonoma County, California.
Cabernet Sauvignon vineyard.

LAUREL GLÈN VINEYARD
Founded 1980, Glen Ellen, Sonoma County, California.
Cases per year: 5,000.
Estate vineyard: Laurel Glen Vineyard.
Wines regularly bottled: Estate bottled, vintage-dated Cabernet Sauvignon.

A simple "bell cap" corkpuller

Laurel Vineyard
Sonoma County, California.
Balverne estate Cabernet Sauvignon, Cabernet Franc and Merlot Chalk Hill vineyard.

LAUREL HILL VINEYARD
Founded 1984, vineyard 1980, Henryville, Lawrence County, Tennessee.
Storage: Stainless steel. Cases per year: 2,200.
Estate vineyard: Laurel Hill Vineyard.
Wines regularly bottled: Estate bottled Vidal Blanc, Marechal Foch 1985, also estate bottled Cabernet Sauvignon, Chardonnay.

LAWRENCE TREVILLE de (See de Treville Lawrence.)

Lawrence Estate Vineyard
Ulster County, New York.
Benmarl estate Hybrids Hudson River Region vineyard.

Layne Vineyards
Josephine County, Oregon.
Cabernet Sauvignon, Chardonnay and Merlot vineyards.

Lazy Creek Estate Vineyard
Mendocino County, California.
Chardonnay, Pinot Noir, Gewurztraminer Anderson Valley vineyard.

LAZY CREEK VINEYARD
Founded 1973, Philo, Mendocino County, California.
Cases per year: 1,000.
Estate vineyard: Lazy Creek Vineyards.
Wines regularly bottled: Estate bottled, vintage-dated Chardonnay, Pinot Noir, Gewurztraminer.

Le Baron Vineyard
Sonoma County, California.
Sonoma/Windsor estate Johannisberg Riesling Russian River Valley vineyard.

L. LE BLANC VINEYARDS (See California Growers Winery.)

Le Blanc Vineyards
Napa County, California.
Sauvignon Blanc and Zinfandel Napa Valley vineyard.

Leaky Lake Vineyard
Napa County, California.
Cabernet Sauvignon, Chardonnay, Gamay Rosé Napa Valley vineyard.

LEE, MICHAEL J.
Michael J. Lee was born and reared in San Francisco. After graduation from California State University, San Francisco, with a degree in marketing, Lee and five partners purchased Kenwood Vineyards in 1970. He has served as general manager of the Sonoma County winery since that time, in addition to working as Cellarmaster and Production Manager. His winemaking education has come through direct experience and extension courses in enology at the University of California, Davis. He was appointed winemaker at Kenwood in 1982.

Lee Vineyard
Napa County, California.
Pinot Noir Carneros vineyard.

LEELANAU PENINSULA VITICULTURAL AREA
The area is a triangular-shaped peninsula, in the northwestern portion of Michigan's lower peninsula, and consists of the mainland portion of Leelanau County, excluding the offshore islands.

LEELANAU WINE CELLARS LTD.
Founded 1975, Omena, Leelanau County, Michigan.
Storage: Stainless steel. Cases per year: 12,000.
Estate vineyard: Leelanau Peninsula.
Wines regularly bottled: Vintage-dated Vignoles, DeChaunac Rosé, Merlot, Cabernet Sauvignon, Chardonnay; also Aurora, Baco Noir; Fruit Wines; and Leelanau Red, White Rosé. Occasionally bottles Festival Cherry.

lees
Yeast sediment, crystalized excess tartaric acid, and pigment deposited by wine in the storage vats after fermentation.

LEEWARD
Founded 1978, Ventura, Ventura County, California.
Storage: Oak and stainless steel. Cases per year: 7,500.
Label indicating non-estate vineyard: MacGregor Vineyard, Ventana Vineyard, Bien Nacido Vineyard.
Wines regularly bottled: Vintage-dated Chardonnay, Cabernet Sauvignon, Zinfandel, Pinot Noir Blanc. Occasionally bottle Sauvignon Blanc.

LEFRANC (See Thée & Lefranc.)

LEGAUX, PIERRE
In 1793, Pierre Legaux planted a vineyard along the banks of the Schuylkill River, a short distance from Philadelphia. His vineyard became a private corporation, The Pennsylvania Vine Company, and many prominent men of the time were shareholders, including Alexander Hamilton and Aaron Burr.

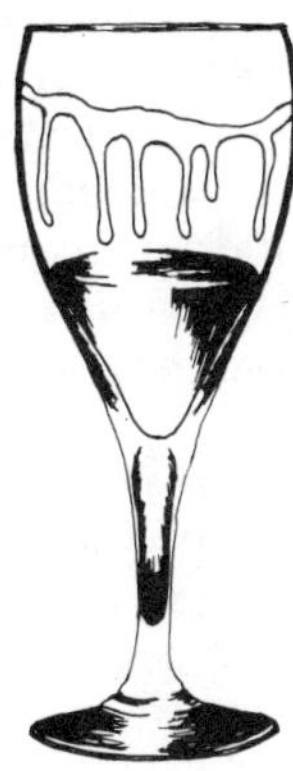

"Legs"

legs
Tasting term.
Legs (also called tears) are caused by a surface tension phenomenon between the alcohol and the glass, not the glycerin. The higher the alcohol the longer the "legs" which run down the inside of the glass after the wine has been swirled.

Lembo Estate Vineyards
Mifflin County, Pennsylvania.
Chelois, DeChaunac and Seyval vineyard.

LEMBO VINEYARDS
Founded 1972, Mifflin County, Pennsylvania.
Storage: Stainless steel. Cases per year: 12,000.
Estate vineyard: Lembo Vineyards.
Wines regularly bottled: Estate bottled Seyval Blanc, Rosé, proprietary Chateau Antoinette, Lembruschini, Country Red.

Lemon Creek Vineyard
Berrien County, Michigan.
Vidal vineyard.

Lemon Ranch
Napa County, California.
Cabernet Sauvignon vineyard.

Lemon/Chabot Vineyard
Napa County, California.
Cabernet Sauvignon vineyard.

LENOIR
A native American grape, of the southeast, producing medium bodied, light red wine.

Lenz Estate Vineyard
Suffolk County, New York.
Gewurztraminer, Merlot, Cabernet Sauvignon, Pinot Noir and Chardonnay vineyard.

LENZ VINEYARDS (VINEYARDS OF PATRICIA AND PETER LENZ)
Founded 1978, Peconic, Suffolk County, New York.
Storage: Oak and stainless steel. Cases per year: 5,000.
Estate vineyard: Lenz Vineyard.
Wines regularly bottled: Estate bottled, vintage-dated Gewurztraminer (dry Alsatian style), Merlot/Cabernet Sauvignon, Pinot Noir, Chardonnay, Sparkling Wine.

LEON MILLOT
A French hybrid grape that produces medium body, fruity red wine similar to the Foch.

LEONETTI CELLAR
Founded 1977, Walla Walla, Washington.
Storage: Oak. Cases per year: 500-700.
Estate vineyard: Leonetti Vineyard.
Label indicating non-estate vineyard: Sagemoor Vineyards.
Wines regularly bottled: Estate bottled, vintage-dated Merlot; vintage-dated Cabernet Sauvignon.

Leonetti Cellar Estate Vineyard
Walla Walla, Washington.
Cabernet Sauvignon, Merlot Walla Walla Valley vineyard.

LES COLOMBES (See Cote des Colombes Vineyard.)

Les Pierres Vineyard
Sonoma County, California.
Sonoma-Cutrer estate Chardonnay Sonoma Valley vineyard.

LES VIGNOBLES CHANTECLER LTEE
Founded 1971, Rougemont, Quebec, Canada.
Storage: Stainless steel. Cases per year: 200,000.
Wines regularly bottled: Proprietary Rossini, Kastle Wien, Plaisir D'Amour, Bellini. Occasionally Vin de France.

LES VIGNOBLES DU QUEBEC (VINIFICATION) INC.
Founded 1973, Hemmingford, Quebec, Canada.
Storage: Stainless steel. Number of cases produced per year: 150,000.
Wines regularly bottled: Proprietary Petit Prince Red and White, Seigneur de Beaujeu.

LETT, DAVID
David Lett was graduated, with a bachelor's degree in philosophy and pre-medicine in 1961, from the University of Utah. After working briefly with Lee Stewart at Souverain, Lett enrolled at the University of California, Davis. He obtained another bachelor's degree in 1963, this time in viticulture, with a minor in enology. While at Davis, Lett developed an interest in the Pinot Noir grape, and, following graduation, spent nine months in Europe investigating the climate and soil requirements of this varietal. He returned to the United States in search of a region appropriate for Pinot Noir growth. In 1965. he and his wife, Diana, established The Eyrie Vineyards in the Willamette Valley of western Oregon. Lett found the southern slope and light soil close to the conditions in French Pinot Noir vineyards, causing the grapes to "struggle" and develop character during their slow maturation. In addition to Pinot Noir, The Eyrie Vineyards also produce Chardonnay, Pinot Gris, Muscatel Ottonel and Pinot Meunier. Lett planted vinifera grapes in the Willamette Valley, and recognized its potential for cool climate varietals. His Pinot Noirs have helped to establish Oregon as a major wine producing region. Lett is a founding member of the Oregon Winegrower's Association, and a member of the Oregon Wine Advisory Board and the American Society of Enologists.

Leveroni Vineyards
Sonoma County, California.
Chardonnay vineyard.

LIBERTY SCHOOL (See Caymus Vineyard.)
Cabernet Sauvignon (vintage-dated).

LICHINE, ALEXIS (1913-)
Wine importer, expert, and writer/encyclopedist.

> "When it comes to wine, I tell people to throw vintage charts out the window and invest in a corkscrew. The best way to learn about wines is in the drinking."

light ruby
A red with a limited blue presence, and of a lighter shade.

light straw
Color of very light yellow.

light wines
Light table wines must be under 10% alcohol. If caloric content is stated, the protein, carbohydrate and fat content must be stated on the label. If the label indicates "reduced in calories", it must be 25% lower in calories than a similar regular product of the same company.

Light wines are produced by two processes, or by a combination of both:

1. Grapes are picked at full ripeness and fermented normally. The wine is then subjected to a flash evaporation process, which evaporates much of the alcohol away. Normal wine can then be added back to bring the alcohol level to any point between zero and normal. The flash evaporation may be done in a centrifuge-vacuum machine, or in a standard "vacuum pan".
2. The grapes are picked before being fully ripe and fermentation is stopped before it is complete so that there is less alcohol, naturally.

There are disadvantages to both methods. Fully ripe grapes usually produce the most natural flavor—but much of this flavor is lost with the alcohol when it is evaporated away. Under-ripe grapes usually have less natural flavor, but more is lost in the process since there is no heating or evaporation. Probably the best light wines are produced from a combination of both processes. Process #1 can be used to produce totally alcohol-free wine.

LIMBERGER/LEMBERGER
A vinifera grape that produces a Gamay style red wine. Used as a varietal in Austria and Hungary. Also Germany (Wurthenberg). Successfully grown in Yakima Valley, Washington State.

Lime Kiln Valley Vineyard
San Benito County, California.
Enz estate Pinot St. George, Sauvignon Blanc, Zinfandel, Orange Muscat and French Colombard vineyard.

LIME KILN VALLEY VITICULTURE AREA
This area is approximately 2,300 acres in San Benito County, California.

LINGANORE VALLEY VITICULTURAL AREA
The viticultural area lies east of the town of Frederick, and encompasses an area of approximately 57,000 acres of Maryland.

LIQUEUR DE TIRAGE
A solution of sugar and wine added to the champagne cuvée to induce second fermentation.

LITTLE AMANA WINERY, INC.
Amana, Iowa County, Iowa.
Storage: Oak.
Wines regularly bottled: Fruit and Berry Wines: Grape, Rhubarb, Cherry, Apricot, Plum, Dandelion, Strawberry, Raspberry, Cranberry, Blueberry, Blackberry.

Little Vineyard Estate Vineyard
Bucks County, Pennsylvania.
Seyval, Vidal, Cayuga, Niagara, Concord, Chambourcin, Chancellor, DeChaunac, Chardonnay vineyard.

THE LITTLE VINEYARD
Founded 1982, Quakertown, Bucks County, Pennsylvania.
Storage: Oak and stainless steel. Cases per year: 400.
Estate vineyard: The Little Vineyard.
Wines regularly bottled: Seyval, Vidal, Cayuga, Niagara, Concord, Chambourcin, Chancellor, DeChaunac, Chardonnay (vintage); also Country Red, White, Rosé.

LIVE OAKS WINERY
Founded 1912, Gilroy, Santa Clara County, California.

Storage: Oak and stainless steel. Cases per year: 50,000.

Label indicating non-estate vineyard: Rampoldi Ranch.

Wines regularly bottled: Grenache Rosé, Chenin Blanc (sweet, medium, extra dry), Haut Sauterne, Sauterne and Burgundy.

LIVERMORE VALLEY CELLARS

Founded 1978, Livermore, Alameda County, California.

Storage: Oak and stainless steel. Cases per year: 2,000.

Estate vineyard: Lagiss Vineyard.

Wines regularly bottled: Estate bottled, vintage-dated Chardonnay, Grey Riesling, Fumé Blanc, French Colombard, Golden Chasselas.

LIVERMORE VALLEY HISTORY (See Concannon and Wente.)

LIVERMORE VALLEY VITICULTURAL AREA

Livermore Valley is one of the coastal intermountain valleys that surround the San Francisco Bay depression. In the east, the watershed area of Livermore Valley is bounded by the Altamont Hills and Crane Ridge, in the south by Cedar Mountain Ridge and Rocky Ridge, in the west by Pleasanton Ridge and in the north by the Black Hills. The valley's geographic location generally is the area covered by the political boundaries of Murray and Pleasanton townships.

Llano Estacado Estate Vineyard

Lubbock County, Texas.

Chenin Blanc vineyard.

LLANO ESTACADO WINERY

Founded 1974, Lubbock, Lubbock County, Texas.

Storage: Oak and stainless steel. Cases per year: 30,000.

Estate vineyard: Llano Estacado Vineyard.

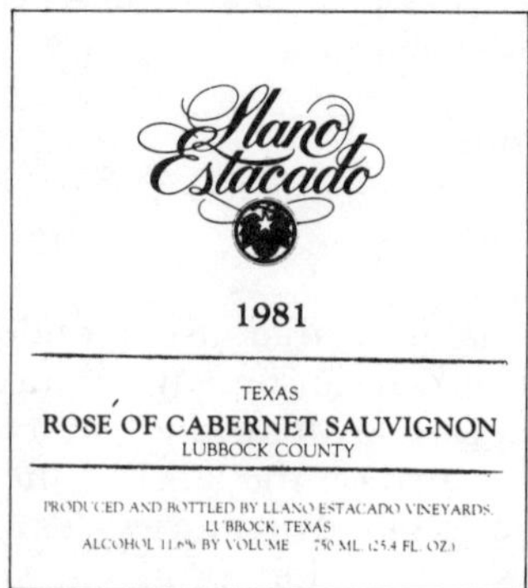

Label indicating non-estate vineyard: Hagens Vineyard; Slaughter Vineyards; Cox Vineyards.

Wines regularly bottled: Estate bottled, vintage-dated Chenin Blanc; vintage-dated Chenin Blanc, Chardonnay, Cabernet Sauvignon, Dry Sauvignon Blanc, Johannisberg Riesling, French Colombard, Zinfandel; and vintage-dated Premium Red and White.

Second label: Staked Plains Winery.

LLORDS & ELWOOD WINERY

Founded 1955, Fremont, Alameda County, California.

Storage: Oak. Cases per year: 15,000.

Wines regularly bottled: Vintage-dated The Rare Chardonnay, Castle Magic Riesling, Rosé of Cabernet, Velvet Hill Pinot Noir; also Great Day D-r-ry Sherry, Dry Wit Sherry, The Judge's Secret Cream Sherry, Ancient Proverb Port (all proprietary wines).

Loewenthal Ranch

Humboldt County, California.

Zinfandel vineyard.

Lolonis Estate Vineyard

Mendocino County, California.

Zinfandel, Chardonnay, Petite Sirah, Sauvignon Blanc, Johannisberg Riesling, Grey Riesling, Cabernet Sauvignon vineyard.

LOLONIS VINEYARDS AND WINERY
Vineyard founded 1920, winery 1982, Mendocino County, California.
Storage: Oak and stainless steel.
History: Tryton Lolonis came to the United States from Greece at age 16 and, through his own hard work, saved enough to start his vineyard in 1920. Two years later he sent for his bride to be, Eugenia, and at age 15, Eugenia came to California to marry Tryton. Today at age 87 she still acts as the force behind this famous vineyard. Up until recently, the vineyard has been providing grapes for award winning wines. In 1982, they also started producing their own wines.
Estate vineyard: Lolonis Vineyards.
Wines regularly bottled: Estate bottled, vintage-dated Chardonnay and Late Harvest Sauvignon Blanc.

Jack London Ranch
Sonoma County, California.
Cabernet Sauvignon and Pinot Noir Sonoma Valley vineyard.

LONDON WINERY LIMITED
Founded 1925, London, Niagara Peninsula, Canada.
Storage: Oak and stainless steel. Cases per year: 300,000.
History: The history of London Winery Limited can be traced to 1871. Canada's first winery, the Hamilton Dunlop winery, was established in Brantford, Ontario by Major J.S. Hamilton, the dean of Canada's wine producers.
In 1874, Major Hamilton became associated with Thaddeus Smith of the Vin Villa Vineyards (established in 1860), located at Point Au Pelee Island, the most southern point in Canada.
In 1888, from the above two companies, the Pelee Island Wine and Vineyard Company was established. The assets of this historic winery were sold to London Winery Limited in the year 1945.
London Winery Limited was established, in 1925, by A.N. Knowles and J.C. Knowles, born in Nassau, Bahamas. The Knowles brothers came to Canada while still in their early twenties.
Estate vineyard: London Vineyard.
Wines regularly bottled: Marechal Foch, Baco Noir; Sparkling wines; Vermouths; Ports and Sherries; Honey wine; Communion Wines; Cider; Table Wines (Red, White, Rosé); proprietary Vinroi Red and White, Londini, Bellevista, DeChaunac, Van Buren, Delaware, Bellevinta Red and White; Chablis.

Lone Oak Vineyard
Monterey County, California.
Pinot Blanc/Late Harvest vineyard.

Lone Oak Vineyard
Sonoma County, California.
Chardonnay vineyard.

LONG ISLAND REGION
The North Fork strip of Long Island, New York is ideally suited for winemaking and grape growing. The growing season is approximately 210 days a year (the same as Bordeaux). It is bounded on three sides by water, which provides temperatures that never go below zero in the winter, and low humidity.

Long Valley Vineyards
Riverside County, California.
Mt. Palomar estate Chardonnay, Sauvignon Blanc, White Riesling, Chenin Blanc, Cabernet Sauvignon, Shiraz, Petite Sirah and Palomino vineyard.

Long Estate Vineyard
Napa County, California.
Chardonnay and Johannisberg Riesling Napa Valley vineyard.

Long Vineyard
Sonoma County, California.
Kalin estate Chardonnay vineyard.

LONG VINEYARDS
Founded 1977, St. Helena, Napa County, California.

Storage: Oak and stainless steel. Cases per year: 1,500.

Estate vineyard: Long Vineyards.

Label indicating non-estate vineyard: University of California, Davis Experimental Vineyards.

Wines regularly bottled: Estate bottled, vintage-dated Chardonnay, Johannisberg Riesling; also vintage-dated Cabernet.

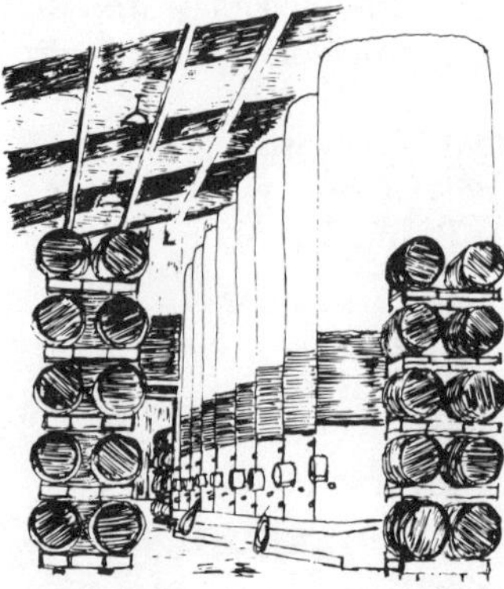

Stainless steel fermentation and oak aging barrels

LONG, ZELMA

Zelma Long was studying for her master's degree at the University of California, Davis, when Mike Grgich asked her to help out during the 1969 harvest at the Robert Mondavi Winery. She found that her undergraduate degree in the sciences, obtained at Oregon State University, was useful in winemaking. She agreed to stay at Mondavi, learning enology under Grgich. When Grgich left in 1972, Long was promoted to head enologist, a position she held for eight years. In 1979, Simi Winery, Inc., hired Long as vice president/winemaker. She is responsible for the selection of grapes, staff, equipment, and winemaking techniques at the winery. She works closely with Simi President Michael Dixon and with Moet Chandon, French owners of the winery, to improve production and style. Simi has recently planted its first vineyards, and plans to expand in the next few years. Long is involved with several wine industry organizations. She is a member and past president of the Napa Valley Wine Technical Group, a former director of the American Society of Enologists, and president of the American Vineyard Association. She is also winemaker at Long Vineyards, a small winery she owns with her former husband.

LONGFELLOW, HENRY WADSWORTH (1807-1882)

American poet.

"This song of mine
Is a song of the vine
To be sung by the glowing embers
Of wayside inns,
When the rain begins
To darken the drear Novembers

For the richest and best
Is the wine of the West
That grows by the Beautiful River;
Whose sweet perfume
Fills all the room
With a benison on the giver

When you ask one friend to dine,
Give him your best wine!
When you ask two,
The second best will do."

LONGORIA, RICK

Rick Longoria became interested in winemaking when he was a student in sociology at the University of California, Berkeley. After graduation in 1973, with a bachelor's degree, Longoria joined the cellar crew of the Buena Vista Winery in Sonoma. While there, he continued his education with viticulture and enology classes at Santa Rosa Junior College. Eighteen months later, he was hired as cellar foreman at the Firestone Vineyards in the Santa Ynez Valley. Then, in 1978, he moved to the Napa Valley, where he worked at Chappellet Vineyards. Longoria returned to the Santa Ynez Valley one year later to become winemaker at the newly established J. Carey Vineyards and Winery, Inc., in Solvang. He has since been producing both white and red table wines, created to enhance good food, for the J. Carey line. In 1982, Longoria experimented with Pinot Noir and Chardonnay from Santa Maria Valley grapes, which he released under his own label. He hopes to slowly expand Longoria Wine Cellars. He is a member of the American Society of Enologists.

RICHARD LONGORIA WINES (See J. Carey Cellars.)

LONGWORTH, NICHOLAS (1782-1863)
Horticulturist.

In the 1850s, the Buckeye State was the leading wine producing area in the United States, thanks to Nicholas Longworth.

Nicholas Longworth started the wine industry in Ohio, around Cincinnati. He discovered that the Catawba and the Isabella were best suited for the soil of these vineyards.

The grapes and the wines he produced were carefully nurtured. The wines from the grapes that he grew became known throughout the world. His wine house in Ohio was visited by many European vintners and winemakers. A successful lawyer and landowner, he was also a noted horticulturist before his interest in grapes and winemaking. Many articles were written by Longworth explaining his experiments with growing crops, what he was attempting to achieve and how he accomplished his purpose.

His best remembered work is on the subject of wine—"A Letter, on the Cultivation of the Grape and Manufacture of Wine" (1846).

Later, disaster struck the Ohio Valley and the thriving vines were wiped out by mildew. Every known remedy was tried to save the vines, and the wine industry of Ohio, but it was to no avail. Ohio is again growing grapes, manufacturing wine, and doing very well with its products. And one and all still talk of the Father of the Ohio Wine Industry—Nicholas Longworth, Horticulturist Extraordinaire.

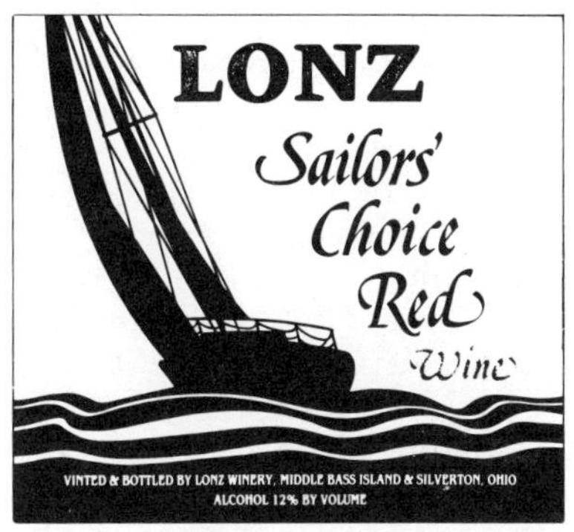

LONZ WINERY

Founded 1860's, Middle Bass Island, Ohio.

Storage: Oak and stainless steel. Cases per year: 20,000.

History: Opened during the Civil War as Golden Eagle Winery. By 1875, it was the largest wine producer in the United States. Winery is located on Middle Bass Island, on Lake Erie, only reachable by boat or small plane. The entrepreneurial president of Lonz, Robert Gottesman, has made the winery an attractive visitors point.

Wines regularly bottled: Baco Noir, Chelois, Blackberry; Champagne; and proprietary Lonzbrusco.

LORAMIE CREEK VITICULTURAL AREA

The area in Shelby County, Ohio is located in west-central Ohio and consists of approximately 3,600 acres in Shelby County, with designated boundaries partly within the townships of Cynthian, Loramie, and Washington.

Los Alamos Vineyard

Santa Barbara County, California.

Pinot Noir, Sauvignon Blanc, Chardonnay vineyard.

LOS ALTOS (See Rappazini.)

LOS HERMANOS (See Beringer.)
Chablis, Burgundy, Vin Rosé, Rhine, Gamay Beaujolais, French Colombard, Chenin Blanc, Cabernet Sauvignon. Light Wines are Chablis, Rosé, French Colombard, Rhine, Chenin Blanc.

Los Ninos Vineyard
Napa Valley, California.
Conn Creek estate Chardonnay and Pinot Noir Napa Valley vineyard.

Los Vignerons Vineyards
Sonoma County, California.
Zinfandel and Semillon Sonoma Valley vineyard.

Los Vinedos Del Rio
Sonoma County, California.
Louis Martini estate Chardonnay, Merlot, Gewurztraminer and Gamay Russian River vineyard.

LOS VINEROS WINERY
Founded 1980, Santa Maria, Santa Barbara County, California.
Storage: Oak and stainless steel. Cases per year: 18,000.
Wines regularly bottled: Vintage-dated Chardonnay, Sauvignon Blanc, Chenin Blanc, Blanc of Cabernet Sauvignon, Gewurztraminer, Pinot Noir, Cabernet Sauvignon. Occasionally bottles Blanc of Pinot Noir.

LOST MOUNTAIN WINERY
Founded 1981, Sequim, Clallam County, Washington.
Storage: Oak. Cases per year: 400.
Label indicating non-estate vineyard: Sagemoor Farms.
Wines regularly bottled: Vintage-dated Cabernet Sauvignon, Pinot Noir, Merlot, Zinfandel, Barbera, Petite Sirah, Red.

Lovin Vineyards
Mendocino County, California.
Johannisberg Riesling Potter Valley vineyard.

LOWDEN SCHOOLHOUSE WINERY
Founded 1983, Lowden, Walla Walla County, Washington.
Storage: Oak and stainless steel. Cases per year: 5,000.
Label indicating non-estate vineyard: Sagemoor Vineyards, Balcolm and Moe Vineyards.
Wines regularly bottled: Vintage-dated Semillon and Merlot. Occasionally bottled Gewurztraminer.

LOWER LAKE WINERY
Founded 1977, Lower Lake, Lake County, California.
Storage: Oak and stainless steel. Cases per year: 5,000.
Label indicating non-estate vineyard: Stromberg Vineyard, Vina Las Lomas Vineyard.
Wines regularly bottled: Vintage-dated Cabernet Sauvignon, Cabernet Sauvignon Reserve, White Cabernet Sauvignon, Fumé Blanc.

LUCAS HOME WINE (See The Lucas Winery.)

Lucas Vineyard
Seneca County, New York.
Lucas estate Cayuga, Ravat, Seyval, Chardonnay, Riesling, Gewurztraminer, DeChaunac, Vidal and Rayon D'Or Finger Lakes vineyard.

LUCAS WINERY
Founded 1980, Interlaken, Seneca County, New York.
Storage: Stainless steel. Cases per year: 600-2,500.
Estate vineyard: Lucas Vineyard.
Wines regularly bottled: Estate bottled Cayuga Semi-Dry, Cayuga Natural; estate bottled proprietary Interlaken Red, Tugboat Red, Tugboat White.

The Lucas Winery Estate Vineyard
San Joaquin County, California.
Zinfandel vineyard.

THE LUCAS WINERY
Founded 1977, Lodi, San Joaquin County, California.
Storage: Oak. Cases per year: 1,000.
Estate vineyard: The Lucas Vineyard.
Wines regularly bottled: Estate bottled, vintage-dated Zinfandel.
Second label: Lucas Home Wine.

LUCIA, DR. SALVATORE P. (1901-1984)
Physician, professor emeritus of medicine, University of California Medical Center (San Francisco), health authority.

"Among the dietary beverages used by man through the centuries, the next in importance to milk has been wine.

"Wine was born, not invented.

"Like any old friend, it continues to surprise us in new and unexpected ways."

Fred Luke Vineyards
Lake Erie County, Pennsylvania.
Penn Shore estate Hybrids and Native American vineyards.

LUPER, JERRY
Jerry Luper began his training as a winemaker while he was a chemistry student at Modesto Junior College. He worked for two years at Gallo as a wine analyst, participating in extensive tastings, routine analysis, and development of experimental winemaking techniques. In 1962, Luper joined the U.S. Army, adding Czechoslovakian to his knowledge of Spanish, French and German by studying at the Army Language School in Monterey, California. After another year at Gallo, he then accepted a position at Shell Oil Research before returning to Fresno State College, where he earned a degree in enology. Luper then went to work for the Louis M. Martini Winery as a cellarman, and, in 1970, joined the new Freemark Abbey Winery as winemaker. In 1976, he left the winery for a year of travel and study in Europe with his family. When he returned to the United States, Luper was named winemaker at Chateau Montelena. In 1982, he became partner and general manager of Chateau Bouchaine in the Napa Valley. He is responsible for all stages of winemaking at the new winery, and is concentrating on the production of Pinot Noir and Chardonnay. The first wines will be released in 1984.

Lyeth Estate Vineyard
Sonoma County, California.
Cabernet Sauvignon, Cabernet Franc, Merlot, Malbec, Sauvignon Blanc, Semillon Alexander Valley vineyard.

LYETH VINEYARD AND WINERY LIMITED

Founded 1981, Geyserville, Sonoma County, California.

Storage: Oak and stainless steel. Cases per year: 30,000.

Estate vineyard: Lyeth Vineyard.

Wines regularly bottled: Estate bottled, vintage-dated Lyeth (blend of Cabernet Sauvignon, Merlot, Malbec and Cabernet Franc), Lyeth Blanc (Sauvignon Blanc and Semillon), and Vintage Port.

LYNFRED WINERY

Founded 1975, Roselle, DuPage County, Illinois.

Storage: Oak and stainless steel.

LYTTON SPRINGS WINERY

Founded 1975, Healdsburg, Sonoma County, California.

Storage: Oak and stainless steel. Cases per year: 8,000-10,000.

Estate vineyard: Valley Vista Vineyard.

Wines regularly bottled: Estate bottled, vintage-dated Zinfandel.

Wines regularly bottled: Vidal Blanc, Cabernet Sauvignon; also Fruit Wines.

Lynne's Vineyard

Riverside County, California.

Mesa Verde Winery estate Gamay Beaujolais and Emerald Riesling Temecula vineyard.

MABRY III, WILLIAM

The Mabry family has been involved in winemaking since William Mabry III was in fourth grade and helped to produce some homemade wine from Concord grapes. When his father retired to Sonoma in 1969, their winemaking interest grew to include viticulture. At age 18, Mabry helped in the care of the family vineyards and winemaking production, started by his father in 1972. Rene Lacasia, of nearby Buena Vista Vineyards, assisted Mabry in his home winemaking efforts. Mabry then enrolled in enology and viticulture courses at the University of California, Davis and helped to plant the newest family vineyard in the Alexander Valley. In 1974, Mabry produced the first commercial Landmark Vineyards wine at Mount Veeder Winery in the Napa Valley. The family purchased land in the Russian River Valley, in 1975, and built the Landmark Vineyards winery in 1976. Mabry manages the entire winemaking operation, with the help of consultant Brad Webb.

Mac Bride Vineyard

San Luis Obispo County, California.
Chardonnay vineyard.

Mac Corquodale Vineyard

Douglas City, Oregon.
Pinot Noir, Chardonnay and Chenin Blanc vineyard.

Mac Gregor Vineyard

San Luis Obispo County, California.
Chardonnay Edna Valley vineyard.

MAC ROSTIE, STEVEN W.

Steven W. MacRostie is a native Californian. After graduation from the Whitman College in 1968, with a bachelor's degree in biology, he served with the United States Army. While stationed in Italy, MacRostie developed an interest in wine and winemaking. Upon his return to the United States, he earned a master's degree in viticulture and enology from the University of California, Davis. Since graduation, in 1974, he has served as vice president and winemaster at Hacienda Wine Cellars in Sonoma County. MacRostie is a member of the American Society of Enologists.

MADDALENA VINEYARD (See San Antonio Winery.)

made by

Unfortunately this term does not mean what it implies. Under Federal Regulations, a bonded premises (not necessarily a winery) may use this term if they "treated the wine" prior to bottling. Treated includes such things as blending, aging or adjusting the wine in one form or another. It does not guarantee that the wine or any part was fermented on the premises.

MADEIRA

A rich fortified wine of the Sherry class, originally from the Island of Madeira.

maderise

When referring to Madeira, it is the process of heating the wine for several months at about 110° degrees F to give it its characteristic flavor, true to type. When referring to other wines, it usually means "oxidize," an unfavorable term, an acquired brownish color.

MADERO (See Casa Madero - Mexico.)

Madonna Vineyard
Napa County, California.
Pinot Noir Carneros vineyard.

Madrona Estate Vineyards
El Dorado County, California.
Cabernet Sauvignon, Merlot, Zinfandel, Chardonnay, White Riesling and Gewurztraminer vineyard.

MADRONA VINEYARDS
Founded 1980, Camino, El Dorado County, California.
Storage: Oak and stainless steel. Cases per year: 10,000.
Estate vineyard: Madrona Vineyards.
Wines regularly bottled: Estate bottled, vintage-dated Cabernet Sauvignon, Merlot, Zinfandel, Chardonnay, White Riesling, Gewurztraminer, White Zinfandel.

magenta
Wine color.
A deep, dark, purplish red.

MAGNANI, ROBERT L.
Robert L. Magnani worked in electronics for fifteen years before becoming professionally involved in the wine field. Although he had been a home winemaker for many years under the guidance of his uncle, an Italian winemaker, he earned an associate of arts degree in Electronics at Diablo Valley College in 1960. He went to work first for the Lawrence Radiation Laboratory, then five years later was hired by Tinsley Optical Laboratories. During his nine years working as a project engineer, Magnani helped to establish Grand Cru Vineyards in the Sonoma Valley. He designed and supervised the construction of all the winery equipment, and has participated in production since 1970. Magnani currently serves as vice president and winemaker at Grand Cru, and is responsible for their line of table, dessert and sparkling wines. He has created special carbonic maceration tanks for the winery, and has been particularly interested in the development of the varietal wines, Gewurztraminer and Chenin Blanc. He feels the Sacramento Delta area is especially well suited to the Chenin Blanc grape. Magnani is a member of the American Society of Enologists and the Brotherhood of the Knights of the Vine.

MAHER, RICHARD L.
Richard L. Maher was graduated from Rensselaer Polytechnic Institute. After three years as an officer in the United States Marine Corps, he enrolled at Stanford University where he earned an M.B.A.
Next came five years of brand management at Proctor and Gamble and three more years in brand management at Ernest & Julio Gallo. In 1968 and 1969, Mr. Maher was assistant director of new product development for Heublein, a post he left for the executive vice presidency of Shakeys Restaurants.
In mid-1972, he was named vice-president of marketing for Heublein. From 1975 to 1983 he achieved an enviable growth record as president of Beringer.
In 1983, he joined Joseph E. Seagram and Sons as President of The Seagram Wine Group.

Mahre Vineyard
Yakima Valley, Washington.
Quail Run estate Riesling vineyard.

MALBEC
One of the three great Bordeaux grapes (Cabernet Sauvignon, Cabernet Franc, Merlot). Long lasting, dry and tannic. Usually blended with Cabernet Sauvignon.

Malinda's Vineyards
Charleston, Tennessee.
Muscadine and Scuppernong vineyard.

MALVASIA BIANCA
A grape that, because of its sweetness, is used for dessert wines or sweet table wine. Also used for Asti Spumante style sparkling wines. Originally imported from the Asti region of Italy.

Mantey Estate Vineyards
Erie County, Ohio.
Concord, Catawba, Chelois, Baco, Seyval and Seibel vineyard.

MANTEY VINEYARDS
Founded 1880, Sandusky, Erie County, Ohio.
Storage: Oak. Cases per year: 25,000.
History: Founded by the Mantey family until acquired by Robert Gottesman in 1980. One of the oldest wineries in Ohio. Today it is a leading producer of wines made from Ohio grapes.
Estate vineyard: Mantey Vineyards.
Wines regularly bottled: Pink Catawba, Dry Catawba, Cream Catawba, Mellow Concord, Dry Concord; generics: Chablis, Haut Sauterne, Burgundy; proprietary Blue Face, Rosey Grape, White Grape; also Vin Rosé; occasionally Johannisberg Riesling, Agawam and Ives Seedling.

MANUKAI (See Fruit Wines of Florida.)
Pineapple Isle.

Maresch Vineyards
Yamhill, Oregon.
Pinot Noir, White Riesling vineyard.

MARIETTA CELLARS
Founded 1980, Healdsburg, Sonoma County, California.
Storage: Oak. Cases per year: 5,000.
Wines regularly bottled: Vintage-dated Cabernet Sauvignon, Zinfandel; and Old Red Wine.

Marina Vineyard
Napa County, California.
Acacia estate Chardonnay Carneros vineyard.

M. MARION AND COMPANY (A negociant Company)
Founded 1981, Saratoga, Santa Clara County, California. Negociant.
Cases per year: 35,000.
Wines regularly bottled: Vintage-dated Cabernet Sauvignon, Chardonnay, White Zinfandel.

Mark West Estate Vineyards
Sonoma County, California.
Chardonnay, Pinot Noir, Gewurztraminer, Johannisberg Riesling Russian River Valley vineyard.

MARK WEST VINEYARDS AND WINERY
Founded 1976, Forestville, Sonoma County, California.
Storage: Stainless steel. Cases per year: 18,000.
Estate vineyard: Mark West Vineyard.
Label indicating non-estate vineyard: Wasson Ranch, Rue Vineyards.
Wines regularly bottled: Estate bottled, vintage-dated Chardonnay, Gewurztraminer, Johannisberg Riesling, Pinot Noir, Pinot Noir Blanc. Occasionally bottles Zinfandel, Fondue Blanc, Russian River Red.

MARKHAM VINEYARDS
Founded 1978, St. Helena, Napa County, California.
Storage: Oak and stainless steel. Cases per year: 18,000.
Estate vineyard: Oak Knoll Ranch, Calistoga Ranch, Yountville Ranch.
Wines regularly bottled: Estate bottled, vintage-dated Johannisberg Riesling, Gamay Blanc, Chenin Blanc, Cabernet Sauvignon; also vintage-dated Muscat de Frontignan, Chardonnay and Merlot.
Second label: Vin Mark.

Markham Vineyards
Lane County, Oregon.
Pinot Noir and Riesling vineyard.

MARKKO VINEYARD
Founded 1968, Conneaut, Ashtabula County, Ohio.
Storage: Oak and stainless steel. Cases per year: 2,000.
History: First winery to use appellation, Lake Erie, on a wine label, in 1973. First to produce Chardonnay, Johannisberg Riesling and Cabernet Sauvignon in Ohio.
Estate vineyard: Markko Vineyards.
Wines regularly bottled: Estate bottled, vintage-dated Chardonnay, Johannisberg Riesling, Cabernet Sauvignon; occasionally Sweet Riesling.
Second label: Underridge.

Markko Estate Vineyard
Ashtabula County, Ohio.
Chardonnay, Riesling and Cabernet Sauvignon Lake Erie vineyard.

Marlo Estate Vineyard
Shelby County, Ohio.
Aurora, Villard Blanc, Seyval Blanc, Marechal Foch, Concord, Delaware and Niagara Laramie Creek vineyard.

MARLO WINERY
Fort Laramie, Shelby County, Ohio.
Storage: Oak and stainless steel.
Estate vineyard: Markko Vineyard.
Wines regularly bottled: Estate bottled Aurora, Villard Blanc, Foch; also Seyval Blanc, Pink Concord, Delaware, Pink Catawba, Niagara, Concord.

Marlstone Vineyard
Sonoma County, California.
Clos du Bois estate Cabernet Sauvignon and Merlot Alexander Valley vineyard.

MARQUAND, JOHN P. (1893-1960)
American writer.

"I hope that I am as broadminded as others, and you have always seen a decanter of wine on my table."

MARSALA WINE
A generic, sweet, dessert and cooking wine. Mission and Grenache grapes are predominantly used. Dark amber in color, with strong raisin flavor. Genuine Marsala from Sicily is flavored with grape concentrate and fortified with grape spirit.

Martha's Vineyard
Napa County, California.
Cabernet Sauvignon Napa Valley vineyard.

MARTHA'S VINEYARD VITICULTURAL AREA
A new viticultural district describing the island off the southeast coast of Massachusetts. Consists of Martha's Vineyard and Chappaquiddick Island, in Dukes County, Massachusetts.

Martin Brothers Estate Vineyard
San Luis Obispo County, California. Martin Bros.
Chardonnay, Sauvignon Blanc, Chenin Blanc, Semillon, Zinfandel, Nebbiolo-Michet vineyard.

MARTIN BROTHERS WINERY
Founded 1981, Paso Robles, San Luis Obispo County, California.
Storage: Oak and stainless steel. Cases per year: 5,000.
Estate vineyard: Martin Brothers Vineyard.
Label indicating non-estate vineyard: Paragon Vineyards, Rancho Tierra Rejada Vineyards.
Wines regularly bottled: Vintage-dated Chardonnay, Zinfandel, Sauvignon Blanc, Dry Chenin Blanc.

Martin Ray Estate Vineyards
Santa Clara County, California.
Chardonnay Santa Cruz Mountains vineyard.

MARTIN RAY VINEYARDS
Founded 1946, Palo Alto, Santa Clara County, California.
Storage: Oak. Cases per year: 4,000.
Estate vineyard: Martin Ray Vineyards.
Label indicating non-estate vineyard: Winery Lake Vineyard, Dutton Ranch, Steltzner Vineyard.
Wines regularly bottled: Estate bottled, vintage-dated Chardonnay. Also vintage-dated Chardonnay, Pinot Noir, Merlot, Cabernet Sauvignon; and vintage-dated Champagne.
Second label: La Montana.

THE MARTIN WINERY
Founded 1970, Culver City, Los Angeles County, California.
Storage: Oak and stainless steel. Cases per year: 500.
Label indicating non-estate vineyard: Radike Vineyards, Jensen Vineyards, Estrella Vineyards.
Wines regularly bottled: Vintage-dated Merlot, Cabernet Sauvignon, Sauvignon Blanc, Chardonnay.

MARTINI, LOUIS
Four generations of the Martini family have been involved in winemaking. As a child in San Francisco, Louis M. Martini helped his father, Agostino, in home winemaking projects. After a six-month course in enology, in Italy, Martini returned to assist his father in the serious production of dry red and white wines. Seven profitless years later, Martini accepted a series of jobs in other wine cellars. In 1920, he was finally able to begin grape and grape product sales under the Martini name, in the San Joaquin Valley. In 1933, he expanded his business and founded the Louis M. Martini Winery, in the Napa Valley. Through the acquisition of several pre-Prohibition vineyards, he was soon able to offer vintage-dated Cabernet Sauvignon, Zinfandel and Pinot Noir. In 1941, his son, Louis P. Martini, joined the winery, after obtaining a bachelor's degree in food technology from the University of California at Berkeley, where he was a member of the famous class of winemakers in

1940. His college career included a year at Davis where he studied viticulture and enology. Louis P. Martini was vice president and production manager from 1946 to 1967, and he has been president and general manager since 1968. Under his direction, the winery has expanded from two to five buildings.

He was a pioneer in the mechanical harvesting of fine wine grapes, and has worked to improve grape varieties through clonal selection. In addition to his scientific and technical contributions, Martini has been actively involved in the development of the wine industry. He is a charter member of the American Society of Enologists, and served as its president in 1956-57. He was chairman of the California Wine Institute in 1977-78. He is a member, and past president, of the Napa Valley Vintners and the Napa Valley Technical Group. Martini has been honored with the Leon B. Adams Achievement Award in 1981, the American Society of Enology Merit Award in 1981, and the American Wine Society Award of Merit, 1983. Louis P. Martini's son, Michael, became winemaker in 1977. He is a Supreme Knight in the Brotherhood of the Knights of the Vine.

LOUIS M. MARTINI

Founded 1933, St. Helena, Napa County, California.

Storage: Oak and stainless steel. Cases per year: 300,000

History: See Martini Biography.

Estate vineyard: Monte Rosso Vineyard, La Loma and Las Amigas Vineyards, Los Vinedos del Rio, Glen Oak Vineyards.

Wines regularly bottled: Estate bottled, vintage-dated Chardonnay, Johannisberg Riesling, Gewurztraminer, Folle Blanche, Dry Chenin Blanc, Cabernet Sauvignon, Pinot Noir, Merlot, Barbera, Zinfandel, Gamay Beaujolais, Sauvignon Blanc. Also semi-generics: Burgundy and Chablis.

Martini and Prati Estate Vineyards

Sonoma County, California.

Grey Riesling, Chenin Blanc, Pinot Noir, Cabernet Sauvignon, Zinfandel, Pinot Chardonnay vineyard.

MARTINI AND PRATI WINES

Founded 1951, Santa Rosa, Sonoma County, California.

Storage: Oak and stainless steel. Cases per year: 25,000. (Bottled under their own label.)

Estate vineyard: Martini and Prati Wines.

Wines regularly bottled: Estate bottled, vintage-dated Cabernet Sauvignon; vintage-dated Pinot Chardonnay. Also Pinot Noir, Zinfandel and Grey Riesling, Cabernet Sauvignon; two generics: Chablis and Sauterne.

Second label: Fountain Grove.

PAUL MASSON VINEYARDS

Founded 1852, Saratoga, Santa Clara and Monterey Counties, California.

Storage: Oak and stainless steel. Cases per year: 8,000,000.

History: The winery dates to 1852, when Etienne Thée, a vigneron from Bordeaux, first planted grapes on the original Narvaez land grant, south of San Jose, and pioneered commercial winegrowing in the region. Thée was succeeded by his son-in-law, Charles Lefranc, and, in turn, by the latter's son-in-law, Paul Masson.

Born in 1859, in Beaune, in the Burgundy district of France, Paul Masson came to California at the age of 19, when the phylloxera wine pest had devastated the vineyard on the Cote d'Or where his family had made wine for three centuries. At that time, the eyes of Old World vintners were attracted to California because of its ideal climate for winegrowing.

The young Burgundian emigré first enrolled at the University of the Pacific, in Santa Clara, to continue the scientific studies begun at the Sorbonne in Paris. While a student, he became acquainted with his compatriot, Charles Lefranc, who was adding to the vineyards inherited from Etienne Thée. Paul Masson became interested in the vineyards and, also, in Lefranc's pretty daughter, Louise. He went to work for Lefranc, married Louise, and planted his own vineyard—the "Vineyard in the Sky"—in the Santa Cruz Mountains above Saratoga. After Lefranc died, the vineyards were merged and Paul Masson gave the operations his name.

On periodic visits to his native France, Paul Masson brought back cuttings of choice European wine grape varieties. Some he planted to make his wines, champagnes and brandy; others, he gave to neighboring growers, who agreed to let him select the best of the crop produced.

He built a great masonry winery at his mountain vineyard, with its foundations deep in the hillside to maintain constantly cool temperatures for aging wine. When the 1906 earthquake destroyed St. Patrick's Church in nearby San Jose, he purchased its 12th century Romanesque portal, originally brought around Cape Horn from Spain, and erected it as part of the winery facade. Today this famous winery, damaged by fire in 1941 and restored as Paul Masson designed it, is officially designated by the State of California as Landmark Number 733. It is still in active use as Bonded Winery #144, holding hundreds of barrels of fine Sherry for aging. Grapes are still harvested each year from the old vineyards: some of the vines are from the original plantings and are of considerable interest to viticulturists.

Paul Masson Vineyards is a unit of the Seagram Wine Company, New York (Joseph E. Seagram, USA).

Estate vineyard: Pinnacles Vineyards.

Wines regularly bottled: Estate bottled, vintage-dated Chardonnay, Riesling,-Gewurztraminer, Fumé Blanc; and vintage-dated Cabernet Sauvignon, Pinot Noir, Chardonnay, Johannisberg Riesling, Zinfandel, Chenin Blanc, French Colombard, Gamay Rosé. Also semi-generic wines: Rosé, Burgundy, Chablis, Rhine, Vin Rosé Sec; proprietary Rhine Castle, Emerald Dry. Dessert wines produced are Sangria, Tawny, Rich Ruby and Rare Souzao Ports, Pale Dry, Medium Dry, Cocktail, Golden Cream and Rare Cream Sherries, Madeira; and Vermouth, Sparkling Wines (Crackling Rosé, Crackling Chablis, Sparkling Burgundy, Very Cold Duck, (vintage) Brut, (vintage) Extra Dry, Pink). Also produce Light Chablis, Light Rosé, Light Rhine.

MASTANTUONO

Founded 1977, Paso Robles, San Luis Obispo County, California.

Storage: Oak and stainless steel. Cases per year: 5,000.

Estate vineyard: Mastantuono Vineyard.

Label indicating non-estate vineyard: Dante Dusi Vineyard, Rancho Tierra Rejada.

Wines regularly bottled: Estate bottled, vintage-dated Zinfandel; vintage-dated White Zinfandel and Zinfandel. Occasionally bottle vintage-dated Cabernet Sauvignon, Sauvignon Blanc, Zinfandel Centennial, Zinfandel Dolce.

Mastantuono Estate Vineyard

San Luis Obispo County, California. Zinfandel vineyard.

Matanzas Creek Estate Vineyards

Sonoma County, Santa Rosa, California. Matanzas Creek.

Chardonnay and Merlot Sonoma Valley vineyard.

MATANZAS CREEK WINERY
Founded 1978, Sonoma County, California.
Storage: Oak and stainless steel. Cases per year: 5,000.
Estate vineyard: Matanzas Creek Vineyard.
Wines regularly bottled: Estate bottled, vintage-dated Pinot Noir and Chardonnay; also bottle Sauvignon Blanc, Chardonnay, Merlot, Cabernet Sauvignon.

Victor Matheu Vineyard
Mendocino County, California.
White Riesling and Chardonnay vineyard.

MATROSE (See Pat Paulsen Vineyards.)
Dry Gewurztraminer.

Matzke Vineyard
La Porte, Indiana.
Ravat vineyard.

L. MAWBY VINEYARDS AND WINERY
Founded 1973, Suttons Bay, Leelanau County, Michigan.
Storage: Oak and stainless steel. Cases per year: 600.
Estate vineyard: Elm Valley, Keswick Vineyard.
Wines regularly bottled: Estate bottled, vintage-dated Vignoles, Red, White and proprietary Turkey Red (100% Foch).
Second label: Elm Valley Vineyards.

MAY WINE
An aromatic wine flavored with woodruff, an herb. Light and fruity, with a scented bouquet and fruity, sweet flavor.

Mayacamas Estate Vineyards
Napa County, California.
Cabernet Sauvignon, Chardonnay and Sauvignon Blanc vineyard.

MAYACAMAS VINEYARDS
Founded 1889, Napa, Napa County, California.
Storage: Oak. Cases per year: 5,000.
History: The Mayacamas Winery was built in 1889 by John Henry Fisher, an immigrant from Stuttgart (where he was a sword engraver). Bulk red and white table wines were made and there was a small distillery. Vineyard plantings were of Zinfandel and "Sweetwater" grapes. Fisher sold the property after the turn of the century and the winery fell into disuse. In 1941, after several changes in ownership, Jack and Mary Taylor (he of England, she of California) acquired the empty stone winery, distillery and the ancient, declining vineyards. The winery was renovated, the distillery made into a home and the old vines torn out. The Taylors replanted the vineyards to Chardonnay and Cabernet Sauvignon and gave the winery the name Mayacamas.
Mayacamas is the name of the mountains which separate the Napa and Sonoma valleys, where the vineyards and winery are located. It is a Spanish adaptation of an Indian word meaning "howl of the mountain lion". Cougars and bobcats still roam this range and the name inspired the Mayacamas label design of two lions rampant.
In 1968, the Mayacamas Vineyards and Winery property was sold again. Now it is owned by Robert and Elinor Travers.
Estate vineyard: Mayacamas Vineyards.
Wines regularly bottled: Estate bottled, vintage-dated Cabernet Sauvignon, Chardonnay, Sauvignon Blanc; Zinfandel and Pinot Noir, in alternate years.

Robert Mazza Estate Vineyards

Lake Erie, Pennsylvania.

DeChaunac, Leon Millot, Chancellor vineyard.

MAZZA VINEYARDS

Founded 1972, North East, Erie County, Pennsylvania.

Storage: Stainless steel. Cases per year: 9,200.

Estate vineyard: Robert Mazza Vineyard.

Label indicating non-estate vineyard: Russ Osen Vineyards.

Wines regularly bottled: Seyval, Vidal, Riesling, Cabernet Sauvignon, Baco Noir, Chelois, Foch, Catawba, Niagara, Dutchess; Cherry, Strawberry.

MAZZEI, PHILIP

Florence, Italy. Born 1730. American Patriot and Vigneron. Mazzei was the vigneron for Thomas Jefferson and became a Revolutionary patriot. In 1773 he brought 1,000 vines and 10 vignerons to Jefferson's Monticello plantation in Virginia and planted a vineyard. Mazzei became friends with Patrick Henry, George Mason, and Benjamin Franklin, who convinced Philip Mazzei to organize the men and the material and come to America and establish vineyards for his friend, Thomas Jefferson, who was most desirous of making his own wine from his own vineyards. Fillippo and Dr. Lapo Mazzei are descendants of Mazzei, the vigneron and American patriot . . . Dr. Mazzei is the chairman of the Chianti Classico Consortium of Florence, Italy.

MC CALL, THOMAS

Founded 1823, first commercial winery in Georgia. Vineyard planted in Laurens County.

MC CLELLAND, JOHN P.

John P. McClelland has had a long and successful career with Almaden Vineyards. He started, in 1957, as a sales supervisor in southern California, and five years later he was promoted to Southwest Division manager. In 1969, he became national sales manager and was transferred to the wineries executive offices. Within three years, McClelland became vice president in charge of sales, and then marketing. In 1976, he was elected president and chief operations officer of Almaden Vineyards. In 1983, he became chief executive officer of Geyser Peak Wineries. McClelland was responsible for many sales, marketing and packaging innovations, including creation of the Charles LeFranc Founder's line of wines and bag-in-box wines. He has also been active in many wine industry organizations serving as President of the North Coast Wine Society of Southern California for two terms and, then, as a director of the Monterey Winegrowers Council. Currently, he is third vice president of the California Wine Institute.

Grapes picked by hand

McCord Vineyards

Lake Erie, Pennsylvania.

Penn Shore estate Hybrid and Native American vineyard.

MC CREA, ELEANOR WHEELER

Eleanor Wheeler McCrea graduated from Wellesley College in 1929. In 1943, she and her husband, Fred, were visiting friends in Napa Valley and fell in love with the area. They decided to look for land to start a vineyard and bought a 160 acre farm which they, along with their children, started planting five acres at a time. In 1952, they started Stony Hill Winery. In 1971, Fred McCrea hired Michael Chelini, a young Californian, as winemaker and passed on his skills to Michael. Since 1977, when Fred died, Eleanor has taken over the running of the winery with the help of her winemaker. Today a lady in her 70's, she is totally involved in this small and excellent producer of bottled-by-hand California wines.

McCrea Vineyards
Sonoma County, California.
Chardonnay Sonoma Valley vineyard.

MC DOUGALL, JAMES A. (1817-1867)
United States Senator from California.

"Do you remember any great poet that ever illustrated the higher fields of humanity that did not dignify the use of wine from Homer on down?"

McDowell Valley Estate Vineyards
Mendocino County, California.
Chenin Blanc, French Colombard, Sauvignon Blanc, Semillon, Riesling, Chardonnay, Grenache, Zinfandel, Cabernet Sauvignon, Syrah McDowell Valley vineyard.

MC DOWELL VALLEY VINEYARDS
Founded 1979, Hopland, Mendocino County, California.
Storage: Oak. Cases per year: 40,000.
History: Grapes have been grown in McDowell Valley since 1890. The original ranch was homesteaded by Paxton McDowell in the early 1800's. Richard and Karen Keehn founded the winery.
Estate vineyard: McDowell Valley Vineyards.
Wines regularly bottled: Estate bottled, vintage-dated Chenin Blanc, French Colombard, Fumé Blanc, Chardonnay, Grenache, Zinfandel, Syrah, Cabernet Sauvignon; and a Red and White Table Wine (vintage).

MC DOWELL VALLEY VITICULTURAL AREA
Mendocino County, California.

McFadden Ranch
Mendocino County, California.
Sauvignon Blanc Potter Valley vineyard.

McGilvery Vineyard
Sonoma County, California.
Cabernet Sauvignon vineyard.

McGregor Estate Vineyard
Yates County, New York.
Johannisberg Riesling, Chardonnay, Gewurztraminer, Pinot Noir, Ravat and Cayuga Finger Lakes vineyard.

MC GREGOR VINEYARD
Founded 1973, Dundee, Yates County, New York.
Storage: Stainless steel. Cases per year: 1,000.
Estate vineyard: McGregor Vineyard.
Wines regularly bottled: Estate bottled, vintage-dated Chardonnay, Riesling, Gewurztraminer, Pinot Noir, Special Select Riesling, Late Harvest Riesling, Chancellor, Cayuga, Ravat.

McHenry Estate Vineyards
Santa Cruz County, California.
Chardonnay, Pinot Noir Santa Cruz Mountains vineyard.

MC HENRY VINEYARDS
Founded 1980, Santa Cruz, Santa Cruz County, California.
Estate vineyard: McHenry Vineyards.
Wines regularly bottled: Estate bottled, vintage-dated Pinot Noir, Chardonnay.

McIntosh Estate Vineyard
Brown County, Ohio.
Concord, Catawba, Niagara, Aurora, Baco Noir and Delaware Ohio River Valley vineyard.

MC INTOSH WINERY

Founded 1965, Bethel, Brown County, Ohio.

Storage: Oak.

Estate vineyard: McIntosh Vineyard.

Wines regularly bottled: Estate bottled Sweet Catawba, Dry Catawba, Pink Catawba, Niagara, Concord; generic: Sauterne; and Rosé.

Second label: McIntosh's Ohio Valley Wine.

MC KENZIE CREEK WINERY (See Roudon-Smith Vineyards.)

McKenzie River Vineyards

Lane County, Oregon.

Gewurztraminer vineyard.

MC LESTER WINERY

Founded 1980, Inglewood, Los Angeles County, California.

Storage: Oak and stainless steel. Cases per year: 1,000.

Label indicating non-estate vineyard: Jensen Vineyard, Radike Vineyard, Cowan Vineyard.

Wines regularly bottled: Vintage-dated Cabernet Sauvignon, Cabernet Sauvignon Rosé, Zinfandel. Occasionally Merlot, French Colombard, White Zinfandel and Sauvignon Blanc.

"Mead" is wine made from honey

MEAD

Honey Wine (Mead) is a wine produced by the fermentation of a solution of pure honey. It has slight honey bouquet and flavor. Honey wines are pale to dark amber in color. Mead can only be made by reducing down its natural sugar to a fermentable level (under 25-30%). It can be dry, sweet, high or low alcohol.

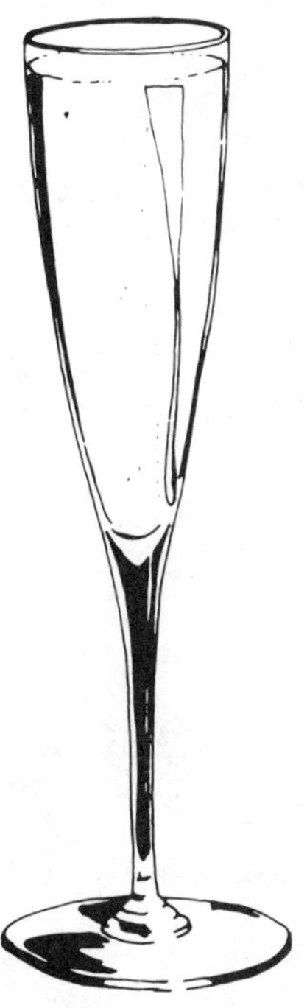

Mead Ranch

Napa County, California.

Zinfandel Atlas Peak vineyard.

MEADOR, DOUG

Doug Meador is the man behind Ventana Vineyards and Ventana Vineyards Winery in Soledad, California. Born in Tacoma, Washington, he was graduated, with a degree in economics, from the University of Washington in 1965. He was an apple grower and a pilot in the United States Navy for seven years before beginning his career in the wine industry. His wife, Shirley, designs the labels for the winery; his daughter, Roni, does deliveries and his son Darren, works in the vineyards and winery when he is not in school. Meador works seven days a week supervising both the grape growing and winemaking operations.

Meadows' Vineyard
Linn County, Oregon.
Marechal Foch, Cabernet Sauvignon, White Riesling, Mueller-Thurgau and Gamay Beaujolais vineyard.

medium
As applied to Sherry, it means slightly sweet, in the middle range between dry and sweet.

medium ruby
Wine color.
Red with still notable quantities of blue.

medium straw
Wine color.
A yellow of medium intensity.

MEIER'S WINE CELLARS
Founded 1895, Cincinnati, Hamilton County, Ohio.
Storage: Oak and stainless steel. Cases per year: 1,300,000.
History: Meier's is one of Ohio's oldest and today is the largest winery. Its vineyard, Isle St. George, lies northwest of Sandusky. Perry's battle of Lake Erie took place off the Isle St. George. Catawba grapes have been grown continuously, for more than 100 years. In 1982, Isle St. George was designated a viticultural area by the United States Government because of its unique climate and grape growing conditions.
Estate vineyard: Isle St. George Vineyard.
Wines regularly bottled: Baco Noir, Gewurztraminer, Riesling, Pinot Rosé, Chardonnay, Steuben, Chelois, Ives Seedling, Soft and Sweet Catawba, Pink and White Catawba, Concord, Blackberry, Red Seedling; generics: Chablis, Rhine, Mellow Burgundy; White and Pink Champagnes, Cold Duck, Spumante; also proprietary: Island Red, White and Rosé, Fu-W Cincinnati Red and White, La Brusca Bianco, Rosato, Rubio; also Spiced Wine.

Melange a Deux Vineyard
Sonoma County, California.
Johannisberg Riesling and Gewurztraminer vineyard.

mellow
Soft, well-balanced wine. Mellowness and velvet richness are the special characteristics of old, well aged reds. Sometimes used to designate well-matured Sherries, wines containing some sweetness.

MENDOCINO VINEYARDS (See Cresta Blanca.)
The label was developed from the post World War II brand of the Mendocino Vineyards Wine Company. Red and White wine, Zinfandel, Chenin Blanc and Cabernet Sauvignon.

MENDOCINO VITICULTURAL AREA
Mendocino County, California, is located in the southernmost part of Mendocino County; consists of about 430 square miles (275,200 acres). It includes the watershed areas and drainage basins of both the Navarro and Russian Rivers. It lies entirely within the North Coastal Viticultural Area.
There are approximately 10,596 acres of grapes in the area. "Mendocino" overlaps with five other viticultural areas. Those areas include McDowell Valley, Cole Ranch, Potter Valley, Anderson Valley and North Coast.

MENDOCINO WINE COMPANY (See Edmeades Vineyards.)

MENDOCINO WINE COUNTRY
By the 1850's, many small vineyards were established along the foothills of the Ukiah Valley. Pioneers such as Samuel Orr and Walter Anderson planted their early vineyards.

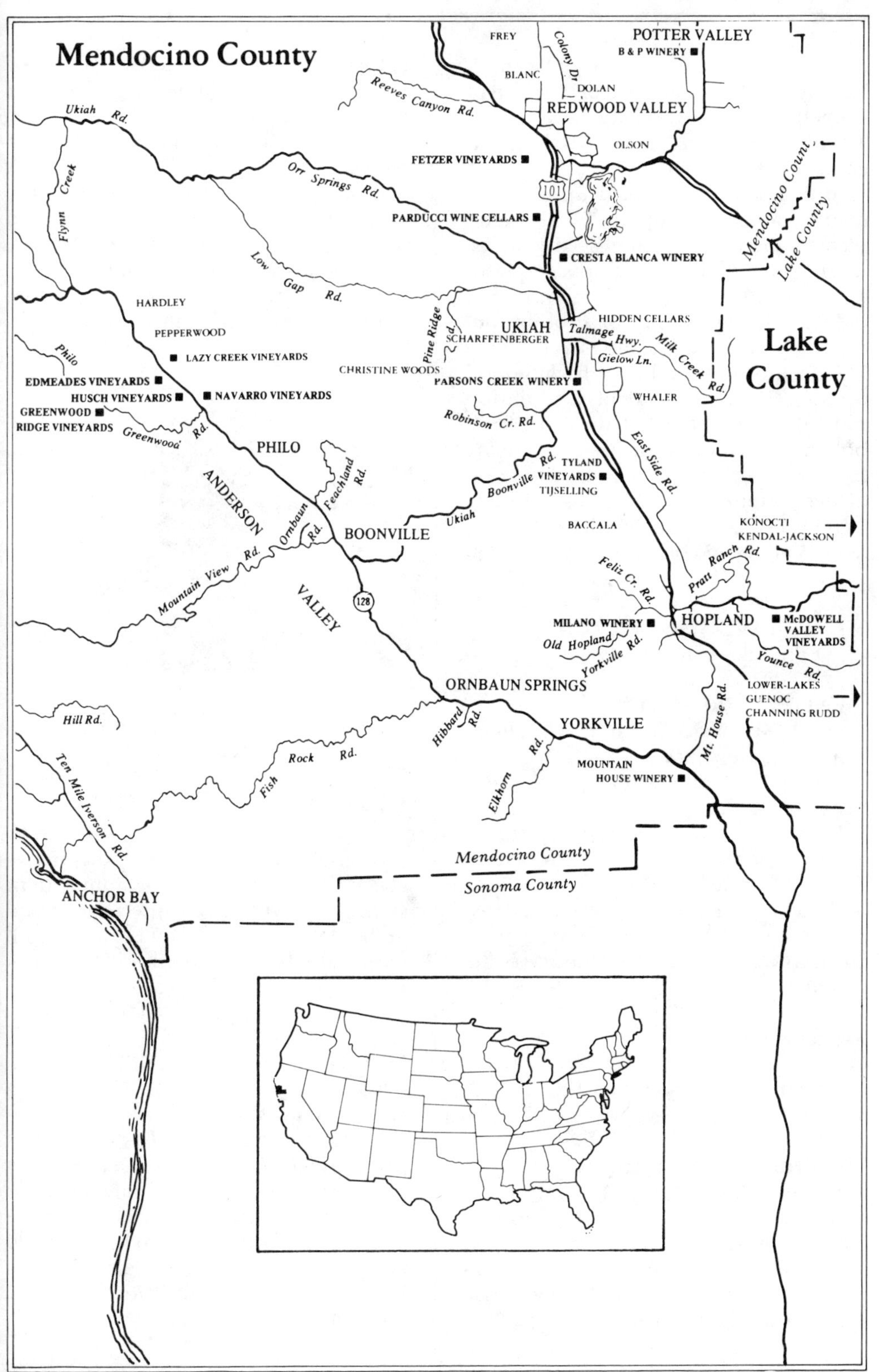
Mendocino County
Lake County
FREY
BLANC
Colony Dr.
DOLAN
POTTER VALLEY
B & P WINERY
REDWOOD VALLEY
OLSON
Reeves Canyon Rd.
Ukiah Rd.
Flynn Creek
Orr Springs Rd.
FETZER VINEYARDS
101
PARDUCCI WINE CELLARS
CRESTA BLANCA WINERY
Mendocino County
Lake County
Low Gap Rd.
HARDLEY
PEPPERWOOD
Pine Ridge Rd.
UKIAH
SCHARFFENBERGER
HIDDEN CELLARS
Talmage Hwy.
Milk Creek Rd.
Gielow Ln.
Philo
LAZY CREEK VINEYARDS
CHRISTINE WOODS
PARSONS CREEK WINERY
WHALER
EDMEADES VINEYARDS
HUSCH VINEYARDS
NAVARRO VINEYARDS
GREENWOOD RIDGE VINEYARDS
Greenwood Rd.
Robinson Cr. Rd.
PHILO
ANDERSON
Peachland Rd.
East Side Rd.
TYLAND VINEYARDS
TIJSELLING
Boonville Rd.
Ukiah
Ornbaun Rd.
BOONVILLE
BACCALA
KONOCTI
KENDAL-JACKSON
Pratt Ranch Rd.
Feliz Cr. Rd.
Mountain View Rd.
VALLEY
128
MILANO WINERY
HOPLAND
McDOWELL VALLEY VINEYARDS
Old Hopland
Yorkville Rd.
Younce Rd.
ORNBAUN SPRINGS
LOWER-LAKES
GUENOC
CHANNING RUDD
Hill Rd.
Hibbard Rd.
YORKVILLE
Mt. House Rd.
Rock Rd.
Fish
Elkhorn Rd.
MOUNTAIN HOUSE WINERY
Ten Mile Iverson Rd.
Mendocino County
Sonoma County
ANCHOR BAY

The first winery was established in 1879 by Louis Finne, north of the town of Calpella. By 1880 there were more than 300 acres of vineyard, mostly in the north end of the Ukiah Valley.

In the early 1900's, until Prohibition, many Italian immigrants settled and planted vineyards and established wineries. There were names like Cuppucci, Masoletti, Luccassi, Fomasero, and Parducci.

After the devastation of Prohibition, only a few wineries remained, setting the Mendocino wine industry back several decades. From the end of Prohibition until the wine boom of the late 1960's only a few wineries survived. There are two major regions in the southern section of Mendocino County. The Russian River Valley, the Potter, Ukiah, McDowell and Sanel Valleys and on the valley floor and in the foothills of the Anderson Valley.

MENDOCINO WINE GUILD

(See Edmeades Vineyards.)

MENGHINI WINERY

Founded 1982, Julian, San Diego County, California.

Storage: Oak and stainless steel. Cases per year: 2,500.

Estate vineyard: Julian Vineyards.

Label indicating non-estate vineyard: Bell Vineyard, Mira Monte.

Wines regularly bottled: Vintage-dated Chardonnay, Chenin Blanc, Sauvignon Blanc, Cabernet. (In the future, estate Chardonnay.)

Mercer Ranch

Washington.

Cabernet Sauvignon and Chenin Blanc vineyard.

Meredyth Estate Vineyards

Fauquier County, Virginia, Meredyth.

Aurora, Seyval, Chardonnay, Riesling, Sauvignon Blanc, Villard Blanc, Rayon d'Or, Vidal, Delaware, Cabernet Sauvignon, Merlot, Rougeon, DeChaunac, Marechal Foch, Leon Millot vineyard.

MEREDYTH VINEYARDS

Founded 1972, Middleburg, Fauquier County, Virginia.

Storage: Oak and stainless steel. Cases per year: 9,000.

Estate vineyard: Meredyth Vineyards.

Wines regularly bottled: Estate bottled Seyval Blanc, Villard Blanc, Aurora, Chardonnay, Riesling, Sauvignon Blanc, Delaware, Rougeon, Cabernet Sauvignon, Marechal Foch, DeChaunac, Villard Noir, Merlot.

Merlot

MERLOT

Wine grape.

A grape that produces a distinctive aromatic, spicy, wine, medium red in color, with some of the green olive and herbaceous odor of Cabernet Sauvignon. Originally from the Bordeaux region of France where it is a major component in the wines of Pommard and St. Emilion.

MERRITT ESTATE WINERY

Founded 1976, Forestville, Chautauqua County, New York.

Storage: Oak and stainless steel. Cases per year: 6,000.

History: The original homestead has been in the family since the late 1800's. It was the first farm winery to be organized after the New York State Farm Winery Law went into effect in 1976.

Estate vineyard: Triple M Farms.

Wines regularly bottled: Estate bottled, vintage-dated Chautauqua Niagara, Aurora, Seyval Blanc, Rosé de Chaunac, Marechal Foch; estate bottled Chautauqua Red, White and Rosé, Pink Catawba, proprietary Sangría de Marguerite and Mereo (mulled wine). Occasionally bottle Vidal Blanc.

Second label: Sheridan Wine Cellars.

MERRITT ISLAND VITICULTURAL AREA

Merritt Island, located in Yolo County, California, is a man-made island consisting of approximately 5,000 acres. The proposed viticultural area is located six miles south of Sacramento and is the first island forming the alluvial fan of the Sacramento Delta.

JOHN B. MERRITT

(See Bandiera Winery.)

MESA VERDE WINERY

Founded Vineyard 1974, Winery 1980, Temecula, Riverside County, California.

Storage: Oak and stainless steel. Cases per year: 20,000.

Estate vineyard: Lynne's Vineyard.

Wines regularly bottled: Estate bottled, vintage-dated Gamay Beaujolais; vintage-dated Chardonnay, Johannisberg Riesling, Sauvignon Blanc, Chenin Blanc, Cabernet Sauvignon and Merlot.

Messina Hof Estate Vineyard

Brazos County, Texas.

Lenoir, Cabernet Sauvignon and Chenin Blanc vineyard.

MESSINA HOF WINE CELLARS

Founded Vineyard 1977, Winery 1983, Bryan, Brazos County, Texas.

Storage: Oak and stainless steel. Cases per year: 4,000.

Estate vineyard: Messina Hof Vineyard.

Wines regularly bottled: Vintage-dated Chenin Blanc, Cabernet Sauvignon and proprietary Papa Paula Porto. Also Vino Di Amore Sweet Bianco and Vino Di Amore Rosso.

metallic

Tasting term.

A defect in wine. A hint of bitterness, a hard finish.

A champagne flute displays "fine perlage"

METHODE CHAMPENOISE

(See Champagne.)

M.E.V. (See Mt. Eden Vineyards.)

Vintage-dated Chardonnay.

MEXICO, HISTORY

The beginnings of the Western wine culture can be traced back more than 450 years to early Mexico.

The Governor of New Spain, from 1521 to 1527, Hernando Cortez, the conqueror of Mexico, ordered vineyards planted in the New World with cuttings brought from Spain. Cortez insisted that all ships coming from Spain to his chief port of Veracruz, Mexico, should bring with them supplies of grape cuttings and seeds.

In 1524, Cortez decreed that all Spaniards holding land grants in Mexico must plant annually, for five years, a thousand grapevines for each hundred Indians located on their ranchos. By 1554, vine-growing was well established on the Mexican mainland.

It is interesting to note that during the Sixteenth Century, the now famous vineyards of Argentina, Chile, and Peru sprang from the wine culture of Mexico, and not that of Europe.

In the Seventeenth Century, Jesuit missionaries extended Spain's frontiers across the Sea of Cortez (Gulf of California) to the 1,000-mile-long peninsula of Lower California to build missions dedicated to the spread of Catholicism. Knowing that the remoteness of the missions made it imperative for each to become self-sufficient, he padres brought with them many fundamental supplies, including seeds and plantings for the establishment of agriculture.

Two commercial wineries of importance were established in Mexico in the 1800's—one in the state of Coahuila, in 1860, and the other, in 1888, at "Bodegas de Santo Tomás", near the present-day port of Ensenada, Baja California.

In 1697, the site selected for the first of the Lower California missions—San Francisco Xavier—was near the small coastal community of Loreto, 735 miles below the California border. Here, the energentic Father Juan Ugarte, a professor of philosophy at the Jesuit College in Mexico City, planted the first vineyards in Lower California. These bluish-black grapes were believed to be the seedlings of the La Mancha grape of Spain. They were called the Criolla, or Mission grapes. The vines prospered and soon were supplying enough grapes for the production of the wine necessary for the padre's Sacramental purposes and to administer to the sick.

Wine production was extremely difficult due to the lack of adequate containers for storing the juice during fermentation. Father Ugarte solved this problem by arranging for a supply of earthen jugs to be brought to his mission, by ship, from the Philippines.

In April of 1791, Mission Santo Tomás de Aquino (Saint Thomas Aquinas) was established by a Dominican friar, Father José Loriente, in beautiful San Solano Valley, 34 miles south of the Mexican coastal city of Ensenada. This rich-loamed area, with its cool Pacific Ocean breezes, warm sun, and occasional morning fogs was renamed Santo Tomás Valley in honor of the mission.

In addition to a favorable climate, a further advantage to Santo Tomás Valley for the growing of vineyards was a blessing in the form of abundant surface water and nearby mountain springs for irrigation of the grapes. The coastal valley also had good drainage for its fields.

Dominican church records disclose that more than 5,000 vines were growing and thriving in Santo Tomás Valley by 1800.

In August of 1825, the historic era of the mission padres in Baja California came to an end and the Mexican government took over all mission properties. The mission in Santo Tomás Valley was abandoned and the vineyards dried and became dormant.

Today, the principal areas are in Northern Baja California, Southern Coahuila. Other limited areas are in Durango, Central Chihuahua, Central Aquascalientes and Southern Queritaro.

MEYER, JUSTIN

Justin Meyer received his bachelor's degree in economics from St. Mary's College in Moraga, then taught high school for three years prior to continuing his education at the University of California, Davis. There he first earned a bachelor's degree in viticulture and enology, then a master's degree in horticulture. Meyer began his winemaking career in 1964, working at The Christian Brothers. In 1972, he formed a partnership with Raymond T. Duncan. They established Silver Oak Cellars in Oakville, California, to produce wines from grapes grown in the North Coast counties. In 1975, Meyer and Duncan purchased the Franciscan Winery. Today, Meyer produces wines with a commitment to excellence. Silver Oak Cellars produces Cabernet Sauvignon exclusively, made from their Alexander Valley vineyard grapes.

MICHAEL'S

(See Artisan Wines)

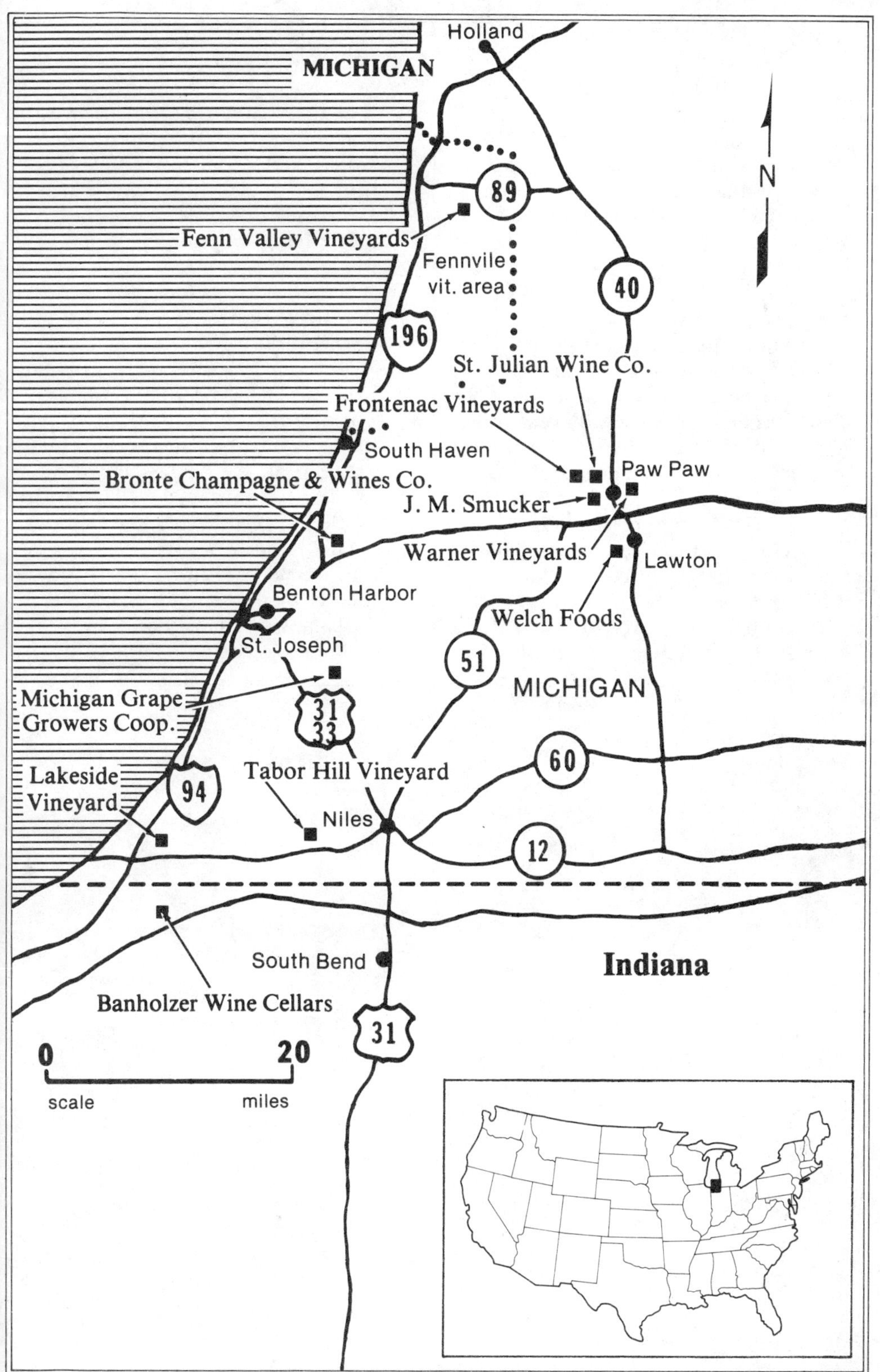

Holland
MICHIGAN
N
89
Fenn Valley Vineyards
Fennvile
vit. area
40
196
St. Julian Wine Co.
Frontenac Vineyards
South Haven
Paw Paw
Bronte Champagne & Wines Co.
J. M. Smucker
Warner Vineyards
Lawton
Benton Harbor
Welch Foods
St. Joseph
51
MICHIGAN
Michigan Grape
Growers Coop.
31
33
60
Lakeside
Vineyard
94
Tabor Hill Vineyard
Niles
12
South Bend
Indiana
Banholzer Wine Cellars
31
0
20
scale
miles

microclimates

Micro-climates are actually climates within climates and are a very important consideration when viticulturists determine what should be planted within a relatively small area. Several micro-climates can exist within one vineyard where the land might change in direction (facing north or south), elevation (valley bottom or slope) and proximity to small or large bodies of water.

When talking about climates and wine it is important to keep in mind that grapes must be planted in the area which is best for a particular variety. For example, the process of ripening is slower under cool conditions, a higher degree of acidity is retained in mature grapes that are planted in cooler regions, and more color remains in the skins of red grapes that are grown in cool as opposed to warm vineyards.

Although there are exceptions, in California here is a general rule of thumb, with districts followed by names of varieties favored in each.

Coldest Climates: Santa Cruz, Santa Maria, Sonoma, Upper Monterey: Pinot Noir, Pinot Blanc, Johannisberg Riesling, Gewurztraminer, Grey Riesling.

Medium Cool: Monterey, Napa Carneros, Santa Barbara, Sonoma: Pinot Blanc, Chardonnay, Sylvaner, Grenache, Grey Riesling, Johannisberg Riesling, Petite Sirah, Zinfandel (sometimes).

Medium Warm: Lower Monterey, San Luis Obispo, Upper Napa Valley: Semillon, Sauvignon Blanc, Chenin Blanc, Zinfandel, Cabernet Sauvignon, Petite Sirah, Grenache, Napa Gamay, French Colombard, Chardonnay (sometimes).

Hot Climates: (Central Valley) Bakersfield, Fresno, Lodi, Modesto: French Colombard, Ruby Cabernet, Barbera, Carignane, Palomino, Tinta Madera, Grenache.

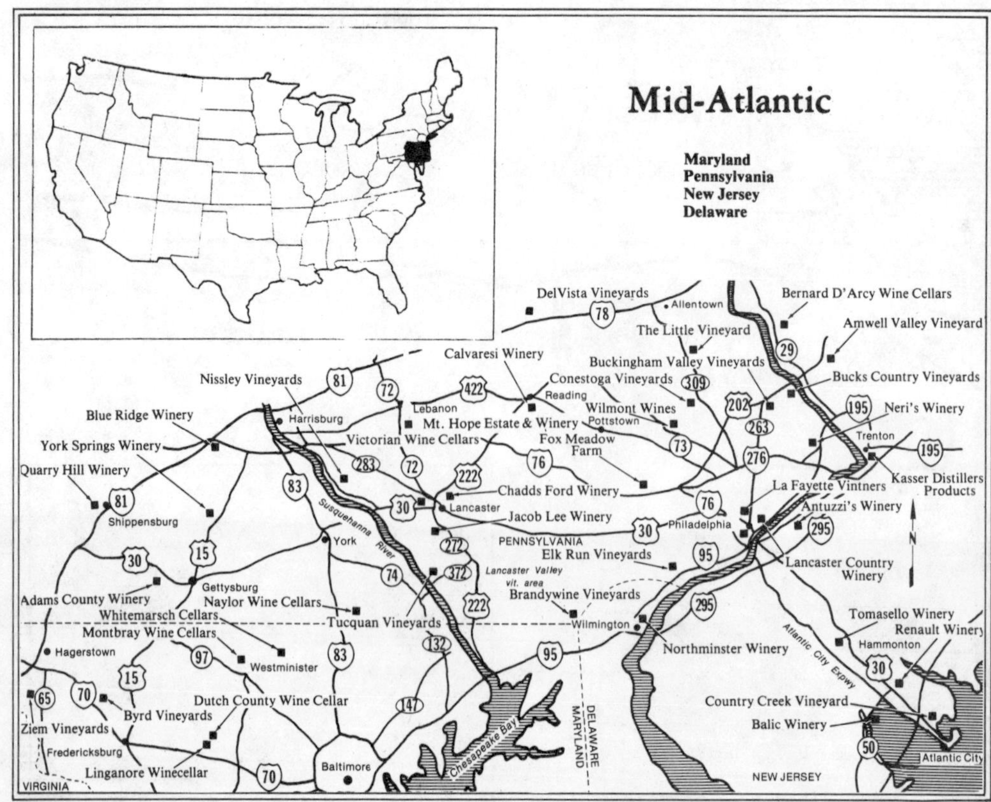

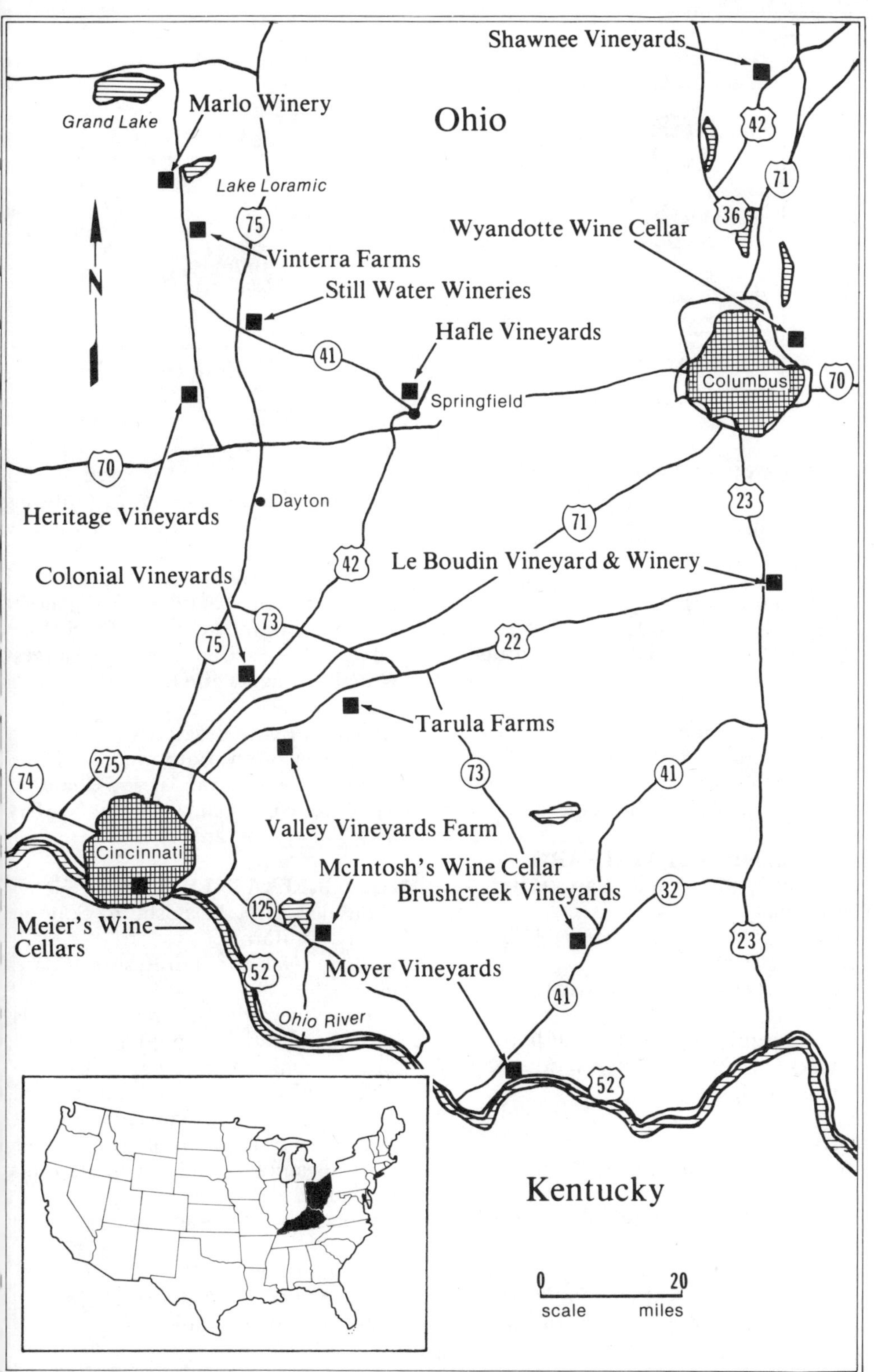
Shawnee Vineyards
Marlo Winery
Grand Lake
Ohio
Lake Loramic
75
42
71
36
Wyandotte Wine Cellar
N
Vinterra Farms
Still Water Wineries
Hafle Vineyards
41
Springfield
Columbus
70
70
Heritage Vineyards
Dayton
23
71
42
Colonial Vineyards
Le Boudin Vineyard & Winery
73
75
22
Tarula Farms
275
74
73
41
Valley Vineyards Farm
Cincinnati
McIntosh's Wine Cellar
Brushcreek Vineyards
32
125
Meier's Wine Cellars
23
52
Moyer Vineyards
41
Ohio River
52
Kentucky
0
20
scale
miles

MIDULLA VINEYARDS (See Fruit Wines of Florida.)

Lorenz Blanc, Jean de Noir, Classic Semi-Dry Red, Classic Semi-Dry White.

Midulla Vineyards

Pasco County, Florida.

Stover, Conquistador and Lake Emerald vineyard.

Mihaly Estate Vineyard

Napa County, California.

Estate Chardonnay, Pinot Noir, Sauvignon Blanc Napa Valley vineyard.

LOUIS K. MIHALY VINEYARD

Founded 1979, Napa, Napa County, California.

Storage: Oak and stainless steel. Cases per year: 10,000.

Estate vineyard: Louis K. Mihaly Vineyard.

Wines regularly bottled: Estate bottled, vintage-dated Pinot Noir, Chardonnay, Sauvignon Blanc.

MILANO WINERY

Founded 1977, Hopland, Mendocino County, California.

Storage: Oak and stainless steel. Cases per year: 10,000.

Label indicating non-estate vineyard: Lolonis Vineyard, Anderson Valley Vineyard, Scharffenberger Vineyard, Keith Vineyard, Sanel Valley Vineyard, Pacini Vineyards, Garzini Vineyards, Ordways Valley Foothill Vineyards, Victor Mathew Vineyards.

Wines regularly bottled: Vintage-dated Sauvignon Blanc, Chardonnay, Cabernet Sauvignon, Pinot Noir, Zinfandel (Late Harvest), Gewurztraminer, Late Harvest White Riesling, White Riesling,

Mill Creek Estate Vineyards

Sonoma County, California.

Cabernet Sauvignon, Merlot, Chardonnay, Pinot Noir, Gamay Beaujolais Dry Creek Valley vineyard.

MILL CREEK VINEYARDS

Founded 1975, Healdsburg, Sonoma County, California.

Storage: Oak and stainless steel. Cases per year: 12,000.

Estate vineyard: Mill Creek Vineyards.

Label indicating non-estate vineyard: Claus Neumann Vineyards.

Wines regularly bottled: Estate bottled, vintage-dated Cabernet Sauvignon, Merlot, Chardonnay, Cabernet Blush, Pinot Noir, Gamay Beaujolais; and Gewurztraminer (vintage-dated).

Second label: Feltz Springs.

Mill Station Vineyard

Sonoma County, California.

Chardonnay vineyard.

Miller Associates
Yolo County, California. R & J Cook.
Estate Cabernet Sauvignon, Sauvignon Blanc and Chenin Blanc vineyard.

J.F. MILLER & COMPANY
Napa, Napa County, California.
History: Produces grape wine carbonated in the bottle without recourse to secondary fermentation. A patented process (Millerway).
Wines regularly bottled: Millerway Carbonated Wine.

James Miller Vineyards
Sonoma County, California.
Gewurztraminer Alexander Valley vineyard.

MILLER, MARK
Mark Miller was raised in Oklahoma. He studied art there, and in California, in preparation for a career in illustration. His first job was as a costume designer for Fox Studios in the early 1940's. After serving four years in the U.S. Army, Miller moved to New York where he pursued a successful career as a free-lance illustrator, working for such magazines as the *Saturday Evening Post, Cosmopolitan,* and *Ladies Home Journal.* In 1957, he and his wife, Dene, purchased a large vineyard on the bank of the Hudson River, just outside Marlboro, New York. They removed the old labrusca vines and replanted the Benmarl Vineyards in vinifera and French-American hybrids. In 1961, Miller's work took him to Europe, where he and his family lived for the next seven years, in England, Paris, Burgundy and the Cote d'Azur. While there, Miller observed the European wine industry, and commuted to Benmarl to maintain the vineyards. In 1967, the family returned to found the Benmarl Wine Company. Miller financed the new winery in a unique manner. He founded the Societe des Vignerons, in which members pay an initial membership fee, then pay yearly dues to sponsor two vines. In return, they receive one case of specially labeled wine each year. Miller has experimented with over 50 grape varieties since the founding of Benmarl, and has increased production to over 24,000 gallons each year.

Mirabelle Vineyard
Sonoma County, California.
Sonoma-Cutrer estate Chardonnay Russian River Valley vineyard.

Miramonte Vineyard
Riverside County, California.
Chenin Blanc and Sauvignon Blanc Temecula vineyard.

Mirande Vineyard
El Dorado County, California.
Chardonnay vineyard.

MIRASSOU
The Mirassou family traces its American winegrowing tradition back five generations to the arrival of Pierre Pellier in California in 1854. He was looking for gold, but recognized the Santa Clara Valley as an ideal location for varietal grapes. He is credited with planting the first French Colombard, Grey Riesling and Pinot Noir, amongst other varietals, from Europe. In 1881, the Mirassou name came into the family when Pellier's daughter, Henrietta, married a neighboring winemaker, Pierre Mirassou, who

joined the family business and, when he died unexpectedly in 1889, left the operation of the vineyards to Henrietta and their three sons, Peter, Herman and John. As the sons married, the holdings were divided, with Peter and his wife settling in the home at the winery's current Aborn Road location. The vineyards were replanted following the phylloxera epidemic of 1894, and Peter continued operation of the winery through Prohibition, producing wine grapes for home winemaking. In 1933, his sons, Edmund and Norbert, replanted the vineyards in premium varietals. They took over operation of the winery in 1937, with Norbert in charge of the vineyards and Edmund directing winemaking and finances. In 1961, as agricultural land became scarce in the Valley, the Mirassou's expanded, with the first vineyard planting in Monterey County. The 5th generation began the marketing and distribution of wine under the Mirassou name in 1966. Today the winery is owned and operated by Peter, Jim and Daniel Mirassou. The Mirassou family celebrated their 130th winemaking anniversary in 1984.

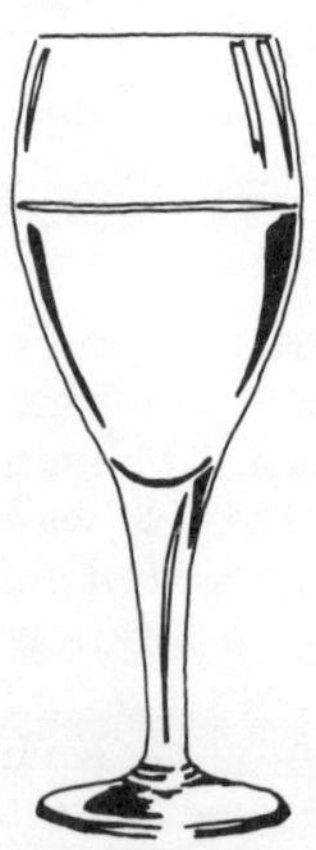

The graceful "Tulip" glass

MIRASSOU

Founded 1854, San Jose, Monterey, Santa Clara Counties, California.

Storage: Oak and stainless steel. Cases per year: 350,000.

History: The history is in the family. See above.

Estate vineyard: San Vicente Vineyard, Mission Ranch, Mirassou Vineyards.

Wines regularly bottled: Estate bottled, vintage-dated Late Harvest Johannisberg Riesling, Chardonnay, White Burgundy (Pinot Blanc), Gewurztraminer, Johannisberg Riesling, Chenin Blanc, Monterey Riesling, Gamay Beaujolais, Petite Sirah, Cabernet Sauvignon. Also vintage-dated Burgundy, Dry Chablis and Petite Rosé. Sparkling wines produced are vintage-dated Blanc de Noir Champagne, Au Naturel Champagne and Brut Champagne. Harvest Reserve Selection includes Monterey Fumé, Cabernet Sauvignon and Chardonnay.

MIRASSOU, E.A.

Edmund Mirassou is one of the pioneers of the American wine industry. He and his brother, Norbert, were the first winegrowers to recognize and develop the potential of Monterey County as an important winegrowing region. Ed and Norb also developed the first permanent vineyard irrigation system. Chairman of the California Wine Advisory Board for 22 years, and the American Society of Enologists' Man-of-the-Year in 1979, E.A. Mirassou is important in the modern history of American wine.

MIRASSOU, HENRIETTA PELLIER

In 1874, when Miss Pellier was in her early teens, her father Pierre Pellier brought her into his Evergreen Winery and taught her how to make his Pellier Santa Clara wines; thus she became one of America's first female winermakers.

MISSION

A black grape introduced by the Jesuit Missionaries who planted the first vinifera grapes at the San Diego Mission in California in the late 1800's. Used for white dessert wines such as Angelica.

MISSION PERIOD

Father Juan Ugarte, about 1697, planted the first vines on California soil, at Mission San Francisco Xavier, in what is now Baja California. Vines were then sent to other missions for plantings; vines were brought to San Diego, in 1769, by Father Serra; the San Gabriel vineyard was planted in 1771. By the end of the century, wines were made at San Juan Capistrano, Santa Barbara, San Diego, San Gabriel and San Buena Ventura. Vineyards were established in the north at San Carlos, Soledad, San Antonio, San Luis Obispo and Santa Clara.

Mission San Jose Vineyard

Alameda County, California.

Weibel estate Pinot Noir, Chardonnay Santa Clara Valley vineyard.

MISSION STREET (See Arbor Crest.)

Cherry wine.

Mission Vineyards

Napa County, California.

Dunn estate Cabernet Sauvignon Napa County vineyard.

MISSISSIPPI DELTA VITICULTURAL AREA

The viticultural area extends for approximately 180 miles mostly in Northwestern Mississippi with very small segments in Tennessee and Louisiana.

MISSOURI RIESLING GRAPE

A labrusca grape that produces a light, aromatic white wine. Originated by Nicholas Grein about 1870 at Hermann, Missouri.

MISSOURI WINE COUNTRY

The tradition of fine winemaking runs deep into the rolling hills and valleys of Missouri.

As early as 1823, French Jesuit missionaries began to produce wines from the wild grapes which flourished in Florissant, "The Valley of the Flowers" near St. Louis.

By 1843, German settlers began to cultivate vineyards on the sunny hillsides overlooking the Missouri River valley, near the town of Hermann, Missouri.

In 1866, Missouri was the second largest wine producing state in the nation, and the city of St. Louis became America's center of wine study and research.

When phylloxera threatened to destroy the vineyards of France, it was resistant root stocks from Missouri that the French chose for grafting to save their vineyards.

Today, a great renaissance is taking place.

MIZNER, WILSON (1872-1933)
writer and commentator on society.

> "A good party is where you enjoy good people, and they taste even better with Champagne."

MJC Estate Vineyard

Montgomery County, Virginia.

Chardonnay, Pinot Noir/Pinot Meunier, White Riesling North Fork of Roanoke vineyard.

MJC VINEYARD

Founded 1981, Blacksburg, Montgomery County, Virginia.

Storage: Oak and stainless steel. Cases per year: 1,500.

History: Vineyard is in the 22 mile viticultural district of North Fork of Roanoke on the Eastern Continental Divide.

Estate vineyard: MJC Vineyard.

Wines regularly bottled: Estate bottled, vintage-dated Johannisberg Riesling, Chardonnay, Pinot Noir. Occasionally bottle Blanc de Landot Noir.

Second labels: Pearis Mountain, Appalachian Harvest.

MISSOURI MAP

M & M Vineyards
Mendocino County, California.
Gamay Beaujolais vineyard.

WINES IN MODERATION

ADE, GEORGE—American humorist and playwright. (1866-1944)

"Good wine is an aid to digestion and a promoter of good cheer. I don't think anyone will find a sound argument against the moderate use of it."

AGASSIZ, LOUIS—American naturalist. (1807-1873)

"I hail with joy—for I am a temperance man and a friend of temperance—I hail with joy the efforts that are being made to raise wine in this country. I believe that when you have everywhere cheap, pure, unadulterated wine, you will no longer have need for either prohibitory or license laws."

CADMAN, CHARLES WAKEFIELD—American musician and composer. (1881-1946)

"I heartily favor the use of wine and its sensible use as a beverage as a help in the digestion of the evening meal. I have found the use of wine in my own life one of great benefit. I truly believe that if the American public were educated in the use of wine both as a mild stimulant and tonic, and as an aid to digestion, the national concern about 'hard liquors' would solve itself."

GATES, NORMAN E.—Founder and Grand Commander of the Knights of the Vine Brotherhood.

"Water separates the people of the world, wine unites them."

HOLMES, OLIVER WENDELL—American man of letters. (1809-1894)

"'Tis but the fool that loves excess;
Has thou a drunken soul?
Thy bane is in thy sallow skull,
Not in my silver bowl."

JACKSON, JUDGE—Formerly of U.S. Court of Customs and Patents Appeal.

"Wine seems to have been a concomitant of mankind as far back as books and tradition reach. It appears to have been what man has always considered the principal and finest product of the grape . . . Wine always has, and probably always will, grace boards of refinement and gentle conduct."

"Ordinary wine is used as a common and usual beverage by multitudes of our people instead of water. The Saviour changed water into wine at the behest of His Virgin Mother at the wedding feast; it was used at the Last Supper and, as a matter of common knowledge, it is a part of the very core of the most sacred religious rites of both Christian and other faiths."

JEFFERSON, THOMAS—Third president of the United States. (1743-1826)

"I rejoice, as a moralist, at the prospect of a reduction of the duties on wine by our national legislature. No nation is drunken where wine is cheap and none sober where the dearness of wine substitutes ardent spirits as the common beverage."

"I think it is a great error to consider a heavy tax on wines as a tax on luxury. On the contrary, it is a tax on the health of our citizens."

POLLOCK, CHANNING—American playwright. (1880-1946)

"Moderation in the drinking of good wine often is conducive to health and happiness. Wine-drinking peoples are never—or seldom—intemperate."

WRIGHT, RICHARDSON—American editor and writer on gardening. (1886-1961)

"The proper tempo for wine drinking is leisurely. It is almost meditative. It aids in digestion, supplies needed vitamins and carbohydrates and suffuses the whole being with warmth and friendliness not to be captured by any other means."

"A people who drink wine, who prefer wine to spirits, are a temperate people."

"Mounting taxes laid on wine drive the temperate man's beverage from his table."

"He who drinks spirits wants to forget; He who drinks wine wants to remember."

"Drinking a little wine with meals dulls the edge of worry and banishes anger. Worry and anger cause more high blood pressure than the general run of doctors realized."

MODESTO

In Stanislaus County, California Modesto is the headquarters of E. & J. Gallo Winery. Here wines are aged, bottled and shipped.

MOFFETT, WILLIAM J.

Following two generations of newspapermen, William J. Moffett began his career as a journalist in Minneapolis in 1963, but then entered book publishing shortly thereafter, with the title of field editor for Charles Scribner's Sons, a position he held until 1974. Four years previously, Moffett had purchased a vineyard on Seneca Lake in New York's Finger Lakes District. With partner/wife, Hope Merletti, he launched the first issue of Eastern Grape Grower Magazine in February, 1975, to create a new feeling of community and camraderie among eastern American and Canadian winegrowers.

The first major development was the trade show, Wineries Unlimited.

Springing from the very first Wineries Unlimited was development of the Eastern Wine Competition for wineries east of the Rocky Mountains.

Also resulting from the interaction of wineries attending Wineries Unlimited, the Association of American Vintners was organized in 1977 as a not-for-profit industry trade association to represent the interests of eastern winemakers at the federal level.

Mr. Moffett was appointed to serve as executive director of the Association of American Vintners at its formation. Mr. Moffett also serves as publisher of Eastern Grape Grower and Winery News.

MON AMI WINE COMPANY

Founded 1870, Port Clinton, Ottawa County, Ohio.

Storage: Oak and stainless steel. Cases per year: 25,000.

History: The winery was constructed of limestone more than 100 years ago.

Label indicating non-estate vineyard: Mon Ami.

Wines regularly bottled: Baco Noir, Spiced Wine, Pink and White Catawba, Delaware, Chardonnay, Johannisberg Riesling; Rosé; LaVin and Brut Champagne.

MONARCH WINE COMPANY, INC.

Founded 1933, Brooklyn, Kings County, New York.

Storage: Oak and stainless steel.

History: Still owned by the original founder. The winery has made a household name of its Manischewitz Kosher wines.

Wines regularly bottled: Under the Manischewitz label are Concord, Catawba, Cream White Concord, Cream Red Concord, Medium-Dry Concord, Blackberry, Cherry, Loganberry, Elderberry; semi-generic Dry Burgundy, Chablis; Pink and White Champagnes, Cold Duck, Sparkling Burgundy; and proprietary Almonetta, Mochanetta, Pina Coconetta, Strawberry Coconetta.

MONARCH WINE COMPANY OF GEORGIA

Founded 1933, Atlanta, Fulton County, Georgia.

Storage: Oak and stainless steel.

Wines regularly bottled: Kosher Concord, Blackberry, Cherry, Niagara, Scuppernong; also "cooking" wines.

C.K. MONDAVI

(See Charles Krug Winery.)

Chablis, Rhine, Vin Rosé, Burgundy, Barberone, Chianti, Zinfandel, Fortissimo, Bravissimo.

MONDAVI, PETER R.

Peter R. Mondavi, the son of Cesare and Rosa Mondavi who immigrated to the United States in 1907 and established residence in Virginia, Minnesota. Peter Sr. was born in Minnesota on November 11, 1914.

In 1922 Cesare, Rosa and the Mondavi family moved to Lodi, California where Cesare continued his grape shipping business which he had started in Minnesota. Peter Sr. attended grade school and high school in Lodi. In 1933, after graduating high school, he attended Stanford University where he received his B.S. degree in general economics with a minor in chemistry. He then attended the University of California at Berkeley where he studied enology with Professor Cruess. Under the supervision of Professor Cruess he completed numerous cold fermentation experiments on both red and white wines. The results were used in his production of wines at the Charles Krug Winery which his father had purchased in 1943. This purchase was a continuation of his winery adventure which had started with Repeal in 1933.

During the early years, Peter purchased French oak barrels for the aging of Chardonnay, Sauvignon Blanc, Cabernet Sauvignon and Pinot Noir. By 1957 he had fully succeeded in the true sterile filtration of wines to remove all of the yeast from wines which still had fermentable sugars. With this technique came the development of Charles Krug Chenin Blanc along with other delicate wines with residual fermentable sugars remaining in the final bottled product.

Today, with the aid of his two sons Marc and Peter, Jr., he is completing his dream, the new highly sophisticated Charles Krug Winery.

MONDAVI, ROBERT G.

In 1913, Robert Gerald Mondavi was born in Virginia, Minnesota, to Italian immigrants Cesare and Rosa Mondavi. Cesare moved his family to Lodi in 1923, where he operated a fruit-shipping firm that sent grapes from the Napa and Central Valleys to the East coast. In 1936, Robert was graduated from Stanford University, with a degree in general economics, and joined his father at the small winery the family had just purchased in Napa. Working in the cellars of Sunnyhill Winery, Robert gained first-hand knowledge of making and marketing wine. He put his education to use when the family bought the defunct Charles Krug Winery in 1943. As general manager, Robert replanted the existing vineyards in quality grape varietals. By 1957, the Mondavis were marketing premium varietal wines throughout the United States. Early success was found with Chenin Blanc and, in 1966, Fume Blanc, made from the Sauvignon Blanc grape. 1966 was the year for Robert to build a winery for the Robert Mondavi family. Michael was ready to enter the business and Tim would eventually join after finishing school. Marcia would also join the men in the Robert Mondavi Winery. In 1978 Michael became president in charge of finance and marketing. Tim soon took over as winemaker in addition to his assignment as co-winemaker for Opus One. In recent years Robert has more time to study trends and innovations in the industry, for Robert Mondavi is an innovator and creator. Slowing down for Robert simply means going from a run to a gallop. He is a great credit to the industry and to Cesare, who started it all.

ROBERT MONDAVI WINERY

Founded 1966, Oakville, Napa County, California.

Storage: Oak and stainless steel.
History: See Robert Mondavi biography.
Estate vineyard: To-Kalon Vineyard, Oak Knoll Vineyard.
Wines regularly bottled: Estate bottled, vintage-dated Cabernet Sauvignon, Pinot Noir, Chardonnay, Johannisberg Riesling, Chenin Blanc, Moscato D'Oro, Fumé Blanc (Sauvignon Blanc). Also produce Robert Mondavi Red, White and Rosé. (See Opus One.)

Monitz Vineyard
Santa Clara County, California.
Cabernet Sauvignon, Morgan Hill vineyard.

Monmartre Vineyard
Santa Clara County, California.
Congress Springs estate Cabernet Sauvignon, Chardonnay and Zinfandel Santa Cruz Mountain vineyard.

MONT EAGLE (See Vichon Winery.)
White Table Wine (Chenin Blanc), Red Table Wine (Pinot Noir).

Mont Elise Estate Vineyard
Klickitat County, Washington.
Pinot Noir, Gewurztraminer, Gamay Beaujolais vineyard.

MONT ELISE VINEYARDS, INC.
Founded 1974, Bingen, Washington.
Storage: Oak and stainless steel. Cases per year: 4,000.
Estate vineyard: Mont Elise Vineyards.
Label indicating non-estate vineyard: Graves Vineyard.
Wines regularly bottled: Estate bottled, vintage-dated Pinot Noir, Gewurztraminer, Pinot Noir Rosé, Gamay Beaujolais; vintage-dated Chenin Blanc.

MONT LA SALLE (See The Christian Brothers.)
Chablis, Haut Sauterne, Rosé, Burgundy, Chateau des Freres, St. Paul, La Salle Special, Port, Tokay, Muscatel, Angelica. Altar Wine (also called Sacramental or Mass Wine).

MONT ST. JOHN CELLARS
Founded 1934, Napa, Napa County, California.
Storage: Oak and stainless steel. Cases per year: 12,000.
Estate vineyard: Mont St. John Vineyards.
Wines regularly bottled: Estate bottled, vintage-dated Chardonnay, Pinot Noir, Johannisberg Riesling, Gewurztraminer, Muscat Canelli; also vintage-dated Cabernet Sauvignon, Petite Sirah, Zinfandel. Produce Wine Country White Wine and Wine Country Red Wine.

Mont St. John Estate Vineyards
Napa County, California.
Chardonnay, Pinot Noir, Johannisberg Riesling, Gewurztraminer, Muscat Canelli Napa Valley vineyard.

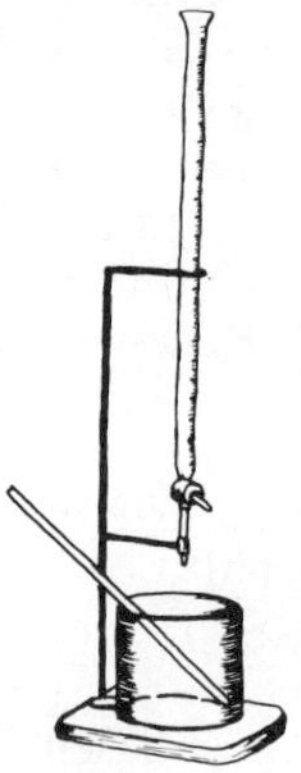

Titrating set to test for total acidity

R. MONTALI WINERY
Founded 1982, Berkeley, Alameda County, California.
Storage: Oak and stainless steel. Cases per year: 50,000.
History: Dr. Richard Carey, one of the most innovative of California's winemakers joined Montali when its owner acquired a great deal of winemaking equipment and inventory from Dr. Carey's former winery. As we say, "a good marriage".
Label indicating non-estate vineyard: French Camp Vineyard, Bien Nacido Vineyard, Bowman Vineyard.

Wines regularly bottled: Vintage-dated Blanc Fumé, Chardonnay, Zinfandel, Cabernet, Gewurztraminer, Pinot Noir; Bel Blanc (Blanc de Noir).
Second labels: Sonoma Mission, Lombard Hill, Jonathan.

Montbray Vineyard
Carroll County, Maryland.
Montbray estate Seyve Villard, Foch, Riesling, Chardonnay, Cabernet Sauvignon, Cabernet Franc and Merlot vineyard.

MONTBRAY WINE CELLARS, LTD.
Founded 1964, Westminister, Carroll County, Maryland.
Storage: Oak. Cases per year: 2,200.
History: Dr. Mowbray has made many important contributions to the wine business by producing a varietally labeled Seyve Villard (Seyval Blanc), introducing commercial wines in Maryland made from Chardonnay, Riesling and Cabernet Sauvignon, and a "clone vineyard" of vines propagated from a single cell of the French Hybrid Seyve Villard.
Estate vineyard: Montbray Vineyards.
Wines regularly bottled: Marechal Foch, Chardonnay, Johannisberg Riesling, Cabernet Sauvignon, Seyve Villard.

MONTCLAIR WINERY
Founded 1975, Piedmont, Alameda County, California.
Storage: Oak. Cases per year: 1,300.
Label indicating non-estate vineyard: Kelley Creek Vineyard, Teldeschi Vineyard, Stags Leap Vineyard.
Wines regularly bottled: Vintage-dated Zinfandel, Petite Sirah, French Colombard.

Montdomaine Estate Vineyard
Albermarle County, Virginia.
Merlot, Cabernet Sauvignon, Monticello Region vineyard.

MONTDOMAINE VINEYARDS
Founded 1980, Charlottesville, Albemarle County, Virginia.
Storage: Oak and stainless steel. Cases per year: 2,000.
History: The first estate winery in Albemarle County since 1914 state Prohibition.
Estate vineyard: Montdomaine Vineyard.
Wines regularly bottled: Estate bottled, vintage-dated Chardonnay, Merlot.
Second label: Monticello Wine Company.

Monte Rosso
Sonoma County, California.
Louis Martini estate Zinfandel, Cabernet Sauvignon, Johannisberg Riesling and Folle Blanche vineyard.

Montelle Estate Vineyard
St. Charles County, Missouri.
Seyval, Vidal, Villard Blanc, Baco Noir, Marechal Foch, Delaware and Millot August Area vineyard.

MONTELLE VINEYARDS, INC.
Founded 1969, Augusta, St. Charles County, Missouri.
Storage: Oak and stainless steel.
Estate vineyard: Montelle Vineyards.
Wines regularly bottled: Varietal and proprietary blends.

MONTEREY COUNTY
Central Coast. (See Monterey Wine Country.)

MONTEREY PENINSULA WINERY
Founded 1974, Monterey, Monterey County, California.
Storage: Oak and stainless steel. Cases per year: 15,000-20,000.
History: Located in the city limits of Monterey, the winery occupies the stately old ranch house of a former Spanish land grant. Known to locals of the Monterey Peninsula as the "Chateau" or the "Old Stonehouse", the property was originally the Rancho Saucito (meaning "little willows"). During the 1920's the Charles Ryan family handbuilt the structure as their home. The uniquely sturdy building was completely constructed of native

limestone found on the property, and its 5-foot-thick walls now protect the wines naturally at a cool 58 degrees F.

Label indicating non-estate vineyard: Cobblestone Vineyard, Ferrera Vineyard, Dusi Vineyards, Willow Creek Vineyard, Junction Vineyards, Lone Oak Vineyard, Arroyo Seco Vineyard, Hacienda Vineyard, Sleepy Hollow Vineyard.

Wines regularly bottled: Vintage-dated Chardonnay, Pinot Blanc, Cabernet Sauvignon, Zinfandel, Barbera, Pinot Noir, Late Harvest Zinfandel, Late Harvest Pinot Blanc, Late Harvest Chardonnay, Late Harvest Johannisberg Riesling. Also bottle Plum, Apricot, Malvasia Bianca, Muscat Canelli and Cream Sherry. Occasionally bottle White Zinfandel, Late Harvest Chenin Blanc.

Second label: Monterey Cellars/Big Sur Wines.

THE MONTEREY VINEYARD

Founded 1974, Gonzales, Monterey County, California.

Storage: Oak and stainless steel. Cases per year: 50,000 cases of vintage varietals and 90,000 cases of vintage classic blends per year.

History: Located between the Gavilan Heights and Santa Lucia mountain ranges, near the town of Gonzales, on portions of land originally granted by the Government of Mexico to Theodore Gonzales in 1836. Founded by a group of Monterey County grape growers, the winery is, today, owned by the Seagram Wine Corporation. Richard G. Peterson is the President and Winemaker (see Peterson biography). Not to be confused with "Monterey Vineyards", which is a giant public limited partnership vineyard operation without a winery.

Label indicating non-estate vineyard: French Camp Vineyard. The Monterey Vineyard indicates vineyard on label whenever the winery feels that the grapes from that vineyard are worthy of so indicating. Because the vineyards in Monterey County are only 10 years old, it is the opinion of the winemaker that more time is needed before specifically earmarking individual vineyards for estate bottling.

Wines regularly bottled: Vintage-dated Chardonnay, Fumé Blanc, Johannisberg Riesling, Chenin Blanc, Pinot Blanc, Gewurztraminer, Rosé of Cabernet Sauvignon; vintage-dated Brut Champagne and Classic California Red, Dry White and Rosé. Occasionally bottles Thanksgiving Harvest Johannisberg Riesling, Botrytis Sauvignon Blanc, Soft White Riesling, December Harvest Zinfandel and Cabernet Sauvignon. The winery has no second label, although Taylor California Cellars originally began as the second label for The Monterey Vineyard.

Barrel fermentation and aging cellar

MONTEREY WINE COUNTRY

In 1935, classifying the best growing areas for wine grapes, staff members of the University of California at Davis examined the climate of Monterey County and found the combination of soil, drainage and temperature ideal. But it wasn't until 1961, with demand for table wines steadily increasing, that the first important plantings of vineyards were made by Mirassou Vineyards of San Jose, Paul Masson Vineyards of Saratoga, and Wente Brothers of Livermore who were the pioneers.

In the early 1970's the real explosion in planting began multiplying the acreage 30 fold. Vineyards were planted on the benchlands of the mountain ranges that define the eastern and western borders of Monterey County. Each vineyard was scientifically laid out, with widely-spaced rows to allow for mechanical harvesting. All are trellised. Only the classic wine grape varieties were planted—those that were in demand in the growing table wine market and were well-suited to the long, cool Monterey season. Pinot Noir, Gewurztraminer, and Chardonnay predominate in the northern-most areas. Mid-way down the valley, others, including Johannisberg Riesling and Cabernet Sauvignon, make their appearance. In the warmer areas in the south, where the cooling effect of the ocean breezes is less influential, there are Chenin Blanc, Zinfandel and Petite Sirah, along with other varieties which thrive better in slightly higher temperatures.

Montevina Estate Vineyard
Amador County, California.
Zinfandel, Cabernet Sauvignon, Barbera, Cabernet Franc, Sauvignon Blanc, Semillon, Orange Muscat Shenandoah Valley vineyard.

Montevina Vineyard
Amador County, California.
Cabernet Sauvignon and Zinfandel vineyard.

MONTEVINA WINES, INC.
Founded 1973, Plymouth, Amador County, California.
Storage: Oak and stainless steel. Cases per year: 40,000.
Estate vineyard: Monteviña Vineyards.
Label indicating non-estate vineyard: Tepusquet Vineyards.
Wines regularly bottled: Estate bottled, vintage-dated Zinfandel, Sauvignon Blanc, Cabernet Sauvignon, White Zinfandel, Semillon.

MONTICELLO CELLARS
Founded 1980, Yountville, Napa County, California.
Storage: Oak. Cases per year: 15,000.
Estate vineyard: Monticello Vineyards, Big Ranch Vineyards.
Wines regularly bottled: Estate bottled, vintage-dated Chardonnay, Sauvignon Blanc, Gewurztraminer; White Dinner Wine. Occasionally produce Chateau "M" Late Harvest Sauvignon Blanc and Cabernet Sauvignon (both vintage-dated).

Monticello Cellars Estate Vineyard
Napa County, California.
Chardonnay, Gewurztraminer, Sauvignon Blanc Napa Valley vineyard.

Monticello Vineyards
Napa County, California.
Chardonnay Napa Valley vineyard.

Monticello Vineyards
Yolo County, California.
R & J Cook estate Cabernet Sauvignon, Merlot, Petite Sirah and Chenin Blanc vineyard.

MONTICELLO VITICULTURAL AREA
The viticultural area in central Virginia called "Monticello" is locally and nationally known as the home of Thomas Jefferson.

Moon Vineyards
Napa County, California.
Pinot Noir and Zinfandel vineyard.

MOORE-DUPONT
Founded 1972, Stoddard County, Missouri.
Storage: Oak and stainless steel. Cases per year: 1,000-5,000.
Estate vineyard: United Ridge Vineyard.
Label indicating non-estate vineyard: Vinland Vineyards.
Wines regularly bottled: Chambourcin, Seyval Blanc, Vidal Blanc, Villard Blanc, White Catawba, Rosé Catawba, Chelois Rosé.

John Morehead Farm/Debevc Vineyards
Lake County, Ohio.

Delaware and Cabernet Sauvignon Lake Erie vineyard.

Morehouse Vineyards
Yates County, New York.
Golden Muscat Finger Lakes vineyard.

Morelli Ranch
Sonoma County, California.
Zinfandel Green Valley vineyard.

Moreman Vineyards
Franklin County, Washington.
Worden's estate Chardonnay, Sauvignon Blanc, White Riesling, Gewurztraminer, Muscat Canelli, Cabernet Sauvignon and Pinot Noir vineyard.

MORGAN HILL CELLARS
(See Pedrizzetti Winery.)

MORGAN WINERY
Founded 1982, Salinas, Monterey County, California.
Storage: Oak and stainless steel. Cases per year: 2,000.
Label indicating non-estate vineyard: Hillside Vineyard, Cobblestone Vineyard.
Wines regularly bottled: Vintage-dated Chardonnay.

MORLEY, CHRISTOPHER (1890-1957)
American writer.

"Wine opens the heart. It warms the shy poet hidden in the cage of the ribs. It melts the wax in the ears that music may be heard.

"It takes the terror from the tongue that truth may be said, or what rhymes marvelously with truth.

"The soft warm sting on the cheekbones that a ripe Burgundy gives is only the thin outward pervasion of a fine heat within, when the cruel secret moulder of the wit leaps into clear flame, flame that consumes the sorry rubbish of precaution and cajolery.

"The mind is full of answers. Then presently, if you have dealt justly with the god, not brutishly, he gives you the completest answer of all—sleep."

J.W. MORRIS WINERY
Founded 1975, Emeryville, California, moved to Healdsburg, Sonoma County, California in 1982.
Storage: Oak and stainless steel. Cases per year: 25,000.
Estate vineyard: Black Mountain Vineyard.
Wines regularly bottled: Vintage-dated Chardonnay, Sauvignon Blanc, Cabernet Sauvignon; and White Private Reserve, Red Private Reserve (vintage-dated), also estate bottled Vintage Port.

MORTON, LUCIE
Lucie Morton is a dynamic young woman who was graduated from the University of Pennsylvania with a degree in 1971, but, after becoming involved with her father's vineyard in Virginia, decided to change her career to viticulture. In 1972 she attended a viticulture course of the Vinifera Wine Growers Association in Virginia. She attended the Cours Superieur International de Viticulture in Montpellier, France and received her degree in 1974. For her degree, she produced a fifty page thesis, written in French, on viticulture in the Eastern United States. Since 1974, she has worked as a writer, lecturer and consultant on viticulture in general, and wine growing East of the Rockies as a specialty. In 1975, with the encouragement of Philip Hiaring, Philip Jackisch and Leon Adams, Lucie set out to translate Professor Galet's Ampelography. This task was completed in 1979 when Cornell University Press published "A Practical Ampelography", by Pierre Galet, translated and adapted by Lucie T. Morton. Her next book will be "Wines and Wineries of Eastern America."

Morton Vineyards
Napa County, California.

Zinfandel and Cabernet Sauvignon Napa Valley vineyard.

MOSCATO
The Italian name for the varietal grape (Muscat) that produces sweet wines. Moscato Amabile, Moscato D'Oro.

MOSCATO CANELLI (MUSCAT)
Wine grape.
The Moscato produces a rich, fruity wine with a delicate muscat character. Sweet enough to be a dessert wine also. Will age very well.

MOSELLE WINE
A generic semi-dry wine. Usually produced from Chenin Blanc and Riesling grapes. Semi-dry, mildly sweet, pale straw in color.

MOTHER LODE (See California Cellar Masters.)
Spice Jubilee and May Wine. Also generics: Chablis, Mellow Burgundy. Only varietal is Zinfandel Rosé.

MOTHER LODE (See Sierra Vista Winery.)
Mother Lode Red, Mother Lode Gold.

MOTHER LODE VINEYARDS
(See Karly Wines.)

MOULTON, EDWARD A.
Edward A. Moulton became interested in wine more than twenty years ago. At first it was an avocation, but gradually it became an occupation when he briefly became a partner with Richard Vine. Moulton has worked for several east coast wineries over the years, including Niagara Falls Winecellar, Buck's Country Vineyards and Gross Highland Wineries. He has been particularly successful with indigenous grape varieties, and has won numerous winemaking awards. Moulton is currently vice-president and winemaker of Meier's Wine Cellars, Inc. in Silverton, Ohio.

Mount Baker Estate Vineyards
Whatcom County, Washington.
Muller Thurgau, Gewurztraminer, Riesling, Foch, Leon Millot vineyard.

MOUNT BAKER VINEYARDS, INC.
Founded 1978, Everson, Whatcom County, Washington.
Storage: Oak and stainless steel. Cases per year: 8,370.
Estate vineyard: Mt. Baker Vineyards.
Wines regularly bottled: Estate bottled, vintage-dated Chardonnay, Pinot Noir, Late Harvest Riesling, Muller Thurgau, Gewurztraminer, Nouveau Foch, Leon Millot, Apple, Plum; estate bottled, vintage-dated proprietary Crystal Rain Blanc, Crystal Rain Rouge, Crystal Rain Rosé, Mountain Spring Muscat, Madeline Angevine, Okanogan Riesling, Precoce de Malingre. Occasionally Late Harvest wines.
Second label: Lucie Kuhlman. (Not to be confused with French hybrid.)

Mount Eden Estate Vineyards
Santa Clara County, California.
Cabernet Sauvignon, Pinot Noir and Chardonnay Santa Cruz Mountains vineyard.

MOUNT EDEN VINEYARDS
Founded 1972, Saratoga, Santa Clara County, California.
Storage: Oak. Cases per year: 5,000.

Estate vineyard: Mt. Eden Vineyard.

Label indicating non-estate vineyard: Ventana Vineyards.

Wines regularly bottled: Estate bottled, vintage-dated Pinot Noir, Cabernet Sauvignon, Chardonnay.

Second label: M.E.V.

MOUNT PALOMAR WINERY

Founded 1975, Temecula, Riverside County, California.

Storage: Oak and stainless steel. Cases per year: 12,000.

History: John Poole planted one of the early vineyards in 1969 which was part of the "new wave" of wineries that rediscovered the Temecula area.

Estate vineyard: Long Valley Vineyards.

Wines regularly bottled: Estate bottled, vintage-dated Chardonnay, Sauvignon Blanc, White Riesling, Chenin Blanc, Shiraz, Petite Sirah, Cabernet Sauvignon, Cabernet Rosé; vintage-dated proprietary Chateau Palomar. Bottles Cocktail, Golden, Cream Sherries. Produces Vintage White and Vintage Rhine (semi-generic), Sangria and more semi-generics: Burgundy, Vin Rosé, Chablis. Occasionally bottles Special Reserve Port, Sweet Cabernet, Zinfandel and White Riesling.

Second labels: Rancho Temecula, Long Valley Vineyards.

The classic "Brandy" snifter

Mount Veeder Estate Vineyards

Napa County, California.

Cabernet Sauvignon, Cabernet Franc, Merlot, Malbec, Petite Verdot vineyard.

MOUNT VEEDER WINERY

Founded 1973, Napa, Napa County, California.

Storage: Oak and stainless steel. Cases per year: 5,000.

History: Formerly known as Bernstein Vineyards.

Estate vineyard: Mount Veeder Vineyards.

Wines regularly bottled: Estate bottled, vintage-dated Cabernet Sauvignon, Zinfandel, Chardonnay, Chenin Blanc; also non-vintage Zinfandel, Pinot Blanc and proprietary Sidehill Blanc.

MOUNTAIN HOUSE WINERY

Founded 1980, Cloverdale, Mendocino County, California.

Storage: Oak and stainless steel. Cases per year: 2,500.

History: In 1859, Alexander McDonald, a former New Yorker who had settled in California after the Mexican War, purchased a tract of land in the Mendocino Mountains. The inland region of Mendocino County was then a wilderness of Redwood forests and grassy valleys. On the coast, the lumber industry was developing, but the coastal towns were accessible only by sea.

In the next few years, the first dirt roads were cut through the rugged mountain terrain to connect the County with San Francisco. One route stretched south from the interior agricultural valleys around Ukiah; another pushed inland for 50 miles from the seacoast. These two great roads crested at McDonald's property, where they converged for their southward journey to San Francisco.

In due course, McDonald built a stage stop and an inn to accommodate the growing stream of travelers and commerce. "Mountain House" soon became such a prominent landmark that the road to

Ukiah was named "Mountain House Road" and the coastal route became known as the "McDonald-to-the-Sea" Highway.

In the late 1930's, present Highway 101 supplanted Mountain House Road as the main route to the Northern California interior.

In 1979, the property became Mountain House Winery.

Wines regularly bottled: Vintage-dated Chardonnay, Cabernet Sauvignon, Mendocino Gold and Vermillion. Occasionally bottled: Vintage-dated Late Harvest Zinfandel.

MOUNTAIN RED CHIANTI

A generic dry wine. Robust and tart, with berry-like bouquet.

MOUNTAIN RED WINE

A generic red table wine. Just another name for semi-generic Burgundy.

MOUNTAIN WHITE WINE

A generic white table wine. Just another name for semi-generic Chablis.

mousseux

Foamy, frothy, sparkling. Also means effervescent wines that are not produced in the Champagne region of France.

MOWBRAY, G. HAMILTON

G. Hamilton Mowbray has had a distinguished career in psychology. He earned a master's degree in physiological and experimental psychology from the Johns Hopkins University, in 1950, and, three years later, obtained a Ph.D. in experimental psychology from Cambridge University, England. He worked first as senior staff research psychologist, from 1953 to 1970, at the Johns Hopkins Applied Physics Laboratory, then on their principal professional staff until 1973. Meanwhile, Mowbray pursued his interest in wine. Since 1966, Mowbray has been founder and co-owner of the Montbray Wine Cellar, Ltd. He is a free-lance wine writer and has lectured on wine at Mt. Vernon College in Washington, D.C., and at the University of Maryland. He has served as a wine consultant to the Bank of America, Time-Life Books, Inc., and wineries in Maryland, Virginia, Pennsylvania and West Virginia. Mowbray has also been involved as a founding member of the American Wine Society, and received their Award of Merit in 1977.

Moyer Estate Vineyards

Adams County, Ohio.

Vidal Blanc, Chambourcin, Villard Noir and Vidal Blanc Ohio River Valley vineyard.

MOYER VINEYARDS

Founded 1973, Manchester, Adams County, Ohio.

Storage: Stainless steel.

Estate vineyard: Moyer Vineyard.

Wines regularly bottled: Estate bottled Vidal Blanc, Chambourcin, Villard Blanc de Noir.

MOYER TEXAS CHAMPAGNE COMPANY

Founded 1980, New Braunfels, Comal County, Texas.

Cases per year: 3,000-4,000.

Wines regularly bottled: Champagnes (Natural, Brut, Extra Dry.)

MR. DUDE 44 (See Fruit Wines of Florida.)

Dude 44 White, Dude 44 Super Red.

Mt. Glenn Vineyard
Mendocino County, California.
Chardonnay Anderson Valley vineyard.

MT. HOPE ESTATE AND WINERY

Founded 1979, Cornwall, Lancaster County, Pennsylvania.

Storage: Stainless steel. Cases per year: 21,000.

History: In 1800, Henry Bates Grubb built the sandstone mansion at Mt. Hope. The mansion is like a feudal English manor. The Victorian elegance of the mansion and its castle walls, turrets, hand-painted ceilings, antiques and furnishings of the Victorian Period make the mansion an exciting visit. The vineyards are on the mansion grounds, and the tasting room is the old billard room.

Estate vineyard: Mt. Hope Vineyard.

Wines regularly bottled: Estate bottled, vintage-dated Seyval Blanc, Vidal Blanc, Cabernet Sauvignon, Chardonnay, Rosé of Chelois, White Riesling, Fruit wines.

Mt. Hope Estate Vineyard
Lancaster County, Pennsylvania.
Vidal Blanc, Riesling vineyard.

Mt. Pleasant Estate Vineyard
St. Charles County, Missouri.
Vidal Blanc, Seyval Blanc, Villard Noir, Cynthiana and Cordon Rouge Augusta Area vineyard.

MT. PLEASANT VINEYARDS

Founded 1881, Augusta, St. Charles County, Missouri.

Storage: Oak and stainless steel. Cases per year: 15,000.

History: The town of Augusta was founded in 1836 and incorporated in 1855. The winery and vineyards of Mount Pleasant were founded less than 25 years later, winning medals at the Columbian Exposition (1893) and the St. Louis World's Fair (1904).

Estate vineyard: Mt. Pleasant Vineyard.

Wines regularly bottled: Estate bottled, vintage-dated Concord, Vidal Blanc, Missouri Riesling, Seyval Blanc, Villard Noir, Cynthiana, Cordon Rouge; also Rosé, Champagne, Red Labrusca; proprietary Stark's Star (after dinner wine), Steuben, Munch, White Munch, Munch Rosé.

Mt. St. Helena Vineyard
Lake County, California.
Cabernet Sauvignon vineyard.

Mt. St. Clare Vineyard
Lake County, California.
Channing Rudd estate Merlot and Cabernet Sauvignon vineyard.

Mt. Veeder Vineyard
Napa County, California.
Malbec vineyard.

Mtn. Cove Vineyard
Nelson County, Virginia.
La Abra estate Villard Blanc, Baco Noir vineyard.

MTN. GOLD (See Lamont Winery.)

MTN. VIEW (See Gemello Winery.)

THE MTN. VIEW WINERY
Founded 1980, Mtn. View, Santa Clara County, California.
Storage: Oak. Cases per year: 1,000.
Label indicating non-estate vineyard: Hughes Vineyard, Hawkes Vineyard.
Wines regularly bottled: Vintage-dated Chardonnay and Cabernet Sauvignon.

Muellers Vineyard
Napa County, California.
Chateau Chevre estate Cabernet Franc and Sauvignon Blanc vineyard.

MULLER-THURGAU
A vinifera grape that is a cross of White Riesling and Sylvaner that produces a fresh, lightly spicy white wine. The floral characteristic comes from its White Riesling ancestry.

Murphy Ranch
Sonoma County, California.
Sauvignon Blanc Alexander Valley vineyard.

MUSCADINE GRAPES
Native to the southeastern United States, they were growing long before the arrival of the first settlers. Noble, Carlos and Welder are examples of its descendants.

MUSCAT
The name for a family of grapes distinguished by their spicy "raisin" flavor. Muscat of Alexandria is grown in California for table grapes, raisins and common sweet wines. Muscat Canelli and Orange Muscat are considered finer wine grapes in California, where they are almost always made sweet.

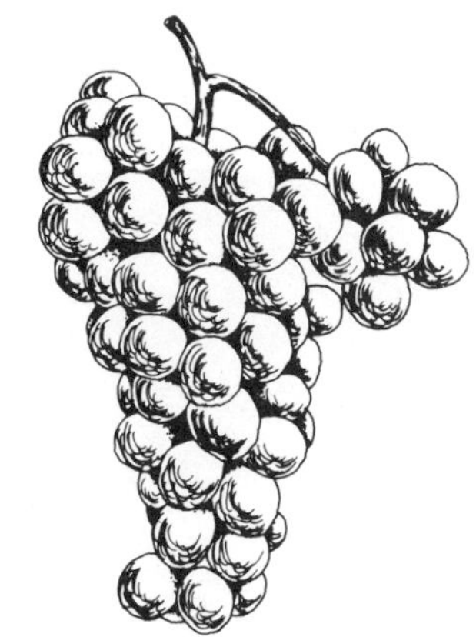

Muscat

MUSCAT FRONTIGNAN
Wine grape.
Primarily used for dessert wines. This is the French name for Muscat Blanc and Muscat Canelli.

MUSCATEL
A rich, flavorful, sweet dessert wine made from Muscat grapes, having their unmistakable flavor and aroma. Its color ranges from golden, or dark, amber to red. While most Muscatel is made from the Muscat of Alexandria grape and is golden in color, several other varieties are used.
Red Muscatel, Black Muscatel: Muscatels which are red or dark red and are sometimes made from black Muscat grapes.
Muscat de Frontignan, Muscat Canelli: Muscatel made from the Muscat Blanc variety of Muscat grapes. Both names refer to the same grape.

must
The term used for the juice of crushed grapes during the fermentation process. After fermentation it becomes wine.

Naismith Vineyards
Yolo County, California.
Petite Sirah vineyard.

NAKED MOUNTAIN VINEYARD
Founded 1976, Markham, Fauquier County, Virginia.
Storage: Oak and stainless steel. Cases per year: 600.
Wines regularly bottled: Vintage-dated Chardonnay, Sauvignon Blanc, Riesling, Cabernet Sauvignon; and Claret.

NAPA CELLARS
Founded 1978, Oakville, Napa County, California.
Storage: Oak and stainless steel. Cases per year: 10,000.
Label indicating non-estate vineyard: Black Mountain Vineyard.
Wines regularly bottled: Vintage-dated Cabernet Sauvignon, Chardonnay, Sauvignon Blanc, Zinfandel.

NAPA COUNTY
Some of the outstanding vineyard areas of California are in this county. Napa Valley, Pope Valley, Gordon Valley, Wooden Valley, Carneros, Calistoga, Napa, Stag's Leap, Rutherford, St. Helena, Mount Veeder, Yountville, Oakville and Pritchard Hills are the major areas.

Napa Creek Estate Vineyard
Napa County, California.
Johannisberg Riesling and Gewurztraminer vineyard.

NAPA CREEK WINERY
Founded 1980, St. Helena, Napa County, California.
Storage: Oak. Cases per year: 12,000.
Estate vineyard: Napa Creek Vineyard.
Wines regularly bottled: Estate bottled, vintage-dated Johannisberg Riesling and Gewurztraminer; vintage-dated Chardonnay, Fumé Blanc, Dry Chenin Blanc, Cabernet Sauvignon.

NAPA GAMAY
This is probably not the grape that produces France's Beaujolais although many say it is. Not to be confused with "Gamay Beaujolais", which is one of the lighter clones of Pinot Noir in California.

NAPA VALLEY
The Napa Valley is relatively small, extending from San Francisco Bay at Napa, some thirty-five miles, to the foothills of Mt. St. Helena. The valley received its name from the "Nappas," Indians who lived there in the early nineteenth century. The first settlers arrived about 1836, but no appreciable growth in the population took place until the 1840's. With the arrival of many immigrant families, considerable agriculture developed in grain and livestock farming. St. Helena was founded in 1853; followed by Calistoga in 1859. The railroad to San Francisco was completed in 1868. The first vineyard for wine making purposes was planted in the valley in 1838 by George C. Yount and started producing wine near what is Yountville in 1844. The 1870's constituted the decade of greatest expansion in the vineyard industry of the valley, reaching a total of 11,000 acres by 1886. This famous wine producing valley is home to some of California's most prestigious wineries.

NAPA VALLEY MAP

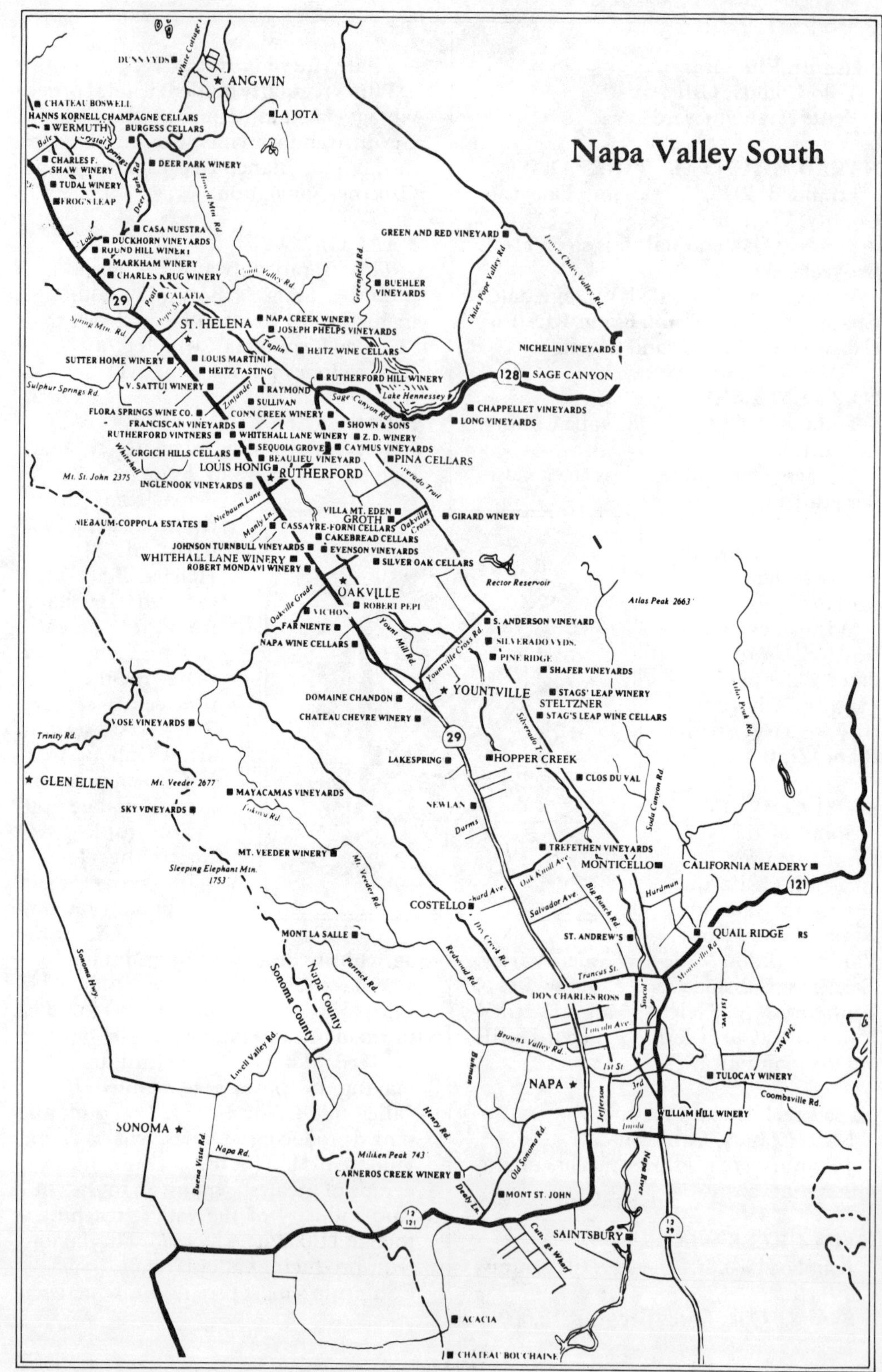

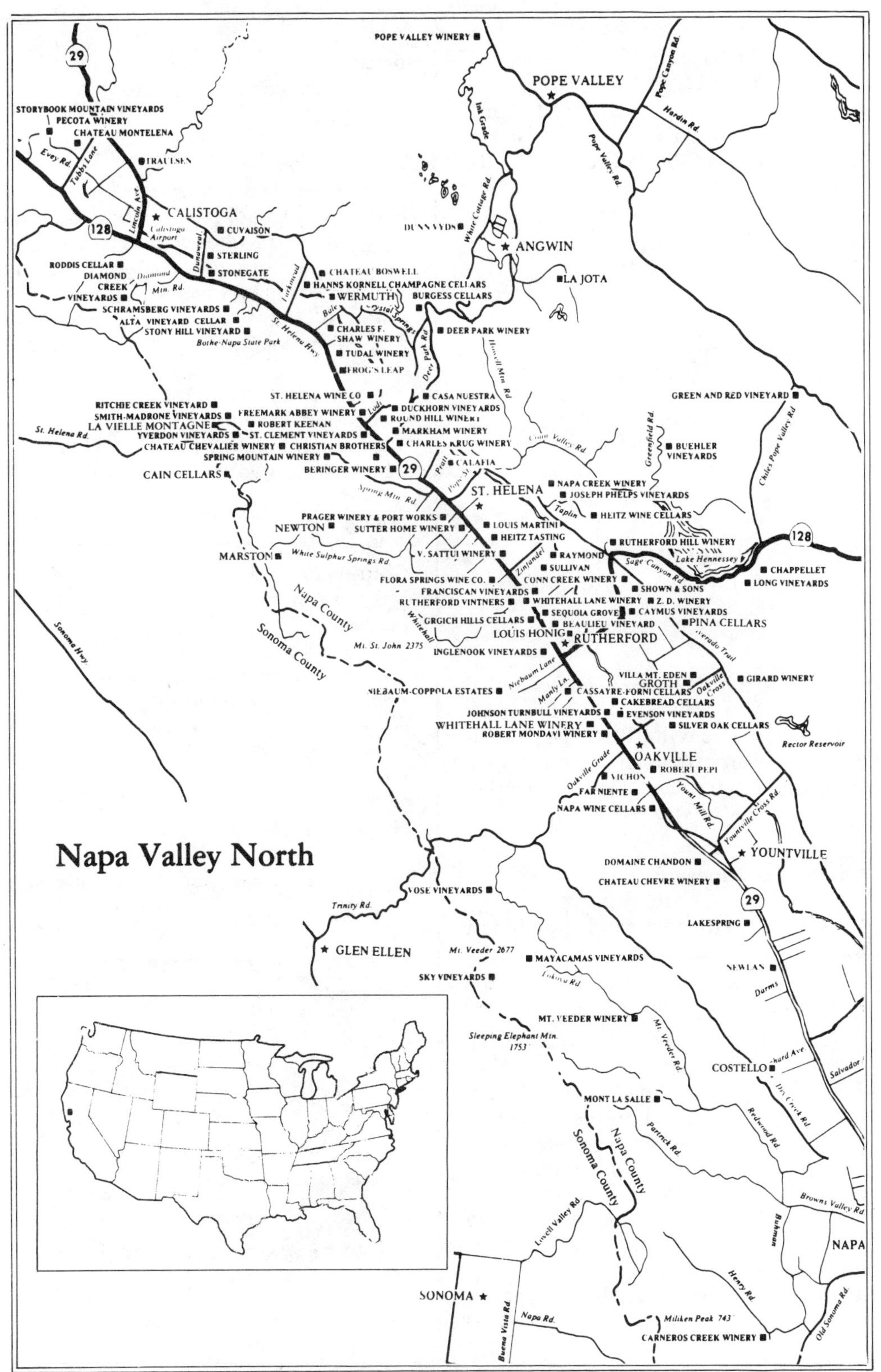
Napa Valley North
POPE VALLEY WINERY
POPE VALLEY
Pope Canyon Rd.
Hardin Rd.
Pope Valley Rd.
Ink Grade
White Cottage Rd.
DUNN VYDS
ANGWIN
LA JOTA
29
STORYBOOK MOUNTAIN VINEYARDS
PECOTA WINERY
CHATEAU MONTELENA
Evey Rd.
Tubbs Lane
TRAULSEN
Lincoln Ave.
CALISTOGA
Calistoga Airport
128
CUVAISON
STERLING
STONEGATE
Dunaweal
Larkmead
RODDIS CELLAR
DIAMOND CREEK VINEYARDS
Diamond Mtn. Rd.
CHATEAU BOSWELL
HANNS KORNELL CHAMPAGNE CELLARS
WERMUTH
BURGESS CELLARS
SCHRAMSBERG VINEYARDS
ALTA VINEYARD CELLAR
STONY HILL VINEYARD
Bothe-Napa State Park
St. Helena Hwy.
CHARLES F. SHAW WINERY
TUDAL WINERY
FROG'S LEAP
DEER PARK WINERY
Deer Park Rd.
Howell Mtn. Rd.
RITCHIE CREEK VINEYARD
SMITH-MADRONE VINEYARDS
LA VIELLE MONTAGNE
YVERDON VINEYARDS
CHATEAU CHEVALIER WINERY
St. Helena Rd.
ST. HELENA WINE CO
FREEMARK ABBEY WINERY
ROBERT KEENAN
ST. CLEMENT VINEYARDS
CHRISTIAN BROTHERS
SPRING MOUNTAIN WINERY
BERINGER WINERY
CAIN CELLARS
Lodi
CASA NUESTRA
DUCKHORN VINEYARDS
ROUND HILL WINERY
MARKHAM WINERY
CHARLES KRUG WINERY
CALAFIA
Pratt
Spring Mtn. Rd.
ST. HELENA
GREEN AND RED VINEYARD
Greenfield Rd.
BUEHLER VINEYARDS
Chiles Pope Valley Rd.
NAPA CREEK WINERY
JOSEPH PHELPS VINEYARDS
Taplin
HEITZ WINE CELLARS
PRAGER WINERY & PORT WORKS
NEWTON
SUTTER HOME WINERY
LOUIS MARTINI
HEITZ TASTING
MARSTON
White Sulphur Springs Rd.
V. SATTUI WINERY
RUTHERFORD HILL WINERY
Sage Canyon Rd.
Lake Hennessey
RAYMOND
SULLIVAN
Zinfandel
CHAPPELLET
LONG VINEYARDS
FLORA SPRINGS WINE CO.
CONN CREEK WINERY
FRANCISCAN VINEYARDS
SHOWN & SONS
RUTHERFORD VINTNERS
WHITEHALL LANE WINERY
Z. D. WINERY
Napa County
Sonoma County
Sonoma Hwy.
Whitehall
GRGICH HILLS CELLARS
SEQUOIA GROVE
CAYMUS VINEYARDS
BEAULIEU VINEYARD
PINA CELLARS
LOUIS HONIG
RUTHERFORD
Mt. St. John 2375
INGLENOOK VINEYARDS
VILLA MT. EDEN
GROTH
GIRARD WINERY
Oakville Cross
NIEBAUM-COPPOLA ESTATES
Niebaum Lane
Manly La.
CASSAYRE-FORNI CELLARS
CAKEBREAD CELLARS
JOHNSON TURNBULL VINEYARDS
EVENSON VINEYARDS
WHITEHALL LANE WINERY
SILVER OAK CELLARS
ROBERT MONDAVI WINERY
Rector Reservoir
OAKVILLE
Oakville Grade
ROBERT PEPI
VICHON
FAR NIENTE
NAPA WINE CELLARS
Yount Mill Rd.
Yountville Cross Rd.
YOUNTVILLE
DOMAINE CHANDON
CHATEAU CHEVRE WINERY
VOSE VINEYARDS
Trinity Rd.
GLEN ELLEN
LAKESPRING
Mt. Veeder 2677
MAYACAMAS VINEYARDS
SKY VINEYARDS
NEWLAN
Durms
MT. VEEDER WINERY
Mt. Veeder Rd.
Sleeping Elephant Mtn. 1753
COSTELLO
Salvador
Dry Creek Rd.
MONT LA SALLE
Redwood Rd.
Partrick Rd.
Sonoma County
Napa County
Browns Valley Rd.
Lovell Valley Rd.
Buhman
NAPA
Henry Rd.
SONOMA
Buena Vista Rd.
Napa Rd.
Miliken Peak 743
CARNEROS CREEK WINERY
Old Sonoma Rd.

NAPA VALLEY VITICULTURAL AREA

The Napa Valley viticultural area includes all of Napa County, except for land northeast of Putah Creek and Lake Berryessa.

NAPA VINTNERS

(See Don Charles Ross Winery.)

NAPLES VALLEY

(See Widmer's Wine Cellars.)
Burgundy, Chablis, Rosé.

Narsai David Vineyard

Napa County, California.
Merlot Napa Valley vineyard.

NASHOBA VALLEY WINERY

Founded 1978, Concord, Middlesex County, Massachusetts.

Storage: Stainless steel. Cases per year: 5,000.

Wines regularly bottled: Estate bottled Fruit and Berry Wines: Dry Semi-Sweet Sparkling Apple Wine (Brut); Dry, Semi-Sweet Blueberry Wine; Cranberry-Apple; Pear; proprietary: Orchard Run (Apple-Pear) and Cyser (Apple-Honey).

Nathan Fay Vineyard

Napa County, California.
Cabernet Sauvignon and Sauvignon Blanc Stags Leap District vineyard.

NATIVE AMERICAN GRAPE

Vitis Muscadinia is not a species. It is a sub-genus which includes the species Rotundifolia, which in turn includes the Muscadine varieties. The other sub-genus is Vitis Euvitis, which includes Vitis Vinifera and a long list of species native to North America, including a few pertinent to winemaking; among these are Labrusca, Aestivalis and Riparia. Of the various species, some 4,000 varieties of American and American hybrid grapes have been identified.

natural

The designation for absolutely dry champagne.

natural wine

When used on a label it means that no brandy or alcohol has been added to the wine, all the alcohol being produced by "natural" fermentation. It does not imply that no sugar or water was added. The "natural" only refers to the fermentation and not to the pureness of the juice.

Navarro Estate Vineyards

Mendocino County, California.
Gewurztraminer, Pinot Noir, Chardonnay Anderson Valley vineyard.

NAVARRO VINEYARDS

Founded 1974, Philo, Mendocino County, California.

Storage: Oak and stainless steel. Cases per year: 8,000.

Estate vineyard: Navarro Vineyards.

Wines regularly bottled: Estate bottled, vintage-dated Gewurztraminer, Late Harvest Gewurztraminer, Chardonnay, Pinot Noir; vintage-dated Chardonnay, White Riesling, Cabernet Sauvignon.

Navoni Vineyard

Napa County, California.
Sauvignon Blanc Napa Valley vineyard.

NAYLOR WINE CELLARS, INC.
Founded 1978, Stewartstown, York County, Pennsylvania.
Storage: Oak and stainless steel. Cases per year: 1,500.
Estate vineyard: Charles Vintners.
Label indicating non-estate vineyard: Seven Valleys Vineyard.
Wines regularly bottled: Estate bottled, vintage-dated Catawba, Chardonnay, Cabernet Sauvignon, Cayuga, Worden, Landal Noir, Baco Noir, Foch, Rose O Chaunac, Millot, Aurora; Golden Grenadier, Ruby Grenadier, All American; also vintage-dated Vidal, Riesling, Chambourcin, Seyval; Niagara, also proprietary: York White Rosé; also Rhinelander; Occasionally proprietary Fragola (Strawberry), Frambois (Raspberry); Merlot, Gewurztraminer, Chelois.

NEBBIOLO
A grape producing a dry, fruity and tart wine. Originally from Italy where it is used to produce Barolo and Gattinara.

NEGOCIANTS (Eleveurs)
The literal meaning is to raise or bring up—as a child. Negociants select and purchase wines from producers and then mature and bottle them as through they were their own. When ready, the wines are offered for sale.

NEHALEM BAY WINE COMPANY
Founded 1973, Nehalem, Tillamook County, Oregon.
Storage: Oak and stainless steel. Cases per year: 24,000.
History: The winery is located on the Oregon coast.
Wines regularly bottled: Vintage-dated Pinot Noir; and Cabernet Sauvignon, Blackberry, Plum; also Red and Rosé Table Wines. Occasionally bottles Cranperé (cranberry and pears).

Nelson Ranch
Mendocino County, California.
Chardonnay vineyard.

Nelson Vineyards
Sonoma County, California.
Chardonnay Sonoma Valley vineyard.

Nepenthe Vineyard
San Luis Obispo County, California.
Cabernet vineyard.

NERI WINE CELLARS
Founded 1979, Langhorne, Bucks County, Pennsylvania.
Storage: Oak and stainless steel. Cases per year: 1,500.
Estate vineyard: Neri's Vineyard.
Wines regularly bottled: Seyval, Aurora, Dutchess, Foch, Baco Noir and proprietary: Northampton Red and White, Red Rouge; occasionally Apple, Dolce Mela (sweet apple).

Neri's Estate Vineyard
Bucks County, Pennsylvania.
Foch, Baco Noir, Seyval, Aurora, Dutchess, Niagara, Concord vineyard.

NERVO (See Geyser Peak Winery.)
Pinot Noir Rosé, Cabernet Sauvignon, Chianti, Country White and Red, Winterchill White.

NEUHARTH, EUGENE
In 1928, Eugene Neuharth was born into a grape growing family, in Lodi,

California. He participated in Future Farmers of America throughout high school and leased vineyards from his father for his projects. After graduation, Neuharth leased more vineyards and became a full-time viticulturist. He gradually purchased vineyards until he owned over 300 acres in the Lodi area. During this period, he was actively involved with the Lodi Wine Growers' Association. In 1970, he decided to sell his vineyards, and three years later, he and his wife, Maria, moved to Sequim, in northwestern Washington. Although they made their living from selling and investing in real estate, they both enjoyed a home winemaking group, and were impressed with the quality of the resulting wines. In 1978, they decided to expand their hobby into a business, and founded the Neuharth Winery. Although their first crush, in 1979, included grapes imported from California, they now specialize in quality table wines made exclusively from eastern Washington grapes. Their own small vineyard will come into maturity soon.

NEUHARTH WINERY

Founded 1979, Sequim, Clallam County, Washington.

Storage: Oak and stainless steel. Cases per year: 1,700.

Label indicating non-estate vineyard: Sagemoor Farms.

Wines regularly bottled: Vintage-dated Cabernet Sauvignon, Merlot, Chardonnay, Johannisberg Riesling; vintage-dated proprietary: Dungeness Red, White Rosé. Occasionally Zinfandel and Chenin Blanc.

Claus Neumann Vineyards

Sonoma County, California.

Gewurztraminer Dry Creek vineyard.

NEVADA CITY CELLARS (See Nevada City Winery.)

Vintage-dated Zinfandel, White Riesling, Victorian White, Mountain White.

NEVADA CITY WINERY

Founded 1980, Nevada City, Nevada County, California.

Storage: Oak and stainless steel. Cases per year: 5,000.

History: The winery resumes a tradition in Nevada City that began with the 49ers. By 1860, the town's miners were turning grapes into a different type of gold and by 1880 there were 450,000 vines in Nevada City alone. Prohibition cut off the growing wine business and it lay dormant until 1980, the year that a group of local grape growers and wine lovers formed Nevada City Winery.

Estate vineyard: Snow Mountain Vineyard, Quail Glen Vineyards.

Wines regularly bottled: Vintage-dated Zinfandel, Douce Noir (Charbono), Late Harvest Riesling, Cabernet Sauvignon, Petite Sirah, Late Harvest Zinfandel, Pinot Noir. Occasionally bottles Bottle Blanc.

Second label: Nevada City Cellars.

NEW MEXICO

Early in the seventeenth century, long before grape growing and wine making became established in California, missions along the Rio Grande River in New Mexico were making wine for use in their religious services.

Travelers passing through this area in the mid 1800's were highly impressed by the quality of the wines being produced and spread the news throughout the southwest.

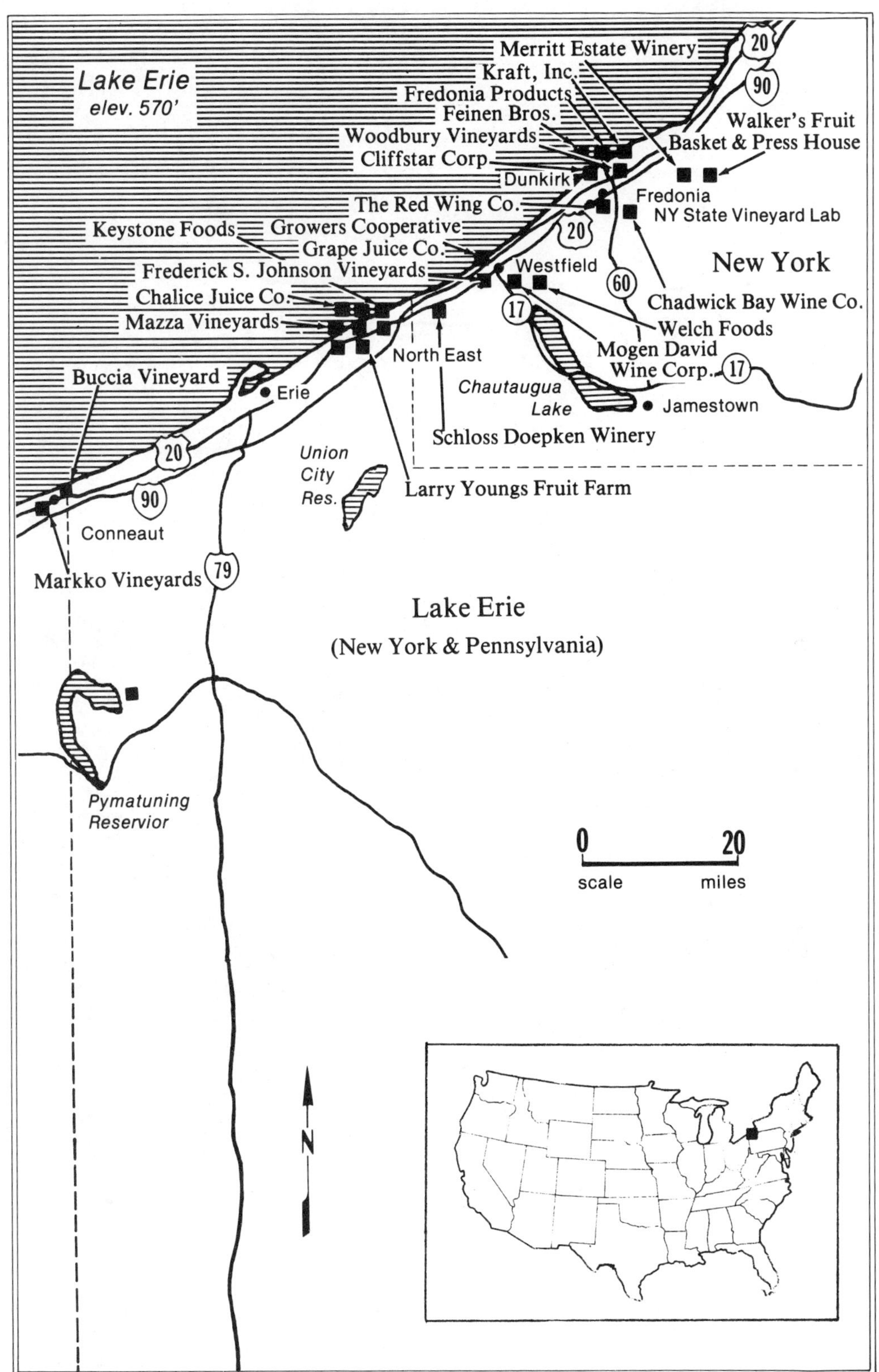
Lake Erie
elev. 570'
Merritt Estate Winery
Kraft, Inc.
Fredonia Products
Feinen Bros.
Woodbury Vineyards
Cliffstar Corp.
Dunkirk
Walker's Fruit
Basket & Press House
Fredonia
NY State Vineyard Lab
The Red Wing Co.
Keystone Foods
Growers Cooperative
Grape Juice Co.
Frederick S. Johnson Vineyards
Westfield
New York
Chalice Juice Co.
Mazza Vineyards
Chadwick Bay Wine Co.
Welch Foods
Mogen David
Wine Corp.
North East
Buccia Vineyard
Erie
Chautaugua
Lake
Jamestown
Schloss Doepken Winery
Union
City
Res.
Larry Youngs Fruit Farm
Conneaut
Markko Vineyards
Lake Erie
(New York & Pennsylvania)
Pymatuning
Reservior
0
20
scale
miles
N
20
90
60
17
79

NEW YORK MAP

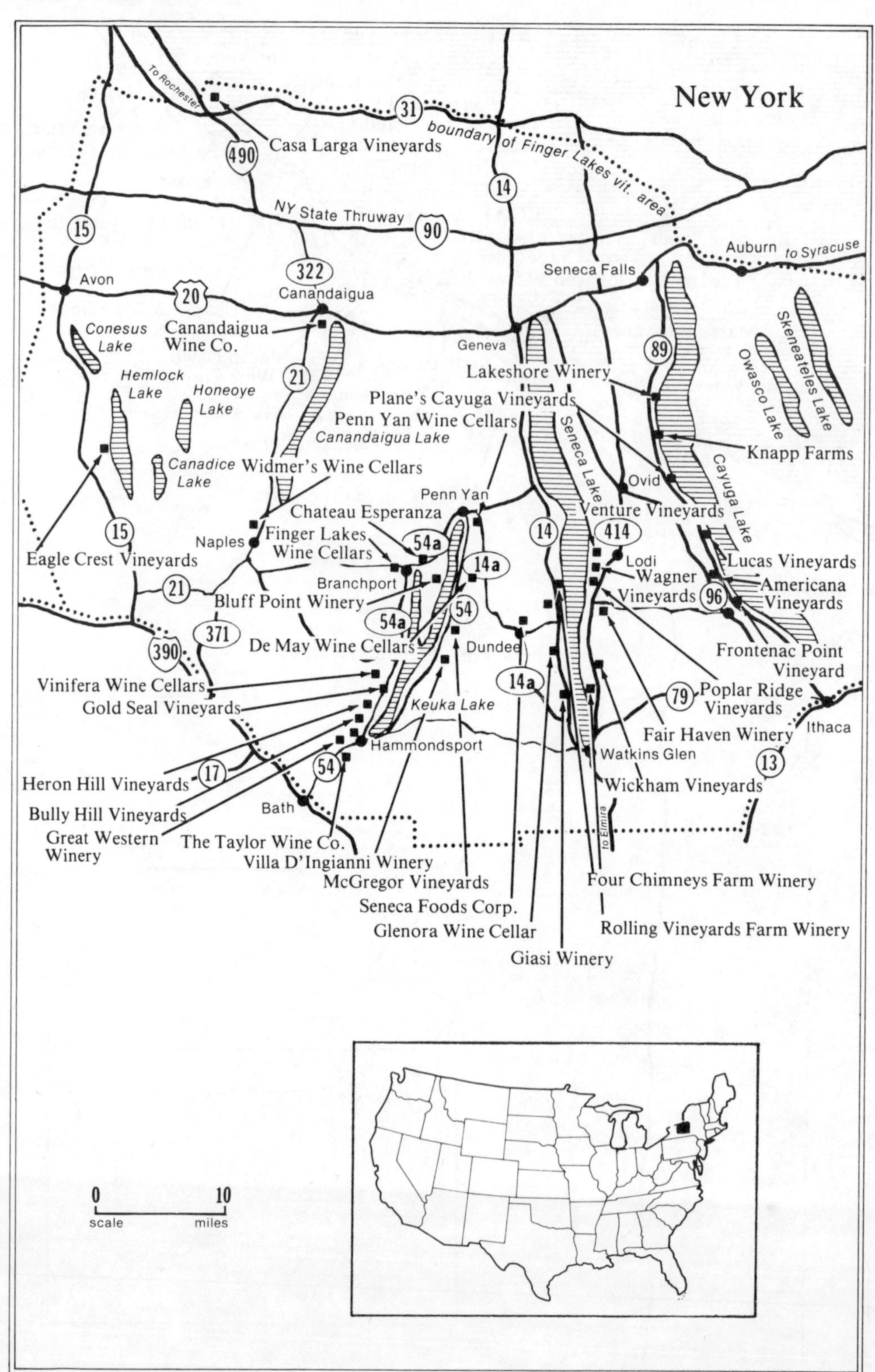

New York
To Rochester
Casa Larga Vineyards
boundary of Finger Lakes vit. area
NY State Thruway
Auburn
to Syracuse
Seneca Falls
Avon
Canandaigua
Conesus Lake
Canandaigua Wine Co.
Geneva
Hemlock Lake
Honeoye Lake
Canadice Lake
Lakeshore Winery
Plane's Cayuga Vineyards
Penn Yan Wine Cellars
Canandaigua Lake
Seneca Lake
Skeneateles Lake
Owasco Lake
Knapp Farms
Cayuga Lake
Widmer's Wine Cellars
Ovid
Penn Yan
Chateau Esperanza
Venture Vineyards
Naples
Finger Lakes Wine Cellars
Eagle Crest Vineyards
Lodi
Lucas Vineyards
Branchport
Wagner Vineyards
Americana Vineyards
Bluff Point Winery
De May Wine Cellars
Dundee
Frontenac Point Vineyard
Vinifera Wine Cellars
Gold Seal Vineyards
Keuka Lake
Poplar Ridge Vineyards
Ithaca
Hammondsport
Fair Haven Winery
Watkins Glen
Heron Hill Vineyards
Wickham Vineyards
Bully Hill Vineyards
Bath
to Elmira
Great Western Winery
The Taylor Wine Co.
Villa D'Ingianni Winery
McGregor Vineyards
Four Chimneys Farm Winery
Seneca Foods Corp.
Glenora Wine Cellar
Rolling Vineyards Farm Winery
Giasi Winery
0
10
scale
miles

In 1880, census figures showed New Mexico wineries were producing more than 900,000 gallons of wine from 3,100 acres of vineyards, and production was steadily increasing. Unfortunately, the Prohibition era forced the closing of these wineries, few of which ever reopened. In 1977, La Vina Winery was built in an area where vineyards once flourished, the fertile and beautiful Mesilla Valley of the Rio Grande.

NEWARK WINES
(See Hildebrand Estates, Canada.)

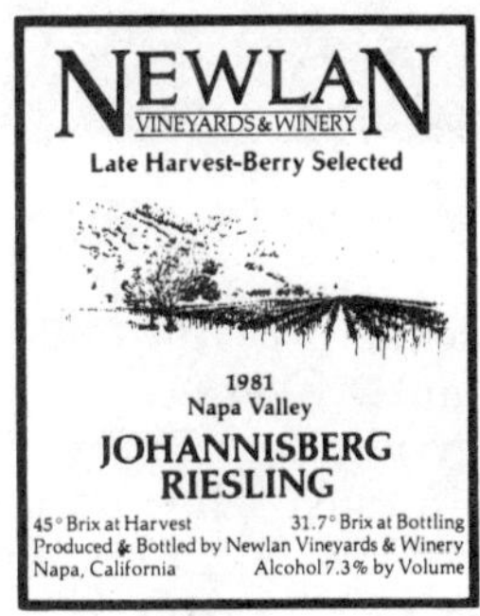

NEWLAN VINEYARDS AND WINERY
Founded 1981, Napa, Napa County, California.
Storage: Oak and stainless steel. Cases per year: 3,000.
Estate vineyard: Dry Creek Vineyard, Yonne Vineyard.
Label indicating non-estate vineyard: Sounder's Vineyard.
Wines regularly bottled: Estate bottled, vintage-dated Cabernet Sauvignon, Pinot Noir, Sauvignon Blanc, Chardonnay; vintage-dated Lake Harvest Johannisberg Riesling and Red Table Wine.

NEWMARK, JOSEPH
In 1857, Joseph Newmark, who was a Los Angeles lay rabbi, certified wine coming from the Los Angeles area and being shipped to San Francisco, as Kosher. He had been trained as a schochet in Brodnick, Poland, and was the Los Angeles Jewish patriarch for years. In 1862, he helped organize and became the first president of Congregation B'nai Brith, which today is known as the Wilshire Boulevard Temple.

Newton Estate Vineyard
Napa County, California.
Cabernet Sauvignon, Merlot, Sauvignon Blanc and Chardonnay vineyard.

NEWTON WINERY
Founded 1978, St. Helena, Napa County, California.
Storage: Oak and stainless steel. Cases per year: 13,000.
Estate vineyard: Newton Vineyard.
Wines regularly bottled: Estate bottled, vintage-dated Cabernet Sauvignon, Merlot, Sauvignon Blanc, Chardonnay.

NIAGARA
An American hybrid grape that produces very aromatic, medium sweet, fruity white wine.

NICASIO VINEYARDS
Founded 1955, Soquel, Santa Cruz County, California.
Storage: Oak. Cases per year: 200.
Wines regularly bottled: Vintage-dated Riesling, Zinfandel, Zinfandel Rosé, Cabernet, Chardonnay, Riesling cuvée Champagne au Natural, Chardonnay cuvée Champagne Au Natural. Occasionally bottles Petite Sirah, Merlot, Grenache.

Nichelini Estate Vineyard
Napa County, California.
Chenin Blanc, Sauvignon Blanc, Gamay, Zinfandel, Petite Sirah, Cabernet Sauvignon vineyard.

NICHELINI VINEYARD
Founded 1890, St. Helena, Napa County, California.
Storage: Oak and stainless steel. Cases per year: 4,000.

Estate vineyard: Nichelini Vineyard.

Wines regularly bottled: Estate bottled Chenin Blanc, Sauvignon Blanc, Gamay, Zinfandel, Petite Sirah, Cabernet Sauvignon.

Nicol Ranch

Solano County, California.

White Riesling, Gamay and Colombard Suisun Valley vineyard.

NIEBAUM-COPPOLA ESTATE

Founded 1978, Rutherford, Napa County, California.

Cases per year: 4,000.

Estate vineyard: Coppola Estate Vineyard.

Wines regularly bottled: Premium Red Table Wine (blend of Cabernet Sauvignon, Cabernet Franc and Merlot).

Nielsen Vineyard

Santa Barbara County, California.

Zaca Mesa estate Chardonnay, Sauvignon Blanc, Riesling and Cabernet Sauvignon Santa Maria Valley vineyard.

NIGHTINGALE, MYRON S.

Myron S. Nightingale Sr.'s introduction to winemaking come in 1940, when, as a member of the famous vintage class of winemakers at the University of California at Berkeley, he received a degree in bacteriology.

In 1949, he became assistant winemaker for Italian Swiss Colony at Asti, California. This was the operations center for his winery, which, in the early fifties, was a premium North Coast vintner. He held this post until 1953 when he became, not only winemaker, but operations manager of the famous Cresta Blanca Winery at Livermore, California.

There his reputation was greatly enhanced through his pioneering work with botrytized wines. The Premier Semillon which climaxed this research, was heralded by admirers within the industry, as well as the consuming public.

In 1971, Myron Nightingale was appointed winemaster of Beringer Vineyards. In December 1983, he retired and is presently acting as winemaster consultant for Beringer.

905

(See Bardenheier's Wine Cellars.)

Sweet and Dry Vermouth, Rhine, Chablis, Burgundy, Rosé, Pink Chablis, Port, Sherry, Muscatel, White Port.

Nissley Estate Vineyards

Lancaster County, Pennsylvania.

Seyval, Aurora, Vidal, Chancellor, DeChaunac, Gewurztraminer, Chardonnay, Johannisberg Riesling vineyard.

NISSLEY VINEYARDS

Founded 1978, Bainbridge, Lancaster County, Pennsylvania.

Storage: Stainless steel. Cases per year: 18,000.

Estate vineyard: Nissley Vineyards.

Wines regularly bottled: Estate bottled, vintage-dated Gewurztraminer, Chardonnay, Johannisberg Riesling; vintage-dated Seyval Blanc, Aurora, Vidal, Chancellor, DeChaunac and proprietary: Greystone Kiss, Naughty Marietta.

NITROGEN GAS

To maintain pressure during filtering and bottling of sparkling wine. To prevent oxidation of wine.

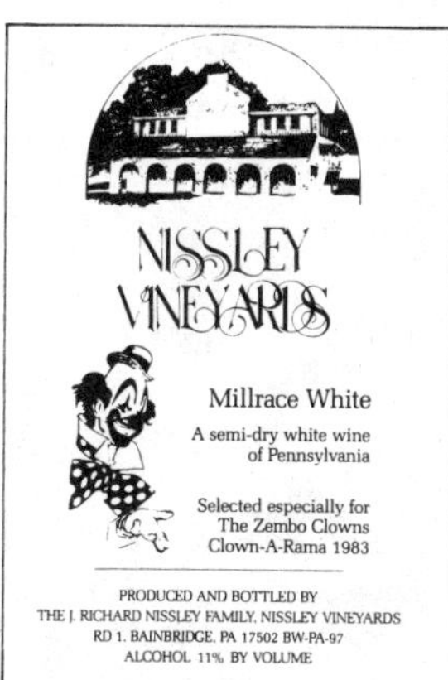

NITTANY VALLEY WINERY
Founded 1980, State College, Centre County, Pennsylvania.
Storage: Stainless steel.
Label indicating non-estate vineyard: Larry Young's Fruit Farm; Warner's Fruit Farm.
Wines regularly bottled: Vintage-dated Seyval Blanc, Vidal Blanc, Chancellors Foch, Apple; vintage-dated Sangría and proprietary: Larmes Du Lion, Ris Du Lion, Coeur Du Lion; occasionally Aurora, Baco Noir.

NOBLE
A red muscadine grape that produces light, fruity Rosé wine, with a fragrant nose.

NOBLE ROT (See Botrytis.)

NOBLE VINEYARDS
Founded 1973, Kerman, Kern County, California.
Storage: Stainless steel.
History: Owned by Pacific Land and Viticulture, the 3,600 acre vineyard produces wines to be used and/or bottled by other wineries.
Estate vineyard: Noble Vineyard.
Wines produced: French Colombard, Chenin Blanc, Semillon, Barbera, Dry White and Dry Red.

Nonini Brothers Estate Vineyard
Fresno County, California.
Barbera, Grenache, Zinfandel vineyard.

A. NONINI WINERY
Founded 1936, Fresno, Fresno County, California.
Storage: Oak. Cases per year: 2,700.
Estate vineyard: Nonini Brothers Vineyard.
Wines regularly bottled: Vintage-dated Barbera, Sweet Zinfandel, Grenache, Zinfandel; vintage-dated semi-generics: Burgundy, Chablis, Chablis Blanc, Chianti, Sauterne, Vin Rosé and Claret.

non-vintage
Wines blended from several vintages to obtain higher standard quality. Non-vintage wines can be the same quality from year to year.

Norman Vineyard
San Luis Obispo County, California.
Zinfandel, Barbera, Chardonnay and Cabernet Sauvignon Paso Robles vineyard.

Norse Vineyard
Sonoma County, California.
Wheeler estate Cabernet Sauvignon, Zinfandel Dry Creek Valley vineyard.

NORTH CAROLINA WINE HISTORY
The earliest written account of the "White Grape", as it was called by the colonists, occurs in Giovanni da Verrazzano's logbook. Verrazzano, the Florentine navigator who explored the Cape Fear River Valley for France in 1524, wrote that he saw "... Many vines growing naturally there ..." and that "... without doubt they would yield excellent wines."
Native grapes (Vitis rotundifolia—now known as Muscadines) were used in wine by the country's first settlers. Sir John Hawkins, in relieving the French at Fort Caroline in 1565, found 20 hogsheads of Muscadine wine. During the 17th century there are instances recorded where Muscadine wine was used as a medium of exchange.

Sir Walter Raleigh's explorers, Captains Phillip Amadas and Arthur Barlowe wrote, in 1584, that the coast of North Carolina was "... so full of grapes as the very beating and surge of the sea overflowed them ... in all the world, like the abundance is not to be found".

In 1585, Governor Ralph Lane stated, in describing North Carolina to Sir Walter Raleigh, "We have discovered the main to be the goodliest soil under the cope of heaven, so abounding with sweet trees that bring rich and most pleasant gummes, grapes of such greatness, yet wild, as France, Spain, nor Italy hath no greater ..."

A Castle Hayne Vineyard Company existed in the 1870s. Far surpassing all rival winemakers in the vicinity was the family of Sol Bear. By 1902, the Bears employed six commercial travellers and had an agent in New York, for their business ranged far beyond the state. A new winery was built about 1902 at Front and Marsteller Streets in Wilmington, that was capable of turning out 200,000 gallons a year. The Bears' plant was one of the largest consumers of muscadines in the nation.

There are many other fine wineries in North Carolina's history, some of which were located at Conover, Eagle Springs, Gibson, Littleton, Louisburg, Manteo, Murphy, Peachland, Pettigrew State Park, Holly Ridge, Samarcand, Tyron, Warrenton, Willard, Edenton and Icard.

The thriving commercial wine industry was brought to a conclusion in January 1909, with the adoption of state-wide Prohibition.

In the early 60's, out-of-state wineries began to be interested in the unique qualities of Muscadines. In response to a vigorous research program at North Carolina State University, several new varieties of Muscadines have been developed.

NORTH COAST VITICULTURAL AREA

"North Coast" California's largest Appelation, contains over 4,700 square miles and incorporates portions of Napa, Sonoma, Mendocino, Solano, Lake and Marin counties.

Within the North Coast viticultural area boundaries are viticultural areas: Napa Valley, Guenoc Valley, Sonoma Valley, McDowell Valley, Suisun Valley, Solano County, Green Valley, Cole Ranch, Dry Creek, Los Carneros and Anderson Valley in addition to several other viticultural areas.

Foothill vineyard

North Del Rio Estate Vineyard
Val Verde County, Texas.
Muller Thurgau and Pinot Noir vineyard.

North Del Rio Vineyard
Val Verde County, Texas.
Val Verde estate Muller Thurgau, Pinot Noir vineyard.

NORTH FORK OF THE ROANOKE VITICULTURAL AREA

North Fork of Roanoke is bounded on the west by the Alleghany Mountain ridges of the Eastern Continental Divide, on the south by the Pedlar Hills, and on the north and east by the Pearis and Ft. Lewis Mountains.

North Salem Estate Vineyard
Westchester County, New York.

Seyval, Foch, Chancellor, DeChaunac, Chelois and Cascade Hudson River Region vineyard.

NORTH SALEM VINEYARD
Founded 1965, North Salem, Westchester County, New York.
Storage: Stainless steel. Cases per year: 1,000.
Estate vineyards: North Salem Vineyard.
Wines regularly bottled: Estate bottled, vintage-dated Seyval, Foch, proprietary: Preview and Red Wine (Chancellor and DeChaunac).

Northeast Estate Vineyard
Dutchess County, New York.
Marechal Foch and Aurora Hudson River Region vineyard.

NORTHEAST VINEYARD
Founded 1972, Millerton, Dutchess County, New York.
Storage: Oak. Cases per year: 300.
Estate vineyard: Northeast Vineyard.
Wines regularly bottled: Foch, Aurora.

NORTHMINSTER WINERY INC.
Founded 1978, New Castle County, Delaware.
Storage: Oak. Cases per year: 1,000.
Wines regularly bottled: White Wines.

THE NORTHWEST WINE COUNTRY

Grape growing in the great Northwest can be traced back at least as far as 1825, when members of the Hudson Bay Company planted vines at Fort Vancouver on the north side of the Columbia River near Portland, Oregon. These vines and most of the plantings by settlers throughout the remainder of the 19th century were of the labrusca variety, a hardy American vine relatively easy to cultivate but not known for producing a grape from which fine wines were made.

As irrigation systems were developed in the Northwest in the early 20th century, vinifera vines, the "wine bearer," the vine that had produced all the great wines of Europe for centuries, were planted in larger numbers by growers. It was not until the mid-1960's, however, that these vinifera varieties were seriously cultivated, leading to the emergence of the Northwest as a major source of quality American wines.

The essential elements in the successful commercial production of fine wines are climate and soil. The Northwest has a variety of each, but, importantly, several areas of this part of the country are ideal for growing certain varieties of grapes.

The cooler climate and relatively short growing season with long hours of sunlight in Oregon, eastern Washington and Idaho, are excellent for the early maturing Gewurztraminer, Chardonnay, Johannisberg Riesling and Pinot Noir.

Soil conditions vary throughout the area but many vineyards of the Northwest are planted in soil containing small rocks and pebbles, which not only aid drainage but help the soil retain heat from the sun. Others are planted on a sandy loam, which, like the slate found in the soil of many German vineyards, reflects light from the sun back to the vines. In certain other areas the iron content in the soil is very high, quite similar to some areas in the Burgundy region of France.

NORTON

An American red wine hybrid growing principally in Missouri and considered promising for this region.

nose

Refers mainly to the qualities of bouquet and aroma.

nouveau

Literally "New" wine. In the tradition of Beaujolais, wine is bottled and released in late November, a few days after fermentation is finished. It is fresh, yeasty; a wine that captures the full fresh essence of the grape. Usually made from Gamay, Pinot Noir and Zinfandel, often using the carbonic maceration technique.

Novitiate

Santa Clara County, California.

Semillon Santa Cruz Mountain vineyard.

NOVITIATE WINES

Founded 1888, Los Gatos, Santa Clara Valley, California.

Cases per year: 30,000.

Wines regularly bottled: Vintage-dated Grenache Rosé, Chenin Blanc, Pinot Blanc, White Malvasia; also Cabernet Sauvignon, Angelica, Black Muscat, Cream Sherry, Muscat Frontignan and Port. Also produce Burgundy, Chablis, Red and White Dinner Wine. Occasionally bottle Black Rosé (Rosé of Black Muscat).

nutty

Tasting term.

The desirable odor and taste of appetizer and dessert wines and particularly well-processed Sherries.

Nyland Vineyards

Solano County, California.

Chardonnay and Barbera vineyard.

OAK BARREL WINECRAFT
Founded 1956, Berkeley, Alameda County, California.
Storage: Oak and stainless steel.
Wines regularly bottled: Moscato Secco, Pinot Noir, Zinfandel, Cabernet Sauvignon, Light Muscat; and Chablis, Vino Bianco and proprietary Chateau Oak Barrel (white wine).

OAK CREEK VINEYARD
Sonoma County, California.
Balverne estate Johannisberg Riesling Chalk Creek vineyard.

OAK KNOLL RANCH
Sonoma County, California.
Chardonnay, Sauvignon Blanc, Chenin Blanc, Pinot Noir Dry Creek vineyard.

Oak Knoll Vineyard
Napa County, California.
Robert Mondavi estate Cabernet Sauvignon, Pinot Noir, Chardonnay, Johannisberg Riesling, Chenin Blanc, Moscato d'Oro and Sauvignon Blanc vineyard.

OAK KNOLL WINERY
Founded 1970, Hillsboro, Washington County, Oregon.
Storage: Oak and stainless steel. Cases per year: 20,000.
Wines regularly bottled: Vintage-dated Chardonnay, Pinot Noir, Pinot Noir Blanc, White Riesling, Muscat of Alexandria, Gewurztraminer, Cabernet Sauvignon; and Gooseberry, Rhubarb, Red Currant, Plum, Strawberry, Blackberry, Loganberry, Raspberry, Boysenberry; occasionally a Niagara, Sauvignon Blanc, Zinfandel.

OAK RIDGE VINEYARDS
Founded 1934, Lodi, San Joaquin County, California.
Cases per year: 365,000.
History: A cooperative of approximately 125 families; formed after the repeal of Prohibition.
Wines regularly bottled: Estate bottled, vintage-dated Chenin Blanc, French Colombard, Dry Semillon, Gamay Rosé, Ruby Cabernet, Petite Sirah, Barbera, Zinfandel (White) and Cabernet Sauvignon. Semi-generics are Chablis, Burgundy, Rhine, Sauterne, Vin Rosé, Chianti. Dessert wines are Muscatel, Tinta Madeira Port, Port, Cream Sherry, Pale Dry Sherry, White Port, Tokay, Tawny Port, Marsala. Also Sweet and Dry Vermouth. Also proprietary Angelica Antigua and Angelica.

Oak Valley Vineyard
Sonoma County, California.
Hacienda estate Cabernet Sauvignon, Sauvignon Blanc, Semillon, Zinfandel vineyard.

Oakcross Vineyards
Napa County, California.
Groth estate Cabernet Sauvignon, Sauvignon Blanc, Chardonnay vineyard.

Oakencroft Estate Vineyard
Albermarle County, Virginia.
Chardonnay, Cabernet Sauvignon, Merlot Monticello vineyard.

OAKENCROFT VINEYARDS
Founded 1978, Charlottesville, Albermarle County, Virginia.
Storage: Stainless steel. Cases per year: 500.
History: Vineyards are in Charlottesville, which was the capital of the Virginia

wine belt in the 1800's. Often referred to as Jefferson County, since nearby Monticello was Thomas Jefferson's home.

Estate vineyard: Oakencroft Vineyards.

Wines regularly bottled: Estate bottled, vintage-dated Chardonnay, Seyval Blanc.

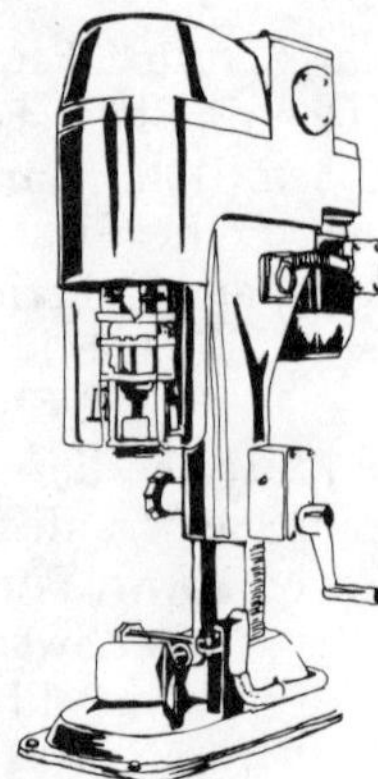

Capsuler machine forms capsules to bottles

OAKVILLE

In Napa County in the middle of the Napa Valley.

Oakville Vineyard

Napa County, California.

Franciscan estate Zinfandel, Merlot, Chenin Blanc, Johannisberg Riesling, Chardonnay and Cabernet Sauvignon Napa Valley vineyard.

Oasis Estate Vineyard

Fauquier County, Virginia.

Semillon, Sauvignon Blanc, Gewurztraminer, Chardonnay, Pinot Noir, Merlot, Cabernet Sauvignon, Chelois, Chancellor, Rayon d'Or, Seyval Blanc, Foch vineyard.

OASIS VINEYARD

Founded 1977, Hume, Fauquier County, Virginia.

Cases per year: 4,000.

Estate vineyard: Oasis Vineyard.

Wines regularly bottled: Estate bottled, vintage-dated proprietary Semillon, Sauvignon Blanc, Gewurztraminer, Chardonnay, Cabernet Sauvignon, Chelois, Chancellor, Seyval Blanc and Champagne (all under Domaine de La Venne label).

Oberhellmann Estate Vineyards

Gillespie County, Texas.

Chardonnay, Johannisberg Riesling, Sauvignon Blanc, Gewurztraminer, Semillon, Pinot Noir, Merlot, Cabernet Sauvignon vineyard.

OBERHELLMANN VINEYARDS

Founded 1972, Fredericksburg, Gillespie County, Texas.

Storage: Oak and stainless steel. Cases per year: 12,000.

History: Vineyards are 14 miles north of historic Fredericksburg in the Bell Mountain region. Several years were spent experimenting with more than 26 varieties before focusing on the nine current varieties.

Estate vineyard: Oberhellmann Vineyard.

Wines regularly bottled: Estate bottled, vintage-dated Johannisberg Riesling, Gewurztraminer, Chardonnay, Cabernet Sauvignon, Pinot Noir; estate bottled, vintage-dated Liebsfand, Federweiss, Edelblume, Shillerwine and proprietary Domane Red and White.

Second label: Oberhof.

OBESTER WINERY

Founded 1977, Half Moon Bay, San Mateo County, California.

Cases per year: 6,000.

History: Co-founder Sandra is granddaughter of John Gemello, who founded Gemello Winery in 1934.

Label indicating non-estate vineyard: Ventana Vineyard, Vinifera Vineyard, Redwood Ranch.

Wines regularly bottled: Vintage-dated Sauvignon Blanc, Johannisberg Riesling, Cabernet Sauvignon, White Cabernet, Zinfandel. Occasionally bottles Gamay.

OBLATE FATHERS

Missionaries who planted vines in British Columbia in 1864.

ODE TO CATAWBA WINE
by Henry Wadsworth Longfellow

"The song of mine
Is the song of the Vine
To sung by the glowing embers
Of wayside inns
When the rain begins
To darken the drear Novembers.

It is not a song
Of the Scuppernong
From Warm Carolian valleys,
Nor the Muscadel
That bask in our garden alleys.

Nor the red Mustang
Whose clusters hand
O'er the waves of the Colorado,
An the firey flood
Of whose purple blood
Has a dash of Spanish bravado.

For the richest and best
Is the wine of the West,
That grows by the Beautiful River,
Whose sweet perfume
Fills all the room
With a bension on the river.

And as hollow trees
Are the haunts of bees,
For ever going and coming;
So this crystal hive
Is all alive
With a swarming and buzzing and humming.

Very good in its way
Is the Verzenay,
Or the Sillery soft and creamy;
But the Catawba wine
Has a taste more divine
More dulcet, delicious and dreamy.

There grows no vine
By the haunted Rhine,
By Danube or Guadalquivir,
Nor island or cape,
That bears such a grape
As grows by the Beautiful River.

Drugged is their juice,
For foreign use,
When shipped o'er the reeling Atlantic,
To rack our brains
With the fever pains,
That have driven the Old World frantic.

To the sewers and sinks
Will all such drinks
And after them tundle the mixer.
For a poison malign
Is such Borgia wine,
Or he best but a Devil's Elixir.

While pure as a spring
Is the wine I sing,
And to praise it, one needs but name it;
For Catawba wine
Has need of no sign,
No tavern-bush proclaim it.

And this Song of the Vine,
This greeting of mine,
The winds and the birds shall deliver.
To the Queen of the West,
In her garlands dressed,
On the banks of the Beautiful River."

odor

The odors of a wine (the way a wine smells) may be divided into two groups, aroma and bouquet odors and off odors.

Aroma and Bouquet Odors: The term aroma is reserved to describe those pleasant and desirable odors which are characteristic of the unfermented grape. Bouquet refers to the odors produced by the interactions of the aroma substances with the container, with a small quantity of oxygen, and with one another. One can distinguish tank aging bouquet from bottle bouquet. It is, particularly, during long bottle aging of suitable red table wines that pronounced bouquet develops.

Off Odors: Odors which are foreign to the normal smell of a clean sound wine, are known as off odors.

Off odors resulting from the extensive use of sulfur compounds to prevent secondary fermentation.

Evaluating the "nose" of the wine

a. Sulfur dioxide—used in excessive amounts has an unpleasant effect upon the tissue of the throat and nose, as well as a pungent smell.

b. Hydrogen sulfide—the smell of rotten eggs results from the reduction of free sulfur, or breakdown of sulfur containing amino acids in nitrogen poor musts.

c. Mercaptans—are compounds with intensely disagreeable smells, similar to that of hydrogen sulfide. In these compounds one of the hydrogen atoms of hydrogen sulfide is replaced by an ethal, or similar, alkyl group. It is often described as a garlic odor. To distinguish between hydrogen sulfide and mercaptan is sometimes difficult.

Off odors resulting from other causes:

a. Oxidized: The odors known as oxidized result from the action of air on the various components of wine. The particular smell produced is dependent upon the grape variety from which the wine was made, the rate at which air is introduced, the length of time the wine is aerated, and the temperature of the wine during aeration. There is some resemblance between an oxidized odor and that of prolonged aging in wood in contact with air. The distinction between the two odors can be established only by experience and by knowledge of the appropriateness of type.

1. Vapid: The smell of a wine that has been mildly aerated in the recent past is known as vapid. Wines usually recover from the vapid condition upon storage in full containers. (Temporary oxidation after bottling is often called bottled sickness.)

2. Over Oxidized: The smell of a wine that is excessively oxidized.

3. Ullaged: The odor acquired by table wines, when left in partly filled casks for a long time, results from slow oxidation. The odor may be complicated by the smells produced by film yeasts or bacteria.

4. Aldehydic: In dessert wines the aldehydic odor may be caused by the addition of wine spirits high in aldehyde. Aldehyde odor in table wine is evidence of over-oxidation.

5. Esterified: Aromatic odor due to excessive esterification.

odor, alcoholic

Young dessert wines may often have an odor of wine spirits. These wines may be clean smelling but the overwhelming portion of the odor is "alcoholic." These wines will lose this odor with some aging.

odor, cooked or burnt

Appetizer or dessert wines which have been subjected to a heating process to extract color or flavor, or which have been baked at an excessive temperature, show the defective odors, cooked or burnt.

odor, corked or corky

The smell called corky might indicate defects in cork extracted by the wines.

odor, filter pad

Wines filtered through asbestos, filter pads, cloths, filter powders or cellulose which were not properly processed have an objectionable special odor.

odor, hot fermentation

This odor may also be called pomacy.

It may be caused by leaving the wine on the pomace too long a period or may result from too high a fermentation temperature in contact with the pomace.

odor, lees
When the wine is left too long in contact with the lees, some unpleasant decomposition products of the lees are absorbed. These may range from smells reminiscent of yeast to, cheese-like smells, and are frequently complicated by the presence of mercaptans.

odor, moldy
The smell of molds may appear in wines made from moldy grapes or in wines which have been stored in moldy cooperage.

odor, overaged
The smell, known as overaged, is most generally found in white table wines which have been aged too long. The odor probably results from oxidative changes and is usually accompanied by darkening in color.

odor, poor or hot brandy
Appetizer and dessert wines sometimes smell of the aldehydes, esters, and fusel oils that are characteristic of poor wine spirits.

odor, stagnant
Wines stored in containers, which have been previously filled with stagnant water, absorb some of the odor from the tank. This smell is the result of the action of micro-organisms carried in the water on the substances in the wood or surface of the tank or cask.

odor, woody
The characteristic odor of wet wood is apparent in wines aged for a long period in wooden tanks or casks. A trace of this smell, particularly that of oak, is desirable in fine red table wines, well aged Dry Sherries, Sherries and Tawny Ports, but, when excessive, is unpleasant, especially in white table wines.

Pinot Noir

OEIL DE PERDRIX
A pale wine made from the free-run juice of Pinot Noir grapes. The French translation is "eye of the partridge." The same color as the bird's pink iris.

OHIO RIVER VALLEY VITICULTURAL AREA
The area covers Indiana, Ohio, West Virginia and Kentucky. It consists of 26,000 square miles, with 570 acres of grape vines.

OHIO WINE COUNTRY
The history of wine production in Ohio began at a time when few white settlers in the area were even concerned with statehood. It began with Moravian missionaries, who had come to work with the Delaware Indians, and with French settlers who had brought native vines with them to the Marietta and Gallipolis areas. It was not until the 1820's and the introduction of the Catawba grape, a domestic variety rugged enough to withstand Ohio's climate, that commercial growing of grapes in Ohio was possible. It was largely due to the efforts of Nicholas Longworth, a New England lawyer who had his practice in the Cincinnati area, that the grape industry began here. Longworth was so successful at growing grapes

and making wine that he soon had to abandon his law practice to keep up with his vineyard labors.

Longworth was a man of vision, and others of the area saw the potential for the greater Cincinnati area to truly become a "Rhineland" in America. It was on the terraced hillside covered with vines that they saw an industry develop which would soon rival European wine imports. From the Catawba, Longworth was able to produce a light, semi-sweet wine, which challenged the trend of the rather strong American wines of that time. By the mid-1800's, Longworth's wine was finding favor in the European, as well as American, markets, and it was this success that encouraged others to begin establishing vineyards. Soon many acres of vines were growing and by 1845 the annual production of wine was 300,000 gallons. The nation was now accepting this area as the wine capital.

As the taste for Ohio wines developed in the 1800's, a problem also developed. It was black rot and mildew in the vineyards, conditions which would soon affect the entire Ohio grape industry. While some of the larger vineyards were able to lose a portion of their crop and still make a profit, many of the smaller growers were forced to abandon their vines. The diseases were devastating; black rot caused young grapes to mummify and turn black; the mildew infected the leaves and grapes, resulting in heavy loss in yield and quality. With the manpower shortage brought on by the Civil War, the disease became permanently established in the vineyards. There was some effort to eradicate the problem after the war but for all practical purposes it was too late.

During the successful period for the southwestern Ohio growers there surfaced an interest in grape production on the Lake Erie Islands and the adjacent southern lake shore. With the decline in the Cincinnati area, the Lake Erie area emerged as a major center. The spread of the industry eastward from Bass Island and northward from Cincinnati was the beginning of the Ohio Grape Belt.

oily

An unfavorable tasting term for a wine with an oily appearance or feel in the mouth.

OK CELLARS

(See Calona Wines, Canada.)

OKANAGAN

This British Columbian valley, running north from the Washington Border for 120 miles, is apple country, where winemaking started in 1933 with the establishment of Calona Wines Ltd. at Kelowna to make fruit wines. Since then a considerable grape wine industry has developed and is still growing. There are now a dozen wineries. The severe winter climate is somewhat moderated by a chain of large, deep lakes. The 50th parallel latitude, with long summer days, and the favorable volcanic soils, provide a satisfactory environment for a number of French-American hybrids and a few vinifera varieties.

OKANAGAN RIESLING

From a Hungarian grape, of unknown heritage, with Riesling characteristics. Produces fresh, fruity, white wine with a flowery aroma and an intense character.

Old Casteel Estate Vineyard
San Luis Obispo County, California. Old Casteel.
Zinfandel, Grenache, Carignane vineyard.

OLD CASTEEL VINEYARDS
Founded 1980, Paso Robles, San Luis Obispo County, California.
Storage: Oak. Cases per year: 2,500.
Estate vineyard: Old Casteel Vineyards.
Label indicating non-estate vineyard: Beckwith Ranch Vineyard, Norman Vineyard, Radeki Vineyard.
Wines regularly bottled: Estate bottled, vintage-dated Zinfandel and Grenache. Also vintage-dated Zinfandel.

OLD CREEK RANCH WINERY
Founded 1981, Oakview, Ventura County, California.
Storage: Oak and stainless steel. Cases per year: 650.
History: Old Creek Ranch lies along the banks of the San Antonio Creek, at the south end of the Ojai Valley. The site is adjacent to an old winery established around 1900.
Label indicating non-estate vineyard: Rancho Sisquoc Vineyards, Vineyard Nepenthe, Bien Nacida Vineyard.
Wines regularly bottled: Vintage-dated Johannisberg Riesling, Cabernet Sauvignon, Gamay Beaujolais, Sauvignon Blanc.

OLD NAUVOO (See Gem City Vineland Company.)
Concord White, Niagara, Rosé, Sauterne, Burgundy.

OLD RANCH (See J. Filippi Vintage Company.)

Old South Estate Vineyard
Adams County, Mississippi.
Carlos, Noble and Magnolia vineyard.

OLD SOUTH WINERY
Founded 1979, Natchez, Adams County, Mississippi.
Storage: Stainless steel. Cases per year: 2,000.
History: The Galbreath ancestors brought vines to Mississippi in 1815, from South Carolina. The making of Muscadine wines is a family heritage of over 100 years.
Estate vineyard: Old South Vineyards.
Wines regularly bottled: Estate bottled Carlos, Noble, Dry Carlos, Dry Noble, Noble Rosé.

Old Vines River West Vineyard
Sonoma County, California.
Sonoma/Windsor estate Zinfandel Russian River Valley vineyard.

OLD WINE CELLAR WINERY
Founded 1962, Amana, Iowa County, Iowa.

Wines regularly bottled: Fruit and Berry Wines.

OLDE NORTH VINEYARDS
(See Duplin Winery.)

OLIVER, WILLIAM W.
William W. Oliver is a professor of law in the School of Law at Indiana University and has served as a law clerk to Chief Justice of the Supreme Court, Earl Warren. He became interested in grape growing after hearing a lecture on French hybrid varieties. Not only was he the first to plant a commercial vineyard of French hybrids in Indiana since the 1800's, he planted the vines on reclaimed coal stripped soil. The Oliver Wine Company, Inc. is the oldest winery in Indiana, producing wines from French hybrids and Concord grapes, as well as Mead from honey. The annual Camelot Wine Festival, held at the winery in Bloomington, Indiana, draws 10,000 wine enthusiasts each spring.

Oliver Estate Vineyard
Monroe County, Indiana.
Baco Noir, Aurora, Chelois, Marechal Foch, Cascade and DeChaunac vineyard.

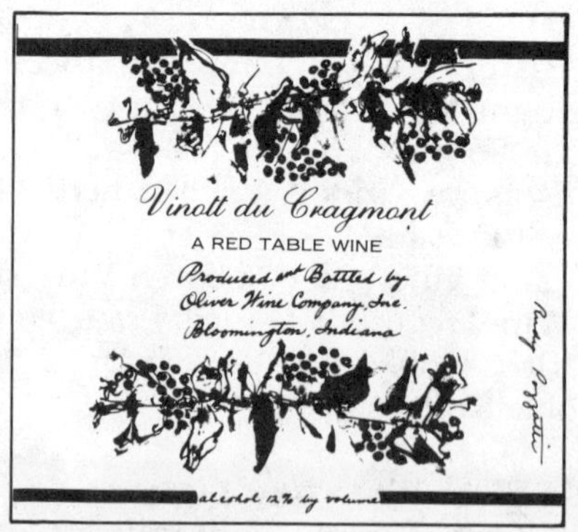

OLIVER WINE COMPANY
Founded 1972, Bloomington, Monroe County, Indiana.
Storage: Oak and stainless steel. Cases per year: 10,000.
History: The oldest winery in Indiana. (See William Oliver biography.)
Estate vineyard: Oliver Vineyard.
Wines regularly bottled: Estate bottled DeChaunac, Vidal; proprietary Oliver Soft Red, Camelot Mead, Vino de Cragmont; also country White and Red.

OLOROSO
A sweet Sherry of medium type, darker and richer than Amontillado.

Olson Estate Vineyards
Mendocino County, California.
Petite Sirah, Zinfandel, Sauvignon Blanc, French Colombard North Coast vineyard.

OLSON VINEYARDS
Founded 1982, Redwood Valley, Mendocino County, California.
Storage: Oak and stainless steel. Cases per year: 3,000.
Estate vineyard: Olson Vineyards.
Label indicating non-estate vineyard: Olson Vineyards West, Quillen Ranch.
Wines regularly bottled: Estate bottled, vintage-dated White Zinfandel, Fumé Blanc, Sauvignon Blanc, Late Harvest Zinfandel, Gamay; also vintage-dated Zinfandel and French Colombard.
Second label: Vista Mendocino.

on the yeast
Period, during second fermentation, when wine and yeast cells are allowed to remain in contact, contributing to traditional "yeasty" nose of Champagne. The "yeastiness" in the Champagne remains as long as the wine is sound. Also practiced when barrel fermenting Chardonnay.

onion skin
A light rose tone with definite copper lights; luminous in the same way an onion is.

ONTARIO WINE STANDARDS
Varietal wine may be identified by the grape variety from which it is made,

providing 75% or more of the wine has been derived from that particular variety. If that variety happens to be from Class 5 or above (i.e., vinifera or hybrid grapes), then the remaining portion, in any blend, must also be from the same classification. Additionally, any varietal wine must derive its predominant aroma and taste characteristic from its designated grape variety.

Vintage wine may be identified by the harvest year of the grapes from which it has been made, providing 85 percent or more of the wine has been derived from that year.

Estate bottled wine must meet the Ontario Superior standard and is made from grapes grown in vineyards owned and operated by the winery bottling the wine.

Ontario Superior wine is a designation reserved for wines that adhere to set standards and have the approval of the official tasting committee as to their superior characteristics:

Still Table Wine

1. must be made entirely from fresh grapes of Class 5 or above,
2. must not exceed 13 percent alcohol by volume,
3. no alcohol may be added in the production,
4. must have cork closure.

Crackling Wine

1. must conform to federal regulations concerning pressure,
2. must be made entirely from fresh grapes of Class 5 or above,
3. must not exceed 13 percent alcohol by volume,
4. no alcohol may be added in the production.

Sparkling Wine

1. must conform to federal regulations concerning pressure,
2. must have undergone a secondary fermentation,
3. must not exceed 13 percent alcohol by volume,
4. no alcohol may be added in the production.

Dessert Wine

1. must be aged at least three years, with a minimum of 51 percent of the blend aged in small oak casks,
2. must contain not less than 17 percent, nor more than 20 percent alcohol by volume,
3. must have cork closure.

Champagne, to be labelled champagne, must undergo a secondary fermentation in the bottle or in a closed tank under CO_2 pressure. Artifically carbonated wine cannot be labelled Champagne. The method of production must be stated on the label, i.e., Charmat Process, fermented in bottles, fermented in this bottle.

OPUS ONE

A collaboration of Robert Mondavi Winery and Chateau Mouton Rothschild of France to produce a California Cabernet Sauvignon. Tim Mondavi is the Mondavi winemaker and Lucien Sionneau is the Mouton winemaker.

orange red

There is more yellow and ochre with the red, and it is rather pale in old wines.

Ordway's Valley Foothill Vineyards

Mendocino County, California.

Gewurztraminer and White Riesling Anderson Valley vineyard.

OREGON MAP

SHALLON WINERY
NAHALEM BAY WINERY
COTE DES COLOMBES
TUALATIN VINEYARDS
SHAFER VINEYARD
ELK COVE VINEYARDS
CHEHALEM MT. WINERY
ADELSHEIM VINEYARD
KNUDSEN ERATH
SOKOL BLOSSER
CHATEAU BENOIT
THE EYRIE VINEYARDS
ARTERBERRY, LTD.
AMITY VINEYARDS
ELLENDALE VINEYARDS
SERENDIPITY CELLARS
26
PORTLAND
HOOD RIVER VINEYARDS
OAK KNOLL
MT. HOOD WINERY
PONZI VINEYARDS
WASSON
BIG FIR WINERY
CENTURY HOMES
99
197
MULHAUSEN VINEYARDS
HIDDEN SPRINGS WINERY
GLEN CREEK WINERY
HONEY WOOD WINERY
I-5
ALPINE VINEYARDS
FORGERON VINEYARD
EUGENE
HINMAN VINEYARDS
N
HENRY WINERY
HILLCREST VINEYARD
ROSEBURG
BJELLAND VINEYARDS
GRANTS PASS
MEDFORD
199
VALLEY VIEW
SISKIYOU VINEYARDS
Oregon

OREGON WINE COUNTRY

Western Oregon has long been established as an outstanding region for growing fruit and berries. Chardonnay and Pinot Noir were planted in Forest Grove at the turn of the century. In the early 60's, the first modern plantings of European varietal grapes, principally Pinot Noir, White Riesling, Chardonnay and Gewurztraminer, took place and Oregon has leaped forward to produce world class wines. The main growing areas are the Willamette Valley, the Roseburg area, the Tualatin Valley, the Applegate River and other parts of South Oregon. Planted in a cool growing zone, similar to the great wine regions of Northern Europe, the vineyards of Oregon are producing a quality of grape that is outstanding. The early settlers, who brought vines with them over the Oregon Trail, would be proud of the wine that is produced today.

OREGON WINE LABELING LAW

A vintage date indicates that at least 95% of the grapes used were harvested during the year stated. Varietally-labeled wine must show an appellation, or statement of the exact geographic origin of the grapes used. The name of a wine can be varietal (containing at least 90% of the grape variety named, while the BATF requires only 75%). No Oregon produced wine may be given a generic name. This means that it cannot be called by European geographic names such as Chablis, Burgundy, etc.

CHARLES ORTMAN (See St. Andrews Winery.)

Vintage-dated Chardonnay, Sauvignon Blanc.

Russ Osen Vineyards

Lake Erie, Pennsylvania.

Riesling vineyard.

ounces and liters

New Size	Equivalent Ounces	Old Size
3 liters	101.4 oz	Gallon (128 oz) or Jereboam (102.4 oz)
1.5 liters	50.7 oz	Half Gallon (64 oz) or Magnum (51.2 oz)
1 liter	33.8 oz	Quart (32 oz)
750 milliliters	25.4 oz	Quart (32 oz) or Fifth (25.6 oz)
375 milliliters	12.7 oz	Pint (16 oz) or Tenth (12.8 oz)
167 milliliters	6.3 oz	Half Pint (8 oz) or Split (6.4 oz)
50 milliliters	1.7 oz	Miniature (1.6 oz)

oxidation

The effect of air upon wine. The character of a wine can be substantially altered by exposure to air. Partially filled bottles will quickly oxidize if not refrigerated. White wines not carefully made or stored will take on a brownish color and "burnt sugar" taste. The production of Sherry and Madeira requires slow oxidation.

OZEKI SAN BENITO, INC.

Hollister, San Benito County, California.

Storage: Oak and stainless steel.

History: The first joint venture Sake brewery in America.

Wines regularly bottled: Sake.

PACHECO PASS VITICULTURAL AREA

Pacheco Pass Viticultural Area is located in Pacheco Pass, near Hollister, California. It is approximately 3,200 acres, distinguished from surrounding area by its terrain, soil and climate.

Pacheco Ranch Estate Vineyard

Marin County, California.

Cabernet Sauvignon North Coast vineyard.

PACHECO RANCH WINERY

Founded 1979, Ignacio, Marin County, California.

Storage: Oak. Cases per year: 900.

History: The Pacheco Ranch is located in the town of Ignacio, named for the Ranch's founder Ignacio Pacheco. Ignacio received a Mexican land grant, in 1836, entitling the Ranch to membership in the 100 year club. It is one of the oldest holdings in California, still in the hands of the original grantees, engaged in an agricultural enterprise. The vineyard was planted in 1970 and was the first modern commercial planting in Marin County. All the labor is provided by the partners, 8th generation Californians.

Estate vineyard: Pacheco Vineyards.

Label indicating non-estate vineyard: Meeken & Mettlen Vineyards.

Wines regularly bottled: Estate bottled, vintage-dated Cabernet Sauvignon. Also vintage-dated Chardonnay. Occasionally Cabernet Rosé.

Second label: RMS Cellars.

Pacini Vineyards/Garzini Vineyards

Mendocino County, California.

Zinfandel vineyard.

PAGE MILL WINERY

Founded 1976, Los Altos Hills, Santa Clara County, California.

Storage: Oak. Cases per year: 2,000.

Label indicating non-estate vineyard: Keene Dimick Vineyards, Volken Eisele.

Wines regularly bottled: Vintage-dated Chardonnay, Cabernet Sauvignon, Zinfandel, Dry Chenin Blanc and Sauvignon Blanc.

Paicines Vineyards

San Benito County, California.

Almaden estate Burger, Chardonnay, Chenin Blanc, Gewurztraminer, Grey Riesling, Pinot Blanc, Sauvignon Blanc, Semillon, Johannisberg Riesling, Folle Blanche, Veltliner, Cabernet Sauvignon, Gamay Beaujolais, Grenache, Pinot Noir, Tinta Madeira vineyard.

PAICINES VITICULTURAL AREA

The Paicines, consisting of 4,500 acres within San Benito County, is located approximately 17 miles north of the Pennacles National Monument and Park and due east of Cienega Valley.

Palace Hill Ranch

Sonoma County, California.

Zinfandel Dry Creek Valley vineyard.

pale straw yellow with greenish tints

A very, very pale yellow, with a slightly green tint.

PALMETTO COUNTRY

(See Fruit Wines of Florida.)

Orange, Tangerine, Grapefruit Wines.

PALOMINO

Known predominantly as the Sherry grape of Spain.

PALOS VERDES WINERY

Founded 1983, Los Angeles County, California.

Storage: French oak. Cases per year: 1,000.

Label indicating non-estate vineyard: Bien Nacido.

Wines regularly bottled: Vintage-dated Chardonnay and Sauvignon Blanc.

PAPAGNI, ANGELO

Angelo Papagni is a first generation Italian wine grower. His father, Demetrio, came from the Bari district of Italy to the San Joaquin Valley in 1912. Born and brought up in Fresno, Angelo began his viticultural education when he was old enough to follow his father into the family vineyards. The vineyards continue to produce table wine grapes and varietal grapes for the home winemaking market today. Angelo, however, dreamed of making premium table wines under his own label. He began construction of a new winery for that purpose in 1973, in Madera; his first wines were released in 1975. He maintains a vigorous program of viticultural experimentation and has served in several wine industry organizations, including membership in the California Grape and Tree Fruit League, president of the San Joaquin Wine Grower's Association, and member of the board of directors of the California Wine Institute. He is also a Supreme Knight in the Brotherhood of the Knights of the Vine.

PAPAGNI VINEYARDS

Founded Vineyards 1920, Winery 1973, Madera, Madera County, California.

History: Demetrio Papagni, the grape grower had left his native Italy and journeyed 8,000 miles directly to Fresno, California. His dream was to plant and cultivate his own vineyard.

Demetrio worked the land and later managed vineyards for several local growers in the San Joaquin Valley. With the money saved over eight years, he purchased his first plot of land in 1920. A young man's life-long dream came true when Demetrio planted his first vine. By the end of the year, 20 acres of young Alicante Bouschet, Muscat Alexandria, and Thompson Seedless table grapes became the nucleus of Papagni Vineyards.

During the early 1920's, however, life was not easy for vineyard owners. Prohibition had been in effect for two years and wineries were closed down. Many California grape growers were pulling grape vines out of the ground and planting other crops.

Despite the restrictions imposed by Prohibition, heads of households could still legally produce up to 200 gallons of wine in their own homes (a law still in effect today) and the demand for fresh grapes did exist. This meant "opportunity" to the father and son, Angelo.

The next decade brought an end to Prohibition and the beginning of the great Depression.

The early 1940's, however, painted a brighter picture for many growers, including Demetrio and Angelo Papagni. The two men were able gradually and methodically to expand their plantings. Experiments to determine which grape varieties produced finer quality fruit were conducted. Angelo applied for registration at the University of California at Davis but the department had closed. The viticultural and enological schools at University of California, Davis and University of California, Fresno (then Fresno State College) had not yet opened as independent divisions.

Experimentation with premium varietal grapes expanded dramatically in the 50's.

The Papagni grapes became highly respected in the marketplace. It was this reputation for quality that caused Angelo

Papagni to formulate plans for someday producing his own vintage-dated bottled wines—a project that would require years of research and devotion.

Construction of the winery was completed in 1973; the first bottle of Angelo Papagni wine was sold in September, 1975.

Estate vineyard: Vallis Vineyard, Clovis Ranch Vineyard, Borita Vineyard.

Wines regularly bottled: Estate bottled, vintage-dated Zinfandel, Rosé of Gamay, Late Harvest Zinfandel, Barbera, Chenin Blanc, Muscat Alexandria, Alicante Bouchet, Moscatop d'Angelo, Fumé Blanc, Chardonnay, Charbono, Late Harvest Emerald Riesling; Madera Rosé, Spumante d'Angelo, Sparkling Chenin Blanc, Chardonnay Au Natural, Brut and Extra Dry Champagne; also Finest Hour Dry and Cream Sherry. And a proprietary wine Fu Jin, Bianca di Madera. All wines are estate bottled and vintage-dated.

paper white

The lightest color in white wines.

Paragon Vineyard

San Luis Obispo County, California.

Chardonnay and other grape varieties, Edna Valley vineyard.

PARDUCCI, JOHN

John Parducci's father, Adolph, purchased 100 acres of vineyards near Ukiah, in Mendocino County, in 1933. He and his three brothers grew up assisting their father in the vineyards and at the primitive Home Ranch winery, which was without electricity until 1939. John became winemaker, in 1944, when Adolph retired. Parducci is considered a pioneer in his use of varietal grapes in the region. He produced early bottlings of Zinfandel, Petite Sirah and French Colombard, and has continued to test different vinifera varieties for suitability to the Mendocino area. He believes the character of a variety can be lost through inappropriate blending, heavy oak flavors or excessive filtration and fining. He recognized the varietal grape-growing potential of the Lake County region, and advised the planting of Zinfandel, Petite Sirah and Cabernet Sauvignon. Consistent with his interest in growing the best grapes appropriate to any region, Parducci is an outspoken advocate of a system of "appellation of origin." He is a Supreme Knight in the Brotherhood of the Knights of the Vine.

Parducci Vineyard

Mendocino County, California.

Chardonnay vineyard.

Wine aging cellar

PARDUCCI WINE CELLARS

Founded 1932, Ukiah, Mendocino County, California.

Storage: Oak and stainless steel. Cases per year: 450,000.

History: Founded by Adolph Parducci, who, in 1932, ended his search for the ideal combination of soil and climate by planting his vineyards in Mendocino County. Adolph carved out his vineyards at what today is called "Home Ranch". At the end of Prohibition, he began creating wines of distinction. Adolph married and had four sons, two of whom, John and George, presently operate Parducci Wine Cellars.

Adolph, the pioneer in bringing finer grapevines to the northernmost reaches of California's wine country, passed on his viticultural and winemaking skills to John. George looks after management duties. Since John's sons joined the winery, a fourth generation of the Parducci family is becoming involved in the winemaking. The Parducci family celebrated 50 years of creating fine varietal wines from Mendocino County in 1982.

Estate vineyard: Home Ranch, Talmage Vineyard, Largo Vineyard.

Wines regularly bottled: Estate bottled, vintage-dated Chenin Blanc, French Colombard, Flora, Gewurztraminer, Muscat Canelli, Sauvignon Blanc, Riesling, Charbono, Barbera, Gamay Beaujolais, Merlot. Semi-generic vintage-dated Chablis and Burgundy. Also estate bottled, vintage-dated Cellar Master Selection Chardonnay, Cabernet Sauvignon, Petite Sirah, Pinot Noir and Zinfandel. This selection is in particular years when only small quantities are produced.

PARKER, DOROTHY (1893-1967)
American writer, humorist.

"Three are the things I shall never attain—
Envy, content and sufficient champagne."

PARRAS (See Casa Madero, Mexico.)

PARSONS CREEK WINERY
Founded 1979, Ukiah, Mendocino County, California.

Storage: Stainless steel. Cases per year: 13,000.

Wines regularly bottled: Vintage-dated Johannisberg Riesling, Chardonnay; occasionally bottle Gewurztraminer, Colombard, Zinfandel (vintage-dated).

PASO ROBLES VITICULTURAL AREA

Wine grapes have been grown in the Paso Robles area since the founding of the California missions. Mission San Miguel Archangel produced wine in 1797. Total vineyard plantings in the area comprise approximately 4,000 acres.

Paso Robles, in San Luis Obispo County, California is bounded on the west and south by the Santa Lucia Mountain range and on the east by the Cholame Hills. The Salinas River has its headwaters at Santa Margarita Lake, just south of the boundary, and flows northward through the viticultural area into the Salinas Valley.

PASO ROBLES WINE COUNTRY

Wine grapevines were introduced into the Paso Robles area by the Franciscan missionaries. All through the 1800's, there were small plantings of winegrapes. Early county assessor records show that, in the period from 1873 to 1883, there were in excess of 80,000 grapevines planted.

In 1882, Andrew York, with the help of his three sons, built a small winery on the eastern slopes of the Santa Lucia Mountains. That winery is still in existence today as York Mountain Winery. In 1890, Rotta Winery (now Las Tablas Winery) was founded.

In 1914, Ignace Paderewski, the famed Polish pianist and statesman, established a vineyard on his 2,500 acre Rancho San Ignacio in the Adelaide area, west of Paso Robles. His San Ignacio Zinfandel was known throughout the state for its quality.

PASTORI WINERY
Founded 1914, Cloverdale, Sonoma County, California.

Paterson Winery
Benton County, Washington.
Chateau Ste. Michelle estate Chardonnay, Sauvignon Blanc, Chenin Blanc, Grenache, Johannisberg Riesling, Semillon, Cabernet Sauvignon Columbia Valley vineyard.

PATRICIA
A native American grape that produces wine similar to Concord. Primarily used for Sparkling, Sweet and blending wines.

Pat Paulsen Estate Vineyard
Sonoma County, California.
Sauvignon Blanc, Chardonnay, Cabernet Sauvignon Alexander Valley vineyard.

PAT PAULSEN VINEYARDS
Founded 1980, Cloverdale, Sonoma County, California.
Storage: Oak and stainless steel. Cases per year: 10,000.
Estate vineyard: Pat Paulsen Vineyards.
Label indicating non-estate vineyard: Lou Preston Vineyards, Long Vineyards.
Wines regularly bottled: Estate bottled, vintage-dated Sauvignon Blanc, Chardonnay, Cabernet Sauvignon. Also vintage-dated Dry Muscat Canelli, Chardonnay.
Second label: Matrose.

PEARIS MOUNTAIN
(See MJC Vineyard.)
Estate bottled, vintage-dated Seyval Blanc, Chambourcin, Marechal Foch.

PEARL OF CSABA
A vinifera grape that produces a pleasant Muscat type white wine.

Peck Ranch
San Luis Obispo County, California.
Sauvignon Blanc vineyard.

Robert Pecota Estate Vineyard
Napa County, California.
Sauvignon Blanc, Cabernet Sauvignon, Colombard, Gamay Beaujolais Napa Valley vineyard.

ROBERT PECOTA WINERY
Founded 1978, Calistoga, Napa County, California.
Storage: Oak and stainless steel. Cases per year: 10,000.
Estate vineyard: Robert Pecota Vineyards.
Wines regularly bottled: Estate bottled, vintage-dated Sauvignon Blanc, Cabernet Sauvignon, Gamay Beaujolais, Colombard.

PEDRIZZETTI WINERY
Founded 1913, Morgan Hill, Santa Clara County, California.
Cases per year: 35,000.
History: Purchased in 1945 from the original owner that built the winery in 1913 and added to it in 1938.
Label indicating non-estate vineyard: Gena's Vineyard, Shell Creek Vineyard.
Wines regularly bottled: Vintage-dated White Zinfandel, Chardonnay, Gewurztraminer, French Colombard, Chenin Blanc, Zinfandel Rosé, Zinfandel, Barbera, Cabernet Sauvignon, Petite Sirah. Also bottle semi-generics. Occasionally bottle Johannisberg Riesling.
Second labels: Golden Hills, Morgan Hill Cellars, Crystal Springs.

Pedro Domecq Estate Vineyards
Baja California, Mexico.
Cabernet Sauvignon, Chenin Blanc, Riesling and Zinfandel Calafia Valley vineyard.

PEDRO DOMECQ WINES

Founded 1958, Baja, Mexico.

Storage: Stainless steel.

History: The House of Pedro Domecq was established in Spain in 1730. In the Jerez region it owns nearly 6,000 acres of vineyards and in the Rioja region another 3,000 acres where the vineyards provide for the famous Domecq Sherries and Rioja wines. In the early 1950's the Domecq family planted vineyards in the Calafia Valley of Baja California region of Mexico.

Estate vineyard: Pedro Domecq Vineyards.

Wines regularly bottled: Estate bottled, vintage-dated Cabernet Sauvignon, Riesling, Blanc de Blanc and Zinfandel. Also Los Reyes (Red, White, Rosé), Calafia (Red, White, Rosé), Padre Kino (Red, White, Rosé) and Chateau Domecq.

PEDRO XIMINEZ

A grape that produces a sweet dessert Sherry.

J. PEDRONCILLI WINERY

Founded 1904, Geyserville, Sonoma County, California.

Storage: Oak and stainless steel. Cases per year: 140,000.

Wines regularly bottled: Vintage-dated Chardonnay, Sauvignon Blanc, French Colombard, Gewurztraminer, Johannisberg Riesling, Chenin Blanc, Zinfandel Rosé, Gamay Beaujolais, Pinot Noir, Zinfandel, Cabernet Sauvignon. Occasionally bottle "Vintage Selection" Zinfandel and Cabernet Sauvignon. Also Sonoma Red, White and Rosé.

Pelee Island Estate Vineyard

Ontario County, Canada.

Johannisberg Riesling, Welsh Riesling, Scheurebe, Gewurztraminer, Pinot Noir, Chardonnay vineyard.

PELEE ISLAND WINERY, INC.

Founded 1983, Kingsville, Ontario, Canada.

Storage: Oak and stainless steel. Number of cases produced per year: 12,000.

History: The first winery, called Vin Villa, was opened on the Pelee Island in 1866.

Estate vineyard: Pelee Island Vineyard.

Wines regularly bottled: Estate bottled, vintage-dated Johannisberg Riesling, Scheurebe, Kerner, Chardonnay, Pinot Noir, Rosé and Late Harvest Johannisberg Riesling.

Pellegrini Estate Ranch

Sonoma County, California. Pellegrini. Chardonnay vineyard.

PELLEGRINI VINEYARDS

Founded 1934, Santa Rosa, Sonoma County, California.

Storage: Oak and stainless steel. Cases per year: 60,000.

Estate vineyard: Pellegrini Ranch.

Wines regularly bottled: Estate bottled Chardonnay; also Cabernet Sauvignon, Fumé Blanc, Zinfandel, Chenin Blanc, French Colombard and a proprietary wine Clos Du Merle. Also Premium Dry Red, Dry White and Rosé.

PENDLETON WINERY, LTD.

Founded 1977, Santa Clara, Santa Clara County, California.

Storage: Oak. Cases per year: 4,000.

Wines regularly bottled: Vintage-dated Chardonnay, Cabernet Sauvignon, Pinot Noir.

Second label: Arroyo.

penetrating
Tasting term.
Insinuates itself; attacks.

Penn Shore Estate Vineyards
Erie County, Pennsylvania. Penn Shore.
Ravat, Vidal, Seyval, Chancellor, Baco Noir, Catawba, Concord Lake Erie vineyards.

PENN SHORE VINEYARDS, INC.
Founded 1969, North East, Erie County, Pennsylvania.
Storage: Oak and stainless steel. Cases per year: 15,000.
History: The owners of Penn Shore pioneered the State Farm Winery Act in 1969.
Estate vineyard: Sceifond Vineyards; Fred Luke Vineyards; McCord Vineyards.
Wines regularly bottled: Estate bottled, vintage-dated Ravat Blanc, Vidal Blanc, Seyval Blanc, Chancellor Noir, Baco Noir; estate bottled Pink and White Catawba, Concord; proprietary Holiday Spice; also semi-generics: Chablis, Burgundy; and Rosé and Kir. Occasionally Seyval Blanc and Champagne.
Second label: Free Spirit Wines.

Robert Pepi Estate Vineyard
Napa County, California.
Chardonnay Napa Valley vineyard.

ROBERT PEPI WINERY
Founded 1981, Oakville, Napa County, California.
Storage: Oak and stainless steel. Cases per year: 15,000.
Estate vineyard: Robert Pepi Vineyard.
Label indicating non-estate vineyard: Vine Hill Ranch Vineyard.
Wines regularly bottled: Estate bottled, vintage-dated Chardonnay; also vintage-dated Cabernet Sauvignon, Sauvignon Blanc, Semillon.

Pepperwood Vineyard
Sonoma County, California.
Balverne estate Dry Gewurztraminer Chalk Hill vineyard.

Perdido Estate Vineyard
Baldwin County, Alabama.
Muscadines: Magnolia, Higgins, Scuppernong vineyard.

PERDIDO VINEYARDS
Founded 1979, Perdido, Baldwin County, Alabama.
Storage: Oak and stainless steel. Cases per year: 13,000.
History: The winery produces wines from native southern Muscadine grapes. Licensed as Alabama Winery N1 in 1979, the first since Prohibition.
Estate vineyard: Perdido Vineyards.
Label indicating non-estate vineyard: The Graham Farm; Still Pond Vineyard.
Wines regularly bottled: Estate bottled, vintage-dated Magnolia; vintage-dated Apple, Noble.

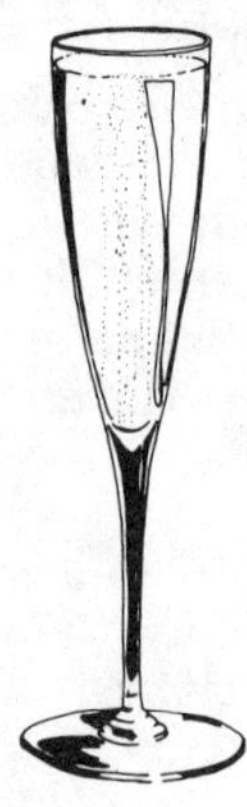

A champagne flute displays "fine perlage"

PERLAGE

Bubbles that are fine, large or medium of natural carbon dioxide which rise in a regular and continuous fountain in a glass of sparkling wine.

PERLE

A vinifera grape; Gewurztraminer crossed with Muller Thurgau. Produces light white wine.

PERSISTENT FOAM

After having poured out the sparkling wine, the foam is full and lasting.

Person Orchards

Yakima, Washington.

Semillon, Chardonnay, Riesling, Chenin Blanc and Gewurztraminer vineyard.

persuasive

Convincing; easily judged; communicative.

Pesenti Estate Vineyard

San Luis Obispo County, California.

Zinfandel, Ruby Cabernet vineyard.

PESENTI WINERY

Founded 1934, Templeton, San Luis Obispo County, California.

Cases per year: 50,000.

History: Founded by Frank Pesenti and still family owned and operated.

Estate vineyard: Pesenti Vineyards.

Wines regularly bottled: Zinfandel, Cabernet Sauvignon, Zinfandel Blanc, Ruby Cabernet, Rosé of Ruby Cabernet, Cabernet Sauvignon Blanc, Rosé of Cabernet Sauvignon, Zinfandel Rosé, Late Harvest Zinfandel, Cabernet Sauvignon (nouveau).

PETALUMA CELLARS (See La Crema Vinera.)

PETERSON, DR. R.G.

Dr. R.G. Peterson began his career in enology as a home winemaker. After obtaining a Ph.D. in agricultural chemistry from the University of California, Berkeley, Peterson joined the E. & J. Gallo Winery. From 1958 to 1968, he worked in research and new product development, becoming research director and assistant production manager in charge of winemaking. For the next 5 years, he was winemaker-production manager at Beaulieu Vineyard. Peterson became winemaker at The Monterey Vineyard in 1973, and assumed the presidency in 1974. He has also been directly responsible for the development of both the generic and varietal table wines marketed under the Taylor California Cellars label, as well as for the line of Taylor California Cellars champagnes. In addition to publishing more than 200 articles in technical journals and writing a periodic newsletter, Peterson has been active in many professional wine organizations. He is a member of the American Society of Enologists, and served as a director from 1973 to 1975, and as its president in 1977-78. He is a charter member of the Society of Wine Educators and was director of the association in 1979-80. He is a director of the California Wine Institute, member of Les Amis du Vin, International Wine and Food Society, W.I.N.O., The Confrerie de la Chaine des Rotisseurs and Supreme Knight of the Brotherhood of the Knights of the Vine. Peterson also continues to teach wine appreciation, technology and home winemaking at Hartnell College in Salinas.

PETERSON, HEIDI

Winemaker at Buehler Vineyards, Heidi was Jerry Luper's assistant at Chateau Bouchaine until 1982. Prior to that, she was a cellar worker at Franciscan Vineyards and, earlier, at The Monterey Vineyard. An enology graduate from University of California at Davis in 1978, she did cellar work for one season in Germany and in Australia prior to working in Napa Valley.

Peterson Vineyard

Seneca County, New York.

Ravat, Chardonnay Finger Lakes vineyard.

petillant

Sparkling as referring to bubbling or fizzy wine. French for "crackling". A wine less fizzy than "cremant" or "mousseux". Perlé is also used.

PETITE SIRAH

A grape that produces a dry, full-bodied, tannic wine. Believed to have originated in the Middle East, as the Shiraz grape, prior to being planted in France. The California Petite Sirah is thought to be the French Duriff and is not the same as the Syrah of the Rhone Valley. Ages well.

PHEASANT RIDGE WINERY

Founded 1982, Lubbock, Lubbock County, Texas.

Storage: Oak and stainless steel. Cases per year: 2,000.

History: Vineyards are on the High Plains of Texas.

Estate vineyard: Cox Family Vineyard.

Label indicating non-estate vineyard: Boepple Vineyards, Hildebrand Vineyards.

Wines regularly bottled: Vintage-dated Cabernet Sauvignon, Chenin Blanc, Sauvignon Blanc and Premium Red.

JOSEPH PHELPS VINEYARDS

Founded 1972, St. Helena, Napa County, California.

Storage: Oak. Cases per year: 55,000.

Estate vineyard: Joseph Phelps Backus Vineyard.

Non-estate vineyard: Eisele Vineyards, San Giacomo Vineyard.

Wines regularly bottled: Vintage-dated Chardonnay, Late Harvest Scheurebe, Cabernet Sauvignon, Johannisberg Riesling (Early Harvest, Regular, Late Harvest, Select Late Harvest and Special Select Late Harvest), Zinfandel, Gewurztraminer, Chardonnay, Syrah, Sauvignon Blanc, Cabernet Sauvignon. Also Vin Blanc, Vin Rouge.

PHENOLS (See Tannin.)

PHINIOTIS, ELIAS G.

Dr. Elias G. Phiniotis was born into a grape growing and winemaking family in Cyprus. As a youth, he worked at the "KEO" winery in Pera-Pedhi, a large commercial operation. After high school, Phiniotis continued his education, first at Kingsway Day College in London, then at the University of Technical Sciences in Budapest. He was graduated with a master's degree in chemical engineering, with a specialty in food chemistry and technology; he then obtained a Ph.D. three years later at the Research Institute of Viticulture and Enology, also in Budapest. Phiniotis began his professional career as Winemaker at the LOEL Winery and Distillers in Cyprus. Two years later, in 1977, he accepted a position as wine consultant for Wine Art in Vancouver. The following year he joined Golden Valley Wines in British Columbia, where he worked consecutively as quality control manager, winemaker, and vice-president plant manager. In 1980, Phiniotis left to become research director and quality control manager at Casabello Wines, also in British Columbia. He was later promoted to chief winemaker. In 1981, he was hired by Calona Wines as research director and quality control manager. He has since become enologist and now serves as chief enologist and research director. Phiniotis is a member of the American Society of Enologists.

Piazza Vineyards
Phelps County, Missouri.
Riesling vineyard.

Piccho Vineyard
San Luis Obispo County, California.
Pressoir Deutz Winery estate Pinot Noir, Pinot Blanc, Chardonnay and Chenin Blanc Central Coast vineyard.

Pickle Canyon Vineyards
Napa County, California.
Merlot, Zinfandel, White Riesling and Chenin Blanc vineyard.

PICONI WINERY, LTD.

Founded 1980, Temecula, Riverside County, California.

Storage: Oak and stainless steel.

Wines regularly bottled: Vintage-dated Chenin Blanc, Fumé Blanc, Chardonnay, Petite Sirah, Cabernet Sauvignon.

PIEDMONT CELLARS

Founded 1980, Oakland, Alameda County, California.

Storage: Oak. Cases per year: 1,000.

Label indicating non-estate vineyard: Freites Ranch, Freiberg Ranch, Horne Ranch.

Wines regularly bottled: Vintage-dated Chardonnay, Pinot Noir, Cabernet Sauvignon.

Piedmont Estate Vineyard
Fauquier County, Virginia.
Semillon, Chardonnay, Seyval vineyard.

PIEDMONT VINEYARDS AND WINERY, INC.

Founded 1973, Middleburg, Fauquier County, Virginia.

Storage: Oak and stainless steel. Cases per year: 3,360.

History: Founded by Mrs. Thomas Furness, the only woman, to date, who has single-handedly started a winery and a winery operation, and at the age of 75. Piedmont is the first commercial vinifera vineyard in Virginia. The original buildings date back to the mid 1700's and the main house is a Virginia Historical Society landmark.

Estate vineyard: Piedmont Vineyards.

Wines regularly bottled: Estate bottled, vintage-dated Virginia Semillon, Seyval Blanc, Chardonnay and Chardonnay Reserve.

PINA CELLARS

Founded 1979, Rutherford, Napa County, California.

Storage: Oak and stainless steel. Number of cases produced per year: 450.

Wines regularly bottled: Vintage-dated Chardonnay and Zinfandel.

PINE RIDGE

Founded 1978, Napa, Napa County, California.

Storage: Oak and stainless steel. Number of cases produced per year: 12,000-14,000.

Estate vineyard: Pine Ridge Winery Vineyard.

Wines regularly bottled: Estate bottled, vintage-dated Chardonnay; also vintage-dated Cabernet Sauvignon, Chenin Blanc, Merlot.

Pine Ridge Estate Vineyard

Napa County, California.

Chardonnay (Stag's Leap District), Cabernet Sauvignon (Rutherford District), Chenin Blanc (Yountville District) vineyard.

PINEAU DE LA LOIRE

Name often used in France for a grape also widely grown in California, where it is best known as Chenin Blanc.

pink champagne

The pink color results from letting the juice of red grapes remain with the grape skins during fermentation, until the desired hue is obtained.

Pinnacles Vineyards

Monterey County, California.

Paul Masson estate Chardonnay, Cabernet, Johannisberg Riesling, Gewurztraminer, Pinot Noir and Sauvignon Blanc vineyards.

PINOT BLANC

A grape, Chardonnayish in character, that produces a light, dry, medium-bodied, moderately tart wine with a pronounced grape flavor and aroma. The better ones are rich and full and age well for 2-3 years.

PINOT CHARDONNAY

(See Chardonnay.)

PINOT GRIS

White variety of Pinot Noir.

Pinot Noir

PINOT NOIR

A grape that produces clear, brilliant, medium to deep red color. Rich, with just a hint of violets; velvety and full of flavor. Originally from France, where it is used in all of the great red Burgundies: Beaune, Pommard and Cote D'Or. Also the principal grape in Champagne making. Age 3-7 years.

PINOT NOIR BLANC

White wine made from Pinot Noir grapes by removing juice from skins immediately after grapes are crushed. It is usually faintly pink.

PINOT ST. GEORGE
A grape that produces a robust, earthy, fruity wine similar to the California Gamay. Relatively rare.

PINSON (See Casa Pinson Hermanos.)

PIPER-SONOMA
Founded 1980, Windsor, Sonoma County, California.
Cases per year: 100,000.
Wines regularly bottled: Champagne, vintage-dated Brut, Blanc de Noirs, Tete du Cuvée.

Pirtle's Weston Estate Vineyard
Platte County, Missouri.
Leon Millot, Seyval, Villard Noir and Baco Noir vineyard.

PIRTLE'S WESTON VINEYARDS
Founded 1980, Weston, Platte County, Missouri.
Cases per year: 1,000.
History: Winery is in the former German Lutheran Evangelical Church, built in 1867 and is in the National Register of Historic Places.
Estate vineyard: Pirtle Weston Vineyards.
Wines regularly bottled: Vintage-dated Villard Noir; Leon Millot, Seyval, Apple; also Mellow Red and Claret. Occasionally Mead (Honey Wine).

Piterra Vineyard
Hampshire County, West Virginia.
Robert Pliska estate Foch, Chancellor, Seyval vineyard.

Plane's Cayuga Vineyard
Seneca County, New York.
Chancellor and Cayuga White Finger Lakes vineyard.

PLANE'S CAYUGA VINEYARD
Founded 1980, Ovid, Seneca County, New York.
Storage: Oak and stainless steel. Cases per year: 3,500.

Estate vineyard: Plane's Cayuga Vineyard.
Wines regularly bottled: Estate bottled, vintage-dated Chardonnay, Riesling, Cayuga White, Chancellor, Ravat Vignoles, Ravat Vignoles Late Harvest.

PLANTATIONS
(See Berrywine Plantations Wine Cellars.)
Foch, Chancellor, Rosé, Seyval, White, Apple.

ROBERT F. PLISKA & COMPANY WINERY
Founded 1975, Purgitsville, Hampshire County, West Virginia.
Storage: Oak and stainless steel. Cases per year: 1,000.
Estate vineyard: Piterra Vineyards.
Wines regularly bottled: Estate bottled, vintage-dated proprietary 101 Piterra Foch, Chancellor, Seyval, Assumption Aurora. Occasionally proprietary Assumption Rosé.

Joseph E. Pohorly Vineyards
Niagara-on-the-Lake, Ontario, Canada.
Gewurztraminer vineyard.

POINT LOMA WINERY
Founded 1980, San Diego, San Diego County, California.
Storage: Oak and stainless steel. Cases per year: 500.

Wines regularly bottled: Gamay Beaujolais, Red Table Wine.

POLITICAL AREA

Term defined in regulations of the U.S. Bureau of Alcohol, Tobacco and Firearms. It may be the entire United States, a single state or county or a multi-state or multi-county area. The main thing is that 75 percent of the wine must originate within the area named.

For multi-state and multi-county areas, there are further stipulations:

1. The states must be adjoining for a multi-state designation.
2. All counties lumped together in a multi-county unit must lie within the same state, though they needn't be adjacent to one another.
3. The percentage of wine coming from each separate county, or state, must be spelled out on the label, and 100 percent of the wine must come from the counties or states named on the label.

Polson Vineyards

Sonoma County, California.

Cabernet Sauvignon, Merlot, Zinfandel, Chardonnay and Chenin Blanc Dry Creek Valley vineyard.

POMACE

The pulp, skins and seeds of grapes remaining after the juice or newly fermented wine has been drawn off or pressed out.

POMMERAIE VINEYARDS

Founded 1979, Sebastopol, Sonoma County, California.

Cases per year: 2,000.

Estate vineyard: Green Valley Vineyards.

Wines regularly bottled: Estate bottled Cabernet Rosé; also vintage-dated Chardonnay, Cabernet Sauvignon.

Ponderosa Valley Vineyard

New Mexico County, New Mexico.

La Chiripada estate French Hybrids and White Riesling vineyard.

Ponnequin Vineyards

Lake County, California.

Chardonnay and Sauvignon Blanc vineyard.

PONZI VINEYARDS

Founded 1970, Beaverton County, Oregon.

Estate vineyard: Ponzi Vineyards.

Wines regularly bottled: Pinot Noir, White Riesling, Pinot Gris and Chardonnay.

POOLE, JOHN P.

The first 30 years of John P. Poole's life were spent in communications. As a young man, in the 1930's, he worked as a wireless operator on ships sailing all over the world. During World War II he served with both the Royal Air Force of Great Britain and the United States Army Signal Corps. After the war he went into radio and television broadcasting and both constructed and managed more than ten radio and television stations in different parts of the country. In 1968, John decided to withdraw from broadcasting and became fascinated with the possibilities of viticulture. He purchased acreage and planted a vineyard in the vicinity of Temecula, California. For the next twelve years John devoted his energies to the establishment of an excellent vineyard providing grapes well suited to the micro-climate of the Temecula region. In 1975, John built the Mount Palomar Winery where excellent wines are produced at affordable prices. Today John is joined by his son, Peter, in the management of both the vineyard and the winery.

POPE VALLEY
Situated in Napa County north of the Napa Valley.

POPE VALLEY WINERY
Founded 1972, Pope Valley, Napa County, California.
Wines regularly bottled: Vintage-dated Chardonnay, Sauvignon Blanc, Cabernet Sauvignon, Zinfandel, Pinot Noir.

POPLAR VINEYARDS (See Chateau Bouchaine.)

Poplar Ridge Estate Vineyards
Seneca County, New York.
Riesling, Chardonnay, Cabernet Sauvignon, Chelois, Seyval Blanc, Cayuga, Vidal Blanc, Ravat and Aurora Finger Lakes vineyard.

POPLAR RIDGE VINEYARDS, INC.
Founded 1980, Valois, Seneca County, New York.
Storage: Stainless steel. Cases per year: 5,500.
History: David Bagley, the owner/winemaker, has accomplished a great deal in a few years. Starting at Bully Hill in 1971 where he "learned basic wine-making and viticulture". He spent two years at the Brotherhood Winery and then went to the Wagner Vineyard where he became the head winemaker. In 1980 he started his own vineyard and winery.
Estate vineyard: Poplar Ridge Vineyards.
Wines regularly bottled: Vintage-dated Riesling, Cayuga White, Ravat, Vidal Blanc, Seyval Blanc, Foch, Aurora, Delaware, Chelois, proprietary: Valois Blanc, Valois Rouge; and Rosé.

PORT
Port is a fortified, rich, fruity, heavy-bodied, sweet wine; usually deep red.

However, there is Tawny Port, and White Port. Port originated in Portugal but may be made and legally called "Port" in the United States. Many grape varieties can be used in making Port, including Carignane, Petite Sirah, Tinta Cao, Tinta Madeira and Zinfandel. Port is not baked (as are Madera and some Sherries), influenced by yeast (as 'Flor' Sherry) or flavored as Marsala. The alcohol content is usually 18-21 percent; but may be 14-24%.

Fermentation. Grapes, fresh from the vineyard, are crushed into a fermentation vat or tank. Cultured yeast is added and, within 12-24 hours the yeast is at work converting the natural sugar of the grape into alcohol.

Fortification. When the yeast has worked to a point pre-determined by the winemaker and the desired amount of natural grape sugar remains unfermented in the must, high proof alcohol is added. This alcohol addition, or fortification, kills the yeast and preserves the natural sweetness of the wine, while raising the alcohol level to the desired percentage.

Pressing. Pressing may take place just before or after fortification. Grapes, seeds and juice are passed through the press and pressure is applied. This squeezes all of the juice out of the grapes and leaves behind the dry 'pomace' of skins and seeds.

Types of Port. Vintage Port is a wine made from one vintage only and is usually bottled within 18-24 months of harvest. When bottled, it is dark, tannic and raw. Its development occurs over many years in the bottle and requires cellaring to achieve peak quality. Late-bottled Port is also a vintage wine, but is aged in a barrel to achieve maximum development. It is usually ready to drink when bottled.

Ruby Port is a blend of vintages. The wine is blended into a hue and fruitiness of a young wine and is enjoyable, ready to drink when bottled.

Tawny Port is a blend of vintages of such age as to mute the youthful purples and reds of the young wine to earthy orange. Tawny Ports are usually drier and, if genuine, more costly than Ruby Ports.

PORT TINTA WINE
Port produced from Tinta Madeira or Tinta Cao grapes.

Porter-Bass Vineyards
Sonoma County, California.
Zinfandel vineyards.

PORTET, BERNARD
Bernard Portet is a sixth generation winemaker. Born in 1944 in Cognac, France, Portet spent much of his childhood on the estate of Chateau Lafite-Rothschild, where his father was regisseur. After graduating as an ingenieur agronome from the Toulouse School of Agronomy, he attended the Montipellier School of Agronomy, where he earned a diploma of enology. He then worked for two years in Morocco, teaching agronomy and serving as assistant manager of a pilot farm at the Institute of Agronomy. In 1970, Portet was hired by John Goelet to search the world for a viticultural region equivalent to the best in France. He traveled Europe, Africa, South America and Australia before recommending the southeastern area of the Napa Valley. He was put in charge of building the Clos Du Val Winery, and planting the 120 acres of Cabernet Sauvignon, Merlot and Zinfandel vines. Since 1972, Portet has produced Cabernet Sauvignons in a subtle, Bordeaux style, designed to complement a good meal. He has recently expanded the vineyards to the Carneros Creek area, where Pinot Noir and Chardonnay have been planted. Portet believes that the winemaking is a business, one that requires good management skills in addition to artistry and devotion. He is active not only in industry events, but gives both his time and wine to cultural and charity organizations. He is a Master Knight in the Brotherhood of the Knights of the Vine.

Portola Valley Vineyards
San Mateo County, California.
Chardonnay vineyard.

Possum Trot Estate Vineyards
Brown County, Indiana.
Marechal Foch, Aurora and Seyval Blanc Ohio River Valley vineyard.

POSSUM TROT VINEYARDS
Founded 1978, Unionville, Brown County, Indiana.
Storage: Oak and stainless steel. Cases per year: 1,000.
Estate vineyard: Possum Trot Vineyards.
Wines regularly bottled: Estate bottled

Aurora, Marechal Foch, Seyval Blanc; proprietary Zaraguega Sangria; also Vignoles, Mulled Wine.

Second label: Benora.

POST, MATHEW JOSEPH

Mathew Joseph Post was born in Altus, Arkansas, in 1925. After serving in the Navy Medical Corps during World War II, Post enrolled at St. Louis University. In 1947, he became his father's partner at the Post Winery, and later assumed proprietorship. He is currently chairman of the board. Post Winery is a family run business, and several of Post's twelve children assist in winery duties: Mathew Jr. is operation manager; Paul directs sales and marketing; Thomas supervises viticulture; and Andrew serves as winemaker. In 1967, the family was named "Farm Family of the Year." Post has also been active in many civic organizations and his church. He was on the City Council and served as Mayor of Altus for twelve years.

Post Estate Vineyards

Franklin County, Arkansas.

Catawba, Delaware, Landot Noir, Seyval Blanc, Villard Blanc, Niagara, Noble and Magnolia Altus vineyard.

POST WINERY

Founded 1880, Altus, Franklin County, Arkansas.

Storage: Stainless steel. Cases per year: 50,000.

History: The Post Family emigrated to the United States from Nederwurtgback, Germany during the Franco Prussian War in 1872. Jacob and Marie Post first settled in Indiana and, in 1880, moved to Altus, Arkansas because of the similarity of the area to their homeland. Land grants from the railroads allowed them to purchase a farm site and plant vineyards. While stopping for wood and water for the steam engines, rail passengers passing through Altus were given the opportunity of purchasing fruit, vegetables and wines.

Jacob's eldest son, Joseph, and his wife, Kathrin, also were farmers and vintners. Kathrin became known to the local residents as "Ma" Post, as their home was both a restaurant and wine shop. Prohibition and the Depression only served to increase her business, as she chose to continue her way of life in this remote Arkansas village. Punishment did finally come. She was sent to a federal prison in New Jersey for her activities. A full pardon came six months later as Prohibition ended and winemaking began a new life in Altus.

Kathrin and Joseph Post's oldest son, James, became one of the leaders in the effort to revitalize the wine community. James Post's first contribution came with his work in formulating rules to govern the Arkansas wine industry, which became law during the 1922 Arkansas General assembly. For three years, he acted as a viticultural consultant with Missouri Pacific Railroad to reestablish fresh market and wine grape production in the Arkansas River Valley. In 1937, he formed the Altus Cooperative Winery in an effort to unite winemakers and growers to increase profits. Eventually, he bought the winery with his son, Mathew, and started the Post Winery. His winemaking career was cut short by a fatal auto accident in 1951. Mathew became the new president and winemaker.

Estate vineyard: Post Winery Vineyard.

Wines regularly bottled: Estate bottled, vintage-dated Seyval Blanc, Seyval Blanc Special Select Sweet, Delaware, Ives Noir, Steuben, Cynthiana, Dry Catawba; vintage-dated Niagara, Catawba, Aurora, Concord; also Red Muscadine (Noble), White Muscadine (Magnolia and Carlos), White Post; Champagne (Aurora and Seyval Blanc), Sherry; semi-generics White Port, Sauterne, Rhine, Chablis, Burgundy; and Vin Rosé, Strawberry.

Second labels: Post Familie, Post, Altus, Subiaco Abbey (alter wines).

Potter-Cowan Vineyards

Amador County, California.

White Zinfandel Shenandoah vineyard.

POTTER VALLEY VITICULTURAL AREA

The area is in east central Mendocino County and consists of approximately 27,500 acres of valley floor surrounded by mountains.

POURRITURE NOBLE

French name for Botrytis Cinerea, a mold which forms on some grapes and concentrates the sugar and the acid content (see Botrytis).

Pratt Vineyard

Napa County, California.

Conn Creek estate Cabernet Sauvignon vineyard.

PRECOCE DE MALINGRE

A vinifera grape introduced to the Northwest by Dr. Robert Norton. Produces a golden-green pleasant-aroma wine.

Presque Isle Estate Vineyards

Erie County, Pennsylvania.

Chardonnay, Riesling, Aligote, Pinot Gris, Vidal, Dutchess, Delaware, Cabernet Sauvignon, Cabernet Franc, Gamay, Chambourcin, Foch, Leon Millot, DeChaunac vineyard.

Presque Isle Vineyards

Lake County, Pennsylvania.

Delaware, Vidal Blanc and Chardonnay Lake Erie vineyard.

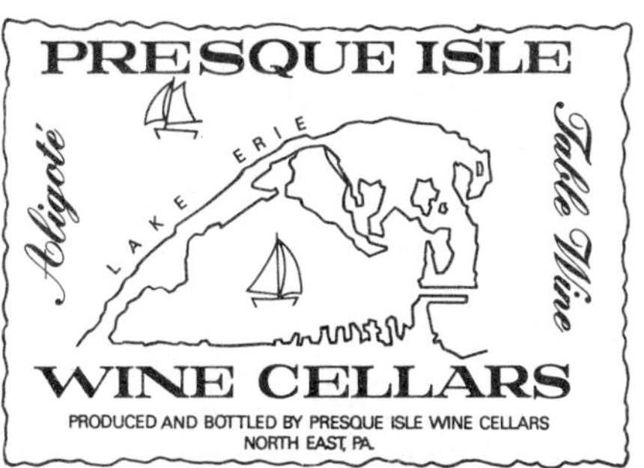

PRESQUE ISLE WINE CELLARS

Founded 1964, North East, Erie County, Pennsylvania.

Storage: Oak and stainless steel. Cases per year: 1,000-1,500.

Estate vineyard: Presque Isle Vineyard.

Wines regularly bottled: Estate bottled, vintage-dated Johannisberg Riesling, Cabernet Sauvignon, Pinot Gris, Chardonnay, Aligoté, Vidal Blanc, Dutchess, Chambourcin, Foch, Delaware, Catawba, Chancellor, Cabernet Franc, Gamay Beaujolais, Petite Sirah, DeChaunac, Leon Millot.

PRESSOIR DEUTZ WINERY

Founded 1983, Arroyo Grande, San Luis Obispo County, California.

Storage: Stainless steel. Cases per year: 1,300 to 30,000 (1990).

History: Deutz is one of the most prestigious Champagne producers in France. Not just another French company that is starting in California. Their first Champagne will be released in 1986.

Estate vineyard: Piccho Vineyard.

Wines regularly bottled: Only Champagne.

PRESTON, BILL

Bill Preston was born and raised in the

Tri-Cities of Washington State (Pasco, Kennewick and Richland). He spent four years in the Air Force and 30 years in the family tractor-implement-irrigation business. Preston Vineyards were started in 1972, with 50 acres; another 130 acres were added in 1979. Preston Wine Cellars had its first crush in 1976.

Preston Wine Cellars is a family owned and operated winery. He is a Master Knight in the Brotherhood of the Knights of the Vine.

Preston Estate Vineyard

Sonoma County, California.

Sauvignon Blanc, Cabernet Sauvignon, Zinfindel, Chenin Blanc, Muscat Canelli, Gamay Dry Creek vineyards.

Preston Ranch Estate Vineyard

Franklin County, Washington.

Johannisberg Riesling, Chenin Blanc, Sauvignon Blanc, Chardonnay, Gewurztraminer, Merlot, Cabernet Sauvignon, Pinot Noir vineyards.

PRESTON VINEYARDS AND WINERY

Founded 1975, Healdsburg, Sonoma County, California.

Cases per year: 12,000.

Estate vineyard: Preston Vineyards

Wines regularly bottled: Estate bottled, vintage-dated Sauvignon Blanc, Zinfandel, Chenin Blanc, Gamay, Cuvée de Fumé . Occasionally estate bottles Dry White Wine.

PRESTON WINE CELLARS

Founded 1976, Pasco, Franklin County, Washtingon.

Storage: Oak and stainless steel. Cases per year: 40,000.

Estate vineyard: Preston Ranch.

Label indicating non-estate vineyard: Balcom and Moe Vineyards, Sagemoor Farms.

Wines regularly bottled: Vintage-dated Chardonnay, Chenin Blanc, Fumé Blanc, Gewurztraminer, Merlot, Pinot Noir, Pinot Noir Blanc, Johannisberg Riesling, Select Harvest White Riesling, Cabernet Sauvignon; non-vintage Gamay Beaujolais Rosé; proprietary Desert Rose and vintage-dated Desert Gold; also White Table Wine; occasionally Late Harvest Riesling, White Riesling Ice Wine, Select Reserve Cabernet Sauvignon, Late Harvest Sauvignon Blanc.

Second label: Columbia River Cellars.

PREVOST, PAUL

In the village of Frenchtown, which overlooks the Delaware River, the Prevost, French Revolutionary refugee, cross-pollinated vines with native grapes and produced the variety of Delaware which is considered to be one of the best of the American varieties.

The classic "Paris goblet"

Priest Ranch

Napa County, California.

Cabernet Sauvignon vineyard.

PRINCE MICHAEL VINEYARDS

Private Stock Estate Vineyard

Boone County, Iowa.

Aurora, Marechal Foch, Baco Noir, Fredonia and Seyval vineyard.

PRIVATE STOCK WINERY

Founded 1977, Boone, Boone County, Iowa.

Storage: Stainless steel. Cases per year: 5,000.

History: Iowa's largest winery.

Estate vineyard: Private Stock Vineyard.

Wines regularly bottled: Estate bottled, vintage-dated Emperor, Concord Sweet and Dry, Niagara Sweet and Dry; proprietary Iowa White Sweet and Dry, Van Buren Sweet and Dry; also Grape White Sweet and Dry, Strawberry Sweet and Dry, Blackberry, Cherry Sweet and Dry, Blueberry, Cranberry. Occasionally Pineapple, Banana.

PRODUCED BY

Under Federal Regulations it certifies that not less than 75% of the wine was fermented at the winery, or in the case of Champagne, the transformation from still wine to Champagne was done at the winery. (See also MADE BY.)

pronounced

Tasting term.

Very strong; persistent.

PROPRIETARY LABELING

Wine name which is the property of the winery; e.g. Thunderbird, Wild Irish Rose.

Provenza Estate Vineyards

Montgomery County, Maryland.

Seyval Blanc, Vidal 256, Vidal Blanc, Rayon D'Or, Foch, Cascade, Chancellor, Villard Noir and Chambourcin vineyard.

PROVENZA VINEYARDS

Founded 1974, Brookeville, Montgomery County, Maryland.

Storage: Oak and stainless steel.

Estate vineyard: Provenza Vineyards.

Wines regularly bottled: Estate bottled Seyval Blanc; also Cascade Rosé; Red, White, Rosé Table Wine.

Prudence Island Estate Vineyards

Newport County, Rhode Island.

Chardonnay, Riesling, Gewurztraminer, Gamay Beaujolais, Cabernet, Merlot vineyard.

PRUDENCE ISLAND VINEYARDS

Founded 1973, Prudence Island, Newport County, Rhode Island.

Storage: Stainless steel. Cases per year: 1,000.

History: Winery is on a small island in Narragansett Bay. The 1783 farmhouse has an underground winery. The first commercial vinifera vineyard ever in Rhode Island.

Estate vineyard: Prudence Island Vineyards.

Wines regularly bottled: Estate bottled, vintage-dated Chardonnay, Gamay Beaujolais, Cabernet, Gewurztraminer, Riesling, Merlot.

John Pun Vineyard

Napa County, California.

Chardonnay vineyard.

punt

The indent in the bottom of the Champagne bottle or any other bottle. Also called a "kick."

PUTNAM, NINA WILCOX (1888-1962)

American writer.

> "The grape absorbs the sun, the wine puts the sunshine into men's hearts; without it the world would begin to look for vices to take the place of conviviality."

QUADY, ANDREW K.

Andrew K. Quady was graduated from California State Polytechnic University, Pomona, in 1969, with a bachelor's degree in chemical engineering. After working for two years in the pyrotechnics industry, Quady continued his education at the University of California, Davis, where he received his master's degree in food science, with a specialization in enology. Following graduation, he worked first at Montcalm Vintners, then at Lodi Vintners. A surplus of red wine and inspiration from wine merchant Darrell Corti led to the production of the first Quady Port in 1975. Made from Amador County Zinfandel, the port was fermented at Lodi Vintners, barrel aged at Oakville Vineyards and bottled at Rutherford Hill Winery. In 1976, Quady accepted a position with United Vintners as a processing engineer, where he worked for five years. During that time, he and his wife purchased a small bonded winery and planted seven acres of Portuguese grape varieties. In 1981, he left United Vintners to operate Quady Winery full-time. Quady continues to produce vintage ports from Amador County Zinfandel. He experiments with other Zinfandels and other varieties, such as Orange Muscat.

QUADY WINERY

Founded 1975, Madera, Madera County, California.

Storage: Oak and stainless steel. Cases per year: 5,000.

Labels indicating non-estate vineyard: Shenandoah School Road Vineyard, Clockspring Vineyard, Frank's Vineyard.

Wines regularly bottled: Vintage-dated proprietary Orange Muscat, Essensia, Port of the Vintage; and Vintage Port.

Quail Glen Vineyards

Nevada County, California.

Pinot Noir, Cabernet Sauvignon vineyard.

Quail Hill Ranch

Sonoma County, California.

Pinot Noir vineyard.

QUAIL RIDGE

Founded 1978, Napa, Napa County, California.

Cases per year: 6,000.

History: The winery cellar is in the historic Hedgeside Caves which were handhewn in 1884. According to the Napa Register in 1885, the cost of the building equipped "will probably not fall far short of $15,000, and it will be as commodious in extent as it is comely in appearance."

Estate vineyard: Quail Ridge Vineyard.

Label indicating non-estate vineyard: Cyril Saviez Vineyard.

Wines regularly bottled: Vintage-dated French Colombard, Chardonnay, Cabernet Sauvignon.

Quail Ridge Estate Vineyard

Napa County, California.

Chardonnay Napa Valley vineyard.

QUAIL RUN VINTNERS

Founded 1982, Zillah, Yakima County, Washington.

Storage: Oak and stainless steel. Cases per year: 14,000-25,000.

Estate vineyard: Whiskey Canyon Vineyards.

Label indicating non-estate vineyard: Willard Farms, Section 1 Vineyard; Don Mahre Vineyard; Emerald Acres; Ed Crawford Vineyard.

Wines regularly bottled: Vintage-dated Johannisberg Riesling, Gewurztraminer, Lemberger, Cabernet Sauvignon, Chardonnay, White Riesling, vintage-dated proprietary Morlo Muscat; White Table Wine.

Quarry Hill Estate Vineyard

Cumberland County, Pennsylvania.

Chelois, Vidal, Delaware, Diamond vineyard.

QUARRY HILL WINERY

Founded 1981, Shippensburg, Cumberland County, Pennsylvania.

Storage: Stainless steel.

Estate vineyard: Quarry Hill Vineyard.

Wines regularly bottled: Vidal, Delaware, Diamond, Pink Diamond, Semi-Sweet and Dry Chelois Rosé; Nectarine and Peach Wines; also Mountain Rosé.

Quartz Ridge Vineyard

Sonoma County, California.

Balverne estate Zinfandel Chalk Hill vineyard.

Quemado Vineyard

Val Verde County, Texas.

Val Verde estate Lenoir, Herbemont vineyard.

Quercus Vineyard

Lake County, California.

Cabernet Sauvignon vineyard.

QUILCEDA CREEK VINTNERS, INC.

Founded 1978, Snohomish, Snohomish County, Washington.

Storage: Oak. Cases per year: 500.

Wines regularly bottled: Vintage-dated Cabernet Sauvignon.

Quillen Ranch

Mendocino County, California.

Sauvignon Blanc vineyard.

QUITZOW, AUGUSTUS

In 1880, Quitzow built a winery and distillery that is today Geyser Peak Winery.

RACKING

The drawing of clean wine off its lees (sediment) into another storage container.

Radeki Vineyard

San Luis Obispo County, California.

Petite Sirah, Zinfandel and Merlot Paso Robles vineyard.

A. RAFANELLI

Founded 1972, Healdsburg, Sonoma County, California.

Storage: Oak. Cases per year: 3,000.

Estate vineyard: A. Rafanelli Vineyard.

Wines regularly bottled: Estate bottled, vintage-dated Zinfandel and Gamay Beaujolais.

A. Rafanelli Vineyard

Sonoma County, California.

Alderbrook estate Chardonnay, Sauvignon Blanc, Semillon vineyard.

RALEIGH, SIR WALTER

Famed in English history as a statesman, soldier, explorer, man of letters; favorite of Queen Elizabeth; proponent of the wine grape.

In American history, Walter Raleigh is known as a colonizer and developer of the winegrape culture.

For his Queen, Raleigh organized an expedition to the New World. The site of Roanoke Island was the destination because his emissaries had reported luscious clusters of grapes there for winemaking.

The colony thrived in the beginning, but, when supply ships finally arrived after years of delay, the settlement and its inhabitants had disappeared, including the first English child born in America, Virginia Dare ... (Captain Paul Garrett made her everlasting when he named his wine in her honor, Virginia Dare ...)

Historians still hope to discover what actually happened to the "Citie of Raleigh", The Lost Colony.

In the historic town of Mantoe, on Roanoke Island, some three hundred years later, the famous Walter Raleigh Vine, or the Mother Vine, is still blooming, still bearing its rich clusters of winegrapes ... as it did for the Raleigh colony. But it has grown to gargantuan size, its branches spread out over half an acre, its trunk in excess of two feet.

Until recently, when the Mother Vineyard and Winery operated and produced wine, its major source of supply came from the grapes of the "Mother Vine".

RAMEY, BERN C.

Bern C. Ramey is well known in the wine industry. He has worked as winemaker, merchandising specialist, sales executive, author, lecturer, and educator. His first experience after completing enology and viticulture courses at the University of California, Davis, was in planting the French-American grapes in Illinois. He also founded a champagne winery. For the next 11 years, he was responsible for sales in the East and Midwest for "21" Brands, Inc. Then in 1970, he joined Paul Masson Vineyards as merchandising specialist; he was appointed vice-president in 1971. He has also held a variety of specialized sales positions with Browne Vintners Company and Chateau & Estate Wines. Most recently, he has served as consultant to Paul Masson. Ramey is a prolific writer: He founded the quarterly publication *Wine Illustrated,* and authored the first *Pocket Dictionary of Wines* in 1970. He has written chapters on wine and champagne for *World Book Encyclopedia,* and published a two-volume Wine Record Album in 1964. He also wrote *The Great Wine Grapes and the Wines They Make* in 1979, and is currently revising it for a second edition. His speaking engagements on winemaking and wine marketing are popular internationally. Ramey is active in many professional wine associations, including membership in the International Wine and Food Society, and the Confrerie de la Chaine des Rotisseurs. He is a charter member of the American Society of Enologists and a Supreme Knight of the Brotherhood of Knights of the Vine.

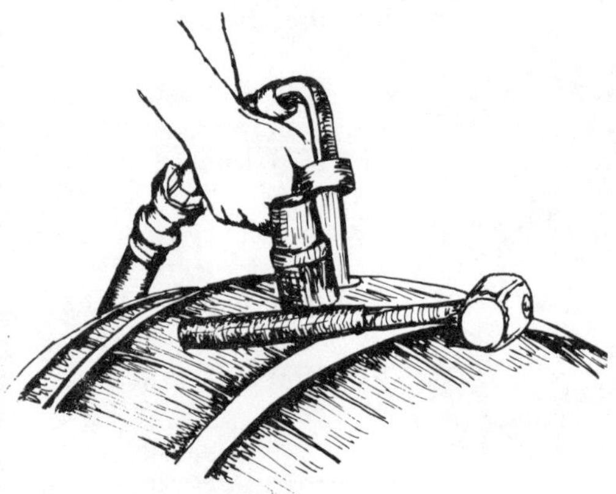

Racking wine

RANCHITA OAKS WINERY, INC.
Founded 1971, San Miguel, San Luis Obispo County, California.
Storage: Oak. Cases per year: 12,000.
Estate vineyard: Bergstrom Vineyards.
Wines regularly bottled: Estate bottled, vintage-dated Zinfandel, White Riesling, Petite Sirah, Cabernet Sauvignon, Chardonnay.
Second label: Cross Canyon Vineyards.

Rancho Alta Vista
Sonoma County, California.
Gewurztraminer Sonoma Valley vineyard.

RANCHO CALIFORNIA
The Riverside County area often called Temecula, and fast becoming one of California's new vineyard areas.

RANCHO DE PHILO
Founded 1975, Alta Loma, San Bernardino County, California.
Storage: Oak and stainless steel. Cases per year: 200.
Estate vineyard: Rancho de Philo.
Wines regularly bottled: Cream Sherry.

Rancho de Philo Estate Vineyard
San Bernardino County, California.
Mission vineyard.

Rancho Dos Amigos
San Luis Obispo County, California.
Barbera vineyard.

Rancho Ferrino
Coahuila, Mexico.
Moscatel, Palomino vineyard.

Rancho Los Dolores
Baja California, Mexico.
Bodegas de Santo Tomás estate Cabernet Sauvignon, Chardonnay, Pinot Noir, Barbera, Chenin Blanc vineyard.

Rancho Regalo del Mar
San Diego County, California.
John Culbertson estate Chardonnay, Semillon and Muscat Canelli vineyard.

Rancho Santa Isabel
San Vicente, Mexico.
Bodegas de Santo Tomás estate Cabernet Sauvignon, Chardonnay, Barbera, Chenin Blanc, French Colombard, Grenache, Carignane, Valdepenas, Palomino, Rubi Red vineyard.

Rancho Sisquoc Vineyards
Santa Barbara County, California.
Cabernet Sauvignon, Merlot, Johannisberg Riesling, Chardonnay and Sauvignon Blanc Santa Maria Valley vineyard.

RANCHO SISQUOC WINERY
Founded 1977, Santa Maria, Santa Barbara County, California.
Storage: Oak and stainless steel. Cases per year: 4,000.
Estate vineyard: Rancho Sisquoc Winery Vineyard.
Wines regularly bottled: Estate bottled, vintage-dated Chardonnay, Sauvignon Blanc, Johannisberg Riesling, Franken Riesling, Cabernet Sauvignon Blanc, Merlot, Cabernet Sauvignon.

Rancho Tierra Rejada
San Luis Obispo County, California.
Sauvignon Blanc, Cabernet Sauvignon, Zinfandel vineyard.

RAPAZZINI WINERY
dba B & R Vineyards, Inc.
Founded 1962, Gilroy, Santa Clara County, California.
Storage: Oak and stainless steel. Cases per year: 25,000.
Estate vineyard: Rapazzini Winery Vineyard; B & R Vineyards, Inc.
Wines regularly bottled: Vintage-dated Muscat Canelli, Gewurztraminer, Johannisberg Riesling, Fumé Blanc, Cabernet Sauvignon.
Second labels: Los Altos; San Juan Bautista.

Rapidan River Estate Vineyards
Orange County, Virginia.
Riesling, Chardonnay, Gewurztraminer, Pinot Noir vineyard.

RAPIDAN RIVER VINEYARDS
Founded 1978, Culpepper, Orange County, Virginia.
Storage: Oak and stainless steel.
History: In 1710 Governor Alexander Spotswood established a colony of German settlers from the Rhine country along the Rapidan River at a place called Germanna. For many years vineyards were successful and then the vines failed due to the phylloxera. This touch of history influenced Dr. Gerhard Guth, a surgeon from

Hamburg, to plant a vineyard along the Rapidan River in 1978.
Estate vineyards: Rapidan River Vineyard.
Wines regularly bottled: Estate bottled, vintage-dated Dry Riesling, Semi-Dry Riesling, Chardonnay, Gewurztraminer.

RAVAT (See Vignoles.)

RAVENSWOOD
Founded 1976, Sonoma, Sonoma County, California.
Storage: Oak. Cases per year: 5,000.
Label indicating non-estate vineyard: Dickerson Vineyard, Dry Creek, Benchland, Batto Ranch, Preston Vineyard.
Wines regularly bottled: Vintage-dated Zinfandel, Cabernet Sauvignon. Occasionally Merlot.

Raymond Estate Vineyard
Napa County, California.
Cabernet Sauvignon, Zinfandel, Chardonnay, Johannisberg Riesling, Chenin Blanc Napa Valley vineyard.

RAYMOND VINEYARD AND CELLAR
Founded 1974, St. Helena, Napa County, California.
Storage: Oak and stainless steel.
History: An old California wine family involved in viticulture dating back to the 1870's.
Estate vineyard: Raymond Vineyards.
Wines regularly bottled: Estate bottled, vintage-dated Caberent Sauvignon, Zinfandel, Chardonnay, Johannisberg Riesling, Chenin Blanc.

RAYON D'OR
A French hybrid which produces excellent white wine. Prized in blends for "interest", and occasionally found bottled as varietal. Spicy flavor.

Rays Ranch
Sonoma County, California.
Seghesio estate Cabernet Sauvignon, Carignane, Golden Chasselas Alexander Valley vineyard.

Red Barn
Napa County, California.
Freemark Abbey estate Caberent Sauvignon, Chardonnay, Sauvignon Blanc, Pinot Noir and Merlot Napa Valley vineyard.

RED DINNER WINES (TABLE WINES)
Red dinner wines usually accompany main course dishes. Most such red wines are dry, sometimes tart and might even be astringent in flavor. They blend well with red meats, spaghetti and highly seasoned foods. Alcoholic content of red dinner wines is from 10 to 14 percent. Some red dinner wines have small amounts of natural sugar and are "mellower"

Red Hill Vineyard
Douglas County, Oregon.
Gewurztraminer, Cabernet Sauvignon, Chardonnay, White Riesling and Pinot Noir Umpqua Valley vineyard.

RED HILLS VINEYARD
(See Arterberry Ltd.)
Vintage-dated Red Hills Vineyard Sparkling Wine.

Red Willow
Yakima County, Washington.
Cabernet Sauvignon Yakima Valley vineyard.

Red Winery Vineyard
Sonoma County, California.
Hafner estate Cabernet Sauvignon, Chardonnay, Gewurztraminer and White Riesling Alexander Valley vineyard.

REDFORD CELLARS
(See Amity Vineyards.)

REDWOOD CANYON CELLARS (See Calafia Cellars.)

Redwood Ranch
Sonoma County, California.
Cabernet Sauvignon Alexander Valley vineyard.

REDWOOD VALLEY
In Mendocino County, this northern valley is by the Russian River.

Redwood Valley Vineyards
Mendocino County, California.
Pinot Blanc vineyard.

Reed Vineyard
San Benito County, California.
Calera estate Pinot Noir vineyard.

Reese Vineyard
Napa County, California.
Cabernet Sauvignon and Chardonnay Edna Valley vineyard.

refined
Tasting term.
Harmonious, composed, elegant.

REFLECTIONS
Found laterally in the glass—in young white, they can be greenish, in old white wines, yellow-gold; in young red wines, violet-blue, in old red wines, orange brick.

Reif Estate Vineyard
Ontario, Canada.
Ortega, Riesling, Dutchess, Foch, Gamay Beaujolais, Kerner, Verdelet, Seyve Villard, Villard Noir, Seyval Blanc, Vidal, Siegfried Rebe, DeChaunac, Pinot Chardonnay, Gewurztraminer vineyard.

REIF WINERY, INC.
Founded 1982, Niagara-on-the-Lake, Ontario, Canada.
Estate vineyard: Reif Vineyard.
Wines regularly bottled: Estate bottled, vintage-dated Riesling, Siegfried Rebe, Vidal, Seyval Blanc; vintage-dated proprietary Rheingold, Rosengarten.

Reiner Mannhardt Vineyards
Okanagan Valley, British Columbia.
White Riesling, Ehrenfelser vineyard.

Reis Estate Vineyards
Texas County, Missouri.
Leon Millot, Seyval Blanc, Villard Blanc, Catawba and Concord Ozark Plateau vineyard.

REIS WINERY
Founded 1978, Texas County, Missouri.
Storage: Oak and stainless steel.
Estate vineyard: Reis Vineyards.
Wines regularly bottled: Estate bottled Leon Millot, Seyval Blanc, Villard Blanc, Vidal, Pink Catawba, Concord; generics: Rhine, Burgundy, also Mountain White and Mountain Red.

Remick Ridge Vineyard
Sonoma County, California.
Smothers estate Chardonnay and Sauvignon Blanc vineyard.

REMUAGE
The riddling or turning of the inverted Champagne bottles to dislodge yeast sediment and allow it to collect on the cork.

Renault Estate Vineyard
Atlantic County, New Jersey.
Johannisberg Riesling, Niagara, Noah, Ives, Elvira, Catawba, Baco Noir, DeChaunac and Seyval Villard Blanc vineyard.

RENAULT WINERY

Founded 1864, Egg Harbor City, Atlantic County, New Jersey.

Storage: Oak and stainless steel. Cases per year: 50,000.

History: The 1,100 acre vineyard and winery were founded by Louis Nicholas Renault. In 1919, John D'Agostino bought the winery and operated it during the 14 years of Prohibition under a government permit. Renault Wine Tonic, with an alcoholic content of 22 percent, was its main product and was sold in drugstores. After Repeal, he bought two California wineries, Montebello at St. Helena and St. George at Fresno and brought the wines to Egg Harbor for blending and bottling. John died in 1948 and, in 1977, Joseph P. Milza, former newspaper owner and publisher, bought the winery and vineyard. The winery is recognized as a historical site, being the oldest winery with its own vineyard in continuous operation in the United States and the largest in New Jersey.

Estate vineyard: Renault Vineyard.

Wines regularly bottled: Estate bottled, vintage-dated Elvira, Pink Catawba, Cabernet Sauvignon, Noah; also estate bottled sparkling wines: Pink, White and Blueberry Champagne, Cold Duck, Blue Duck, Spumante, Sparkling Burgundy; generics: Chablis, Rhine; some proprietary: Pink Lady, Royal Rouge.

RENICK WINERY

Founded 1982, Sturgeon Bay, Door County, Wisconsin.

Storage: Stainless steel. Cases per year: 3,000.

Estate vineyard: Hillside Orchards.

Wines regularly bottled: Cherry Wine; proprietary Pomme Blanc (Apple-Grape); non-alcoholic Cherry-Grape Fitz, Sparkling Apple Cider. Occasionally bottle Cabernet Sauvignon.

RESERVA DE LA CASA (See Casa Madero, Mexico.)

RESIDUAL SUGAR

Natural sugar that has remained unfermented throughout the fermentation process. It is possible to stop a fermentation by addition of alcohol, chilling or by yeast removal (centrifuge, filter etc.) leaving some residual sugar behind for natural sweetness in the wine. The following is a guideline often used for residual sugar.

0.5% or less	Dry
0.6 to 1.5%	Lightly Sweet
1.6 to 3.0%	Medium Sweet
3.1 to 5.0%	Sweet
More than 5%	Very Sweet

Rest & Be Thankful Vineyard

San Luis Obispo County, California. Merlot vineyard.

RESTAURANT RESERVE (See Bardenheier's Wine Cellars.)

Rhine, Chablis, Burgundy, Rosé, Pink Chablis.

RETSINA

The ancient Greeks were involved in wine making, dating back between 1,000 and 800 B.C.

They used pine barrels in which to store their wine, as pine trees grew in abundance in Greece. They became so accustomed to the pine flavor that later they added resin (the pitch of pine trees) to their wine in its processing stage.

This resin-flavored wine grew in popularity, even to the extent of becoming the favorite drink of all Greece. The philosophers, Socrates, Aristotle, Plato and the general, Alexander the Great, all reveled in its taste.

When Western Europe was still in an uncivilized stage, men from Western Europe were sent to the Near East to be educated in Greek culture. One of the things they brought back to Western Europe was the art of wine making, but they made their wine without the addition of resin, processing it in oak barrels instead, as oak trees were common in Western Europe.

To this day, the Greeks still use resin and pine barrels in their wine making, while Western Europe uses oak barrels.

The only producer of Retsina in America is Nicholas G. Verry of California who imports resin from the pine forests of Greece.

RHINE WINE

From the name of the Rhine River in Germany. Used to describe a white dinner or table wine, lightly to moderately sweet, pleasantly tart, pale golden or slightly green gold in color. Medium bodied, fresh and fruity.

Rhinefarm Vineyards

Sonoma County, California.

Gundlach-Bundschu estate Cabernet, Merlot, Zinfandel, Pinot Noir, Chardonnay, Gewurztraminer, Kleinberger, Sylvaner and Johannisberg Riesling Sonoma Valley vineyard.

RICE, ELMER (1892-1967)

American playwright.

"You can have too much champagne to drink but you can never have enough."

Ricetti Vineyard

Mendocino County, California.

Zinfandel vineyard.

RICHBURG, JOHN

Although John Richburg's experience in the wine industry dates back to picking grapes in the Napa Valley as a child, he was not always sure that he wanted to become a winemaker. After graduating from Napa Valley College, he joined the U.S. Air Force. When he married into a winemaking family, in 1968, he began to consider a career in enology, and, in 1971, he enrolled in the Department of Viticulture and Enology at the University of California, Davis. During his studies, Richburg began to work three days a week at Inglenook Vineyards in the laboratory. After graduation in 1973, he joined Inglenook full-time as an assistant winemaker. In 1976, Richburg was promoted to cellarmaster.

Riddling by experienced hands

RIDDLING

The slight turning or shaking of bottles of sparkling wine or champagne, that have been placed in a special rack, neck down. By turning the bottles on a daily basis, the sediment in the bottle settles in the neck. This is done prior to disgorging to remove the sediment.

Ridge Estate Vineyards

Santa Clara County, California.

Cabernet Sauvignon Monte Bello Ridge vineyard.

RIDGE VINEYARDS
Founded 1962, Cupertino, Santa Clara County, California.
Storage: Oak.
Estate vineyard: Monte Bello (Monte Bello Ridge) Vineyard.
Label indicating non-estate vineyard: Dusi Vineyard, Esola Vineyard, Eschen Vineyard, Howell Mountain Vineyard.
Wines regularly bottled: Cabernet Sauvignon, Zinfandel, Petite Sirah.

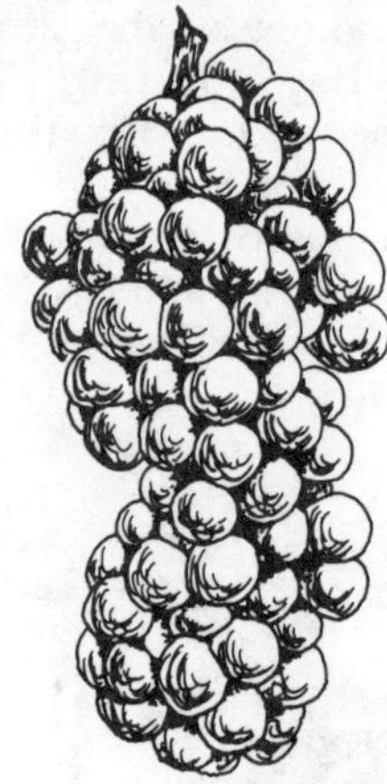

Riesling

RIESLING (See White Riesling.)

RINK, BERNIE
Following in the family tradition of winemaking, which includes bootlegging by his father, Bernie Rink began planting French hybrid vines on the Leelanau Peninsula, Michigan, in 1964. While working as dean of library service at Northwestern Michigan College, he experimented with different grapes to determine which were best suited to the microclimate of his lakeside site. After choosing six varietals, Rink built the Boskydel Vineyards winery in 1975. Rink and his sons now produce dry and semi-dry table wines from 5 hybrid and White Riesling grapes. Other wineries have benefited from Rink's early pioneering efforts, and the Leelanau Peninsula has recently been designated a viticultural district.

Rios Rancho Vineyard
Santa Ynez County, California.
Vega estate Pinot Noir, Riesling and Johannisberg Riesling vineyard.

RIPARIA, VITIS
A native American blue-black grape species abundantly found in the woods and river banks of the East and Midwest. One of the "American Roots" upon which many of the European vines are grown. Descendants of this grape are used for wine making in the upper Midwest. Primarily used for blending. Two hybrid varieties with Riparia parentage are the white grapes, Elvira and Missouri Riesling.

ripe
The term used to describe a wine which has attained maturity, mellowness, perfection. When the term "ripe for bottling" is used, it means the wine has improved in the cask to the highest point possible.

Ripley Vineyard
Litchfield County, Connecticut.
Chardonnay and Seyval vineyard.

Ritchie Creek Estate Vineyard
Napa County, California.
Cabernet Sauvignon and Chardonnay Ritchie Creek vineyard.

RITCHIE CREEK VINEYARDS
Founded 1974, St. Helena, Napa County, California.
Storage: Oak. Cases per year: 800.
Estate vineyard: Ritchie Creek Vineyard.
Wines regularly bottled: Estate bottled, vintage-dated Cabernet, Chardonnay.
Second label: Vineyard 1967.

RIVER BEND (See Davis Bynum Winery.)

RIVER COUNTRY (See Wollersheim Winery.)
Red, White, Rosé.

River East Vineyards
Sonoma County, California.

Sonoma/Windsor estate Pinot Noir and Chardonnay vineyard.

RIVER OAKS VINEYARD (See Clos du Bois.)

River Oaks Estate Vineyards
Sonoma County, California.
Chardonnay, Pinot Noir, Gamay Beaujolais, White Riesling, Gewurztraminer, Cabernet Sauvignon, French Colombard, Chenin Blanc, Sauvignon Vert Alexander Valley vineyard.

RIVER OAKS VINEYARDS
Founded 1977, Healdsburg, Sonoma County, California.
Storage: Oak and stainless steel. Cases per year: 20,000.
Estate vineyard: River Oaks Vineyard.
Wines regularly bottled: Estate bottled, vintage-dated Gewurztraminer, Johannisberg Riesling, Gamay Beaujolais, Cabernet Sauvignon, Zinfandel, Pinot Noir; also vintage-dated Dry White Table Wine, Premium Red and White Table Wine and a non-vintage Rosé.

River Ridge Estate Vineyard/Chateau Ste. Michelle
Benton County, Washington.
White Riesling, Chenin Blanc, Chardonnay, Sauvignon Blanc, Grenache, Cabernet Sauvignon and Semillon Columbia Valley vineyard.

RIVER RIDGE WINERY
Founded 1981, Paterson, Benton County, Washington.
Storage: Oak and stainless steel. Cases per year: 600,000.
History: The newest winery owned by Chateau Ste. Michelle.
Estate vineyard: Cold Creek, Hahn Hill, River Ridge.
Wines regularly bottled: Estate bottled, vintage-dated Cabernet Sauvignon, Blanc de Noir and Late Harvest White Riesling; estate bottled Grenache Rosé, Chenin Blanc, Rosé of Cabernet, Johannisberg Riesling, Gewurztraminer, Merlot, White Riesling, Muscat Canelli, Fumé Blanc and Chardonnay. Wines occasionally bottled are Ice Wine, Botrytised Riesling, Blanc de Noir Sparkling wine and Chateau Reserve wines.
Second label: Farron Ridge.

River Road North Vineyard
Sonoma County, California.
Toyon estate Sauvignon Blanc, Zinfandel Alexander Valley vineyard.

RIVER ROAD VINEYARDS
Founded 1979, Forestville, Sonoma County, California.
Storage: Oak. Cases per year: 4,700.
Estate vineyard: Forestville Ranch, River Road North Vineyard.
Wines regularly bottled: Vintage-dated Chardonnay, Johannisberg Riesling, Fumé Blanc, Zinfandel.

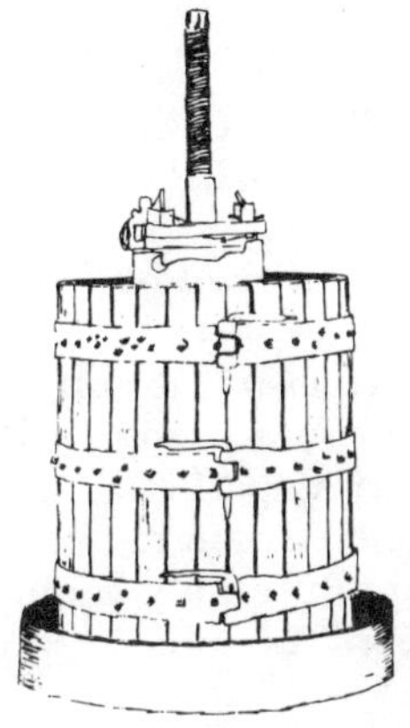

Wooden grape "basket press"

RIVER RUN VINTNERS
Founded 1978, Watsonville, Santa Cruz County, California.
Storage: French, American oak, stainless steel. Cases per year: 1,000-2,000.
Label indicating non-estate vineyard: Ventana Vineyards, Penny Smith Vineyard, Shandon Valley Vineyards.
Wines regularly bottled: Vintage-dated Chardonnay, Zinfandel, Cabernet Sauvignon, Petite Sirah.

River West Vineyard
Sonoma County, California.
Sonoma/Windsor estate Chardonnay, Cabernet Sauvignon, Merlot and Johannisberg Riesling Russian River Valley vineyard.

RIVERSIDE FARM (See Foppiano Wine Company.)
Zinfandel, Chenin Blanc, French Colombard. Premium Dry Red, Premium Dry Rosé, Premium Dry White.

Riverview Vineyard
Santa Ynez County, California.
Vega estate Gewurztraminer vineyard.

RKATSITELI
Russian grape that produces a white wine. A few acres are grown in California.

RMS CELLARS (See Pacheco Ranch Winery.)

ROBIN FILS & CIE, LTD.
Founded 1939, Batavia, Genesee County, New York.
Storage: Redwood. Cases per year: 500,000.
Estate vineyard: Robin Fils Vineyard.
Wines regularly bottled: Estate bottled, proprietary: Capri Pink Catawba, Niagara, Cream Sherry, Sangria, Vin Rosé, Sherry, Sparkling Burgundy, Spumante, Champagne, Pink Chablis, Chianti, Sauterne.
Second label: Imperator.

Robin Fils Estate Vineyard
Genesee County, New York.
Catawba, DeChaunac and Niagara vineyard.

Rock Hill Vineyard
Amador County, California.
Zinfandel Shenandoah Valley vineyard.

Rocking Chair
Yakima County, Washington.
Riesling vineyard.

ROCKY KNOB VITICULTURAL AREA
The Rocky Knob viticultural area is approximately 9,000 acres located in portions of Floyd and Patrick Counties in Virginia.

RODDIS CELLARS
Founded 1979, Napa County, California.
Cases per year: 500.
Estate vineyard: Diamond Mountain Vineyard.
Wines regularly bottled: Estate bottled, vintage-dated Cabernet Sauvignon.

Rodeno Vineyards
Napa County, California.
Chardonnay, French Colombard and Sauvignon Blanc vineyard.

ROLLING HILLS VINEYARDS
Founded 1980, Camarillo, Ventura County, California.
Storage: Oak and stainless steel. Cases per year: 800.
Estate vineyard: Rolling Hills Vineyard.
Label indicating non-estate vineyard: Tepusquet Vineyards, Miramonte Vineyards.
Wines regularly bottled: Vintage-dated Chardonnay, Merlot, Cabernet Sauvignon, Zinfandel.

Rolling Vineyards Farm Estate Vineyard
Schuyler County, New York.
Johannisberg Riesling, Chardonnay, Gewurztraminer, Pinot Noir, Seyval Blanc,

Vidal, Ravat, Cayuga, Villard, Foch, Chelois, Baco Noir, Landot, Catawba and Concord Finger Lakes vineyard.

ROLLING VINEYARDS FARM WINERY

Founded 1981, Hector, Schuyler County, New York.

Storage: Oak and stainless steel. Cases per year: 1,000.

Estate vineyard: Rolling Vineyards.

Wines regularly bottled: Estate bottled, vintage-dated Aurora, Seyval, Johannisberg Riesling, Vidal, Foch, Chelois, Chardonnay; also proprietary Seneca Lake Red and Rosé.

ROMA WINE COMPANY

Founded in the 1890's. Currently owned by Guild Wineries.

ROMBAUER VINEYARDS

Founded 1982, St. Helena, Napa County, California.

Storage: Oak and stainless steel. Cases per year: 3,000.

Wines regularly bottled: Vintage-dated Cabernet Sauvignon, Chardonnay.

room temperature

The temperature at which red wines are usually served, between 60 degrees and 65 degrees F. Older and great wines are better at room temperature. Other red wines (particularly young ones) are often preferred slightly cool. Wine is never brought to room temperature by abrupt "warming," which would spoil it, but it is left to stand for a few hours before serving. There are other opinions which indicate that half an hour is sufficient.

ROSATI WINERY

Founded 1934, St. James, Phelps County, Missouri.

Storage: Oak and stainless steel. Cases per year: 10,000.

History: Established by a local group of Italian grape growers; originally called Knobview Fruitgrowers Co-op. The name was changed in honor of Bishop Rosati, the first bishop of St. Louis.

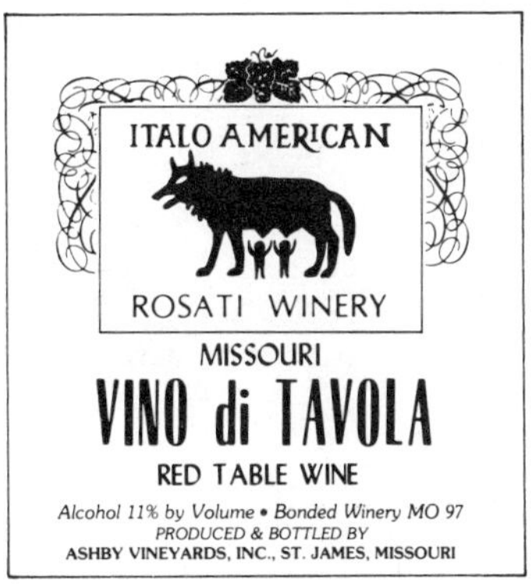

Label indicating non-estate vineyard: Stoltz Vineyards, Bardenheier Vineyard, Piazza Vineyards; Ashby Vineyard; Abby Vineyard.

Wines regularly bottled: Vintage-dated Chelois, Marechal Foch, Missouri Riesling, Old Fashioned Concord, White Concord, Champagne, Spice Wine, Peach, Apple, Sangria, Red, White, Vin Rosé, proprietary Blackberry Hill, Cherry Hill, Dolce Bianco.

ROSÉ

The French word for pink dinner wine, sometimes called a luncheon wine. Rosés range from dry to slightly sweet and are usually fruity-flavored light-bodied and made from Cabernet, Gamay, Grenache, Grignolino or Zinfandel grapes. Alcohol content is 10 to 14 percent, usually about 12 percent by volume. The pink or pale red color is obtained by removing the red grape skins as soon as the required amount of color has been attained by the wine.

Rosé table wines: Cabernet Sauvignon Rosé, Gamay Rosé, Grenache Rosé, Grignolino Rosé, Petite Sirah Rosé, Pinot Noir Rosé, Zinfandel Rosé.

Rose Bower Estate Vineyard

Prince Edward County, Virginia.

Seyval, Rayon D'Or, Chardonnay,

Chancellor Noir, Cabernet Sauvignon, Johannisberg Riesling, Vidal vineyard.

ROSE BOWER VINEYARD AND WINERY

Founded 1974, Hampden Sydney, Prince Edward County, Virginia.

Storage: Oak and stainless steel. Cases per year: 2,000.

Estate vineyard: Rose Bower Vineyard.

Label indicating non-estate vineyard: Allegheny Mountains Vineyard.

Wines regularly bottled: Estate bottled, vintage-dated Seyval; estate bottled Nouveau Foch, Johannisberg Riesling, Chardonnay, Cabernet Sauvignon, proprietary Hampden Forest, Briery Lake.

Rosebud Vineyards

Sacramento County, California.

Chenin Blanc vineyard.

ROSENBLUM CELLARS

Founded 1978, Oakland, Alameda County, California.

Storage: Oak and stainless steel.

Label indicating non-estate vineyard: Cohn Vineyard, Cullinan Vineyard, Lone Oak Vineyard, St. George Vineyard, McGilvery Vineyard.

Wines regularly bottled: Vintage-dated Cabernet Sauvignon, Zinfandel, Chardonnay, Petite Sirah, Sparkling Gewurztraminer. Occasionally bottles Chenin Blanc and Johannisberg Riesling.

ROSETTE

A French hybrid that produces a Rosé type wine.

DON CHARLES ROSS WINERY

Founded 1976, Napa, Napa County, California.

Cases per year: 2,000.

Wines regularly bottled: Cabernet Sauvignon.

Second label: Napa Vintners.

B & B ROSSER WINERY

Founded 1979, Oconee County, Georgia.

Storage: Oak and stainless steel.

Wines regularly bottled: Cabernet Franc, Carmine, White Riesling and Chardonnay.

ROSS-KELLEREI WINERY

Founded 1980, Nipomo, San Luis Obispo County, California.

Storage: Oak and stainless steel. Cases per year: 2,000.

Label indicating non-estate vineyard: Los Alamos Vineyard.

Estate vineyard: Vintage-dated Pinot Noir, Chardonnay, Cabernet Sauvignon, Johannisberg Riesling/White Riesling, Sauvignon Blanc, Pinot Noir Blanc, Cabernet Blanc. Occasionally bottle Red and White Burgundy.

ROTOLO AND ROMEO WINES

Founded 1980, Avon, Livingston County, New York.

Cases per year: 250.

Wines regularly bottled: Vintage-dated DeChaunac.

Second label: Rotolo & Romeo Wine Cellars.

ROTUNDIFOLIA (VITIS)

Native American species of grapes grown in the South from Louisiana to North Carolina. The muscadine varieties, Scuppernong, Magnolia and Carlos are examples.

ROUDON, ROBERT

Born into a teetotaling environment in

Texas during the Great Depression, Robert Roudon did not discover wine until he joined the U.S. Army and was stationed in the Pfalz region of Germany. Following his return, he attended the University of New Mexico, where he earned a degree in fine arts-music in 1960. He enjoyed his first California and French wines during this time. He became an engineer in southern California, and was able to attend short courses in viticulture and enology at the University of California, Davis. By 1967, Roudon decided to become a winemaker and winery owner. After an attempt to establish a winery in southern California failed, he moved to the Santa Clara Valley to accept a position with the Amdahl Corporation. There he met James Smith, who joined him in forming the Roudon-Smith Vineyards, Inc., in 1971. A small vineyard was established in the Santa Cruz Mountain area; the first crush took place in a rented building, and the following year the winery was moved to the basement of Roudon's home. In 1978, a 10,000 case winery was completed near Scott's Valley, and both Roudon and Smith became full time winemakers. Roudon produces wine influenced by his direct experience with German and French wines, adapted to California conditions.

Roudon-Smith Estate Vineyards

Santa Cruz County, California.

Chardonnay Santa Cruz Mountains vineyard.

ROUDON-SMITH VINEYARDS, INC.

Founded 1972, Santa Cruz, Santa Cruz County, California.

Storage: Oak. Cases per year: 10,000.

Estate vineyard: Roudon-Smith Vineyards.

Label indicating non-estate vineyards: Nelson Vineyard, Nelson Ranch, Chauret Vineyard, Burgstrom Ranch, Steiner Vineyard.

Wines regularly bottled: Estate bottled, vintage-dated Chardonnay, vintage-dated Zinfandel, Cabernet Sauvignon, Chardonnay, Petite Sirah, Pinot Noir.

Second label: McKenzie Creek Winery.

ROUGEON

A French hybrid that produces a fruity, light body red wine.

Grape vine bud break

round

Tasting term.

Soft, velvety in the mouth, with plenty of glycerine.

ROUND HILL CELLARS

Founded 1977, St. Helena, Napa County, California.

Storage: Stainless steel. Cases per year: 65,000.

Wines regularly bottled: Vintage-dated Chardonnay, Fumé Blanc, Chenin Blanc, Johannisberg Riesling, Gewurztraminer, Cabernet Sauvignon, Zinfandel, Petite Sirah, Pinot Noir, Gamay Rosé, Muscat Canelli and non-vintage Chardonnay "House", and semi-generics: Chablis, Burgundy.

Second label: Rutherford Ranch.

rounded

A wine with all the vinous elements; well-balanced.

Rowan Vineyards

Santa Barbara County, California.

Chardonnay vineyard.

Royal Kedem Estate Winery

Ulster County, New York.

Seyval Blanc, Aurora, DeChaunac, Tokay, Concord, Chenin Blanc and Zinfandel Hudson River Region vineyard.

ROYAL KEDEM WINERY

Founded 1948, Milton, Ulster County, New York.

Storage: Oak and stainless steel. Cases per year: 350,000.

History: Kedem was founded in a small Czechoslavakian village. In 1948 the Herzog family moved to the United States to continue in the wine business. The winery is in a 120 year old railroad station.

Estate vineyard: Royal Kedem Vineyards.

Wines regularly bottled: Estate bottled Seyval Blanc, Aurora Blanc, DeChaunac, Tokay, Concord, Chenin Blanc, Zinfandel, Fruit, Dry Champagne, Honey Wine, Chablis and Rosé.

ROYALTY

A hybrid grape that was developed at University of California to grow in San Joaquin Valley as a base for Port-type wines.

Roza Estates Vineyard

Yakima Valley, Washington.

The Hogue Cellars. Estate Riesling and Chardonnay vineyard.

RUBIRED

A hybrid red grape that is used in blending and for Port wine.

RUBY

A Port of very deep-red color, usually quite young, as opposed to one which has been aged for some time in wood and has become "tawny", which is pale in color.

RUBY CABERNET

A grape developed by the University of California to grow under the warm conditions of the San Joaquin Valley. It is the child of the Cabernet Sauvignon and the Carignane. The wine is dry, has a Cabernet-like aroma, good acidity and a fruity flavor.

Ruby Hill Vineyard

Alameda County, California.

Semillon and Zinfandel Livermore Valley vineyard.

CHANNING RUDD CELLARS

Founded 1976, Middletown, Lake County, California.

Storage: Oak. Cases per year: 1,000.

Estate vineyard: Mount St. Clare Vineyard.

Label indicating non-estate vineyard: Bella Oaks Vineyard; Morton Adamson Vineyards, Fay Vineyard, Mt. Veeder Vineyard, Mt. St. Helena Vineyard, Espinosa Vineyard, Preston Vineyards, Guenoc Vineyard.

Wines regularly bottled: Vintage-dated Cabernet Sauvignon, Zinfandel; occasionally vintage-dated Petite Sirah, Chardonnay, Chenin Blanc, Sauvignon Blanc, Semillon, Malbec, Petite Vendot, Cabernet Franc, and Port (Zinfandel and Bordeaux blends).

Rue Vineyards

Sonoma County, California.

Zinfandel vineyard.

RUNQUIST, JEFF

The first experience Jeff Runquist had in the wine industry was in 1977. On leave from his studies at the University of California, Davis, he worked at Paul Masson, in Madera, as a lab technician. After returning to school, he spent the following summer at Montevina Wines under Cary Gott, establishing a lab. Following graduation in 1980, Runquist joined Montevina as assistant winemaker/enologist. He was promoted to winemaker in 1982.

RUSHING, SAMUEL HORTON

Samuel Horton Rushing was born in Starkville, Mississippi. He became inter-

ested in making wine while in high school, with his father O.W. Rushing, Jr. After graduating high school, he entered Mississippi State University where he became more involved with fermenting different types of fruits and malts. At the time, Mississippi State was located in a dry county. He was asked to leave one dormitory because of his "hobby" of making wine and home brew but found another dorm that appreciated his contribution to campus life.

After discharge from the U.S. Army, in 1974, he and his wife Diane returned to Mississippi State University, where both graduated in 1976. They moved to the family farm that is located in the heart of the "Mississippi Delta", Merigold, Mississippi. It was here that they began exploring the possibilities of beginning the first commercial winery in Mississippi since the repeal of Prohibition. They pushed for the Native Wine Act of 1976 to be passed and began building and planting. On September 16, 1977 they were issued the license for Mississippi Bonded Winery #1.

Rushing Vineyard

Bolivar County, Mississippi.

The Winery Rushing estate Noble, Carlos and Magnolia Mississippi Delta vineyard.

Russian River Valley Vineyard

Sonoma County, California.

Landmark estate Chardonnay vineyard.

RUSSIAN RIVER VALLEY VITICULTURAL AREA

The viticultural area is comprised for approximately 150 square miles in Sonoma County—one mile north-west of Santa Rosa. There are more than 8,000 acres of producing vineyard area. The area includes those areas where the Russian River and some of its tributaries flow. The principal distinctive characteristic of the Russian River Valley viticultural area is the significant climate effect from coastal fogs which give cooler growing temperatures and distinguish the area from the warmer neighboring valleys such as Dry Creek Valley, Alexander Valley and Sonoma Valley. The Alexander Valley is a bend of the Russian River; Dry Creek is a tributary on the west. Much of Sonoma Valley is cooler than the Russian because of the air flow from San Francisco Bay.

Russian River Vineyards

Sonoma County, California.

Topolos estate Chardonnay, Merlot, Cabernet Sauvignon vineyard.

RUSSIAN RIVER VINEYARDS

(See Topolos at Russian River.)

Rutherford Hill Estate Vineyard

Napa County, California

Cabernet Sauvignon, Gewurztraminer, Merlot, Sauvignon Blanc, Pinot Noir, Johannisberg Riesling Napa Valley vineyard.

RUTHERFORD HILL WINERY

Founded 1976, Rutherford, Napa County, California.

Storage: Oak and stainless steel. Cases per year: 100,000.

Estate vineyard: Rutherford Hills Vineyards.

Label indicating non-estate vineyard: Jaeger Vineyards, Mead Ranch.

Wines regularly bottled: Estate bottled, vintage-dated Chardonnay, Cabernet Sauvignon, Gewurztraminer, Merlot, Sauvignon Blanc, Zinfandel, Pinot Noir, Johannisberg Riesling. Also vintage-dated Zinfandel.

RUTHERFORD RANCH

(See Round Hill Cellars.)

Chardonnay, Sauvignon Blanc, Cabernet Sauvignon, Zinfandel.

Rutherford Vineyards

Napa County, California.

Rutherford Vintners estate Cabernet Sauvignon, Johannisberg Riesling Napa Valley vineyard.

RUTHERFORD VINTNERS

Founded 1976, Rutherford, Napa County, California.

Storage: Oak. Cases per year: 12,000.

Estate vineyard: Rutherford Vineyards.

Wines regularly bottled: Vintage-dated Cabernet Sauvignon, Pinot Noir, Merlot, Johannisberg Riesling, Chardonnay; Muscat of Alexandria; and vintage-dated Chateau Rutherford Special Reserve Cabernet Sauvignon.

SAGARGNAC (See Casa Madero, Mexico.)

SAGE CANYON WINERY
Founded 1981, Rutherford, Napa County, California.
Storage: Oak and stainless steel. Cases per year: 1,200.
Label indicating non-estate vineyard: Oakville Ranch.
Wines regularly bottled: Vintage-dated Chenin Blanc.

Sagemoor Farms
Franklin County, Washington.
Cabernet Sauvignon, Chardonnay, Pinot Noir, Gewurztraminer, Riesling Columbia Valley vineyard.

SAINT MACAIRE
Dry, red table wine grape variety.

St. Amant Vineyard
Amador County, California.
Zinfandel vineyard.

St. Andrew's Estate Vineyard
Napa County, California.
Chardonnay vineyard.

ST. ANDREWS WINERY
Napa, Napa County, California
Storage: Oak and stainless steel.
Estate vineyard: St. Andrews Vineyard.
Wines regularly bottled: Estate bottled, vintage-dated Chardonnay.

ST. CARL (See The Brander Vineyard.)
Merlot Blanc, Cabernet Blanc, Cabernet Franc Blanc.

St. Charles Vineyard
Santa Clara County, California.
Pinot Blanc and Pinot Noir Santa Cruz Mountains vineyard.

St. Clair Vineyard
Napa County, California.
Pinot Noir Carneros vineyard.

St. Clement Estate Vineyard
Napa County, California.

ST. CLEMENT VINEYARDS
Founded 1975, St. Helena, Napa County, California.
Storage: Oak and stainless steel. Cases per year: 10,000.
Estate vineyard: St. Clement Vineyards.
Wines regularly bottled: Vintage-dated Cabernet Sauvignon, Chardonnay, Sauvignon Blanc.

ST. CROIX WINERY, INC.
Founded 1981, Prescott, Pierce County, Wisconsin.
Storage: Oak and stainless steel. Cases per year: 2,500.
Label indicating non-estate vineyard: Eisert Vineyard.
Wines regularly bottled: Vintage-dated DeChaunac, Seyval Blanc, Millot, Rosé Sec; semi-enerics: Chablis, Rhine, Burgundy; also Apple and Rosé. Occasionally Foch and Cherry.

St. Francis Vineyard
Sonoma County, California.
Chardonnay, Johannisberg Riesling, Gewurztraminer, Pinot Noir, Merlot Sonoma Valley vineyard.

ST. FRANCIS WINERY AND VINEYARD
Founded 1973, Kenwood, Sonoma County, California.
Storage: Oak and stainless steel. Cases per year: 20,000.

Estate vineyard: St. Francis Vineyard.

Label indicating non-estate vineyard: Terra Pulchra Vineyard, Jacobs Vineyard, Potter Valley Vineyard.

Wines regularly bottled: Estate bottled Chardonnay, Johannisberg Riesling, Gewurztraminer, Pinot Noir, Merlot; also Muscat Canelli, Chardonnay.

St. George Vineyard

Napa County, California

Petite Sirah vineyard.

St. Gertrudis Vineyard

Riverside County, California.

Cabernet Sauvignon Temecula vineyard.

ST. HELENA

This town in Napa County is home to many prestigious wineries and vineyards.

St. Helena Home Vineyard

Napa County, California.

Cabernet Sauvignon and Petite Sirah vineyard. This is the initial vineyard that Jacob Beringer started in the 1870's.

St. James Estate Vineyard

Phelps County, Missouri.

Concord, Catawba, Niagara, Delaware, Manson, Cynthiana, Vidal Blanc, Villard Noir, Villard Blanc, Chancellor, Seyval Blanc, Rougeon and Isabella vineyard.

ST. JAMES WINERY

Founded 1970, St. James, Phelps County, Missouri.

Storage: Stainless steel. Cases per year: 15,000.

Estate vineyard: St. James Vineyards.

Wines regularly bottled: Estate bottled Sweet Catawba, Vidal Blanc, Munson, Villard Noir, Cynthiana; also Velvet Red and White, Pink and Sweet Catawba, Mellow Red and White, Country Red and White, Fruit Wines, Mead. Occasionally bottles Chancellor.

St. Jean Vineyards

Sonoma County, California.

Chateau St. Jean estate Chardonnay, Sauvignon Blanc, Pinot Blanc, Semillon and Gewurztraminer Sonoma Valley vineyard.

ST. JOHNS (See Bardenheier's Wine Cellars.)

Port, Sherry, Muscatel, White Port.

ST. JULIAN WINE COMPANY, INC.

Founded 1921, Paw Paw, Van Buren County, Michigan.

Storage: Stainless steel. Cases per year: 100,000.

History: St. Julian was established by Mariano Meconi in 1921 in Windsor, Ontario as the Italian Wine Company. After Prohibition was repealed, in 1933, Meconi moved his operations across the river to Detroit. Three years later, in order to be near the grapes, he relocated his winery in Paw Paw, the heart of Michigan's grape growing district. There the area's hilly terrain and Lake Michigan's weather-tempering influence proved ideal for grape cultivation.

Throughout the later 1930's, the Italian Wine Company flourished, bottling sweet wines under a variety of labels. During World War II, when America fought Fascist Italy and anti-Italian sentiment developed, Meconi sought a less conspicuous name for his winery. He chose St. Julian, the patron saint of his birthplace, Faleria, Italy, a small village north of Rome.

After the war, the business passed into the hands of Eugene and Robert Meconi, Meconi's sons, and Apollo Braganini, Meconi's son-in-law. Apollo's son David is now president of the winery.

Wines regularly bottled: Vintage-dated Vidal Blanc Reserve, Seyval Blanc Reserve, Vignoles Reserve, Chancellor Noir, Seyval Blanc (Dry), Sparkling Wines: Cold Duck, Vidal Champagne, Brut, White and Pink, Spumante; vintage-dated generic: Burgundy, non-vintage generics Chablis, Rhine; proprietary Sholom, Frankenmuth White and Rosé, Friars Blanc, Noir Rosé; Van Buren Dry and Cream Sherries, Van Buren Port, a Solera Cream Sherry; also Red Rosé and Sparkling Juices.

St. Regis Vineyard

Napa County, California.

Chardonnay and Gewurztraminer Napa Valley vineyard.

STE. CHAPELLE VINEYARDS, INC.

Founded 1976, Caldwell, Canyon County, Idaho.

Storage: Oak and stainless steel. Cases per year: 110,000.

History: The name of the winery, and building itself, were inspired by the Ste. Chapelle chapel in Paris.

Estate vineyard: Symms Family Vineyard.

Label indicating non-estate vineyard: Mercer Ranch.

Wines regularly bottled: Estate bottled, vintage-dated Johannisberg Riesling, Special Harvest Johannisberg Riesling, Idaho Chenin Blanc, Idaho Chardonnay, Blanc de Noir, vintage-dated Washington Chardonnay, Rosé of Cabernet Sauvignon, Cabernet Sauvignon, Gewurztraminer, Special Harvest Gewurztraminer, Merlot; also Muscat of Alexandria and vintage-dated Blanc de Blanc.

Second label: Sunny Slope Vineyards.

STE. MICHELLE CELLARS, LTD.

Founded Ontario 1921, British Columbia, Ontario and Alberta.

Storage: Oak and stainless steel.

History: The original Jordan winery was established in the village of Jordan, Ontario. Ste. Michelle Wines is western Canada's oldest winery, with origins dating back to 1923. In that year, the Growers Wine Company was formed in Victoria, to produce wine from the loganberries which grew profusely on the Saanich Peninsula of Vancouver Island. This early company quickly expanded, and was instrumental in developing the first vineyards in the Okanagan Valley. The first wine from Okanagan grapes was produced by Growers in 1932. By 1966, the company had changed its name from Growers Wines to Castle Wines. Then, in 1974, the corporate name became Ste. Michelle Wines: a wholly-owned subsidiary of Jordan and Ste. Michelle Cellars Ltd., with associated wineries in British Columbia, Alberta and Ontario.

Wines regularly bottled: Maria Christina (White, White Light, Red, Red Light), Toscano (White, White Light, Red), Falkenberg (White), Interlude (White), Spumante Bambino (White 7% Sparkling Cuvée), Lonesome Charlie (Red 7% Sparkling Cuvée), Gold Seal Champagne, Classic Cream Sherry, Classic Cream Port, Milano Vermouth, Grand Cuvée (Red and White), Lieberstein (White), Branvin Sherry, Growers Cider (Medium and Extra Dry, 6% alcohol), Johannisberg Riesling. Occasionally: Seyval Blanc, Pinot Chardonnay, Marechal Foch and Johannisberg Riesling Special Reserve.

SAINTSBURY

Founded 1981, Napa County, California.

Cases per year: 15,000.

Wines regularly bottled: Vintage-dated Chardonnay and Pinot Noir.

Sakonnet Estate Vineyards

Newport County, Rhode Island.

Estate Riesling, Chardonnay, Pinot Noir, Aurora, Seyval, Vidal, Chancellor, Foch, Millot vineyard.

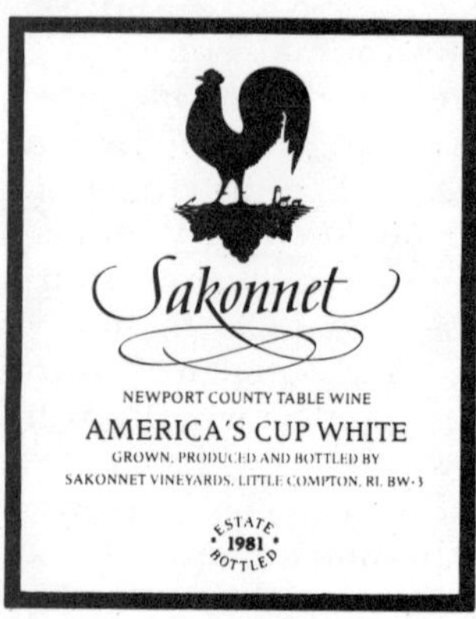

SAKONNET VINEYARDS

Founded 1975, Little Compton, Newport County, Rhode Island.

Storage: Oak and stainless steel. Cases per year: 10,000.

History: Winery is in the seaside town of Little Compton. The vineyards are on a ridge bordered by the Sakonnet River and the Patchet Reservoir.

Estate vineyard: Sakonnet Vineyards.

Wines regularly bottled: Estate bottled, vintage-dated Riesling, Pinot Noir, Americas Cup White and Rhode Island Red. Vintage-dated Spinnaker White, Vidal and Chardonnay. Also bottled is Compass Rosé. Occasionally estate bottled, vintage-dated Late Harvest Riesling and Vidal Reserve.

SALISHAN VINEYARDS

Founded 1982, La Center, Clark County, Washington.

Storage: Oak and stainless steel. Cases per year: 1,600-2,000.

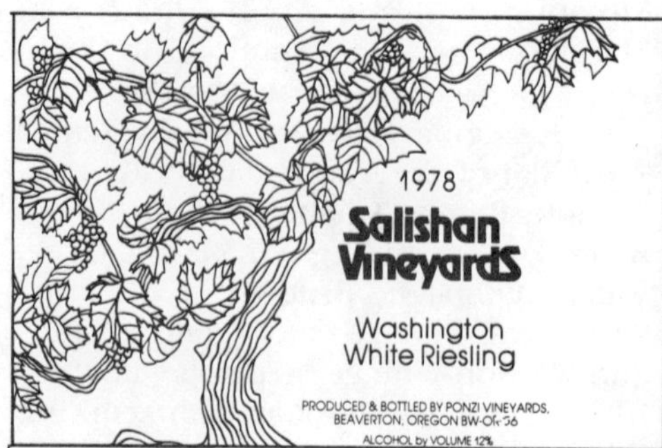

History: Lincoln Wolverton prepared 50 years of weather data comparing the vineyard area with Dijon, France and put all on a computer, day by day, month by month and discovered an incredible similarity in latitude, rainfall, minimum and maximum temperatures. Joan Wolverton, the winemaker, runs the winery with a nearly all-woman crew.

Estate vineyard: Salishan Vineyards.

Wines regularly bottled: Estate bottled, vintage-dated Pinot Noir, Chardonnay, White Riesling, Chenin Blanc, Cabernet Sauvignon Rosé.

Second label: Timmen's Landing.

SALVADOR

A red grape used in blending.

Salzbeger/Chan Vineyard

Dry Creek, California.

Chardonnay and Sauvignon Blanc vineyard.

SAN ANTONIO WINERY

Founded 1917, Los Angeles, Los Angeles County, California.

Cases per year: 400,000.

History: Still in operation by fourth generation of the Riboli family.

Wines regularly bottled: Vintage-dated Chardonnay, Sauvignon Blanc, Light Barbera, Chenin Blanc, French Colombard, White Zinfandel, Gewurztraminer,

Johannisberg Riesling; vintage-dated semi-generics: Velvet Chablis, Rosé and Burgundy. Also proprietary Almondoro.
Second label: Maddalena Vineyard.

SAN BENITO COUNTY
Paicines, Cienega, Hollister.

SAN BENITO VINEYARDS
Founded 1971, Hollister, San Benito County, California.
Storage: Stainless steel. Cases per year: 30,000.
Wines regularly bottled: Fruit wines: Apricot, Pomegranate, Plum, Strawberry, Blackberry, Raspberry, Boysenberry, Pineapple.
Second label: California Fruit and Berry Wine Company.

SAN GABRIEL (See Casa Bello, Canada.)

SAN JOAQUIN COUNTY
Central Valley.

SAN JOAQUIN VALLEY
Most of America's jug wines come from this valley. It runs from Bakersfield north to Sacramento.

San Jose Vineyards
Santa Clara County, California.
Almaden estate Johannisberg Riesling and Gamay Beaujolais vineyard.

SAN JUAN (See Rapazzini Winery.)

SAN JUAN BAUTISTA (See Rapazzini Winery.)

SAN LORENZO
(See Casa Madero, Mexico.)

San Lorenzo Vineyard, Santa Barbara Vineyard, San Judas Vineyard
Parras, Mexico.
Casa Madero estate Chenin Blanc, Colombard, Pinot Blanc, White Riesling, Zinfandel, Petite Sirah, Cabernet Sauvignon, Merlot, Malbec, Ruby Cabernet vineyard.

San Lucas Vineyard
Monterey County, California.
Almaden estate Chardonnay, Chenin Blanc, Sauvignon Blanc, Semillon, Johannisberg Riesling, Folle Blanche, Veltliner, Cabernet Sauvignon, Napa Gamay, Merlot, Petite Sirah, Pinot St. George, Ruby Cabernet, Tinta Madeira and Zinfandel vineyard.

SAN LUIS OBISPO COUNTY
Paso Robles, Edna Valley, Templeton, York Mountain are wine districts in this Central California coastal county.

SAN MARTIN WINERY
Founded 1905, San Martin, Santa Clara County, California.
Storage: Oak and stainless steel.
History: The first vineyards for San Martin Winery were planted in 1905. When the vines came into bearing in 1908, the winery was incorporated.
Wines regularly bottled: Vintage-dated Chenin Blanc, Chardonnay, Cabernet Sauvignon, Sauvignon Blanc, Soft Chenin Blanc, Soft Gamay Beaujolais, Soft Johannisberg Riesling, Emerald Riesling, Late Harvest Johannisberg Riesling, Zinfandel. Also bottles semi-generics: vintage-dated Chablis, Burgundy, Rosé, Rhine and non-vintage Table Wines (Red, White, Rosé).

San Pasqual Valley Estate Vineyard

San Diego County, California.

Sauvignon Blanc, Chenin Blanc, Muscat Canelli, Gamay, San Pasqual Valley vineyard.

SAN PASQUAL VINEYARDS

Founded 1976, San Diego, San Diego County, California.

Storage: Oak and stainless steel. Cases per year: 20,000.

Estate vineyard: San Pasqual Valley Vineyards.

Label indicating non-estate vineyard: Sierra Madre Vineyard.

Wines regularly bottled: Estate bottled, vintage-dated Fumé Blanc, Dry Chenin Blanc, Chenin Blanc, Muscat Canelli, Gamay Nouveau (maceration carbonique), Blanc de Noir. Also some vintage-dated Sparkling Wines. Occasionally bottles Semillon.

SAN PASQUAL VALLEY VITICULTURAL AREA

San Diego County. A relatively small valley located in the Santa Ysabel watershed, 10 to 15 miles east of the Pacific Ocean. It is enclosed by a line which begins on the valley floor and rises to a point 500 feet above sea level, and a manmade western boundary—Interstate 15.

San Saba Vineyard

Monterey County, California.

Cabernet Sauvignon vineyard.

San Vicente/Mission Ranches

Monterey County, California.

Mirassou estate Chardonnay, Pinot Blanc, Gewurztraminer, Chenin Blanc, Monterey Riesling, Gamay Beaujolais, Petite Sirah, Cabernet Sauvignon, Chablis and Petite Rosé vineyards on opposite sides of the valley.

Sanchez Creek Estate Vineyards

Parker County, Texas.

Cabernet Sauvignon, Merlot, Syrah, Zinfandel, Chambourcin, Bellendais, Malbec, Landot, Seyval Blanc, Vidal Blanc, Grenache, Rayon d'Or, Florentil, Villard Noir, Ravat Noir vineyard.

SANCHEZ CREEK VINEYARDS

Founded 1975, Weatherford, Parker County, Texas.

Storage: Oak and stainless steel. Cases per year: 2,000.

Estate vineyard: Sanchez Creek Vineyards.

Wines regularly bottled: Vintage-dated Cabernet Sauvignon, Ruby Cabernet, Rouge.

SANDSTONE WINERY

Founded 1960, Amana, Iowa County, Iowa.

History: Winery is in one of the oldest homes in the Amana Colonies; built in 1855; a National Historic Landmark.

Wines regularly bottled: Fruit and Berry Wines: Grape, Rhubarb, Cherry, Plum, Strawberry, Raspberry, Blackberry, Cranberry.

Sanel Valley Vineyard

Mendocino County, California.

Cabernet Sauvignon and Chenin Blanc vineyard.

SANFORD & BENEDICT

Founded 1972, Lompoc, Santa Barbara County, California.

Cases per year: 10,000.

Estate vineyard: Sanford & Benedict Vineyards.

Wines regularly bottled: Estate bottled Pinot Noir, Cabernet, Chardonnay, Merlot, Riesling; estate bottled proprietary wines: La Purisma (White Table Wine) and Tintillon (Light Pinot Noir).

Sanford & Benedict Estate Vineyard

Santa Barbara County, California.

Pinot Noir, Chardonnay, Merlot, Riesling, Cabernet Sauvignon, Cabernet Franc Santa Ynez Valley vineyard.

Sanford Estate Ranch Vineyards
Santa Barbara County, California.
Sauvignon Blanc, Chardonnay, Pinot Noir Santa Ynez Valley vineyard.

SANFORD WINERY
Santa Barbara, Santa Barbara County, California.
Storage: Oak and stainless steel.
Estate vineyard: Sanford Ranch Vineyards.
Label indicating non-estate vineyard: Sierra Madre Vineyards, Paragon Vineyards.
Wines regularly bottled: Vintage-dated Chardonnay, Sauvignon Blanc, Pinot Noir Vin Gris, Pinot Noir.

Sangiacomo Vineyards
Sonoma County, California.
Chardonnay Sonoma Valley vineyard.

SANTA BARBARA COUNTY
Santa Maria and Santa Ynez.

SANTA BARBARA WINERY
Founded 1962, Santa Barbara, Santa Barbara County, California.
Storage: Oak and stainless steel. Cases per year: 7,500.
History: The oldest of the modern day wineries in Santa Barbara County.
Estate vineyard: Santa Ynez Vineyards.
Label indicating non-estate vineyard: Los Alamos Vineyard.

Wines regularly bottled: Estate bottled, vintage-dated Cabernet Sauvignon, Gris, Zinfandel, White Zinfandel, Chenin Blanc, Cabernet Sauvignon Blanc, Johannisberg Riesling. And a vintage-dated Sauvignon Blanc.
Second label: Solvang Fruit Wine.

SANTA CLARA COUNTY
Santa Clara Valley, Santa Cruz Mountains and Gilroy—Hecker Pass.

SANTA CRUZ MOUNTAIN VINEYARD
Founded 1974, Santa Cruz, Santa Cruz County, California.
Storage: Oak. Cases per year: 3,000.
History: The original John Jarvis vineyard was established on the site in 1863.
Estate vineyard: Jarvis Vineyard.
Label indicating non-estate vineyard: Bates Ranch, Jones Ranch.
Wines regularly bottled: Estate bottled, vintage-dated Pinot Noir. Also vintage-dated Cabernet Sauvignon and Duriff.

SANTA CRUZ MOUNTAINS VITICULTURAL AREA
The geographic boundaries established for the Santa Cruz Mountains include portions of San Mateo, Santa Clara and Santa Cruz Counties.

Santa Gertrudice Vineyard
Riverside County, California.
Johannisberg Riesling vineyard.

Santa Maria Hills Vineyards
Santa Barbara County, California.
Chardonnay vineyard.

SANTA MARIA VITICULTURAL AREA
The geographic boundaries established for the Santa Maria Valley viticultural area include portions of Santa Barbara and San Luis Obispo Counties.

SANTA SOFIA
(See Vinificacion y Tecnologia, S.A.)

SANTA YNEZ VALLEY VITICULTURAL AREA
The new viticultural area is a valley bordering the Santa Ynez River. It is comprised of 285 square miles, with 1,200 acres of vineyards.

Santa Ynez Valley Estate Vineyards
Santa Barbara County, California.
Cabernet Sauvignon, Zinfandel, Chardonnay, Chenin Blanc, Johannisberg Riesling, Sauvignon Blanc, Pinot Noir Santa Ynez Valley vineyard.

Counterscrew type corkpuller

SANTA YNEZ VALLEY WINERY
Founded 1976, Santa Ynez, Santa Barbara County, California.
Storage: Oak and stainless steel. Cases per year: 10,000.
History: Winery planted the first wine grapes in the valley in 1969.
Estate vineyard: Viña de Santa Ynez (Home Vineyard), Brandez Vineyard, Los Olivos Vineyard.
Label indicating non-estate vineyard: Rancho Tierra Rejada.
Wines regularly bottled: Vintage-dated Sauvignon Blanc Reserve, Blanc de Cabernet, White Riesling, Gewurztraminer, Chardonnay, Merlot.

Santa Ynez Vineyard
Santa Barbara County, California.
Sauvignon Blanc, Cabernet Sauvignon, Zinfandel, Chardonnay, Chenin Blanc, Johannisberg Riesling Santa Ynez Valley vineyard.

SANTINO WINES
Founded 1979, Plymouth, Amador County, California.
Storage: Oak and stainless steel. Cases per year: 15,000.
Label indicating non-estate vineyard: Downing Vineyard, D'Agostini Brothers Vineyard, Cowan Family Vineyard, Eschen Vineyards, Baldinelli Vineyard.
Wines regularly bottled: Vintage-dated Zinfandel, Cabernet Sauvignon, Sauvignon Blanc and White Harvest Riesling. Also bottle Late Harvest Riesling; and Red and White Table Wines.

Sarah's Estate Vineyard
Santa Clara County, California.
Chardonnay vineyard.

SARAH'S VINEYARD
Founded 1978, Gilroy, Santa Clara County, California.
Storage: Oak and stainless steel. Cases per year: 1,700.
Estate vineyard: Sarah's Vineyard.
Label indicating non-estate vineyard: Ventana Vineyard, Paragon Vineyard.
Wines regularly bottled: Vintage-dated Chardonnay, Riesling, Cabernet Sauvignon.

SARATOGA CELLARS (See Kathryn Kennedy Winery.)

Varietal, vintage-dated Dry White Wines.

SATIETY

Founded 1981, Davis, Yolo County, California.

Storage: Stainless steel. Cases per year: 400.

Estate vineyard: Satiety Vineyard.

Wines regularly bottled: Satiety.

Satiety Estate Vineyard

Yolo County, California.

Chenin Blanc, French Colombard, Sauvignon Blanc, Semillon, Muscat Blanc, Orange Muscat, Cabernet Sauvignon.

SATTUI, DARYL

Vittorio Sattui founded the V. Sattui Wine Company, in 1894, in San Francisco. Although he was forced out of business during Prohibition, Vittorio's great-grandson, Daryl Sattui, resumed the family enterprise in the Napa Valley. While obtaining a bachelor's degree in business from San Jose State University, Sattui became involved in several entrepreneurial businesses. Following a year in Europe, he returned to the University of California, Berkeley, and earned a master's degree in business. In 1975, he opened V. Sattui Winery, just outside of St. Helena. Sattui believes that the region is best suited to the growth of red wine grape varieties, and he has limited his wine production to reds and rosés. Although he eventually plans to plant his own vineyards, Sattui currently buys all of his grapes and sells his wine through the winery.

V. SATTUI WINERY

Founded 1885, St. Helena, Napa County, California.

Storage: Oak and stainless steel. Cases per year: 9,000.

History: Daryl is the great grandson of the winery's founder.

Label indicating non-estate vineyard: Preston Vineyard.

Wines regularly bottled: Vintage-dated Chardonnay, Muscato Dry Johannisberg Riesling, Off Dry Johannisberg Riesling, Dry Gamay Rouge, Muscat Zinfandel, Cabernet Sauvignon; a non-vintage Madeira; and a vintage-dated Napa Valley Red Wine.

Saucelito Canyon Estate Vineyard

San Luis Obispo County, California.

Zinfandel, Cabernet Sauvignon vineyard.

SAUCELITO CANYON WINERY

Founded 1982, Arroyo Grande, San Luis Obispo County, California.

Storage: Oak and stainless steel. Cases per year: 1,000.

History: The original vineyards at Saucelito Canyon were planted in 1880 by Henry Ditmas, who homesteaded the land. Grapes were grown and wine was produced until the 1940's when the vineyard and homestead were abandoned. The original vines survived until 1974 when the property was purchased and the original three acre Zinfandel vineyard was restored.

Estate vineyard: Saucelito Canyon Vineyard.

Wines regularly bottled: Estate bottled White Zinfandel, Zinfandel, Cabernet Sauvignon.

Sausal Estate Vineyard

Sonoma County, California.

Zinfandel, Cabernet Sauvignon, French Colombard, Petite Sirah, Napa Gamay vineyard.

SAUSAL WINERY

Founded 1973, Healdsburg, Sonoma County, California.

Storage: Oak and stainless steel. Cases per year: 8,000.

Estate vineyard: Sausal Vineyard.

Wines regularly bottled: Estate bottled, vintage-dated Zinfandel, Cabernet Sauvignon, Sausal Blanc; vintage-dated Chardonnay. Occasionally bottles Pinot Noir Blanc.

SAUTERNE

A name derived from the Bordeaux term, Sauternes for its sweet table wine. Sauterne wines are golden-hued, fragrant, full-bodied, white dinner or table wines. They vary greatly because the sweetness and varietal content of Sauterne are not defined by regulations. Generally, American Sauternes are less sweet than those of France.

Cyril Saviez Vineyard

Napa County, California.

French Colombard vineyard.

Sauvignon Blanc

SAUVIGNON BLANC

A grape producing white table wines that range from dry to sweet. When produced "dry", the wine is excellent with seafood and poultry. When sweet, it is delightful to sip after being well chilled.

One of the principal grapes of the Sauternes district of France. The Sauvignon Blanc also has several aliases—using the Fumé name before, or after, the Blanc: Fumé Blanc, Blanc Fumé Sauvignon and Blanc de Sauvignon. The wines carrying the Fumé tend to be drier, with a grassy or smoky taste. Sauvignon Blanc is richer, sweeter and fruitier, unless indicated as "dry".

SAUVIGNON VERT

Wine Grape.

A grape that is mainly used for blending because of its low acidity. Not a true Sauvignon.

savory

Tasting term.

Alive, fresh with pleasant acidity.

Sax Estate Farms

Franklin County, Arkansas.

Delaware, Cynthiana, Niagara, Campbells Early vineyard.

HENRY J. SAX WINERY

Founded 1923, Altus, Franklin County, Arkansas.

Storage: Oak. Cases per year: 750.

History: The Sax winery has been in the family since 1882. The original log cabin in on the premises.

Estate vineyard: Sax Farms.

Wines regularly bottled: Campbells Early, Delaware, Cynthiana, Niagara; occasionally Muscadine.

Sceiford Vineyards

Erie County, Pennsylvania.

Penn Shore estate Hybrids and Native American Lake Erie vineyard.

SCHAPIRO'S WINE COMPANY, LTD.

Founded 1899, New York County, New York.

Storage: Oak. Cases per year: 100,000.

History: Producer of Kosher wines.

Only winery in Manhattan, it is owned and operated by the fourth generation.

Wines regularly bottled: Extra Heavy Cream White, Red and Pink, Naturally Sweet Concord; Blackberry, Cherry, Sweet Tokay, Plum Cream, Cream Honey, Cream Cherry and Blackberry; Sparkling Wines; semi-generics: Chablis, Burgundy, Rhine, Sauterne, Malaga; Also Sangria, Vin Rosé and Proprietary Cream Almondetta, Pina Cocatina.

SCHARFFENBERGER CELLARS

Founded 1980, Ukiah, Mendocino County, California.

Storage: Stainless steel. Cases per year: 15,000.

Estate vineyard: Eagle Point Ranch.

Label indicating non-estate vineyards: Valley Foothills Vineyards, Wiley Ranch, Dennison Ranch, Day Ranch, Lolonis Vineyards.

Wines regularly bottled: Vintage-dated Sparkling Wines (Brut and Reserve), Chardonnay and Blanc de Noir.

Second label: Eaglepoint.

Scharffenberger Vineyard

Mendocino County, California.

Zinfandel vineyard.

SCHEUREBE

In 1916, German botanist George Scheu developed a new wine grape by crossing Sylvaner and Riesling. Riesling-like, but strongly floral. In 1956, the grape was officially named Scheu Rebe.

SCHLOSS DOEPKIN WINERY

Founded 1980, Ripley, Chautauqua County, New York.

Storage: Oak and stainless steel. Cases per year: 1,500.

Estate vineyard: R.D. Watso Farms, Schloss Doepken Winery.

Wines regularly bottled: Estate bottled, vintage-dated Johannisberg Riesling, Chardonnay, Gewurztraminer; estate bottled, vintage-dated proprietary: Roxann Rouge, Ripley Red, September Rouge, Checktowaga White, Chautauquablumchen; and Schloss Blanc.

Schloss Tucker Estate Vineyard

Loudoun County, Virginia.

Pinot Chardonnay, Pinot Noir, White Riesling, Nebbiolo and Sangiovese vineyard.

SCHLOSS TUCKER NURSERY AND VINEYARD.

Founded 1967, Loudoun County, Virginia.

Cases per year: 2,500. (Proposed)

History: First pressing in 1985.

Estate vineyard: Schloss Tucker Vineyard.

Wines regularly bottled: Estate bottled Champagne Blanc de Blanc and Blanc de Noir.

SCHRAM, JACOB

Jacob Schram was an early winemaking pioneer in the Napa Valley. He came to the United States, from Germany, at the age of 14 and moved to California in 1857, where he bought some hillside property. He and his new wife, Annie Christine Weber, cleared and planted the land in vineyards, and built a large stone house and winery. Two sets of tunnels, between 200 and 400 feet each, were dug into the hill for aging the wines. Both Schrams worked actively in production and promotion of their wines, and, by the 1880's, were well known for their Schramsberger Riesling, Hock and Burgundy.

SCHRAMSBERG VINEYARDS

Founded 1962, Calistoga, Napa County, California.

Cases per year: 25,000.

History: The first winery on the hillsides of the Napa Valley. See biographies of Jack Davies, the present owner, and Jacob Schram, the founder.

Wines regularly bottled: Vintage-dated sparkling wines (Reserve Blanc de Noirs, Blanc de Blancs, Cuvée de Pinot, Cremant).

SCHUG CELLARS

(See Storybook Mountain Vintners.)

Vintage-dated Chardonnay, Pinot Noir. Occasionally Sauvignon Blanc, Cabernet Franc.

SCHUG, WALTER

Walter Schug's involvement with wine is the natural result of a long family tradition of grape growing and wine making. Born in Germany he grew up on the state-owned vineyard and winery estate of Assmannshausen in the Rheingau. Staatsweingut Assmannshausen is well known for specializing in the production of Pinot Noir.

He served a six-year work and study apprenticeship in several well-known German vineyards and wineries and, in 1959, received a degree in viticulture and enology from the renowned college of Geisenheim, Germany.

Following the completion of his degree, he spent a year in California to further his study of winemaking. In 1961 he and his wife Gertrud, also from a well-known German wine family, immigrated to the United States and settled in the San Joaquin Valley where he was employed by California Grape Products Corporation.

In 1966, Schug joined the E. & J. Gallo Winery where he became responsible for grower relations and quality control in the North Coast Counties of Napa, Sonoma and Mendocino.

He participated in the awakeing premium wine production and, in 1973, was chosen to help plan and develop the vineyards and winery of Joseph Phelps in St. Helena, Napa Valley. For 10 years Schug served as Phelps' vice president, winemaker and vineyard manager and with great versatility combined old-world skills with new found knowledge in a wide style of wines, many of which became industry benchmarks.

In 1980, the Schug family founded Schug Cellars, and Walter Schug became a partner to Dr. J. Bernard Seps to rebuild the historic Grimm Winery in Calistoga, now called Storybook Mountain Vintners. To provided more time for the development of Schug Cellars, Walter Schug relinquised his duties at Phelps in the summer of 1983 while remaining in the capacity of consultant.

With 30 years of technical experience in the trade, Schug has established himself as a speaker, judge of wines, and friend and advisor to many of his colleagues. As industry spokesman, he helped introduce premium California wines to such markets as Great Britain, Germany, Switzerland and Austria.

Schug is a member of the American Society of Enologists, the California Wine Institute, Napa Valley Vintners Association, Napa Valley Wine Technical Group, Bund Der Ingenieure Des Weinbaus Geisenheim and honorary member of the Rheingauer Weinkonvent in Germany.

In addition to publishing many articles in both English and German language journals, Schug is also a contributor to the University of California Sotheby Book of California Wine. He is a docent at the Napa Valley Wine Library Association, the Napa Valley Wine Symposium, and has been guest lecturer at the University of California at Davis, Napa Junior College and the Institute of Masters of Wine, in Oxford, England.

Schwartzman

Yakima Valley, Washington.

The Hogue Cellars estate Riesling vineyard.

SCRIMSHAW
(See Crosswoods Vineyards, Connecticut.)

SCUPPERNONG

The oldest cultivated native American grape (Muscadine). Produces medium to sweet, grapey red and white wine.

SEA RIDGE WINERY

Founded 1980, Cazadero, Sonoma County, California.

Storage: Oak. Cases per year: 5,000.

Label indicating non-estate vineyard: Searby Vineyards, Mill Station Vineyards, Bohan Vineyards, Summa Vineyards, Porter-Bass Vineyards.

Wines regularly bottled: Vintage-dated Chardonnay, Pinot Noir.

Second label: Wild Boar Cellars.

Seabrook Vineyards

Hunterdon County, New Jersey.
French Hybrids vineyard.

Searby Vineyards

Sonoma County, California.
Chardonnay vineyard.

SEBASTIANI, AUGUST

August Sebastiani was the son of Samuele who had arrived from Tuscany as a teenager, to work as a laborer who hauled cobblestones from the Sonoma quarries to San Francisco for street paving.

In 1904, Samuele bought a horse barn in Sonoma and founded the Sebastiani Vineyards. August began working at an early age and was put to the test while still in his pre-teens. His father was a task master who believed that the way to learn about life and work was to start almost as soon as one was old enough to hold a rake. Samuele instilled in young August the old Italian tradition that love of family and hard work were the way to live one's life.

August started college during Prohibition and left before he finished because Repeal came and his father needed him at the winery. He was a man of great determination, a shrewd winemaker, farmer and businessman whose constant uniform of overalls was deceiving to those who were of the "shirt and tie" school and thought that he was just a "farmer".

August Sebastiani brought "nouveau" to America, in 1972, and was a leader in many wine marketing methods.

A force in the development of Sonoma wines and one who was given a great deal of respect for his abilities. August's son, Sam, now operates the winery that his grandfather and father worked so hard to establish.

Barrel aging in caves

SEBASTIANI VINEYARDS

Founded 1904, Sonoma, Sonoma County, California.

Storage: Oak and stainless steel. Cases per year: 2,500,000.

History: The winery has historical status as the site of Mission Vineyards established in 1825.

Estate vineyard: Vigna del Lago Vineyard, Green Acres Vineyard, Wilson Ranch, Town Vineyard.

Wines regularly bottled: Vintage-dated Barbera, Cabernet Sauvignon, Chardonnay, Zinfandel, Chenin Blanc, Gamay Beaujolais, Gewurztraminer, Green Hungarian, Johannisberg Riesling, Pinot Noir, Pinot Noir Blanc, Rosa Traminer, Grenache, French Colombard, Gamay Rosé; and Brut sparkling wine. Occasionally bottle vintage-dated Gamay Beaujolais

Nouveau, Pinot Noir Tres Rouge, Muscat Canelli, Eagle Cabernet, Black Beauty Zinfandel.

Second label: August Sebastiani.

sec

French word for "dry". When applied to Champagne, it actually means medium sweet.

sediment

Deposit which results from aging in the bottle. Sediment does not harm the wine if it is not disturbed; it is very often an indication that the wine is a better and older one.

A bottle of wine showing sediment should be left to stand until the sediment has dropped to the bottom of the bottle. It should then either be decanted, or poured carefully, in order to allow only clear wine to pass into the glass.

Seeger Vineyard

Ontario, Canada.

Chardonnay vineyard.

Segas Vineyard

Napa County, California.

Chardonnay and Johannisberg Riesling vineyard.

SEGHESIO

Founded 1902, Cloverdale, Sonoma County, California.

Storage: Oak and stainless steel. Cases per year: 11,000.

History: Founded by Edoardo Seghesio in 1902 at Chianti, California. In 1886, Edoardo was asked to come to the United States by Pietro C. Rossi, a leading winemaker, of that era, for Italian Swiss Colony. Edoardo and Rossi had their origins in the same area of Piedmont, Italy.

Edoardo received no pay for the first three years. He stayed in the cookhouse where he received free meals, as was the policy of the Colony in Asti. When Edoardo did get paid, it was in one lump sum which, along with all his earnings for the next five years, made possible his own venture into grape growing.

In 1893, Edoardo, then 32, married Angela Dionisia Vasconi, the niece of Asti superintendent Vasconi. Angela was fifteen.

In 1894, they purchased the original Seghesio grape ranch between Asti and Geyservile.

In 1902, the first Seghesio winery was built by the family.

By 1905, Angela and Edoardo had five children: Ida, Frank, Arthur, Inez, and Eugenio. The family was complete in 1919 with the birth of Pio Eugene (Pete).

The arrival of Prohibition, in 1919, brought changes for America. The Colony went on the market, a victim of Prohibition's restrictions. The Seghesio family bought the Colony and were given credit for saving it for the Sonoma Country wine industry. Asti was about to become a sheep ranch.

In 1920 the Colony's previous owners, among them the Rossi brothers, re-entered the partnership, which lasted until 1933. Edoardo passed away in November 1934, after a brief illness.

In 1949, they purchased the 1,000,000 gallon plant of the Alta Vineyards Company, which is now the Healdsburg winery.

Estate vineyard: Hopland Ranch, Frank's Ranch, Ray's Ranch, Ellis Ranch, Home Ranch, Dry Creek Ranch, Keyhole Ranch.

Wines regularly bottled: Estate bottled, vintage-dated Cabernet Sauvignon, Zinfandel, Chenin Blanc, French Colombard and Marian's Reserve (Petite Sirah, Zinfandel). Occasionally bottled White Zinfandel, Chianti and Carignane.

select late harvest

Equivalent to the Auslese and Beerenauslese wines of Germany. Picked only from grapes that have been infected by Botrytis. By law must have a minimum 28 degrees Brix at harvest.

THOMAS SELLARDS WINERY ·
Founded 1980, Sebastopol, Sonoma County, California.
Storage: Oak. Cases per year: 7,500.
Wines regularly bottled: Vintage-dated Cabernet Sauvignon, Zinfandel, Merlot. Occasionally bottle Gewurztraminer, Petite Sirah, Gamay Beaujolais (all vintage-dated.)

Selleck Vineyard
San Benito County, California.
Calera estate Pinot Noir vineyard.

Sémillon

SEMILLON
A grape that is a companion to the Sauvignon Blanc in the Sauternes region of France. Records show that the Semillon has been planted in the Sauternes district of France since the first century. The wine is made in both dry and sweet versions. Dry it has a perfumey, aromatic flavor. Goes well with poultry and cream sauces. Sweet, it is rich and full.

Sena Vineyard
Sandova City, New Mexico.
Black Malvoise vineyard.

Seneca East Vineyard
Steuben County, New York
Gold Seal estate Chardonnay, White Riesling, Elvira, Niagara, Ravat, Seyval Blanc and Colobel Finger Lakes vineyard.

Sequoia Grove Estate Vineyards
Napa County, California.
Chardonnay, Cabernet Sauvignon, Merlot, Cabernet Franc Napa Valley vineyard.

SEQUOIA GROVE VINEYARDS
Founded 1980, Napa, Napa County, California.
Storage: Oak. Cases per year: 7,500.
History: Winery building dates back to the 19th Century and stands in the shade of a circle of 100 year old Sequoia Redwood trees.
Estate vineyard: Sequoia Grove Vineyards.
Label indicating non-estate vineyard: Cutrer Vineyard, Fay Vineyards, Redwood Vineyards.
Wines regularly bottled: Estate bottled, vintage-dated Chardonnay; vintage-dated Chardonnay, Cabernet Sauvignon Cask 1, Cabernet Sauvignon Cask 2.

Serena Vineyard
Napa County, California.
Pinot Noir vineyard.

SERENDIPITY CELLARS WINERY
Founded 1981, Monmouth, Polk County, Oregon.
Storage: Oak and stainless steel. Cases per year: 750.
Estate vineyard: Serendipity Vineyard.
Label indicating non-estate vineyard: Bethel Heights Vineyard, Mac Corquodale Vineyard, Meadows Vineyard.
Wines regularly bottled: Vintage-dated Pinot Noir, Pinot Noir Blanc, Pinot Noir Fruite, Marechal Foch, Muller Thurgau/White Riesling, Chenin Blanc, Chardonnay. Occasionally Pinot Noir Rosé.

Serendipity Estate Vineyard
Polk County, Oregon.
Pinot Noir, Chardonnay, Pinot Gris, Cabernet Sauvignon vineyard.

serving temperatures
Champagne: 50 degrees; Red Table Wine: 65-70 degrees; Sherry: 65-70 degrees; Port: 65-70 degrees; Rosé: 50 degrees; White Table Wine: 55-65 degrees.

servings

ml = milliliter, l = liter	Dinner Wines Champagne (servings)	Appetizer-Dessert Wines (servings)
187 ml	2	3
375 ml	2-3	4-6
750 ml	4-6	8-12
1 l	6-8	10-14
1.5 l	9-12	15-21
3 l	18-24	30-40

SETTLERS CREEK
(See Delicato Vineyards.)

Seven Lakes Estate Vineyard
Oakland County, Michigan.
DeChaunac, Chancellor, Cascade, Seneca, Aurora, Vidal Blanc, Seyval Blanc and Ravat vineyard.

SEVEN LAKES VINEYARD
Founded 1982, Oakland County, Michigan.
Storage: Oak and stainless steel. Cases per year: 1,000.
Estate vineyard: Seven Lakes Vineyard.
Wines regularly bottled: Estate bottled Chancellor Red, DeChaunac Rosé.

Seven Valleys Vineyard
York County, Pennsylvania.
Chambourcin vineyard.

SEYVAL BLANC
A French hybrid, white, some Chardonnay parentage, widely grown east of the Rockies, producing dry to medium dry—even late harvest style wines for table and dessert use. Dry examples tend to improve for several years in bottle; sweeter (Germanic) style versions are often best when consumed young. Dry versions are excellent with seafoods, and especially, shellfish. Sweeter versions are for sipping and light dishes.

SHADOW CREEK CHAMPAGNE CELLARS
San Luis Obispo County, California.
Label indicating non-estate vineyard: Robert Young Vineyard (Pinot Noir), plus three Sonoma Vineyards (Chardonnay).
Wines regularly bottled: Vintage-dated Sparkling Wines: Blanc de Blanc, Blanc de Noir, Brut and non-vintage Brut.

Shafer Estate Vineyard
Washington County, Oregon.
Pinot Noir, Chardonnay, Riesling, Sauvignon Blanc, Gewurztraminer Willamette Valley vineyard.

SHAFER VINEYARD CELLARS
Founded 1980, Washington County, Oregon.
Cases per year: 5,000.
Estate vineyard: Shafer Vineyard.
Wines regularly bottled: Estate bottled, vintage-dated Pinot Noir, Pinot Noir Blanc, Riesling Reserve, Riesling Willamette Valley, Dry Riesling, Chardonnay, Sauvignon Blanc, Gewurztraminer.

SHAFER VINEYARDS
Founded 1979, Napa, Napa County, California.
Cases per year: 13,000.
Estate vineyard: B/J Ranch.
Wines regularly bottled: Estate bottled, vintage-dated Cabernet Sauvignon, Zinfandel, vintage-dated Chardonnay.
Second label: Chase Creek.

Shaffer Vineyard
Sandoval County, New Mexico.
Zinfandel and Muscat of Alexandria.

SHALLON WINERY

Founded 1978, Astoria, Clatsop County, Oregon.

Cases per year: 1,000.

Wines regularly bottled: Wild Evergreen Blackberry, Peach, "Cran du Lait" (from Cranberries and Whey). Occasionally Spiced Apple, Rhubarb and Red Table Wine.

Second label: Fort Astoria, Gillnetter's Delight, Under the Bridge.

Shandon Valley Vineyards

San Luis Obispo County, California.

Zinfandel vineyard.

SHANKEN, MARVIN R.

Marvin R. Shanken is a leading spokesperson for the wine industry. After obtaining a master's degree in business administration from the American University, Shanken became a respected New York City research analyst and investment banker. It wasn't until 1972 that Shanken became involved in the wine field. In that year he purchased Impact, a small industry newsletter. Under his direction, the publication has developed into an important trade journal, offering in-depth articles on industry issues. Then, in 1979, Shanken bought the Wine Spectator, tripling circulation of the consumer newspaper. The most recent addition to his wine publications is the Market Watch, which analyzes trends and forecasts major growth areas in the industry. Annual industry-related studies are also issued by M. Shanken Communications, Inc. Shanken conducts regular seminars in connection with his publications, including the California Wine Experience and the Impact Marketing Seminar.

CHARLES F. SHAW VINEYARD

Founded 1979, St. Helena, Napa County, California.

Storage: Oak and stainless steel. Cases per year: 20,000.

Estate vineyard: Domaine Elucia Vineyard.

Wines regularly bottled: Estate bottled, vintage-dated Napa Valley Gamay; vintage-dated Napa Valley Gamay, Chardonnay, Fumé Blanc and Gamay Nouveau.

Second label: Bale Mill Cellars.

Shell Creek Vineyards

San Luis Obispo County, California.

Petite Sirah, Gamay, Cabernet Sauvignon and Barbera vineyard.

Shenandoah School Road

Amador County, California.

Zinfandel Shenandoah Valley vineyard.

SHENANDOAH SPRINGS VINEYARD

Founded 1980, Fiddletown, Amador County, California.

Wines regularly bottled: Vintage-dated Zinfandel and Sauvignon Blanc.

SHENANDOAH: TWO VITICULTURAL AREAS

Two new viticultural areas, one in California and the other in Virginia and West Virginia, will be known as Shenandoah Valley. Evidence established two points clearly: 1) that the Shenandoah Valley in Virginia and West Virginia is nationally known; 2) that California's Shenandoah Valley also is nationally known to wine consumers as a specific grape growing area in Amador County, California.

The use of the name "Shenandoah Valley, California" allows consumers to identify where grapes used to make that wine are grown.

The name "Shenandoah Valley" for the Virginia and West Virginia viticultural area is well known and does not require adding the names of the two states.

Shenandoah Estate Vineyards
Amador County, California.
Sauvignon Blanc, Cabernet Sauvignon Shenandoah Valley vineyard.

SHENANDOAH VINEYARDS
Founded 1977, Plymouth, Amador County, California.
Cases per year: 8,000.
Estate vineyard: Home Vineyard.
Label indicating non-estate vineyard: Dal Porto Vineyard, Eschen Vineyard.
Wines regularly bottled: Estate bottled, vintage-dated Sauvignon Blanc; vintage-dated White Zinfandel, Zinfandel, Black Muscat. Cabernet Sauvignon; Zinfandel Port, non-vintage. Occasionally bottles Late Harvest Zinfandel and proprietary Mission del Sol.
Second label: SV.

SHENANDOAH VINEYARDS
Founded 1977, Edinburg, Shenandoah County, Virginia.
Storage: Oak and stainless steel. Cases per year: 7,000-11,000.
Estate vineyard: (Shenandoah Vineyards) Crystal Hill, Cedar Lane, Willow Run.
Wines regularly bottled: Estate bottled, vintage-dated Vidal Blanc, Chambourcin; vintage-dated Vidal Blanc, Seyval Blanc, Chambourcin, Johannisberg Riesling, Chardonnay, Pinot Noir, Cabernet Sauvignon; also vintage-dated Blanc, Rosé.
Second label: Stoney Creek.

SHERIDAN
A native American grape that produces wines similar to Concord. Used primarily for sweet wines and blending.

SHERIDAN (See Merritt Estate Winery.)
White, Red.

shermat
Short for "Sherry Material." Young wine, adjusted by the addition of wine spirits to the desired alcohol content, destined to be made into Sherry by one of several different methods.

SHERRILL CELLARS
Founded 1973, Woodside, Santa Clara County, California.
Storage: Oak and stainless steel. Cases per year: 2,500.
Label indicating non-estate vineyard: Shell Creek Vineyards, Dusi Vineyard, Wiedeman Vineyard.
Wines regularly bottled: Vintage-dated Cabernet Sauvignon, Petite Sirah, Gamay Blanc, Zinfandel.
Second label: Skyline.

SHERRY
The most popular appetizer wine of all; always higher in alcohol content than table wine, is often made from Palomino, Mission or Pedro Ximenez. Sherry has a characteristic "nutty" flavor. Its color ranges from pale gold to dark amber, and it is either dry, medium dry or sweet. The sweet is often called "cream" Sherry. Under United States labeling regulation, the unqualified word "Sherry"

means an alcohol content of not less than 17%. Between 14% and 17% the wine may be labeled "Light Sherry".

The sweeter Sherries are usually served with dessert, or as between meals refreshment.

SHERRY, DRY

Light straw to light amber in color, with a nutty Sherry character. Light in body, but alcoholic and mellow in flavor. Sugar content should be lower than 2.5%.

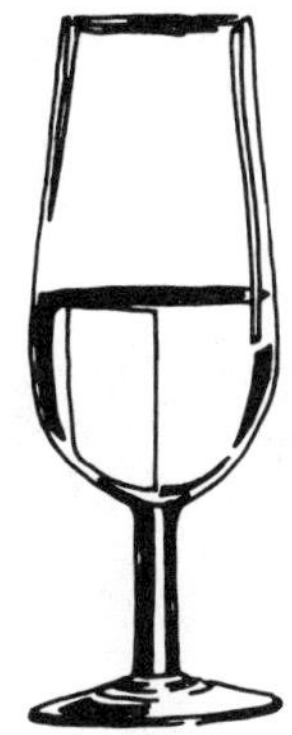

The traditional "Sherry" glass

SHERRY, DRY FLOR AND MEDIUM FLOR

Same characteristics and sugar content of corresponding sherries with the exception of a pronounced flor or mild yeasty flavor.

sherry, making

Although many different grapes are used to make Sherry, many winemakers use Mission, Palomino, Thompson Seedless and Pedro Ximenez. After fermentation of the juice has reached the desired stage—when the wine is as dry as the producers style requires—brandy is added to stop fermentation. The new wine is called shermat. Then many wineries age the wine at a warm temperature in lined or stainless steel or concrete tanks or in redwood containers. This process, at temperatures anywhere between 100 to 140 degrees F (38 to 60 degrees C), continues for from three months to a year. Sometimes it is done in a heated room, sometimes in tanks heated by coils, and sometimes by the heat of the sun. Later, the Sherry is allowed to cool gradually to cellar temperature and it is then aged like other wines. The heating, the oxidation due to the prolonged contact of the warm wine with air, and the aging in wood barrels all combine to develop the pleasant "nutty" flavor characteristic of Sherry. In addition, others produce a "flor" Sherry, using either the Spanish method which allows a film—yeast growth called "flor"—to form on the surface of the wine in partially-filled containers—or the "submerged flor" process. These also impart a distinctive flavor to the wine. Some other wineries offer blends of baked and "flor" Sherries.

Some wineries operate Sherry Soleras. A solera consists of barrels lying one on top of another four or five tiers high, the oldest at the bottom and the youngest at the top. At periodic intervals, the matured Sherry is drawn from the bottom barrel to be bottled. This barrel is then replenished from the one above, and so on. The top barrel is filled with new wine. By this method the young wine mixes with the older to provide a uniform product of high quality year after year.

SHERRY, MEDIUM

Light golden amber to medium golden amber in color. Medium bodied, nutty character. Sugar content should be between 2.5 and 4.0%. Alcoholic content between 16% and 20%.

SHERRY, SWEET (CREAM)

Medium to dark amber in color. Full bodied, rich and nutty with well developed Sherry character. Sugar content should not be less than 4.0%, and alcoholic content usually above 18%.

short

Tasting term.

Yielding, not persistent, fleeting.

Shown & Sons Estate Vineyards
Napa County, California.
Cabernet Sauvignon, Zinfandel, Johannisberg Riesling, Chenin Blanc, Chardonnay Napa Valley vineyard.

SHOWN & SONS VINEYARDS
Founded 1978, Rutherford, Napa County, California.
Storage: Oak and stainless steel. Cases per year: 15,000.
Estate vineyard: Shown & Sons Vineyards.
Wines regularly bottled: Estate bottled, vintage-dated Cabernet Sauvignon, Johannisberg Riesling, Chenin Blanc, Late Harvest Johannisberg Riesling, Zinfandel. Occasionally bottles "Ricardo's Robust 1975 Zinfandel Vinegar."

SIERRA FOOTHILLS WINE COUNTRY
One of the oldest California wine regions; has roots deep in California history. Many who came seeking their fortune in gold turned to a more settled way of life, planting vineyards and orchards, as the rush for gold subsided. By 1890, grape growing and winemaking had become a major industry, with many hundreds of acres of vineyards and dozens of small wineries scattered throughout the Mother Lode. Closing of the mines, followed by population decline, phylloxera vine disease and Prohibition contributed to the eventual abandonment of all but a few vineyards. Shenandoah Valley still possesses the oldest producing vineyard and the D'Agostini Winery, fourth oldest in California.

During the 1960's, rediscovery of Amador County's rich and intense Zinfandels, and experimental plantings in El Dorado, Placer and Calaveras Counties assisted by the University of California, proved the area suited to the production of premium varietal wine grapes, and attention was again focused on the Sierra Foothills.

Sierra Madre Vineyard
Santa Barbara County, California.
Sauvignon Blanc, Pinot Noir vineyard.

Sierra Vista Estate Vineyard
El Dorado County, California.
Cabernet Sauvignon, Chardonnay, Sauvignon Blanc, Zinfandel, Semillon vineyard.

SIERRA VISTA WINERY
Founded 1977, Placerville, El Dorado County, California.
Storage: Oak and stainless steel. Cases per year: 2,000.
Estate vineyard: Sierra Vista Vineyard.
Wines regularly bottled: Estate bottled, vintage-dated Late Harvest Zinfandel, Cabernet Sauvignon, Fumé Blanc, vintage-dated Zinfandel, White Zinfandel, non-vintage Zinfandel.
Second label: Mother Lode, Higgins Vineyard.

SILKWOOD CELLARS
Founded 1979, Yountville, Napa County, California.
Wines regularly bottled: Vintage-dated Chardonnay, Sauvignon Blanc.

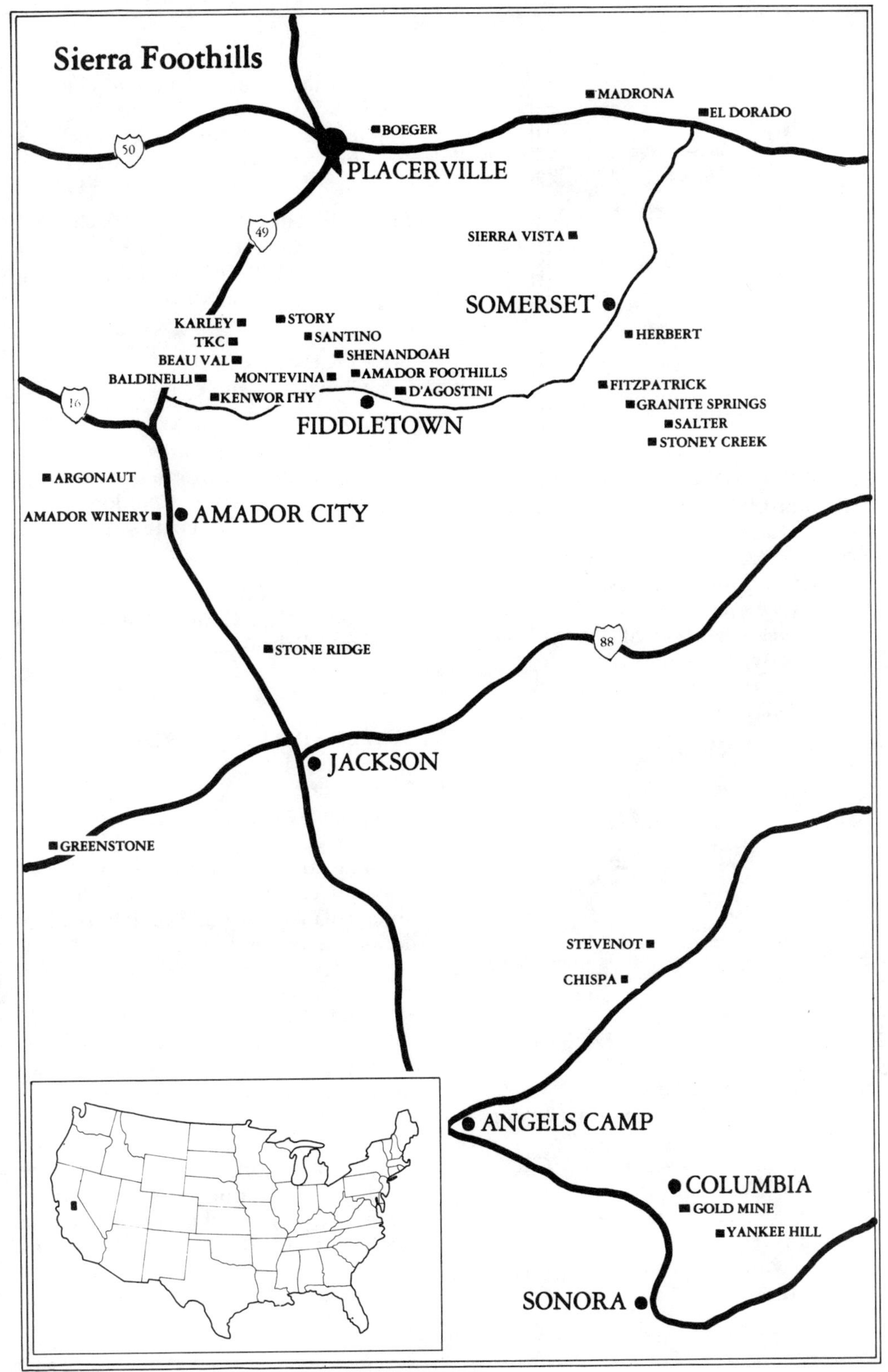
Sierra Foothills
MADRONA
EL DORADO
BOEGER
50
PLACERVILLE
49
SIERRA VISTA
SOMERSET
KARLEY
STORY
TKC
SANTINO
HERBERT
BEAU VAL
SHENANDOAH
BALDINELLI
MONTEVINA
AMADOR FOOTHILLS
FITZPATRICK
KENWORTHY
D'AGOSTINI
16
GRANITE SPRINGS
FIDDLETOWN
SALTER
STONEY CREEK
ARGONAUT
AMADOR WINERY
AMADOR CITY
STONE RIDGE
88
JACKSON
GREENSTONE
STEVENOT
CHISPA
ANGELS CAMP
COLUMBIA
GOLD MINE
YANKEE HILL
SONORA

SILVER MOUNTAIN VINEYARDS
Founded 1979, Los Gatos, Santa Clara County, California.
Storage: Oak. Cases per year: 1,500.
Label indicating non-estate vineyard: Ventana Vineyards, Polson Vineyards.
Wines regularly bottled: Vintage-dated Chardonnay, Zinfandel.

SILVER OAK CELLARS
Founded 1972, Oakville, Napa County, California.
Storage: Oak. Cases per year: 15,000.
History: Winery only produces one wine; it is cellared and bottled 5 years before it is released.
Estate vineyard: Silver Oak Cellars Vineyard.
Label indicating non-estate vineyard: Bonny's Vineyard.
Wines regularly bottled: Vintage-dated Cabernet Sauvignon.

Silverado-Atlas Ranch
Napa County, California.
William Hill estate Chardonnay and Cabernet Sauvignon Napa Valley vineyard.

SILVERADO CELLARS
(See Chateau Montelena.)

Silverado Estate Vineyards
Napa County, California.
Cabernet Sauvignon, Chardonnay and Sauvignon Blanc Silverado vineyards.

SILVERADO TRAIL
A road in Napa County that is between Napa and Calistoga.

SILVERADO VINEYARDS
Founded 1981, Napa, Napa County, California.
Storage: Oak and stainless steel. Cases per year: 20,000.
Estate vineyard: Silverado Vineyards.
Wines regularly bottled: Estate bottled, vintage-dated Cabernet Sauvignon, Chardonnay, Sauvignon Blanc.

SIMI WINERY
Founded 1881, Healdsburg, Sonoma County, California.
Storage: Oak and stainless steel. Cases per year: 130,000.
History: On December 6, 1881, two Italian immigrant brothers, Guiseppe and Pietro Simi, purchased a winery on Front Street near the train depot in Healdsburg for $2,250 in gold coin and named it "Simi Winery". In 1883 they purchased more land and started a second

winery and called it "Montepulciano Winery" in honor of the district in Italy where they were born. The original Simi ceased production after the 1906 earthquake. After Repeal it was decided to drop the name "Montepulciano" because it was difficult to pronounce and it was decided to go back to the "Simi" label. In 1981 Moët-Hennessy became the new owners of Simi.

Wines regularly bottled: Vintage-dated Chardonnay, Chenin Blanc, Gewurztraminer, Rosé of Cabernet Sauvignon, Pinot Noir, Sauvignon Blanc, Zinfandel, Cabernet Sauvignon. Occasionally bottle Burgundy.

Simms Vineyard
El Dorado County, California.
Cabernet Sauvignon vineyard.

SIN ZIN (See Alexander Valley Vineyards.)
Zinfandel.

SINGLETON, VERNON L.

After four years of active duty in World War II, Vernon L. Singleton completed his Ph.D. in protein biochemistry at Purdue University in 1951. He was first employed as a research chemist in antibiotics at the Lederle Division of American Cyanamid Company for four years, then as a biochemist at the Pineapple Research Institute in Honolulu. He served concurrently as an associate professor of chemistry at the University of Hawaii. Beginning in 1958, he joined the staff in the Department of Viticulture and Enology at the University of California, Davis, as assistant enologist. His current position is as professor of enology and chemist in the Agricultural Experiment Station. Singleton is well-known throughout the wine industry for his authoritative writings on wine and wine chemistry. He has published, with his students and colleagues, over 125 books, articles, papers and patents. His *Wine, an Introduction for Americans,* written with Maynard A. Amerine, is considered a classic in the field. His recent research has included further studies of phenols and tannins, chemical reactions of wine aging and storage, and the chemical effects of harvesting and processing. Singleton has been awarded the first André Simon Literary Prize (with Amerine) for the best book in English on a gourmet subject; the first Walter and Carew Reynell Fellowship; and the Eighth Biennial Wine Research Award of the Society of Medical Friends of Wine. He has also been active in the American Society of Enologists and served as its president in 1975-76.

Siskiyou Estate Vineyards
Josephine County, Oregon.
Cabernet Sauvignon, Gamay, Pinot Noir, Gewurztraminer, Chenin Blanc, Merlot vineyard.

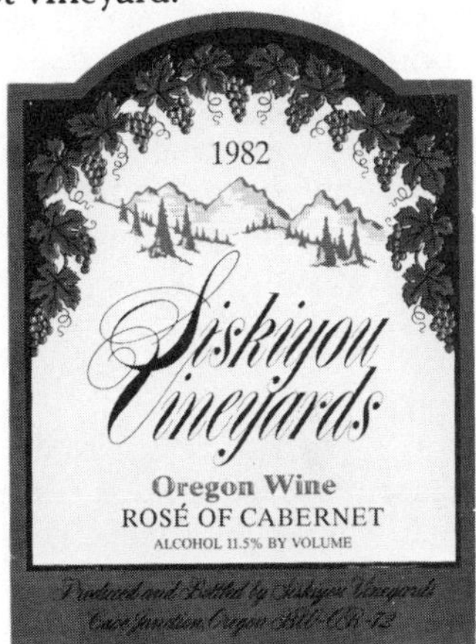

SISKIYOU VINEYARDS

Founded 1978, Cave Junction, Josephine County, Oregon.

Cases per year: 6,000.

Estate vineyard: Siskiyou Vineyards.

Wines regularly bottled: Estate bottled, vintage-dated Semillon, White Riesling, Chenin Blanc, Sauvignon Blanc, Gewurztraminer, Chardonnay, Cabernet Sauvignon, Pinot Noir, Zinfandel, Rosé of Cabernet.

size

Present Liter Size (Amount) ml = milliliter L = liter	Ounces	Old U.S. Bottle Size (Name)	Ounces
100 ml	3.8	Miniature	2.0
187 ml	6.34	Split	6.4
375 ml	12.68	10th	12.8
750 ml	25.36	5th	25.6
1 L	33.81	Quart	32.0
1.5 L	50.72	Magnum	51.2
3 L	101.44	Jeroboam	102.4

SKODA, BERNARD

Bernard Skoda studied winemaking in the Alsace region of France before coming to America with his family in 1954. He worked first at a large department store, setting up a wine section, then at Parrot & Co., the Louis M. Martini and Wente Bros. distributor. In 1961, Skoda accepted the position of sales director at the Martini Winery and worked there for 15 years. During that time, he began to search the area for suitable vineyard property that he might tend in his impending retirement. In 1967, he acquired 25 acres north of Rutherford, and planted them in Cabernet Sauvignon. Further vineyards were purchased and planted in Johannisberg Riesling in 1972. He also began to clear the land for a winery, which was completed in 1977. After he resigned from Martini in 1976, Skoda established Rutherford Vintners. It is a family-run operation, with Skoda serving as owner, winemaker, viticulturist, and tour guide. His wife, Evelyn, works in the tasting room and office, and their son, Louis, helps out on weekends. Skoda specializes in the Cabernet Sauvignon and the Johannisberg Riesling, grown in his vineyards, and also produces Pinot Noir and Merlot made from grapes purchased from friends.

Sky Estate Vineyards

Napa, Napa County, California.
Zinfandel vineyard.

SKY VINEYARDS

Founded 1973, Napa County, California.
Storage: Oak. Cases per year: 2,000.
Estate vineyard: Sky Vineyards.
Wines regularly bottled: Estate bottled, vintage-dated Zinfandel.

SKYLINE (See Sherrill Cellars.)

Skyline Vineyard

Marion County, Oregon.
Riesling, Chardonnay, Pinot Noir and Gewurztraminer Willamette Valley vineyard.

Slaughter Vineyards

Lubbock County, Texas.
Chenin Blanc, Chardonnay and Cabernet Sauvignon vineyard.

Sleepy Hollow Vineyards

Monterey County, California.
Zinfandel, Pinot Noir and Chardonnay vineyard.

Scott Knight Smith Vineyard

Santa Clara County, California.
Cabernet Sauvignon vineyard.

SMITH, ARCHIE M.

Archie M. Smith, Jr. enlisted as a naval aviation flight trainee in 1941. He served overseas as a second lieutenant, USMCR, during World War II. When he was retired for disability, Smith enrolled in the University of Virginia, where he was graduated with a degree in psychology in 1949. After graduate work at both Tulane University and Florida State University, he worked briefly for Brown-Forman Distillers Corporation. In 1953, he moved to a farm in Fauquier County, Virginia, where he raised cattle and crop farmed for eighteen years. He attended some of the early viticulture and enology courses of the Vinifera Wine Growers Association. In 1971, Smith replanted his land in vineyards, and now grows over fifty acres of vinifera wine grapes. He is president of Meredyth Vin yards. Smith is also founder and chairman of the Virginia Wineries Association, and a director of the American Association of Vintners.

Smith & Hook Estate Vineyard
Monterey County, California.
Cabernet Sauvignon vineyard.

SMITH & HOOK VINEYARD
Founded 1974, Gonzales, Monterey County, California.
Storage: Oak and stainless steel. Cases per year: 10,000.
Estate vineyard: Smith & Hook Vineyard.
Wines regularly bottled: Estate bottled, vintage-dated Cabernet Sauvignon.

Smith-Madrone Estate Vineyard
Napa County, California.
Chardonnay, Johannisberg Riesling, Pinot Noir, Cabernet Sauvignon vineyard.

SMITH-MADRONE VINEYARDS
Founded 1977, St. Helena, Napa County, California.
Storage: Oak and stainless steel. Cases per year: 4,000.
Estate vineyard: Smith-Madrone Vineyard.
Wines regularly bottled: Estate bottled, vintage-dated Chardonnay, Pinot Noir, Cabernet Sauvignon; also vintage-dated Johannisberg Riesling.

smoky
Tasting term.
A noticeable odor of smoke.

SMOKY MOUNTAIN WINERY
Founded 1981, Gatlinburg, Tennessee.
Storage: Stainless steel. Cases per year: 5,000.
Wines regularly bottled: Red, White, Rosé.

SMOTHERS-VINE HILL WINES
Founded 1977, Santa Cruz, Santa Cruz County, California.
Cases per year: 4,000.
Estate vineyard: Remick Ridge Vineyard.
Label indicating non-estate vineyard: Green Pastures Vineyard.
Wines regularly bottled: Estate bottled, vintage-dated Chardonnay, Sauvignon Blanc; vintage-dated Gewurztraminer (sweet and dry), Zinfandel, Cabernet Sauvignon, Chardonnay, White Riesling (off dry). Occasionally bottles White Riesling and vintage-dated Pinot Noir.

SNOQUALMIE WINERY
Founded 1983, Issaquah, King County, Washington.
Storage: Oak and stainless steel. Cases per year: To be determined.
History: Joel K. Klein who is owner and winemaker came to the Northwest to be the winemaker at Chateau Ste. Michelle. Prior to that he helped to build and

design Geyser Peak in California where he was also winemaker. He is associated with David Wyckoff of Yakima Valley whose family has been growing vinifera for many years.

Wines regularly bottled: Vintage-dated Chenin Blanc, Gewurztraminer, Muscat Canelli, Semillon, White Riesling, Chardonnay, Cabernet Sauvignon and Merlot.

Snow Mountain Vineyard

Nevada County, California.

Pinot Noir and Chardonnay vineyard.

soft

Tasting term.

Yielding, low in acidity.

soft wines (See light wines.)

New low alcohol (7-10% wines).

Sokol Blosser Estate Vineyard

Yamhill County, Oregon.

Pinot Noir, White Riesling, Chardonnay vineyard.

SOKOL BLOSSER WINERY

Founded 1971, Dundee, Yamhill County, Oregon.

Storage: Oak and stainless steel. Cases per year: 25,000.

Estate vineyard: Sokol Blosser Vineyard.

Label indicating non-estate vineyards: Durant Vineyards, Sagemoor Farms, Fuqua Vineyards.

Wines regularly bottled: Estate bottled, vintage-dated White Riesling, Gewurztraminer, Pinot Noir, Chardonnay, Mueller-Thurgau. Also vintage-dated Sauvignon Blanc, Merlot, Pinot Noir Rosé.

solera

The Spanish system of progressively blending Sherries in tiers of small casks—to blend Sherries of the same type but varying ages.

SOLANO WINERY

(See Wooden Valley Winery.)

Chardonnay.

Solano Winery Vineyard

Solano County, California.

Chardonnay Suisun Valley vineyard.

SOLVANG FRUIT WINE

(See Santa Barbara Winery.)

Fruit wines: Olallieberry, Strawberry, Raspberry.

SONOMA COUNTY

One of California's leading wine producing areas. Dry Creek Valley, Alexander Valley, Russian River Valley, Sonoma Valley and Carneros are the outstanding vineyard areas of the county.

Cutrer

Sonoma County, California.

Chardonnay vineyard.

SONOMA-CUTRER VINEYARDS

Founded 1980, Windsor, Sonoma County, California.

Storage: Oak and stainless steel.

Estate vineyard: Cutrer Vineyards, Mirabelle Vineyard, Les Pierres Vineyard.

Wines regularly bottled: Vintage-dated Chardonnay.

SONOMA MISSION

(See R. Montali Winery.)

SONOMA VALLEY CELLARS

Founded 1981, Sonoma County, California.

Storage: Stainless steel. Cases per year: 8,000.

History: A partnership to produce Champagne combining the Duckhorn's involvement in the wine industry, the Hunter's who are growers with the technical supervision of consultant Dimitri Tchelistcheff.

Estate vineyard: Hunter Farms Vineyard.

Wines regularly bottled: Brut de Noir Champagne.

SONOMA VALLEY VITICULTURAL AREA

Area bordered by the ridge called

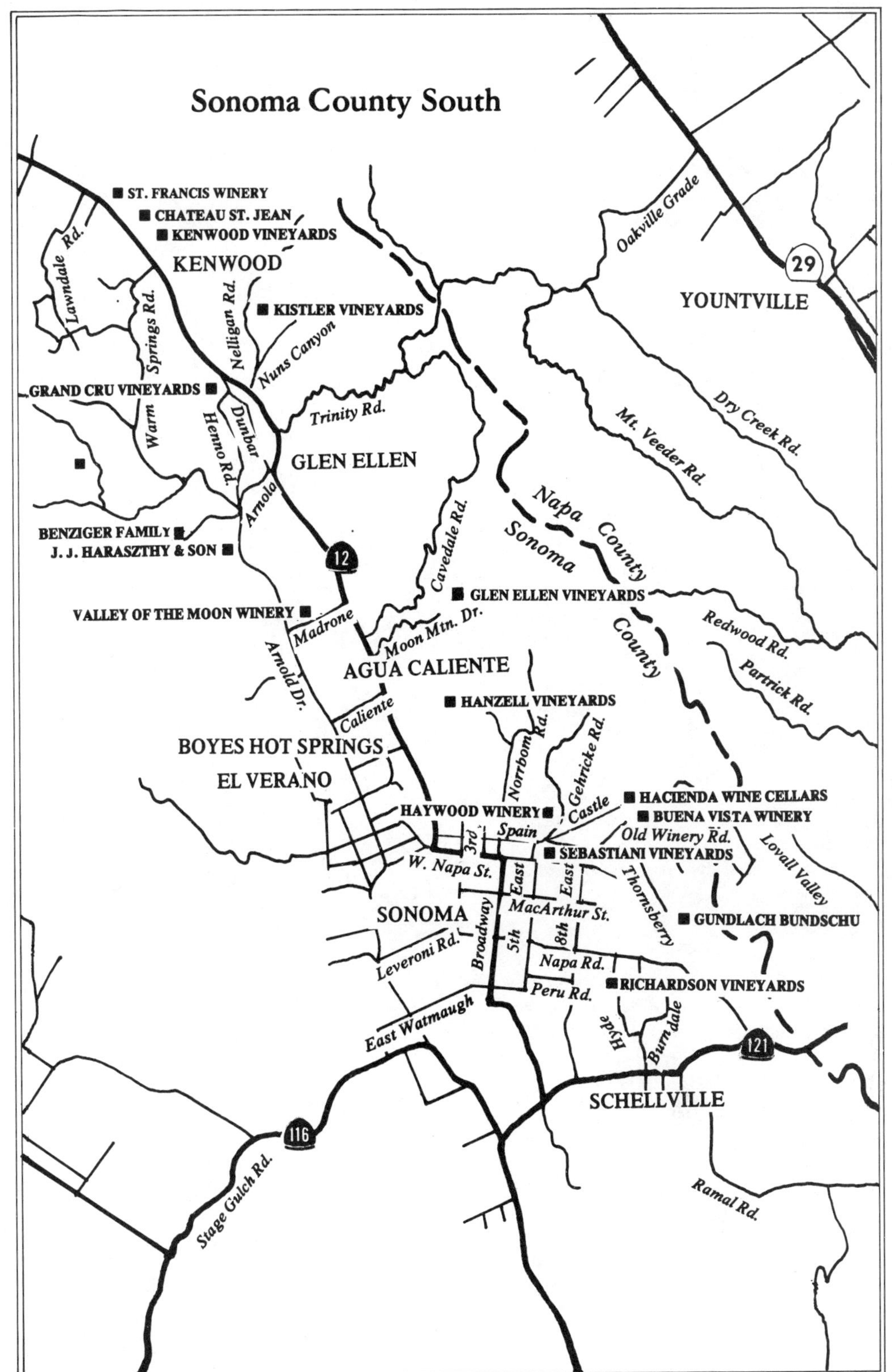
Sonoma County South
ST. FRANCIS WINERY
CHATEAU ST. JEAN
KENWOOD VINEYARDS
KENWOOD
Lawndale Rd.
Warm Springs Rd.
Nelligan Rd.
KISTLER VINEYARDS
Nuns Canyon
GRAND CRU VINEYARDS
Henno Rd.
Dunbar
Trinity Rd.
GLEN ELLEN
Arnold
BENZIGER FAMILY
J. J. HARASZTHY & SON
12
VALLEY OF THE MOON WINERY
Madrone
Arnold Dr.
Cavedale Rd.
Moon Mtn. Dr.
GLEN ELLEN VINEYARDS
AGUA CALIENTE
Caliente
HANZELL VINEYARDS
BOYES HOT SPRINGS
EL VERANO
Norrbom Rd.
Gehricke Rd.
HAYWOOD WINERY
Castle
Spain
3rd
HACIENDA WINE CELLARS
BUENA VISTA WINERY
Old Winery Rd.
SEBASTIANI VINEYARDS
W. Napa St.
East
SONOMA
Broadway
MacArthur St.
5th
8th
Thornsberry
Lovall Valley
GUNDLACH BUNDSCHU
Leveroni Rd.
Napa Rd.
Peru Rd.
RICHARDSON VINEYARDS
Hyde
Burndale
East Watmaugh
116
121
SCHELLVILLE
Stage Gulch Rd.
Ramal Rd.
Oakville Grade
29
YOUNTVILLE
Dry Creek Rd.
Mt. Veeder Rd.
Napa County
Sonoma County
Redwood Rd.
Partrick Rd.

SONOMA MAP

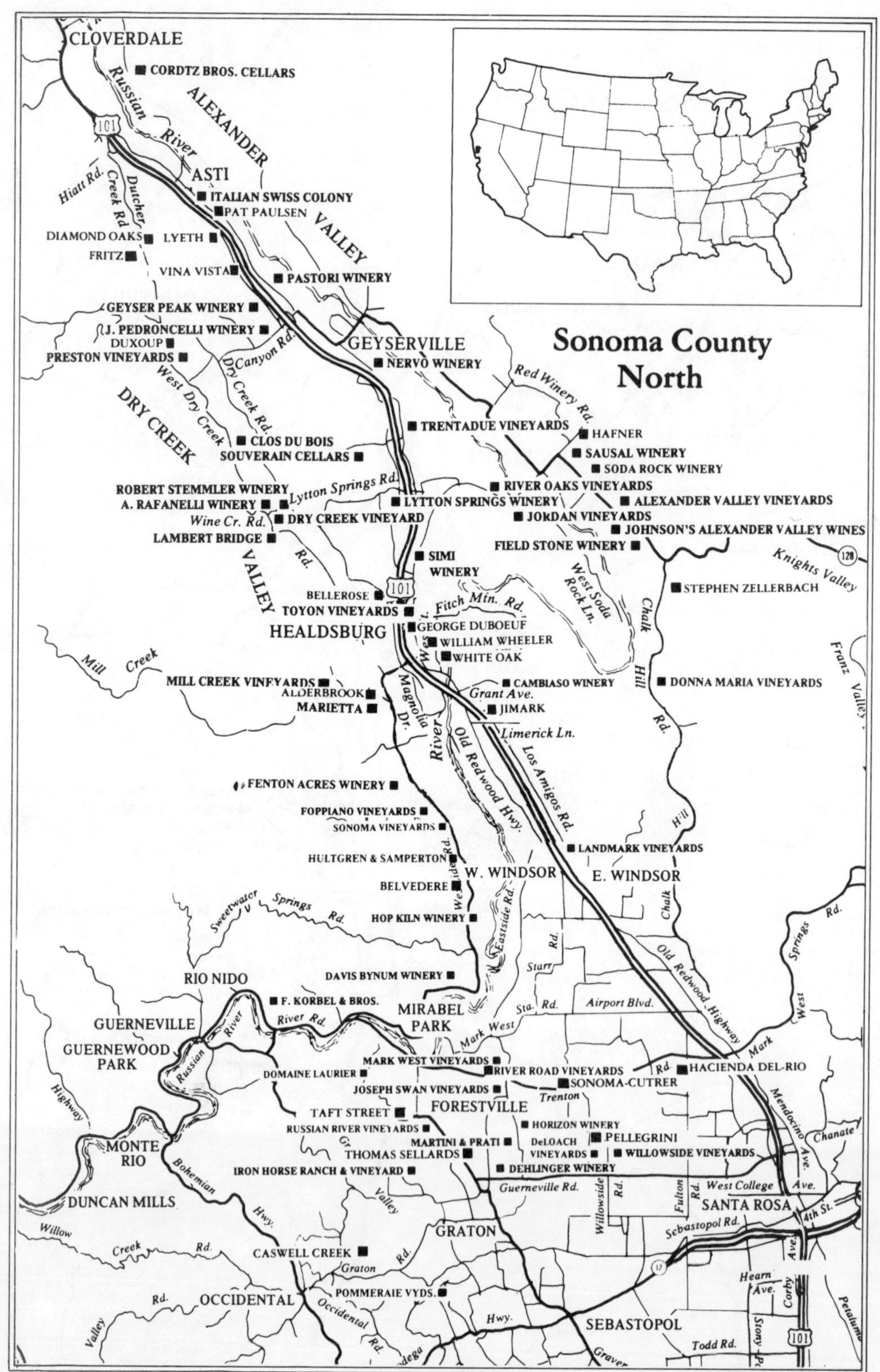
Sonoma County North
CLOVERDALE
CORDTZ BROS. CELLARS
Russian River
ALEXANDER VALLEY
101
Hiatt Rd.
Dutcher Creek Rd
ASTI
ITALIAN SWISS COLONY
PAT PAULSEN
DIAMOND OAKS
LYETH
FRITZ
VINA VISTA
PASTORI WINERY
GEYSER PEAK WINERY
J. PEDRONCELLI WINERY
DUXOUP
PRESTON VINEYARDS
Canyon Rd.
GEYSERVILLE
NERVO WINERY
Red Winery Rd.
West Dry Creek
Dry Creek Rd.
DRY CREEK VALLEY
TRENTADUE VINEYARDS
HAFNER
CLOS DU BOIS
SOUVERAIN CELLARS
SAUSAL WINERY
SODA ROCK WINERY
ROBERT STEMMLER WINERY
Lytton Springs Rd.
RIVER OAKS VINEYARDS
A. RAFANELLI WINERY
LYTTON SPRINGS WINERY
ALEXANDER VALLEY VINEYARDS
Wine Cr. Rd.
DRY CREEK VINEYARD
JORDAN VINEYARDS
JOHNSON'S ALEXANDER VALLEY WINES
LAMBERT BRIDGE
FIELD STONE WINERY
128
Knights Valley
SIMI WINERY
West Soda Rock Ln.
Chalk Hill Rd.
STEPHEN ZELLERBACH
BELLEROSE
West Fitch Mtn. Rd.
TOYON VINEYARDS
HEALDSBURG
GEORGE DUBOEUF
WILLIAM WHEELER
WHITE OAK
Mill Creek
Franz Valley
MILL CREEK VINEYARDS
CAMBIASO WINERY
DONNA MARIA VINEYARDS
ALDERBROOK
Grant Ave.
Magnolia Dr.
MARIETTA
JIMARK
Limerick Ln.
River
Old Redwood Hwy.
Los Amigos Rd.
FENTON ACRES WINERY
FOPPIANO VINEYARDS
SONOMA VINEYARDS
Westside Rd.
HULTGREN & SAMPERTON
LANDMARK VINEYARDS
W. WINDSOR
E. WINDSOR
BELVEDERE
Sweetwater Springs Rd.
HOP KILN WINERY
Eastside Rd.
Starr Rd.
Old Redwood Highway
West Springs Rd.
DAVIS BYNUM WINERY
RIO NIDO
F. KORBEL & BROS.
Airport Blvd.
MIRABEL PARK
GUERNEVILLE
River Rd.
Mark West Sta. Rd.
GUERNEWOOD PARK
Russian River
MARK WEST VINEYARDS
Mark West Rd
DOMAINE LAURIER
RIVER ROAD VINEYARDS
HACIENDA DEL-RIO
JOSEPH SWAN VINEYARDS
SONOMA-CUTRER
Trenton
Highway
TAFT STREET
FORESTVILLE
Mendocino Ave.
HORIZON WINERY
RUSSIAN RIVER VINEYARDS
MARTINI & PRATI
DeLOACH VINEYARDS
PELLEGRINI
Chanate
MONTE RIO
THOMAS SELLARDS
WILLOWSIDE VINEYARDS
Bohemian Hwy.
IRON HORSE RANCH & VINEYARD
DEHLINGER WINERY
Green Valley
Guerneville Rd.
Willowside Rd.
Fulton Rd.
West College Ave.
DUNCAN MILLS
SANTA ROSA
4th St.
Willow Creek Rd.
GRATON
Sebastopol Rd.
CASWELL CREEK
Graton Rd.
12
Hearn Ave.
Corby Ave.
POMMERAIE VYDS.
OCCIDENTAL
Occidental Rd.
Valley
Hwy.
SEBASTOPOL
Petaluma
Todd Rd.
Stony Pt.
101

Sonoma Mountain on the west and by the Mayacamas Mountains on the east. Includes the city of Sonoma, in Sonoma County, California. The area extends from San Francisco Bay, north 14 miles and is approximately 3.5 miles wide at the Sonoma-Napa counties line.

Sonoma Vineyard
Sonoma County, California.
Landmark estate Chardonnay vineyard.

SONOMA VINEYARDS (WINDSOR)
Founded 1970, Windsor, Sonoma County, California.

Storage: Oak and stainless steel. Cases per year: 500,000.

History: Rodney D. Strong was the original founder, starting with Tiburon Vintners in Marin in 1957. Then, in 1961, a move was made to Windsor, adding the Windsor Vineyards label. In 1973, the name was changed to Sonoma Vineyards, a corporation was formed and the winery was built. Rod is still the winemaker.

Estate vineyard: Alexander's Crown Vineyard, River West Vineyard, Chalk Hill Vineyard, River East Vineyard, LeBaron Vineyard.

Wines regularly bottled: Estate bottled, vintage-dated Cabernet Sauvignon, Chardonnay, Zinfandel, Pinot Noir, Johannisberg Riesling. Also vintage-dated Cabernet Sauvignon, Zinfandel, Chardonnay, Johannisberg Riesling, Gewurztraminer, Fumé Blanc, Chenin Blanc.

Second label: Pacific Coast Vineyards, Cellar Select, Healdsburg Cellars.

SONOMA WINE COUNTRY
In 1823, the Franciscan Fathers laid the foundation for this viticultural district at their northernmost Mission, San Francisco de Solano Sonoma. By 1824, Padre Jose Altimira recorded more than 1,000 vines of the mission variety in his Sonoma Vineyards. When the Mexican government, in 1835, forced the abandonment of the missions, the military governor of Mexican California, General Vallejo, continued his grape growing. In 1846, the Bear Flag was raised by Americans who deposed Vallejo, in the Sonoma Plaza, across the street from Mission San Francisco de Solano. By 1854, the General's vineyards boasted some 5,000 vines.

In 1862, Colonel Agoston Haraszthy returned to Sonoma with over 100,000 cuttings of European grape varieties.

The wine industry expanded North to Healdsburg, where the first commercial winery in Northern Sonoma County was established during the Civil War. With the completion of the railroad from San Francisco, in 1871, the four areas of Alexander Valley, Dry Creek Valley, Russian River Valley, and the Sonoma Valley made up the Sonoma County Wine Country.

"Head pruned" grape vine

SOTOYOME WINERY
Founded 1974, Healdsburg, Sonoma County, California.

Storage: Oak and stainless steel. Cases per year: 2,500.

History: The winery was established on 10 acres of the old rancho Sotoyome, a Mexican land grant of 1840 which covered 48,000 acres. The Rancho originally included the Alexander Valley, the Dry Creek Valley and the middle course of the Russian River. The town of Healdsburg lies in the approximate center of the old Rancho.

Wines regularly bottled: Vintage-dated Chardonnay, Cabernet Sauvignon, Petite Sirah, Zinfandel.

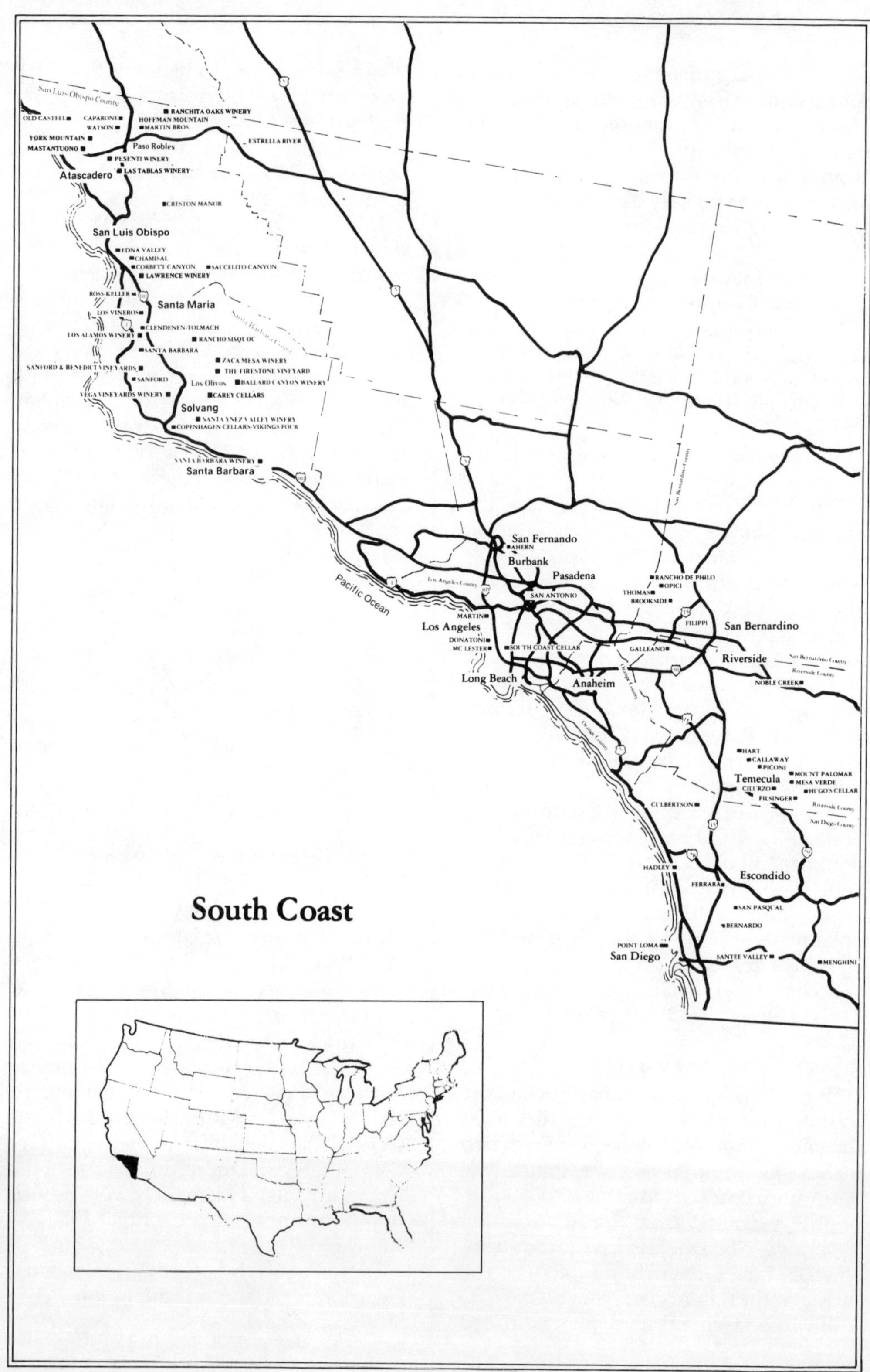

San Luis Obispo County
RANCHITA OAKS WINERY
OLD CASTEEL
CAPARONE
WATSON
HOFFMAN MOUNTAIN
MARTIN BROS.
YORK MOUNTAIN
MASTANTUONO
Paso Robles
ESTRELLA RIVER
PESENTI WINERY
LAS TABLAS WINERY
Atascadero
CRESTON MANOR
San Luis Obispo
EDNA VALLEY
CHAMISAL
CORBETT CANYON
SAUCELITO CANYON
LAWRENCE WINERY
ROSS-KELLER
Santa Maria
LOS VINEROS
CLENDENEN-TOLMACH
LOS ALAMOS WINERY
RANCHO SISQUOC
SANTA BARBARA
ZACA MESA WINERY
THE FIRESTONE VINEYARD
SANFORD & BENEDICT VINEYARDS
SANFORD
Los Olivos
BALLARD CANYON WINERY
VEGA VINEYARDS WINERY
CAREY CELLARS
Solvang
SANTA YNEZ VALLEY WINERY
COPENHAGEN CELLARS-VIKINGS FOUR
SANTA BARBARA WINERY
Santa Barbara
San Fernando
AHERN
Burbank
Pasadena
Los Angeles County
Pacific Ocean
SAN ANTONIO
RANCHO DE PHILO
OPICI
THOMAS
BROOKSIDE
MARTIN
Los Angeles
FILIPPI
San Bernardino
DONATONI
MC LESTER
SOUTH COAST CELLAR
GALLEANO
Riverside
San Bernardino County
Riverside County
Long Beach
Anaheim
NOBLE CREEK
Orange County
HART
CALLAWAY
PICONI
Temecula
MOUNT PALOMAR
MESA VERDE
CILURZO
HUGO'S CELLAR
FILSINGER
CULBERTSON
San Diego County
HADLEY
Escondido
FERRARA
SAN PASQUAL
BERNARDO
POINT LOMA
San Diego
SANTEE VALLEY
MENGHINI
South Coast

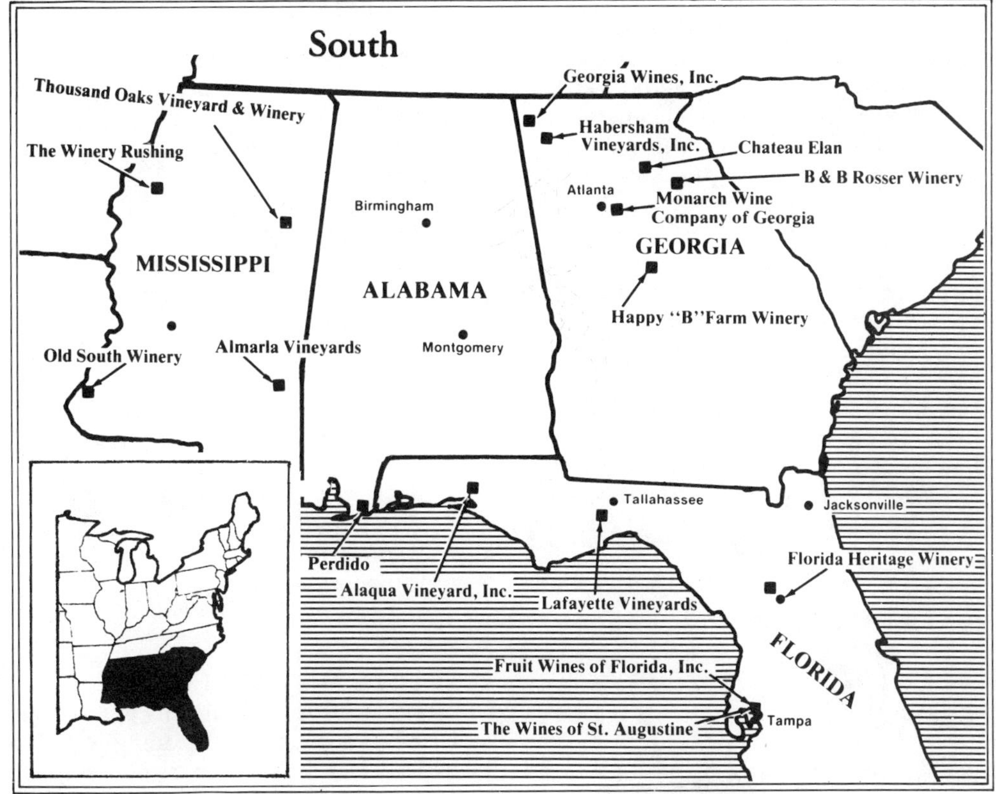

SOUTH CENTRAL COAST VITICULTURAL AREA

Includes the areas of Santa Barbara and San Luis Obispo.

sour

A sour wine is not necessarily a spoiled wine. It is not proper to call a dry, or tart wine "sour", although that is precisely the taste of acids. Those who are wine knowledgeable use "tart", "crisp", or "acidic", rather than the word "sour". There is no valid reason for this snobbism, however.

Acid = Sour = Tart (a good connotation)
Astringent = Tannin = Bitter = mouth drying or puckery
Volatile Acid = Spoiled Wine = "fruity sour" = Estery (bad connotation).

SOUTH COAST CELLAR

Founded 1977, Gardena, Los Angeles County, California.

Label indicating non-estate vineyard: Tepusquet Vineyards, Rancho Mission Viejo.

Wines regularly bottled: Cabernet Sauvignon, Merlot, Petite Sirah, Zinfandel.

SOUTHEASTERN NEW ENGLAND VITICULTURAL AREA

Located in Connecticut, Rhode Island and Massachusetts—includes New London County, Connecticut east of the Mystic River, all of Rhode Island except most of Kent and Providence counties, and all of southeastern Massachusetts east and south of the Norfolk-Bristol County boundary, the Amtrak mainline, and the Neponset River. All offshore islands between Boston and the Mystic River are included.

SOUVERAIN CELLARS

Founded 1943, Geyserville, Sonoma County, California.

Storage: Oak and stainless steel. Cases per year: 150,000.

History: The original Souverain winery was founded in 1943 by J. Leland Steward. In the early 1970's, under the auspices of the Pillsbury Company, the winery on Howell Mountain in the Napa Valley was sold and a new winery was built in Rutherford in 1973.

Plans were also made, that year, to build a second winery in Sonoma County. In 1976, the Rutherford winery was sold to a group of investors and the Sonoma facility and the Souverain name were purchased by a group of vineyardists from Napa, Sonoma and Mendocino. Under this ownership, Souverain has access to the best grapes in the tri-county viticultural areas.

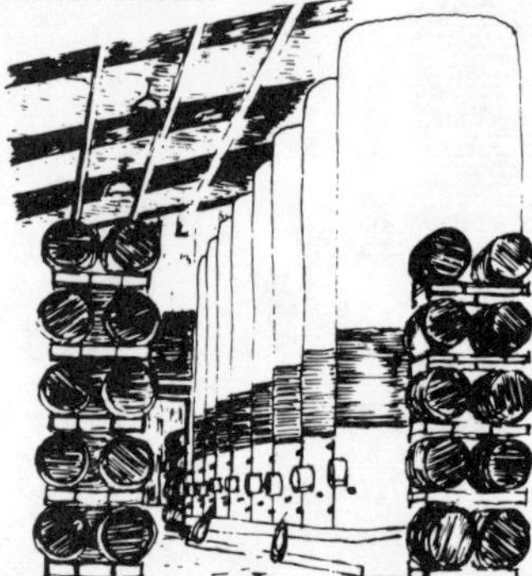

Stainless steel fermentation and oak aging barrels

Estate vineyard: Over 300 limited partners of the North Coast Grape Growers Association.

Wines regularly bottled: Estate bottled, vintage-dated Fumé Blanc, Chardonnay, Cabernet Sauvignon; vintage-dated Chardonnay, Fumé Blanc, Johannisberg Riesling, Gewurztraminer, Chenin Blanc, Grey Riesling, Colombard Blanc, Muscat Canelli, Cabernet Sauvignon, Merlot, Pinot Noir, Zinfandel, Charbono, Petite Sirah, Gamay Beaujolais, Piñot Noir Rosé. Also produce White, Red, Rosé Table Wines.

Second label: North Coast Cellars.

SOUZAO

A Portuguese grape used for Port. Produces a deep color wine.

SPANISH WORDS

Age—edad, epoca, periodo
Alcohol—alcohol
American—Americano
Aperitif—aperitivo
Barrel—barril, cuba
Bottle—botella, frasco
Bouquet—ramo, ramillete, aroma
Brandy—conac, aguardiente
Fermentation—fermentacion
Fine—fino, refinado
Grape—uva, vid
Label—marbete, rotulo, etiqueta, marca
Vine—vain, enredadera
Vineyard—vina, vinedo
Vintage—vendimia
Vintner—vinatero
Wine—vino
Wine Cellar—bodega
Wine Skin—bota o pollejo de vino

SPARKLING BURGUNDY

A red wine made sparkling by secondary fermentation in closed containers. It is usually semi-sweet or sweet. Barbera, Carignane, Petite Sirah and Pinot Noir are the grapes most used for its production.

sparkling wines

Sparkling wines are wines which have been made naturally effervescent by a second fermentation, in closed containers. Sparkling wines can be red, pink or white, with an alcohol content of 10-14 percent.

Sparkling wines: Champagne, Cold Duck, Sparkling Burgundy, Sparkling Muscat, Sparkling Rosé.

spatlese

The German word for "late picking". This word is not used by American vineyards thus, in its place, the wines are called "Late Harvest." Grapes that are picked in an over-ripe condition, with some incidence of Botrytis, and have a minimum of 24 degrees Brix at harvest. Harvest sugar and residual sugar must be on label.

Spaulding Vineyard
Napa County, California.
Stonegate estate Merlot, Chardonnay Napa Valley vineyard.

special select late harvest
Equivalent to the Trockenbeerenauslese wines of Germany. Wines produced only from grapes totally affected with Botrytis, and in some cases the berries are fully raisined and have a minimum of 35 degrees Brix at time of harvest. Harvest sugar and residual sugar must be on label.

Spill Pond Vineyard
Arlington, Georgia.
Muscadines, Carlos vineyard.

Spires/Sicler Vineyard
Valencia County, New Mexico.
Chancellor, Rougeon, Vidal, Baco Noir vineyard.

Spottswoode Estate Vineyards
Napa County, California.
Cabernet Sauvignon, Sauvignon Blanc, Cabernet Franc, Merlot, Semillon Spring Mountain vineyard.

SPOTTSWOODE WINERY
Founded 1982, St. Helena, Napa County, California.
Storage: Oak. Cases per year: 2,500.
History: Vineyard first planted in 1879, the main house and aging cellars constructed in 1882, one of the four wineries located in the city limits of St. Helena.
Estate vineyard: Spottswoode Vineyards.
Wines regularly bottled: Estate bottled, vintage-dated Cabernet Sauvignon, Sauvignon Blanc.

Spring Lake Vineyard
Napa County, California.
Chardonnay and Cabernet Sauvignon Napa Valley vineyard.

Spring Ledge Vineyard
Yates County, New York.
Glenora estate Johannisberg Riesling and Seyval Blanc Finger Lakes vineyard.

SPRING MOUNTAIN
The area in Napa county that is west of St. Helena in the Napa Valley.

Spring Mountain Estate Vineyard
Napa County, California.
Chardonnay, Cabernet Sauvignon, Sauvignon Blanc, Pinot Noir Napa Valley vineyard.

SPRING MOUNTAIN VINEYARDS
Founded 1968, St. Helena, Napa County, California.
Storage: Oak and stainless steel. Cases per year: 25,000.
Estate vineyard: Spring Mountain Vineyards.
Wines regularly bottled: Vintage-dated Chardonnay, Cabernet Sauvignon, Sauvignon Blanc, Pinot Noir (Les Trois Cuvées).
Second label: Falcon Crest.

Spring Valley Vineyard
Polk County.
Hidden Springs estate Chardonnay, Pinot Noir and Riesling vineyard.

Springdale Vineyard
Washington County, Oregon.
Pinot Noir Willamette Valley vineyard.

SPRINGWOOD
(See Barnes Wines, Ltd.)
Bottle Springwood Canadian Sauterne, Rubiwein, Still Cold Duck, Crackling Cold Duck.

Spurgeon Estate Vineyards
Grant County, Wisconsin.
Labrusca vineyard.

SPURGEON VINEYARDS AND WINERY
Founded 1981, Highland, Grant County, Wisconsin.

Storage: Oak and stainless steel. Cases per year: 2,000.

Estate vineyard: Spurgeon Vineyards.

Wines regularly bottled: Cherry, Cranberry-Apple; proprietary Wisconsin Red, Big Spring Red, White, Pink and Rosé; occasionally Strawberry, Apple and Mead.

Spurlock-Ashby Vineyard

Missouri.

Concord vineyard.

ST, STE

(listed under Saint.)

STAG'S LEAP

The vineyard area in Napa County, east of Yountville known mainly for the growing of outstanding Cabernet Sauvignon grapes.

The classic "Paris goblet"

STAG'S LEAP WINE CELLARS

Founded 1972, Napa, Napa County, California.

Storage: Oak and stainless steel. Cases per year: 20,000.

Estate vineyard: Stag's Leap Vineyards.

Label indicating non-estate vineyard: Birkmyer Vineyards.

Wines regularly bottled: Estate bottled, vintage-dated Cabernet Sauvignon, Merlot; vintage-dated Chardonnay, Gamay Beaujolais, White Riesling, Sauvignon Blanc, Petite Sirah.

Second label: Hawk Crest.

Stag's Leap Wine Cellars Estate Vineyards

Napa County, California.

Cabernet Sauvignon and Merlot vineyards.

STAGS' LEAP WINERY

Founded 1972, Napa, Napa County, California.

Storage: Oak and stainless steel. Cases per year: 10,000.

History: Stags' Leap is an historic 19th century estate in Yountville, in the Napa Valley, near the Silverado Trail. Founded in the 1880's by Horace Blanchard Chase of Chicago. The Chases supposedly called their wine Stags' Leap in reference to an Indian legend, which held that a magical stag had once escaped a hunter by leaping a great distance from one crag in the mountain to another. The focal point of the estate was the manor house complete with 40 foot stone tower. Chase built a winery and planted about 200 acres of vineyards. The Chases sold the property and it became an inn. During World War II, it was a Navy billet and then fell into disrepair until its current owners bought the property, in 1970, and started to restore the vineyards, house and winery.

Estate vineyard: Stags' Leap Vineyards.

Label indicating non-estate vineyard: Pedregal Vineyard.

Wines regularly bottled: Estate bottled, vintage-dated Petite Sirah, Chenin Blanc, Merlot, Cabernet Sauvignon, Pinot Noir. Occasionally bottle non-vintage Burgundy.

Second label: Pedregal.

Stags' Leap Winery Estate Vineyard

Napa County, California.

Petite Sirah, Chenin Blanc, Merlot, Cabernet Sauvignon and Pinot Noir Napa Valley vineyard.

P & M STAIGER

Founded 1973, Boulder Creek, Santa Cruz County, California.

Storage: Oak and stainless steel. Cases per year: 450.

Estate vineyard: P & M Staiger Vineyard.

Wines regularly bottled: Estate bottled, vintage-dated Chardonnay, Cabernet Sauvignon.

P & M Staiger Estate Vineyard
Santa Cruz County, California.
Chardonnay, Cabernet Sauvignon, Merlot Santa Cruz Mountain vineyard.

STANFORD
(See Weibel Vineyards.)

STANFORD, LELAND

In addition to building railroads, founding Stanford University and serving as Governor of California, Leland Stanford was a vineyard owner and wine producer. In 1869, he acquired a square mile of land in the Mission San Jose area, which his brother planted in 350 acres of vines. A large winery was also built, and by 1881, wines and brandies were being sold under the Stanford label. Sometime before 1889, Stanford deeded the property to his brother. By that time, however, he was involved in a much larger winery operation. In 1881, he purchased 9,000 acres of the Vina grant in Tehama County, including the vineyards and winery. Stanford imported not only European grape cuttings, but hired experienced French vineyard workers to plant and care for the vines. He continued to expand, and by 1888, Vina was the largest vineyard in the world with 3,575 acres and 2,600,000 vines. The winery and distillery, supplied with the latest in winemaking technology, produced mostly brandy due to the poor quality of the grapes. Stanford attempted to produce table wines at another vineyard, in San Mateo County, but died, in 1893, before any of the prize-winning vintages were produced.

Stanton's Pinot Patch
Napa County, California.
Pinot Noir vineyard.

STAR OF ISRAEL
(See Honeywood Winery.)

Blackberry, Loganberry, Creme Currant, Concord.

Stargazers Vineyard
Chester County, Pennsylvania.
Chardonnay vineyard.

State Lane Vineyard
Napa County, California.
Beringer estate Cabernet vineyard.

STEARN'S WHARF VINTNERS

Founded 1982, Santa Barbara, Santa Barbara County, California.

Wines regularly bottled: Vintage-dated Johannisberg Riesling, Chenin Blanc, Cabernet Sauvignon Blanc. Also Chardonnay, Fumé Blanc, Cabernet Sauvignon, Petite Sirah, Barbera.

Steiner Vineyard
Sonoma County, California.
Cabernet Sauvignon vineyard.

Stelling Vineyard
Napa County, California.
Far Niente estate Chardonnay, Cabernet Sauvignon, Cabernet Franc and Merlot Halter Valley vineyard.

Steltzner Estate Vineyard
Napa County, California.
Chenin Blanc, Cabernet Sauvignon, Cabernet Franc and Chardonnay Napa Valley vineyard.

STELTZNER VINEYARDS
Founded 1983, Napa County, California.
Storage: Oak.
Estate vineyard: Steltzner Vineyard.
Wines regularly bottled: Estate bottled, vintage-dated Cabernet Sauvignon and Chardonnay.

Robert Stemmler Estate Vineyard
Sonoma County, California.
Chardonnay Dry Creek Valley vineyard.

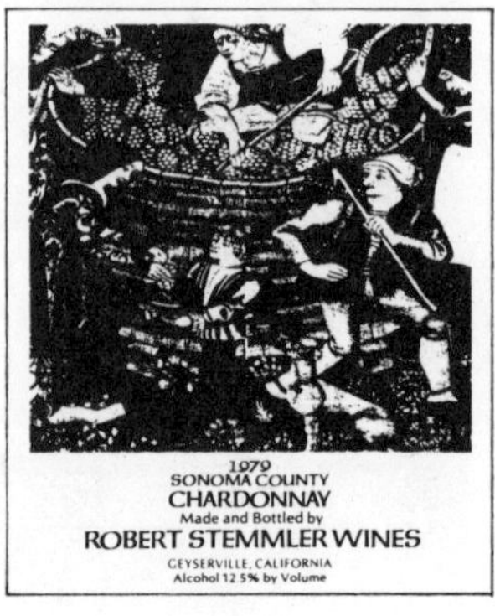

ROBERT STEMMLER WINERY
Founded 1977, Healdsburg, Sonoma County, California.
Storage: Oak and stainless steel. Cases per year: 7,000-8,000.
Estate vineyard: Robert Stemmler Vineyards.
Wines regularly bottled: Estate bottled, vintage-dated Chardonnay; vintage-dated Chardonnay, Cabernet, Sauvignon Blanc, Pinot Noir.
Second label: Bel Canto.

stemmy
Tasting term.
The term stemmy is used to describe the flavor of those wines which have been fermented too long or pressed too hard in the presence of stems.

STEPHENS WINERY
(See Girard Winery.)

STERLING
(See Sterling Vineyards.)
Vintage-dated Chardonnay, Sauvignon Blanc, Light Zinfandel Port; and Red and White.

STERLING, GEORGE (1869-1926)
American poet.

"Into a crystal cup the dusky wine
I pour, and, musing at so rich a shrine,
I watch the star that haunts its ruddy gloom.

"Wine gives all and gives forever.

"He who clinks his cup with mine,
Adds a glory to the wine."

Sterling Estate Vineyards
Napa County, California.
Chardonnay, Sauvignon Blanc, Cabernet Sauvignon, Merlot, Semillon, Cabernet Franc, Petite Verdot vineyard.

STERLING VINEYARDS
Founded 1964, Calistoga, Napa County, California.
Storage: Oak and stainless steel.
Estate vineyard: Sterling Vineyards.
Wines regularly bottled: Estate bottled, vintage-dated Chardonnay, Sauvignon Blanc, Cabernet Sauvignon, Reserve Cabernet Sauvignon, Merlot, Cabernet Blanc.
Second label: Sterling.

Steuk Estate Vineyard
Erie County, Ohio.
Black Pearl, Elvira, Montefiore, Clinton and Riesling Lake Erie Area vineyard.

STEUK, WILLIAM CHARLES
William Charles Steuk is the fifth generation William Steuk to be involved in the wine industry. He is a practicing attorney with the firm of Flynn, Py and Kruse in Sandusky, Ohio, and has managed the operation of the Steuk Wine Company since his father's death in 1975. The winery is part of the family farm business, which includes 20 acres of apple orchards. They specialize in table wines and sparkling wines made from native local grapes such as Concord, Delaware, Niagara and Black Pearl. The winemaking tradition may continue to a sixth generation with Steuk's son, William Robert Sommer Steuk.

THE STEUK WINE COMPANY
Founded 1855, Sandusky, Erie County, Ohio.
Storage: Oak and stainless steel.
History: The Steuk Wine Company is one of the oldest wineries in the United States still owned and operated by the founding family. In the 1850's, two great-grandfathers of the present owner commenced their grape growing and wine making operations. They were Louis Harms and William Leopold Steuk.
Louis Harms was one of the pioneer growers on Put-in-Bay Island who purchased land from J.D. Rivera St. Jurgo and engaged in the grape and wine business until 1876. At that time, he sold his holdings and bought a farm in Euclid, Ohio and resumed the same occupation.
William Leopold Steuk arrived in Sandusky about 1850 and very soon became interested in grapes and wine making. He purchased property on the corner of Market and Decatur Streets, which became his home and winery. He also purchased land on Venice Road, which became an important grape vine nursery. At a later time, vineyards and land were purchased in Venice.
A daughter of Louis Harms, Julia, and a son of William Leopold Steuk, Edward L., were married thus uniting two families of similar tradition. In time, Edward L. Steuk became the owner of the vineyards and winery. His entire life was devoted to this pursuit. In the course of his long career, he developed a grape variety that was unique in that the sugar content was sucrose rather than the usual dextrose and levulose.
Edward L. Steuk died in 1917, thus was spared the anguish of dissolving his lifelong heritage as decreed by the adoption of the Prohibition Act. His son, William Louis, long associated with his father, became the operator. It was his lot to comply with the Prohibition law and close the winery. However, the vineyards were continued.
After repeal of Prohibition, he returned to wine making. One son, William Kark Steuk became owner and winemaker.
Estate vineyard: Steuk Vineyard.
Wines regularly bottled: Sweet Concord, Sweet Niagara, Pink Catawba, Delaware Rosé, Dry Concord, Dry Catawba, Dry Delaware, Elvira, Seyval Blanc, Beta, Marechal Foch, Dry Black Pearl, Sweet Black Pearl; Sparkling wines: Brut, Extra Dry and Pink Champagnes, Sparkling Catawba; Dry and Mellow Burgundy; and Labrusca Rosé.

STEVENOT

Founded 1974, Murphys, Calaveras County, California.

Storage: Oak and stainless steel. Cases per year: 13,000.

History: The original ranch was built in the 1880's by the Shaw family from the Falkland Islands.

Estate vineyard: Stevenot Vineyards.

Wines regularly bottled: Estate bottled, vintage-dated Zinfandel and Cabernet Sauvignon. Vintage-dated Chenin Blanc, Sauvignon Blanc, Chardonnay, Zinfandel Blanc.

Stevenot Estate Vineyards

Calaveras County, California.

Zinfandel, Cabernet Sauvignon and Chardonnay vineyard.

STEVENSON, ROBERT LOUIS
(1850-1894)

Scottish essayist, novelist and poet.

"A bottle of good wine, like a good act, shines ever in the retrospect.

"Vines, and the vats and bottles in the cavern, made a pleasant music for the mind."

(Re—California vineyards)

"Those lodes and pockets of earth, more precious than the precious ores, that yield inimitable fragrance and soft fire, those virtuous Bonanzas where the soil has sublimated under sun and stars to something finer, and the wine is bottled poetry ..."

STEWART VINEYARD

Founded 1983, Granger, Yakima County, Washington.

Storage: Oak and stainless steel. Cases per year: 5,500.

Estate vineyards: Sunnyside Vineyard, Wahluke Slope Vineyard.

Wines regularly bottled: Estate bottled, vintage-dated Chardonnay, White Riesling, Muscat Canelli, Gewurztraminer, Late Harvest Gewurztraminer and Cabernet Sauvignon.

Stewarts Sunnyside Vineyard

Yakima County, Washington.

Riesling, Cabernet Sauvignon, Gewurztraminer Yakima Valley vineyard.

Stewarts Wahluke Slope Vineyard

Grant County, Washington.

Riesling, Muscat Canelli, Chardonnay, Gewurztraminer Columbia Valley vineyard.

Stillwater Estate Vineyards

Miami County, Ohio.

Aurora, Chelois, Bellandais, Baco Noir, Villard Blanc, Niagara, Villard Noir, Catawba, Concord, Cayuga and Foch vineyard.

STILLWATER WINERIES, INC.

Founded 1973, Troy, Miami County, Ohio.

Storage: Stainless steel.

Estate vineyard: Stillwater Vineyards.

Wines regularly bottled: Aurora, Baco Noir, Cayuga, Chelois, Villard Blanc, Catawba, Foch, Bellandais, Niagara, Concord, Sparkling Catawba.

STODDARD, RICHARD HENRY
(1825-1903)

American poet.

"Day and night my thoughts incline
To the blandishments of wine;
Jars were made to drain, I think,
Wine, I know, was made to drink.

"When I die, (the day be far)
Should the potter make a jar
Out of this poor clay of mine,
Let the jar be filled with wine."

Stoltz Vineyards

Philips County, Missouri.

French Hybrids vineyard.

Stone Church Vineyards
Franklin County, Missouri.
Edelweiss estate Golden September, Chancellor, Seyval Blanc, Vidal and Villard Blanc vineyard.

Stone Hill Estate Vineyards
Gasconade County, Missouri.
Vidal, Missouri Riesling, Villard Noir, Ravat, Chancellor, Norton, Niagara, Seyval, Catawba and Colobel vineyard.

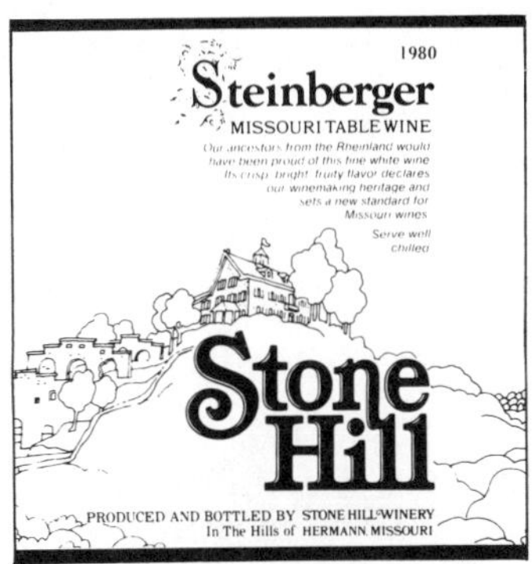

STONE HILL WINERY
Founded 1847, Hermann, Gasconade County, Missouri.
Storage: Oak and stainless steel. Cases per year: 36,431.
History: The winery, a national historic site, grew to be the 3rd largest winery, second in the nation by the turn of the century. Closed during Prohibition and re-opened in 1965. Gold medals were awarded in Vienna (1873), Philadelphia (1876), Paris (1878), New Orleans (1885) and the Pan American Exposition (1901).
Estate vineyard: Stone Hill Vineyard.
Wines regularly bottled: Pink Catawba, Concord, Missouri Riesling, Norton, Harvest Peach; also proprietary Festive Red, White and Rosé, Rosé Montaigue, Montaigue Blanc; also Steinberger and generic Golden Rhine. And Champagne.

STONE MILL WINERY
Founded 1973, Ozaukee County, Wisconsin.

Storage: Stainless steel. Cases per year: 4,500.
History: In 1864, the mill was built by hand from stones removed from the creek bed and nearby quarries and the rushing waters of the creek supplied the power. The mill was converted to a winery in 1972. Stone Mill wines come in authentic hand cast stoneware crocks and are considered collectors items.
Wines regularly bottled: Aurora, DeChaunac, Seyval, Niagara, Catawba, Natural Cherry, Cranberry-Apple, proprietary American Red and White; and Colonial Spice Cherry.

Stonecrest Vineyard
Sonoma County, California.
Balverne estate Sauvignon Blanc and Semillon Chalk Hill vineyard.

Stonecrop Estate Vineyard
New London County, Connecticut.
Seyval, Chardonnay, Vidal Blanc, Marechal Foch vineyard.

STONECROP VINEYARD
Founded 1979, Calistoga, New London County, Connecticut.
Storage: Stainless steel.
History: Vineyards are planted on the historic homestead of Paul Wheeler, built in 1750.
Estate vineyard: Stonecrop Vineyards.

Wines regularly bottled: Estate bottled, vintage-dated Marechal Foch; also Vidal Blanc, Seyval Blanc, Marechal Foch; estate bottled, vintage-dated White Table Wine (Rayon D'Or grapes).

STONEGATE WINERY
Founded 1973, Sutter Creek, Napa County, California.
Storage: Oak and stainless steel. Cases per year: 20,000.
Estate vineyard: Spaulding Vineyard.
Wines regularly bottled: Estate bottled, vintage-dated Chardonnay, Merlot, vintage-dated Cabernet Sauvignon, Chardonnay, Sauvignon Blanc.

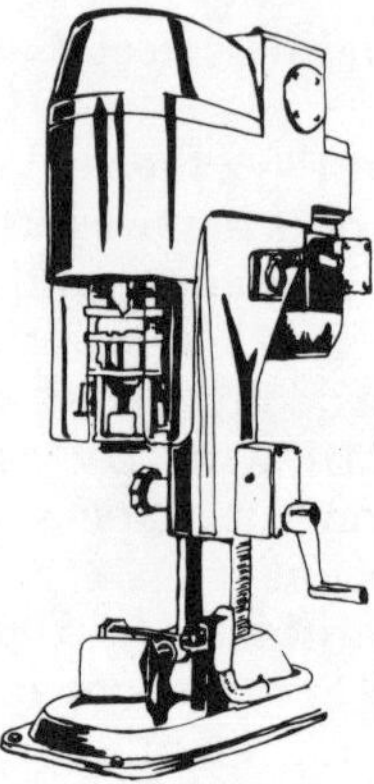

Capsuler machine forms capsules to bottles

STONERIDGE
Founded 1975, Amador County, California.
Storage: Oak. Cases per year: 1,000.
Label indicating non-estate vineyard: Milanovich-Cassinelli Vineyards.
Wines regularly bottled: Vintage-dated Zinfandel, Ruby Cabernet, White Zinfandel.

STONEY CREEK
(See Shenandoah Vineyards, Virginia.)

STONEY CREEK VINEYARDS
Founded 1979, Somerset, El Dorado County, California.
Storage: Oak and stainless steel. Cases per year: 1,000.
Estate vineyard: Stoney Creek Vineyards, Simms Vineyard.
Wines regularly bottled: Estate bottled, vintage-dated Zinfandel, Zinfandel Blanc; and vintage-dated Fairplay Red.
Second label: Gerwer.

Stony Hill Estate Vineyard
Napa County, California.
Chardonnay, Riesling, Gewurztramier, Semillon vineyard.

STONY HILL VINEYARD
Founded 1953, St. Helena, Napa County, California.
Storage: Oak. Cases per year: 4,000.
Estate vineyard: Stony Hill Vineyard.
Wines regularly bottled: Estate bottled, vintage-dated Chardonnay, White Riesling, Gewurztraminer, Semillon de Soleil.

Stony Ridge Vineyard
Alameda County, California.
Kalin estate Semillon and Zinfandel Livermore Valley vineyard.

STONY RIDGE WINERY
Founded 1975, Pleasanton, Alameda County, California.
Storage: Oak and stainless steel. Cases per year: 150,000.
Estate vineyard: Stony Ridge Vineyard.
Label indicating non-estate vineyard: Smith & Hook Vineyard; La Reina Vineyard.
Wines regularly bottled: Estate bottled, vintage-dated Chardonnay, Zinfandel, Sauvignon Blanc, proprietary Chevrier (Dry Semillon) and Crescent Gold; also vintage-dated Chardonnay, Cabernet Sauvignon, White Zinfandel and non-vintage Champagnes (Blanc de Noir; Malvasia Bianca); bottle also vintage-dated White Zinfandel and vintage-dated proprietary Chevrier (Dry Semillon).

Stony Ridge Winery Estate Vineyard
Alameda County, California.
Chardonnay, Zinfandel, Semillon, Pinot Noir and Sauvignon Blanc Livermore Valley vineyard.

stop fermentation

The term that describes how, in winemaking, pure grape brandy is added to a fermenting dessert wine to check the fermentation. This prevents complete conversion of the natural grape sugar into wine alcohol and carbon dioxide so that the wine is sweeter than if fermentation had run its course. Some other ways are adding CO_2

Story Estate Vineyard

Amador County, California.

Zinfandel, Mission, Chenin Blanc, Shenandoah Valley vineyard.

STORY VINEYARD

Founded 1973, Plymouth, Amador County, California.

Storage: Oak and stainless steel. Cases per year: 20,000.

Estate vineyard: Story Vineyard.

Wines regularly bottled: Estate bottled, vintage-dated Zinfandel and non-vintage Zinfandel; also vintage-dated Premier White.

Second label: Cosumnes River Vineyard.

Storybook Mountain Estate Vineyards

Napa County, California.

Zinfandel Napa Valley vineyard.

STORYBOOK MOUNTAIN VINTNERS

Founded 1980, Calistoga, Napa County, California.

Storage: Oak. Cases per year: 15,000.

History: In 1883, 100 years ago, Jacob and Adam Grimm, immigrants from Germany like their fellow vintners Jacob Schram and Jacob Beringer, carved three deep caves into the Mayacamas Range in the farthest northern part of Napa Valley and planted the surrounding acreage to Zinfandel grapes.

There followed the disasters of nature and Prohibition and the caves and vineyards lay unused for decades.

In 1976, Dr. J. Bernard Seps acquired the forlorn estate and appropriately named it Storybook Mountain. With great effort he and his family fought back the forest that had re-invaded the hills and planted 36 acres of Zinfandel vines which now supply the grapes for the Storybook Mountain Estate Vineyards label.

Walter Schug, a winemaker of Joseph Phelps Vineyards for 10 years, joined J. Bernard Seps in 1980 to rebuild the historic winery.

Storybook Mountain Vintners Winery is now the home to the individual winemaking skills of its two partners, who are marketing their wines under the Storybook Mountain Vineyards and Schug Cellars labels.

Estate vineyard: Storybook Mountain Vineyards, Schug Cellars.

Wines regularly bottled: Estate bottled, vintage-dated Zinfandel, vintage-dated vineyard labeled Pinot Noir and Chardonnay.

STOVER

A Florida hybrid grape developed by Loren H. Stover at University of Florida agricultural research center. A cross between Florida wild grapes and cultivated varieties. Produces a white fruity wine.

Strawberry Ridge Vineyard

Connecticut.

Hopkins estate Cayuga White and Seyval Blanc vineyard.

Stretch Island Vineyards

Mason County, Washington.

Hoodsport state Island Belle Grape vineyard.

Stromberg Vineyard

Lake County, California.

Cabernet Sauvignon vineyard.

STRONG, RODNEY D.

Rodney D. Strong's long career in the wine industry has included many different jobs. He has held apprenticeships in France and in Germany, in the homes of relatives involved in winemaking in the Rheingau region; in 1960, he opened a business of bottling wines and selling them under the label of Tiburon Vintners. In 1961, he founded Sonoma Vineyards, through acquisition of a small vineyard and winery in Windsor, Sonoma County. He began careful plantings of varietal grapes and production of quality table wines. He has recently collaborated with Piper-Heidseick to create the new Piper-Sonoma line of California sparkling wines. In addition to his duties as chairman and winemaster at Windsor Winery, Strong has served as past president of the Sonoma County Wine Grower's Association.

He is currently on the Board of Directors of the California Wine Institute and is a member of the American Society of Enologists, and is a Supreme Knight in the Brotherhood of the Knights of the Vine.

stuck fermentation

This occurs when the temperature of fermentation becomes too high for the yeast to grow and ferment.

sugar content

The following are average sugar percentage contents: Aperitif: 0.5-3.5%; Red: 0-1/5%; White: 0-4.0%; Rosé: 0-2.0%; Dessert: 5.0-14%; Champagne: 0.5-5.0%.

Note: Late Harvest white wines might go from 4% to 5% or more.

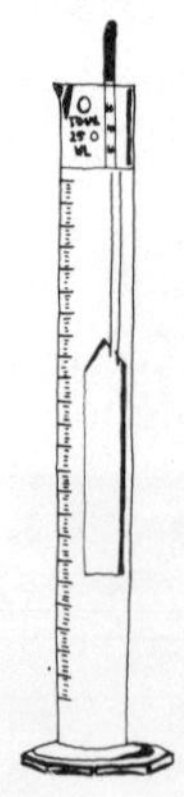

A saccharometer for testing sugar content

sugars

Ripe grapes often contain more than 20 per cent sugar, measured by winemakers as "20 degrees Balling," or "20 degrees Brix," which are similar laboratory calibrations. The percentage of sugar multiplied by 0.55 calculates the alcohol a wine will contain if fermented completely dry. (20 Brix equals 11 per cent alcohol.)

Most of the sugars in a grape are fructose and glucose. They are produced in the vine by the leaves, energized by sunlight to convert carbon dioxide and water into carbohydrates. In intense, enduring sunlight, this process, called photosynthesis, tends to produce high sugar levels in the fruit.

Yeast, feeding and multiplying in the grape juice ("must") produces alcohol, carbon dioxide and other substances which interact among themselves in the natural but very complex processes which convert the juice to wine. Traces of other sugars than fructose and glucose in the juice are thought to be resistant to yeast action and to be the reason that even a "bone dry" wine can be shown by analysis to retain them, even though they are below the taste threshhold. For most tasters that threshhold is at about one per cent, although a fine-tuned palate can detect sugar in still drier wines.

Sugarloaf Hill

Wisconsin County, Wisconsin.

Wollersheim estate Seyval Blanc vineyard.

SUISUN VALLEY VITICULTURAL AREA

Suisun Valley is located in the southwestern portion of Solano County, adjacent to the Napa County line, and east of Green Valley. Suisun Valley has about 800 acres of grapes within its three-mile-wide, eight-mile-long area. It lies within the southern end of two ranges of the Coast Range—the Vaca Mountains on the east and the Mount George Range on the west. The valley ends in the south at the Suisun Bay marshlands.

Sullivan Estate Vineyards
Napa County, California.
Chenin Blanc, Chardonnay, Merlot, Zinfandel, Cabernet Sauvignon vineyard.

SULLIVAN VINEYARD WINERY
Founded 1979, Rutherford, Napa County, California.
Storage: Oak and stainless steel. Cases per year: 6,000.
Estate vineyard: Sullivan Vineyards.
Wines regularly bottled: Estate bottled, vintage-dated Cabernet Sauvignon, Chardonnay, Chenin Blanc, Merlot, Zinfandel.

SUMAC RIDGE ESTATE WINERY, LTD.
Founded 1979, Summerland, British Columbia, Canada.
Storage: Oak.
Wines regularly bottled: Estate bottled, vintage-dated Verdelet, Okanagan Riesling, Chancellor, Riesling/Chancellor, Gewurztraminer, Chardonnay, Chenin Blanc. Also proprietary Summerland Rosé.

Summa Vineyards
Sonoma County, California.
Pinot Noir vineyard.

Summerhill Estate Vineyards
Santa Clara County, California.
Cabernet Sauvignon vineyard.

SUMMERHILL VINEYARDS
Founded 1917, Gilroy, Santa Clara County, California.
Storage: Oak and stainless steel. Cases per year: 30,000.
Estate vineyard: Summerhill Vineyards.
Wines regularly bottled: Chenin Blanc, French Colombard, Riesling, Cabernet Sauvignon, Zinfandel; and semi-generics: Chablis and Rhine. Also produce Rosé Table Wine, Loganberry, Plum and Apricot. Occasionally Port Soleil, Sherry Soleil, Grignolino, Aleatico.
Second label: Gavilan Creek, Thigpen Reserve.

Tying cane to grape stake

SUMMIT (See Geyser Peak) (Wine-in-Box)
Rhine, Rosé, Chablis, Burgundy, Winterchill.

SUMMUM WINERY
Founded 1975, Salt Lake City, Salt Lake County, Utah.
Storage: Oak. Cases per year: 5,000.
History: Summum is the only winery currently licensed by the State of Utah and is only the second winery ever so licensed, the first such winery having been established and operated by Brigham Young.
Wines regularly bottled: Proprietary Nectar of Meditation, Nectar of Cause and Effect, Nectar of Neutralization.

Sumner Vineyards
Solano County, California.
Cabernet Sauvignon, Sauvignon Blanc, Gamay and French Colombard Suisun Valley vineyard.

Sundial Ranch
Mendocino County, California.
Fetzer estate Chardonnay, Pinot Noir, Chenin Blanc and Johannisberg Riesling vineyard.

Sunny Slope Vineyard
Franklin County, Missouri.
Eckert's Sunny Slope estate Chancellor, DeChaunac, Vidal, Seyval, Foch and Baco Noir vineyard.

SUNNY SLOPE VINEYARDS
(See Ste. Chapelle Vineyards, Inc.)
White table wine.

Sunnyside Vineyard
Yamhill County, Oregon.
Riesling and Pinot Noir Willamette Valley vineyard.

Sunol Valley Vineyard
Alameda County, California.
Villa Armando estate Chardonnay, Cabernet Sauvignon, Malvasia Bianca, Petite Sirah vineyard.

SUNRISE WINERY
Founded 1976, Santa Cruz, Santa Cruz County, California.
Storage: Oak and stainless steel. Cases per year: 2,000.
Label indicating non-estate vineyard: Arata Vineyard, Wasson Vineyard.
Wines regularly bottled: Vintage-dated Chardonnay, Pinot Noir, Cabernet Sauvignon. Occasionally bottles Sonoma Sauvignon Blanc, Sonoma Zinfandel.

supple
Tasting term.
Softness and balance coming together, a certain harmony giving the sensation of roundness.

SUSINE CELLARS
Founded 1981, Suisun City, Solano County, California.
Storage: Oak and stainless steel. Cases per year: 800.
Label indicating non-estate vineyard: Sumner Vineyards, Nicol Ranch.
Wines regularly bottled: Vintage-dated Johannisberg Riesling, Sauvignon Blanc, Cabernet Sauvignon; also vintage-dated Susine White, Susine Bouquet and non-vintage Susine Red. Occasionally bottle Susine Rosé, Late Harvest Johannisberg Riesling.

Sutter Basin Vineyard
Yolo County, California.
Zinfandel vineyard.

SUTTER HOME WINERY, INC.
Founded 1874, St. Helena, Napa County, California.
Storage: Oak and stainless steel. Cases per year: 500,000.
History: The winery was originally built in 1874, and "Bob" Trinchero purchased the property in 1946.
Wines regularly bottled: Vintage-dated White Zinfandel, Zinfandel, Muscat Amabile, Dessert Zinfandel.

SV (See Shenandoah Vineyards.)
Red Table Wine.

JOSEPH SWAN VINEYARDS
Founded 1968, Forestville, Sonoma County, California.
Storage: Oak. Cases per year: 1,000.
Estate vineyard: Joseph Swan Vineyards.
Wines regularly bottled: Estate bottled, vintage-dated Pinot Noir, Chardonnay and Cabernet Sauvignon. Also vintage-dated Zinfandel.

sweet
Pleasant taste characteristic of sugar. The sweetness comes from the fermentation of the sweet grape juice and how it is balanced with the acids, not from addition of sugar.
The taste of sweetness is on the tongue. Comes from sugar remaining after fermentation or blending of sweet juice.

sweet dessert wines

Full bodied wines served with desserts, and as refreshments. The alcoholic content, not less than 17% for Sherry, 18% for all other dessert wines. They range from medium sweet to sweet and in color from pale gold to red.

Sycamore Creek Estate Vineyards

Santa Clara County, California.

Zinfandel, Cabernet Sauvignon, Chardonnay Uvas Valley vineyard.

SYCAMORE CREEK VINEYARDS

Founded 1975, Morgan Hill, Santa Clara County, California.

Storage: Oak and stainless steel.

History: The vineyards site was a pre-Prohibition family winery known as the Marachetti Ranch. The present owners purchased and reestablished the winery in 1975.

Estate vineyard: Sycamore Creek Vineyards.

Label indicating non-estate vineyard: La Reina Vineyards, River Road Vineyards, Smith & Hook Vineyards.

Wines regularly bottled: Estate bottled, vintage-dated Carignane, Zinfandel, Chardonnay; also vintage-dated Johannisberg Riesling, Chardonnay, Cabernet Sauvignon, Pinot Noir and White Burgundy (100% Chardonnay), Summer Chardonnay.

SYLVANER

A German grape that produces dry to semi-dry wines, tart, clean and fruity. Not a Riesling, although sometimes labeled as Riesling or Franken Riesling.

Symms Family Vineyard

Canyon County, Idaho.

Ste. Chapelle estate Johannisberg Riesling, Chardonnay, Chenin Blanc, Gewurztraminer, Pinot Noir vineyard.

SYRAH

Red wine grape, grown in the Rhone Valley of France. It is rare in California also known as Shiraz. It is not Petite Sirah.

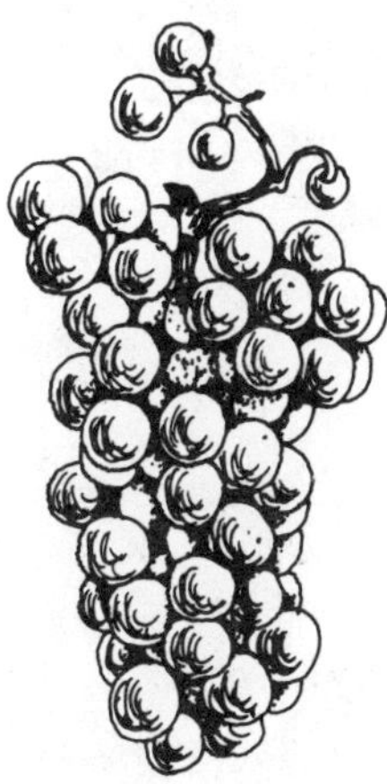

Sylvaner

table or dinner wine

The "right" name for all still wines with 7% to 14% alcohol content by volume. Most table or dinner wines are dry, but it is wrong to call all of them "dry" wines. That was formerly the practice, but it has been discontinued because many dinner wines, like Sweet Sauterne, are actually semi-sweet or sweet, while some wines of the dessert or appetizer class, like Sherry, are nearly dry. "Table" or "dinner wine" is the "right" term because most wines of that class are used with meals and also because the term guides the consumer in selecting wines of this class for mealtime use. The class includes the wines sometimes referred to as "light wines," "dry wines," or "natural wines."

Tabor Hill Estate Vineyard

Berrien County, Michigan.

Chardonnay, Riesling, Vidal, Seyval, Ravat and Baco Noir vineyard.

TABOR HILL VINEYARDS

Founded 1970, Buchanan, Berrien County, Michigan.

Storage: Oak and stainless steel. Cases per year: 20,000.

Estate vineyard: Tabor Hill.

Label indicating non-estate vineyard: Sagemoor Vineyards, Matzke Vineyards, Lemon Creek Vineyard, Dick Tropp Lazy Acre Vineyard.

Wines regularly bottled: Estate bottled, vintage-dated Seyval Blanc; vintage-dated Chardonnay, Sweet Harvest Ravat, Vidal Blanc Demi-Sec, Cabernet Sauvignon, Johannisberg Riesling, Vidal Blanc Sec; also Baco Noir Reserve. Occasionally bottles Seyval Late Harvest Ice Wine.

TAFT STREET WINERY

Founded 1982, Forestville, Sonoma County, California.

Storage: Oak and stainless steel. Number of cases produced per year: 4,000.

Wines regularly bottled: Vintage-dated Chardonnay, Pinot Noir; vintage-dated proprietary Cabernet blends and White House Wine.

ROBERT TALBOTT VINEYARD AND WINERY

Founded 1983, Monterey County, California.

Storage: Oak and stainless steel. Cases per year: 10,000—15,000.

History: Founded by the direct descendants of Edward and Elizabeth Talbott who owned Talbott's Vineyard, 1,021 acres, on Elk Ridge in Anne Arundel County, Maryland in 1689, and the same family for which Talbott County, Maryland is named.

Estate vineyard: Diamond T, Sara's Acre, Chalk Flats Vineyards.

Wines regularly bottled: Estate bottled, vintage-dated Chardonnay, Pinot Noir and Sauvignon Blanc.

Talmage Vineyard

Mendocino County, California.

Parducci estate Riesling, Chenin Blanc, Cabernet Sauvignon, Pinot Noir and Gamay Beaujolais vineyard.

TANCER, FORREST

Forrest Tancer's agricultural beginnings can be traced to summers spent on his family's ranch in Sonoma County. After graduation from the University of California, Berkeley, in 1969, Tancer took viticulture classes at California State University, Fresno. He then worked, 18 months, for the Peace Corps as an agricultural extension agent in Brazil. When he and his wife returned to the United States in 1971, Tancer was hired by Rod Strong of Sonoma Vineyards to manage the Iron Horse ranch vineyards. He also worked in the Sonoma Vineyards cellars and, in 1977, was promoted to cellarmaster. When the Iron Horse ranch was sold in 1976, the new owners, Audrey and Barry Sterling, asked Tancer to remain as manager. In 1979, he and the Sterlings formed a partnership to establish Iron Horse Vineyards. The 1978 Chardonnay was the first wine released under the Iron Horse label. In addition to the Iron Horse ranch vineyards, Tancer uses grapes from his family's ranch, including Sauvignon Blanc, Cabernet Sauvignon, Zinfandel and Cabernet Franc. Tancer has recently added sparkling wines to the Iron Horse Vineyards line.

tannic

Tasting term.

High tannin content, astringent or bitter or mouth-dry or puckery sensation.

tannins or phenols

Give red wine its color, and red and white wines their astringent or bitter taste (but not "tart") and much of what the tongue senses as "body" is alcohol or glycerin in the wine.

Tannin in wine comes from the grape skins, stems (even seeds if they happen to get crushed) and also, important to the eventual wine flavor, from barrels the wine was aged in at the winery. Most white wines are much lower in tannin than most red wines, but no grape wine is completely free of it. However, white wines aged in wood (Chardonnay, Sauvignon Blanc and a few others) can contain considerable of tannin, one of the reasons that these wines live longer in the bottle than others. Tannins are natural antioxidants and, since oxygen is the greatest enemy of aging wine, tannins are responsible for extending the life of bottled wine. "Fresh and Fruity" white wines, not aged in wood and not fermented in contact with skins or stems, don't contain much tannin and don't taste bitter or astringent and don't have long lives in the bottle.

tart

Possessing agreeable acidity; in wine, tartness reflects the content of agreeable fruit acids.

tartaric

The presence of tartaric acid. The most prevalent natural acid in grapes and wine.

tartrates

Clear, harmless crystals of potassium bitartrate that form as wine is aged; very common in white wines, where tartrates form a granular sediment in the tank or bottle when the wine is chilled.

TARULA FARMS

Founded 1967, Clarksville, Clinton County, Ohio.

Storage: Oak and stainless steel. Cases per year: 600.

Estate vineyard: Tarula Farms.

Wines regularly bottled: Estate bottled Tarula White Seyval; also Tarula White Aurora, Tarula Red DeChaunac, Tarula Red Foch, proprietary Country White, Red and Rosé. Occasionally Baco Rosé, Vidal, Foch Nouveau.

Tarula Farms Estate Vineyard
Clinton County, Ohio.
Seyval Blanc, Aurora, Baco Noir, Marechal Foch, Catawba, Concord Ohio Valley vineyard.

taste
True taste sensations are probably limited to the four classes: sour or acid; sweet; bitter; and salt, with the tactile ability to discern viscosity also of importance. In the examination of wines, however, there are a number of sensations—due perhaps to complex interactions of the odor receptors and the taste receptors—which only become apparent when the wine is taken into the mouth. Warming up the wine in the mouth may also release odors. For this reason a number of terms are listed under tastes which could also be classified as odors.

ACID—The acid taste in wines results from the presence of tartaric and malic acids principally, although citric, succinic, lactic and acetic acids also contribute. Terms used to designate acid concentrations are listed.

Flat—This and similar terms describe wines with low acidities. The terms soft and mellow are frequently applied to wines of low acidity.

Tart—Wines of a pleasing freshness and balance, with higher acidity, usually are described as tart.

Green, Acidulous or Unripe—These terms are used to describe wines unbalanced because of excessive acidity.

SWEETNESS—The sensation of sweetness in wines is derived mainly from the presence of the sugars, glucose and fructose. The term "dry" is used to indicate the lack of sweetness. Glycerol also contributes to the sweet taste, while acidity and astringency counteract the sweet impression.

Dry—Wines which give no impression of sweetness on tasting usually have less than 1.0% reducing sugar.

Low Sugar—This term describes wines containing between 0.2 and 1.0% sugar.

Medium Sugar—This term describes wines containing between 1.0 and 4.0% sugar.

High Sugar—Wines containing over 4.0-15% sugar are described as having high sugar content.

The sweetness of Sparkling wines is *usually* considered in the following order of increasing sweetness for any one producer:

1. Brut or Nature.
2. Demi Sec or Extra Dry.
3. Sec or Sweet.

"Taste vin" used by restaurant sommelier

tasting wine
In tasting wines the color, clarity, aroma, bouquet, tartness, flavor, astringency, degree of sweetness and balance are all to be considered.

Look of color and clarity.
Smell for aroma and bouquet.
Taste for flavor and body.

Taurian Vineyards
Sonoma County, California.
Petite Sirah and Zinfandel vineyard.

tawny
Term applied to Ports and other red wines which have a brownish or golden tinge instead of the customary ruby, resulting from the loss of pigment by oxidation and condensation during long aging, filtering or fining.

TAYLOR, WALTER S. ("The winemaker without a name")

Walter S. Taylor, grandson of the founder of Taylor Wine Company of New York, was heir to the multimillion dollar business. That was before 1970, when he upset the company's board of directors by publicly criticizing the adulteration of New York State wines with chemicals and out-of-state grapes. He was kicked out of the company in a secret meeting. Determined to make wine that is responsible to the public, Taylor began production at Bully Hill Vineyards on land he had purchased in 1958, in the Finger Lakes Region. Seven years later, when Coca-Cola purchased the Taylor Wine Company, he was sued for his use of the Taylor name. When he lost in court, he asked visitors to the winery to help eliminate the Taylor name from all his labels with a black marker pen. Since that time, in 1977, he has used a variety of names on his hand-drawn labels, including Walter St. Bully, Count Cyclops, Mr. George C. Raccon and the Bully Hill Billy Goat. Taylor is fond of saying "They got my name, but they didn't get my goat." Today, Taylor produces 100% New York State wines and champagne under the "wine without guilt" motto.

TAYLOR CALIFORNIA CELLARS
(See The Monterey Vineyard.)

Founded 1979, Gonzales, Monterey County, California.

History: Owner is the Seagram Wine Company. The success of the Taylor California Cellars wines are due to the masterful blending and selection of California varietal grapes by master winemaker Dr. Richard G. Peterson.

Wines regularly bottled: Chardonnay, Cabernet Sauvignon, Sauvignon Blanc, Johannisberg Riesling, Chenin Blanc, Zinfandel, French Colombard; semi-generics: Chablis, Burgundy, Rhine, Light Chablis, Light Rhine; also Light Rosé, Rosé, Dry White, Dry Red, and Champagne (Brut, Extra Dry, Pink).

GREYTON H. TAYLOR WINE MUSEUM

The Greyton H. Taylor Wine Museum is a non-profit, educational institution provisionally chartered by the Educational Department of the State of New York. It occupies three buildings, one of which dates back to 1880, and contains displays of early wine and champagne making equipment (some from as long ago as 1828), as well as the museum gift shop and the Champagne Country Cafe. A second building was added, in 1969, to house the museum's extensive collection of old cooper's tools and horse-drawn vineyard equipment. And in 1973, the Walter Taylor Memorial Building was built to accommodate a gallery of original art work by Walter S. Taylor and the museum's unique collection of White House glassware, including goblets and champagne glasses ordered by Mary Todd Lincoln at the beginning of her husband's presidential administration.

TAYLOR, ROD

Rod Taylor received his bachelor's degree in microbiology from the University of British Columbia in 1968. In 1970, he was hired by Andres Wines (British Columbia) Ltd. as assistant production manager. Four years later, he was promoted to winemaker. Taylor has been involved in the development of several popular Andres wines, including Baby Duck and Hochtaler. He has also been involved in the expansion of the wine industry in British Columbia, and he helped to establish German clone vinifera vineyards of Riesling, Ehrenfelser and Scheurebe.

THE TAYLOR WINE COMPANY, INC.

Founded 1880, Hammondsport, Steuben County, New York.

Storage: Oak and stainless steel.

History: The Taylor Wine Company was founded in 1880 on a small plot of vineyards overlooking Keuka Lake, at Hammondsport. While several wineries closed during Prohibition, Taylor continued by producing Sacramental wines and grape juice. Today, Taylor is owned by the Seagram Wine Company.

Estate vineyard: Taylor Wine Vineyards.

Wines regularly bottled: Estate bottled Burgundy, Chablis, Rhine, Sauterne; Pink Catawba; Sparkling wines: Brut, Extra Dry, Pink Champagne, Sparkling Burgundy and Cold Duck; Sherries: Dry, Golden, Cream, Empire Cream, Cooking; Port, Tawny Port; Sangria, Rosé, Vermouth; Lake Country Red, White, Pink, Gold, Soft Red, Soft White, Soft Pink and Chablis.

TCHELISTCHEFF, ANDRE

Andre Tchelistcheff was born into an aristocratic Moscow family in 1901. During the Bolshevik revolution, his family fled across the border to the Crimea, where Tchelistcheff was given a small vineyard by his godfather. It was his first experience with viticulture, which remains to this day his favorite occupation.

Tchelistcheff has been involved in the California wine industry for over forty years. He started at Beaulieu Vineyards, where he worked until 1973, establishing his renowned reputation as an enologist. During that period, he began to accept freelance consulting jobs. His first customer was Frank Bartholomew, who revived the Buena Vista Winery at the end of World War II. In 1947, Tchelistcheff founded the Napa Valley Research Laboratory and Enological Research Center. He worked with the Associated Vintners in Washington state, beginning in 1966, to establish the premium grape-growing region of the Yakima Valley. He closed his business in 1950, but continues his consulting.

Tchelistcheff is an advocate of the California microclimate concept. He considers the central Napa Valley as ideal for Cabernet Sauvignon, the Carneros Creek area for Pinot Noir and Chardonnay, and the Santa Ynez region very promising for white wine grapes. He praises Central Coast microclimates such as Temecula, where Callaway produces German-style wines.

A simple "bell cap" corkpuller

Tchelistcheff was an early advocate of the enological profession. He was a founding member of the American Society of Enologists in 1949. He is a Supreme Knight in the Brotherhood of the Knights of the Vine.

Today, Tchelistcheff continues to lend his expertise to many wineries.

TEDESCHI VINEYARD AND WINERY

Founded 1974, Ulupalakua, Maui County, Hawaii.

Storage: Oak and stainless steel. Cases per year: 2,000.

History: Tedeschi's winery and vineyard is on the grounds of history-rich Ulupalakua Ranch, which dates back to 1856 when a retired sea captain, James Makee, bought the ranch. The converted century-old lava rock and plaster "Jail house" at the old estate is today's wine tasting room. This unique vineyard and winery are at the 2,000 foot level overlooking the coastline of Maui.

Wines regularly bottled: Proprietary Maui Blanc (pineapple wine); Carnelian Champagne.

Teldeschi Vineyard

Sonoma County, California.

Zinfandel, Petite Sirah and French Colombard vineyard.

Templeton Vineyard
San Luis Obispo County, California.
Zinfandel vineyard.

Tepusquet Vineyard
Santa Barbara County, California.
Chardonnay, Gewurztraminer, Johannisberg Riesling, Chenin Blanc, Cabernet Sauvignon, Merlot, Pinot Noir, Gamay Beaujolais and Sauvignon Blanc vineyard.

Terra Point
Clayton County, Illinois.
Marechal Foch vineyard.

Terra Pulchra Vineyard
Sonoma County, California.
Muscat Canelli Sonoma Valley vineyard.

Terra Rosa
Sonoma County, California.
Fieldstone estate Cabernet Sauvignon, Petite Sirah, Johannisberg Riesling and Gewurztraminer Alexander Valley vineyard.

Terry Mountain Vineyard
Napa County, California.
Pinot Noir vineyard.

Richard Tennis Vineyard
Lake County, California.
Konocti estate Cabernet Sauvignon Clear Lake vineyard.

Tewksbury Estate Vineyards
Hunterdon County, New Jersey.
Chardonnay, Gewurztraminer, Chambourcin, Gamay Beaujolais, Johannisberg Riesling and Rayon D'Or vineyard.

TEWKSBURY WINE CELLARS

Founded 1979, Lebanon, Hunterdon County, New Jersey.

Storage: Oak and stainless steel. Cases per year: 4,500.

Estate vineyard: Tewksbury Vineyards.

Wines regularly bottled: Estate bottled, vintage-dated Chardonnay, Gewurztraminer, Johannisberg Riesling, Gamay Beaujolais, Chambourcin, Rayon D'Or; also vintage-dated Seyval Blanc. Occasionally bottles Delaware, Apple Wine.

TEXAS

Grapes can grow in Texas, but for many years the idea of Texas grapes and wine was just a dream of many—and a remembrance for others.

Texas has many native species of grapes, many of which are lumped under the name "Mustang grape." Texas also has a significant grape and wine history. Early Spanish missionaries brought "Mission" grapes to the Rio Grande River Valley, near El Paso, for making Sacramental wines. Numerous vineyards continued in that area for many years. Texas has had a winery called Val Verde Winery located at Del Rio since 1883.

A Texan helped save the French grape and wine industry. T.V. Munson, from Denison, conducted grape research in the early 1900's and developed rootstocks that were resistant to diseases brought from America to France. He received the French Legion of Merit Award for his efforts. Mr. Munson also developed many new grape varieties that were used for table and wine grapes throughout the state.

At the 1904 World's Fair in St. Louis, a wine from grapes grown near Barstow in West Texas won a blue ribbon, so Texas and Texans have a long and proud grape heritage.

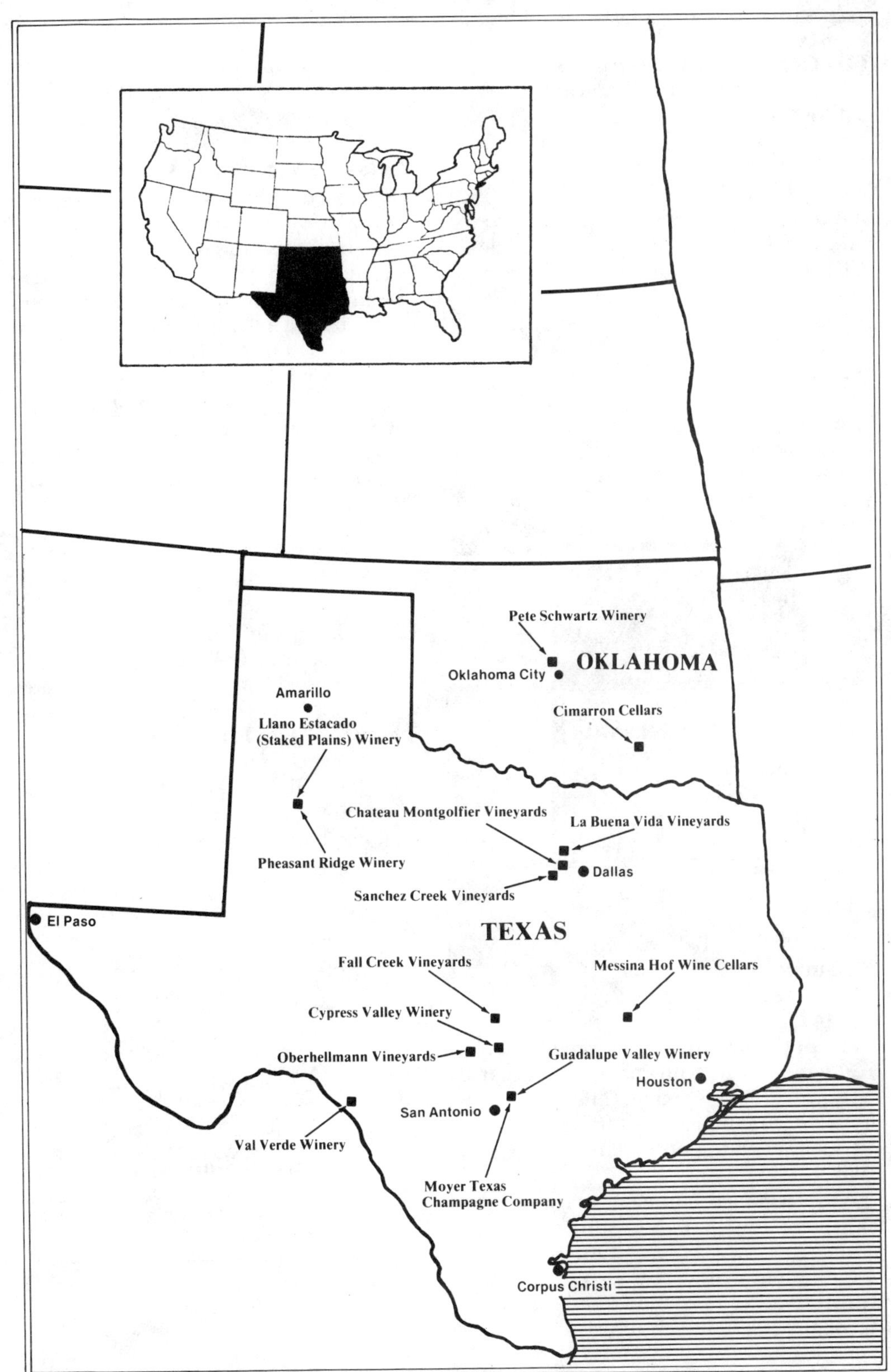
Pete Schwartz Winery
OKLAHOMA
Oklahoma City
Cimarron Cellars
Amarillo
Llano Estacado
(Staked Plains) Winery
Chateau Montgolfier Vineyards
La Buena Vida Vineyards
Pheasant Ridge Winery
Dallas
Sanchez Creek Vineyards
El Paso
TEXAS
Fall Creek Vineyards
Messina Hof Wine Cellars
Cypress Valley Winery
Oberhellmann Vineyards
Guadalupe Valley Winery
Houston
San Antonio
Val Verde Winery
Moyer Texas
Champagne Company
Corpus Christi

THACKREY AND COMPANY

Founded 1980, Bolinas, Marin County, California.

Storage: Oak and stainless steel.

Label indicating non-estate vineyard: Winery Lake Vineyard.

Wines regularly bottled: Proprietary wines: Orion (Pinot Noir), Pleiades (Chardonnay), Aquila (Merlot/Cabernet Sauvignon/Cabernet Franc).

THIGPEN RESERVE

(See Summerhill Vineyards.)

thin

Tasting term.

Unbalanced, lacking desired body. Watery mouth feeling.

PAUL THOMAS

Founded 1979, Bellevue, King County, Washington.

Storage: Stainless steel. Cases per year: 14,000.

History: The winery has proven that fruit wines do age and change from year to year, as do "grape" wines. They produce dry table wines made from fruits.

Label indicating non-estate vineyards: Sagemoor Farms.

Wines regularly bottled: Vintage-dated Cabernet Sauvignon, Sauvignon Blanc, White Riesling, Muscat Canelli; also Crimson Rhubarb, Dry Bartlett Pear, Raspberry, Elegance (Bing Cherries). Occasionally Nectarine and Apricot.

Thomas Vineyard

Riverside County, California.

Chardonnay Temecula vineyard.

THOMAS VINEYARDS

Founded 1839, Cucamonga, San Bernardino County, California.

Storage: Oak. Cases per year: 20,000.

History: On March 3, 1839, Tiburcio Tapia was given the Cucamonga land grant by Juan Alvarado, governor of Mexico. Tapia built an adobe home, planted a vineyard and started a winery. Thomas is now owned and operated by the Joseph Filippi family.

Wines regularly bottled: Zinfandel, Grenache Rosé, Chablis Blanc, Burgundy, Rhine.

THOMAS WINES

(See J. Filippi Vintage Company.)

H.T. Thompson Vineyard

Napa County, California.

Sauvignon Blanc Napa Valley vineyard.

Thousand Oaks Estate Vineyards

Oktibbeha County, Mississippi.

Villard Blanc, Baco Noir, Niagara, Magnolia, Noble, Verdelet Blanc, Marechal Foch and Landot Noir vineyard.

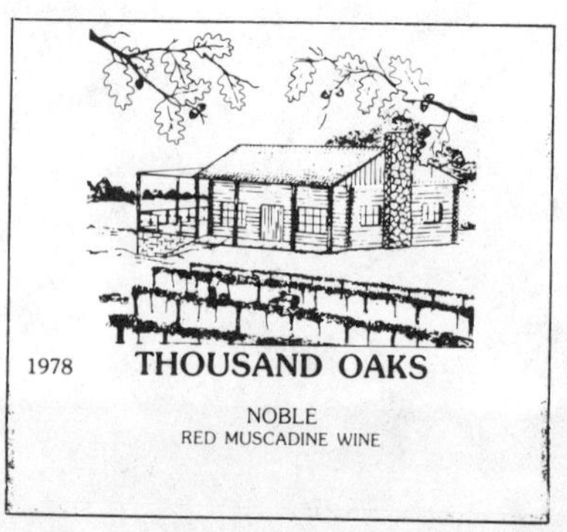

THOUSAND OAKS VINEYARD AND WINERY
Founded 1972, Starkville, Oktibbeha County, Mississippi.
Storage: Stainless steel. Cases per year: 7,500.
Estate vineyard: Thousand Oaks Vineyard.
Wines regularly bottled: Estate bottled, vintage-dated Villard Blanc, Baco Noir, Niagara; vintage-dated Noble; also Magnolia and Sweet White Muscadine.

THREE MOUNTAIN WINES
(See JFJ Bronco Winery.)

Three Palms Vineyard
Napa County, California.
Merlot Napa Valley vineyard.

TIJSSELING VINEYARDS
Founded 1981, Ukiah, Mendocino County, California.
Storage: Oak and stainless steel.
Wines regularly bottled: Chardonnay, Sauvignon Blanc, Petite Sirah; Red and White Table Wines; also Champagnes.

TIMMEN'S LANDING
(See Salishan Vineyards.)
Red and White Table Wine.

BROTHER TIMOTHY
Born Anthony George Diener, Brother Timothy, F.S.C., entered the Christian Brothers novitiate in 1928. As a student, his first experience in the wine industry involved moving large casks from the old Martinez winery to the new Christian Brothers winery location at Mont La Salle, in the Napa Valley. After additional study at St. Mary's College, Moraga, Brother Timothy taught chemistry at local Catholic high schools until he was assigned as wine chemist at the Mont La Salle in 1935. He has worked in all aspects of the wine business at the Christian Brothers winery, and currently holds the position of vice-president and cellarmaster. He supervises all winemaking activities. He is also curator of an extensive corkscrew collection, comprised of more than 1,500 pieces. Brother Timothy is a Supreme Knight in the Brotherhood of the Knights of the Vine.

TINTA MADEIRA
Wine grape famous in Portugal and grown in the warm districts of California. Used in the best of California Ports.

TINTA RUBY PORT WINE
A wine produced from Tinta Madeira grapes. Ruby red in color and rich, with a fruity bouquet. Full bodied and sweet.

tirage (or liqueur de tirage)
The sugar or sweetener added to still wine to induce yeast cells to begin secondary fermentation for Champagne production.

tired
Tasting term.
The term tired is used to describe the impression made by wines which have been excessively processed in the cellar or are too old. Such wines usually show lack of freshness, fruitiness and aroma.

TOBIAS VINEYARDS
Founded 1980, Paso Robles, San Luis Obispo County, California.
Label indicating non-estate vineyard: Dusi Ranch, Radike Ranch, Jones Ranch.

Wines regularly bottled: Zinfandel and Petite Sirah.

To-Kalon Vineyard

Napa County, California.

Robert Mondavi estate Pinot Noir, Chardonnay, Cabernet Sauvignon, Johannisberg Riesling, Chenin Blanc, Moscato D'Oro and Sauvignon Blanc Napa Valley vineyard.

TOKAY

Tokay is midway in sweetness between Sherry and Port. It is amber-colored with a slightly "nutty" or Sherry-like flavor. It is a blend of dessert wines, usually Angelica, Port and Sherry. American Tokay is not to be confused with Tokaji wines from Hungary or with any Tokay grape, which may or may not be used in its production. Flame Tokay is a seedless red table grape, not used for wine and unrelated to the Hungarian Tokay.

TONIO CONTI SPARKLING WINES

Founded 1983, San Luis Obispo County, California.

Storage: Oak and stainless steel. Cases per year: 3,000.

Wines regularly bottled: Blanc de Blanc (Chardonnay) Champagne.

Tomasello Estate Vineyard

Atlantic County, New Jersey.

Villard Blanc, DeChaunac, Seyval Blanc, Vidal Blanc, Catawba, Concord and Noah vineyard.

TOMASELLO WINERY

Founded 1933, Hammondton, Atlantic County, New Jersey.

Storage: Oak and stainless steel. Cases per year: 33,000.

Estate vineyard: Tomasello Vineyard.

Wines regularly bottled: Estate bottled Dry Villard Blanc, Dry DeChaunac; also Spumante, Champagnes (Blanc de Blancs, Natural, Brut, Extra Dry, Pink), Sparkling Burgundy; and proprietary Moderately Sweet (Red, White, Rosé).

TOPOLOS AT RUSSIAN RIVER

Founded 1978, North Forestville, Sonoma County, California.

Storage: Oak and stainless steel. Cases per year: 5,000.

Wines regularly bottled: Russian River Vineyards.

Label indicating non-estate vineyard: Taurian Vineyards, Quail Ridge Ranch, Arnesberg Vineyard.

Wines regularly bottled: Vintage-dated Pinot Noir, Zinfandel, Petite Sirah, Cabernet Sauvignon, Chardonnay, White Riesling; vintage-dated Sonoma Blanc. Occasionally bottles Gravenstein Blanc (dry Apple Wine).

Topolos Estate Vineyard

Sonoma County, California.

Chardonnay, Merlot and Cabernet Sauvignon vineyard located at winery. Zinfandel, White Riesling and Cabernet Sauvignon Sonoma Mountain vineyard.

Town Vineyard

Sonoma County, California.

Sebastiani estate Cabernet Sauvignon, Barbera, Cabernet Blanc and Sylvaner Sonoma Valley vineyard.

Toyon Estate Vineyards

Sonoma County, California.

Cabernet Sauvignon Alexander Valley vineyard.

TOYON VINEYARDS

Founded 1979, Healdsburg, Sonoma County, California.

Cases per year: 4,000.

Estate vineyard: Toyon Vineyard.

Label indicating non-estate vineyard: River Road Vineyard; Handley Ranch; Edna Valley Vineyards.

Wines regularly bottled: Estate bottled, vintage-dated Cabernet Sauvignon, vintage-dated Sauvignon Blanc, Chardonnay, Cabernet Sauvignon.

Second label: Charis Vineyards, Herrera Cellars, Osprey Vineyards, Rose Family Wines.

Gewurztraminer

TRAMINER

A German grape that produces a light, fruity, semi-dry, soft white wine. Commonly grown in the Alsace province of France. This was the name of the grape until the Alsatian winemakers discovered that its best clones and best vintages were spicier in flavor and brought spicier profits when labeled Gewurztraminer. (See Gewurztraminer.)

transfer method

A method used to produce Champagne. After secondary fermentation is completed in the bottle, wine is passed through a filter to remove sediment, and transferred to another bottle saving having to manually remove sediment.

TREE POINT

(See Arbor Crest.)

Trefethen Estate Vineyard

Napa County, California.

Chardonnay, White Riesling, Gewurztraminer, Pinot Noir, Merlot, ZInfandel and Cabernet Sauvignon Napa Valley vineyards.

TREFETHEN VINEYARDS

Founded 1886, Napa, Napa County, California.

Storage: Oak and stainless steel. Cases per year: 40,000.

History: Throughout its history, what is today called the Trefethen Winery and was originally known as Eshcol has used grapes which were grown and owned by its proprietor. In 1968, Gene and Kate Trefethen bought the original land and acreage adjoining the property.

It all began in 1886, when James and George Goodman, prominent banking brothers, built the winery in the middle of their 280 acre parcel at the southern end of the Napa Valley. The designer was the leading winery architect of that time, Captain Hamden McIntyre. For $1,500 he planned and constructed what has become one of the oldest wooden wineries in the Valley.

The Goodmans named their property "Eshcol" after the Eshcol of Biblical fame. In 1888, the Goodman's "Eshcol" Cabernet Sauvignon made wine history by winning first place at the San Francisco Viticultural Fair.

At the turn of the century, J. Clark Fawver bought Eshcol. Why he bought the winery remains a mystery. Fawver did not like wine and he never served a drop in his home. Perhaps his reason for buying Eshcol was the land. He was a farmer and it is said that he loved the earth and took pride in his vineyard. Fawver died in 1940, leaving no one behind to tend the vines or make wine. This, combined with a depressed grape and wine market, led to the closing of the old wooden winery and disinterest in the surrounding acreage.

Estate vineyards: Trefethen Vineyards.

Wines regularly bottled: Estate bottled, vintage-dated Chardonnay, Cabernet Sauvignon, Pinot Noir, White Riesling; and proprietary Eshcol Red Wine, Eshcol White Wine.

Trentadue Estate Vineyards

Sonoma County, California.

Aleatico, Carignane, Cabernet Sauvignon, Gamay, Merlot, Petite Sirah, Zinfandel, Chardonnay, Chenin Blanc, French Colombard, Johannisberg Riesling, Semillon vineyard.

TRENTADUE WINERY

Founded 1969, Geyserville, Sonoma County, California.

Storage: Oak and stainless steel. Cases per year: 25,000.

Estate vineyard: Trentadue Vineyards.

Wines regularly bottled: Estate bottled, vintage-dated Aleatico, Carignane, Cabernet Sauvignon, Gamay, Merlot, Petite Sirah, Zinfandel, Late Harvest Zinfandel, Chardonnay, Chenin Blanc, French Colombard, Johannisberg Riesling, Semillon, White Riesling, Sauvignon Blanc; also Alexander Valley Red and White Wines; and an Early Burgundy.

Tres Ninos Vineyards

Napa County, California.

Villa Mt. Eden estate Pinot Noir Napa Valley vineyard.

Tri-Mountain Estate Vineyard

Frederick County, Virginia.

Cabernet Sauvignon, Riesling, Sauvignon Blanc, Chardonnay, Aligoté, Vidal, Seyval, Chancellor, Ravat, Chelois, Baco, Villard Blanc, Verdelet, Villard Noir Shenandoah Valley vineyard.

TRI-MOUNTAIN WINERY AND VINEYARDS, INC.

Founded 1981, Middletown, Frederick County, Virginia.

Storage: Stainless steel. Cases per year: 4,000.

History: In the cradle of three mountains—the Blue Ridge to the East, the Massanutten to the South, and the Great North Mountain to the West—in the heart of the Shenandoah Valley.

Estate vineyard: Tri-Mountain Vineyard.

Wines regularly bottled: Estate bottled, vintage-dated Late Harvest Vidal, Cabernet Sauvignon, Tri-Mountain Red; vintage-dated Seyval/Aligoté; vintage-dated proprietary Massanutten White, Blue Ridge Rosé, Great North Mountain Concord; and vintage Apfelwein.

TRINCHERO, LOUIS

Six generations of winemakers, in Italy and the United States, preceded Louis "Bob" Trinchero. His family moved to the Napa Valley, to join his uncle in operating the Sutter Home Winery. Everyone in the family participated. Trinchero worked after school and during the summers until he joined the U.S. Air Force. Upon his return, he learned the art of winemaking under the direction of his uncle. When his uncle retired two years later, he sold his interest in the winery to Trinchero and his father. Trinchero took over as winemaker, gradually changing it from the production of jug wines to premium table wines, as the public's demand for good wines increased. His goal was to produce a great red wine. In 1970, he realized his dream with the release of the 1968 Zinfandel, made from Amador County grapes. It sold out quickly, and Trinchero recognized the potential of wines made from the Gold Country grapes. Trinchero specializes in one varietal, Zinfandel, producing both a red and a white wine from the grape.

Triple M Farms

Chautauqua County, New York.

Merritt estate Seyval Blanc, Marechal Foch, Aurora, Baco Noir, Niagara and Vidal Blanc Lake Erie vineyards.

trockenbeerenauslese
(special select late harvest)
The German word for "dried berry selection". Refers to the overripe grapes as a result of Botrytis, that are allowed to stay on the vine until they shrink and the water evaporates thus giving the appearance of raisins. They must be picked raisin by raisin. Minimum Brix: 35 degrees in both Germany and the United States.

Dick Tropp Lazy Acre Vineyard
Berrien County, Michigan.
Vidal vineyard.

TROWBRIDGE, JOHN TOWNSEND
(1827-1916)
American writer.

"What age a richer life begins,
The spirit mellows:
Ripe age gives tone to violins,
Wine, and good fellows."

Truehard Vineyards
Napa County, California.
Merlot Napa Valley vineyard.

Truluck Estate Vineyards
Clarendon County, South Carolina.
Chambourcin, Villard Blanc, Verdelet, Cayuga White, Ravat, Cabernet Sauvignon, Chardonnay, Sauvignon Blanc, Merlot, Barbera vineyard.

TRULUCK, JIM
Jim Truluck grew up in Lake City, South Carolina. After graduation from the University of Louisville School of Dentistry, Truluck served for two years in the U.S. Dental Corps in the Loire region of France. It was during this time that he developed an interest in wine and enology. He returned to Lake City to open a dental practice, but also began to experiment with growing grapes on what was formerly his grandfather's tobacco farm. In 1976, he opened Truluck Vineyards and Winery. Although Truluck continues to work full-time as a dentist, the family operation produces over 18,000 gallons of wine each year, from both French hybrid and vinifera grapes. Popular selections include Carolina Rosé, Villard Blanc, Blanc de Chambourcin and, following the suggestion of noted French viticulturist M. Pierre Galet, Rosé de Chambourcin. Truluck has also produced a Cabernet Sauvignon, Riesling, Chardonnay and a Gamay Beaujolais.

TRULUCK VINEYARDS
Founded 1976, Lake City, Florence County, South Carolina.
Storage: Oak and stainless steel. Cases per year: 8,000-10,000.
Estate vineyard: Truluck Vineyards.
Wines regularly bottled: Estate bottled, vintage-dated Blanc de Chambourcin, Chambourcin, Rosé de Chambourcin, Villard Blanc, Cayuga White, Golden Muscat, Munson Red, Verdelet, Vidal Blanc, Ravat Blanc, Cabernet Sauvignon, Gamay, Riesling; vintage-dated Carlos; proprietary Carolina Red, White and Rosé.

Tsugwale Vineyard
King City, Washington.
Vierthaler estate Goldener Gutedel vineyard.

Tualatin Estate Vineyard
Washington County, Oregon.
White Riesling, Pinot Noir, Gewurztraminer, Chardonnay, Muscat, Flora Willamette Valley vineyard.

TUALATIN VINEYARDS
Founded 1973, Forest Grove, Washington County, Oregon.
Storage: Oak and stainless steel. Cases per year: 25,000.
Estate vineyard: Tualatin Vineyard.
Wines regularly bottled: Estate bottled, vintage-dated White Riesling, Early Muscat, Pinot Noir, Chardonnay, Pinot Noir Blanc, Sauvignon Blanc; also Red and White Table Wine.

TUCKERS CELLARS
Founded 1981, Sunnyside, Yakima County, Washington.
Storage: Oak and stainless steel. Cases per year: 4,500.
Estate vineyard: Tucker Cellars
Wines regularly bottled: Estate bottled, vintage-dated Johannisberg Riesling, White Riesling, Chenin Blanc, Gewurztraminer, Chardonnay; vintage-dated Cabernet Sauvignon and Muscat Canelli.

Tucker Cellars Estate Vineyards
Yakima County, Washington.
Riesling, Chardonnay, Chenin Blanc, Gewurztraminer, Muscat Yakima Valley vineyard.

Tucquan Estate Vineyard
Lancaster County, Pennsylvania.
Seyval Blanc, Foch, Chancellor, Concord, Steuben, Niagara, Cascade vineyard.

TUCQUAN VINEYARD
Founded 1968, Holtwood, Lancaster County, Pennsylvania.
Storage: Oak and stainless steel. Cases per year: 1,000.
History: The oldest commercial winery in Lancaster County, Pennsylvania. The first successful vineyards were planted in Lancaster County in 1790 and thrived until the mid-1800's. The Hamptons were responsible for vineyards being planted once again.
Estate vineyard: Tucquan Vineyard.
Wines regularly bottled: Estate bottled Seyval Blanc, Foch, Chancellor, Concord, Steuben, Niagara, Cascade. Occasionally Peach.

Tudal Estate Vineyard
Napa County, California.
Cabernet Sauvignon vineyard.

TUDAL WINERY
Founded 1979, St. Helena, Napa County, California.
Storage: Oak and stainless steel. Cases per year: 2,500.
Estate vineyard: Tudal Vineyard.
Wines regularly bottled: Estate bottled, vintage-dated Cabernet Sauvignon, Chardonnay.

TULOCAY WINERY
Founded 1975, Oakville, Napa County,

California.

Storage: Oak and stainless steel. Cases per year: 2,000.

Wines regularly bottled: Vintage-dated Cabernet Sauvignon, Pinot Noir, Zinfandel, Chardonnay. Occasionally bottle Pinot Blanc.

Label indicating non-estate vineyards: Mayer Vineyard, Decelles Vineyard, Haynes Vineyard, Phillips Vineyards, Egan Vineyard, Stags' Leap Vineyard.

turbid

Tasting term.

Not luminous; muddy; unclear.

TURGEON & LOHR WINERY

Founded 1974, San Jose, Santa Clara, Monterey County, California.

Storage: Oak and stainless steel. Cases per year: 110,000.

Estate vineyard: Greenfield Vineyards.

Label indicating non-estate vineyard: Rosebud Vineyards.

Wines regularly bottled: Estate bottled, vintage-dated Chardonnay, Pinot Blanc, Johannisberg Riesling, Fumé Blanc, Monterey Gamay, Pinot Noir; vintage-dated Petite Sirah, Chenin Blanc, Late Harvest Johannisberg Riesling. Occasionally bottles Late Harvest Sauvignon Blanc, Chardonnay Reserve, Cabernet Sauvignon.

Turkey Hill Vineyard/Hoot Owl Creek

Sonoma County, California.

Fieldstone estate Cabernet Sauvignon Alexander Valley vineyard.

Turnbull/Fay Vineyard

Napa County, California.

Cabernet Sauvignon Stags Leap District vineyard.

Turner Estate Vineyards

Lake County, California.

Cabernet Sauvignon, Zinfandel, Chardonnay, Sauvignon Blanc, Chenin Blanc, Johannisberg Riesling, Napa Gamay, Petite Sirah, Ruby Cabernet vineyard.

TURNER WINERY

Founded 1979, Woodbridge, Lake County, California.

Storage: Oak and stainless steel. Cases per year: 100,000.

History: Originally known as the Urgon Winery, now Turner, was built at the turn of the century by Adolph Bauer. The name Urgon came from "Urgon Station" a railroad switching station terminal that is now the town of Woodbridge. The Turner family acquired the winery in 1979.

Estate vineyard: Turner Vineyard.

Wines regularly bottled: Estate bottled, vintage-dated Cabernet Sauvignon, Zinfandel, Chenin Blanc, Johannisberg Riesling, White Zinfandel; vintage-dated Merlot, Chardonnay, Fumé Blanc, Cabernet Sauvignon; also semi-generic Burgundy, Chablis, Rhine and Rosé Table Wine.

Twin Rivers Vineyards

El Dorado County, California.

Zinfandel vineyard.

214

(See Bardenheier's Wine Cellars.)

Tawny Port, Dry Cocktail Sherry, Cream Sherry.

Tyland Estate Vineyards

Mendocino County, California.

Chardonnay, Chenin Blanc, Gewurztraminer, Gamay Beaujolais, Zinfandel and Caberent Sauvignon vineyard.

TYLAND VINEYARDS

Founded 1979, Ukiah, Mendocino County, California.

Storage: Oak and stainless steel. Cases per year: 20,000.

Estate vineyard: Tyland Vineyard.

Wines regularly bottled: Estate bottled, vintage-dated Cabernet Sauvignon, Zinfandel, Chardonnay, Pinot Noir Blanc, Chenin Blanc, Gamay Beaujolais, vintage-dated Gewurztraminer.

U.C. Davis Experimental Vineyard

Napa County, California.

Napa Valley vineyard used for growing and various experiments including spacing and trellising.

ullage

The amount of air-space above the wine in a bottle, or cask, which is no longer full. Excessive ullage leads to spoilage.

UMPQUA VALLEY VITICULTURAL AREA

Umpqua Valley in Douglas County, Oregon is located entirely within Douglas County, in the southwest part of Oregon, and consists of approximately 1,200 square miles.

unbalanced

Tasting term.

Excessive or inadequate amounts of one constituent or another cause disharmony and wines so constituted are described as unbalanced.

uncorking

Carefully cut away the capsule just under the bottle lip. You might find, between the cork and the capsule, a mold with a color which can vary between white and black, through brown or green. Don't worry about it, for it is rather a good sign—your cellar has sufficient humidity.

Wipe the neck of the bottle carefully and use a really good corkscrew, which permits easy, steady and progressive withdrawal of the cork. The thread of the corkscrew, by the way, should be long enough to penetrate the cork completely but without coming out of the other side. Smell the extreme end of the cork—it should smell of wine. Perhaps once in 5,000 bottles, there might be a smell of rotten cork. If it is very weak, it may go off after the decanting but, if it is strong, then it is probably hopeless.

unharmonious

Tasting term.

Lacking balance.

Uniacke Estate Farms

British Columbia, Canada.

Gewurztraminer, Johannisberg Riesling, Pinot Noir, Semillon, Merlot, Verdelet, Chelois, DeChaunac Okanagan District vineyard.

UNIACKE WINES LIMITED

Founded 1980, Okanagan Mission, British Columbia, Canada.

Storage: Oak and stainless steel. Cases per year: 6,000.

History: The family name, Uniacke, dates back to 14th Century Ireland when a member of the Fitzgerald family demonstrated outstanding duty to his king and was recognized as being 'without peer', or 'unique'. Thereafter, Fitzgerald and his family took the name Uniacke with them until present day times.

Estate vineyards: Uniacke Farms.

Wines regularly bottled: Estate bottled, vintage-dated Johannisberg Riesling, Okanagan Riesling, Gewurztraminer, Chelois Rosé, Merlot; vintage-dated Chasselas, Pinot Noir; occasionally Late Harvest Johannisberg Riesling.

United Ridge Vineyard

Scott County, Missouri.

Moore-Dupont estate Catawba, French

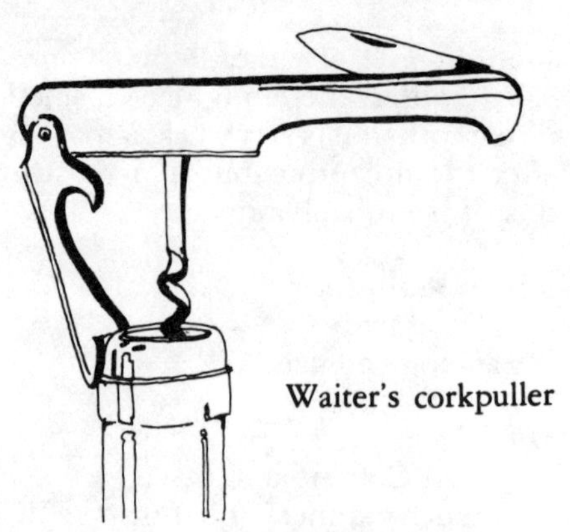
Waiter's corkpuller

Hybrids and Chardonnay vineyard.

Upton Ranch,
Amador County, California.
Zinfandel Shenandoah Valley vineyard.

Uvas Valley Vineyard
Santa Clara County, California.
Kirigin estate Pinot Noir, Cabernet Sauvignon and Malvasia Bianca vineyard.

Vail Vista
Sonoma County, California.
Chardonnay and Cabernet Sauvignon Alexander Valley vineyard.

VAL VERDE WINERY
Founded 1883, Del Rio, Val Verde County, Texas.
Storage: Oak and stainless steel. Cases per year: 2,000.
History: Founded by Frank Qualia and operated today by third generation Texas winemaker. Was the only licensed winery in state from 1949 to 1976.
Estate vineyard: Del Rio Vineyard, Quemado Vineyard, North Del Rio Vineyard.
Wines regularly bottled: Lenoir, Herbemont; Rosé, Tawny Port; and proprietary Sweet Amber.

Valle de Guadalupe
Baja, Mexico.
Casa Pinson estate Cabernet Sauvignon, Ruby Cabernet, Pinot Noir, Gamay, Petite Sirah, Nebiolo, Merlot, Chenin Blanc, Colombard, Ugni Blanc, Grenache, Riesling, Chardonnay vineyard.

VALLEJO, GENERAL DON MARIANO GUADALUPE
Commandante of the California Army, upon orders from the governor of the Mexican State, California, took over the Sonoma Missions vineyards and replanted them in 1836. He dominated Northern California's wine production from 1834 to 1859.

VALLEY OF THE MOON
Founded 1941, Glen Ellen, Sonoma County, California.
Storage: Oak and stainless steel. Cases per year: 35,000-40,000.
History: Named by the Wappo, Miwok and Pomo Indians as the "Valley of the Seven Moons", Jack London, the writer, shortened the name. Originally a part of the Agua Caliente Rancho granted by the Mexican Government to Lazaro Pena, the land was purchased by General M.G. Vallejo. In 1851, Joseph Hooker, known as "Fighting Joe Hooker" of the Union Army bought 640 acres and planted the vineyard. The Harry Parducci family bought the winery in 1941.
Estate vineyard: Valley of the Moon.
Label indicating non-estate vineyard: Leveroni Vineyards.
Wines regularly bottled: Estate bottled, vintage-dated French Colombard, Dry French Colombard, Semillon, Zinfandel, Zinfandel Rosé, White Zinfandel; vintage-dated Pinot Noir, Pinot Noir Blanc, Chardonnay, a non-vintage Zinfandel. Also Vin Rosé, Claret, Chablis and Burgundy.

Valley of the Moon Winery Estate Vineyard
Sonoma County, California.
Zinfandel, French Colombard, Semillon, Alicante Bouschet vineyards.

Valley View Estate Vineyard
Jackson County, Oregon.
Cabernet Sauvignon, Chardonnay, Pinot Noir, Gewurztraminer, Merlot vineyard.

VALLEY VIEW VINEYARD
Founded 1979, Jacksonville, Jackson County, Oregon.
Storage: Oak and stainless steel. Cases

per year: 6,000.

History: The original Valley View was founded in 1854 by Peter Britt. The Wisnousky family, current owners, planted a new vineyard in 1972.

Estate vineyard: Valley View Vineyard.

Label indicating non-estate vineyard: Layne Vineyards, Gerber Vineyards, Carpenter Vineyard.

Wines regularly bottled: Estate bottled, vintage-dated Cabernet Sauvignon, Chardonnay, Pinot Noir, Merlot, Gewurztraminer.

Second label: Jefferson State Brand.

Valley Estate Vineyards

Warren County, Ohio.

Baco Noir, DeChaunac, Niagara, Concord and Catawba Ohio River Valley vineyard.

VALLEY VINEYARDS

Founded 1969, Morrow, Warren County, Ohio.

Storage: Oak and stainless steel. Cases per year: 10,000.

Estate vineyard: Valley Vineyards.

Wines regularly bottled: Estate bottled, vintage-dated Baco Noir, DeChaunac, Niagara, Pink Concord, Pink Catawba, Honey (Mead); proprietary Blue Eye, Hillside Red and Valley Rosé, Sangria, Ohio Sauterne.

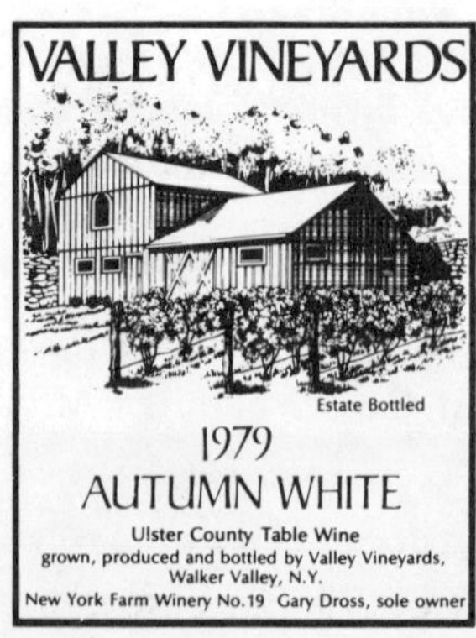

Valley Estate Vineyards

Ulster County, New York.

Riesling, Chardonnay, Cayuga, Ravat, Foch, Baco Noir and Seyval Hudson River Region vineyard.

VALLEY VINEYARDS

Founded 1978, Walker Valley, Ulster County, New York.

Cases per year: 2,000.

Estate vineyard: Valley Vineyards.

Wines regularly bottled: Estate bottled, vintage-dated Riesling, Cayuga White, Seyval; vintage-dated Chardonnay; also estate bottled, vintage Rosé; proprietary Autumn Red, Autumn White and Nouveau. Occasionally estate bottles Pinot Noir.

Valley Vista Vineyard

Sonoma County, California.

Lytton Springs estate Zinfandel vineyard.

Vallis Vineyard

Madera County, California.

Papagni estate Late Harvest Zinfandel vineyard.

varietal

When a wine is named for the principal grape variety from which it is made, it is said to have a varietal name. Cabernet Sauvignon, Chardonnay, Zinfandel, Pinot Noir are some of the varietal names for wine types in the United States. Under Federal regulations, when any grape varietal name is used on a wine label, 75% of the resulting wine must be from that variety, except in the case of American hybrids with Vitis Labrusca parentage, in which case only 51% is necessary.

Veeder Hills Vineyard

Napa County, California.

Cabernet Sauvignon and Chardonnay Veeder Peak vineyard.

Veeder Park Vineyard

Napa County, California.

William Hill estate Chardonnay and Cabernet Sauvignon Napa Valley vineyard.

VEGA VINEYARD WINERY

Founded 1979, Buellton, Santa Barbara County, California.

Storage: Oak and stainless steel. Cases per year: 4,000.

History: Rancho la Vega became a reality when the marriage of Micaela Cota and Dr. Ramon de la Cuesta combined a portion of Rancho Santa Rosa and Rancho Nojoqui—an eight thousand acre portion. Besides the generous portion of land, Micaela's dowry included 10 cows, a bull, and 10,000 pesos. The primary task at the new Rancho la Vega was establishing a vineyard. Vines from this original vineyard are producing grapes to this day.

The de la Cuesta home was finished in 1853 and consisted of nine rooms and a large courtyard. The arrival of children necessitated the enclosure of the courtyard to be used as a schoolroom.

The house is adobe and redwood timber. The walls are three feet thick. The timber was carried by oxen and mule trains over Gaviota Pass on trails barely passable by foot, much less by carts.

Today this lovely old house is home to Mike Mosby, vineyard manager for Vega Vineyards Winery. William Mosby acquired the property and planted vines in 1971.

Estate vineyard: Riverview Vineyard, Rios Ranch Vineyard.

Label indicating non-estate vineyard: Caldwell Vineyard, Rancho Sisquoc.

Wines regularly bottled: Estate bottled, vintage-dated Gewurztraminer, Johannisberg Riesling, White Riesling, Pinot Noir. Occasionally Cabernet Sauvignon, Chardonnay.

Second label: Los Padres.

veiled

Tasting term.

Appears veiled or hazy.

velvety

Tasting term for a wine which feels soft, round and rich in the mouth.

VENGE, NILS

In 1970, after completing the viticulture and enology program at the University of California, Davis, Nils served as vineyard manager for Charles Krug before joining Sterling as co-vineyard manager. During his years with Sterling and Krug, Nils developed literally hundreds of acres of vineyards for the wineries. At the time he joined Villa Mt. Eden, in February of 1973, his job was to develop the vineyards surrounding the old winery, including one plot adjacent to what is now the Groth's Oakcross Vineyard (where he is now winemaker). During this time, Villa Mt. Eden was reactivated as a winery and its first commerical crush, 47 tons, was in 1974, with Nils as winemaker.

Ventana Estate Vineyards

Monterey County, California.

Chardonnay, Chenin Blanc, Sauvignon Blanc, Pinot Blanc, White Riesling, Pinot Noir, Cabernet Sauvignon, Merlot, Cabernet Franc, Petite Sirah vineyard.

VENTANA VINEYARDS WINERY

Founded 1978, Soledad, Monterey County, California.

Storage: Oak and stainless steel. Cases per year: 37,000.

Estate vineyard: Ventana Vineyards.

Wines regularly bottled: Estate bottled Chardonnay, Sauvignon Blanc, White Riesling, Chenin Blanc, Dry Chenin Blanc, Petite Sirah, Rosé of Petite Sirah, Syrah, Cabernet Sauvignon; also Pinot Noir and vintage-dated Sparkling Wine. Occasionally bottles Botrytis White Riesling and Sauvignon Blanc.

Second label: Los Coches Cellars; Douglas Meador.

VENTURA

A native American grape that produces a foxy white wine.

Venture Vineyards
Seneca County, New York.
Cayuga White, Riesling Finger Lakes vineyard.

Verdekal Vineyard
Cumberland County, Pennsylvania.
Blue Ridge estate Aurora, Baco, DeChaunac, Chelois and Foch vineyard.

VERDELET
A French hybrid grape that produces dry to medium sweet, fruity white wines.

VERGENNES
An American hybrid that produces dry, delicate white wine.

VERMOUTH
A wine flavored with herbs and other aromatic substances. The two principal types are dry (pale) and sweet (dark), usually fortified with Brandy.

For Vermouth, neutral white wines are first selected and aged. Then they are flavored by an infusion of herbs, and more aging follows. Vermouth ranges from 15-20 percent alcohol content.

VERNIER WINES
Founded 1981, Seattle, Washington.
Storage: Oak and stainless steel. Cases per year: 3,500.
Wines regularly bottled: Sauvignon Blanc, White Riesling, Chardonnay, Cabernet Sauvignon and Semillon.

NICHOLAS G. VERRY
Parlier, Fresno County, California.
Wines regularly bottled: Retsina.

CONRAD VIANO WINERY
Founded 1946, Contra Costa County, California.
Storage: Oak.
Estate vineyard: Viano Vineyard.
Wines regularly bottled: Cabernet Sauvignon, Zinfandel, French Colombard, Chardonnay, Sauvignon Blanc, Gamay, Zinfandel Rosé; also generics Burgundy and Chablis.

VICHON WINERY
Founded 1980, Oakville, Napa County, California.
Storage: Oak and stainless steel. Cases per year: 40,000-50,000.
Label indicating non-estate vineyard: Fay Vineyard, Volker Eisele Vineyard.
Wines regularly bottled: Vintage-dated Cabernet Sauvignon, Chardonnay and a proprietary Chevrier Blanc (50% Semillon, 50% Sauvignon Blanc).
Second label: Mont Eagle.

VIDAL BLANC
A popular French hybrid grape that produces dry to medium, fresh, fruity white wine. In exceptional years, a late harvest is produced with rich honey flavor.

MANFRED VIERTHALER WINERY
Founded 1976, Sumner, Pierce County, Washington.
Storage: Oak and stainless steel. Cases per year: 7,500.
Estate vineyard: Tsugwale Vineyard; EL-HI Hill Vineyard.
Wines regularly bottled: Cabernet Sauvignon, Chenin Blanc, Select Harvest White Riesling, White Riesling, Cream Sherry and Port; semi-generics: Rhine, Rosé, Burgundy, Harvest Burgundy, Chablis, Moselle, Sylvaner; also Barberone Red, Angelica, Banquet White.

Vigna del Lago
Sonoma County, California.
Sebastiani estate Chardonnay, Sauvignon Blanc and Traminer Sonoma Valley vineyard.

VIGNES, JEAN LOUIS
When Jean Louis Vignes arrived in California, in 1831, from France, the Los Angeles area had already been established as a grape growing area. There were six wine growers and close to 100 acres of mission grape vines. Vignes, trained as a cooper and distiller, was the first professional enologist to recognize the potential for California wines. By 1833, he had purchased 104 acres, in what is now the heart of downtown Los Angeles, and had planted the land with the first imported European grape varieties in California. The cuttings were sent from France to Boston, then to the west coast on trading vessels. Although the date of his first vintage is unknown, it is thought to have been about 1837. By 1840, Vignes was producing many wines and brandies, from 40,000 vines, at his Aliso Vineyard. Many were shipped to the ports of Santa Barbara, Monterey and San Francisco, and were sold for $2 a gallon for white wines, $4 a gallon for brandies. Vignes was the first to age his wines in quantity, and believed that sea travel aided in aging. He kept an extensive cellar, and advertised some of his wines as 20 years old. Between 1840 and 1850, over a dozen new wine producers joined Vignes in the Los Angeles area. It became the largest grape growing region in the United States, producing 57,355 gallons of wine in 1850. Upon his retirement in 1855, at the age of 76, Vignes sold Aliso to one of the eight relatives he had persuaded to emigrate to California.

VIGNOLES
A French hybrid that produces a crisp white dinner wine. Also known as Ravat.

Pruning back the vine canes

vigorous
Tasting term.
Well structured in body and character.

VILLA ARMANDO WINERY
Founded 1903, Pleasanton, Alameda County, California.
Storage: Oak and stainless steel.
Estate vineyard: Sunol Valley Vineyard.
Wines regularly bottled: Estate bottled Chardonnay, Cabernet Sauvignon, Malvasia Bianca, Petite Sirah; also Cabernet Sauvignon, Napa Gamay Rosé, Chenin Blanc, Valdepenas, Muscat Canelli; proprietary wines: Rustico Red, Rustico Mellow White, Rustico Mellow Red, Rustico Mellow Rosé, Orobianco, Rubinello.

VILLA D'INGIANNI WINERY, INC.
Founded 1972, Dundee, New York.
Cases per year: 35,000.
Wines regularly bottled: Riesling, Chardonnay, Niagara, Concord, Pink Catawba, DeChaunac, Dutchess, Delaware, Ravat, Isabella, Baco Noir, Ottonel; semi-generics: Rhine, Sauterne, Pink Chablis, Chablis, Burgundy; also Rosé.
Second label: Fox Lane, Sunshine, Steuven Valley, Golden Valley, Autumn Harvest, Crooked Lake.

VILLA MT. EDEN
Founded 1881, Oakville, Napa County, California.

Storage: Oak and stainless steel. Cases per year: 20,000.

History: Villa Mt. Eden's winery building was built in 1886, and the hand sawn planks and stone foundation are still visible in the winery. Zinfandel and Riesling were among the first varieties planted in 1881.

Estate vineyard: Villa Mt. Eden.

Wines regularly bottled: Estate bottled, vintage-dated Cabernet Sauvignon, Chardonnay, Pinot Noir, Chenin Blanc, Gewurztraminer; also estate bottled proprietary Ranch Red and White Table Wines.

Villa Mt. Eden Estate Vineyard

Napa County, California.

Cabernet Sauvignon, Chardonnay, Pinot Noir, Chenin Blanc, Gewurztraminer Napa Valley vineyard.

Villa Paradiso Vineyards

Founded 1981, Morgan Hill, Santa Clara County, California.

Storage: Oak and stainless steel. Cases per year: 250/500.

Wines regularly bottled: Vintage-dated Zinfandel.

VILLA SORRENTO

(See Bianchi Winery.)

VILLARD BLANC

A French hybrid grape that produces a fresh, fruity white wine produced mainly in Virginia and Texas.

VIN MARK (See Markham Vineyards.)

Merlot.

Vina de Santa Ynez Vineyard

Santa Ynez County, California.

Johannisberg Riesling Santa Ynez Valley vineyard.

Vina de Santa Ynez Vineyard

Santa Barbara County, California.

Santa Ynez estate Sauvignon Blanc, Chardonnay, Semillon, White Riesling, Gewurztraminer, Cabernet Sauvignon, Merlot, Cabernet Franc Santa Ynez Valley vineyard.

Vina Las Lomas

Lake County, California.

Cabernet Sauvignon vineyard.

Vina Madre Estate Vineyard

Chaves County, New Mexico.

Chardonnay, Cabernet Sauvignon, Gamay, Ruby Cabernet, Zinfandel, Riesling and French Colombard Vina Madre vineyards.

VINA MADRE WINERY

Founded 1978, Dexter, Chaves County, New Mexico.

Storage: Stainless steel.

History: The first winery in the Pecos Valley, southeastern New Mexico.

Estate vineyard: Vina Madre

Wines regularly bottled: Estate bottled, vintage-dated Chardonnay, Cabernet Sauvignon, Gamay, Ruby Cabernet, Zinfandel, Riesling, French Colombard; also Burgundy.

VINA VISTA VINEYARDS

Founded 1971, Geyserville, Sonoma County, California.

Cases per year: 3,000.

Produces Zinfandel and other table wines.

vine and wine years

As climates change, so does the vine and wine year. There are no hard and fast rules.

January and February

A very important time of year, for now comes the pruning, a difficult and laborious task that takes much practice to learn. The object of the exercise is to reduce the number of fruitbearing buds in order that the sap when it rises, concentrates its efforts on producing quality fruit in good quantity.

March, April, May

The time growth starts, leaves are produced, budding takes place, new shoots

Smudge pot in winter vineyard

start growing—and weeds. Therefore, three basic jobs are done during this period:

1. The soil that had been banked up around the base of the vines is now drawn away and ploughed in with a light fertilizer.
2. Weeding is carried out in order that the weeds do not use the natural ingredients in the soil, thus reserving them for the precious grapes.
3. Unwanted shoots and leaves have to be pruned off to thin out the vegetation in order to concentrate the sap in more needed areas and to expose the budding grapes to more sunlight (light and other constituents produce sugar).

June, July, August

Buds grow and then comes the flowering, after that the grapes start to form and grow in size and maturity. However, it is also a problem time for the vine can be attacked by diseases. Sprayed or dusted against, mildew, botrytis and insects.

September-October

When the grapes are ripe, they are then picked.

1. The grapes are pressed in large, specially made presses.
2. White wines—the juice is run off into vats or barrels and fermentation begins.
3. Fermentation: depending on the temperatute of the juice, it will ferment faster as the temperature rises and vice-versa. Ideally a good fermentation needs to be fairly slow. Fermentation is the interaction of sugar and yeast, which produce alcohol, CO_2 gas and heat.
4. Red Wines—the grapes are destemmed and crushed and the resultant juice and skins are put into fermentation vats. During fermentation, apart from converting sugar to alcohol, alcohol and temperature attack the skins and extract the coloring matter.
5. After the fermentation, the wine has in it dead yeasts and other solids that sink to the bottom; occasionally some solids stay in suspension making the wine cloudy. The wine is drawn off the deposit and put into tanks and barrels to start maturing and lose its cloudiness.

Treatments

Racking: Transferring wine from one barrel to another, leaving behind the deposit.

Fining: A treatment used to take out cloudiness (e.g. egg whites, etc.) or to take out unwanted minerals, bacteria, etc. (e.g. iron, copper, protein, yeast, etc.)

Filtering: To take out any small solids that may still be in the wine.

October, November, December

The harvesting having been completed, in cold climates, the earth is built up around the base of the vines to protect them from winter frosts. The vines are then lightly pruned in preparation for the main pruning in January/February.

Vine Hill Ranch

Napa County, California.

Cabernet Sauvignon Napa Valley vineyard.

VINE, RICHARD

Richard Vine began his winemaking career in 1961 as a cellarworker for the Pleasant Valley Wine Company, makers of Great Western Champagnes and wines. He soon became an apprentice winemaker, and after completion of an evening study program at Corning Community College, was appointed winemaker. By 1971, Vine was promoted to executive vice-president. In 1973, he became vice president/winemaker at Warner Vineyards, Inc. Vine is currently cellarmaster at the A.B. McKay Food and Enology Laboratory at Mississippi State University, where he received his Ph.D. in agricultural economics. In addition, Vine serves as consultant to thirty-one wineries, throughout the United States, and is author of numerous articles and three books, including *Commercial Winemaking*. His newest book, *Wine Appreciation* is scheduled to be released in 1985. Vine is a Supreme Knight in the Brotherhood of the Knights of the Vine, a member of the Alpha Zeta Agricultural Fraternity, the Sigma Xi Research Fraternity, and the American Society of Enologists.

Vineberg Vineyards
Sonoma County, California.
Pinot Noir and Chardonnay Carneros vineyard.

Vinedos Don Luis
Ojocaliente, Mexico.
Industrias de la Fermentacion estate Cabernet Sauvignon, Chardonnay vineyard.

Vinedos El Queretano
Ezequiel Montes, Mexico.
Cabernet Sauvignon, Ugni Blanc vineyard.

Vinedos Hidalgo
San Clemente, Mexico.
Cavas de San Juan estate Cabernet Sauvignon, Malbec, Ugni Blanc, Grenache, Carignane, Pinot Noir vineyard.

Vinedos La Escondida
Luis Moya, Mexico.
Industrias de la Fermentacion estate Pinot Noir, Malbec, Merlot vineyard.

Vinedos La Esperanza
Luis Maya, Mexico.
Industrias de la Fermentacion estate Merlot, Semillon vineyard.

Vinedos La Estancia
Luis Maya, Mexico.
Industrias de la Fermentacion estate Chenin Blanc, Sylvaner vineyard.

Vinedos Los Lobos
Tequisquiapan, Mexico.
Ugni Blanc, Chenin Blanc vineyard.

Vinedos Queretanos
San Clemente, Mexico.
Cavas de San Juan estate Cabernet Sauvignon, Chenin Blanc, Verdona, Alicante Bouschet, Traminer, Feher Szagos, Merlot, Sauvignon vineyard.

Vinedos San Isidro
San Isidro, Mexico.
Cavas de San Juan estate Traminer, Riesling, Macabeo, Feher Szagos, Ugni-Blanc, Merlot, Chenin Blanc vineyard.

vinegar, wine
The production of vinegar requires two fermentation processes. The first transforms the sugar of the fruit or juice into alcohol. This is brought about by yeast, a microscopic organism of the plant kingdom. The second changes the alcohol into acetic acid, caused by acetic acid bacteria. The alcoholic fermentation must be complete before the transformation to vinegar is allowed to start, otherwise the yeast fermentation will be stopped by the acetic acid and unfermented sugar will remain in the vinegar.

1. Crush Grapes. This may be done by hand or with a potato masher. Press the crushed fruit through a double thickness of cheesecloth. To each quart of juice add about F cake of fresh yeast, which should be well broken up and thoroughly mixed with the juice. The juice must not be above 90 degrees F. when the yeast is added. Allow it to stand in a stone or glass jar with the lid removed and with the jar

covered with a cloth until gas formation ceases; or allow the juice to ferment in a gallon jug or bottle of any suitable size and plugged with cotton or covered with cloth. The fermentation usually requires 2 weeks.

2. When gas formation has ceased, separate the fermented liquid from the sediment. To each quart of this liquid add about half a pint of good unpasteurized vinegar. Cover the jar or bottle with a cloth, to exclude insects and allow it to stand in a warm place until the vinegar is strong enough to use. Separate it from the "mother of vinegar" and sediment, bottle it, and cork tightly. Mother of vinegar is the white, rubbery mass of vinegar bacteria that often forms in vinegar. Heating the bottled vinegar to between 140 and 160 degrees will pasteurize it and preserve it.

Note: Step One may be eliminated by using a commercial wine and proceeding with Step Two.

The vinegar or acetic acid formation usually takes six to eight weeks.

Vineyard Hill

Santa Clara County, California.

Zinfandel Santa Cruz Mountain vineyard.

VINEYARD HILL FARM

VINEYARD 1976

(See Ritchie Creek Vineyards.)

Vineyard View

Contra Costa County, California.

Barbera vineyard.

Vineyards of Piterra

Hampshire County, West Virginia.

Robert F. Pliska estate Aurora, Foch, Chancellor, Seyval vineyard.

vinifera grapes

The European grape species of which such varieties as Chardonnay, Riesling, Cabernet Sauvignon and Pinot Noir, among others are part. Vinifera vines are subject to winter damage and costly to grow in colder parts of the United States east of the Continental Divide. Crosses of vinifera and other species (see Hybrids) were developed to compensate for the lack of winter hardiness of such varieties. Also susceptible to phylloxera, so require American or Hybrid rootstock.

Vinifera Vineyard

Mendocino County, California.

Sauvignon Blanc, Gewurztraminer, Chenin Blanc vineyard.

VINIFERA WINE CELLARS

Founded 1962, Hammondsport, Steuben County, New York.

Cases per year: 50,000.

History: Dr. Konstantin Frank is one of the most outstanding men in the American wine industry. In 1925, as a professor of viticulture and enology in Russia, he was ordered by the Russian Government to "fill" the cellar and restore a vineyard destroyed by phylloxera. He restored the vineyard—and with the same grapes that the Catholic monks had planted 200 years earlier. In 1941 he escaped from Russia to Austria. Since 1945 he has been an American and German advisor on confiscated Nazi properties. In 1951, he came to America and his mission in life has been to introduce, promote, grow and recommend grape varieties that were planted 200 years ago for the Russian Czar's family. "What was good enough 200 years ago for a Czar's family must be good for American people too."

Estate vineyard: Vinifera Wine Cellars Vineyard.

Wines regularly bottled: Estate bottled, vintage-dated Rkaziteli, Saperavi, Mzwani, Furmint, Riesling, Pinot Noir, Aligoté, Gewurztraminer, Muscat, Ruby Cabernet, Gamay Beaujolais, Chardonnay and Cabernet Sauvignon.

Vinifera Wine Cellars Estate Vineyard

Steuben County, New York.

Riesling, Gewurztraminer, Pinot Gris, Chardonnay, Pinot Noir, Cabernet Sauvignon, Muscat Frontignan, Pedro Ximènez, Fermint, Sereksia, Rkatsiteli, Saperavi and Mzwani Finger Lake vineyard.

VINIFICACION Y TECNOLOGIA, S.A.

Vinland Vineyards

Stoddard County, Missouri.

Moore-Dupont estate French Hybrids and Vinifera vineyard.

vinosity

Tasting term.

The grape character.

vinous

Tasting term.

This term is used to describe the smell of wine when no varietal or distinct aroma is detectable.

vintage

The gathering of grapes and their fermentation into wine; also the crop of grapes or wine of one season. A vintage wine produced in the United States is one labeled with the year in which at least 95% of its grapes were gathered and crushed and the juice therefrom fermented. A vintage year is one in which grapes reach full maturity. Particularly applicable in Europe, where growing conditions vary greatly from year to year.

"Head pruned" grape vine

vintage-dated

95% of the wine must have been grown and produced from grapes harvested from the stated year, and the wine must be labeled with a county, state, or viticultural area designation. Consequently an American made wine can never have a vintage date where the grapes were grown in one state and the wine made in another.

vintage-year

The year of the harvest from which the wine is made.

Sometimes, for many, "vintage" has come to be synonymous with "quality". It is true that wines made by expert winemakers, from grapes ripened under exceptionally favorable climatic conditions, may have a higher degree of pleasant characteristics, but three factors are of prime importance in connection with vintage:

1. It is impossible, at the time of harvest, to predict with certainty the quality of the wine being made and how it will develop through fermentation and aging.

2. The estimation of the general quality of a vintage may be correct, as a whole, yet be entirely wrong for the wines coming from any specific vineyard or district. Much depends, also, on the experience and integrity of the winemaker. Different quantities of wine are made every year—some of which are better and some worse than the rating vintage would indicate.

3. Great vintage wines take a very long time to mature and, until then, are less palatable than so-called "off vintage" wines which meanwhile have reached their peaks.

Vintage Vineyard
Santa Ynez County, California.
Ballard Canyon estate Johannisberg Riesling, Cabernet Sauvignon, Muscat Canelli, Chardonnay and Zinfandel Santa Ynez Valley vineyards.

Vinterra Farm Estate Vineyard
Shelby County, Ohio.
Baco, Vidal, Villard, Aurora, DeChaunac, Catawba, Niagara and Chancellor Loramie Creek vineyard.

VINTERRA FARM WINERY
Founded 1972, Houston, Shelby County, Ohio.
Storage: Oak and stainless steel.
Estate vineyard: Vinterra Farm.
Wines regularly bottled: Estate bottled Baco Noir, Aurora, Concord, Vidal, Pink Catawba, Rosé, proprietary Rouge Blanc; generics: Rhine, Chablis, Sauterne, Burgundy and Pink Chablis.

Checking vines in dormant vineyard

vintner
One who makes wine from grapes: the winemaker.

violet-blue
Wine color.
The color of the must of very young (a few months) wines with a notable blue tone.

VIRGINIA ROSE
(See Ingleside Plantation Winery.)

VIRGINIA WINE COUNTRY
Viticulture, or winegrowing, has been intertwined with the history of Virginia ever since the Jamestown colonists made the first wine in the New World.

Several prominent Virginians made important contributions to grapegrowing. The state's own original wine pioneer, Thomas Jefferson, also known as "the father of American wines", labored long in his Monticello vineyard to prove Virginia could produce fine wines. Colonel Robert Bolling of Buckingham County became the first American writer on grapes.

But it was Dr. D.N. Norton of Richmond who, in 1835, brought Virginia a major viticultural victory when he developed the first "non-foxy" wine grape. Norton claret wine won international acclaim and stimulated the planting of 4,000 acres of grapes statewide. With vineyards concentrated in Albemarle County, Charlottesville was christened "the capital of the Virginia wine belt" and the Rivanna River, "the Rhine of America".

Until the early 1970's, the idea of Virginia again producing fine wines was merely a gleam in the eyes of a few visionary Virginians. Today Virginia has vineyards and wineries budding everywhere. There are four viticultural areas: Monticello, Rocky Knob, Shenandoah Valley and North Fork of the Roanoke.

Viridian Vineyard
Napa County, California.
Girard estate Cabernet Franc, Cabernet Sauvignon, Sauvignon Blanc and Semillon Napa Valley vineyard.

VISTA MENDOCINO
(See Olson Vineyards.)

VIRGINIA MAP

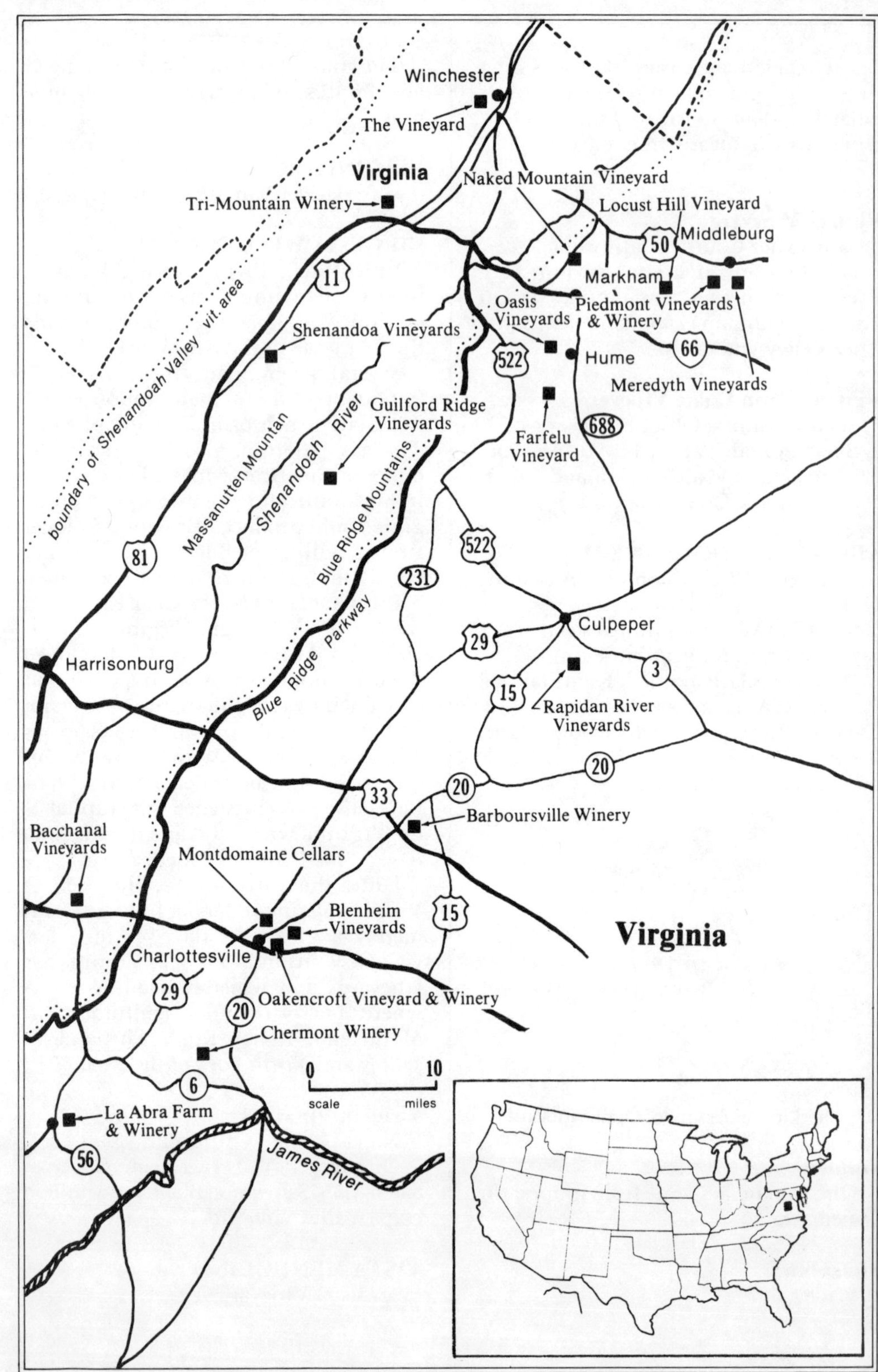

Wind machine for frost control

viticulture

The cultivation of the vine, also the science of grape production.

viticultural area

A grape growing region distinguished by unique characteristics such as climate, soil, elevation and/or other geographic features which distinguish it from surrounding areas. Designation as a viticultural area by the BATF permits wineries to use its designation in labeling and advertising. If BATF-approved, the name can appear when 75% of the wine comes from grapes grown in that area or state.

Vogensen Vineyard

Sonoma County, California.
Zinfandel vineyard.

Volker Eisle Vineyard

Napa County, California.
Zinfandel Chiles Valley vineyard.

Volkhardt Vineyards

Solano County, California.

Chateau de Leu estate Chardonnay, Sauvignon Blanc, French Colombard and De Leu Blanc Green Valley vineyard.

VON STIEHL WINERY

Founded 1961, Kewaunee County, Wisconsin.

Storage: Stainless steel. Cases per year: 6,000.

History: The winery is housed in a limestone and brick building that was built in 1853. Unique to the winery is the Von Stiehl aging wrap. Each bottle of wine is wrapped in gauze and covered with a mixture of paint, flour and water to form a cast which keeps the wine in complete darkness and allows it to age in the bottle.

Wines regularly bottled: Fruit wines: Natural Sweet Cherry, Dry Cherry, Sweet Apple, Dry Apple.

Vose Estate Vineyard

Napa County, California.

Chardonnay, Cabernet, Zinfandel vineyard.

VOSE VINEYARDS

Founded 1970, Napa, Napa County, California.

Storage: Oak and stainless steel. Cases per year: 15,000.

Estate vineyard: Vose Vineyard.

Wines regularly bottled: Estate bottled, vintage-dated Chardonnay, Cabernet Sauvignon, Zinfandel; vintage-dated Fumé Blanc, Gewurztraminer.

Second label: Redwood Ridge.

VUYLSTEKE, RON

Ron Vuylsteke's winemaking experience began in the 1960's when his wife, Marj, suggested making wine from a bountiful crop of blackberries. After much experimentation and self-education through books and short courses at the University of California, Davis, Vuylsteke founded Oak Knoll Winery in Oregon. His winemaking goal is to produce a wine for every taste and occasion. Oak Knoll produces both fruit and berry wines, as well as those made from vinifera grapes. Vuylsteke has worked closely with Oregon State University on research projects and has taught a course on fruit and berry winemaking at the University of California, Davis. He has also been active in professional wine organizations, including the American Society of Enologists and membership on the board of directors of the Oregon Winegrower's Assocation.

WAGNER, PHILIP
Philip Wagner began his career as a journalist. He was editor first of the Evening Sun, then The Sun, in Baltimore. An early interest in wine and viticulture prompted him, in 1935, to begin extensive experimentation with American, vinifera and French hybrid vines. He found the Maryland climate perfect for research: he could test for disease resistance, frost and hail resistance and damage from wind, birds, ripe rot and splitting—all at the same site. He gave the grapes only routine care. He was convinced that any varieties that did not survive were not suitable for growth in most East Coast regions. His rigorous experimentation led to his strong support of French hybrids for wine production east of the Rockies. He and his wife began a grape nursery with successful vines, then, in 1945, established a winery. He published *American Wines and Winemaking* in 1933, followed by *A Wine-Grower's Guide* in 1945. Both books have been recently revised, and are an important source for amateur and small commercial winemakers. Wagner has twice held the position of Regent's Lecturer at the University of California and was honored as an Officier du Merite Agricole by the French Government.

Wagner Estate Vineyards
Seneca County, New York.
Chardonnay, Johannisberg Riesling, Gewurztraminer, Seyval, Aurora, Cayuga White, Ravat, Delaware, DeChaunac and Rougeon Finger Lakes vineyard.

WAGNER VINEYARDS
Founded 1978, Lodi, Seneca County, New York.
Storage: Oak and stainless steel. Cases per year: 50,000.
History: Although the winery was started in 1978, Bill Wagner the owner, was no stranger to vineyards having been a grape grower for over 30 years. He is recognized as one of the leaders in Eastern Viticulture.
Estate vineyard: Wagner Vineyards.
Wines regularly bottled: Estate bottled, vintage-dated Seyval, Seyval Blanc, Aurora, Chardonnay, Johannisberg Riesling, Delaware, DeChaunac, Champagne, DeChaunac Lot #1 & Lot #2, Rougeon, Rougeon Light; also estate bottled DeChaunac Rosé, Sparkling Rosé; estate bottled proprietary: Alta B, Senlaka Rosé, Capital Red and White, Caywood Red; Red Wine; vintage-dated Port.

Wahl Vineyards
Yamhill County, Oregon.
Riesling Willamette Valley vineyard.

WAIT, FRONA EUNICE
In 1889, Wait wrote *Wines and Vines of California, or a Treatise on the Ethics of Wine Drinking*.

Walker Fruit Press
Lake County, New York.
Catawba and Dutchess Lake Erie vineyard.

Walker Valley Estate Vineyards
Ulster County, New York.
Riesling, Cayuga, Seyval, Rayon D'Or, Millot Hudson River Region vineyard.

WALKER VALLEY VINEYARDS
Founded 1978, Ulster County, New

York.

Storage: Oak and stainless steel. Cases per year: 2,000.

Estate vineyard: Walker Valley Vineyards.

Wines regularly bottled: Estate bottled, vintage-dated Riesling, Cayuga White, Seyval, Rayon D'Or, Leon Millot; vintage-dated Ravat, Chardonnay. Occasionally vintage-dated Pinot Noir, Marechal Foch, Chancellor Noir.

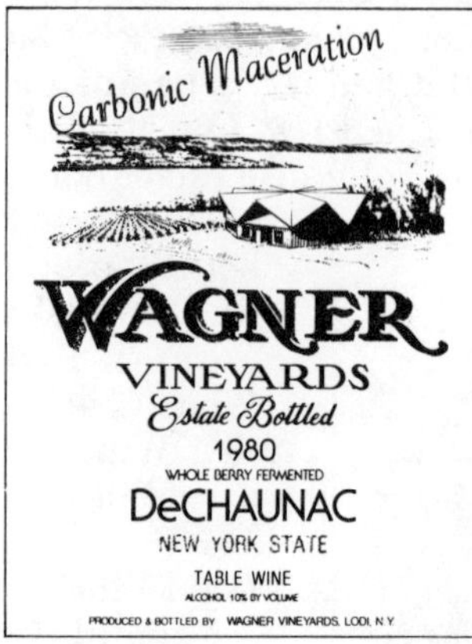

Walker Vineyard

El Dorado County, California.

Chenin Blanc and Zinfandel vineyard.

WALKER WINES

Founded 1979, Palo Alto, Santa Clara County, California.

Cases per year: 1,000.

Label indicating non-estate vineyard: Shell Creek Vineyards, Bowman Vineyards, Nyland Vineyards.

Wines regularly bottled: Vintage-dated Chardonnay, Petite Sirah, Gamay, Barbera, White Zinfandel. Occasionally Cabernet Sauvignon.

WALLA WALLA VALLEY VITICULTURAL AREA

A viticultural area in southeast Washington and northeast Oregon. Walla Walla Valley is approximately 178,560 acres and is bounded by three physical features: The Touchet Slope, Horse Heaven Ridge, and the Blue Mountains.

WALLACE, MIKE

Mike Wallace is a native of Seattle, Washington. After graduation from high school, Wallace joined the United States Air Force and was stationed at Hamilton Air Force Base in California. While there, he became interested in wine, and visited many of the nearby Napa and Sonoma wineries. Following his discharge, Wallace enrolled in Western Washington University, and received his bachelor's degree in biology and chemistry in 1968. For the next three years, he worked as a research technician in microwaves and physical medicine at the University of Washington. At the time, he viewed wine as an avocation. Wallace became professionally involved when he read a local newspaper article on vineyard research being conducted at Washington State University and he contacted Dr. Walter Clore at the research station in Prosser. Clore convinced him that fine wine grapes could be grown in the region. Wallace undertook a viticultural and enological education, first through reading, then as a graduate student at the University of California, Davis. In 1971, he returned to Washington, and planted 25 acres of grapes in the Yakima Valley, just north of Prosser. While the vines matured, Wallace worked as a research technician with Walter Clore. In 1976, Hinzerling Vineyards produced its first wine in a converted truck garage. In addition to his research work and winemaking, Wallace has been actively involved in the development of the Washington State wine industry. He is chairman of the board for the Washington Wine Institute, and was a founding member of the Enological Society of the Pacific Northwest. He has worked towards legislative changes beneficial to wine production, is vice-chairman of the Advisory Committee for the Washington State Department of Agriculture Wine Marketing Program, and is a member of the Wine and Winegrape Research Advisory Committee of the University of Washington. Wallace also established the enology and viticulture collection of the Prosser Library.

Walsh Vineyards
Napa County, California.
Cabernet Franc, Cabernet Sauvignon, Chardonnay, White Riesling, Muscat Canelli, Sauvignon Blanc and Zinfandel Napa Valley vineyard.

John Carl Warneke
Sonoma County, California.
Chardonnay Alexander Valley vineyard.

WARNELIUS VINEYARDS (See Vina Vista Vineyards.)

WARNER, JAMES J.
James J. Warner is the third generation in his family to be involved in winemaking and viticulture. After graduation from the Michigan State University in 1965, he founded Warner Vineyards and Warner Vineyards Supplies in Paw Paw, Michigan. He has served as president and director of both organizations since that time. In addition, he has held the position of vice-president and director of the First National Bank of Lawton, for 11 years, and most recently became a director of Peninsular Products. Warner has been active in many professional wine associations, including the American Society of Enologists and Les Amis du Vin. He has been a director of the American Association of Vintners, the Concord Grape Association and the Michigan Wine Institute.

Warner Estate Vineyards
Van Buren County, Michigan.
Seyval, Aurora, Vidal, Ravat, Chelois, Marechal Foch, Chancellor Noir, Chardonnay and Riesling Lake Michigan Shores vineyard.

WARNER VINEYARDS
Founded 1938, Paw Paw, Van Buren County, Michigan.
Storage: Stainless steel. Cases per year: 150,000.
History: Winery is now operated by third generation of the Warner family.
Estate vineyard: Warner Vineyards.

Wines regularly bottled: Estate bottled, vintage-dated Aurora, Chancellor, Riesling, Chardonnay; vintage-dated Brut Champagne, proprietary Pol Pereaux Brut, Extra Dry, Pink and Cold Duck Champagnes; and proprietary L'Aurore Superior, Jacks or Better, Jokers Wild; also produce Seyval, Currant; also bottle Spumante, Sherry, Port; and semi-generics: Mountain Chablis, Mountain Rosé, Mountain Burgundy, Rhine, Sauterne, Pink Chablis, Burgundy, Chablis.
Second label: Cork.

WARNER WEST
Cabernet Sauvignon Blanc produced by a partnership of Warner Vineyards, Michigan and Copenhagen Cellars, California.

WASHINGTON, GEORGE (1732-1799)
American statesman.

"My manner of living is plain and I do not mean to be put out of it. A glass of wine and a bit of mutton are always ready, and such as are content to partake of that are always welcome. Those who expect more will be disappointed."

WASHINGTON WINE COUNTRY
The first of the European or vinifera varieties were brought to Washington's

WASHINGTON MAP

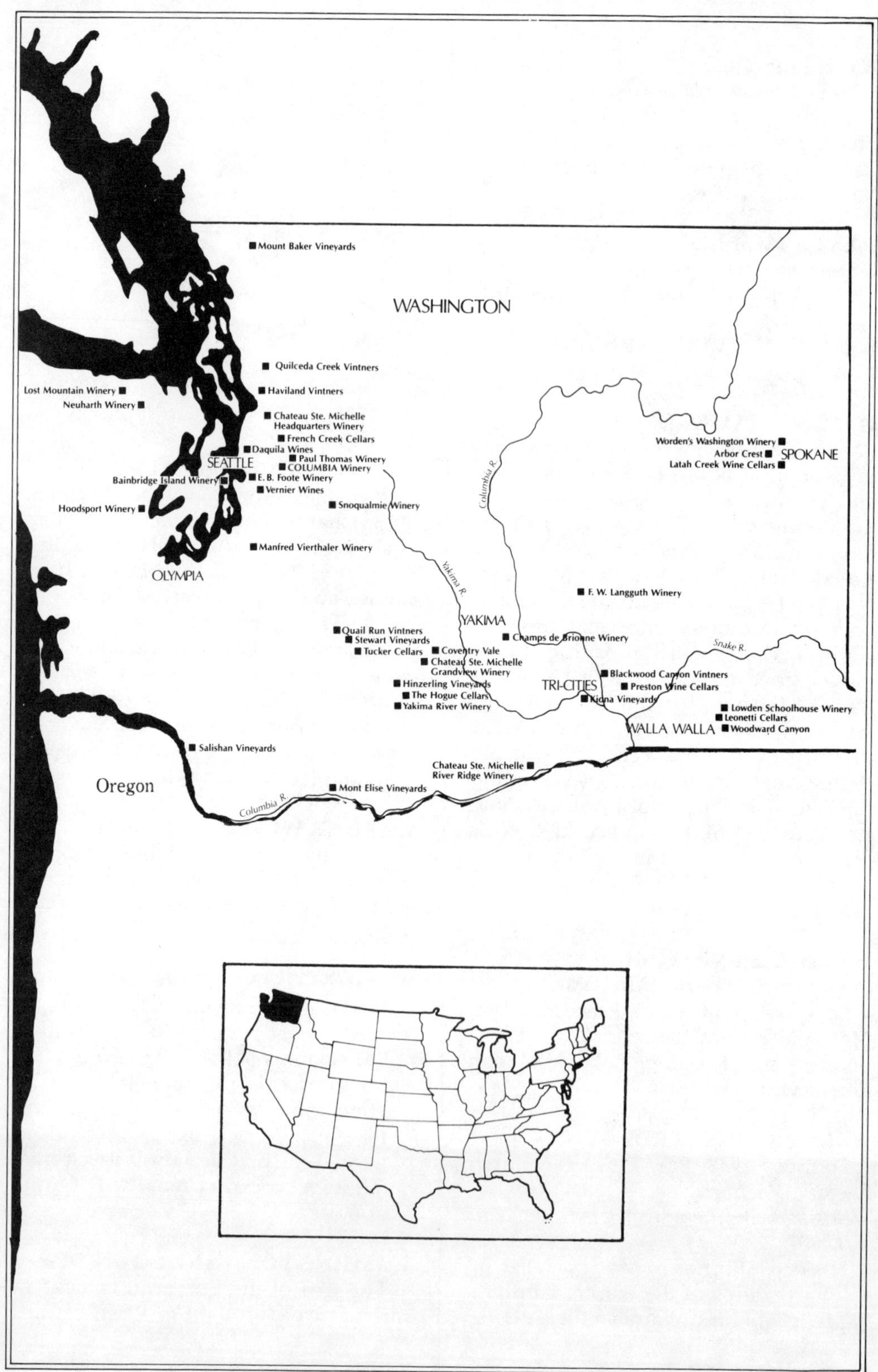

Columbia Basin around 1900, as the semi-desert was coming under irrigation. Due to limited demand, the earliest attempts at varietal wines weren't extremely successful, and most of Washington's winemakers were content producing ordinary dessert wines such as Port and Sherry. These could be made from the more common North American, of Labrusca, variety grapes that had long been established in the State.

However, in 1966, world renowned winemaker Andre Tchelistcheff sampled a homemade Gewurztraminer made from grapes grown in the Yakima Valley. Tchelistcheff was astounded, calling it the finest Gewurztraminer yet produced in the United States.

The next year, under Tchelistcheff's expert direction, the first of Washington's truly premium wines—Cabernet Sauvignon, Pinot Noir, Semillon and Grenache—were "challenging the quality supremacy of California varietal wines." Thus began Washington's reputation as a producer of fine varietal wines. In the years since, the size and number of Washington's wineries and vineyards have grown considerably.

Today there are over 8,000 acres, with an estimated 200,000 additional acres that are suitable for vinifera varieties.

Wasson Brothers Estate Vineyard

Clackamas County, Oregon.

Pinot Noir, Gewurztraminer, Chardonnay, Muscat vineyard.

WASSON BROTHERS WINERY

Founded 1981, Oregon City, Clackamas County, Oregon.

Storage: Oak and stainless steel. Cases per year: 2,000.

Estate vineyard: Wasson Vineyard.

Label indicating non-estate vineyard: La Gae Vineyard, Wertz Vineyard.

Wines regularly bottled: Vintage-dated Pinot Noir, Chardonnay, White Riesling and Red Table Wine; also Fruit wines: Rhubarb, Loganberry, Boysenberry, Plum, Bing Cherry, Blackberry.

Wasson Vineyard

Sonoma County, California.

Chardonnay vineyard.

WATERBROOK WINERY

Founded 1983, Lowden, Walla Walla County, Washington.

Storage: Oak and stainless steel. Cases per year: 1,000.

Estate vineyard: Gold Hill Vineyard to be planted in 1985.

Non-estate vineyard: Balcom and Moe Vineyard.

Wines regularly bottled: Vintage-dated Cabernet Sauvignon, Sauvignon Blanc and Merlot.

R.D. Watso Farms

Chautauqua County, New York.

Schloss Doepken estate Gewurztraminer, Concord, Johannisberg Riesling, Chardonnay vineyard.

Watson Estate Vineyard

San Luis Obispo County, California.

Johannisberg Riesling, Pinot Noir vineyard.

WATSON VINEYARDS

Founded 1981, Paso Robles, San Luis Obispo County, California.

Storage: Oak and stainless steel. Cases per year: 1,000.

History: The vineyards are part of the ranch that belonged to Ignace Paderewski the great pianist. Paderewski produced award winning Zinfandels around 1914.

Estate vineyard: Watson Vineyards, Mac Bride Vineyards.

Wines regularly bottled: Estate bottled, vintage-dated Johannisberg Riesling, Pinot Noir; vintage-dated Chardonnay.

Waverly Farm

Fauquier County, Virginia.

Seyval Blanc Appalachian Mountains vineyard.

Weather Vane Vineyards, Rocky Knob

Floyd County, Virginia.

Chateau Morrisette/Woolwine estate Seyval Blanc, Riesling, Chardonnay, Merlot, Gamay, Pinot Noir, Foch and Baco Noir vineyard.

R.W. WEBB WINERY
Founded 1980, Tucson, Pima County, Arizona.
Storage: Oak and stainless steel. Cases per year: 2,500.
Estate vineyard: B & W Vineyards.
Wines regularly bottled: Estate bottled, vintage-dated Cabernet Sauvignon, Arizona Chardonnay, Johannisberg Riesling, Arizona Zinfandel, Arizona Petite Sirah, Arizona French Colombard; vintage-dated Emerald Gold (Emerald Riesling), Arizona Blanc de Blanc.
Second label: Arizona Territory.

Weber Vineyards
Yamhill County, Oregon.
White Riesling, Pinot Noir, Chardonnay, Gewurztraminer Willamette Valley vineyard.

The Wedding Vineyard
Sonoma County, California.
Cabernet Sauvignon vineyard.

WEIBEL, FRED E.
Fred E. Weibel was born in Switzerland, attended the City College of Bern, and completed coursework in wine chemistry and enology in Altenburg, Thuringia, Germany, before coming to the United States. In 1945, Weibel and his father, Rudolf, purchased property in the Santa Clara Valley, at Mission San Jose, where they founded Weibel Champagne Vineyards. Weibel is president of the family operated winery. His son, Fred Jr., serves as executive vice-president/secretary-treasurer, daughter, Linda, works as plant operations manager, and daughter, Diana, is involved part-time in public relations. Weibel's wife, Marlene, manages the office and tasting rooms. Weibel became a citizen in 1963 and has been very active in community affairs. He is a member of many local educational and service organizations and has received many awards of recognition for his leadership. In addition, he is also a member of the Confrerie de la Chaine des Rotisseurs.

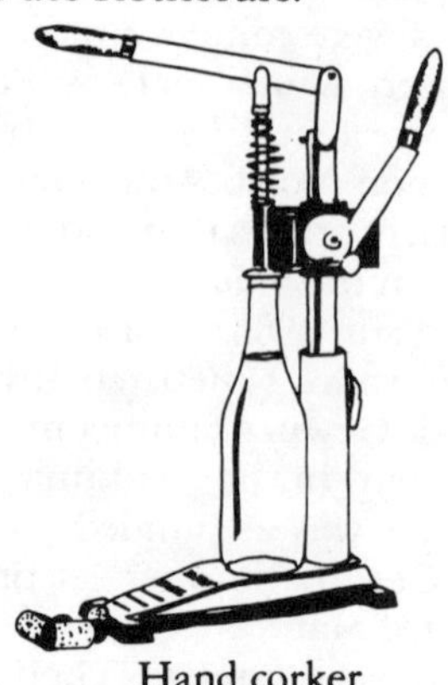

Hand corker

WEIBEL VINEYARDS
Founded 1869, Mission San Jose, Mendocino, Alameda Counties, California.
Storage: Oak and stainless steel. Cases per year: 1,000,000.
History: California's first governor and founder of Stanford University, Leland Stanford, established the winery in 1869. (See Stanford biography.)
Estate vineyard: Mission San Jose.
Wines regularly bottled: Estate bottled, vintage-dated Chardonnay, Pinot Noir; vintage-dated Chardonnay, Grey Riesling, Pinot Noir Blanc, White Cabernet Sauvignon, White Zinfandel, Johannisberg Riesling, Pinot Noir, Cabernet Sauvignon Reserve, Cabernet Sauvignon, Petite Sirah, non-vintage Green Hungarian; also generics: Chablis, Burgundy and Vin Rosé. Also dessert wines: Dry Bin Sherry, Medium Sherry, Amber Cream Sherry, Rare Port, Cream of Black Muscat; and Sparkling wines: Blanc de Noir, Stanford Champagne, Sparkling Green Hungarian, Crackling Rosé, Sparkling Burgundy; also Hoffberg May Wine and a proprietary Tangor; and Sweet and Dry Vermouth. Occasionally bottles Cabernet Sauvignon Rosé, White Pinot Noir and White Cabernet Sauvignon.

Second labels: Chateau Lafayette, Chateau Napoleon, Stanford.

Weidman Vineyard
Santa Clara County, California.
Zinfandel vineyard.

Weinbau Vineyards
Grant County, Washington.
F.W. Langguth estate Riesling, Muscat, Chardonnay and Gewurztraminer Columbia Valley vineyard.

WELDER
A native American white grape which is the only commercially grown Muscadine that originated in Florida. Produces a light-bodied wine with a distinct aroma.

well balanced
When the many odor and flavor substances of wines are present in quantities such that the concerted impression is pleasant, the wine is described as balanced.

WELSCH, WILLIAM
William Welsch was trained as a research chemist, and worked twenty-five years in the retail building supply business prior to becoming a viticulturist. A two year search for suitable land led him to western Michigan, where in 1973, he purchased 230 acres of continuous farms. Welsch and his son, Douglas, cleared the land and planted the Fenn Valley Vineyards—primarily Riesling, Chardonnay, Gewurztraminer and several different hybrid varieties of grapes. The winery was built in 1975. Until the vineyards became productive, Douglas, serving as winemaker, made wines from purchased grapes. When many of the hybrids proved unsatisfactory, they were replaced. Fenn Valley now specializes in fruity, German-style wines made from White Riesling, Vidal Blanc, Seyval Blanc, Foch, Chardonnay and Gewurztraminer. His goal is to produce the best wines possible in the difficult Michigan climate, recognizing that superior quality will mean lower yield.

WENTE BROS.
Founded 1883, Livermore, Alameda County, California.
Storage: Oak and stainless steel. Cases per year: 700,000.
History: (See Wente biography.)
Estate vineyard: Wente Bros. Vineyard.
Wines regularly bottled: Estate bottled, vintage-dated wines include Sauvignon Blanc, Dry Semillon, Petite Sirah, and proprietary Chateau Wente. Wines bearing the vintner grown designation from Monterey County, also vintage-dated, are Pinot Blanc, Pinot Noir, Johannisberg Riesling, Gewurztraminer, Arroyo Seco Riesling and Blanc de Noir. Wines with a California appellation, vintage-dated include Le Blanc de Blancs, Chablis, Rosé Wente, Grey Riesling, Pinot Chardonnay, Zinfandel and Gamay Beaujolais.

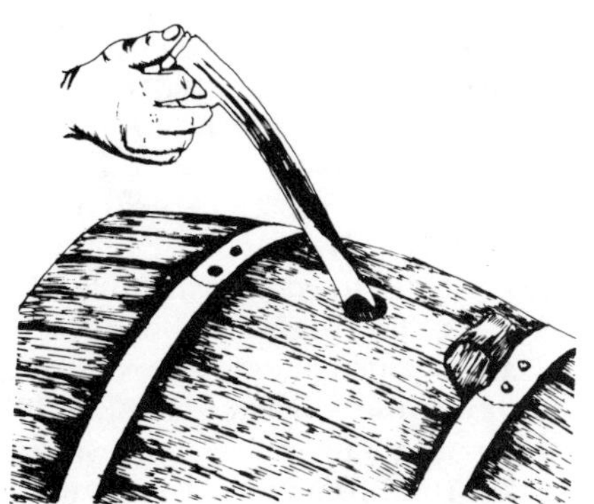

Extracting wine with a glass "wine thief"

Wente Bros. Estate Vineyard
Alameda County, California.
Sauvignon Blanc, Semillon, Petite Sirah Livermore Valley vineyard; Monterey County estate Pinot Blanc, Pinot Noir, Johannisberg Riesling vineyard.

WENTE, FAMILY HISTORY
Carl Heinrich Wente arrived in Northern California from Germany in the 1870's; he worked as a laborer and was hired by winemaker Charles Krug. He soon became cellarman, and developed both his winemaking skills and ambitions. He founded his own vineyards and winery in the

Livermore Valley. Two of his sons, Herman and Ernest, became the Wente Bros., Inc., and maintained the vineyards through Prohibition. After Repeal, they began to produce wines using a varietal label, an innovative departure from the European tradition of generic names in marketing their wines throughout the United States. In 1949, after receiving a bachelor's degree in microbiology from Stanford University, Ernest's son, Karl, joined the family business. Karl expanded the winery at the original Livermore site. His contributions to the science and art of winemaking enriched the wine industry and encouraged its growth in California and throughout the United States. Karl was instrumental in the development of pneumatic pruning shears for vineyard pruning, machine harvesting and field crushing of grapes for better juice quality, and he pioneered the use of centrifuges in winemaking. Karl managed the family winery until his death in 1977. Wente Bros. is now operated by the fourth generation: Eric, Phil and Carolyn, along with their mother, Jean. Eric obtained a master's degree in enology from the University of California, Davis, in 1974, and worked as winemaker for several years before becoming winery president in 1977. Phil also graduated from the University of California, Davis, in 1974, with a bachelor's degree in agricultural science and management. He manages the winery's vineyards and farms in the Livermore Valley and in Monterey County. Carolyn joined the brothers as vice-president in charge of public relations for Wente Bros. after graduating from Stanford University. Vineyard holdings and wine production have doubled under the direction of the younger Wentes, and they have developed such industry innovations as the "California bottle" design, and the tank press used for immediate treatment of juice for champagne cuvée.

WERMUTH WINERY

Founded 1982, Napa County, California.

Storage: Oak. Cases per year: 550.

Wines regularly bottled: Estate bottled, vintage-dated French Colombard, vintage-dated Sauvignon Blanc.

West Park Estate Vineyards

Ulster County, New York.

Chardonnay Hudson River Region vineyard.

WEST PARK VINEYARDS

Founded 1980, West Park, Ulster County, New York.

Storage: Oak and stainless steel. Cases per year: 10,000.

History: The winery was started by an interesting combination of four with a love for wine. Louis Fiore, president of Cybernetics and Kevin Zraly, wine director and consultant who are owners with Ms. O'Connell from the cosmetics industry and Velda Bennett, a horse woman.

Estate vineyard: West Park Vineyards.

Wines regularly bottled: Estate bottled, vintage-dated Chardonnay.

Westside Vineyards

Sonoma County, California.

Fenton Acres estate Chardonnay, Pinot Noir, Sauvignon Blanc, French Colombard and Gamay (Napa), Zinfandel vineyard.

WESTWIND WINERY

Founded 1983, Bernalillo, Sandoval County, New Mexico.

Cases per year: 5,000.

Wines regularly bottled: Vintage-dated Vidal Blanc, Chancellor, French Colombard, Ruby Cabernet, Zinfandel, Baco; also vintage-dated proprietary El Viejo, Ojo de Perdiz, Rio Grande Rojo. Occasionally bottles proprietary Lincoln County White.

WEST-WHITEHILL WINERY LTD.

Founded 1981, Keyser, Mineral County, West Virginia.

Storage: Stainless steel. Cases per year: 800.

Estate vineyard: Whitehills Vineyard.

Wines regularly bottled: Vintage-dated Aurora, Foch, Seyval Blanc, and proprietary Highland Red and Rosé.

WETZEL, HANK

Hank Wetzel was born in Dayton, Ohio on February 19, 1951. He was raised in Palos Verdes Estates, a suburb of Los Angeles. In 1963 his father purchased 240 acres of pastureland in Alexander Valley in Sonoma County. In 1964, plantings of premium wine grapes began. Hank received his bachelor's degree in fermentation science from the University of California, Davis, in 1974. While attending Davis he worked at several wineries in Sonoma and Napa counties. In 1975, Hank formed a partnership with his father, other family members and three friends, and began to produce his own wine under the label Alexander Valley Vineyards. Hank is a director and past-president of the sixty-four winery member Sonoma County Winegrowers. He is a member of the American Society of Enologists.

Whaler Estate Vineyard

Mendocino County, California.

Zinfandel vineyard.

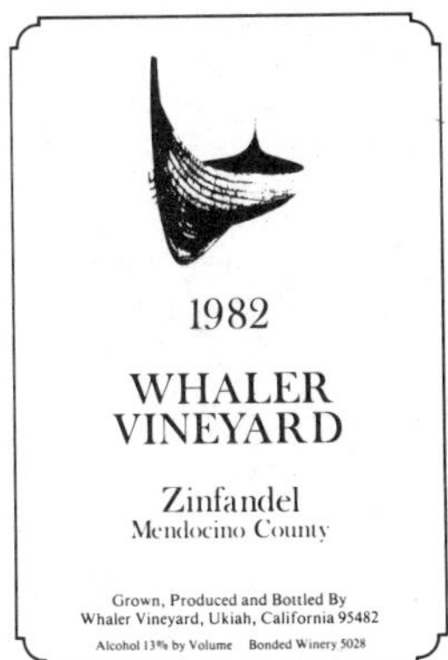

WHALER VINEYARD

Founded 1981, Ukiah, Mendocino County, California.

Storage: Oak. Cases per year: 3,000.

Estate vineyard: Whaler Vineyard.

Wines regularly bottled: Estate bottled, vintage-dated Zinfandel, White Zinfandel.

Wheeler Vineyards

Fentress County, Tennessee.

Highland Manor estate Alwood, Bath, Catawba, Concord, Delaware, Aurora, Baco Noir, Foch, DeChaunac and Cabernet Sauvignon vineyard.

WILLIAM WHEELER WINERY

Founded 1981, Healdsburg, Sonoma County, California.

Storage: Oak and stainless steel. Cases per year: 13,000-14,000.

Estate vineyard: Norse Vineyard.

Wines regularly bottled: Estate bottled, vintage-dated Cabernet Sauvignon; vintage-dated Chardonnay, Sauvignon Blanc. Occasionally Pinot Noir.

Second label: Healdsburg Wine Company.

Whiskey Canyon Vineyard

Yakima County, Washington.

Quail Run estate Riesling, Chardonnay, Gewurztraminer, Merlot, Cabernet Sauvignon, Semillon and Muscat Blanc Yakima Valley vineyard.

White Mountain Estate Vineyard

Belknap County, New Hampshire.

Marechal Foch, Aurora, Chelois, Diamond and Delaware vineyard.

WHITE MOUNTAIN VINEYARD

Founded 1972, Laconia, Belknap County, New Hampshire.

Cases per year: 120,000.

History: The winery overlooks the Belknap Mountain Range.

Estate vineyard: White Mountain Vineyard.

Wines regularly bottled: Estate bottled, proprietary semi-generics: Mont Blanc Chablis and Burgundy, Winnepesaukee Chablis and Burgundy, Mont Blanc Rosé, Winnepesaukee Rosé, Chablis, Burgundy and Rosé.

WHITE OAK VINEYARDS

Founded 1981, Healdsburg, Sonoma County, California.

Storage: Oak and stainless steel. Cases per year: 6,000.

Wines regularly bottled: Vintage-dated Chardonnay, Chenin Blanc, Sauvignon Blanc, Johannisberg Riesling, Zinfandel, Cabernet Sauvignon; a proprietary Fitch Mountain Red.

Riesling

WHITE RIESLING (Riesling)

A vinifera grape that is used in producing the great white German wines. The wines produced are fruity, delicate and range from medium dry to medium sweet. In California the name Johannisberg is often used.

white table wines

White dinner or table wines vary from extremely dry and tart to sweet and full-bodied. Their color ranges from pale straw to deep gold and their alcohol content from 10 to 14 percent. Most popular white dinner wines fall into three semi-generic types: Chablis, Rhine or Sauterne. The varietal white wines are all discussed under their varietal names.

White table wines—semi-generic: Chablis, Moselle, Rhine, Sauterne, Vino Bianco.

White table wines—varietal: Chardonnay (Pinot Chardonnay), Chenin Blanc, Emerald Riesling, French Colombard, Gewurztraminer, Pinot Blanc, Pinot Noir Blanc, Sauvignon Blanc, Semillon, Sylvaner, White Riesling (Johannisberg), Zinfandel Blanc, Seyval Blanc, Verdelet Blanc, Vidal Blanc, Villard Blanc.

WHITEHALL LANE WINERY

Founded 1980, St. Helena, Napa County, California.

Cases per year: 12,000.

Estate vineyard: Whitehall Lane Winery Vineyard.

Label indicating non-estate vineyard: Cerro Vista Vineyard; Serena Vineyard.

Wines regularly bottled: Estate bottled, vintage-dated Chardonnay, Sauvignon Blanc, Chenin Blanc; vintage-dated Chardonnay, Blanc de Pinot Noir, Sauvignon Blanc, Cabernet Sauvignon; a proprietary Fleur d'Helene; and White Table Wine. Occasionally vintage-dated Merlot.

Second label: Jacabels Cellars.

Whitehall Lane Winery Estate Vineyard

Napa County, California.

Chardonnay, Chenin Blanc, Sauvignon Blanc, Merlot, Cabernet Sauvignon vineyard.

Whitehills Vineyard

Mineval County, West Virginia.

West Whitehill estate Aurora, Foch, Chardonnay, Seyval Blanc vineyard.

Whitney's Vineyard

Sonoma County, California.

Fisher estate Chardonnay vineyard.

Don J. Wickham Vineyard

Schuyler County, New York.

Aurora, Baco Noir, DeChaunac, Seyval, Cayuga White, Ravat and Delaware Finger Lakes Region vineyard.

WICKHAM VINEYARDS LTD.

Founded 1981, Hector, Schuyler County, New York.

Storage: Stainless steel. Cases per year: 7,000-17,000.

History: Winery is owned and operated by three generations of the Wickham family who have been fruit farmers for over 100 years. Grape growing has been a specialty for the last 100 years.

Estate vineyard: Don J. Wickham Vineyard.

Label indicating non-estate vineyard: Mark Wagner Vineyard, Cameron Hosmer Vineyard, Plane's Cayuga Vineyard, Dick Peterson Vineyard, Venture Vineyards.

Wines regularly bottled: Vintage-dated Aurora, Baco Noir, Cayuga White, Ravat 51, Seyval Blanc, Johannisberg Riesling, Chardonnay. Occasionally vintage Delaware, Marechal Foch and Rosé.

Widmer Estate Vineyards

Ontario Country, New York.

Foch, Ravat, Vidal Blanc, Seyval Blanc and Cayuga White Finger Lakes vineyard.

WIDMER'S WINE CELLARS, INC.

Founded 1888, Naples, Ontario County, New York.

Storage: Oak and stainless steel. Cases per year: 380,000.

History: John Jacob Widmer went to Naples, New York, from Switzerland in 1882 at the suggestion of his brother who had preceded him to the Finger Lakes Region. In 1882 he bought his first plot. He cleared it and planted vines in the spring of 1883. John Jacob built his homestead in 1885 and started his winery in 1888. Until his death in 1930 he was active in the winery and vineyards.

Estate vineyard: Widmer Vineyards.

Wines regularly bottled: Pink Catawba, Cayuga White, Wood Age Foch; estate bottled, vintage-dated Vidal Blanc, Nouveau Foch; Sherry, Cream and Pale Dry Sherry; Extra Dry Champagne; semi-generics: Select American Burgundy, Sauterne, Haut Sauterne, Chablis Blanc, Rhine; proprietary Lake Niagara Light, Pink, Red, Light Champagne and Crackling Lake Niagara, Lake Roselle; also Port.

Second label: Naples Valley.

Wiedeman's Vineyard

Santa Clara County, California.

Cabernet Sauvignon and Zinfandel vineyard.

WIEDERKEHR, ALCUIN (AL)

Alcuin Wiederkehr's grandfather emigrated to Altus, Arkansas in the 1880's. He established a wine cellar and a tradition of winemaking in the Arkansas region. Although Al received his bachelor of science degree from the University of Notre Dame, he studied law at the University of Arkansas Law School and studied enology and viticulture at the University of California at Davis in 1960. In 1961 to 1962, he participated in the exchange student scholarship program with the Centre Nationale de Jeunes Agriculteurs in France. There, he studied winemaking at major research centers throughout the country, and lived with ten different winemaking families. Upon his return, he became chairman of the board and chief executive officer of Wiederkehr Wine Cellars, Inc. Under his direction, storage has increased and the three labels in the product line have expanded to 55 labels. He started an annual European-style grape festival and opened the winery to tours. Wiederkehr has been instrumental in the development of the Arkansas wine industry. He was a founding member of the Arkansas Grape Grower's Association, and the Arkansas Wine Producer's Association, of which he is president. Al has served on the board of directors of the Arkansas State Horticultural Society since 1966, and was president in 1966-1967. He founded *The Arkansas Horticulturist,* and has been involved in the passage of eleven bills, beneficial to all wines, in the state legislature. One of the most significant was the bill he authored that allowed wines to be served in restaurants. He is a Supreme Knight in the Brotherhood of the Knights of the Vine.

WIEDERKEHR WINE CELLARS

Founded 1880, Altus, Franklin County, Arkansas.

Storage: Oak and stainless steel.

History: Winery was founded by Swiss immigrant Johann Andreas Wiederkehr. His first wine cellar, now the Weinkeller Restaurant, is on the National Register of Historic Places.

Estate vineyard: Wiederkehr Vineyards.

Wines regularly bottled: Estate bottled, vintage-dated Johannisberg Riesling, Sauvignon Blanc, Rosé de Cabernet Sauvignon, Late Harvest Johannisberg Riesling, Johannisberg Riesling, Blanc de Noirs of Pinot Noir, Cabernet Sauvignon, Gewurztraminer, Pinot Chardonnay; also Cynthiana, Pink Catawba, Verdelet Blanc, Niagara; a proprietary Di Tanta Maria (vintage-dated); Cream Sherry, Dry Sherry, Cocktail Sherry, Sherry; semi-generics: Alpine Rosé, Burgundy, Chablis Blanc; Sparkling wines: Chateau du Monte Pink Champagne, Cold Duck, Chateau du Monte White Champagne, Hans Wiederkehr Extra Dry Champagne; also Vin Rosé Sec, Del Rosé, Edelweiss.

Wiederkehr Wine Cellars Estate Vineyards

Franklin County, Arkansas.

Johannisberg Riesling, Cabernet Sauvignon, Chardonnay, Pinot Noir, Sauvignon Blanc, Verdelet, Gewurztraminer, Cynthiana vineyard.

Hermann J. Wiemer Estate Vineyards

Yates County, New York.

Johannisberg Riesling, Chardonnay, Pinot Noir and Gewurztraminer Finger Lakes vineyard.

HERMANN J. WIEMER VINEYARD, INC.

Founded 1979, Dundee, Yates County, New York.

Storage: Oak and stainless steel. Cases per year: 4,000.

History: The vineyards are all vinifera plantings. Over 300,000 grafted vinifera vines are produced annually and are sold to leading vineyards throughout the United States. Born in West Germany, Hermann Wiemer has been doing grafting, nursery and vineyard work since the age of 10. Upon graduating from the Geisenheim Institute, he came to the United States in 1968.

Estate vineyard: Hermann J. Wiemer Vineyard.

Wines regularly bottled: Vintage-dated Johannisberg Riesling (Dry), Late Harvest

Johannisberg Riesling, Chardonnay; occasionally Gewurztraminer, Champagne and Trockenbeerenauslese.

WILD BOAR CELLARS
(See Sea Ridge Winery.)

Wildwood Vineyard
Sonoma County, California.
Cabernet Sauvignon, Muscat Canelli and Chardonnay Sonoma Valley vineyard.

WILLAMETTE VALLEY HOMESTEAD
(See The Honey House Winery.)

WILLAMETTE VALLEY VITICULTURAL AREA
In northwest Oregon, Willamette Valley is enclosed by natural boundaries—the Columbia River to the north, the Coast Range Mountains on the west, the Calapooya Mountains on the south, and the Cascade Mountains to the east.

Willard Farms Vineyard
Yakima County, Washington.
Quail Run estate Riesling, Chardonnay, Gewurztraminer, Cabernet Sauvignon, Muscat Blanc, Chenin Blanc Yakima Valley vineyard.

Willjen Hill Vineyard
Litchfield County, Connecticut.
Hopkins estate Ravat 51 (Vignoles), Leon Millot, Seyval Blanc and Aurora vineyard.

WILLOUGHBY WINERY
Founded 1935, Willowick, Lake County, Ohio.
Storage: Oak and stainless steel.
Wines regularly bottled: Pink Catawba, Vino D'Oro, Chancellor Noir, Delaware, Niagara, Concord; also Vin Rosé.

Willow Creek Vineyard
Amador County, California.
Argonaut estate Barbera Shenandoah Valley vineyard.

Willow Creek Vineyards
San Luis Obispo County, California.
Zinfandel vineyard.

WILLOW CREEK VITICULTURAL AREA
Willow Creek in California's Humboldt and Trinity counties is situated in and around the confluence of the Trinity River and the South Fork of the Trinity River, approximately 31 miles inland from the Pacific Ocean. The area surrounding the viticultural area is mountainous, at times rising sharply to high elevations. Very few grapes are grown.

Milk carton protects young vine from animals

Willow Run
Shenandoah County, Virginia.
Shenandoah estate Vidal, Seyval, Riesling, Chardonnay, Chambourcin, Cabernet Sauvignon, Pinot Noir Shenandoah Valley vineyard.

WILSON DANIELS CELLARS
Founded 1978, St. Helena, California.
Storage: Oak and stainless steel. Cases per year: 6,000.
Wines regularly bottled: Vintage-dated Sauvignon Blanc, Chardonnay, Cabernet Sauvignon.

Wilson Ranch
Sonoma County, California.
Sebastiani estate Johannisberg Riesling, Chardonnay and Pinot Noir Sonoma County vineyard.

Wind Hill Vineyard
Washington County, Oregon.
Pinot Noir, Riesling and Chardonnay Willamette Valley vineyard.

WINDSOR VINEYARDS
Founded 1959, Windsor, Sonoma County, California.
Storage: Oak.
History: This is a second label and mail order operation of Sonoma Vineyards, founded by Rod Strong and Peter Friedman at the same time and place as Sonoma Vineyards.
Estate vineyard: Windsor Vineyards.
Wines regularly bottled: Estate bottled, vintage-dated Johannisberg Riesling, Chardonnay, Pinot Noir, Cabernet Sauvignon, French Colombard, Merlot, Zinfandel; vintage-dated Gewurztraminer, Cabernet Sauvignon, Chardonnay, Chenin Blanc, Petite Sirah, Blanc de Noir, Grey Riesling, Rosé of Cabernet, Grenache Rosé, Sauvignon Blanc, Gamay Beaujolais; also bottle Cream and Pale Dry Sherry, Tawny Port, Brut Champagne, Red and White Table Wines, and proprietary Adequate Red, Adequate White; and semi-generics: Burgundy, Chablis and Vin Rosé.

WINDSOR VINEYARDS, INC.
(dba Great River Vineyards.)
Founded 1944, Marlboro, Ulster County, New York.
Storage: Stainless steel.
Wines regularly bottled: Vincent Noir, Baco Noir, Aurora, Seyval Blanc, and New York State Champagne.

wine
In the United States wine may be, and is, produced by the fermentation of any fruit or other agricultural product, resulting in an alcoholic content in excess of 7%. However, when used on a label without qualification it means that the contents have been made from 100% grapes. If a product other than grapes is used, the word "wine" must be qualified by such product, e.g. Apple Wine, Honey Wine, etc.

WINE AND THE PEOPLE
(See Berkeley Wine Cellars.)

Wine and the People
Lake County, Ohio.
Cabernet Sauvignon vineyard.

Wine Creek Vineyard
Sonoma County, California.
Belvedere estate Cabernet Sauvignon vineyard.

wine tasting and scoring
These are the quality factors and score weighting as recommended by University of California, Davis.

Appearance—This refers to clarity, not color. Wine should be free of cloud or sediment (unless it is very old red). "Unfiltered" wines don't often rate high in appearance. Give the wine 2 points if brilliant, only 1 if not crystal clear, and 0 if dull or cloudy.

Color—You need an idea of what color is right for each wine. Golden or amber is unacceptable for white or pink table wines, but exactly right for many dessert wines. Rosé should be distinctly pink, with only a suggestion of orange or red. Whites can be yellow, gold or straw color; flaws are amber or too "water white". Red can have violet tints if young, amber tints if aged. But brown is a flaw, as is too little red color. Perfect color for type: 2 points; reduce points if unacceptable tints or tones begin to predominate.

Aroma and Bouquet—Aroma may be a "varietal" or simply wine-like, "vinous". Intensity of aroma and bouquet may be light, medium or high. Give the wine 4 points for having a clean and unmistakeable aroma, 3 points for good varietal overtones, 2 for being pleasantly vinous, etc. Subtract points for off odors, such as moldy, mousey, corked, sulfur dioxide, sewer-like, excess wood, bacterial, carmel, raisiny, rubbery, etc.

Acescence—Volatile acidity (vinegar) is an index of aerobic bacterial spoilage in wine. If you smell no vinegar, give the wine 2 points; a faint vinegar aroma rates 1 point. If the smell is strong and interfers with the normal wine aroma and bouquet, give the wine 0.

Total Acid—Finally, for the first time, you taste the wine. With too high acidity, wine tastes excessively sharp or sour. Low acid makes the wine seem flat, flabby or even soapy. Rate the acidity level 0, 1 or 2 points.

Sugar—The taste of sugar (sweetness) and acid are evaluated together. Good balance: 1 point; too high or too low in sugar, 0.

Body—Evaluate the mouth-feel of the wine as it is swished about in the mouth. Too thin or too heavy are defects: 0 points. Good balance: 1 point.

Flavor—Is the flavor pleasant, and appropriate for the wine type? Does it correspond well with the smell? 2 points is perfect, 1 if the flavor is slightly marred by the off-tastes, and 0 if off tastes are distinct.

Astringency—This is the taste of tannin (sometimes bitter or mouth puckering) and it should be in balance with the other taste components. Score 0, 1 or 2, depending upon your judgment of the balance.

General Quality—If your overall impression is high, give it 2 points; average 1 point; poor 0 points.

20 is maximum, 17-20 is an outstanding wine, 13-16 sound commercial wine, below 9 is commercially unacceptable wine.

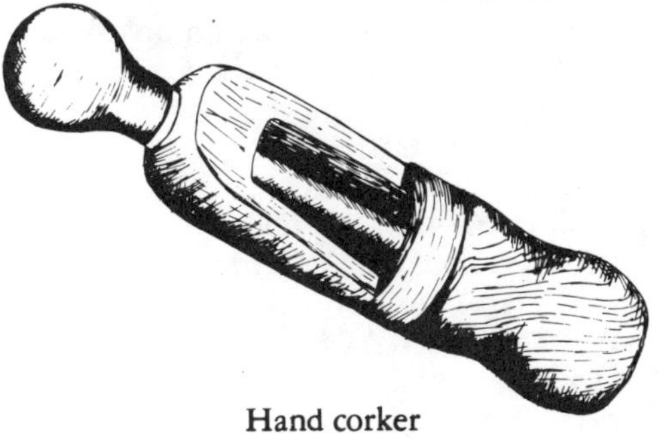

Hand corker

wine thief

A long glass tube used to extract samples of wine from the barrel.

WINEMASTER'S SELECTION (See Calona Wines, Canada.)

WINERY OF THE LITTLE HILLS

Founded 1860, St. Charles, St. Charles County, Missouri.

Storage: Oak and stainless steel. Cases per year: 2,000.

History: The winery was originally called the Wepprich Winery and is one of Missouri's historic wineries.

Estate vineyard: Hercules Vineyard.

Wines regularly bottled: Estate bottled Seyval Blanc; also Villard Blanc, Vidal Blanc, DeChaunac, Chelois, Chancellor, Concord; and Missouri Valley White.

WINERY OF THE ROSES (See California Meadery.)

Produce Chardonnay, Cabernet Sauvignon and Champagne.

Winery Lake Vineyard

Napa County, California.

Pinot Noir, Chardonnay, Merlot, Cabernet Sauvignon, Johannisberg Riesling and Cabernet Franc Carneros vineyard.

THE WINERY RUSHING

Founded 1977, Merigold, Boliver County, Mississippi.

Storage: Stainless steel. Cases per year: 6,000.

History: On September 16, 1977, Mississippi history was made when the winery was issued Bonded Winery Permit Number One, thus establishing Mississippi's first winery since Prohibition.

Estate vineyard: Rushing Vineyard.

Wines regularly bottled: Estate bottled, vintage-dated White Muscadine; also Noble, Carlos, Red, Sweet White, Rosé.

WINIARSKI, WARREN

Although the name Winiarski means "son of the wine (or vine)", Warren

Winiarski began his career as an academician. He obtained a master's degree in the study of Italian Renaissance political theory at the University of Chicago, and planned to earn a doctorate. An interest in wine, however, prompted him to begin making wine at home. In 1964, he moved his family to California to pursue a winemaking education. Winiarski apprenticed first with Lee Stewart at Souverain Cellars, then with Robert Mondavi, from 1966 to 1968. He began work as a winemaking consultant, while searching for his own vineyard property. In 1970, he purchased the site, that was to become Stag's Leap Wine Cellars, in the Napa Valley, where he planted forty-one acres of Cabernet Sauvignon and Merlot. The winery was completed in 1973. Winiarski admired the winemaking style of Andre Tchelistcheff, and Stag's Leap is best known for the production of graceful, elegant Cabernet Sauvignon.

W.I.N.O.
Wine Investigation for Novices and Oenophiles: America's largest consumer wine organization; founded by Jerry D. and Linda Mead in 1969, Orange County California. Now over 50 chapters and over 3,000 regular members; 100s of wine events annually. Membership information: W.I.N.O., Box 7244, San Francisco Ca 94120

WINTER CREEK
(See Caswell Vineyards.)

Winter Creek
Sonoma County, California.
Caswell estate Zinfandel and Pinot Noir Green Valley vineyard.

WINTERS WINERY
Founded 1980, Winters, Yolo County, California.
Storage: Oak and stainless steel.
Label indicating non-estate vineyard: Gordon Ranch Vineyards; Naismith Vineyards, Thompson Vineyards, Brown Vineyards.
Wines regularly bottled: Vintage-dated Sauvignon Blanc, Chenin Blanc, Petite Sirah, Zinfandel, Pinot Noir. Occasionally bottles Gamay.

Wirtz Vineyard
Washington County, Oregon.
Gewurztraminer, Pinot Noir, White Riesling Willamette Valley vineyard.

WISCONSIN WINERY
Founded 1979, Lake Geneva, Wallworth County, Wisconsin.
Storage: Oak.
Wines regularly bottled: Fruit and Berry Wines: Pear, Plum, Apricot, Strawberry, Cranberry, Raspberry, Elderberry.

WITTWER WINERY
Founded 1969, Eureka, Humboldt County, California.
Storage: Oak and stainless steel. Cases per year: 2,100.
Label indicating non-estate vineyard: Conn Ranch, Parducci Vineyards.
Wines regularly bottled: Cabernet Sauvignon, Chardonnay.

Wolf Hill Vineyard
Montgomery County, Maryland.
Catoctin estate Chardonnay, Cabernet Sauvignon, Riesling, Sauvignon Blanc, Seyval, Vidal, Chambourcin, Chancellor vineyard.

Wolf Creek Vineyards
Illinois County, Illinois.
Foch, Seyval and Villard vineyard.

WOLFE, DR. WADE
Dr. Wade Wolfe is director of agriculture operations of Chateau Ste. Michelle. Wolfe

joined the winery staff as its viticulturist in 1978, advancing to his present position in June, 1983. He earned a bachelor of science in biochemistry in 1971 from the University of California at Davis. Wolfe's doctoral research focused on grape genetics. He subsequently served as viticulturist and enologist for a wine grape feasibility study funded by the Four-Corners Regional Commission to explore the potential for wine production in four southwestern states. In 1978 his exploration of emerging wine producing areas led Wolfe to Washington and Chateau Ste. Michelle. His educational background and viticultural experience have earned him a courtesy faculty appointment at Washington State University, the State's agricultural land grant university. He is also a professional member of the American Society of Enologists and Viticulturists.

WOLFSKILL, WILLIAM
Planted 40 acre vineyard in Los Angeles in 1838.

WOLLERSHEIM WINERY, INC.
Founded 1847, Prairie Du Sac, Dane County, Wisconsin.
Storage: Oak and stainless steel. Cases per year: 6,000.
History: The only winery in Wisconsin that grows its own grapes. The vineyard site was selected by European vintners to grow white grapes over 125 years ago. Agoston Haraszthy originally planted vineyards on the same site in the 1840's.
Estate vineyard: Sugarloaf Hill Vineyard, Wollersheim Vineyard.
Label indicating non-estate vineyard: Wolfcreek Vineyards, Terra Vineyards.
Wines regularly bottled: Estate bottled, vintage-dated proprietary Sugarloaf White, Domaine Reserve, Domaine du Sac; vintage-dated Dry Ravat, Foch Reserve, Seyval Blanc Reserve; also Baco Noir, Saukskeller; proprietary River Gold, Harvest Gold, Illinois Red and White. Occasionally Seyval Blanc Ice Wine.
Second label: River Country.

WOOD HILL CELLARS (See Glen Creek Winery.)

Woodbury Estate Vineyards
Chautauqua County, New York.
Chardonnay, Riesling, Gewurztraminer, Cabernet Sauvignon, Pinot Noir, Merlot, Niagara, DeChaunac and Catawba vineyard.

Grapes picked by hand

WOODBURY VINEYARDS
Founded 1971, Dunkirk, Chautauqua County, New York.
Storage: Oak and stainless steel. Cases per year: 8,400.
Estate vineyard: Woodbury Vineyards.
Label indicating non-estate vineyard: Falkenhausen Farms.
Wines regularly bottled: Estate bottled, vintage-dated Chardonnay; vintage-dated Chardonnay, Aurora Blanc, Dutchess, Riesling, Seyval Blanc, Marechal Foch, DeChaunac, Late Harvest Chardonnay; also Chardonnay, Blanc de Blancs, Extra Dry Champagne, Spumante; and proprietary Glacier Ridge White, Red and Rosé.

WOODBURY WINERY
Founded 1977, San Rafael, Marin County, California.
Storage: Oak and stainless steel.
Wines regularly bottled: Vintage Port; and pot distilled Brandy.

Wooden Valley Estate Farms
Solano County, California.

Pinot Chardonnay, French Colombard, Johannisberg Riesling, Green Hungarian, Gewurztraminer, Sauvignon Blanc, Chenin Blanc, Gamay Beaujolais, Cabernet Sauvignon, Zinfandel, Riesling, Pinot Noir, Napa Gamay Windsor Valley vineyard.

WOODEN VALLEY WINERY

Founded 1932, Suisun, Solano County, California.

Storage: Oak and stainless steel. Cases per year: 85,000.

Estate vineyard: Wooden Valley Farms.

Wines regularly bottled: Table and Dessert Wines.

WOODHALL CELLARS

Founded 1983, Sparks, Baltimore County, Maryland.

Storage: Oak. Cases per year: 800.

Estate vineyard: Woodhall Vineyards.

Wines regularly bottled: Seyval Blanc, Vidal Blanc, Chardonnay, Riesling, Cabernet Sauvignon; also Rosé and Red Blend.

Woodhall Estate Vineyards

Baltimore County, Maryland.

Seyval Blanc, Vidal Blanc, Foch, Chambourcin, Chardonnay, Riesling and Cabernet Sauvignon vineyard.

The graceful "Tulip" glass

WOODS, FRANK M.

Frank M. Woods did not begin his career in the wine industry until he had spent many years in advertising and marketing. After graduating from Cornell University's School of Hotel Administration, Woods accepted a job in advertising with Proctor and Gamble. He began his own marketing firm, then several successful years later, formed a partnership with Thomas C. Reed and Dennis Malone, fellow Cornell alumni. They purchased several hundred acres of land in the Sonoma Valley, chosen to suit the varietal to be grown. The vineyards were planted, and for many years, the grapes were sold to nearby wineries. Woods continued his formal education during this period, studying at the University of California, Davis, School of Enology and Viticulture. In the mid-1970's, he and his partners invested in a winery operation. Wines produced from their well-established vineyards were released under the Clos Du Bois label. Woods currently serves as president and marketing director of Clos Du Bois and is actively involved in many wine industry organizations. These include the California Wine Institute, the Sonoma County Wine Grower's Association, the Brotherhood of the Knights of the Vine, Commanderie de Bordeaux, the Vintners Club, Les Amis du Vin, W.I.N.O., and the American Society of Wine Educators.

WOODSIDE VINEYARDS

Founded 1960, Woodside, San Mateo County, California.

Storage: Oak.

Wines regularly bottled: Vintage-dated Cabernet Sauvignon, Pinot Noir, Chardonnay.

WOODWARD CANYON WINERY

Founded 1981, Lowden, Walla Walla County, Washington.

Storage: Oak and stainless steel. Cases per year: 2,000.

Estate vineyard: Woodward Canyon Winery.

Label indicating non-estate vineyard: Sagemoor Farms.

Wines regularly bottled: Estate bottled, vintage-dated Chardonnay; vintage-dated Chardonnay, Cabernet Sauvignon, White Riesling.

Woodward Canyon Winery Estate Vineyard

Walla Walla County, Washington.

Pinot Noir and Chardonnay vineyard.

woody

Tasting term.

The characteristic odor of "wet oak" is apparent in wine aged too long or in faulty wood. The term comes into use when this characteristic is excessive.

WORDEN'S WASHINGTON WINERY

Founded 1979, Spokane, Spokane County, Washington.

Storage: Oak and stainless steel. Cases per year: 15,000.

Label indicating non-estate vineyard: Moreman Vineyards.

Wines regularly bottled: Vintage-dated Johannisberg Riesling, Gewurztraminer, Sauvignon Blanc, Chardonnay, Gamay Beaujolais Rosé, Cabernet Sauvignon.

WRIGHT, JOHN H.

Following graduation from Wesleyan College, with a degree in chemistry, John H. Wright became interested in fine wines while stationed in the Rheingau and Moselle regions of Germany with the United States Army. Upon his return to the United States, Wright began a career as a market development specialist, first with the American Viscose Company, then with Arthur D. Little, Inc. During this time, he continued his wine education through home winemaking, experimental vineyards and bottling of barreled wines from the Burgundy region of France. In 1972, he convinced Arthur D. Little to let him undertake a marketing study on the future of wine in America. The report came to the attention of the Banque Nationale de Paris and Moet-Hennessy. In 1973, Moet decided to enter the American wine market, and asked Wright to be president and chairman of the board of the new company, M & H Vineyards. The winery, specializing in the production of sparkling wines, was completed in 1977 and named Domaine Chandon. Wright, who still considers himself an "enthusiastic amateur", continues leadership of the winery to this day. He is a Supreme Knight in the Brotherhood of the Knights of the Vine.

WYANDOTTE WINE CELLARS

Founded 1977, Gahanna, Franklin County, Ohio.

Storage: Oak and stainless steel.

Wines regularly bottled: Niagara, Pink Catawba, Concord, Delaware, Vidal. Occasionally bottle Berry wines, Dandelion, Rhubarb.

YAKIMA RIVER WINERY

Founded 1978, Prosser, Benton County, Washington.

Storage: Oak and stainless steel. Cases per year: 4,000.

Label indicating non-estate vineyard: Ciel du Cheval Vineyard; Fairacre Nursery; Mercer Ranches.

Wines regularly bottled: Vintage-dated Chenin Blanc, Johannisberg Riesling, Late Harvest White Riesling, Muscat de Frontignan, Gewurztraminer, Gewurztraminer Dry Berry Selection, Cabernet Sauvignon, Merlot, Chardonnay, Gewurztraminer Ice Wine; also proprietary Valley Rosé, Valley White.

YAKIMA VALLEY, WASHINGTON VITICULTURAL AREA

Yakima Valley in south central Washington. The boundaries of the Yakima Valley viticultural area are the mountain ranges surrounding the valley. The boundary follows the crest of the Ahtanum Ridge and the Rattlesnake Hills on the north, crosses the top of Rattlesnake Mountain, Red Mountain, and Badger Mountain on the east, and follows the contour line of Horse Heaven Hills and the crest of the Toppenish Ridge on the south. The western boundary is the lower foothills of the Cascade Mountains.

Yakima is the name of the Yakima Nation, a loose confederacy of Indian tribes which once controlled a vast portion of eastern Washington.

Although Yakima Valley only recently has become recognized as a wine producing region, it has been known as an important agricultural region since the early 1900's when river water was first used to irrigate the valley.

yeast

A micro-organism that ferments sugars into alcohol. Often found on the bloom of grapes.

yeasty

Tasting term.

Wines containing materials which have a flavor or smell of yeast are described as yeasty.

YERBA BUENA

Founded 1977, San Francisco, San Francisco County, California.

Wines regularly bottled: Pinot Noir, Red, White, Rosé, Gewurztraminer.

Yonne Vineyard

Napa County, California.

Sauvignon Blanc and Chardonnay vineyard.

York Creek Vineyard

Napa County, California.

Petite Sirah, Zinfandel, and Cabernet Sauvignon Spring Mountain vineyard.

York Mountain Estate Vineyard

San Luis Obispo County, California.

Pinot Noir, Cabernet Sauvignon, Chardonnay, Zinfandel vineyard.

YORK MOUNTAIN VITICULTURAL AREA

The area is located in the western part of San Luis Obispo County approximately seven miles from the Pacific Ocean. There are about 300 acres of vineyards scattered throughout the proposed 10,000 acres of coastal mountain terrain.

YORK MOUNTAIN WINERY

Founded 1882, Templeton, San Luis Obispo County, California.

Storage: Oak and stainless steel. Cases per year: 3,000.

Estate vineyard: York Mountain Vineyard.

Label indicating non-estate vineyard: McBride Vineyard, Watson Vineyard, Bailey Vineyards.

Wines regularly bottled: Estate bottled, vintage-dated Pinot Noir, Cabernet Sauvignon, Chardonnay, Zinfandel; vintage-dated Chardonnay, Merlot, Johannisberg Riesling; and Dry Sherry, Red and White Wine. Occasionally bottles Chenin Blanc.

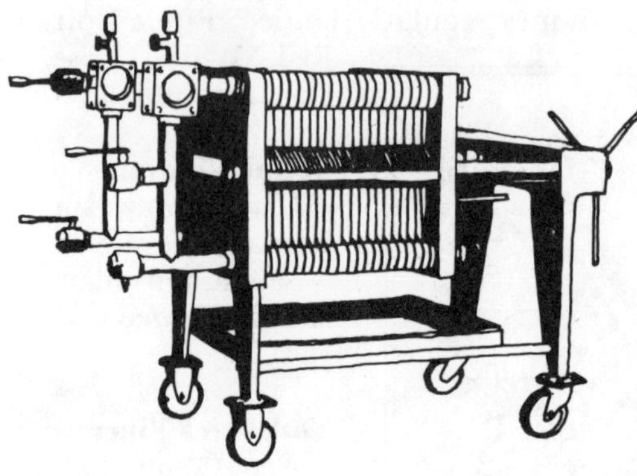

A plate filter is used to improve the clarity of wine

YORK SPRINGS

Founded 1978, York Springs, Adams County, Pennsylvania.

Storage: Oak and stainless steel. Cases per year: 2,000.

History: The vineyards are on the eastern slopes of the Appalachian Mountains in what is called the Piedmont area.

Estate vineyard: York Springs Vineyard.

Wines regularly bottled: Estate bottled, vintage-dated Cabernet Sauvignon, Chardonnay, Johannisberg Riesling, SV 5276, Marechal Foch; also estate bottled Rouge, Susquehanna; and Strawberry.

Larry Young's Fruit Farm

North East County, Pennsylvania.

Seyval, Vidal, Barrat, DeChaunac, Chelois, Chancellor, Foch vineyards.

Robert Young Vineyards

Sonoma County, California.

Sauvignon Blanc, Chardonnay, Gewurztraminer, Johannisberg Riesling, Pinot Blanc and Merlot Alexander Valley vineyard.

Yountville Island

Napa County, California.

Inglenook estate Semillon, Sauvignon Blanc, Chardonnay, Gewurztraminer Napa Valley vineyard.

Yountville Ranch

Napa County, California.

Inglenook estate Chardonnay, Gamay Beaujolais, Pinot Noir, Chenin Blanc and Grey Riesling vineyard.

Yountville Vineyard

Napa County, California.

Beringer estate Grignolino, Chardonnay, Sauvignon Blanc and Johannisberg Riesling vineyard.

YVERDON VINEYARDS

Founded 1970, St. Helena, Napa County, California.

Wines regularly bottled: Vintage-dated Cabernet Sauvignon, Gamay, Chenin Blanc, Johannisberg Riesling, Gewurztraminer.

Zaca Mesa Estate Ranch/Zaca Vineyards
Santa Barbara County, California.
Chardonnay, Riesling, Sauvignon Blanc, Semillon, Cabernet Sauvignon, Merlot, Pinot Noir, Zinfandel and Syrah Santa Ynez Valley vineyard.

ZACA MESA WINERY
Founded 1978, Los Olivos, Santa Barbara County, California.
Cases per year: 90,000.
Estate vineyard: Zaca Mesa Ranch.
Label indicating non-estate vineyard: Tepusquet Vineyard, Rowan Vineyard, Los Alamos Vineyard, Rancho Tierra Rejada, Douglas Vineyard, Rancho Sisquoc.
Wines regularly bottled: Chardonnay, Sauvignon Blanc, Johannisberg Riesling, Cabernet Sauvignon, Zinfandel, Pinot Noir; and proprietary Toyon Blanc, Toyon Noir.

ZD WINES
Founded 1969, Napa, Napa County, California.
Storage: Oak. Cases per year: 10,000.
Label indicating non-estate vineyard: Tepusquet Vineyard, Pickle Canyon Vineyard.
Wines regularly bottled: Vintage-dated Chardonnay, Pinot Noir, Cabernet Sauvignon, Zinfandel.

STEPHEN ZELLERBACH VINEYARD
Founded 1970, Geyserville, Sonoma County, California.
Storage: Oak and stainless steel. Cases per year: 22,000.
Estate vineyard: Elysian Vineyards, Stephen Zellerbach Vineyard.
Label indicating non-estate vineyard: John Carl Warneke Vineyards.

Wines regularly bottled: Vintage-dated Cabernet Sauvignon, Merlot, Chardonnay.

Ziem Estate Vineyards
Washington County, Maryland.
Chancellor, Chelois, Marechal Foch, Seyval, Dutchess and Vidal Blanc vineyard.

ZIEM VINEYARDS
Founded 1972, Fairplay, Washington County, Maryland.
Cases per year: 1,000.
Estate vineyard: Ziem Vineyards.
Wines regularly bottled: Estate bottled Chancellor, Chelois, Marechal Foch, Seyval Blanc, Dutchess, Vidal Blanc.

ZINFANDEL
Wine grape.
A grape that produces three styles of wine: Light—spicy flavor with a berry-like aroma and a tang. It should be consumed in one to three years. Oak aged—intense, berry-like and spicy aroma, full bodied with some tannin and dark in color. Should be aged 3-8 years. Late Harvest—rich, dark, full-bodied, lots of alcohol and tannin. May be aged 8 to 16 years.

Zinfandel

Dutchess

1979 to 1983
EASTERN WINE COMPETITIONS

Wineries eligible to enter the competition were those who produce wines from grapes grown east of the Rocky Mountains in the United States and Canada. The Eastern Wine Competition is the largest and most competitive test for wineries in this region. Judges for the Eastern Wine Competition are members of the wine retail and wholesale industry, restauranteurs, wine researchers and winemakers. Medals are awarded according to an anchored point system and are the result of consensus of the judging panel. Large classes are first screened by the panels to eliminate non-competitive entries and then judged finally for medal. All wines are blind tasted in random order by a minimum of five judges. In 1979, only 82 wineries entered. For 1983, 106 wineries entered a total of 675 wines.

The Eastern Wine Competition is sponsored annually by Eastern Grape Grower and Winery News. Competition Chairman is Dean Thomas Lee Hayes, a former winemaker who is currently a consulting Episcopal minister. Proceeds of the Eastern Wine Competition are donated to the Eastern Section of the American Society of Enologists for award to graduate students pursuing career training in enology and viticulture.

The following is a summary of the judging results of the five competitions from 1979 to 1983. The results for each variety or wine type are grouped together so that you may gather a five year perspective of the best wines judged in any single category. There are a total of 13 different groups and the respective varieties or types are each presented alphabetically within the group.

The wine groups are:

I VINIFERA VARIETALS from Aligote to Semillon.

II NATIVE AMERICAN VARIETALS from Carlos to Scuppernong.

III FRENCH-AMERICAN HYBRID VARIETALS from Aurora to Villard Noir.

IV NATIVE AMERICAN GENERICS.

V FRENCH-AMERICAN HYBRID GENERICS.

VI NATIVE AMERICAN SPARKLING WINES.

VII FRENCH-AMERICAN HYBRID AND VINIFERA SPARKLING WINES.

VIII SPECIALTY SPARKLING WINES.

IX FRUIT WINES from Apple to Mead.

X APERITIF AND FLAVORED SPECIALTY WINES.

XI SHERRY.

XII PORT.

XIII MISCELLANEOUS.

This is the largest summary of ratings of Eastern American Wines available; over 3,300 wines were evaluated for these five years of judgings.

Note: Absence of listing indicates no medal awarded.

LIST OF JUDGES

THE EASTERN WINE COMPETITION 1983

John Chambers, *wine retailer, Peter Morell & Company, New York, NY*

Scott Cook, *restauranteur, Pierce's 1894 Restaurant, Elmira Heights, NY*
Tom Cottrell, *wine researcher, Cornell University, Geneva, NY*
Karen Dumas, *restauranteur, The Greystone Restaurant, Ithaca, NY*
Larry Fuller-Perrine, *wine researcher, Cornell University, Geneva, NY*
James Gallander, *wine researcher, OARDC, Wooster, OH*
Jim Gifford, *winemaker, Gold Seal Vineyards, Hammondsport, NY*
Craig Goldwyn, *president, Beverage Testing Institute, Ithaca, NY*
Vicki Gray, *wine researcher, HRIO, Vineland, Ontario, Canada*
Julie Green, *winemaker, Andres Wines Ltd., Winona, Ontario, Canada*
Patty Herron, *winemaker, The Taylor Wine Company, Hammondsport, NY*
Jim Holsing, *wine consultant, Longmeadow, MA*
William Lamberton, *winemaker, Wickham Vineyards, Hector, NY*
Chuck Mara, *wine retailer, Green Hills Wine & Liquor, Syracuse, NY*
Douglas Moorhead, *winemaker, Presque Isle Wine Cellars, Northeast, PA*
Roger del Nero, *wine retailer, Barbara's World, Albany, NY*
C. Joseph Pierce, *restauranteur, Pierce's 1894 Restaurant, Elmira Heights, NY*
Christina Reynolds, *wine retailer, Northside Wine & Liquors, Ithaca, NY*
Dan Robinson, *winemaker, Widmer's Wine Cellars, Naples, NY*
Jim Rossback, *wine wholesaler, Chateau & Estates Wines, New York, NY*
Alex Sebastian, *restauranteur, The Wooden Angel, Beaver, PA*
Robert Simonelli, *wine retailer, The World of Wine & Spirits, Suffern, NY*
Archie Smith III, *winemaker, Meredyth Vineyards, Middleburg, VA*
Richard Vine, *wine researcher, A.B. McKay Lab, Mississippi State University*
Philip Wagner, *winemaker (ret.), Boordy Nursery, Riderwood, MD*
Bruce Zoecklin, *wine researcher, University of Missouri, Columbia, MO*

I VINIFERA VARIETALS

ALIGOTE, *1980 Rating*
(10 entries)
BRONZE
Aligote/Vinifera Cellars, NY

CABERNET SAUVIGNON, *1979 Rating*
(5 entries)
BEST OF CLASS
Cabernet Sauvignon, Wiederkehr Wine Cellars, Altus, AR
(No medals awarded in class.)

CABERNET SAUVIGNON, *1980 Rating*
(5 entries)
GOLD
Cabernet Sauvignon 1978, Wiederkehr Wine Cellars, AR
SILVER
Cabernet Sauvignon 1978, Markko Vineyard, OH
BRONZE
Cabernet Sauvignon 1977, Wiederkehr Wine Cellars, AR

CABERNET SAUVIGNON, *1981 Rating*
(16 entries)
*GOLD
Allegro Vineyards, PA, 1980
SILVER
Four Chimneys Farm Winery, NY, 1980
BRONZE
Presque Isle Wine Cellars, PA, 1978

CABERNET SAUVIGNON, *1982 Rating*
(15 entries)
BRONZE
*Cabernet Sauvignon 1979, Byrd Vineyards, MD
Cabernet Sauvignon 1980, Byrd Vineyards, MD

CABERNET SAUVIGNON, *1983 Rating*

(14 entries)
BRONZE
*Cabernet Sauvignon 1981, Allegro Vineyards, PA
Cabernet Sauvignon 1980, Ingleside Plantation, VA
Cabernet Sauvignon 1981, Meredyth Vineyards, VA
Cabernet Sauvignon 1981, Byrd Vineyards, MD
HONORABLE MENTION
Cabernet Sauvignon 1982, Ingleside Plantation, VA

CABERNET FRANC, *1983 Rating*
(1 entry)
BRONZE
Cabernet Franc 1980, Presque Isle Wine Cellars, PA

CHARDONNAY, *1979 Rating*
(23 entries)
BEST OF CLASS
Chardonnay, Tabor Hill/Chi Winery, Buchanan, MI
GOLD
Chardonnay, Tabor Hill/Chi Winery, Buchanan, MI
SILVER
Chardonnay, Markko Vineyards, Conneaut, OH
BRONZE
Chardonnay, Haight Vineyards, Litchfield, CT
Chardonnay 1978, Heron Hill Winery, Hammondsport, NY
McGregor Vineyard 1978, Glenora Wine Cellars, Dundee, NY
Chardonnay 1979, Wiederkehr Wine Cellars, Altus, AR
Chardonnay, Byrd Vineyards, Myersville, MD

CHARDONNAY, *1980 Rating*
(22 entries)
GOLD
Chardonnay, Sakonnet Vineyards, RI
SILVER
Chardonnay 1978, Markko Vineyards, OH
Chardonnay 1979, Byrd Vineyards, MD
Chardonnay 1979, Gold Seal Vineyards, NY
Chardonnay 1976, Vinifera Cellars, NY
BRONZE
Chardonnay 1975, Vinifera Cellars, NY
Chardonnay, Casa Larga Vineyards, NY
Chardonnay 1978, Charal Winery & Vineyard, Ontario
Chardonnay 1979, Brotherhood Corp., NY
Chardonnay 1979, Glenora Wine Cellars, NY

CHARDONNAY, *1981 Rating*
(44 entries)
*SILVER
Markko Vineyard, OH, 1979 Lot 921
Wagner Vineyards, NY, 1980
Markko Vineyard, OH, 1979 Lot 923
BRONZE
Vinifera Wine Cellars, NY, 1977
Woodbury Vineyards, NY, von Falkenhausen Vineyard 1980

JUDGES' SPECIAL SELECTION: LATE HARVEST STYLE CHARDONNAY CLASS
GOLD
Cedar Hill Wine Co., OH, 1977
BRONZE
Woodbury Vineyards, NY, Late Harvest 1980

CHARDONNAY, *1982 Rating*
(38 entries)
GOLD
*Chardonnay 1981, Wagner Vineyards, NY
SILVER
Chardonnay 1980, Robert Plane Vineyards, Glenora Wine Cellars, NY

Chardonnay 1981, Prudence Island Vineyards, RI
Chardonnay 1979, Byrd Vineyards, MD
Chardonnay 1980 Estate Bottled, Gold Seal Vineyards, NY

BRONZE
Chardonnay 1981, Byrd Vineyards, MD
Chardonnay 1981, Naked Mountain Vineyard, VA
Chardonnay 1980 Lot 029, Markko Vineyards, OH
Chardonnay 1980 Lot 025, Markko Vineyards, OH

CHARDONNAY, LATE HARVEST STYLE
(1 entry)

BRONZE
Chardonnay Late Harvest 1980, Woodbury Vineyards, NY

CHARDONNAY, *1983 Rating*
(38 entries)

GOLD
*Chardonnay 1981, Gold Seal Vineyards, NY

SILVER
Chardonnay 1982, Sakonnet Vineyards, RI
Chardonnay 1982, Wagner Vineyards, NY
Chardonnay 1982, Plane's Cayuga Vineyards, NY
Chardonnay 1982 Lot 2, Meredyth Vineyards, VA
Chardonnay 1982, Seeger Vineyards (limousin), Inniskillin Wines, Ontario

BRONZE
Chardonnay 1982, Barboursville Winery, VA
Chardonnay 1980, Byrd Vineyards, MD
Chardonnay 1982, Anchor Acres, Glenora Wine Cellars, NY
Chardonnay 1982, Rolling Vineyards, NY
Chardonnay 1982 Lot 1, Meredyth Vineyards, VA
Chardonnay 1982 Lot 3, Meredyth Vineyards, VA
Chardonnay 1982 Lot 4, Meredyth Vineyards, VA
Chardonnay 1982, Commonwealth Winery, MA
Chardonnay 1978, Benmarl Wine Company, NY
Chardonnay 1982, Wickham Vineyards, NY
Chardonnay 1982, Seeger Vineyards (Nevers), Inniskillin Wines, Ontario
Chardonnay 1981, Schloss Doepken Winery, NY
Chardonnay 1982, Nissley Vineyards, PA

HONORABLE MENTION
Chardonnay 1982, Casa Larga Vineyards, NY
Chardonnay 1982, DelVista Vineyards, NJ
Chardonnay 1982, Chateau Gai Wines, Ontario

GAMAY BEAUJOLAIS, *1979 Rating*
(1 entry)

SILVER
Gamay 1978, Inniskillin Wines, Niagara-on-the-Lake, Ontario

GAMAY BEAUJOLAIS, *1980 Rating*
(4 entries)

BRONZE
Gamay Noir 1978, Inniskillin Wines, Ontario

GAMAY BEAUJOLAIS, *1981 Rating*
(6 entries)

*Inniskillin Wines, Ontario, Gamay Noir
(Best of class only awarded.)

GAMAY, *1982 Rating*
(1 entry)

BRONZE
Gamay Noir 1980, Inniskillin Wines, Ontario

GAMAY, *1983 Rating*
(4 entries)
BRONZE

GEWURZTRAMINER, *1979 Rating*
(5 entries)
BEST OF CLASS
Gewurztraminer 1979, Wiederkehr Wine Cellars, Altus, AR
GOLD
Gewurztraminer 1979, Wiederkehr Wine Cellars, Altus, AR
BRONZE
Gewurztraminer 1978, Wiederkehr Wine Cellars, Altus, AR

GEWURZTRAMINER, *1980 Rating*
(4 entries)
GOLD
Gewurztraminer 1978, Vinifera Cellars, NY
SILVER
Gewurztraminer 1979, Casa Larga Vineyards, NY

GEWURZTRAMINER, DRY, (under 1% sugar) *1981 Rating*
(10 entries)
*BRONZE
Gold Seal Vineyards, NY, 1980

GEWURZTRAMINER, MEDIUM SWEET, (1-3% sugar) *1981 Rating*
(6 entries)
*BRONZE
Casa Larga Vineyards, NY, 1980 (Tied for best of class.)

GEWURZTRAMINER, SWEET, (over 3% sugar) *1981 Rating*
(1 entry)
BRONZE
McGregor Vineyards, NY, Late Harvest 1980

GEWURZTRAMINER, *1982 Rating*
(12 entries)
BRONZE
*Gewurztraminer 1982, Casa Larga Vineyards, NY
Gewurztraminer 1982, Hillebrand Estates, Ontario
Gewurztraminer 1980, Newark Wines, Ontario
Gewurztraminer 1982 (medium dry), McGregor Vineyards, NY
HONORABLE MENTION
Gewurztraminer 1982 (medium dry), Great Western Winery, NY

GEWURZTRAMINER, *1983 Rating*
(13 entries)
No Awards

MERLOT, *1981 Rating*
(3 entries)
*Inniskillin Wines, Ontario, 1980 (Best of class only awarded.)

MERLOT, *1982 Rating*
(3 entries)
BRONZE
Merlot 1980, Meredyth Vineyards, VA

MERLOT, *1983 Rating*
(3 entries)
BRONZE
Merlot 1982, Barboursville Winery, VA
HONORABLE MENTION
Merlot 1981, Meredyth Vineyards, VA

MUSCAT, *1980 Rating*
(5 entries)
SILVER
Muscato di Tanta Maria 1980, Wiederkehr Wine Cellars, AR
Muscato di Tanta Maria 1979, Wiederkehr Wine Cellars, AR

MUSCAT, *1981 Rating*
(5 entries)
*SILVER
Canandaigua Wine Company, NY, Canada Muscat 1981

MUSCAT, *1983 Rating*
(3 entries)

GOLD

*Di Tanta Maria 1981 (sweet), Wiederkehr Wine Cellars, AR

SILVER

Di Tanta Maria 1982 (sweet), Wiederkehr Wine Cellars, AR

PINOT NOIR, *1980 Rating*
(3 entries)

No awards.

PINOT NOIR, *1981 Rating*
(8 entries)

BRONZE

*McGregor Vineyards, NY, 1980

Four Chimneys Farm Winery, NY, 1980

PINOT NOIR, *1982 Rating*
(1 entry)

HONORABLE MENTION

Pinot Noir 1981, Shenandoah Vineyards, VA

PINOT NOIR BLANC, *1981 Rating*
(1 entry)

BRONZE

Wiederkehr Wine Cellars, AR, Blancs de Noir 1981

BLANC DE NOIR, *1983 Rating*
(6 entries)

SILVER

*Blanc de Noir 1983 (dry), Wiederkehr Wine Cellars, AR

HONORABLE MENTION

Steuben Blanc de Noir 1982 (medium dry), Cedar Hill Wine Co., OH

RIESLING, *1979 Rating*
(20 entries)

BEST OF CLASS

Riesling, Haight Vineyards, Litchfield, CT

GOLD

Riesling, Haight Vineyards, Litchfield, CT

Late Harvest Riesling 1979, Wiederkehr Wine Cellars, Altus, AR

Riesling 1978, Glenora Wine Cellars, Dundee, NY

BRONZE

Riesling 1979, Wiederkehr Wine Cellars, Altus, AR

Riesling, Casa Larga Vineyards, Fairport, NY

Riesling 1978, Markko Vineyards, Conneaut, OH

JOHANNISBERG RIESLING, DRY, *1980 Rating*
(16 entries)

GOLD

(Tied for top score of competition.)

Johannisberg Riesling Reserve 1979, Markko Vineyards, OH

Johannisberg Riesling Dry 1979, Heron Hill Vineyard, NY

SILVER

Johannisberg Riesling 1979, Wiederkehr Wine Cellars, AR

Johannisberg Riesling 1979, Gold Seal Vineyards, NY

Johannisberg Riesling 1979, Glenora Wine Cellars, NY

BRONZE

White Riesling 1979, Fenn Valley Vineyards, MI

Johannisberg Riesling 1979, Naylor Wine Cellars, PA

JOHANNISBERG RIESLING, SWEET, *1980 Rating*
(6 entries)

GOLD

Johannisberg Riesling Sweet 1979, Haight Vineyards, CT

SILVER

Johannisberg Riesling Sweet 1979 Late Harvest, Wiederkehr Wine Cellars, AR

BRONZE

Johannisberg Riesling Sweet 1979, Sakonnet Vineyards, RI

Johannisberg Riesling Sweet 1980 Late Harvest, Wiederkehr Wine Cellars, AR

RIESLING, DRY, (under 1% sugar) *1981 Rating*
(13 entries)

GOLD

*Casa Larga, NY, Johannisberg Riesling 1980

Sakonnet Vineyards, RI, 1980

SILVER

Chateau Esperanza, NY, Grow Vineyards 1980

BRONZE

Wiederkehr Wine Cellars, AR, Johannisberg Riesling 1981

RIESLING, MEDIUM SWEET, (1-3% sugar) *1981 Rating*
(18 entries)

GOLD

(Tied for best of class.)

*Gold Seal, NY, Estate Bottled 1980

Glenora Wine Cellars, NY Johannisberg Riesling 1980

SILVER

Woodbury Vineyards, NY, Johannisberg Riesling 1980

Rapidan River Vineyards, VA, White Riesling (medium dry) 1981

BRONZE

Rapidan River Vineyards, VA, White Riesling (dry) 1981

RIESLING, SWEET, (over 3% sugar) *1981 Rating*
(5 entries)

GOLD

Gold Seal Vineyards, NY, Estate Bottled Late Harvest 1980

SILVER

Wiederkehr Wine Cellars, AR, Johannisberg Riesling Late Harvest 1981

Wiederkehr Wine Cellars, AR, Johannisberg Riesling Late Harvest 1979

BRONZE

Gold Seal Vineyards, NY, Estate Bottled 1979

RIESLING, DRY, (under .7% sugar) *1982 Rating*
(6 entries)

BRONZE

*Johannisberg Riesling 1980, Wiederkehr Wine Cellars, AR

RIESLING, MEDIUM SWEET, (.7%-3% sugar)
(22 entries)

SILVER

Johannisberg Riesling Springledge 1981, Glenora Wine Cellars, NY

BRONZE

Johannisberg Riesling 1981, Fenn Valley Vineyards, MI

Riesling 1980, McGregor Vineyards, NY

Johannisberg Riesling Lot 114 1981, Markko Vineyards, OH

Johannisberg Riesling 1981, Wagner Vineyards, NY

RIESLING, SWEET, (over 3% sugar) *1982 Rating*
(5 entries)

SILVER

**Late Harvest Riesling 1981, Sakonnet Vineyards, RI

Johannisberg Riesling 1981, Gold Seal Vineyards, NY

Johannisberg Riesling Lot 112 1981, Markko Vineyards, OH

RIESLING, *1983 Rating*
(42 entries)

RIESLING, DRY, (under 0.7% residual sugar)
(9 entries)

GOLD

*J. Riesling 1982, Casa Larga Vineyards, NY

SILVER

J. Riesling 1982, Colio Wines Ltd., Ontario

BRONZE

White Riesling 1983, Rapidan River Vineyards, VA

J. Riesling 1982, Naylor Wine Cellars, PA

HONORABLE MENTION

Riesling Domaine de la Venne 1983, Oasis Vineyards, VA
J. Riesling (nv), Mon Ami Champagne Co., OH

RIESLING, MEDIUM DRY, (0.7 to 3.0% residual sugar)
(23 entries)
GOLD
*J. Riesling 1982, Wagner Vineyards, NY
SILVER
Riesling 1982, Haight Vineyards, CT
J. Riesling 1982, Glenora Wine Cellars, NY
White Riesling (semi-dry) 1983, Rapidan River Vineyards, VA
J. Riesling 1982, Shenandoah Vineyards, VA
J. Riesling 1982, Chateau Esperanza Winery, NY
BRONZE
J. Riesling 1982, Chateau Gai Wines, Ontario
Dry Riesling 1982, Gold Seal Vineyards, NY
Riesling 1982, McGregor Vineyards, NY
Riesling 1982, Sakonnet Vineyards, RI
J. Riesling 1980, Wiederkehr Wine Cellars, AR
J. Riesling 1982, Rolling Vineyards, NY
Riesling Domaine de la Venne 1983, Oasis Vineyards, VA
Riesling 1983, Ingleside Plantation Winery, VA
Riesling 1982, Meredyth Vineyards, VA
Riesling 1982, Commonwealth Winery, MA
HONORABLE MENTION
J. Riesling Berrien County 1982, Tabor Hill Winery, MI

RIESLING, SWEET, (over 3% residual sugar)
(10 entries)
GOLD
*J. Riesling Select Late Harvest 1982, Glenora Wine Cellars, NY
SILVER
J. Riesling Bunch Select 1982, Gold Seal Vineyards, NY
J. Riesling Individual Bunch Selected 1982, Wagner Vineyards, NY
BRONZE
J. Riesling 1982, Wickham Vineyards, NY
J. Riesling Late Harvest 1981, Wiederkehr Wine Cellars, AR
J. Riesling 1981, Gold Seal Vineyards, NY

SAUVIGNON BLANC, *1980 Rating*
(7 entries)
BRONZE
Sauvignon Blanc 1979, Wiederkehr Wine Cellars, AR

SAUVIGNON BLANC, *1981 Rating*
(8 entries)
SILVER
*Oasis Vineyards, VA, Domaine de la Venne 1980
BRONZE
Byrd Vineyards, MD, 1980
Meredyth Vineyards, VA, 1980

SAUVIGNON BLANC, *1982 Rating*
(9 entries)
SILVER
Sauvignon Blanc 1981, Wiederkehr Wine Cellars, AR
BRONZE
Sauvignon Blanc 1979, Byrd Vineyards, MD
Sauvignon Blanc 1980, Byrd Vineyards, MD
Sauvignon Blanc 1981, Naked Mountain Vineyard, VA
Sauvignon Blanc 1981, Meredyth Vineyards, VA

SAUVIGNON BLANC, *1983 Rating*
(5 entries)
BRONZE

*Sauvignon Blanc 1981 (dry), Byrd Vineyards, MD
Sauvignon Blanc 1981 (dry), Wiederkehr Wine Cellars, AR
HONORABLE MENTION
Sauvignon Blanc 1983, Domaine de la Venne (dry), Oasis Vineyards, VA

SEMILLON, *1980 Rating*
(1 entry)
BRONZE
Semillon 1978, Piedmont Vineyards, VA

SEMILLON, *1981 Rating*
(3 entries)
SILVER
*Meredyth Vineyards, VA, 1980
BRONZE
Piedmont Vineyard, VA, 1980 Lot 136

SEMILLON, *1982 Rating*
(3 entries)
SILVER
Domaine de la Venne Semillon 1982, Oasis Vineyards, VA

SEMILLON, *1983 Rating*
(3 entries)
BRONZE
Semillon Domaine de la Venne 1983 (medium dry), Oasis Vineyards, MD

II NATIVE AMERICAN VARIETALS

CARLOS, *1979 Rating*
(4 entries)
BEST OF CLASS
Sweet White, The Winery Rushing, Merigold, MS
BRONZE
Sweet White, The Winery Rushing, Merigold, MS
Rushing White, The Winery Rushing, Merigold, MS

CARLOS, *1981 Rating*
(3 entries)
No awards.

CARLOS, *1982 Rating*
(2 entries)
No awards.

CARLOS, *1983 Rating*
(2 entries)
BRONZE
Carlos 1982 (medium dry), Treuluck Vineyards, SC
HONORABLE MENTION
Carlos Semi-Dry (medium dry) 1982, Alaqua Vineyards, FL

WHITE CATAWBA, *1979 Rating*
(1 entry)
BRONZE
Sweet Catawba, Heineman Winery, Put-in-Bay, OH

WHITE CATAWBA, *1980 Rating*
(1 entry)
SILVER
Sweet Catawba, Heineman Winery, OH

CATAWBA, DRY, (under 1% sugar) *1981 Rating*
(3 entries)
BRONZE
Four Chimneys Farm Winery, NY, Catawba

PINK CATAWBA, *1979 Rating*
(17 entries)
BEST OF CLASS
NY State Pink Catawba, The Taylor Wine Company, Hammondsport, NY

PINK CATAWBA, *1980 Rating*
(10 entries)
GOLD
D'Arcy Pink Catawba, Gross' Highland Winery, NJ
SILVER
Pink Catawba, Heineman Winery, OH
BRONZE
Catawba Rosé, Conestoga Vineyards, PA
Red Catawba, Heritage OH

CATAWBA, MEDIUM SWEET, (1-3% sugar) *1981 Rating*
(5 entries)
SILVER
*Wiederkehr Wine Cellars, AR, Pink Catawba

CATAWBA, SWEET, (over 3% sugar) *1981 Rating*
(10 entries)
SILVER
(Tied for Best of Class)
*Heineman Winery, OH, Sweet Catawba
Heineman Winery, OH, Pink Catawba
BRONZE
Gross' Highland Winery, NJ, Pink Catawba

SWEET CATAWBA, (over 3% sugar) *1982 Rating*
(12 entries)
BRONZE
*Pink Catawba, Gross' Highland Winery, NJ
Sweet Catawba, Heineman Winery, OH

CATAWBA, *1983 Rating*
(13 entries)
BRONZE
*Pink Catawba (sweet), Bernard D'Arcy Wine Cellars, NJ
Pink Catawba 1981 (sweet), Heineman Winery, OH
Catawba 1982 (sweet), Post Winery, AR
HONORABLE MENTION
Catawba (medium dry), Chalet Debonne Vineyards, OH
Pink Catawba 1981 (sweet), Nissley Vineyards, PA
Sweet Catawba 1982 (sweet), Heineman Winery, OH
Catawba 1982 (sweet), Naylor Wine Cellars, PA

CAYUGA WHITE, *1979 Rating*
(6 entries)
BEST OF CLASS
Cayuga White, Brotherhood Winery, Washingtonville, NY
SILVER
Cayuga White, Brotherhood Winery, Washingtonville, NY

CAYUGA, *1980 Rating*
(8 entries)
BRONZE
Cayuga, Commonwealth Winery, MA
Cayuga 1979, Woodbury Vineyards, NY
Cayuga 1979, Chateau Esperanza, NY
Cayuga 1979, Glenora Wine Cellars, NY
Cayuga, Antuzzi's Winery, NJ

CAYUGA WHITE, DRY, (under 1% sugar)
(7 entries)
No awards.

CAYUGA WHITE, MEDIUM SWEET, (1-3% sugar) *1981 Rating*
(3 entries)
BRONZE
*Lucas Vineyards, NY, semi-dry 1980

CAYUGA WHITE, SWEET, (over 3% sugar)
(2 entries)
No awards.

CAYUGA WHITE, DRY, (under .7% sugar) *1982 Rating*
(7 entries)
BRONZE
**Cayuga 1981, Glenora Wine Cellars, NY

CAYUGA WHITE, MEDIUM SWEET, (.7%-3% sugar)
(3 entries)
BRONZE
Cayuga White 1981, Wickham Vineyards, NY

CAYUGA WHITE, SWEET
(1 entry)
No award.

CAYUGA WHITE, ***1983 Rating***
(10 entries)
SILVER
*Cayuga 1982 (dry), Glenora Wine Cellars, NY
BRONZE
Cayuga White 1982 (dry), Widmer's Wine Cellars, NY
Cayuga White 1982 (dry), Commonwealth Wines, MA
Cayuga 1982 (medium dry), Valley Vineyards, NY
Cayuga White 1982 (medium dry), Wickham Vineyards, NY
HONORABLE MENTION
Cayuga 1982 (medium dry), Finger Lakes Wine Cellars, NY

RED CONCORD, ***1979 Rating***
(6 entries)
BEST OF CLASS
Cream Red Concord, Monarch Wine Company, Brooklyn, NY
(No medals awarded in class.)

RED CONCORD, ***1980 Rating***
(5 entries)
GOLD
Concord, Antuzzi's Winery, NJ

WHITE CONCORD, ***1979 Rating***
(3 entries)
BEST OF CLASS
Cream White, Royal Wine Company, Milton, NY
SILVER
Cream White, Royal Wine Company, Milton, NY

WHITE CONCORD, ***1980 Rating***
(1 entry)
SILVER
Cream White Concord, Royal Wine Corp., NY

CONCORD, ***1981 Rating***
(5 entries)
BRONZE
*Huber Orchard Winery, IN, 1980
Bucks Country Vineyards, PA, 1980
Heritage Wine Cellars, PA, (nv)

CONCORD, ***1982 Rating***
(9 entries)
SILVER
*Concord 1981, Bucks Country Vineyards, PA
Concord 1981, Calvaresi Winery, PA
Vinott du Cragmont, Oliver Wine Co., IN

CONCORD, ***1983 Rating***
(5 entries)
BRONZE
Concord 1982 (sweet), Nissley Vineyards, PA

CYNTHIANA, ***1982 Rating***
(3 entries)
SILVER
*Cynthiana 1981, Cowie Wine Cellars, AR
BRONZE
Cynthiana, Wiederkehr Wine Cellars, AR

CYNTHIANA, ***1983 Rating***
(5 entries)
BRONZE
Cynthiana 1983 (medium dry), Wiederkehr Wine Cellars, AR

DELAWARE, ***1979 Rating***
(19 entries)
BEST OF CLASS
Delaware, Naylor Wine Cellars, York, PA
BRONZE
Delaware, Naylor Wine Cellars, York, PA
Delaware 1978, Meredyth Vineyards, Middleburg, VA
Delaware 1979, Post Winery, Altus, AR
Delaware 1979, Wagner Vineyards,

Lodi, NY
Delaware, Chalet Debonne, Madison, OH

DELAWARE, DRY, *1980 Rating*
(4 entries)
BRONZE
Bellevinta 1978, London Winery, Ontario
Delaware Moselle, Gross' Highland Winery, NJ
Delaware 1978, Frederick S. Johnson Vineyards, NY

DELAWARE, SWEET, *1980 Rating*
(10 entries)
GOLD
Pink Delaware 1979, Frederick S. Johnson Vineyards, NY
SILVER
Delaware, Chalet Debonne Vineyards, OH
Liebestropfchen, Frederick S. Johnson Vineyards, NY
BRONZE
Golden Grenadier, Naylor Wine Cellars, PA
Golden Delaware, Brotherhood Corp., NY

DELAWARE, MEDIUM SWEET, (1-3% sugar) *1981 Rating*
(7 entries)
BRONZE
Wagner Vineyards, NY, 1980

DELAWARE, SWEET, (over 3% sugar)
(3 entries)
GOLD
*Johnson Estate, NY, Liebestropfchen 1980

DELAWARE, DRY, (under .7% sugar) *1982 Rating*
(3 entries)
BRONZE
*Delaware 1981, Chalet Debonne Vineyards, OH

DELAWARE, MEDIUM SWEET, (.7%-3% sugar)
(8 entries)
BRONZE
*Delaware 1981, Glenora Wine Cellars, NY
Delaware 1981, Poplar Ridge Vineyards, NY
Country White (Delaware), Gross' Highland Winery, NJ

DELAWARE, SWEET, (over 3% sugar) *1982 Rating*
(8 entries)
SILVER
**Delaware 1981, Wagner Vineyards, NY
BRONZE
Delaware 1981, Wickham Vineyards, NY
Delaware 1981, Meredyth Vineyards, VA

DELAWARE, *1983 Rating*
(8 entries)
SILVER
*Delaware 1982 (sweet), Wagner Vineyards, NY
*Delaware 1982 (sweet), Meredyth Vineyards, VA
BRONZE
Delaware 1982 (sweet), Finger Lakes Wine Cellars, NY
Delaware 1982 Estate Bottled (medium dry), Post Winery, AR
Delaware 1982 Appalachian Harvest (dry), MJC Vineyards, VA
HONORABLE MENTION
Delaware 1982 (medium dry), Calvaresi Winery, PA
Wolf Island (medium dry), Conneaut Cellars Winery, PA

DUTCHESS, *1979 Rating*
(9 entries)
BEST OF CLASS
Great Western Dutchess Rhine, Pleasant Valley Wine Company, Hammondsport, NY
BRONZE
Great Western Dutchess Rhine,

Pleasant Valley Wine Company, Hammondsport, NY

DUTCHESS, *1980 Rating*
(8 entries)
SILVER
Dutchess Rhine, Gross' Highland Winery, NJ
BRONZE
Pennsylvania Dutchess, Conestoga Vineyards, PA

DUTCHESS, DRY, (under 1% sugar) *1981 Rating*
(4 entries)
BRONZE
Gross' Highland Winery, NJ
Charal Winery & Vineyards, Ontario, 1980

DUTCHESS, MEDIUM SWEET, (1-3% sugar)
(3 entries)
BRONZE
Brotherhood Winery, NY

DUTCHESS, MEDIUM SWEET, (.7%-3% sugar) *1982 Rating*
(4 entries)
SILVER
*Dutchess Rhine, Gross' Highland Winery, NJ
BRONZE
Dutchess 1981, Bucks Country Vineyards, PA

DUTCHESS, *1983 Rating*
(2 entries)
HONORABLE MENTION
Dutchess Rhine (medium dry), Bernard D'Arcy Wine Cellars, NJ

MOORE'S DIAMOND, *1980 Rating*
(1 entry)
SILVER
1979 Diamond, Chateau Esperanza, NY

DIAMOND, *1983 Rating*
(1 entry)
BRONZE
Diamond 1981 (dry), Chateau Esperanza Winery, NY

ELVIRA, *1983 Rating*
(1 entry)
BRONZE
Lakeside White 1982 (sweet), Hopkins Vineyard, CT

ISABELLA, *1981 Rating*
(1 entry)
BRONZE
Heritage Wine Cellars, PA

IVES, *1981 Rating*
(4 entries)
*Gross' Highland Winery, NJ
(Best in class award only given.)

IVES, *1982 Rating*
(3 entries)
SILVER
*Ives Noir, Frederick S. Johnson Vineyards, NY
BRONZE
Ives Noir 1981, Post Winery, AR

IVES, *1983 Rating*
(1 entry)
BRONZE
Ives Noir 1982 (medium dry), Post Winery, AR

MAGNOLIA, *1980 Rating*
(2 entries)
BRONZE
Magnolia, Thousand Oaks Vineyard & Winery, MS

ELVIRA, *1983 Rating*
(1 entry)
BRONZE
Lakeside White 1982 (sweet), Hopkins Vineyard, CT

ISABELLA, *1981 Rating*
(1 entry)
BRONZE
Heritage Wine Cellars, PA

IVES, *1981 Rating*
(4 entries)
*Gross' Highland Winery, NJ
(Best in class award only given.)

IVES, *1982 Rating*
(3 entries)
SILVER
*Ives Noir, Frederick S. Johnson Vineyards, NY
BRONZE
Ives Noir 1981, Post Winery, AR

IVES, *1983 Rating*
(1 entry)
BRONZE
Ives Noir 1982 (medium dry), Post Winery, AR

MAGNOLIA, *1980 Rating*
(2 entries)
BRONZE
Magnolia, Thousand Oaks Vineyard & Winery, MS

MAGNOLIA, *1981 Rating*
(1 entry)
SILVER
Thousand Oaks Vineyard & Winery, MS, (nv)

MAGNOLIA, *1982 Rating*
(1 entry)
BRONZE
Magnolia, Thousand Oaks Vineyard Winery, MS

MAGNOLIA, *1983 Rating*
(2 entries)
BRONZE
Magnolia 1981 (sweet), Duplin Wine Cellars, NC

NIAGARA, *1979 Rating*
(16 entries)
BEST OF CLASS
Lake Niagara, Widmer's Wine Cellars, Naples, NY
GOLD
Lake Niagara, Widmer's Wine Cellars, Naples, NY
Niagara, Gross' Highland Winery, Absecon, NJ
SILVER
Cream Niagara, Royal Wine Company, Milton, NY
Niagara, Naylor Wine Cellars, York, PA

NIAGARA, *1980 Rating*
(12 entries)
GOLD
York County Niagara, Naylor Wine Cellars, PA
Niagara, Woodbury Vineyards, NY
SILVER
American Niagara, Gross' Highland Winery, NJ
BRONZE
Cream Niagara, Royal Wine Corp., NY
Niagara, Wiederkehr Wine Cellars, AR

NIAGARA, MEDIUM SWEET, (1-3% sugar) *1981 Rating*
(1 entry)
BRONZE
Gross' Highland Winery, NJ

NIAGARA, SWEET, (over 3% sugar)
(6 entries)
SILVER
*Naylor Wine Cellars, PA, 1979

NIAGARA, MEDIUM SWEET, *1982 Rating*
(3 entries)
No awards.

NIAGARA, SWEET
(6 entries)
BRONZE
*Niagara 1981, Naylor Wine Cellars, PA

NIAGARA, *1983 Rating*
(6 entries)
BRONZE
*Niagara 1982 (sweet), Naylor Wine

NOBLE, *1979 Rating*
(5 entries)
BEST OF CLASS
Sweet Red Muscadine, Thousand Oaks Vineyard/Winery, Starkville, MS
SILVER
Sweet Red Muscadine, Thousand Oaks Vineyard/Winery, Starkville, MS

NOBLE, *1980 Rating*
(2 entries)
SILVER
Noble, Post Winery, AR

NOBLE, *1983 Rating*
BRONZE
Noble Dry 1982 (medium dry), Alaqua Vineyards, FL

SCUPPERNONG, *1979 Rating*
(4 entries)
BEST OF CLASS
Magnolia, Thousand Oaks Vineyard/Winery, Starkville, MS
SILVER
Magnolia, Thousand Oaks Vineyard/Winery, Starkville, MS

SCUPPERNONG, *1981 Rating*
(1 entry)
No award.

SCUPPERNONG, *1983 Rating*
(1 entry)
HONORABLE MENTION
Scuppernong 1981 (sweet), Duplin Wine Cellars, NC

MUSCADINE, *1981 Rating*
(1 entry)
BRONZE
Thousand Oaks, MS, Sweet White Muscadine

Cellars, PA
Niagara (medium dry), Bernard D'Arcy Wine Cellars, NJ
Niagara 1982 (sweet), Post Winery, AR

MUSCADINE, *1982 Rating*
(3 entries)
BRONZE
*Muscadine 1981, Highland Manor Winery, WV

MUSCADINE, *1983 Rating*
(4 entries)
SILVER
*Muscadine 1982 (sweet), Post Winery, AR

III FRENCH-AMERICAN HYBRID VARIETALS

AURORA, *1979 Rating*
(18 entries)
BEST OF CLASS
Aurora 1979, Wagner Vineyards, Lodi, NY
GOLD
Aurora 1979, Wagner Vineyards, Lodi, NY
SILVER
Aurora, Commonwealth Winery, Plymouth, MA
Aurora 1978, Heron Hill Winery, Hammondsport, NY
Aurora 1978, Glenora Wine Cellars, Dundee, NY
BRONZE
Aurora Blanc 1978, Meredyth Vineyards, Middleburg, VA
Chandelle Blanc Aurora, Charal Winery, Blenheim, Ontario

AURORA, DRY, *1980 Rating*
(10 entries)
SILVER
Chandelle Blanc, Charal Winery & Vineyard, Ontario
BRONZE
Aurora 1979, Glenora Wine Cellars, NY
Aurora 1979, Chateau Esperanza, NY

AURORA, SWEET, *1980 Rating*
(4 entries)
SILVER
Aurora 1979, Nissley Vineyards, PA

AURORA, DRY, (under 1% sugar) *1981 Rating*
(4 entries)
BRONZE
Chateau Esperanza, NY, Smith Vineyards 1980

AURORA, MEDIUM SWEET, (1-3% sugar) *1981 Rating*
(7 entries)
SILVER
(Tied for best of class.)
*Leelanau Wine Cellars, MI, (nv)
*Wagner Vineyards, NY, 1980
BRONZE
Naylor Wine Cellars, PA, 1981

AURORA, DRY, (under .7% sugar) *1982 Rating*
(9 entries)
BRONZE
**Waramaug White 1981, Hopkins Vineyard, CT
Aurora 1981, Finger Lakes Wine Cellars, NY

AURORA, MEDIUM SWEET, (.7%-3% sugar)
(7 entries)
BRONZE
*Aurora Blanc-John Henry Vineyards 1981, Chateau Esperanza, NY
Aurora 1981, Nissley Vineyards, PA
Aurora Blanc 1981, Meredyth Vineyards, VA

AURORA, SWEET, (over 3% sugar)
(2 entries)
No awards.

AURORA, *1983 Rating*
(13 entries)
SILVER
*Aurora 1982 (sweet), Andres Wines Ltd., Ontario
BRONZE
Aurora 1983 (medium dry), Hamlet Hill Vineyards, CT
HONCRABLE MENTION
Aurora Special Reserve (dry), Finger Lakes Wine Cellars, NY

BACO NOIR, *1979 Rating*
(14 entries)
BEST OF CLASS
Late Vintage Baco 1978, T.G. Bright, Niagara Falls, Ontario
GOLD
Late Vintage Baco 1978, T.G. Bright, Niagara Falls, Ontario
SILVER
Baco Noir Neruveo, Tabor Hill/Chi Vineyards, Buchanan, MI
BRONZE
1978 Baco Noir, T.G. Bright, Niagara Falls, Ontario
Baco Noir Rosé 1978, Glenora Wine Cellars, Dundee, NY

BACO, *1980 Rating*
(15 entries)
GOLD
Baco Noir, Wollersheim Winery, WI
SILVER
Baco Noir 1975, Tabor Hill Vineyards, MI
Baco, Leelanau Wine Cellars, MI
BRONZE
Baco, Naylor Wine Cellars, PA
Late Harvest Baco, T.G. Bright & Company, Ontario

BACO NOIR, *1981 Rating*
(16 entries)
SILVER
*Canandaigua Wine Co., NY, Special Selection
Charal Winery & Vineyards, Ontario, 1979
BRONZE
Tabor Hill Wine Cellars, MI, Nouveau 1980

BACO NOIR, *1982 Rating*
(10 entries)
SILVER
*Baco Noir 1980, T.G. Bright and Co., Ontario

Late Harvest Baco Noir 1979, T.G. Bright and Co., Ontario

BRONZE

Baco 749 Berrien County, Tabor Hill Vineyard, MI

London Baco Noir 1981, London Winery, Ontario

Premium NYS Baco Noir, Great Western Winery, NY

BACO NOIR, *1983 Rating*
(7 entries)

GOLD

*Baco Noir Estate Bottled 1982, Great Western, NY

HONORABLE MENTION

Baco Noir (nv), Wollersheim Winery, WI

CHAMBOURCIN, *1979 Rating*
(5 entries)

BEST OF CLASS

Chambourcin, La Buena Vida Vineyards, Springtown, TX

CHAMBOURCIN, *1980 Rating*
(3 entries)

No awards.

CHAMBOURCIN, *1981 Rating*
(7 entries)

BRONZE

*Naylor Wine Cellars, PA, 1980

CHAMBOURCIN, *1982 Rating*
(7 entries)

BRONZE

(Tied for best in class.)

*Chambourcin 1981, Naylor Wine Cellars, PA

*Chambourcin 1980, Chadwick Bay Wine Co., NY

CHAMBOURCIN, BLANC DE NOIR
(1 entry)

SILVER

Chambourcin Blanc 1981, Truluck Vineyards, SC

CHAMBOURCIN, *1983 Rating*
(9 entries)

SILVER

*Chambourcin 1981, Shenandoah Vineyards, VA

BRONZE

Chambourcin 1981, Tewksbury Wine Cellars, NJ

Chambourcin 1980 Special Reserve, Shenandoah Vineyards, VA

Chambourcin 1981, Naylor Wine Cellars, PA

HONORABLE MENTION

Chambourcin 1982, Tewksbury Wine Cellars, NJ

Chambourcin Rosé 1982 (medium dry), Calveresi Winery, PA

CHANCELLOR NOIR, *1979 Rating*
(7 entries)

BEST OF CLASS

Chancellor Noir, Country Creek Vineyards, Telford, PA

CHANCELLOR, *1980 Rating*
(6 entries)

SILVER

Chancellor 1979, Chateau Esperanza, NY

BRONZE

Chancellor, Buffalo Valley Winery, PA

CHANCELLOR NOIR, *1981 Rating*
(8 entries)

BRONZE

(Tied for best of class.)

*Chateau Esperanza, NY, Plane Vineyard 1979 Reserve

*Chateau Esperanza, NY, Plane Vineyard 1979

CHANCELLOR NOIR, *1982 Rating*
(8 entries)

BRONZE

*Chancellor Noir 1979, St. Julian Wine Co., MI

Debonne Red-Chancellor 1979, Chalet Debonne Vineyards, OH

Chancellor-1981, The Barry Wine Co., NY

CHANCELLOR, *1983 Rating*
(6 entries)
HONORABLE MENTION
Chancellor 1982, Buffalo Valley Winery, PA

CHELOIS, *1979 Rating*
(8 entries)
BEST OF CLASS
Chelois, Wollersheim Winery, Prairie du Sac, WI
GOLD
Chelois, Wollersheim Winery, Prairie du Sac, WI

CHELOIS, *1980 Rating*
(6 entries)
SILVER
Chelois, Bucks Country Vineyards, PA
BRONZE
Chelois 1979, Nissley Vineyards, PA
Chelois 1978, Inniskillin Wines, Ontario

CHELOIS, *1981 Rating*
(7 entries)
BRONZE
*Inniskillin Wines, Ontario, 1980
Meier's Wine Cellars, OH, (nv)

CHELOIS
(6 entries)
No awards.

CHELOIS, *1983 Rating*
(7 entries)
BRONZE
*Chelois 1982, Cedar Hill Wine Co., OH
Chelois 1982, Oasis Vineyards, VA
HONORABLE MENTION
Chelois 1980, Inniskillin Wines Ltd., Ontario

DE CHAUNAC, *1979 Rating*
(20 entries)
BEST OF CLASS
DeChaunac, Royal Wine Company, Milton, NY
BRONZE
DeChaunac, Royal Wine Company, Milton, NY

DE CHAUNAC, *1980 Rating*
(19 entries)
SILVER
DeChaunac, Brotherhood Corp., NY
BRONZE
Bellevinta 1979, London Winery, Ontario
DeChaunac, Commonwealth Winery, MA
DeChaunac 1979, Inniskillin Wines, Ontario
DeChaunac, Great Western Winery, NY

DE CHAUNAC, *1981 Rating*
(9 entries)
SILVER
*Wagner Vineyards, NY, 1980 Lot 1
BRONZE
Lucas Winery, NY, Interlaken Red 1980
Woodbury Vineyards, NY, 1980

DE CHAUNAC, *1982 Rating*
(11 entries)
BRONZE
*DeChaunac, Commonwealth Winery, MA
DeChaunac Lot 2 1980, Wagner Vineyards, NY

DE CHAUNAC, *1983 Rating*
(6 entries)
BRONZE
*DeChaunac 1980 Lot 2, Wagner Vineyards, NY
Rosé O'DeChaunac 1982, Naylor Wine Cellars, PA
HONORABLE MENTION
DeChaunac, Finger Lakes Wine Cellars, NY

FOCH, *1979 Rating*
(23 entries)

BEST OF CLASS
Foch 1977, Northeast Vineyards, Millerton, NY
GOLD
Foch 1977, Northeast Vineyards, Millerton, NY
SILVER
Foch, Haight Vineyards, Litchfield, CT
Foch 1977, Barnes Wines, St. Catharines, Ontario
Foch 1978, Cedar Hill Wine Company, Cleveland Heights, OH
Foch 1978, Wollersheim Winery, Prairie du Sac, WI
BRONZE
Foch 1977, Meredyth Vineyards, Middleburg, VA
1979 "Nouveau", Glenora Wine Cellars, Dundee, NY

MARECHAL FOCH, *1980 Rating*
(17 entries)
GOLD
1978 Table Wine, Northeast Vineyard, NY
SILVER
Marechal Foch 1979, Inniskillin Wines, Ontario
BRONZE
Marechal Foch 1978, Inniskillin Wines, Ontario
Marechal Foch, Tewksbury Wine Company, NJ

MARECHAL FOCH, *1981 Rating*
(19 entries)
*Northeast Vineyard, NY, 1979
(Best of class only awarded.)

MARECHAL FOCH, *1982 Rating*
(24 entries)
SILVER
*Foch, John Henry Vineyards, Chateau Esperanza, NY
Foch Reserve Cuvée 802, Wollersheim Winery, WI
BRONZE
Marechal Foch 1980, Inniskillin Wines, Ontario
Marechal Foch 1981, Haight Vineyards, CT
Barn Red 1981, Hopkins Vineyards, CT

MARECHAL FOCH, *1983 Rating*
(18 entries)
SILVER
*Marechal Foch 1982, Finger Lakes Wine Cellars, NY
BRONZE
Marechal Foch 1981, Colio Wines Ltd., Ontario
Marechal Foch 1980, Inniskillin Wines, Ontario
Marechal Foch 1982, Inniskillin Wines, Ontario
Marechal Foch 1982, Valley Vineyards, NY
HONORABLE MENTION
Marechal Foch 1982, Alexis Bailly Vineyard, MN

LANDAL, *1982 Rating*
(1 entry)
BRONZE
Landal 244 1981, Naylor Wine Cellars, PA

LANDOT, *1980 Rating*
(1 entry)
BRONZE
Landot, Wollersheim Winery, WI

LANDOT, *1982 Rating*
(1 entry)
No award.

LANDOT, *1983 Rating*
(1 entry)
HONORABLE MENTION
Landot Noir 1982, Baldwin Vineyards, NY

MILLOT, *1979 Rating*
(1 entry)
GOLD
Leon Millot, Alexis Bailly Vineyards, Hastings, MN

LEON MILLOT, *1980 Rating*
(4 entries)
BRONZE
Leon Millot 1979, Alexis Bailly Vineyards, MN

LEON MILLOT, *1981 Rating*
(6 entries)
SILVER
*Alexis Bailly Vineyard, MN, 1980
BRONZE
Woodbury Vineyards, NY, 1980
Tewksbury Wine Cellars, NJ, 1980

LEON MILLOT, *1982 Rating*
(2 entries)
BRONZE
Leon Millot 1980, Inniskillin Wines, Ontario

LEON MILLOT, *1983 Rating*
(4 entries)
SILVER
*Leon Millot 1982, Meredyth Vineyards, VA
BRONZE
Leon Millot 1982, Alexis Bailly Vineyard, MN
HONORABLE MENTION
Leon Millot 1982, Valley Vineyards, NY

RAYON D'OR, *1980 Rating*
(1 entry)
GOLD
Rayon d'Or, La Buena Vida Vineyards, TX

RAYON D'OR, *1981 Rating*
(1 entry)
No award.

RAYON D'OR, *1982 Rating*
(2 entries)
*Rayon d'Or 1981, La Buena Vida Vineyards, TX
No medals.

RAYON D'OR, *1983 Rating*
(3 entries)
SILVER
*Rayon d'Or 1982, Valley Vineyards, NY
Rayon d'Or 1982, La Buena Vida Vineyards, TX

RAVAT (VIGNOLES), *1980 Rating*
(6 entries)
SILVER
Vignoles 1979 Late Harvest, Fenn Valley Vineyards, MI
Ravat 1979, Chateau Esperanza, NY
BRONZE
Ravat 1979, Glenora Wine Cellars, NY

RAVAT BLANC, DRY, (under 1% sugar) *1981 Rating*
(1 entry)
No award.

RAVAT BLANC, MEDIUM SWEET, (1-3% sugar) *1981 Rating*
(3 entries)
SILVER
Glenora Wine Cellars, NY, 1980

RAVAT BLANC, SWEET, (over 3% sugar) *1981 Rating*
(5 entries)
SILVER
Glenora Wine Cellars, NY, Special Selection 1980

RAVAT, DRY, (under .7% sugar) *1982 Rating*
(1 entry)
GOLD
**Ravat Blanc 1981, Penn-Shore Vineyards, PA

RAVAT, MEDIUM SWEET, (.7%-3% sugar)
SILVER
*Ravat 1981 Martini Vineyards, Chateau Esperanza, NY
BRONZE
Ravat Blanc 1981, Glenora Wine Cellars, NY

RAVAT, SWEET, (over 3% sugar)
(5 entries)
BRONZE
(Tied for best in class.)
*Ravat Matzke Vineyard 1981, Tabor Hill Vineyards, MI
*Vignoles Reserve 1981, St. Julian Wine Co., MI
Ravat Matzke Vineyard 1980, Tabor Hill Vineyards, MI
Vignoles 1981, Fenn Valley Vineyards, MI
Ravat Late Harvest Plane's Vineyard 1980, Chateau Esperanza, NY

VIGNOLES (RAVAT), *1983 Rating*
(14 entries)
GOLD
*Ravat Select Harvest 1982 (sweet), Chateau Esperanza Winery, NY
BRONZE
Vignoles Select Harvest 1982 (sweet), Fenn Valley Vineyards, MI
Ravat Selected Sweet Harvest 1981 (sweet), Tabor Hill Vineyards, MI
Ravat Blanc 1982 (medium dry), Glenora Wine Cellars, NY
Ravat 51 (medium dry), Baldwin Vineyards, NY
HONORABLE MENTION
Vignoles 1982, Debevc Vineyards, Chalet Debonne Vineyards, OH

ROUGEON NOIR, *1980 Rating*
(1 entry)
BRONZE
Rougeon, Wagner Vineyards, NY

ROUGEON, *1983 Rating*
(1 entry)
BRONZE
Rougeon 1982, Meredyth Vineyards, VA

SEREKSIA, *1980 Rating*
(1 entry)
SILVER
Sereksia, Vinifera Cellars, NY

SEREKSIA, *1981 Rating*
(1 entry)
SILVER
Vinifera Wine Cellars, NY

SEYVAL BLANC, *1979 Rating*
(32 entries)
BEST OF CLASS
Seyval Blanc 1978, Heron Hill Winery, Hammondsport, NY
SILVER
Seyval Blanc 1978, Heron Hill Winery, Hammondsport, NY
Meredyth 1978 Seyval Blanc, Meredyth Vineyards, Middleburg, VA
Seyval Blanc, Wagner Vineyards, Lodi, NY
BRONZE
1978 Seyval Blanc, Glenora Wine Cellars, Dundee, NY
Seyval Blanc, Mazza Vineyards, North East, PA

SEYVAL BLANC, DRY, *1980 Rating*
(37 entries)
GOLD
Seyval 1979, Glenora Wine Cellars, NY
Seyval 1980, Mazza Vineyards, PA
SILVER
Seyval Blanc Dry, Naylor Wine Cellars, PA
BRONZE
Susquehanna Wine 1979, Nissley Vineyards, PA
Colonnade 1979, Nissley Vineyards, PA
Seyval Blanc 1980, Mount Hope Vineyards & Winery, PA

SEYVAL BLANC, SWEET, *1980 Rating*
(10 entries)
BRONZE
(Tied for best in class.)
Classic White 1979, Nissley Vineyards, PA
Seyval Ice Wine 1979, Tabor Hill Vineyards, MI
Bainbridge White 1979, Nissley Vineyards, PA

SEYVAL BLANC, DRY, (under 1% sugar) *1981 Rating*
(35 entries)
SILVER
*Commonwealth Winery, MA, 1980
Country Creek Winery, PA, (nv)
Inniskillin Wines, Ontario, (nv)
Chateau Esperanza, NY, Hosmer Vineyard (nv)
BRONZE
Tewksbury Wine Cellars, NJ, 1980
Glenora Wine Cellars, NY, Springledge 1980
Wagner Vineyards, NY, 1980

SEYVAL BLANC, MEDIUM SWEET, (1-3% sugar) *1981 Rating*
(5 entries)
No awards.

SEYVAL BLANC, SWEET, (over 3% sugar) *1981 Rating*
(3 entries)
SILVER
Tabor Hill Wine Cellars, MI, Late Harvest 1979

SEYVAL BLANC, DRY, (under .7% sugar) *1982 Rating*
(35 entries)
GOLD
*Seyval Blanc 1981, Wagner Vineyards, NY
SILVER
Seyval Blanc Winemaker's Reserve, Commonwealth Winery, MA
BRONZE
Seyval Blanc 1980, Tewksbury Wine Cellars, NJ
Seyval Blanc 1979, Fenn Valley Vineyards, MI
American Seyval Blanc 1981, Byrd Vineyards, MD

SEYVAL BLANC, MEDIUM SWEET, (.7%-3% sugar)
(9 entries)
GOLD
*October Harvest Seyval 1981, Meredyth Vineyards, VA
SILVER
Wagner's Seyval 1981, Wagner Vineyards, NY
BRONZE
Seyval 1981, Rolling Vineyards, Farm Winery, NY
Seyval Blanc 1981, Penn-Shore Vineyards, PA

SEYVAL BLANC, SWEET, (over 3% sugar)
(3 entries)
GOLD
**Seyval Blanc Dry Berry 1981, Tabor Hill Vineyard, MI

SEYVAL, *1983 Rating*
(68 entries)

SEYVAL, DRY, (under 0.7% residual sugar)
(48 entries)
GOLD
*Seyval 1982, Tewksbury Wine Cellars, NJ
SILVER
Seyval Blanc 1982 Lot 1, Meredyth Vineyards, VA
Seyval Blanc 1982, Fenn Valley Vineyards, MI
Seyval Blanc 1982, Franklin Hill Vineyards, PA
Seyval Blanc Extra Dry Reserve, Hamlet Hill Vineyard, CT
Seyval Blanc 1982, Wickham Vineyards, NY
BRONZE
Seyval Blanc 1981, Fenn Valley Vineyards, MI
Seyval Blanc 1982, Wagner Vineyards, NY
Warner Seyval 1982, Warner Vineyards, MI
Seyval Blanc 1982, Boordy Vineyards, MD
Pearis Mountain Seyval Blanc, MJC Vineyards, VA
Seyval 1983, Oakencroft Vineyard, VA

Seyval Blanc 1983, Ingleside Plantation Winery, VA
Seyval Blanc 1982, Shenandoah Vineyards, VA
Seyval Blanc 1982, Good Harbor Vineyards, MI
HONORABLE MENTION
Seyval 1982, Colio Wines Ltd., Ontario
Seyval Blanc 1981, Piedmont Vineyards, VA
Seyval Blanc 1981 Limited Release, Penn Shore Vineyards, PA
Seyval Blanc 1982, Cedar Hill Wine Co., OH

SEYVAL, MEDIUM DRY, (0.7 to 3.0% residual sugar)
(13 entries)
SILVER
*Seyval Blanc 1982 Estate Bottled, Hamlet Hill Vineyards, CT
Seyval Blanc 1982, Woodbury Vineyards, NY
BRONZE
October Harvest 1982, Meredyth Vineyards, VA
Seyval Blanc 1982, Glenora Wine Cellars, NY
Seyval Blanc 1982, Lynfred Winery, IL
Seyval 1982, Valley Vineyards, NY
Wagner's Seyval 1982, Wagner Vineyards, NY
Seyval 1982, Rolling Vineyards, NY
HONORABLE MENTION
Seyval 1982, Inniskillin Wines, Ontario

SEYVAL, SWEET, (over 3.0% residual sugar)
(7 entries)
GOLD
*Late Harvest Dry Berry Seyval 1981, Tabor Hill Vineyards, MI
BRONZE
Seyval Blanc 1982, Andres Wines Ltd., Ontario
HONORABLE MENTION
Seyval Blanc 1982, Post Winery, AR
Seyval Blanc Reserve 1982, St. Julian Wine Co., MI

VERDELET, *1980 Rating*
(3 entries)
BRONZE
Verdelet, Wiederkehr Wine Cellars, AR

VERDELET, *1981 Rating*
(3 entries)
SILVER
*Great Western, NY, Special Selection 1980

VERDELET, *1982 Rating*
(2 entries)
No awards.

VIDAL BLANC, *1979 Rating*
(18 entries)
BEST OF CLASS
1978 Michigan Reserve, Fenn Valley Vineyards, Fennville, MI
SILVER
1978 Michigan Reserve, Fenn Valley Vineyards, Fennville, MI
1978 Vidal Blanc, Presque Isle Wine Cellars, North East, PA

VIDAL, DRY, *1980 Rating*
(14 entries)
SILVER
(Tied for best in class.)
Vidal-Riesling, Tabor Hill Vineyards, MI
Vidal Blanc, Penn-Shore Vineyards, PA
BRONZE
Vidal Sec, Tabor Hill Vineyards, MI
Vidal Demi-Sec, Tabor Hill Vineyards, MI
Vidal Dry, Byrd Vineyards, MD

VIDAL, SWEET, *1980 Rating*
(6 entries)
SILVER
Vidal Blanc-Reserve 1978, Fenn Valley Vineyards, MI

BRONZE
Vidal Reserve 1979, Sakonnet Vineyards, RI
Vidal Bunch Select 1979, Tabor Hill Vineyards, MI

VIDAL BLANC, DRY, (under 1% sugar) *1981 Rating*
(10 entries)
GOLD
*Chadwick Bay Winery, NY, 1980
SILVER
Byrd Vineyards, MD, (nv)
Tewksbury Wine Cellars, NJ, 1980
BRONZE
Naylor Wine Cellars, PA, 1980

VIDAL BLANC, MEDIUM SWEET, (1-3% sugar) *1981 Rating*
(11 entries)
SILVER
Commonwealth Winery, MA, 1980
BRONZE
Inniskillin Wines, Ontario, 1980
Jordan & Ste. Michelle, Ontario, 1980

VIDAL, DRY, (under .7% sugar) *1982 Rating*
(10 entries)
SILVER
*Vidal Blanc 1981 Debevc Vineyards, Chalet Debonne Vineyards, OH
BRONZE
Vidal 1981, Naylor Wine Cellars, PA
Vidal 1981, Sakonnet Vineyards, RI

VIDAL, MEDIUM SWEET, (.7%-3% sugar)
(9 entries)
No awards.

VIDAL, SWEET, (over 3% sugar)
(4 entries)
SILVER
**Vidal Blanc Reserve 1981, Fenn Valley Vineyards, MI

VIDAL BLANC, ***1983 Rating***
(19 entries)

VIDAL BLANC, DRY, (under 0.7% residual sugar)
(11 entries)
BRONZE
Vidal 1982, Cascade Mountain Vineyards, NY
Vidal Blanc, Chalet Debonne Vineyards, OH
HONORABLE MENTION
Vidal, Conneaut Cellars, PA
Vidal Blanc 1982, Del Vista Vineyards, NJ
Vidal 1982, Sakonnet Vineyards, RI
Vidal 1981, Naylor Wine Cellars, PA

VIDAL BLANC, MEDIUM DRY, (0.7 to 3.0% residual sugar)
(10 entries)
SILVER
*Vidal 1982, Rolling Vineyards, NY
BRONZE
Vidal Blanc 1982, Debevc Vineyards, Chalet Debonne Vineyards, OH
HONORABLE MENTION
Vidal Blanc Demi-Sec Lemon Creek 1981, Tabor Hill Vineyards, MI
Vidal 1982, Inniskillin Wines, Ontario

VIDAL BLANC, SWEET, (over 3.0% residual sugar)
(8 entries)
BRONZE
Vidal Blanc 1980, Landey Vineyards, Conestoga Vineyards, PA
HONORABLE MENTION
Vidal Blanc 1982 Late Harvest, Great Western, NY
Vidal Blanc 1982 Reserve, St. Julian Wine Co., MI
Vidal Blanc 1982 Reserve, Fenn

Valley Vineyards, MI
Vidal Blanc 1981 Landey Vineyards, Conestoga Vineyards, PA
Vidal Blanc 1982 Landey Vineyards, Conestoga Vineyards, PA

VILLARD BLANC, *1979 Rating*
(3 entries)
BEST OF CLASS
Villard Blanc, Thousand Oaks Vineyard/Winery, Starkville, MS
(No medals awarded in class.)

VILLARD BLANC, *1980 Rating*
(2 entries)
BRONZE
Villard Blanc 1979, Truluck Vineyards, SC

VILLARD BLANC, *1982 Rating*
(5 entries)
No awards.

VILLARD NOIR, *1979 Rating*
(4 entries)
BEST OF CLASS
Villard Noir, Laine Vineyards, Fulton, KY
(No medals awarded in class.)

VILLARD NOIR, *1980 Rating*
(4 entries)
SILVER
Villard Noir 1978, Meredyth Vineyards, VA
La Vida del Sol, La Buena Vida Vineyards, TX
BRONZE
Blanc de Noir, La Buena Vida Vineyards, TX
Villard Noir, T.G. Bright & Company, Ontario

VILLARD NOIR, *1981 Rating*
(1 entry)
No awards.

IV NATIVE AMERICAN, GENERIC AND PROPRIETARY TABLE WINES

AMERICAN DRY RED, *1979 Rating*
(10 entries)
BEST OF CLASS
Noble, The Winery Rushing, Merigold, MS
BRONZE
Noble, The Winery Rushing, Merigold, MS

AMERICAN DRY RED, *1980 Rating*
(13 entries)
SILVER
Red Wine, Antuzzi's Winery, NJ
BRONZE
Princiere Red, Chateau Gai Wines, Ontario
Bellevinta, London Winery, Ontario
D'Arch Country Red, Gross' Highland Winery, NJ

AMERICAN DRY RED, (under 1% sugar) *1981 Rating*
(6 entries)
SILVER
Gross' Highland Winery, NJ, Country Red

AMERICAN DRY RED, (under .7% sugar) *1982 Rating*
(6 entries)
*Claret, Heineman Winery, OH
No medals.

AMERICAN SWEET RED, *1979 Rating*
(7 entries)
BEST OF CLASS
Rosario, Brotherhood Winery, Washingtonville, NY
SILVER
Rosario, Brotherhood Winery, Washingtonville, NY
BRONZE
Concord, Heritage Wine Cellars, North East, PA

AMERICAN SWEET RED, *1980 Rating*
(9 entries)
BRONZE
Sweet Claret, Gross' Highland Winery, NJ
Ives Noir 1978, Frederick S. Johnson Vineyards, NY

MEDIUM SWEET RED, (1-3% sugar) *1981 Rating*
(5 entries)
SILVER
Heineman Winery, OH, Burgundy

SWEET RED, (over 3% sugar) *1981 Rating*
(8 entries)
GOLD
*Brotherhood Winery, NJ, Rosario
BRONZE
Gross' Highland Winery, NJ, Sweet Claret
T.G. Bright Wine Co., Ontario, Riuscita Red

MEDIUM SWEET RED, (.7%-3% sugar) *1982 Rating*
(6 entries)
BRONZE
*Burgundy, Heineman Winery, OH

SWEET RED, (over 3% sugar)
(4 entries)
BRONZE
**Labrusca 1981, Bucks Country Vineyards, PA

NATIVE AMERICAN GENERIC RED TABLE WINE, *1983 Rating*
(12 entries)
BRONZE
*American Burgundy (medium dry) Manischewitz, Monarch Wine Co., NY
HONORABLE MENTION
Old Fashioned Blue Face (sweet), Mantey Vineyards, OH

AMERICAN DRY ROSÉ, *1979 Rating*
(7 entries)
BEST OF CLASS
Chandelle Rosé, Charal Winery, Blenheim, Ontario
BRONZE
Chandelle Rosé, Charal Winery, Blenheim, Ontario

AMERICAN DRY ROSÉ, *1980 Rating*
(5 entries)
SILVER
Chandelle Rosé, Charal Winery & Vineyard, Ontario
BRONZE
Rosé, Kolin Winery, PA

DRY ROSÉ, (under 1% sugar) *1981 Rating*
(12 entries)
SILVER
Charal Winery & Vineyards, NY, Chandelle Rosé
BRONZE
Wagner Vineyards, NY, Senlaka Rosé

DRY ROSÉ, (under .7% sugar) *1982 Rating*
(1 entry)
No award.

AMERICAN SWEET ROSÉ, *1979 Rating*
(12 entries)
BEST OF CLASS
Tokay, Royal Wine Company, Milton, NY
BRONZE
Tokay, Royal Wine Company, Milton, NY
Carolina Rosé, Truluck Winery, Lake City, SC
Lake Niagara Pink, Widmer's Wine Cellars, Naples, NY
Rosé, London Winery, London, Ontario

AMERICAN SWEET ROSÉ, *1980 Rating*
(13 entries) BRONZE

Sweet Rosé, Heritage Vineyards, OH
Glacier Ridge Rosé, Woodbury Vineyards, NY
Isabella Rosé, Great Western Winery, NY

MEDIUM SWEET ROSÉ, (1-3% sugar) *1981 Rating*
(11 entries)
BRONZE
Johnson Estate Vineyards, NY, Cascade Rosé 1980

SWEET ROSÉ, (over 3% sugar) *1981 Rating*
(7 entries)
SILVER
*Heritage Wine Cellars, PA, Rosé (nv)
Byrd Vineyards, VA, Church Hill Rosé (nv)

MEDIUM SWEET ROSÉ, (.7%-3% sugar) *1982 Rating*
(4 entries)
BRONZE
*Seneca Lake Rosé 1981, Rolling Vineyards Farm Winery, NY

SWEET ROSÉ, (over 3% sugar)
(9 entries)
BRONZE
**Cracker Ridge Rosé 1981, Chadwick Bay Wine Co., NY
Catawba, Brotherhood Corporation, NY

NATIVE AMERICAN GENERIC ROSÉ TABLE WINE, *1983 Rating*
(3 entries)
BRONZE
Semi-Sweet Rosé 1982 (medium dry), Kolln Vineyards. PA

AMERICAN DRY WHITE, *1979 Rating*
(23 entries)
BEST OF CLASS
Rhine NY State, The Taylor Wine Company, Hammondsport, NY
BRONZE
Rhine NY State, The Taylor Wine Company, Hammondsport, NY

AMERICAN DRY WHITE, *1980 Rating*
(24 entries)
GOLD
Glacier Ridge White, Woodbury Vineyards, NY
SILVER
D'Arch Country White, Gross' Highland Winery, NJ
Alpenweiss, Chateau Gai Wines, Ontario
Dry White, Chateau Gai Wines, Ontario
N.Y.S. Rhine, The Taylor Wine Co., NY
1979 White, Frederick S. Johnson Vineyards, NY
BRONZE
Chablis, St. Julian Wine Co., MI
Edelwein, Chateau Gai Wines, Ontario
Chablis Blanc, Wiederkehr Wine Cellars, AR
Rhine, Warner Vineyards, MI

DRY WHITE, (under 1% sugar) *1981 Rating*
(22 entries)
BRONZE
Brotherhood Winery, NY, Jubilee Chablis
Nissley Vineyards, PA, Millrace White 1979

DRY WHITE, (under .7% sugar) *1982 Rating*
(4 entries)
SILVER
*White Table 1981, Kolln Vineyards & Winery, PA

NATIVE AMERICAN GENERIC WHITE TABLE WINE, *1983 Rating*
(18 entries)
BRONZE
*Maria Christina White 1982 (medium dry), Jordan & Ste.

Michelle Cellars, Ontario
Interlude 1982 (medium dry), Jordan & Ste. Michelle Cellars, Ontario
Semi-Sweet White 1982 (medium dry), Kolln Vineyards, PA
Lake Country Chablis (medium dry), The Taylor Wine Company, NY
Eastern White 1982 (sweet), Buffalo Valley Winery, PA
Sweet White Wine 1982 (sweet), Germanton Vineyard, NC

HONORABLE MENTION
Lonzbrusco White Wine (medium dry), Lonz Winery, OH
Chablis (medium dry), Penn Shore Vineyards, PA

AMERICAN SWEET WHITE, *1979 Rating*
(19 entries)

BEST OF CLASS
Country Gold, Gross' Highland Winery, Absecon, NJ

GOLD
Country Gold, Gross' Highland Winery, Absecon, NJ

BRONZE
Carlos, The Winery Rushing, Merigold, MS
Edelweiss, Wiederkehr Wine Cellars, Altus, AR

AMERICAN SWEET WHITE, *1980 Rating*
(12 entries)

GOLD
D'Arcy Country Gold, Gross' Highland Winery, NJ

SILVER
Sauterne, Brotherhood Corp., NY
Mississippi Niagara, Thousand Oaks Vineyard & Winery, MS

BRONZE
Lake County Gold, The Taylor Wine Co., NY
Matuk Royale Blanc, Royal Wine Corp., NY

MEDIUM SWEET WHITE, (1-3% sugar) *1981 Rating*
(28 entries)

SILVER
*Brotherhood Winery, NY, Rhineling

BRONZE
Nissley Vineyards, PA, Cellar Reserve 1979
Wiederkehr Wine Cellars, AR, Edelweiss

SWEET WHITE, (over 3% sugar) *1981 Rating*
(9 entries)

No awards.

JUDGES' SPECIAL SELECTION: LATE HARVEST STYLE SWEET WHITE CLASS

GOLD
Wiederkehr Wine Cellars, AR, Di Tanta Maria 1979

SILVER
Jordan & Ste. Michelle, Ontario, Pinot Muscato 1980

MEDIUM SWEET WHITE, (.7%-3% sugar) *1982 Rating*
(18 entries)

SILVER
*Chautauqua Blanc 1981, Frederick S. Johnson Vineyards, NY
Rhine Castle 1981, Jordan & Ste. Michelle Cellars, Ontario

BRONZE
Maria Christina White 1981, Jordan & Ste. Michelle Cellars, Ontario
Lake Country Chablis, The Taylor Wine Company, NY
Baron Ludwig, T.G. Bright & Co., Ontario

SWEET WHITE, (over 3% sugar)
(11 entries)

SILVER
**Virginia Dare White NYS, Canandaigua Wine Co., NY

BRONZE

Cracker Ridge White 1981, Chadwick Bay Wine Co., NY

Haut Sauterne 1981, Mantey Vineyards, OH

Manor St. David's Medium Dry, T.G. Bright & Co., Ontario

Country Gold, Gross' Highland Winery, NJ

V FRENCH-AMERICAN HYBRID GENERICS

DRY RED, *1979 Rating*
(35 entries)

BEST OF CLASS

Rhode Island Red, Sakonnet Vineyards, Little Compton, RI

BRONZE

Rhode Island Red, Sakonnet Vineyards, Little Compton, RI

Vin Nouveau 1978, Inniskillin Wines, Niagara-on-the-Lake, Ontario

DRY RED, *1980 Rating*
(37 entries)

SILVER

Rhode Island Red 1979, Sakonnet Vineyards, RI

BRONZE

1978 Table Wine, Northeast Vineyard, NY

Vin Nouveau, Inniskillin Wines, Ontario

DeChaunac, Commonwealth Winery, MA

DRY RED, (under 1% sugar) *1981 Rating*
(42 entries)

GOLD

*Benmarl Wine Co., NY, Cuvée du Vigneron (17A) 1979

SILVER

St. Julian Wine Co., MI, Friar's Noir

Brotherhood Winery, NY, St. Vincent Burgundy

Inniskillin Wines, Ontario, Millot-Chambourcin 1979

Tabor Hill Wine Cellars, MI, Cuvée Rouge (nv)

BRONZE

T.G. Bright Wine Co., Ontario, House Red

T.G. Bright Wine Co., Ontario, Entre Lacs Red

Cascade Mountain Vineyards, NY, Reserve Red 1980

Chateau Gai Wines, Ontario, Seibel

DRY RED, (under .7% sugar) *1982 Rating*
(33 entries)

SILVER

**Blend #10001 1980, Chadwick Bay Wine Co., NY

Reserved Red 1981, Cascade Mountain Vineyards, NY

Northeast Vineyards 1979, Northeast Vineyard, NY

Friar's Noir, St. Julian Wine Co., MI

BRONZE

Dry Red, Fenn Valley Vineyards, MI

Baco-DeChaunac, Wollersheim Winery, WI

Rhode Island Red 1980, Sakonnet Vineyard, RI

Charter Oak Red 1980, Hamlet Hill Vineyards, CT

Cuvée 37B-Estate Blend 1979, Benmarl Wine Co., NY

DRY RED, (under 0.7% residual sugar) *1982 Rating*
(39 entries)

BRONZE

*Burgundy, Wiederkehr Wine Cellars, AR

Classic Red, Finger Lakes Wine Cellars, NY

Illinois Red 1981, Wollersheim Winery, WI

Capital Red, Wagner Vineyards, NY

Domaine Reserve 1981, Wollersheim Winery, WI

Friar's Noir, St. Julian, MI

Northeast Vineyard 1980, NY
Country Red, Alexis Bailly Vineyard, MN
Midland Country Red, Cypress Valley Winery, TX
Entre Lacs Red, T.G. Bright Wines, Ontario
Riserva Rosso 1981, Colino Wines Ltd., Ontario

MEDIUM DRY, (0.7 to 3.0% residual sugar)
(7 entries)
BRONZE
Susquehanna Valley Red 1980, Nissley Vineyards, PA
First Capital (oak aged), Naylor Wine Cellars, PA
Harvest Red 1982, Meredyth Vineyards, VA
HONORABLE MENTION
Mellow Burgundy, Eagle Crest Vineyards, NY

SWEET RED, 1979 Rating
(5 entries)
BEST OF CLASS
"St. John's", Huber Winery, Borden, IN
(No medals awarded in class.)

SWEET RED, *1980 Rating*
(9 entries)
BRONZE
St. John's Red, Huber Orchard Winery, IN
Harvest Red, Commonwealth Winery, MA
Alta B., Wagner Vineyards, NY

MEDIUM SWEET RED, (1-3% sugar) *1981 Rating*
(7 entries)
No awards.

SWEET RED, (over 3% sugar) *1981 Rating*
(7 entries)
No awards.

MEDIUM SWEET RED, (.7%-3% sugar) *1982 Rating*
(8 entries)
BRONZE
*Country Red, Gross' Highland Winery, NJ
First Capital 1981, Naylor Wine Cellars, PA

SWEET RED, (over 3% sugar)
(6 entries)
BRONZE
*Bainbridge Red 1980, Nissley Vineyards, PA

SWEET RED, (over 3% residual sugar) *1983 Rating*
(7 entries)
BRONZE
Naughty Marietta 1982, Nissley Vineyards, PA
Bainbridge Red 1982, Nissley Vineyards, PA
St. John's Red 1982, Huber Orchards Winery, IN
HONORABLE MENTION
Alta B, Wagner Vineyards, NY

SWEET ROSÉ, *1979 Rating*
(7 entries)
BEST OF CLASS
Country Pink, Bucks Country Vineyards, New Hope, PA
SILVER
Country Pink, Bucks Country Vineyards, New Hope, PA

SWEET ROSÉ, (over 3% sugar) *1981 Rating*
(10 entries)
BRONZE
Chalet Debonne, OH, Rosé

MEDIUM SWEET ROSÉ, (.7%-3% sugar) *1982 Rating*
(14 entries)
SILVER
*Rosé, Commonwealth Winery, MA

BRONZE
Country Pink 1981, Bucks Country Vineyards, PA

SWEET ROSÉ, (over 3% sugar) *1982 Rating*
(5 entries)
SILVER
*Holiday Rosé, Nissley Vineyards, PA
BRONZE
Church Hill Rosé 1981, Byrd Vineyards, MD

SWEET ROSÉ, (over 3% residual sugar) *1983 Rating*
(7 entries)
BRONZE
Rosé 1983, Nissley Vineyards, PA
HONORABLE MENTION
DeChaunac Rosé 1982, Good Harbor Vineyards, MI

DRY ROSÉ, *1979 Rating*
(19 entries)
BEST OF CLASS
Vin Rosé Sec, Wiederkehr Wine Cellars, Altus, AR
SILVER
Vin Rosé Sec, Wiederkehr Wine Cellars, Altus, AR

DRY ROSÉ, *1980 Rating*
(15 entries)
BRONZE
Vin Rosé, Fenn Valley Vineyards, MI
Chancellor Rosé, Tewksbury Wine Company, NJ
Rosé, The Taylor Wine Company, NY
Keystone Kiss 1979, Nissley Vineyards, PA

DRY ROSÉ, (under 1% sugar) *1981 Rating*
(11 entries)
BRONZE
*Wiederkehr Wine Cellars, AR, Alpine Rosé

DRY ROSÉ, (under .7% sugar) *1982 Rating*
(12 entries)
SILVER
**Rosé de Cabernet Sauvignon, Wiederkehr Wine Cellars, AR
BRONZE
Vin Rosé Sec, Wiederkehr Wine Cellars, AR

DRY ROSÉ, (under 0.7% residual sugar) *1983 Rating*
(4 entries)
BRONZE
Rosé de Chardonnay 1982, Tewksbury Wine Cellars, NJ
Rosé 1982, Casa Larga Vineyards, NY
HONORABLE MENTION
Elizabeth Rosé 1981, Newark/Hillebrand Cellars, Ontario

MEDIUM DRY ROSÉ, (0.7 to 3.0% residual sugar)
(10 entries)
BRONZE
*Harvest Rosé 1982, Cascade Mountain Vineyards, NY
Renaissance Rosé, Hamlet Hill Vineyards, CT
Plymouth Rock Rosé 1982, Commonwealth Winery, MA

DRY WHITE, *1979 Rating*
(45 entries)
BEST OF CLASS
American Chablis, Golden Rain Tree Winery, Wadesville, IN
GOLD
American Chablis, Golden Rain Tree Winery, Wadesville, IN
America's Cup White, Sakonnet Vineyards, Little Compton, RI
SILVER
Brae Blanc 1978, Inniskillin Wines, Niagara-on-the-Lake, Ontario
Ravat 1978, Glenora Wine Cellars, Dundee, NY
Rhine, Widmer's Wine Cellars, Naples, NY

Chablis White, Brotherhood Winery, Washingtonville, NY
Ravat "51", Antuzzi's Winery, Riverside, NJ

BRONZE
Chablis Nature, Gold Seal Vineyards, Hammondsport, NY
Vin Blanc, Fenn Valley Vineyards, Fennville, MI
Colonade 1978, Nissley Vineyards, Bainbridge, PA
Seyval, Wagner Vineyards, Lodi, NY

DRY WHITE, *1980 Rating*
(32 entries)

GOLD
(Tied for top score of competition.)
Criterion White, Golden Rain Tree Winery, IN
Vidal Blanc, Commonwealth Winery, MA
Aurora 1979, Nissley Vineyards, PA

SILVER
Vidal-Riesling, Tabor Hill Vineyards, MI
Susquehanna White 1979, Nissley Vineyards, PA
Vin Blanc, Fenn Valley Vineyards, MI
Cayuga, Commonwealth Winery, MA
Chablis, Leelanau Wine Cellars, MI

BRONZE
Chablis, Brotherhood Corp., NY

DRY WHITE, (under 1% sugar) *1981 Rating*
(12 entries)

BRONZE
Jordan & Ste. Michelle, Ontario, Rhine Castle 1980

DRY WHITE, (under .7% sugar) *1982 Rating*
(13 entries)

SILVER
(Tied for best of class.)
*Regal White 1981, Fenn Valley Vineyards, MI
*Little White 1981, Cascade Mountain Vineyards, NY

BRONZE
Seyval-Aligote 1981, Tri-Mountain Winery, VA
Sugarloaf 1981, Wollersheim Winery, WI
America's Cup White 1981, Sakonnet Vineyards, RI
Dry White House, T.G. Bright Wine Co., Ontario

DRY WHITE, (under 0.7% residual sugar) *1983 Rating*
(14 entries)

SILVER
*Chablis Blanc, Widmer's Wine Cellars, NY
Covertside White, Haight Vineyards, CT

BRONZE
Ronay Blanc, Bernard D'Arcy Wine Cellars, NJ
America's Cup White, Sakonnet Vineyards, RI
Debonne White 1981, Chalet Debonne Vineyards, OH
Chateau White, Chateau Esperanza, NY

HONORABLE MENTION
Summertide 1983, Cascade Mountain Vineyards, NY
Riserva Bianco 1981, Colio Wines Ltd., Ontario
Casa Larga White 1982, Casa Larga Vineyards, NY

MEDIUM SWEET WHITE, (1-3% sugar) *1981 Rating*
(10 entries)

SILVER
Jordan & Ste. Michelle, Ontario, Maria Christina White 1980

BRONZE
Penn Shore Winery, PA, Free Spirit Chablis, nv

MEDIUM SWEET WHITE, (.7%-3% sugar) *1982 Rating*

(21 entries)
BRONZE
*Vin Blanc 1981, Allegro Vineyards, PA
Frankenmuth White 1981, St. Julian Wine Co., MI
Chablis, Brotherhood Corporation, NY

MEDIUM DRY WHITE, (0.7 to 3.0% residual sugar) *1983 Rating*
(23 entries)
SILVER
Schloss Hillebrand, Hillebrand Estates, Ontario
BRONZE
Spinnaker White 1982, Sakonnet Vineyards, RI
York White Rosé 1982, Naylor Wine Cellars, PA
Keuka Highlands White, McGregor Vineyards, NY
Criterion White, Golden Rain Tree Winery, IN
Ingleside Fraulein 1982, Ingleside Plantation Winery, VA
Rhine Wine, Widmer's Wine Cellars, NY
HONORABLE MENTION
Steller White, Nissley Vineyards, PA
Meadow White, Nissley Vineyards, PA

SWEET WHITE, *1979 Rating*
(11 entries)
BEST OF CLASS
Rayon D'Or, La Buena Vida Vineyards, Springtown, TX
GOLD
Rayon D'Or, La Buena Vida Vineyards, Springtown, TX
SILVER
Church Hill White Wine, Byrd Vineyards, Myersville, MD

SWEET WHITE, *1980 Rating*
(15 entries)
SILVER
Pinot-Moscato, Jordan & Ste. Michelle Cellars, Ontario
Bainbridge White 1979, Nissley Vineyards, PA
BRONZE
Starlite White, Huber Orchard Winery, IN
Bainbridge White 1978, Nissley Vineyards, PA
Church Hill White, Byrd Vineyards, MD

SWEET WHITE, (over 3% sugar) *1981 Rating*
(11 entries)
GOLD
*Gross' Highland Winery, NJ, Country Gold

SWEET WHITE, (over 3% sugar) *1982 Rating*
(7 entries)
SILVER
**Di Tanta Maria 1981, Wiederkehr Wine Cellars, AR
BRONZE
Bainbridge White 1981, Nissley Vineyards, PA
Di Tanta Maria 1980, Wiederkehr Wine Cellars, AR

SWEET WHITE, (over 3% residual sugar) *1983 Rating*
(10 entries)
BRONZE
Millrace White 1982, Nissley Vineyards, PA
Millrace White 1981, Nissley Vineyards, PA
HONORABLE MENTION
Country Gold, Bernard D'Arcy Wine Cellars, NJ
Classic White 1980, Nissley Vineyards, PA

VI NATIVE AMERICAN SPARKLING WINES

BRUT SPARKLING, *1979 Rating*
(8 entries)
BEST OF CLASS

Blanc de Blanc, Gold Seal Vineyards, Hammondsport, NY
GOLD
Blanc de Blanc, Gold Seal Vineyards, Hammondsport, NY
SILVER
Great Western, Pleasant Valley Wine Company, Hammondsport, NY

BRUT SPARKLING, *1980 Rating*
(7 entries)
SILVER
Brut-Very Dry, The Taylor Wine Company, NJ
Brut-Dry, Gross' Highland Winery, NJ
Brut, St. Julian Wine Company, MI
Hans Wiederkehr, Wiederkehr Wine Cellars, AR
BRONZE
Brut, Gold Seal Vineyards, NY
President, T.G. Bright & Company, Ontario

BRUT, *1981 Rating*
(8 entries)
GOLD
*T.G. Bright Wine Co., Ontario, Brut Champagne
SILVER
Gold Seal Vineyards, NY, Brut Champagne
BRONZE
Great Western, NY, Natural Champagne
Chateau Gai Wines, Ontario, Brut Champagne
Gross' Highland Winery, NJ, Champagne

EXTRA DRY & DRY, *1981 Rating*
(11 entries)
SILVER
*Woodbury Vineyards, NY, Extra Dry Champagne, nv
St. Julian Winery, MI, White Champagne
BRONZE
London Wines Ltd., Ontario, Jubilee Champagne
Gold Seal Vineyards, NY, Extra Dry Champagne

BRUT, 1982 Rating
(6 entries)
SILVER
*Brut Champagne, Gross' Highland Winery, NJ
BRONZE
Brut President Champagne, T.G. Bright & Co., ONT
White Champagne, Renault Winery, NJ
Brut Champagne 1980, Mon Ami Champagne Co., OH

EXTRA DRY, *1982 Rating*
(8 entries)
SILVER
*J. Roget American Dry Champagne, Canandaigua Wine Co., NY
President Champagne, T.G. Bright & Co., Ontario
BRONZE
North Carolina Champagne 1981, Duplin Wine Cellars, NC
Extra Dry Champagne 1980, Mon Ami Champagne Co., OH

DRY, *1982 Rating*
(4 entries)
SILVER
*Alpenweiss Sparkling, Chateau Gai Wines, Ontario
BRONZE
Sparkling Scuppernong 1981, Duplin Wine Cellars, NC
Jubilee Champagne, London Winery, Ontario

DRY, *1983 Rating*
(3 entries)
BRONZE
*Extra Dry Champagne (sweet), Gold Seal Vineyards, NY
Champagne Canadian 1982 (sweet), Montravin Cellars, Ontario

EXTRA DRY, *1983 Rating*
(10 entries)
SILVER
*Extra Dry Champagne (medium dry), Great Western, NY
BRONZE
Extra Dry Ohio Champagne (medium dry), Meier's Wine Cellars, OH
Gold Label Extra Dry (medium dry), Canandaigua Wine Company, NY
Podamer Champagne Special Reserve (medium dry), Montravin Cellars, Ontario

BRUT, *1983 Rating*
(9 entries)
BRONZE
*La Mont (medium dry), Chateau Gai Wines, Ontario
HONORABLE MENTION
Brut Champagne (medium dry), Gold Seal Vineyards, NY
Podamer Champagne Brut 1980/1981 (dry), Montravin Cellars, Ontario
Naturel (dry), Great Western, NY
President Brut Champagne (medium dry), T.G. Bright Wines, Ontario

VII FRENCH-AMERICAN & VINIFERA SPARKLING WINES

DRY SPARKLING, *1979 Rating*
(22 entries)
BEST OF CLASS
Hans Wiederkehr White Champagne, Wiederkehr Wine Cellars, Altus, AR
GOLD
Hans Wiederkehr White Champagne, Wiederkehr Wine Cellars, Altus, AR
Vidal Champagne, St. Julian Wine Company, Paw Paw, MI
SILVER
White Champagne, Royal Wine Company, Milton, NY
Spumante Classica, Chateau Gai Wines, Niagara Falls, Ontario
BRONZE
President, T.G. Bright, Niagara Falls, Ontario

DRY SPARKLING, *1980 Rating*
(10 entries)
SILVER
Extra Dry, Great Western Winery, NY
BRONZE
Imperial, Chateau Gai Wines, Ontario

BRUT, *1981 Rating*
(4 entries)
GOLD
*Cedar Hill Wine Co., OH, Natur Methode Champenoise
SILVER
Fenn Valley Vineyards, MI, Blanc de Blancs 1978
BRONZE
Woodbury Vineyards, NY, Blanc de Blancs 1980

EXTRA DRY AND DRY, *1981 Rating*
(8 entries)
SILVER
*Gold Seal, NY, Blanc de Blancs
BRONZE
Ingleside Plantation, VA, Champagne 1980

BRUT, *1982 Rating*
(7 entries)
BRONZE
*Brut Champagne, Gold Seal Vineyards, NY
Brut Blanc de Blancs 1980, Podamer Champagne Co., Ontario

EXTRA DRY, *1982 Rating*
(4 entries)
BRONZE
*Extra Dry Champagne, Gold Seal Vineyards, NY

DRY, *1982 Rating*

(3 entries)
*Hans Wiederkehr Champagne, Wiederkehr Wine Cellars, AR
No awards.

EXTRA DRY, *1983 Rating*
(8 entries)
BRONZE
*Pinot Chardonnay Champagne (medium dry), T.G. Bright Wines, Ontario
HONORABLE MENTION
Champagne (medium dry), Leelanau Wine Cellars, MI

SPARKLING BRUT, *1983 Rating*
(6 entries)
HONORABLE MENTION
Richelieu Brut Champagne (dry), Andres Wines Ltd., Ontario

VIII SPECIALTY SPARKLING WINES

SPECIALTY SPARKLING, *1979 Rating*
(21 entries)
BEST OF CLASS
Sparkling Burgundy, Gold Seal Vineyards, Hammondsport, NY
GOLD
Sparkling Burgundy, Gold Seal Vineyards, Hammondsport, NY
SILVER
Crackling Lake Niagara, Widmer's Wine Cellars, Naples, NY
New York State Pink Champagne, The Taylor Wine Company, Hammondsport, NY
BRONZE
Pink Champagne, Gold Seal Vineyards, Hammondsport, NY
Sparkling Pink Catawba, Gross' Highland Winery, Absecon, NJ

SPUMANTE SPARKLING, *1980 Rating*
(6 entries)
BRONZE
Du Barry Spumante, T.G. Bright & Company, NJ

SPECIALTY SPARKLING, RED, *1981 Rating*
(7 entries)
BRONZE
*Andres' Wine Nova Scotia, Baby Duck
Gold Seal Vineyards, NY, Sparkling Burgundy

SPECIALTY SPARKLING, ROSÉ, *1981 Rating*
(5 entries)
GOLD
*Gross' Highland Winery, NJ, Sparkling Pink Catawba
SILVER
Wiederkehr Wine Cellars, AR, Chateau du Monte Pink
BRONZE
Chateau Gai Wines, Ontario, Chevente Rosé

SPECIALTY SPARKLING, SPUMANTE, *1981 Rating*
(11 entries)
GOLD
*Chateau Gai Wines, Ontario, Spumante Classico
SILVER
Barnes Wines Ltd., Ontario, Spumante Bianco
BRONZE
Gross' Highland Winery, NJ, Spumante
Andres' Wines, Ontario, Spumante
Woodbury Vineyards, NY, Spumante

SPECIALTY SPARKLING WHITE AND PINK, *1982 Rating*
(6 entries)
BRONZE
*Sparkling Pink Catawba, Gross' Highland Winery, NJ
Chateau du Monte Pink Champagne, Wiederkehr Wine Cellars, AR
J. Roget Almante, Canandaigua Wine Co., NY

SPECIALTY SPARKLING SPUMANTE
SILVER
*Spumante Classico, Chateau Gai Wines, Ontario

SPECIALTY SPARKLING RED, *1982 Rating*
(7 entries)
BRONZE
*Sparkling Burgundy, Gold Seal Vineyards, NY
Capri Sparkling Burgundy 1981, Robin Fils et Cie, NY

SPECIALTY SPARKLING RED, *1983 Rating*
(1 entry)
BRONZE
Sparkling Burgundy (sweet), Gold Seal Vineyards, NY

SPECIALTY SPARKLING ROSÉ, *1983 Rating*
(3 entries)
BRONZE
*Pink Champagne (sweet), Gold Seal Vineyards, NY

SPECIALTY SPARKLING SPUMANTE, *1983 Rating*
(8 entries)
SILVER
*Spumante Reiems (sweet), Meier's Wine Cellars, OH
BRONZE
Spumante (sweet), Canandaigua Wine Company, NY
Spumante (medium dry), Bernard D'Arcy Wine Cellars, NJ
Spumante Bambino 1982 (sweet), Jordan & Ste. Michelle Cellars, Ontario
Spumante (sweet), T.G. Bright Wines, Ontario

SPARKLING APPLE, *1983 Rating*
(1 entry)
BRONZE
Sparkling Apple Wine (dry), Nashoba Valley Winery, MA

IX FRUIT WINES

GENERAL, *1980 Rating*
(13 entries)
BRONZE
(Tied for best in class.)
Pear, Leelanau Wine Cellars, MI
Raspberry, Antuzzi's Winery, NJ

APPLE, *1980 Rating*
(7 entries)
BRONZE
Apple, Woodbury Vineyards, NY
Apple, Kolln Vineyards and Winery, PA
Country Roads, Chateau Gai Wines, Ontario

CHERRY, *1980 Rating*
(7 entries)
SILVER
Cherry, Woodbury Vineyards, NY
BRONZE
Festival Cherry, Leelanau Wine Cellars, MI
Cherry, Country Creek Vineyard and Winery, PA

STRAWBERRY, *1980 Rating*
(7 entries)
SILVER
Strawberry, Mt. Hope Vineyards & Winery, PA
BRONZE
Strawberry, Huber Orchard Winery, IN

MEAD, *1980 Rating*
(4 entries)
BRONZE
Honey Royale, Royal Wine Corp., NY
Honey Wine, Valley Vineyards Farm, OH

FRUIT WINES, *1981 Rating*
(37 entries)
BRONZE
Fruit Wines of Florida, Palmetto Country Orange Wine

Nashoba Valley Winery, MA, Cranberry-Apple
Nashoba Valley Winery, MA, Peach
Leelanau Wine Cellars, MI, Morello Cherry
Buffalo Valley Winery, PA, Strawberry
Chateau Gai Wines, Ontario, Country Roads Apple

MEAD (HONEY WINE), ***1981 Rating***
(4 entries)
BRONZE
Valley Vineyards, OH, Honey Wine 1979

FRUIT WINES, ***1982 Rating***
(29 entries)
SILVER
*After Dinner Peach, Nashoba Valley Winery, MA
Festival Cherry, Leelanau Wine Cellars, MI
Semi-Sweet Apple, Nashoba Valley Winery, MA
BRONZE
Blackberry, Lonz Winery, OH
Cranberry Apple, Nashoba Valley Winery, MA
Semi-Sweet Blueberry, Nashoba Valley Winery, MA
Apfelwein, Tri-Mountain Winery, VA
Dutch Apple, Bucks Country Vineyards, PA
Apple, La Abra Farm and Winery, VA
Strawberry Sweet 1982, Private Stock Winery, IA
Strawberry 1982, Buffalo Valley Winery, PA

MEAD (HONEY WINE)
(2 entries)
No awards.

FRUIT WINES, ***1983 Rating***
(18 entries)
SILVER
*After Dinner Peach Wine (sweet), Nashoba Valley Winery, MA
BRONZE
Waterworks Current (sweet), Warner Vineyards, MI
Opus 1 (Peach/Seyval; sweet), Allegro Vineyards, PA
Peach Wine (sweet), La Abra Farm Winery, VA
Dry Blueberry (dry), Nashoba Valley Winery, MA
Semi-Sweet Blueberry (sweet), Nashoba Valley Winery, MA
Raspberry 1982 (sweet), Huber Orchards Winery, IN
Fragola Strawberry 1983 (sweet), Naylor Wine Cellars, PA
Strawberry 1983 (sweet), Buffalo Valley Winery, PA
HONORABLE MENTION
Cranberry Apple Wine Special Dry (medium dry), Nashoba Valley Winery, MA
Nectarine 1983 (sweet), Buffalo Valley Winery, PA

CHERRY
(5 entries)
GOLD
*Cherry 1983 (sweet), Buffalo Valley Winery, PA
SILVER
Cherry 1982 (sweet), Good Harbor Vineyards, MI
HONORABLE MENTION
Wisconsin Cherry 1982 (sweet), Spurgeon Vineyards Winery, WI
Puesta del Sol Cherry, Colorado Mountain Vineyards, CO

APPLE
(11 entries)
SILVER
*Apple 1982 (sweet), Tewksbury Wine Cellars, NJ
BRONZE
Apple 1982 (sweet), Lynfred Winery, IL
Apple Wine (sweet), Conestoga Vineyards, PA
Pomfret White (apple) (medium

dry), Hamlet Hill Vineyards, CT
HONORABLE MENTION
Dry Apple Wine (dry), Nashoba Valley Winery, MA

MEAD, *1983 Rating*
(2 entries)
BRONZE
Ancient Mead (golden), London Winery Ltd., Ontario
HONORABLE MENTION
Camelot Mead (sweet), Oliver Wine Company, IN

X APERITIF AND FLAVORED SPECIALTY WINES

DESSERT SPECIALTY, *1979 Rating*
(38 entries)
BEST OF CLASS
Holiday, Brotherhood Winery, Washingtonville, NY
GOLD
Holiday, Brotherhood Winery, Washingtonville, NY
Manischewicz Cream Anasetta, Monarch Wine Company, Brooklyn, NY
Cream Almonique, Gross' Highland Winery, Absecon, NJ
Strawberry, Huber Winery, Borden, IN
SILVER
Stone's Ginger, Andres Wines, Winona, Ontario
Honey, Royal Wine Company, Milton, NY
BRONZE
Strawberry, Private Stock Winery, Boone, IA
Nashoba Valley Winery, Sommerville, MA
Manischewicz Cream Almonetta, Monarch Wine Company, Brooklyn, NY
Plum, Royal Wine Company, Milton, NY
Pear, Leelanau Wine Cellars, Omena, MI

FLAVORED DESSERT, *1980 Rating*
(18 entries)
SILVER
Cream Almonique, Gross' Highland Winery, NJ
May Wine, Brotherhood Corp., NY
Plum Royale, Royal Wine Corp., NY
Sweet Vermouth, London Winery, Ontario
BRONZE
May Wine, St. Julian Wine Company, MI
Holiday, Brotherhood Corp., NY
Dry Vermouth, London Winery, Ontario
Iowa Sweet White Rhubarb, Private Stock Winery, IA

FLAVORED WINES, (under 14% alcohol) *1981 Rating*
(10 entries)
SILVER
*Gross' Highland Winery, NJ, Cream Almonique
Monarch Wine Co., NY, Pina Coconetta
BRONZE
St. Julian Wine Co., MI, May Wine
Brotherhood Winery, NY, May Wine

FLAVORED WINES, (over 14% alcohol) *1981 Rating*
(12 entries)
GOLD
*Canandaigua Wine Co., NY, Chateau Martin Sweet Vermouth
SILVER
Fruit Wines of Florida, Floriana Almond Tropicale
BRONZE
London Winery Ltd., Ontario, London Sweet Vermouth
Brotherhood Winery, NY, Brother O'Brien
Fruit Wines of Florida, Floriana Cafe Tropicale

FLAVORED WINES, (under 14% alcohol) *1982 Rating*
(14 entries)
SILVER
*Cream d'Or, St. Julian Wine Co., MI
BRONZE
Old Fashioned Spiced Wine, Meier's Wine Cellars, OH
May Wine, Brotherhood Corporation, NY
Cream Almonique, Gross' Highland Winery, NJ
Holiday Wine, Brotherhood Corporation, NY

FLAVORED WINES, (over 14% alcohol) *1982 Rating*
(15 entries)
BRONZE
*Durouget, T.G. Bright & Co., Ontario
Chateau Martin Sweet Vermouth, Canandaigua, NY
Pina Coconetta, Monarch Wine Co., NY
London Sweet Vermouth, London Winery, Ontario
Ancient Mead, Golden, London Winery, Ontario

APERITIF AND FLAVORED WINES, (under 14% alcohol) *1983 Rating*
SILVER
*Frankenmuth May Wine (sweet), St. Julian Wine Company, MI
BRONZE
Cream Almonique (sweet), Bernard D'Arcy Wine Cellars, NJ
Old Fashioned Spiced Wine, Meier's Wine Cellars, OH
Sweet Cyser Honey Apple Wine (sweet), Nashoba Valley Winery, MA

APERITIF AND FLAVORED WINES, (over 14% alcohol) *1983 Rating*
BRONZE
*DuRouget Red (sweet), T.G. Bright Wines, Ontario
HONORABLE MENTION
London Sweet Vermouth (sweet), London Winery Ltd., Ontario

XI SHERRY

COCKTAIL SHERRY, *1979 Rating*
(8 entries)
BEST OF CLASS
Great Western Dry Sherry, Pleasant Valley Wine Company, Hammondsport, NY
SILVER
Great Western Dry Sherry, Pleasant Valley Wine Company, Hammondsport, NY
BRONZE
Special Selection Pale Dry, Widmer's Wine Cellars, Naples, NY

COCKTAIL SHERRY, *1980 Rating*
(8 entries)
SILVER
Fino Sherry, Brotherhood Corp., NY
Cocktail, Wiederkehr Wine Cellars, AR
BRONZE
Hallmark Dry, Chateau Gai Wines, Ontario
Pale Dry-Light, St. Julian Wine Company, MI

COCKTAIL SHERRY, *1981 Rating*
(8 entries)
BRONZE
*St. Julian Wine Co., MI, Van Buren Dry Sherry
Chateau Gai Wines, Ontario, Hallmark Dry Sherry

STRAIGHT SHERRY, *1981 Rating*
(7 entries)
SILVER
*Brotherhood Winery, NY, Golden Mellow Sherry

COCKTAIL SHERRY, *1982 Rating*
(9 entries)

SILVER

**Cocktail Sherry, Gold Seal Vineyards, NY

BRONZE

NYS Solera Dry Sherry, Great Western Winery, NY

STRAIGHT SHERRY, *1982 Rating*
(7 entries)

BRONZE

*No. 22 Sherry, Meier's Wine Cellars, OH

Sherry, Gold Seal Vineyards, NY

Golden Mellow Sherry, Brotherhood Corporation, NY

SHERRY, COCKTAIL DRY, *1983 Rating*
(2 entries)

BRONZE

Solera Dry Sherry (medium dry), Great Western, NY

Pale Dry Sherry, Widmer's Wine Cellars, NY

SHERRY, STRAIGHT, *1983 Rating*
(5 entries)

GOLD

*Sherry, Widmer's Wine Cellars, NY

SWEET SHERRY, *1979 Rating*
(20 entries)

BEST OF CLASS

Cream Sherry, Brotherhood Winery, Washingtonville, NY

GOLD

Cream Sherry, Brotherhood Winery, Washingtonville, NY

Solera Sherry, St. Julian Wine Company, Paw Paw, MI

Richelieu Golden Cream, Andres Wines, Winona, Ontario

Candlelight Sherry, London Winery, London, Ontario

SILVER

Special Selection Cream Sherry, Widmer's Wine Cellars, Naples, NY

Great Western Cream Sherry, Pleasant Valley Wine Company, Hammondsport, NY

Special Selection Sherry, Widmer's Wine Cellars, Naples, NY

BRONZE

Cream Sherry, Widmer's Wine Cellars, Naples, NY

Sweet Sherry, Renault Winery, Egg Harbor City, NJ

Classic Cream, Jordan Wines, St. Catharines, Ontario

CREAM SHERRY, *1980 Rating*
(14 entries)

GOLD

Cream Sherry, Brotherhood Corp., NY

SILVER

Hallmark Cream, Chateau Gai Wines, Ontario

BRONZE

Solera Cream, St. Julian Wine Company, MI

Classic Cream, Jordan & Ste. Michelle Cellars, Ontario

Cream, Great Western Winery, NY

Creme d'Or, London Winery, Ontario

Cream, Gold Seal Vineyards, NY

CREAM SHERRY, *1981 Rating*
(17 entries)

GOLD

*Taylor Wine Company, NY, Empire Cream Sherry

SILVER

St. Julian Wine Co., MI, Solera Cream

BRONZE

London Winery Ltd., Ontario, London Cream

Gold Seal Vineyards, NY, Cream Sherry

CREAM SHERRY, *1982 Rating*
(17 entries)

SILVER

**Cream Sherry, Wiederkehr Wine Cellars, AR

BRONZE

Cream Sherry 1980, Mantey Vineyards, OH
Solera Cream Sherry, St. Julian Wine Co., MI
Empire Cream Sherry, The Taylor Wine Co., NY

SHERRY, CREAM, *1983 Rating*
(10 entries)
BRONZE
*Special Selection Cream Sherry, Widmer's Wine Cellars, NY
HONORABLE MENTION
Cream Sherry, Widmer's Wine Cellars, NY

XII PORTS

RUBY PORT, *1979 Rating*
(7 entries)
BEST OF CLASS
1976 Vintage Port, Brotherhood Winery, Washingtonville, NY
GOLD
1976 Vintage Port, Brotherhood Winery, Washingtonville, NY
Ruby Port, Brotherhood Winery, Washingtonville, NY

RUBY PORT, *1980 Rating*
(5 entries)
SILVER
President, T.G. Bright & Company, Ontario
Ruby, Brotherhood Corp., NY
BRONZE
Port, Royal Wine Corp., NY
Celebration, Brotherhood Corp., NY

RUBY PORT, *1981 Rating*
(13 entries)
SILVER
*Brotherhood Winery, NY
Brotherhood Winery, NY, 1978 Vintage
BRONZE
Brotherhood Winery, NY, 1976 Vintage

RUBY PORT, *1982 Rating*
(8 entries)
GOLD
**Vintage Texas Port 1980, La Buena Vida Vineyards, TX
BRONZE
Port, Gold Seal Vineyards, NY
President Ruby Port, T.G. Bright & Co., Ontario

PORT, RUBY, *1983 Rating*
(8 entries)
BRONZE
*Port, Widmer's Wine Cellars, NY
HONORABLE MENTION
Port, Great Western, NY

TAWNY PORT, *1979 Rating*
(8 entries)
BEST OF CLASS
Celebration Port, Brotherhood Winery, Washingtonville, NY
BRONZE
Celebration Port, Brotherhood Winery, Washingtonville, NY

TAWNY PORT, *1980 Rating*
(4 entries)
No awards.

TAWNY PORT, *1981 Rating*
(3 entries)
BRONZE
*Brotherhood Winery, NY, Celebration Port

TAWNY PORT, *1982 Rating*
(1 entry)
SILVER
*Celebration Port, Brotherhood Corporation, NY

PORT, TAWNY, *1983 Rating*
(1 entry)
BRONZE
Tawny Port, Widmer's Wine Cellars, NY

WHITE PORT, *1982 Rating*
(2 entries)
BRONZE
*White Port, Post Winery, AR

PORT, WHITE, *1983 Rating*
No awards.

MADEIRA, *1983 Rating*
(1 entry)
HONORABLE MENTION
3 Islands Madeira, Long Winery, OH

XIII MISCELLANEOUS CLASSES

NOUVEAU RED, *1981 Rating*
(8 entries)
BRONZE
*Benmarl, NY, Nouveau 1981
Tewksbury Wine Cellars, NJ, Harvest Nouveau

SOFT WINES, *1981 Rating*
(5 entries)
SILVER
(Tied for best of class.)
*Taylor Wine Co., NY, Soft White
*Chateau Gai Wines, Ontario, Capistro 1 Soft Wine

NOUVEAU RED, *1982 Rating*
(2 entries)
BRONZE
Nouveau Red 1982, Ingleside Plantation Winery, VA

SOFT WINES
(5 entries)
BRONZE
Lake County Soft White, The Taylor Wine Co., NY

NOUVEAU, *1983 Rating*
(5 entries)
BRONZE
*Nouveau 1983 (dry), Ingleside Plantation Winery, VA
Nouveau 1983 (dry), Benmarl Wine Co., NY
HONORABLE MENTION
Nouveau 1983 (dry), Oasis Vineyards, VA

SOFT WINES
(8 entries)
HONORABLE MENTION
Lake Country Soft White (sweet), The Taylor Wine Co., NY

Chardonnay

ORANGE COUNTY FAIR CALIFORNIA WINE AWARDS 1979-1984

CALIFORNIA WINE RATINGS

Wine, like music and art is very subjective when it comes to personal taste. When it comes to judging the best, the sum total of a large panel of experts is the closest one can get when seeking an objective professional opinion. In my effort to be as objective as possible, with the cooperation of the Orange County Wine Society, I bring to you the results of the 1979 through 1984 Orange County Fair Commercial Wine competitions.

The following provides a summary of the judging of a total of 7918 different wines. Results are presented alphabetically, by variety. All winners of the same variety, for the six judgings are grouped together. Note: All varieties were not judged in all years. Generic and proprietary wines were judged only in 1984 and are at the end following Sparkling Wines.

Every attempt was made to place each wine in the appropriate price category at the time of judging. Because of possible changes in wholesale prices, coupled with the free market conditions, consumers may find wines marked slightly above or below the price ranges shown. 1979 Competition included nine varieties and 507 entries.

1979 JUDGING PANEL

Richard Arrowhead, *Chateau St. Jean*
David Bennion, *Ridge Vineyards*
Ken Brown, *Zaca Mesa Winery*
Ken Burnap, *Santa Cruz Mountain Vineyard*
Charles Crawford, *E & J Gallo Winery*
Albert Cribari, *B. Cribari & Sons*
Richard Elwood, *Llords & Elwood Winery*
Tom Ferrell, *Inglenook Vineyards*
John Hoffman, *Christian Brothers*
Marty Lee, *Kenwood Vineyards*
John Merritt, *Gundlach-Bundschu Winery*
Bonny Meyer, *Silver Oak Cellars*
Steve Mirassou, *Mirassou Vineyards*
Tim Mondavi, *Robert Mondavi Winery*
Myron Nightingale, *Beringer Vineyards*
Steve O'Donnell, *Callaway Vineyards*
Rosemary Papagni, *Papagni Vineyards*
Phyllis Pedrizzetti, *Pedrizzetti Winery*
Don Sebastiani, *Sebastiani Winery*
Peter Stern, *Turgeon-Lohr Winery*
Rodney Strong, *Sonoma Vineyards*
Janet Trefethen, *Trefethen Vineyards*
Warren Winiarski, *Stag's Leap Wine Cellars*
Frank Woods, *Clos Du Bois*

The 1980 wine competition included thirteen different, popular varietal wine classifications, from 733 entries; including Dry Chenin Blanc, Chenin Blanc—Off Dry, Gewurztraminer, Sauvignon Blanc, Chardonnay, Blanc de Noir, Petite Sirah, Cabernet Sauvignon, Pinot Noir, Zinfandel, Dry Late Harvest Zinfandel, Sweet Late Harvest Zinfandel, and Dry Sherry.

1980 JUDGING PANEL

Ray Krause, *Rancho Yerba Buena Cellars*
Jim Carter, *Sebastiani Vineyards*
Ed Pedrizzetti, *Pedrizzetti Winery*
Albert Cribari, *B. Cribari & Sons*
Bill Jekel, *Jekel Vineyards*
Nils Venge, *Villa Mt. Eden Winery*
J. Michael Rowan, *Jordan Vineyard and Winery*
John A. Parducci, *Parducci Winery, Ltd.*
Myron Nightingale, *Beringer Vineyards*
Charles Crawford, *Ernest & Julio Gallo Winery*
Bonny Meyer, *Silver Oak Cellars*
James Ahern, *Ahern Winery*
Dawnine Sample, *Domaine Chandon*
Richard Arrowood, *Chateau St. Jean*
Richard Elwood, *Llords & Elwood Winery*
Warren Winiarski, *Stag's Leap Wine Cellars*
Zelma Long, *Long Vineyard & Simi Winery*
Steve Mirassou, *Mirassou Vineyards*
Ken Brown, *Zaca Mesa Winery*
James Lawrence, *Lawrence Winery*
Tim Mondavi, *Robert Mondavi Winery*
Frank Woods, *Clos du Bois*
Phyllis Pedrizzetti, *Pedrizzetti Winery*
John Hoffman, *The Christian Brothers*
Robert Kozlowski, *Kenwood Vineyards*
Jim Prager, *Prager Port Works*
George Kolarovich, *Perelli-Minetti Winery*
Peter Stern, *Turgeon-Lohr Winery*
Rodney D. Strong, *Sonoma Vineyards*
David Bennion, *Ridge Vineyards*
John Merritt, *Gundlach-Bundschu Winery*
Leon Sobon, *Shenandoah Vineyards*
John Kenworthy, *Kenworthy Vineyards*
Angelo Papagni, *Papagni Vineyards*
David Stare, *Dry Creek Vineyards*
Steve O'Donnell, *Callaway Vineyard & Winery*

1981 COMPETITION

TOTAL ENTRIES—1130; GOLD MEDALS—112; SILVER MEDALS—139; BRONZE MEDALS—155

1981 PANEL OF JUDGES

Jerry D. Mead, *Chairman*
Jim Ahern, *Ahern Winery*
Richard Arrowood, *Chateau St. Jean*
Dave Bennion, *Ridge Vineyards*
Ken Brown, *Zaca Mesa Vineyards*
Ken Burnap, *Santa Cruz Mountain Vineyard*
Jim Carter, *Sebastiani Vineyards*
Charles M. Crawford, *E & J Gallo*
Albert Cribari, *Cribari & Sons*
Richard Elwood, *Llords & Elwood Winery*
John Hoffman, *The Christian Brothers*
Bill Jekel, *Jekel Vineyards*
Tor Kenward, *Prager Port Works*
John Kenworthy, *Kenworthy Vineyards*
Bob Kozlowski, *Kenwood Vineyards*
Ray Krause, *Farview Farm Vineyard*
James Lawrence, *Lawrence Winery*
Zelma Long, *Long Vineyards & Simi Winery*
Steve MacRostie, *Hacienda Wine Cellars*
John Merritt, *Gundlach-Bundschu Vineyards*
Bonny Meyer, *Silver Oak Cellars*
Tim Mondavi, *Robert Mondavi Winery*
Myron Nightingale, *Beringer Vineyards*
Angelo Papagni, *Papagni Vineyards*
John A. Parducci, *Parducci Wine Cellars*
Phyllis Pedrizzetti, *Pedrizzetti Winery*
Jim Prager, *Prager Port Works*
Dawnine Sample, *Domaine Chandon*
Leon Sobon, *Shenandoah Cellars*
David Stare, *Dry Creek Vineyards*
George Starke, *Napa Wine Cellars*
Peter Stern, *J. Lohr Wines*
John Turner, *Turner Winery*
Janet Trefethen, *Trefethen Vineyards*
Nils Venge, *Villa Mt. Eden*
Warren Winiarski, *Stag's Leap Wine Cellars*

The 1982 competition included seventeen (17) different wine varieties ... Barbera, Cabernet Sauvignon, Charbono, Chardonnay, Chenin Blanc, Gamay Beaujolais, Gewurztraminer, Merlot, Muscat, Petite Sirah, Pinot Blanc, Pinot Noir, Sauvignon (Fumé) Blanc, White (Johannisberg) Riesling, Zinfandel, and Sparkling Wine (all methods). TOTAL ENTRIES—1387; GOLD MEDALS—91; SILVER MEDALS—149; BRONZE MEDALS—204.

1982 PANEL OF JUDGES

Jerry D. Mead, *Chairman*
Jim Ahern, *Ahern Winery*
Bill Arnold, *Smothers Vine Hill Wines*
Richard Arrowood, *Chateau St. Jean*
Fred Brander, *Santa Ynez Winery & Brander Vineyards*
Ken Brown, *Zaca Mesa Vineyards*
Jim Bundschu, *Gundlach-Bundschu Winery*
Jim Carter, *Sebastiani Vineyards*
Jim Concannon, *Concannon Vineyards*
David Cordtz, *Cordtz Brothers Cellars*
Charles M. Crawford, *E. & J. Gallo*
Kerry Damsky, *San Pasqual Vineyards*
Joseph Franzia, *JFJ Bronco Winery*
Don Harrison, *Buena Vista Winery and Vineyards*
Bill Jekel, *Jekel Vineyards*
Josh Jensen, *Calera Wine Company*
Tor Kenward, *Prager Winery & Port Works*
John Kenworthy, *Kenworthy Vineyards*
Bob Kozlowski, *Kenwood Vineyards*
Ray Krause, *Farview Farm Vineyard*
James Lawrence, *Lawrence Winery*
Jerry Lohr, *J. Lohr Wines*
Zelma Long, *Long Vineyards & Simi Winery*
Richard Longoria, *J. Carey Cellars*
John Merritt, *Bandiera Winery*
Bonny Meyer, *Silver Oak Cellars*
Daniel Mirassou, *Mirassou Vineyards*
Myron Nightingale, *Beringer Vineyards*
Paul Obester, *Obester Winery & Gemello Winery*
Steve O'Donnell, *Callaway Vineyards and Winery*
Angelo Papagni, *Angelo Papagni Vineyards*
John A. Parducci, *Parducci Wine Cellars*

Phyllis Pedrizzetti, *Pedrizzetti Winery*
Brian Pendleton, *Pendleton Winery*
Richard G. Peterson, *The Monterey Vineyard*
Jim Prager, *Prager Winery & Port Works*
Dawnine Sample-Dyer, *Domaine Chandon*
David Stare, *Dry Creek Vineyards*
George Starke, *Napa Cellars*
Rodney D. Strong, *Sonoma Vineyards*
John Turner, *Turner Winery*
Nils Venge, *Villa Mt. Eden*
Carolyn Wente, *Wente Bros.*
Warren Winiarski, *Stag's Leap Wine Cellars*

The 1983 competition, for the first time, included every wine varietal made in California, including Rosé and White wines labeled with the varietal from which they were made. In addition, all Sherries and all Sparkling wines were judged. There were 55 different classes, further broken down into sub-categories by price and residual sugar. TOTAL ENTRIES—1854; GOLD MEDALS—128; SILVER MEDALS—180; BRONZE MEDALS—296.

1983 PANEL OF JUDGES

Jerry D. Mead, *Chairman*
Jim Ahern, *Ahern Winery*
Bill Arnold, *Smothers Vine Hill Wines*
Richard Arrowood, *Chateau St. Jean*
David Bennion, *Ridge Vineyards*
Rick Boyer, *Ventana Vineyards*
Fred Brander, *Santa Ynez Valley*
Ken Brown, *Zaca Mesa Winery*
Ken Burnap, *Santa Cruz Mountain Vineyards*
Richard Carey, *R. Montali Winery*
Jim Carter, *Sebastiani Vineyards*
David Cordtz, *Cordtz Brothers Cellars*
Mitch Cosentino, *Crystal Valley Cellars*
Charles M. Crawford, *E. & J. Gallo*
Al Cribari, *Cribari & Sons*
Kerry Damsky, *San Pasqual Vineyards*
Jill Davis, *Buena Vista Winery*
Norman C. de Leuze, *ZD Wines*
Hal Doran, *Parsons Creek Winery*
John Fetzer, *Fetzer Vineyards*
Rod Foppiano, *Louis J. Foppiano*
Joseph Franzia, *JFJ Bronco*
Alison Green, *The Firestone Vineyard*
Gary B. Heck, *Korbel & Bros.*
Rick Jekel, *Jekel Vineyards*
Josh Jensen, *Calera Wine Company*
Tor Kenward, *Prager Winery & Port Works*
John Kenworthy, *Kenworthy Vineyards*
George Kolarovich, *Fairmont Vineyards*
Bob Kozlowski, *Kenwood Vineyards*
Ray Krause, *Farview Farms*
James Lawrence, *Giumarra Vineyards*
Jerry Lohr, *Turgeon & Lohr Winery*
Richard Longoria, *J. Carey Vineyards*
Munro Lyeth, Jr., *Lyeth Vineyard*
Steve MacRostie, *Hacienda Wine Cellars*
John Merritt, *Bandiera Winery*
Bonny Meyer, *Silver Oak Cellars*
Daniel Mirassou, *Mirassou Vineyards*
Tim Mondavi, *Robert Mondavi Winery*
Myron Nightingale, *Beringer Vineyards*
Paul Obester, *Obester & Gemello Wineries*
Steve O'Donnell, *Callaway Vineyards*
Angelo Papagni, *Papagni Vineyards*
John A. Parducci, *Parducci Wine Cellars*
Phyllis Pedrizzetti, *Pedrizzetti Winery*
Brian Pendleton, *Pendleton Winery*
Richard Peterson, *The Monterey Vineyard*
Jim Prager, *Prager Winery & Port Works*
Robert Roudon, *Roudon-Smith Vineyards*
Dawnine Sample-Dyer, *Domaine Chandon*
Leon Sobon, *Shenandoah Vineyards*
George Starke, *Napa Cellars*
Rodney D. Strong, *Sonoma Vineyards*
Brother Timothy, *The Christian Brothers*
John Turner, *Turner Winery*
Nils Venge, *Groth Vineyard & Winery*
Tom Wiggington, *Franciscan Vineyards*
Warren Winiarski, *Stag's Leap Wine Cellars*

1984 WORLD'S LARGEST WINE COMPETITION

In 1984, cork finished generic and proprietary table wines in 750ml bottles were added to the competition, as were all fortified wines. A total of 2307 wines were judged, making the Competition the largest such competition in the world, for the second year in a row. TOTAL ENTRIES—2307; GOLD MEDALS—178; SILVER MEDALS—271; BRONZE MEDALS—422.

1984 PANEL OF JUDGES

Jerry D. Mead, *Chairman*
Jim Ahern, *Ahern Winery*
Bill Arnold, *Smothers Vine Hill Wines*
Richard Arrowood, *Chateau St. Jean Vineyards & Winery*
Robert S. Atkinson, *Stony Ridge*
Dave Bennion, *Ridge Vineyards*
Rick Boyer, *Ventana Vineyards*
Fred Brander, *The Brander Vineyard*
Robert L. Broman, *Stag's Leap Wine Cellars*
Ken Brown, *Zaca Mesa Winery*
David H. Bruce, *David Bruce Winery*
George Bursick, *McDowell Valley Cellars*
Richard Carey, *R. Montali Winery*
Jim Concannon, *Concannon Vineyards*
David Cordtz, *Cordtz Brothers Cellars*
Mitch Consentino, *Crystal Valley Cellars*
Charles M. Crawford, *E. & J. Gallo Winery*
Al Cribari, *Cribari and Sons*
John Culbertson, *John Culbertson Winery*
Kerry Damsky, *San Pasqual Vineyards*
John E. Daumé, *The Daumé* Winery
Jill Davis, *Buena Vista Winery & Vineyards*
Duane de Boer, *Smith & Hook Winery*
Norman C. de Leuze, *ZD Wines*
Hal Doran, *Parsons Creek Winery*
Tom Eddy, *The Christian Brothers*
Richard Elwood, *Llords & Elwood*
Tom Ferrell, *Franciscan Vineyards*
John Fetzer, *Fetzer Vineyards*
Joseph Franzia, *JFJ Bronco Winery*
Cary Gott, *Corbett Canyon Vineyard*
Alison Green, *The Firestone Vineyard*
William T. Harper IV, *Sebastiani Vineyards*
Scott Harvey, *Santino Winery*
Dwayne Helmuth, *Callaway Vineyards & Winery*
Bill Jekel, *Jekel Vineyards*
Josh Jensen, *Calera Wine Company*
Bob Kozlowski, *Kenwood Vineyards*
Ray Krause, *Farview Farm Vineyards*
Jerry Lohr, *Turgeon and Lohr Winery*
Zelma Long, *Long Vineyards & Simi Winery*
Richard Longoria, *J. Carey Vineyards & Cellars*
Steve MacRostie, *Hacienda Wine Cellars*
Michael Martini, *Louis Martini Winery*
Ed Masciana, *HMR Winery*
John Merritt, *Bandiera Winery*
James A. Milone, *Milano Winery*
Daniel Mirassou, *Mirassou Vineyards*
Marc C. Mondavi, *Charles Krug Winery*
Sandra L. Obester, *Gemello Winery*
Paul Obester, *Obester Winery*
Angelo Papagni, *Papagni Vineyards*
John A. Parducci, *Parducci Wine Cellars*
Phyllis Pedrizzetti, *Pedrizzetti Winery*
Brian Pendleton, *Pendleton Winery*
Heidi Peterson, *Buehler Winery*
Richard Peterson, *The Monterey Vineyard*
Jim Prager, *Prager Winery & Port Works*
Walter Raymond, *Raymond Vineyards*
Robert Roudon, *Roudon-Smith Vineyards*
Jack Ryno, *Mesa Verde*
Dawnine Sample-Dyer, *Domaine Chandon*
Leon Sobon, *Shenandoah Vineyards*
Robert M. Stashak, *Korbel Bros.*
David Stare, *Dry Creek Vineyards*
Rodney D. Strong, *Rodney Strongs Wine & Piper Sonoma*
John Turner, *Turner Winery*
Nils Venge, *Groth Vineyard & Winery*
Carolyn Wente, *Wente Bros.*

ALICANTE BOUSCHET, *1983 Rating*
($5.01 to $7.00—1 entry)
BRONZE
Angelo Papagni, 1978, California, Clovis Vineyard

BARBERA, *1982 Rating*
($4.51 to $5.50—6 entries)
BRONZE
Parducci Wine Cellars, 1977, North Coast
Walker Wines, 1979, Solano County, Nyland Vineyard
Borra's Cellar, 1978, California, Estate Bottled

BARBERA, *1982 Rating*
($5.51 and Up—3 entries)
SILVER
Monterey Peninsula Winery, 1979, Northern California, Vineyard View
BRONZE
Sebastiani Vineyards, 1976, Northern California, Proprietors Reserve

BARBERA, *1983 Rating*
(Up to $5.00—5 entries)
GOLD
Louis M. Martini, 1979, California
BRONZE
Parducci Wine Cellars, 1977, Mendocino County

BARBERA, *1983 Rating*
($5.01 to $6.50—2 entries)
GOLD
Gemello Winery, 1977, California

BARBERA, *1983 Rating*
($6.51 and Up—3 entries)
GOLD
Montevina Wines, 1980, Amador County, Estate Bottled, Special Selection
BRONZE
Sebastiani Vineyards, 1977, Northern California, Proprietor's Reserve

BARBERA, *1984 Rating*
(Up to $5.00—4 entries)
GOLD
Estrella River, 1979, San Luis Obispo County, Estate Bottled
SILVER
Papagni Vineyards, 1978, Madera, Clovis Vineyard, Estate Bottled
BRONZE
Louis M. Martini, 1980, California

BARBERA, *1984 Rating*
($5.01 to $6.50—2 entries)
BRONZE
Montevina, 1981, Shenandoah Valley, Estate Bottled

BARBERA, *1984 Rating*
($6.51 and Up—2 entries)
GOLD
Sebastiani Vineyards, 1978, California, Proprietor's Reserve
BRONZE
Pedrizzetti, 1974, California

BARBERA BLANC, *1983 Rating*
(Up to $5.00—1 entry)
BRONZE
San Antonio Winery, 1982, California, Light

BLANC DE NOIR, *1980 Rating*
(Up to $4.00—23 entries)
GOLD
Trader Joe's, NV, "Daybreak" Blanc de Noir, Napa Valley
SILVER
Weibel, 1979, Pinot Noir Blanc, Mendocino "Eye of the Partridge"
Mill Creek, 1979, Burgundy Blanc, Sonoma County
BRONZE
Santa Ynez Valley, 1979, Blanc de Cabernet Sauvignon, California

BLANC DE NOIR, *1980 Rating*
($4.01-$6.00—24 entries)
SILVER
Grand Cru Vineyards, 1979, Pinot Noir Blanc, Alexander Valley,

Garden Creek Ranch
Johnson's Alexander Valley, 1979, Pinot Noir Blanc, Alexander Valley

BRONZE

Hacienda Wine Cellars, 1979, Pinot Noir Blanc, Sonoma County
Geyser Peak, 1979, Pinot Noir Blanc, Sonoma County
Kenwood, 1979, Pinot Noir Blanc, Sonoma Valley
Franciscan, 1978, Pinot Noir Blanc, Napa Valley

BLANC DE NOIR, *1983 Rating*
(Up to $4.50—2 entries)

SILVER

Fritz Cellars, 1982, Sonoma County, Dry Creek Valley, Estate Bottled

BLANC DE NOIR, *1983 Rating*
($4.51 to $6.00—13 entries)

GOLD

Whitehall Lane Winery, 1982, Serena Vineyards

SILVER

Mark West Vineyards, 1982, Russian River Valley, Estate Bottled

BRONZE

Leeward, 1982, Santa Maria Valley, "Coral"

BLANC DE NOIR, *1983 Rating*
($6.01 and Up—2 entries)

SILVER

Ross-Keller, 1982, Santa Barbara County, Los Alamos Vineyard

CABERNET SAUVIGNON, *1979 Rating*
(Up to $4.00—27 entries)

GOLD

Vache, 1976, (Brookside)

SILVER

Cresta Blanca, NV
Guasti, 1977
Stone Creek, 1975, North Coast, Special Selection
Stony Ridge, NV, Lot 7477

BRONZE

Beaulieu, 1976, Beau Tor
Bel Arbres, NV
Franzia, NV
Rancho Yerba Buena, 1976
River Oaks, 1977, Sonoma County

CABERNET SAUVIGNON, *1979 Rating*
($4.01-$8.00—80 entries)

GOLD

Gundlach-Bundschu, 1977, Batto Ranch
Robert Mondavi, 1975
Mountainside, 1976
Parducci, 1975, Cellarmaster's Selection

SILVER

Clos du Val, 1975, Grandval
Fetzer Mendocino, 1976, Estate Bottled
Sterling, 1975, Estate Bottled
Stonegate, 1976, Sonoma County
Trefethen, 1975
Veedercrest, 1977, North Branch Vineyard

BRONZE

Arroyo, 1977
Beaulieu, 1975, Estate Bottled
Dry Creek, 1976, Second Bottling
Franciscan, 1975, Alexander Valley
Freemark Abbey, 1974
Liberty School, Lot 5
Mount Palomar, NV, Temecula
San Martin, 1975
Simi, 1975
Smothers, 1977, Robert Young Vineyard
Sonoma Vineyards, 1975
Souverain, 1976
Stone Creek, 1974, North Coast, Bin 89

CABERNET SAUVIGNON, *1979 Rating*
($8.01 and Up—34 entries)

GOLD

Chateau Montelena, 1974, Sonoma
Chateau St. Jean, 1976, Glen Ellen
The Rothschild Brothers of

California, 1973
Villa Mt. Eden, 1976, Estate Bottled

SILVER

Burgess, 1976, Napa Valley
Inglenook, 1974, Cask D8, Estate Bottled
Robert Mondavi, 1973, Reserve
Spring Mountain, (Les Trois Cuvées), N.V.
Stag's Leap Wine Cellars, 1976
Sterling, 1974, Reserve

BRONZE

Arroyo, 1975, Sonoma
Richard Carey, 1975, Special Reserve Lake County
Freemark Abbey, 1975, Caberent Bosche
Hacienda, 1976
Hoffman Mountain Ranch, 1975, Estate Bottled
Monterey Peninsula, 1976, Monterey
Ridge, 1975, Monte Bello

CABERNET SAUVIGNON, *1980 Rating*
(Up to $5.00—29 entries)

GOLD

Bel Arbres, N.V., California
Fetzer, 1978, Lake County
Almaden, 1977, Monterey
Fetzer, 1977, Mendocino

SILVER

Barengo Vineyards, 1976, Lake County
Christian Brothers, N.V., Napa Valley

BRONZE

Franciscan, 1976, Sonoma County
San Martin, 1977, California
Raymond Hill, 1976, Sonoma County (Trader Joe's Winery)
Giumarra, N.V.
R & J Cook, 1978, Clarksburg Red Table Wine (Made entirely from Cabernet Sauvignon Grapes)

CABERNET SAUVIGNON, *1980 Rating*
($5.01-$10.00—97 entries)

GOLD

Trefethen Vineyards, 1976, Napa Valley
Sebastiani Vineyards, N.V., Northern California
Rutherford Hill, 1976, Napa Valley
Caymus Vineyards, 1976, Napa Valley, Estate Bottled
Lower Lake Winery, 1977, Lake County, Stromberg's Hummel, Lane Vineyards
Raymond, 1977, Napa Valley, Estate Bottled
Freemark Abbey, 1975, Napa Valley
Monterey Peninsula Winery, 1977, Monterey
Boeger, 1977, El Dorado County

SILVER

Dehlinger Winery, 1977, Sonoma County
Beaulieu Vineyards, 1977, Rutherford, Estate Bottled
Diamond Creek, 1977, Red Rock Terrace, First Pick
Estrella River Winery, 1977, San Luis Obispo, Estate Bottled
Souverain, 1976, North Coast
Richard Carey Winery, 1975, Lake County, Cask 714
Sterling, 1976, Napa Valley, Estate Bottled

BRONZE

Stag's Leap Wine Cellars, 1977, Napa Valley
Rancho Yerba Buena, 1978, Alexander Valley
Fieldstone, 1977, Alexander Valley, Estate Bottled
Hacienda Wine Cellars, 1977, Sonoma Valley
Chateau Montelena, 1975, North Coast
Harbor Winery, 1977, Amador County, Deaver Vineyard, (Unfined-Unfiltered)
Beringer, 1976, Sonoma, Knight's Valley Estate
Sebastiani Vineyards, 1973, Proprietor's Reserve, Northern California

Robert Mondavi, 1977
Callaway, 1977, Temecula
Sommelier, 1977, San Luis Obispo
Mario Perelli-Minetti, 1977, Napa Valley
Lambert Bridge, 1976, Sonoma County
Parducci, 1977, Mendocino County
HMR, 1976, Paso Robles, Estate Bottled
Chateau Chevalier, 1978, Napa Valley
Clos du Val, 1977, Napa Valley
Rutherford Vintners, 1977, Napa Valley

CABERNET SAUVIGNON, *1980 Rating*
($10.00 and Up—17 entries)
GOLD
Ridge, 1977, York Creek
Heitz Cellar, 1975, Martha's Vineyard
SILVER
Conn Creek, 1976, Napa
Villa Mt. Eden, 1977, Napa
Buena Vista, 1974, Sonoma, Cask 25
Joseph Phelps, 1977, Napa
BRONZE
Charles Krug, 1974, Lot F-1, Napa Vintage Select

CABERNET SAUVIGNON, *1981 Rating*
(Up to $6.00—45 entries)
GOLD
Almaden Vineyards, 1978, Monterey
Stony Ridge Winery, 1975, North Coast
The Monterey Vineyard, 1978, Central Coast Counties, "Classic Red"
Rancho Yerba Buena, 1978, Alexander Valley
SILVER
Bel Arbres Vineyards, 1978, Lake County
Richard Carey Winery, 1977, Santa Barbara
Franciscan Vineyards, 1978, Alexander Valley
BRONZE
Fetzer Vineyards, 1978, Mendocino
J. Pedroncelli Winery, 1978, Sonoma County
Sebastiani Vineyards, N.V., Northern California
Toyon Vineyards, 1978, North Coast

CABERNET SAUVIGNON, *1981 Rating*
($6.01 to $10.00—92 entries)
GOLD
Hacienda Wine Cellars, 1978, Sonoma Valley, Buena Vista Vineyards
Rutherford Ranch Brand, 1978, Napa Valley
Sunrise Winery, 1978, Mendocino, Frey Vineyard
Cassayre-Forni Cellars, 1978, Napa Valley
Conn Creek Winery, 1977, Napa Valley
Round Hill Vineyards, 1978, Napa Valley
Burgess Cellars, 1978, Napa Valley
McDowell Valley Vineyards, 1978, Mendocino County, Estate Bottled
SILVER
Sycamore Creek Vineyards, 1979, San Luis Obispo County
Trefethen Vineyards, 1977, Napa Valley
Raymond Vineyard & Cellar, 1978, Napa Valley, Estate Bottled
Sotoyome Winery, 1978, Sonoma County
Alexander Valley Vineyards, 1978, Alexander Valley, Estate Bottled
BRONZE
Clos Du Bois, 1978, Alexander Valley
Domaine Laurier, 1978, Sonoma County, Estate Bottled
HMR Vineyards, 1978, Central Coast Counties, Hoffman Vineyards

Pine Ridge, 1978, Napa Valley, Rutherford District
Jekel Vineyard, 1977, Monterey County
Beaulieu Vineyard, 1977, Napa Valley, Rutherford, Estate Bottled
Field Stone Winery, 1978, Alexander Valley
Kenwood Vineyards, 1978, Sonoma Valley
Zaca Mesa, 1978, Santa Ynez Valley, Estate Bottled
Edmeades Vineyards, 1978, Mendocino, Anderson Valley, Estate Bottled
Landmark Vineyards, 1978, Sonoma County (75%), Napa County (25%)
Gundlach-Bundschu Winery, 1979, Sonoma Valley, Olive Hill Vineyard
Roudon-Smith Vineyards, 1978, Sonoma County
San Martin Winery, 1977, California

CABERNET SAUVIGNON, *1981*
Rating
($10.01 and Up—50 entries)

GOLD
Santa Cruz Mountain Vineyard, 1978, Santa Cruz Mountains, John Bates Ranch
Durney Vineyard, 1978, Monterey County, Carmel Valley
Beringer Vineyards, 1977, Napa Valley, Private Reserve, Lemon Ranch
Cakebread Cellars, 1978, Napa Valley
Grand Cru Vineyards, 1978, Alexander Valley, Garden Creek Ranch, "Collector's Series"
Buena Vista Winery, 1977, Sonoma, Haraszthy Cellars, Cask 34

SILVER
Burgess Cellars, 1977, Napa Valley, "Vintage Selection"
Milano Winery, 1978, Mendocino County, Sanel Valley, Sanel Valley Vineyard
Heitz Cellar, 1976, Napa Valley, Martha's Vineyard
Duckhorn Vineyards, 1978, Napa Valley
HMR Vineyards, 1975, Paso Robles, Hoffman Vineyards, "Doctor's Reserve"
Ridge Vineyards, 1978, Napa County, York Creek
William Hill, 1978, Napa Valley, Mount Veeder

BRONZE
Shafer Vineyards, 1978, Napa Valley
Villa Mt. Eden Winery, 1978, Napa Valley, Estate Bottled
The Firestone Vineyard, 1977, Santa Ynez Valley, Vintage Reserve
Clos Du Val, 1978, Napa Valley
Gundlach-Bundschu Winery, 1979, Sonoma Valley, Batto Ranch
Matanzas Creek Winery, 1978, Sonoma County, Estate Bottled
Chateau Montelena Winery, 1977, Sonoma
Pendleton Winery, 1978, Napa Valley
Stag's Leap Wine Cellars, 1978, Napa Valley, Stag's Leap Vineyards
HMR Vineyards, 1977, Paso Robles, Hoffman Vineyards
Diamond Creek Vineyards, 1978, Napa Valley, Red Rock Terrace
Robert Mondavi Winery, 1975, Napa Valley, "Reserve"
Simi Winery, 1974, Alexander Valley, "Reserve Vintage"

CABERNET SAUVIGNON, *1982*
Rating
(Up to $7.00—60 entries)

GOLD
Bel Arbres Vineyards, 1979, Lake County
Geyser Peak Winery, 1977, California

SILVER
Alexander Valley Vineyards, 1979, Alexander Valley, Estate Bottled

Sierra Vista Winery, 1979, El Dorado County
Taylor California Cellars, N.V., California
J. Pedroncelli Winery, 1979, Sonoma

BRONZE

Cambiaso Winery and Vineyards, 1978, Sonoma
Fortino Winery, 1978, Santa Clara County
Fetzer Vineyards, 1980, Lake County
River Oaks Vineyards, 1979, Alexander Valley
Montevina Wines, 1979, Amador County, Estate Bottled
Fetzer Vineyards, 1979, Mendocino County
Beaulieu Vineyard, 1979, Napa Valley, Beau Tour, Estate Bottled
Mount Eden Vineyards, 1978, Santa Cruz Mountains, Estate Bottled

CABERNET SAUVIGNON, *1982 Rating*
($7.01 to $11.00—101 entries)

GOLD

Stony Ridge Winery, 1979, Monterey County, Smith & Hook Vineyard
Field Stone Winery, 1978, Alexander Valley, Estate Bottled
Carmel Bay Winery, 1978, Monterey County, D & M Junction Vineyards

SILVER

HMR Hoffman Vineyards, 1979, Central Coast
McDowell Valley Vineyards, 1979, Mendocino County, Estate Bottled
Fenestra, 1979, Monterey
Pine Ridge Winery, 1979, Napa Valley, Rutherford District
Zaca Mesa Winery, 1978, Santa Ynez Valley, Special Selection

BRONZE

Cordtz Brothers Cellars, 1980, Upper Alexander Valley
Grand Cru Vineyards, 1980, California
Davis Bynum Winery, 1979, Sonoma
Beaulieu Vineyard, 1978, Napa Valley, Rutherford, Estate Bottled
Obester Winery, 1979, Sonoma County, Batto Ranch
Dry Creek Vineyard, 1979, Sonoma County
Trefethen Vineyards, 1978, Napa Valley, Trefethen Vineyards
Girard Winery, 1980, Napa Valley, Estate Bottled
Clos du Bois, 1978, Alexander Valley
Estrella River Winery, 1978, San Luis Obispo, Estate Bottled
Raymond Vineyard & Cellar, 1979, Napa Valley, Estate Bottled
Marietta Cellars, 1978, Dry Creek
Ahlgren Vineyard, 1979, Napa Valley, Rutherford
J. Pedroncelli Winery, 1978, Sonoma, Vintage Selection
Jekel Vineyard, 1978, Monterey, Estate Bottled
Mirassou Vineyards, 1979, Monterey County, Harvest Reserve
Veedercrest Vineyards, 1979, Napa Valley, North Ranch Vineyard
Pommeraie Vineyards, 1979, Sonoma County
Boeger Winery, 1979, El Dorado County, Estate Bottled
Rutherford Hill Winery, 1978, Napa Valley

CABERNET SAUVIGNON, *1982 Rating*
($11.01 and Up—59 entries)

GOLD

Monterey Peninsula Winery, 1979, Monterey, Arroyo Seco

SILVER

Chateau Montelena Winery, 1978, Sonoma
Conn Creek, 1978, Napa Valley, Lot 1
Clos du Bois, 1978, Alexander Valley, Marlstone Vineyard

Stonegate, 1978, Napa Valley
Joseph Phelps Vineyards, 1978, Napa Valley, Backus
Kenwood Vineyards, 1979, Sonoma Valley, Jack London Vineyard
Heitz Wine Cellars, 1977, Napa Valley, Martha's Vineyard
ZD Wines, 1979, California

BRONZE

Beaulieu Vineyard, 1977, Napa Valley, Private Reserve
Charles Krug, 1977, Napa Valley, Vintage Select
Jekel Vineyard, 1978, Monterey, Private Reserve, Estate Bottled
Shown & Sons Vineyards, 1979, Rutherford, Napa Valley, Estate Bottled
Robert Mondavi Winery, 1977, Napa Valley, Reserve
Buena Vista Winery & Vineyards, 1978, Sonoma Valley, Special Selection, Carneros, Estate Bottled

CABERNET SAUVIGNON, *1982*
Rating
($11.51 and Up—83 entries)

GOLD

Buena Vista Winery, 1979, Sonoma Valley, Special Selection
Shenandoah Vineyards, 1980, Amador County
Kenwood Vineyards, 1979, Sonoma Valley, Artist Series
Sycamore Creek, 1981, Central Coast
Sequoia Grove, 1980, Napa Valley, Cask One
Shafer Vineyards, 1979, Napa Valley
Monterey Peninsula, 1980, Monterey County

SILVER

Pine Ridge, 1980, Napa Valley, Rutherford District
Santa Cruz Mountain Vineyard, 1979, Santa Cruz Mountains, Bates Ranch
Rothschild Brothers of California, 1979, Napa Valley, Private Reserve
H. Coturri & Sons, 1980, Sonoma Valley, Horn Vineyards
Santa Cruz Mountain Vineyard, 1980, Napa Valley, Gamble Ranch
Robert Keenan, 1980, Napa Valley, Spring Mountain
Burgess Cellars, 1979, Napa Valley, Vintage Selection
Villa Mt. Eden, 1979, Napa Valley, Estate Bottled, Reserve
Zaca Mesa, 1979, Santa Ynez Valley, American Estate
Durney Vineyard, 1980, Monterey County, Carmel Valley, Estate Grown and Bottled
Kenwood Vineyards, 1980, Sonoma Valley, Jack London Ranch Vineyard

BRONZE

Kathryn Kennedy Winery, 1979, Saratoga, Estate Bottled
Lower Lake Winery, 1979, Lake County, Reserve
Napa Cellars, 1980, Napa Valley
Cassayre Forni, 1980, Napa Valley
Cakebread Cellars, 1980, Napa Valley
Jordan Vineyard, 1979, Alexander Valley, Estate Bottled
HMR, Ltd., 1979, Paso Robles
ZD Wines, 1980, California
Gundlach-Bundschu, 1980, Sonoma Valley, Batto Ranch
Smothers, 1979, Alexander Valley
Napa Creek Winery, 1980, Napa Valley
Iron Horse, 1980, Alexander Valley, Estate Grown
Napa Cellars, 1980, Alexander Valley
William Hill Winery, 1979, Napa Valley, Mt. Veeder
Robert Mondavi, 1978, Napa Valley, Reserve

CABERNET SAUVIGNON, *1983*
Rating
(Up to $7.00—83 entries)

GOLD

J. Lohr, N.V., California

Almaden Vineyards, 1980, Monterey

SILVER

Maddalena Vineyard, 1980, Sonoma County, Reserve

Fetzer Vineyards, 1981, Lake County

Coast Range, 1981, Napa Valley

Geyser Peak, 1979, California

Mirassou Vineyards, 1979, Monterey County

BRONZE

Fetzer Vineyards, 1980, Mendocino

Parducci Wine Cellars, 1980, Mendocino County

The Firestone Vineyard, 1978, Santa Ynez Valley, Arroyo Perdido Vineyard

Vincelli, 1980, San Luis Obispo

Diamond Oaks Vineyard, 1980, North Coast, Thomas Knight

Conn Creek, 1979, Napa Valley

ZD Wines, 1980, Napa Valley

CABERNET SAUVIGNON, *1983 Rating*
($7.01 to $11.50—113 entries)

GOLD

Raymond Vineyard and Cellar, 1979, Napa Valley, Estate Bottled

Lower Lake Winery, 1979, Lake County

The Firestone Vineyard, 1978, Santa Ynez Valley

Fetzer Vineyards, 1979, Mendocino, Cole Ranch

Round Hill Cellars, 1979, Napa Valley

SILVER

Chateau Chevalier, 1980, Napa Valley

Tudal, 1980, Napa Valley, Tudal Vineyard, Estate Bottled

Braren & Pauli, 1980, Mendocino County, Redwood Valley

Buena Vista Winery, 1980, Sonoma County

Trefethen Vineyards, 1979, Napa Valley, Trefethen Vineyards, Estate Bottled

Dehlinger Winery, 1979, Sonoma County

Mirassou Vineyards, 1979, Monterey County, Harvest Reserve

Marietta Cellars, 1980, Sonoma, Dry Creek Valley

Rosenblum Cellars, 1980, Napa Valley

Beaulieu Vineyard, 1979, Napa Valley, Rutherford

Mastantuono, 1981, San Luis Obispo County, Paso Robles, Rancho Tierra Rejada

BRONZE

Monticello Cellars, 1980, Napa Valley

Page Mill, 1980, Napa Valley, Volker Eisele Vineyard

Obester, 1979, Sonoma County, Batto Ranch

Joseph Phelps Vineyards, 1980, Napa Valley

V. Sattui, 1980, Napa Valley, Preston Vineyards

Parducci Wine Cellars, 1980, Mendocino County, Estate Bottled, 50th Anniversary

Simi Winery, 1978, Alexander Valley

Eberle, 1979, San Luis Obispo

Grand Cru Vineyards, 1981, California, Cook's Delta Vineyards

The Monterey Vineyard, 1979, San Luis Obispo, French Camp Vineyard, Special Selection

Flora Springs Wine Co., 1980, Napa Valley, Estate Bottled

J. Pedroncelli, 1978, Sonoma County, Vintage Selection

Beringer, 1978, Sonoma, Knights Valley Estate

Freemark Abbey, 1978, Napa Valley, Cabernet Bosche

Guenoc, 1980, Lake County and Napa County

Rutherford Hill, 1978, Napa Valley

Devlin Wine Cellars, 1980, Sonoma

Stephen Zellerbach, 1979, Sonoma County, Alexander Valley

Sierra Vista Winery, 1980, El

Dorado County, Estate Bottled
Rutherford Vintners, 1978, Napa Valley
Fitzpatrick, 1980, El Dorado County

CABERNET SAUVIGNON, *1984 Rating*
(Up to $7.00—81 entries)

GOLD
- ****Estrella River Winery, N.V., San Luis Obispo County
- Taylor California Cellars, N.V., California
- Grand Cru Vineyards, 1982, California (Vin Maison)

SILVER
- Liberty School, N.V., Napa Valley, Lot 10
- Paul Masson Vineyards, 1982, Sonoma County
- Maddalena Vineyard, 1981, Sonoma County
- Konocti, 1980, Lake County
- Stone Creek, 1980, North Coast
- Bandiera, 1981, North Coast
- Foppiano Wine Company, 1980, Sonoma County
- Bel Arbres, 1980, Sonoma
- Crystal Valley Cellars, 1980, Lake County
- Charles Krug, 1979, Napa Valley
- Round Hill Cellars, N.V., Napa Valley, "House"
- August Sebastiani Country, N.V., California
- Sebastiani Vineyards, 1980, North Coast

BRONZE
- Cribari & Sons, N.V., California
- Almaden Vineyards, 1981, Monterey County
- Sonoma Mission, N.V., California
- Colony, N.V., California, "Classic"
- R. and J. Cook, 1980, Northern California, Clarksburg, Estate Bottled
- Glen Ellen, 1982, Sonoma County, Proprietor's Reserve
- Louis M. Martini, 1980, North Coast
- Parducci Wine Cellars, 1980, Mendocino County
- Inglenook Vineyards, 1981, Napa Valley, Cabinet Selection
- Fetzer Vineyards, 1982, Lake County
- Turner Winery, N.V., California
- Seghesio, 1976, Northern Sonoma, Estate Bottled
- Vincelli Cellars, 1982, Alexander Valley
- J. Pedroncelli, 1981, Sonoma County
- Cordtz Brothers Cellars, 1981, Alexander Valley
- Turner Winery, 1980, Lake County
- Thomas Knight, 1980, Napa Valley
- Piconi, 1981, Temecula, "House Cabernet"
- Sotoyome, 1980, Sonoma County, Dry Creek Valley

CABERNET SAUVIGNON, *1984 Rating*
($7.01 to $12.00—135 entries)

GOLD
- ****Whitehall Lane, 1981, Napa Valley
- Adelaida Cellars, 1981, Paso Robles
- Glen Ellen, 1981, Sonoma Valley, Estate Bottled
- Woodside Vineyards, 1979, Santa Cruz Mountains, La Questa Vineyards, Estate Bottled

SILVER
- Hop Kiln, 1981, Alexander Valley
- Fieldstone, 1980, Alexander Valley, Estate Bottled
- Husch Vineyards, 1981, Mendocino, La Ribera Ranch, Estate Bottled
- Robert Mondavi, 1981, Napa Valley
- Marietta Cellars, 1981, Sonoma County
- Buena Vista, 1981, Napa Valley-Carneros, Estate Bottled
- Adler Fels, 1980, Napa Valley
- Fisher Vineyards, 1980, Sonoma County
- Rutherford Hill, 1979, Napa Valley

Estrella River Winery, 1979, San Luis Obispo County, Estate Bottled
Sierra Vista, 1981, El Dorado, Estate Bottled
Page Mill, 1981, Napa Valley, Volker Eisele Vineyard
HMR, 1979, Paso Robles
Los Vineros, 1981, Santa Barbara County

BRONZE
Ridge, 1981, Tepusquet
J. Carey Vineyards, 1981, Santa Ynez Valley, Alamo Pintado Vineyard, Estate Bottled
Pacheco Ranch, 1980, Marin County
Chalk Hill, 1981, Sonoma County
Inglenook Vineyards, 1980, Napa Valley, Estate Bottled
Dehlinger, 1980, Sonoma County
Rosenblum Cellars, 1982, Napa County, Sonoma County
J.W. Morris, 1981, California Select
Richardson, 1982, Sonoma Valley
Pat Paulsen Vineyards, 1981, Sonoma County
John B. Merritt, 1981, Sonoma County, Dry Creek Valley
Round Hill Cellars, 1980, Napa Valley
Alexander Valley Vineyards, 1981, Alexander Valley, Estate Bottled
Parducci Wine Cellars, 1980, Mendocino, 50th Anniversary Bottling
Edmeades Vineyards, 1981, Mendocino, Anderson Valley
Shown & Sons, 1980, Napa Valley, Shown Family Vineyards, Estate Bottled
Sequoia Grove Vineyards, 1981, Napa Valley
Rolling Hills Vineyards, 1981, Temecula
The Firestone Vineyard, 1979, Santa Ynez Valley
Stag's Leap Wine Cellars, 1980, Napa Valley
Windsor Vineyards, 1980, Sonoma County, River Estates, Vineyard Selection

CABERNET SAUVIGNON, *1984 Rating*
($12.01 and Up—75 entries)

GOLD
V. Sattui, 1980, Napa Valley, Preston Vineyard, Centennial
Robert Stemmler, 1979, Sonoma
Bargetto, 1981, Napa Valley, Lawrence J. Bargetto Dedication
Pine Ridge, 1981, Napa Valley, Rutherford Cuvée
Jekel Vineyards, 1979, Monterey County, Home Vineyard, Private Reserve, Estate Bottled

SILVER
Caymus Vineyards, 1978, Napa Valley, Estate Bottled, Special Selection
Chappellet, 1980, Napa Valley
Kenwood Vineyards, 1981, Sonoma Valley, Jack London Ranch
Milano, 1980, Mendocino County, Sanel Valley Vineyard
Caymus Vineyards, 1980, Napa Valley, Estate Bottled
Souverain, 1978, North Coast, Estate Bottled, Vintage Selection
Rodney Strong, 1979, Alexander Valley, Alexander's Crown Vineyard

BRONZE
Robert Mondavi, 1979, Napa Valley, Reserve
Grand Cru Vineyards, 1980, Alexander Valley, Collector Series
Mount Veeder, 1980, Napa County, Mount Veeder, Bernstein Vineyards
Beringer Vineyards, 1980, Napa Valley, Estate, Lemmon-Chabot Vineyard, Private Reserve
Martin Ray, 1980, Napa Valley, Seltzner Vineyard
William Hill, 1980, Napa Valley-Mt. Veeder
Mayacamas, 1979, California
Monticello Cellars, 1981, Napa Valley
St. Clement, 1981, Napa Valley
Smothers, 1980, Alexander Valley
Heitz Cellars, 1979, Napa Valley, Martha's Vineyard

Pine Ridge, 1981, Napa Valley, Stags Leap Cuvée
Cakebread Cellars, 1980, Napa Valley
Charles Krug, 1978, Napa Valley, Vintage Selection
Calafia Cellars, 1980, Napa Valley, Kitty Hawk Vineyard, Reserve
Freemark Abbey, 1980, Napa Valley, "Bosche"

ROSÉ OF CABERNET, *1979 Rating*
(Up to $3.00—2 entries)
No awards.

ROSÉ OF CABERNET, *1979 Rating*
($3.01 to $5.00—15 entries)
GOLD
Mill Creek, 1978, Cabernet Blush
Simi, 1978, Rosé of Cabernet Sauvignon
SILVER
Firestone, 1978, Rosé of Cabernet Sauvignon
BRONZE
Buena Vista, 1977, Cabernet Rosé, (Rosebrook)
Dry Creek, 1978

ROSÉ OF CABERNET, *1979 Rating*
($5.01 and Up—no entries)

ROSÉ OF CABERNET, *1983 Rating*
($4.51 to $6.00—6 entries)
BRONZE
Simi Winery, 1982, Sonoma County
Fieldstone, 1982, Sonoma County, Alexander Valley, "Spring Cabernet," Estate Bottled

CABERNET ROSÉ, *1984 Rating*
(Up to $4.50—8 entries)
GOLD
The Firestone Vineyard, 1983, Santa Ynez Valley
SILVER
Inglenook Vineyards, 1982, Napa Valley, Estate Bottled
BRONZE
Delicato Vineyards, 1983, California

CABERNET ROSÉ, *1984 Rating*
($4.51 to $6.00—5 entries)
GOLD
Fieldstone, 1983, Alexander Valley, Estate Bottled, (Spring Cabernet)
SILVER
Simi, 1983, North Coast
BRONZE
Windsor Vineyards, 1983, Mendocino County
Mill Creek Vineyards, 1983, Dry Creek Valley, Estate Bottled, "Cabernet Blush"

CABERNET BLANC, *1983 Rating*
(Up to $4.50—4 entries)
GOLD
The Konocti Winery, 1982, Lake County
SILVER
Santa Ynez Valley Winery, 1982, Santa Ynez Valley

CABERNET BLANC, *1983 Rating*
($4.51 to $6.00—10 entries)
SILVER
Los Vineros, 1982, Santa Barbara County
Boeger Winery, 1982, El Dorado County
BRONZE
Obester Winery, 1982, Sonoma County
Ballard Canyon Winery, 1982, Estate Bottled, "Rosalie"

CABERNET BLANC, *1983 Rating*
($6.01 and Up—1 entry)
SILVER
Sterling Vineyards, 1982, Napa Valley, Estate Bottled

CABERNET BLANC, *1984 Rating*
(Up to $4.50—5 entries)
SILVER
Bel Arbres, 1983, North Coast
BRONZE
Saddleback Cellars, N.V., Napa Valley, Estate Bottled

CABERNET BLANC, *1984 Rating*

($4.51 to $6.00—15 entries)
SILVER
Rancho Sisquoc, 1983, Santa Maria Valley, Estate Bottled
Ballard Canyon, 1983, Santa Ynez Valley, Estate Bottled
BRONZE
Santa Ynez Valley, 1983, Santa Ynez Valley
Konocti, 1983, Lake County

CARIGNANE, *1983 Rating*
(Up to $4.50—4 entries)
SILVER
Stony Ridge, N.V., California

CARIGNANE, *1984 Rating*
(Up to $4.50—2 entries)
GOLD
Stony Ridge, N.V., California
BRONZE
Cordtz Brothers Cellars, 1982, Upper Alexander Valley, "Mama Tasca's"

CARIGNANE, *1984 Rating*
($4.51 to $6.00—1 entry)
SILVER
White Oak Vineyards, 1981, Dry Creek Valley

CARNELIAN, *1983 Rating*
(Up to $5.00—1 entry)
SILVER
Giumarra Vineyards, 1981, California

CHARBONO, *1982 Rating*
($4.51 to $6.00—3 entries)
BRONZE
Fortino Winery, 1978, Santa Clara County

CHARBONO, *1982 Rating*
($6.01 and Up—3 entries)
BRONZE
Souverain Winery, 1978, North Coast

CHARBONO, *1983 Rating*
(Up to $4.50—1 entry)
BRONZE
Stone Creek, 1980, North Coast, Limited Release

CHARBONO, *1983 Rating*
($4.51 to $6.50—3 entries)
BRONZE
Parducci Wine Cellars, 1980, Mendocino County

CHARBONO, *1983 Rating*
($6.51 and Up—3 entries)
SILVER
Souverain, 1978, North Coast

CHARBONO, *1984 Rating*
($4.51 to $6.50—2 entries)
BRONZE
Parducci Wine Cellars, 1980, Mendocino County

CHARBONO *1984 Rating*
($6.51 and Up—3 entries)
GOLD
Inglenook Vineyards, 1978, Napa Valley, Estate Bottled
SILVER
Franciscan Vineyards, 1980, Napa Valley
BRONZE
Fortino, 1980, Santa Clara County

CHARDONNAY, *1979 Rating*
(Up to $5.00—24 entries)
SILVER
J. Pedroncelli, 1977
San Martin, 1978
BRONZE
Bel Arbres, 1977
Angelo Papagni, 1978, Madera
Paul Masson, N.V.
River Oaks, 1978
Vache, 1978, (Brookside)

CHARDONNAY, *1979 Rating*
($5.01 to $9.00—63 entries)
GOLD
Robert Mondavi, 1977
Sterling, 1976

SILVER
Chateau St. Jean, 1977, Belle Terre
Chateau St. Jean, 1977, Sonoma County
Clos du Bois, 1977, Second Release
Dry Creek, 1977
Stonegate, 1977, Vail Vista Vineyard

BRONZE
Alexander Valley Vineyards, 1977, Estate Bottled
Franciscan, 1977, Napa Valley
Parducci, 1978
Paul Masson, 1977, Monterey County, Estate Bottled
Rutherford Hill, 1977
Stag's Leap Wine Cellars, 1977, Haynes Vineyard

CHARDONNAY, *1979 Rating*
($9.01 and Up—11 entries)

GOLD
Burgess, 1977, Preston Vineyards
Chateau Montelena, 1976, Napa and Alexander Valleys
Chateau St. Jean, 1977, Robert Young Vineyard
Monterey Peninsula, 1977, Arroyo Seco Vineyard

SILVER
Chappellet, 1976, Napa Valley
Chateau St. Jean, 1977, Les Pierres
Spring Mountain, 1977
Villa Mt. Eden, 1977, Estate Bottled

BRONZE
Burgess, 1977, Sonoma County

CHARDONNAY, *1980 Rating*
(Up to $6.00—24 entries)

GOLD
Louis M. Martini, 1978, California
Sonoma Vineyards, 1978, Sonoma County
River Oaks Vineyards, 1978, Alexander Valley

SILVER
Parducci, 1979, Mendocino County

BRONZE
Paul Masson, 1978, Monterey County, Estate Bottled
Geyser Peak, 1978, Sonoma County

CHARDONNAY, *1980 Rating*
($6.01 to $11.00—66 entries)

GOLD
Trefethen, 1977, Napa Valley
Chateau St. Jean, 1979, Sonoma
Zaca Mesa, 1978, Santa Ynez Valley, Barrel Fermented
Estrella River Winery, 1978, San Luis Obispo
Navarro, 1978, Mendocino
Smothers, 1979, California

SILVER
Landmark, 1978, Sonoma County
Zaca Mesa, 1978, Santa Ynez Valley
Alexander Valley Vineyard, 1978, Alexander Valley
Chateau Montelena, 1978, California
The Firestone Vineyard, 1978, Santa Ynez Valley
Sonoma Vineyards, 1978, River West Vineyard, Estate Bottled
Rutherford Hill, 1978, Napa Valley
HMR, 1978, San Luis Obispo

BRONZE
Dry Creek Vineyard, 1978, Sonoma County
Cuvaison, 1977, Napa Valley
Parducci, 1979, Cellar Master's, Mendocino County
Jekel Vineyard, 1978, Monterey County
Stonegate, 1978, North Coast
Buena Vista, 1978, Sonoma
Conn Creek, 1978
HMR, 1977, Paso Robles, Estate Bottled
Pendleton, 1978, Monterey
Carneros Creek, 1978, Sonoma County
Franciscan, 1978, Temecula
Burgess, 1978, Napa Valley
Clos du Bois, 1978, Second Release, Alexander Valley

CHARDONNAY, *1980 Rating*
($11.01 and Up—9 entries)

GOLD

Chateau St. Jean, 1978, Robert Young Vineyard
Grgich Hills Cellar, 1977, Napa Valley

SILVER
Robert Mondavi, 1978, Napa Valley
Villa Mt. Eden, 1978, Napa Valley

CHARDONNAY, *1981 Rating*
(Up to $7.00—38 entries)

GOLD
Tyland Vineyards, 1979, Mendocino, Estate Bottled
San Martin Winery, 1979, California

SILVER
Parducci Wine Cellars, 1980, Mendocino County

BRONZE
Fetzer Vineyards, 1980, Mendocino

CHARDONNAY, *1981 Rating*
($7.01 to $12.00—110 entries)

GOLD
Chateau Chevalier, 1979, California, Edna Valley Vineyard
Napa Wine Cellars, 1979, Lot 2, Napa Valley
Round Hill Vineyards, 1979, Napa Valley
Trefethen Vineyards, 1978, Napa Valley
Leeward Winery, 1979, Monterey, Ventana Vineyard
Mill Creek Vineyards, 1979, Sonoma County
David Bruce Winery, 1979, San Luis Obispo County
Kenwood, 1979, Sonoma Valley, Beltane Ranch

SILVER
Domaine Laurier, 1979, Sonoma County, Russian River Valley
Smothers, 1980, California
Leeward Winery, 1979, San Luis Obispo, MacGregor Vineyard
Jekel Vineyard, 1979, Monterey County, Estate Bottled
Franciscan Vineyards, 1979, Napa Valley
Beringer Vineyards, 1979, Gamble Ranch Vineyard, Barrel Fermented
Clos du Bois, 1979, Sonoma County, Dry Creek Valley Vineyard
Clos du Bois, 1979, Alexander Valley
Chateau St. Jean, 1979, Alexander Valley, Belle Terre Vineyard

BRONZE
Stony Ridge Winery, 1979, Monterey County, Limousin Oak Aged, Limited Bottling
Parson's Creek Winery, 1980, Mendocino County, Anderson Valley
Chamisal Vineyard, 1979, California, Edna Valley
Santa Ynez Valley Winery, 1979, Napa Valley
Burgess Cellars, 1979, Napa Valley
St. Francis Vineyards, 1979, Sonoma Valley, Estate Bottled
Conn Creek Winery, 1978, Napa Valley
Zaca Mesa, 1979, Santa Ynez Valley, Estate Grown, Special Select
Chateau Montelena Winery, 1979, California
Hacienda Wine Cellars, 1979, Sonoma County, Claire de Lune
Husch Vineyards, 1979, Mendocino, Estate Bottled
Stevenot Vineyards, 1979, El Dorado County
Sunrise Winery, 1979, San Luis Obispo, Edna Valley, MacGregor and Paragon Vineyards
Santa Ynez Valley Winery, 1979, California
ZD Wines, 1979, Santa Barbara
Acacia Winery, 1979, Napa Valley
Ahern Winery, 1979, California

CHARDONNAY, *1981 Rating*
($12.01 and Up—24 entries)

GOLD
Chateau St. Jean, 1979, Gauer Ranch, Alexander Valley
Robert Mondavi Winery, 1978, Napa Valley, Reserve

Beringer Vineyards, 1978, Napa Valley, Estate Bottled, "Private Reserve"
Ventana Vineyards, 1979, Monterey County

SILVER
Robert Mondavi Winery, 1979, Napa Valley
Davis Bynum Winery, 1979, Allen-Hafner, Reserve
Monterey Peninsula Winery, 1979, Monterey, Hacienda Vineyards, Late Harvest
Chateau St. Jean, 1979, Alexander Valley, Robert Young Vineyards
David Bruce Winery, 1979, Santa Cruz County, Estate Bottled
Grgich Hills, 1978, Napa Valley
Stonegate, 1979, Spaulding Vineyards

BRONZE
Alta Vineyard Cellar, 1979, North Coast, Commemorative Bottling
Far Niente Winery, 1979, Napa Valley
Martin Ray Vineyards, 1979, California
Villa Mt. Eden Winery, 1979, Napa Valley, Estate Bottled

CHARDONNAY, *1982 Rating*
(Up to $8.00—47 entries)

GOLD
Filsinger Vineyards & Winery, 1980, Temecula, Estate Bottled
Geyser Peak Winery, 1980, Sonoma County
Clos du Bois, 1980, Alexander Valley

SILVER
Conn Creek, 1980, North Coast, Chateau Maja
San Martin Winery, 1980, California
Husch Vineyards, 1980, Anderson Valley, Mendocino, Estate Bottled

BRONZE
Fetzer Vineyards, 1980, Mendocino County, Barrel Aged
Fetzer Vineyards, 1981, Mendocino County, Early Bottling
J. Pedroncelli Winery, 1980, Sonoma
The Monterey Vineyard, 1979, Monterey County
River Oaks Vineyards, 1981, Alexander Valley

CHARDONNAY, *1982 Rating*
($8.01 to $12.00—103 entries)

GOLD
Ventana Vineyards Winery, 1980, Monterey, Estate Bottled
Raymond Vineyard & Cellar, 1980, Napa Valley, Estate Bottled
Alexander Valley Vineyards, 1980, Alexander Valley, Estate Bottled
Ballard Canyon Winery, 1981, Santa Barbara
Sonoma Vineyards, 1980, Sonoma County, Chalk Hill Vineyard, Estate Bottled, Second Release

SILVER
J. Carey Cellars, 1980, Santa Ynez Valley
Jekel Vineyard, 1980, Monterey
Conn Creek, 1979, Napa Valley
Chateau St. Jean, 1980, Sonoma County
Napa Cellars, 1980, Alexander Valley, Black Mountain Vineyard
Shafer Winery, 1980, Napa Valley, Spicer Ranch, Monticello Vineyards
DeLoach Vineyards, 1980, Russian River Valley
Stevenot Winery, 1980, Mendocino County
Trefethen Vineyards, 1979, Napa Valley, Trefethen Vineyards
Flora Springs Wine Co., 1980, Napa Valley
Parsons Creek Winery, 1980, Sonoma, Dry Creek, Vogenson Ranch
Parsons Creek Winery, 1981, Mendocino
Bargetto Winery, 1980, Santa Barbara County, Tepusquet Vineyard

Santa Ynez Valley Winery, 1980, California
Ahern Winery, 1981, Edna Valley, San Luis Obispo, MacGregor Vineyard
BRONZE
Rutherford Hill Winery, 1980, Napa Valley
Dry Creek Vineyard, 1980, Sonoma
Mill Creek Vineyards, 1980, Sonoma County, Dry Creek Valley, Estate Bottled
David Bruce Winery, 1980, California
McDowell Valley Vineyards, 1980, Mendocino County, Estate Bottled
Lakespring Winery, 1980, Napa Valley
Hacienda Wine Cellars (Inc.), 1980, Sonoma County, Clair de Lune
Cordtz Brothers Cellars, 1980, Upper Alexander Valley
Souverain Winery, 1981, North Coast
Parducci Wine Cellars, 1980, North Coast, Cellar Masters
Conn Creek, 1979, Napa Valley, Yountville
Stonegate, 1980, Napa Valley
Christian Brothers Winery, 1979, Napa Valley
Simi Winery, 1981, Mendocino County
Charles Ortman, 1980, Napa Valley

CHARDONNAY, *1982 Rating*
($12.01 and Up—58 entries)
GOLD
Stag's Leap Wine Cellars, 1979, Napa Valley, Haynes
Chateau St. Jean, 1980, Alexander Valley, Robert Young Vineyards
Veedercrest Vineyards, 1980, Napa Valley, Winery Lake Vineyards
Dry Creek Vineyard, 1980, Sonoma, Vintners Reserve
Milano, 1980, Mendocino, Redwood Valley, Lolonis Vineyards
SILVER
Chateau Chevalier Winery, 1980, Edna Valley
Kenwood Vineyards, 1980, Sonoma Valley, Beltane Ranch
Clos du Bois, 1980, Alexander Valley, Barrel Fermented, Reserve
Smothers, 1980, Sonoma, Green Pastures Vineyard
Burgess Cellars, 1980, Napa
Robert Mondavi Winery, 1980, Napa Valley
BRONZE
Chateau St. Jean, 1980, Alexander Valley, Jimtown Ranch
ZD Wines, 1980, California
Robert Keenan Winery, 1980, Napa Valley
Edmeades, 1980, Mendocino County, Anderson Valley, Reserve

CHARDONNAY, *1983 Rating*
(Up to $8.00—77 entries)
GOLD
Clos du Bois, 1982, Alexander Valley
Fetzer Vineyards, 1981, Mendocino, Barrel Select
J. Pedroncelli, 1981, Sonoma County
SILVER
Weibel Vineyards, 1981, Mendocino County
Fetzer Vineyards, 1982, Mendocino County, "Sundial"
The Christian Brothers, N.V., Napa Valley
BRONZE
Vincelli, 1981, Sonoma
Concannon Vineyard, 1981, Monterey County
River Road Vineyards, 1981, Sonoma County, Russian River Valley
Paul Masson Vineyards, 1982, Monterey County
Sonoma Vineyards, 1981, Sonoma County
Maddalena Vineyard, 1982, San Luis Obispo County, Reserve
Round Hill Cellars, N.V.,

California, House
Madrona Vineyards, 1981, El Dorado County, Estate Bottled

CHARDONNAY, *1983 Rating*
($8.01 to $12.00—141 entries)

GOLD

Cache Cellars, 1981, Monterey County, Ventana Vineyards
Ventana Vineyards, 1981, Monterey County, Estate Bottled
Simi Winery, 1980, Mendocino County
Guenoc, 1981, Lake, Sonoma, and Mendocino Counties
Raymond Vineyard & Cellar, 1981, Napa Valley
Congress Springs, 1982, Santa Clara County
Martin Brothers Winery, 1981, Edna Valley

SILVER

Haywood, 1981, Sonoma County, Chamizal Vineyard, Estate Bottled
Mark West Vineyards, 1980, Russian River Valley, Estate Bottled
Longoria, 1982, Santa Maria Valley
Tulocay, 1981, Napa Valley

BRONZE

Santa Barbara, 1982, Santa Ynez Valley
Estrella River Winery, 1978, San Luis Obispo County, Estate Bottled, Reserve
Fetzer Vineyards, 1981, California, Special Reserve
Sanford, 1981, Santa Maria Valley
Donna Maria Vineyards, 1981, Sonoma County, Chalk Hill Region, Estate Bottled
Donatoni Winery, 1982, Paso Robles
Stevenot, 1981, Sonoma County
Cloudstone Vineyards, 1981, Santa Barbara County
Husch Vineyards, 1982, Mendocino County, La Ribera Ranch, Estate Bottled
Roudon Smith Vineyards, 1981, Sonoma County
V. Sattui, 1980, Napa Valley, Vintage Vineyards
Bel Arbres, 1981, Monterey
J. Lohr, 1981, Monterey
Calera Wine Company, 1982, Santa Barbara County
Windsor Vineyards, 1981, Sonoma County, Russian River Valley, River East Vineyards
The Firestone Vineyard, 1981, Santa Ynez Valley
Mirassou Vineyards, 1981, Monterey County, Harvest Reserve
DeLoach Vineyards, 1981, Sonoma County, Russian River Valley
Tijsseling Vineyards, 1981, Mendocino County

CHARDONNAY, *1983 Rating*
($12.01 and Up—71 entries)

GOLD

Grgich Hills Cellar, 1980, Napa Valley
Chateau St. Jean, 1981, Alexander Valley, Jimtown Ranch
Pine Ridge Winery, 1981, Napa Valley, Stag's Leap District
Chalone, 1981, California, Estate Bottled
David Bruce, 1981, Santa Cruz Mountains, Estate Bottled
Beringer, 1981, Napa Valley, Gamble Ranch, Barrel Fermented, Estate Bottled

SILVER

Pendleton, 1981, Monterey County, Reserve
Jordan Vineyards, 1980, Alexander Valley, Estate Bottled
ZD Wines, 1981, California
Flora Springs Wine Co., 1981, Napa Valley, Estate Bottled, Barrel Fermented
Robert Mondavi, 1981, Napa Valley
Milano, 1981, Mendocino County, Redwood Valley, Lolonis Vineyards

BRONZE

Rodney D. Strong, 1981, Sonoma

County
Alta Vineyard Cellar, 1981, Napa Valley
Zaca Mesa, 1981, Santa Ynez Valley, American
Dry Creek Vineyard, 1981, Vintner's Reserve
Freemark Abbey, 1980, Napa Valley
Fisher Vineyard, 1980, Sonoma County
Smothers, 1981, Sonoma County, Green Pastures Vineyard
Clos du Val, 1981, California
Quail Ridge, 1981, Napa Valley
Girard, 1981, Napa Valley, Estate Bottled
Acacia Winery, 1981, Napa Valley, Carneros District, Winery Lake Vineyard
Trefethen Vineyards, 1980, Napa Valley, Trefethen Vineyards, Estate Bottled
Kolarovich Wines, 1981, Sonoma County, Art Release
Clos du Bois, 1981, Alexander Valley, "Calcaire"

CHARDONNAY, *1984 Rating*
(Up to $8.50—104 entries)

GOLD

Concannon Vineyard, 1983, California, Selected Vineyards
J. Patrick Doré, 1983, Santa Maria Valley, Signature Selection

SILVER

Windsor Vineyards, 1982, Mendocino County
Weibel, 1983, Mendocino County
San Martin, 1982, San Luis Obispo County
Windsor Vineyards, 1983, Sonoma County
Maddalena Vineyard, 1982, San Luis Obispo County
Fetzer Vineyards, 1983, Mendocino, "Sundial"

BRONZE

Parducci Wine Cellars, 1983, Mendocino County
Cambiaso, 1982, Sonoma County
August Sebastiani Country, 1983, California
Bargetto, 1982, California, "Cypress"
Almaden Vineyards, 1983, San Benito County
Braren Pauli, 1981, Mendocino County, Potter Valley
California Hillside Cellars, 1982, Central Coast
J. Pedroncelli, 1982, Sonoma County
Devlin Wine Cellars, 1983, Sonoma
Mirassou, 1983, Monterey County
Inglenook Vineyards, 1982, Napa Valley, Cabinet Selection
Stearns Wharf Vintners, 1982, San Luis Obispo County

CHARDONNAY, *1984 Rating*
($8.51 to $12.00—166 entries)

GOLD

Handley Cellars, 1982, North Coast
Sanford Winery, 1982, Santa Maria Valley
Franciscan Vineyards, 1982, Napa Valley, Carneros Vineyard, Reserve, Estate Bottled
Roudon-Smith Vineyards, 1982, Mendocino, Nelson Ranch
Simi, 1981, Mendocino County
Frick Winery, 1981, Monterey County
Kendall-Jackson, 1983, California
Sycamore Creek, 1983, Monterey County, Sleepy Hollow Vineyard

SILVER

Shafer Vineyards, 1982, Napa Valley
Tyland Vineyards, 1982, Mendocino County, Estate Bottled
Tijsseling Vineyards, 1982, Mendocino County, Estate Bottled
Landmark Vineyard, 1982, Sonoma County
Husch Vineyards, 1982, Anderson Valley, Estate Bottled
Karly, 1983, Santa Maria Valley, Tepusquet Vineyard
Silverado Vineyards, 1982, Napa

Valley
Fetzer Vineyards, 1982, California, Special Reserve
Navarro Vineyards, 1982, Mendocino
Los Vineros, 1981, Santa Barbara County
J. Lohr, 1982, Monterey County, Greenfield Vineyards
Ahern, 1982, Edna Valley, MacGregor Vineyard
Daumé, 1982, Central Coast
Longoria Wine Cellars, 1983, Santa Maria Valley

BRONZE

Parsons Creek, 1982, Sonoma County
Donatoni, 1983, Paso Robles
Chateau Du Lac, 1982, Clear Lake
Leeward Winery, 1982, Central Coast
Iron Horse Vineyard, 1982, Sonoma County, Green Valley
White Oak Vineyards, 1982, Sonoma County
Silver Mountain Vineyards, 1982, Monterey, Ventana Vineyards
DeLoach Vineyards, 1982, Sonoma County, Russian River Valley
Sycamore Creek, 1982, Monterey County, La Reina Vineyards
Saintsbury, 1982, Sonoma County
Congress Springs, 1983, Santa Clara County, Barrel Fermented
Bonny Doon Vineyard, 1983, Monterey County, La Reina Vineyard
William Wheeler, 1982, Sonoma County
V. Sattui, 1982, Napa Valley
Ventana Vineyards, 1982, Monterey County
Markham Vineyards, 1981, Napa Valley
Round Hill Cellars, 1982, Napa Valley
Stonegate Winery, 1982, Alexander Valley
Olson Vineyards, 1983, Mendocino County, Jeff Box Ranch
Souverain, 1982, North Coast
Edmeades Vineyards, 1982, Mendocino, Anderson Valley
David Bruce Winery, 1982, California
Guenoc, 1982, Lake and Mendocino Counties
Qupe, 1982, Santa Maria Valley
Franciscan Vineyards, 1982, Alexander Valley
Sequoia Grove Vineyards, 1982, Sonoma County
J. Carey Vineyards, 1983, Santa Ynez Valley, Adobe Canyon Vineyard, Estate Bottled
Deer Park, 1982, Napa Valley
Mill Creek Vineyards, 1981, Sonoma County, Estate Bottled
Parsons Creek, 1982, Mendocino County
Raymond Vineyard and Cellar, 1982, Napa Valley
Navarro Vineyards, 1981, Mendocino, Premiere Reserve
Alexander Valley Vineyards, 1982, Alexander Valley, Estate Bottled

CHARDONNAY, *1984 Rating*
($12.01 and Up—82 entries)

GOLD

Groth Vineyards, 1982, Napa Valley
Martin Ray, 1982, Sonoma County, Dutton Ranch
Perret Vineyards, 1980, Napa Valley, Carneros District
Chateau St. Jean, 1982, Alexander Valley, Belle Terre Vineyards
Pepperwood Springs, 1983, Mendocino, Botrytis (11.0% Residual Sugar)

SILVER

Stag's Leap Wine Cellars, 1982, Napa Valley
Chateau Montelena, 1982, Alexander Valley
Rodney Strong, 1982, Sonoma County, Chalk Hill Vineyard
Domaine Laurier, 1982, Sonoma County

Chalone Vineyard, 1982, "Chalone", Estate Bottled
Rutherford Hill, 1980, Napa Valley, Jaeger Vineyards, Cellar Reserve
Kenwood Vineyards, 1982, Sonoma Valley, Beltane Ranch
Congress Springs, 1982, Santa Cruz Mountains, Private Reserve
Lolonis, 1982, Mendocino County, Lolonis Vineyards

BRONZE
Ventana Vineyards, 1982, Monterey County, Barrel Fermented
Grgich Hills Cellar, 1981, Napa Valley
Perret Vineyards, 1981, Napa Valley, Carneros District, Perret Vineyard
Ballard Canyon, 1982, Santa Barbara County
J. Lohr, 1981, Monterey, Reserve
Far Niente, 1982, Napa Valley
Manzanita Cellars, 1982, Napa Valley
Chateau Julien, 1982, Monterey County
Trefethen Vineyards, 1981, Napa Valley, Estate Bottled
Cuvaison Vineyard, 1982, Napa Valley
Milano, 1982, Mendocino County, Lolonis Vineyard
ZD, 1982, California
Fisher Vineyards, 1982, Sonoma County
Girard, 1982, Napa Valley, Estate Bottled
Rombauer Vineyards, 1982, Napa Valley
Pine Ridge, 1982, Napa Valley, Stag's Leap Cuvée
St. Clement, 1982, Napa Valley
Smothers, 1983, Sonoma Valley, Remick Ridge Vineyard

CHENIN BLANC, DRY, *1980 Rating*
($3.51 to $5.00—16 entries)

GOLD
Stevenot, 1977, Northern California
McDowell Valley Vineyards, 1979, Mendocino County

SILVER
Callaway, 1979, Temecula

BRONZE
Louis J. Foppiano, 1979, Northern California
Pope Valley, 1979, Napa Valley

CHENIN BLANC, (0.0% to 0.70% residual sugar) *1981 Rating*
(Up to $4.00—7 entries)

GOLD
R & J Cook, 1980, Northern California, Estate Bottled, "Very Dry"

BRONZE
Boeger Winery, 1980, El Dorado County

CHENIN BLANC, (0.0% to 0.70% residual sugar) *1981 Rating*
($4.01 to $6.00—22 entries)

GOLD
Kenwood Vineyards, 1980, California

SILVER
Grand Cru Vineyards, 1980, Clarksburg-Yolo County, Cook's Delta Vineyard

BRONZE
Hacienda Wine Cellars, 1980, California
Preston Vineyards, 1980, Sonoma County, Dry Creek Valley, Estate Bottled
Milano Winery, 1980, Mendocino County
Vache, 1980, Temecula
Louis J. Foppiano Vineyards, 1980, Northern California

CHENIN BLANC, (0.0% to 0.70% residual sugar) *1981 Rating*
($6.01 and Up—4 entries)

SILVER
San Pasqual Vineyards, 1980, San Diego

CHENIN BLANC, (0.0% to 0.70%

residual sugar) *1982 Rating*
(Up to $4.00—6 entries)
SILVER
Cambiaso Winery and Vineyards, 1981, Northern California
California Hillside Cellars, 1980, California

CHENIN BLANC, (0.0% to 0.70% residual sugar) *1982 Rating*
($4.01 to $7.00—32 entries)
GOLD
Ronald Lamb Winery, 1981, Monterey, Ventana Vineyards
Bel Arbres Vineyards, 1981, North Coast
Congress Springs Vineyards, 1980, Santa Cruz Mountains, St. Charles Vineyard
Raymond Vineyard & Cellar, 1980, Napa Valley
SILVER
Sierra Vista Winery, 1981, Clarksburg, Northern California
BRONZE
Martin Brothers, 1981, Central Coast, Sisquoc River Vineyards
Fenestra, 1981, Monterey
L. Foppiano Wine Co., 1980, Northern California

CHENIN BLANC, (0.0% to 0.70% residual sugar) *1982 Rating*
($7.01 and Up—3 entries)
GOLD
R & J Cook, 1980, Northern California, Estate Bottled, Extra Dry
SILVER
Stag's Leap Winery, 1981, Napa, Stag's Leap
BRONZE
Lawrence Winery, 1980, California

CHENIN BLANC, (0.0% to 0.70% residual sugar) *1983 Rating*
(Up to $4.00—12 entries)
GOLD
The Wine Cellars of Ernest & Julio Gallo, N.V., California
Geyser Peak, 1982, Sonoma County, Alexander Valley
Cilurzo Vineyard and Winery, 1982, Temecula
SILVER
Cambiaso Vineyards, 1982, Northern California
BRONZE
Bianchi Winery, N.V., California

CHENIN BLANC, (0.0% to 0.70% residual sugar) *1983 Rating*
($4.01 to $6.00—23 entries)
SILVER
Guenoc, 1982, Lake County, Guenoc Valley
Hart Winery, 1982, Temecula, Miramonte Vineyard
Ronald Lamb, 1982, Monterey County, Ventana Vineyards
Boeger Winery, 1982, El Dorado County
BRONZE
Callaway Vineyard and Winery, 1981, California, Estate Bottled
R & J Cook, 1982, Clarksburg, Very Dry, Estate Bottled
White Oak Vineyards, 1982, Alexander Valley
Miramonte Vineyards, 1982, Temecula, California

CHENIN BLANC, (0.0% to 0.70% residual sugar) *1983 Rating*
($6.01 and Up—10 entries)
SILVER
Chappellet Vineyard, 1981, Napa Valley
BRONZE
Girard, 1982, Napa Valley, Estate Bottled
Napa Creek Winery, 1982, Napa Valley
Cassayre Forni, 1982, Lot 1, Napa Valley

CHENIN BLANC, (0.0% to 0.70% residual sugar) *1984 Rating*
(Up to $4.00—7 entries)
GOLD

Cambiaso, 1983, Yolo County, Clarksburg

BRONZE

Tyland Vineyards, 1982, Mendocino County, Estate Bottled

St. Helena Viticultural Society, 1983, California

St. Helena Viticultural Society, 1983, Napa Valley

CHENIN BLANC, (0.0% to 0.70% residual sugar) *1984 Rating*

($4.01 to $5.50—15 entries)

GOLD

Guenoc, 1983, Guenoc Valley

SILVER

Ross-Keller, 1983, Santa Barbara County, Los Alamos Vineyard

Martin Brothers, 1983, Paso Robles

R.H. Phillips Vineyard, 1983, Yolo County, Dunnigan Hills, "Dry"

BRONZE

Santa Barbara, 1983, Santa Ynez Valley

Corbett Canyon Vineyards, 1983, San Luis Obispo County

Bogle Vineyards, 1983, Clarksburg, (Dry)

CHENIN BLANC, (0.0% to 0.70% residual sugar) *1984 Rating*

($5.51 and Up—16 entries)

GOLD

Shown & Sons, 1983, Napa Valley, Shown Family Vineyards, Estate Bottled

SILVER

Ventana Vineyards, 1982, Monterey County, Barrel Fermented

Bel Arbres, 1982, California

BRONZE

Preston Vineyards, 1982, Sonoma County, Dry Creek Valley, Estate Bottled

Lakespring, 1983, Napa Valley

Raymond Vineyard and Cellar, 1982, Napa Valley

CHENIN BLANC, OFF-DRY, *1980 Rating*

(Up to $3.50—18 entries)

GOLD

Giumarra, N.V., California

Trader Joe's, 1979, California

SILVER

Pedroncelli, 1979, Sonoma County

BRONZE

Richard Carey, N.V., Lot 781, California

Taylor California Cellars, N.V., California

Inglenook, N.V., Navalle, California

J. Lohr, 1979, Northern California

Yverdon, 1978, Napa Valley

CHENIN BLANC, OFF-DRY, *1980 Rating*

($3.51 to $5.00—40 entries)

GOLD

Charles Krug, N.V., Napa Valley

Beringer, 1979, Napa Valley

HMR, 1979, Central Coast Counties, Demi Sec

SILVER

San Pasqual, 1979, San Diego County, Estate Bottled

Parducci, 1979, Mendocino County

San Martin, 1978, California

Lawrence, N.V., California

BRONZE

Paul Masson, N.V., California

Congress Springs, 1979, Santa Cruz Mountains, St. Charles Vineyard

Almaden, 1978, Monterey

The Monterey Vineyard, 1978, Monterey County

CHENIN BLANC, OFF-DRY, *1980 Rating*

($5.01 and Up—9 entries)

GOLD

Robert Mondavi, 1979, Napa Valley

SILVER

Simi, 1978, Mendocino County

Stags' Leap Vintners, 1979, Napa Valley

BRONZE

Mount Veeder, 1979, Napa County, Bernstein Vineyard

CHENIN BLANC, (0.71% and above residual sugar) *1981 Rating*
(Up to $4.00—17 entries)

GOLD

J. Pedroncelli Winery, 1980, Sonoma County

SILVER

Weibel, 1980, North Coast, Mendocino

R & J Cook, 1980, Northern California, Estate Bottled

BRONZE

Taylor California Cellars, N.V., California

Mt. Palomar Winery, 1980, California

CHENIN BLANC, (0.71% and above residual sugar) *1981 Rating*
($4.01 to $6.00—36 entries)

GOLD

Bargetto, 1980, California

Fetzer Vineyards, 1980, North Coast

SILVER

Bel Arbres, 1980, North Coast

Shown & Sons, 1980, Napa Valley, "Dry"

Concannon Vineyard, 1980, Northern California, Noble Vineyards

Sebastiani Vineyards, 1980, Northern California

Parducci Wine Cellars, 1980, Mendocino County

Charles Krug Winery, N.V., Napa Valley

Durney Vineyard, 1980, Monterey County, Carmel Valley

BRONZE

The Monterey Vineyard, 1980, Monterey County

Geyser Peak Winery, 1980, Sonoma County

San Pasqual Vineyards, 1980, San Diego County, Estate Bottled

J. Lohr, 1980, Sacramento County, Rosebud Vineyard

Wente Bros., N.V., California, "Le Blanc de Blanc"

CHENIN BLANC, (0.71% and above residual sugar) *1981 Rating*
($6.01 and Up—4 entries)

GOLD

Callaway Vineyards & Winery, 1978, Temecula, Estate Bottled, "Sweet Nancy"

Robert Mondavi Winery, 1980, Napa Valley

SILVER

The Christian Brothers, 1979, Pineau de la Loire, Estate Bottled

Simi Winery, 1979, Mendocino County

CHENIN BLANC, (0.71% and above residual sugar) *1982 Rating*
(Up to $4.00—29 entries)

GOLD

San Antonio Winery, 1981, California

SILVER

Colony Wines, N.V., California, Colony Classic

The Wine Cellars of Ernest & Julio Gallo, California

BRONZE

Bogle Vineyards, 1981, Clarksburg

Cresta Blanca Vineyards, 1980, Santa Barbara

Lost Hills Vineyards, 1981, California

Giumarra Vineyards, 1981, California

CHENIN BLANC, (0.71% and above residual sugar) *1982 Rating*
($4.01 to $7.00—29 entries)

GOLD

Sonoma Vineyards, 1981, Sonoma County

Beringer Vineyards, 1981, Napa Valley

Souverain Winery, 1981, North Coast

SILVER

Parducci Wine Cellars, 1981, North Coast

Congress Springs Vineyards, 1981,

Santa Cruz Mountains
BRONZE
Ventana Vineyards Winery, 1981, Monterey, Estate Bottled
Bargetto Winery, 1981, Santa Barbara County, Tepusquet Vineyard

CHENIN BLANC, (0.71% and above residual sugar) *1982 Rating*
($7.01 and Up—3 entries)
BRONZE
Fenestra, 1981, Monterey, Select Harvest, Ventana Vineyards

CHENIN BLANC, (0.71% to 3.0% residual sugar) *1983 Rating*
(Up to $4.00—21 entries)
GOLD
Geyser Peak, 1982, Alexander Valley, Nervo Ranch, Estate Bottled, Soft
Delicato Vineyards, 1982, California
SILVER
Paul Masson Vineyards, 1982, California
San Antonio Winery, 1982, California
Mount Palomar Winery, 1982, Temecula, Estate Bottled
Bogle Vineyards Winery, 1982, Clarksburg
BRONZE
Brookside Vineyard Company, N.V., California
Settler's Creek, N.V., California
Cresta Blanca, 1981, Santa Maria Valley
Almaden Vineyards, 1981, Monterey

CHENIN BLANC, (0.71% to 3.0% residual sugar) *1983 Rating*
($4.01 to $6.00—39 entries)
GOLD
Ventana Vineyards, 1982, Monterey County
SILVER
San Martin, 1982, California, Soft
Round Hill Cellars, 1982, California
Fetzer Vineyards, 1982, North Coast
BRONZE
Souverain, 1981, North Coast
J. Pedroncelli, 1982, Sonoma County
Beringer, 1982, Napa Valley
Bargetto Winery, 1982, Santa Barbara County, Tepusquet Vineyard
Mirassou Vineyards, 1982, Monterey County
San Martin, 1982, San Luis Obispo County
Kenwood Vineyards, 1982, California
Estrella River Winery, 1982, San Luis Obispo County, Estate Bottled
Weibel Vineyards, 1982, Mendocino County
Wente Brothers, 1981, California, "Le Blanc de Blancs"
Turner Winery, 1982, Lake County

CHENIN BLANC, (0.71% to 3.0% residual sugar) *1983 Rating*
($6.01 and Up—8 entries)
GOLD
Simi Winery, 1982, North Coast
Grand Cru Vineyards, 1982, Clarksburg, California, Cook's Delta Vineyards
SILVER
Shown & Sons Vineyards, 1981, Napa Valley, Rutherford, Estate Bottled
Robert Mondavi, 1982, Napa Valley
Pine Ridge Winery, 1982, Napa Valley, Yountville District
BRONZE
Felton-Empire Vineyards, 1982, California, Talmadge Town

CHENIN BLANC, (0.71% to 3.0% residual sugar) *1984 Rating*
(Up to $4.00—24 entries)
GOLD
Oak Ridge Vineyards, N.V., California, (Blanc de Blanc)

Colony, N.V., California, "Classic"
SILVER
Oak Ridge Vineyards, 1982, Northern California, Ryer Island Ranch
Paul Masson Vineyards, 1983, California
Estrella River, N.V, San Luis Obispo County
North Coast Cellars, N.V., North Coast
Ernest and Julio Gallo, N.V., California
BRONZE
The Monterey Vineyard, 1983, Monterey County
Papagni Vineyards, 1981, California
Cribari & Sons, N.V., California
Giumarra Vineyards, 1982, California
Breckenridge Cellars, N.V., California

CHENIN BLANC, (0.71% to 3.0% residual sugar) *1984 Rating*
($4.01 to $5.50—34 entries)
GOLD
Winterbrook Vineyards, 1983, Amador County
Ventana Vineyards, 1982, Monterey County
SILVER
Christian Brothers, N.V., Napa Valley
Bogle Vineyards, 1983, Clarksburg
Turner Winery, 1983, Lake County
Weibel, 1983, Mendocino County
Windsor Vineyards, 1983, California
Piconi, 1983, Temecula
Foppiano, 1983, Sonoma County
Parducci Wine Cellars, 1983, Mendocino County
Granite Springs, 1983, El Dorado
BRONZE
White Oak Vineyards, 1983, Napa Valley
Hacienda, 1983, California
Inglenook Vineyards, 1983, Napa Valley, Estate Bottled
Estrella River Winery, 1983, Paso Robles, Estate Bottled

CHENIN BLANC, (0.71% to 3.0% residual sugar) *1984 Rating*
($5.51 and Up—11 entries)
SILVER
Simi, 1983, Mendocino County
BRONZE
Robert Mondavi, 1983, Napa Valley

CHENIN BLANC, (12.0% and up residual sugar) *1983 Rating*
($6.01 and Up—1 entry)
SILVER
Callaway Vineyard and Winery, 1978, California, Estate Bottled, Sweet Nancy

CHENIN BLANC, (3.0% to 6.0% residual sugar) *1984 Rating*
(All Price Ranges—1 entry)
No award.

CHENIN BLANC, (6.0% to 12.0% residual sugar) *1984 Rating*
(All Price Ranges—1 entry)
No award.

CHENIN BLANC, (12.0% and up residual sugar) *1984 Rating*
(All Price Ranges—1 entry)
No award.

EARLY BURGUNDY, *1983 Rating*
(Up to $5.00—1 entry)
BRONZE
Trentadue, 1979, Alexander Valley, Estate Bottled

EMERALD RIESLING, *1983 Rating*
($3.51 to $5.00—2 entries)
BRONZE
San Martin, 1981, Santa Barbara County

FLORA, *1983 Rating*
($4.51 to $6.50—1 entry)
BRONZE

FLORA, (3.0% to 6.0% residual sugar) *1984 Rating*
($3.51 to $5.50—1 entry)
BRONZE
Parducci Wine Cellars, 1983, Mendocino County

FOLLE BLANCHE, *1983 Rating*
($4.51 to $6.50—1 entry)
BRONZE
Louis M. Martini, 1982, Sonoma Valley

FRENCH COLOMBARD, *1979 Rating*
(Up to $2.50—12 entries)
SILVER
Ernest & Julio Gallo, N.V.
Guasti, 1978
Inglenook, Navalle, N.V.
BRONZE
Almaden, N.V.
Colony, N.V.
Giumarra, N.V.
Los Hermanos, N.V.
Setrakian, N.V.

FRENCH COLOMBARD, *1979 Rating*
($2.51 to $4.00—16 entries)
GOLD
Stonegate, 1978, Estate Bottled
SILVER
Parducci, 1978
Souverain, 1978, (Colombard Blanc)
BRONZE
Gavilan, 1978
Paul Masson, N.V.
Sebastiani, N.V.
Sonoma Vineyards, 1977

FRENCH COLOMBARD, *1979 Rating*
($4.01 and Up—1 entry)
SILVER
Hop Kiln, 1977, Russian River Valley

FRENCH COLOMBARD, (0.0% to
Parducci Wine Cellars, 1981, Mendocino County

0.70% residual sugar) *1983 Rating*
($3.51 to $5.50—9 entries)
GOLD
Villa Baccla, 1981, North Coast
BRONZE
Windsor Vineyards, 1981, Sonoma County, River West Vineyard
Windsor Vineyards, 1982, California

FRENCH COLOMBARD, (0.0% to 0.70% residual sugar) *1984 Rating*
($3.51 to $5.50—5 entries)
GOLD
Greenstone, 1983, Amador County
SILVER
Windsor Vineyards, 1983, California

FRENCH COLOMBARD, (0.71% to 3.0% residual sugar) *1983 Rating*
(Up to $3.50—10 entries)
SILVER
Paul Masson Vineyards, 1982, California
The Wine Cellars of Ernest & Julio Gallo, N.V., California
San Antonio Winery, 1982, California

FRENCH COLOMBARD, (0.71% to 3.0% residual sugar) *1983 Rating*
($3.51 to $5.50—10 entries)
SILVER
Fetzer Vineyards, 1982, North Coast
Souverain, 1982, North Coast, Colombard Blanc
BRONZE
Greenstone Winery, 1982, Amador County, Estate Bottled
McDowell Valley Vineyards, 1982, McDowell Valley, Estate Bottled
August Sebastiani, 1982, California, Country
J. Pedroncelli, 1982, Sonoma County
Stone Creek, 1981, North Coast, Limited Release

FRENCH COLOMBARD, (0.71% to

3.0% residual sugar) *1984 Rating*
(Up to $3.50—16 entries)
GOLD
Ernest and Julio Gallo, N.V., California
Giumarra Vineyards, 1983, California
SILVER
Paul Masson Vineyards, 1983, California
San Antonio Winery, N.V., California
August Sebastiani Country, 1982, California
BRONZE
Almaden Vineyards, N.V., California
North Coast Cellars, N.V., North Coast

FRENCH COLOMBARD, (0.71% to 3.0% residual sugar) *1984 Rating*
($3.51 to $5.50—9 entries)
GOLD
Greenstone, 1983, Amador County, "California Colombard"
J. Pedroncelli, 1982, Sonoma County
SILVER
McDowell Valley Vineyards, 1983, McDowell Valley, Estate Bottled
Fetzer Vineyards, 1983, North Coast
Souverain, 1982, North Coast
Parducci Wine Cellars, 1983, Mendocino County

GAMAY, *1982 Rating*
(Up to $5.00—8 entries)
SILVER
Preston Vineyards, 1981, Sonoma, Dry Creek Valley, Estate Bottled
J. Lohr, 1981, Monterey, Greenfield Vineyards

GAMAY, *1982 Rating*
($5.01 to $7.00—7 entries)
SILVER
Alta Vineyard Cellar, 1978, Napa Valley, Commemorative Bottling
San Pasqual Vineyards, 1979, San Pasqual Valley, Indian Summer Harvest
BRONZE
Hop Kiln, 1980, Russian River Valley

GAMAY, *1982 Rating*
($7.01 and Up—2 entries)
SILVER
Lawrence Winery, 1981, San Luis Obispo, French Camp Vineyard Selection

GAMAY, *1983 Rating*
($4.01 to $6.00—7 entries)
GOLD
Chateau De Leu, 1982, Solano County, Green Valley, Estate Bottled
SILVER
San Pasqual Vineyards, 1979, San Diego County, Indian Summer Harvest
BRONZE
Trentadue, 1980, Alexander Valley, Estate Bottled

GAMAY BLANC, *1983 Rating*
($5.01 to $7.00—1 entry)
BRONZE
Kendall-Jackson, 1981, Clear Lake

GAMAY, *1984 Rating*
(Up to $4.00—4 entries)
GOLD
Ventana Vineyards, 1981, Monterey County, Estate Bottled
Charles F. Shaw Winery and Vineyard, 1983, Napa Valley, (Nouveau)
Trentadue, 1980, Alexander Valley

GAMAY, *1984 Rating*
($4.01 to $6.00—4 entries)
GOLD
Rosenblum Cellars, 1982, Napa Valley
SILVER
Charles F. Shaw Winery and Vineyard, 1983, Napa Valley

GAMAY ROSÉ, *1983 Rating*
($4.01 to $6.00—2 entries)
SILVER
Inglenook Vineyards, 1981, Napa Valley, Estate Bottled

GAMAY ROSÉ, *1984 Rating*
(Up to $4.00—4 entries)
GOLD
Cribari & Sons, 1983, California, (Napa Gamay Rosé), "Nuovo"

GAMAY ROSÉ, *1984 Rating*
($4.01 to $6.00—1 entry)
SILVER
V. Sattui, 1983, Sonoma County

GAMAY BEAUJOLAIS, *1982 Rating*
(Up to $4.00—17 entries)
GOLD
Fetzer Vineyards, 1981, Mendocino County
SILVER
Davis Bynum Winery, 1981, Sonoma, Vineyard of Joe Rochioli, Jr., Nouveau
BRONZE
Round Hill Cellars, 1981, Napa, Round Hill
Parducci Wine Cellars, 1981, North Coast
Giumarra Vineyards, 1981, California

GAMAY BEAUJOLAIS, *1982 Rating*
($4.51 to $6.00—17 entries)
GOLD
Buena Vista Winery & Vineyards, 1981, Sonoma Valley, Carneros, Estate Bottled
SILVER
Sonoma Vineyards, 1981, Sonoma County, Prime
Robert Pecota, 1981, Napa Valley
BRONZE
Paul Masson Vineyards, 1980, California
Chateau Nouveau, 1981, Napa Valley

GAMAY BEAUJOLAIS BLANC, *1983 Rating*
(Up to $5.00—1 entry)
SILVER

GAMAY BEAUJOLAIS, *1983 Rating*
(Up to $4.50—11 entries)
SILVER
Tyland Vineyards, 1982, Mendocino County
BRONZE
North Coast Cellars, N.V., North Coast
J. Pedroncelli, 1982, Sonoma County

GAMAY BEAUJOLAIS, *1983 Rating*
($4.51 to $6.50—15 entries)
GOLD
Souverain, 1982, North Coast
Almaden Vineyards, 1982, San Benito County
BRONZE
Inglenook Vineyards, 1981, Napa Valley, Estate Bottled
Buena Vista Winery, 1982, Sonoma Valley, Carneros

GAMAY BEAUJOLAIS, *1984 Rating*
(Up to $4.50—10 entries)
GOLD
Preston Vineyards, 1983, Sonoma County, Dry Creek Valley, Estate Bottled
BRONZE
Bale Mill Cellars, 1983, Napa Valley

GAMAY BEAUJOLAIS, *1984 Rating*
($4.51 to $6.50—18 entries)
GOLD
Almaden Vineyards, 1983, San Benito County
SILVER
Souverain, 1983, North Coast
BRONZE
Buena Vista, 1983, Sonoma Valley, Carneros, Estate Bottled
Robert Pecota, 1983, Napa Valley

Wente Bros., 1982, Livermore Valley

GEWURZTRAMINER, *1979 Rating*
(Up to $4.00—5 entries)
BRONZE
M. LaMont, N.V.
J. Pedroncelli, 1978

GEWURZTRAMINER, *1979 Rating*
($4.01 to $6.50—34 entries)
GOLD
Fetzer, 1978
Gundlach-Bundschu, 1977
SILVER
Mark West, 1977, Ellis Vineyards
BRONZE
Richard Carey, 1978, San Luis Obispo County
Chateau St. Jean, 1978, Sonoma County, (Dry)
Fieldstone, 1978, Estate Bottled
Rutherford Hill, 1977
Simi, 1978
Souverain, 1978

GEWURZTRAMINER, *1979 Rating*
($6.51 and Up—5 entries)
GOLD
Joseph Phelps, 1977, Selected Late Harvest
SILVER
Chateau St. Jean, 1978, Sonoma County
BRONZE
Dry Creek, 1978, Diana's Favorite
Winery Lake Vineyards, Veedercrest, 1978, Late Harvest

GEWURZTRAMINER, *1980 Rating*
(Up to $4.50—8 entries)
BRONZE
San Martin, 1978, Santa Barbara County

GEWURZTRAMINER, *1980 Rating*
($4.51 to $7.00—38 entries)
GOLD
Parducci, 1979, Mendocino County
SILVER
Grand Cru, 1979, Garden Creek Ranch
Wente Brothers, 1978, Arroyo Seco Vineyards, Monterey
Buena Vista, 1978, Sonoma
Souverain, 1978, North Coast
Kenwood, 1979, Sonoma Valley
BRONZE
Felton-Empire, 1979, California, Maritime Vineyard Series
The Monterey Vineyard, 1978, Monterey County

GEWURZTRAMINER, *1980 Rating*
($7.01 and Up—6 entries)
GOLD
Smothers, 1979, Alexander Valley, Late Harvest
Felton-Empire, 1979, Santa Barbara, Tepusquet Vineyards
SILVER
Chateau St. Jean, 1979, Belle Terre and Robert Young Vineyards
BRONZE
Richard Carey, 1979, Santa Barbara County, Late Harvest

GEWURZTRAMINER, (0.0% to 0.70% residual sugar) *1981 Rating*
(Up to $5.00—2 entries)
BRONZE
Louis M. Martini, 1979, California

GEWURZTRAMINER, (0.0% to 0.70% residual sugar) *1982 Rating*
(Up to $5.50—1 entry)
BRONZE
Louis M. Martini Winery, 1980, California

GEWURZTRAMINER, (0.0% to 0.70% residual sugar) *1982 Rating*
($5.51 to $7.50—12 entries)
SILVER
Navarro Vineyards, 1980, Mendocino, Estate Bottled
Sebastiani Vineyards, 1980, Sonoma Valley
BRONZE
Gundlach-Bundschu Winery, 1980,

Sonoma Valley, Estate Bottled, Rhinefarm
Hacienda Wine Cellars, 1981, Sonoma Valley
Evensen Vineyards & Winery, 1980, Napa Valley, Estate Bottled
Monticello Cellars, 1981, Napa Valley

GEWURZTRAMINER, (0.0% to 0.70% residual sugar) *1982 Rating*
($7.51 and Up—1 entry)
BRONZE
Chateau St. Jean, 1981, Alexander Valley, Robert Young Vineyards

GEWURZTRAMINER, (0.0% to 0.70% residual sugar) *1983 Rating*
(Up to $5.50—6 entries)
BRONZE
Cordtz Brothers Cellars, 1982, Upper Alexander Valley, Dry
Edmeades Vineyards, 1982, Mendocino, Anderson Valley

GEWURZTRAMINER, (0.0% to 0.70% residual sugar) *1983 Rating*
($5.51 to $7.50—11 entries)
SILVER
Hacienda Wine Cellars, 1982, Sonoma County
Matrose, 1982, Alexander Valley, Traminer
BRONZE
Monticello Cellars, 1982, Napa Valley

GEWURZTRAMINER, (0.0% to 0.70% residual sugar) *1983 Rating*
($7.51 and Up—3 entries)
SILVER
Balverne, 1981, Sonoma County, Pepperwood Vineyard
BRONZE
Hidden Cellars, 1982, Mendocino County, Anderson Valley Vineyards

GEWURZTRAMINER, (0.0% to 0.70% residual sugar) *1984 Rating*
(Up to $5.50—3 entries)
SILVER
Susiné Cellars, 1983, Napa Valley
BRONZE
Tyland Vineyards, 1982, Mendocino County, Anderson Valley

GEWURZTRAMINER, (0.0% to 0.70% residual sugar) *1984 Rating*
($5.51 to $7.50—14 entries)
GOLD
Matrose, 1983, Alexander Valley, (Traminer)
SILVER
Rutherford Hill, 1982, Napa Valley
Hop Kiln, 1983, Russian River Valley
BRONZE
Smothers, 1983, Sonoma County
Vega Vineyards, 1983, Santa Ynez Valley, Estate Bottled
Fieldstone, 1983, Alexander Valley, Estate Bottled
Gundlach-Bundschu, 1983, Sonoma Valley, Rhinefarm Vineyards, Estate Bottled

GEWURZTRAMINER, (0.0% to 0.70% residual sugar) *1984 Rating*
($7.51 and Up—5 entries)
GOLD
Chateau St. Jean, 1983, Alexander Valley
SILVER
Monticello Cellars, 1983, Napa Valley

GEWURZTRAMINER, (0.71% to 3.0% residual sugar) *1981 Rating*
(Up to $5.00—9 entries)
BRONZE
Toyon Vineyards, 1980, Sonoma County

GEWURZTRAMINER, (0.71% to 3.0% residual sugar) *1981 Rating*
($5.01 to $7.00—35 entries)
SILVER
Joseph Phelps Vineyards, 1979, Napa Valley

Buena Vista Winery, 1979, Sonoma
Rutherford Hill Winery, 1979, Napa Valley

BRONZE

Clos du Bois, 1979, Alexander Valley, Early Harvest
Kenwood Vineyards, 1980, Sonoma Valley
Husch Vineyards, 1980, Mendocino County, Estate Bottled

GEWURZTRAMINER, (0.71% to 3.0% residual sugar) *1981 Rating*
($7.01 and Up—7 entries)

GOLD

Chateau St. Jean, 1980, Sonoma County, Frank Johnson Vineyard

BRONZE

Chateau St. Jean, 1980, Alexander Valley, Belle Terre Vineyards
Beringer Vineyards, 1980, Napa Valley
Gundlach-Bundschu Winery, 1979, Sonoma County
Grand Cru Vineyards, 1980, Alexander Valley, Garden Creek Ranch
Veedercrest, 1979, Napa Valley, Winery Lake Vineyard

GEWURZTRAMINER, (0.71% to 3.0% residual sugar) *1982 Rating*
(Up to $5.50—9 entries)

GOLD

Husch Vineyards, 1981, Anderson Valley, Mendocino, Estate Bottled

SILVER

J. Pedroncelli Winery, 1981, Sonoma

BRONZE

Almaden Vineyards, 1980, San Benito
The Monterey Vineyard, 1981, Monterey County

GEWURZTRAMINER, (0.71% to 3.0% residual sugar) *1982 Rating*
($5.51 to $7.50—38 entries)

GOLD

Sonoma Vineyards, 1981, Sonoma County
St. Francis Vineyards, 1981, Sonoma Valley, Estate Bottled

SILVER

Veedercrest Vineyards, 1980, Napa Valley, Winery Lake Vineyards
Souverain Winery, 1980, North Coast
Parsons Creek Winery, 1981, Mendocino, Anderson Valley, Philo Foothills Ranch
Donna Maria Vineyards, 1981, Chalk Hill, Sonoma County
Beringer Vineyards, 1980, Napa Valley

BRONZE

Clos du Bois, 1981, Alexander Valley, Early Harvest
Joseph Phelps Vineyards, 1980, Napa Valley

GEWURZTRAMINER, (0.71% to 3.0% residual sugar) *1982 Rating*
($7.51 and Up—5 entries)

SILVER

Chateau St. Jean, 1981, Sonoma Valley, Frank Johnson Vineyards

BRONZE

Grand Cru Vineyards, 1981, Alexander Valley, Garden Creek Ranch
Milano, 1981, Mendocino County, Anderson Valley Vineyard, Late Harvest

GEWURZTRAMINER, (0.71% to 3.0% residual sugar) *1983 Rating*
(Up to $5.50—9 entries)

GOLD

Windsor Vineyards, 1982, California
The Wine Cellars of Ernest & Julio Gallo, N.V., California, Limited Release

SILVER

J. Pedroncelli, 1982, Sonoma County

GEWURZTRAMINER, (0.71% to 3.0% residual sugar) *1983 Rating*

($5.51 to $7.50—28 entries)

GOLD

St. Francis, 1982, Sonoma Valley, Estate Bottled

Milano, 1982, Mendocino County, Anderson Valley, Ordways Valley Foothill Ranch

DeLoach Vineyards, 1982, Sonoma County, Russian River Valley

SILVER

Los Vineros, 1982, Santa Maria Valley

BRONZE

Charles Krug, 1981, Napa Valley

Parducci Wine Cellars, 1981, Mendocino County

GEWURZTRAMINER, (0.71% to 3.0% residual sugar) *1983 Rating*

($7.51 and Up—6 entries)

SILVER

Grand Cru Vineyards, 1982, Alexander Valley, Garden Creek Ranch

BRONZE

Chateau St. Jean, 1982, Alexander Valley, Belle Terre and Robert Young Vineyards

GEWURZTRAMINER, (0.71% to 3.0% residual sugar) *1984 Rating*

(Up to $5.50—11 entries)

SILVER

Souverain, 1983, North Coast

Cranbrook Cellars, 1983, Napa Valley

BRONZE

Round Hill Cellars, 1983, Napa Valley

Wine Discovery, 1983, Napa Valley

Davis Bynum, 1983, Sonoma County, Westside Road

GEWURZTRAMINER, (0.71% to 3.0% residual sugar) *1984 Rating*

($5.51 to $7.50—31 entries)

GOLD

Fetzer Vineyards, 1983, North Coast

Charles Krug, 1982, Napa Valley

Parducci Wine Cellars, 1983,

GEWURZTRAMINER, (3.1% to 6.0% residual sugar) *1981 Rating*

($5.01 to $7.00—1 entry)

SILVER

Fetzer Vineyards, 1980, California

GEWURZTRAMINER, (3.1% to 6.0% residual sugar) *1982 Rating*

(Up to $5.50—2 entries)

SILVER

River Oaks Vineyards, 1980, Alexander Valley, Early Harvest

BRONZE

Lost Hills Vineyards, 1980, Alexander Valley

GEWURZTRAMINER, (3.1% to 6.0% residual sugar) *1982 Rating*

($5.51 to $7.50—1 entry)

BRONZE

Fetzer Vineyards, 1981, California Mendocino County

GEWURZTRAMINER, (3.0% to 6.0% residual sugar)

SILVER

Hacienda, 1983, Sonoma County

Mark West, 1983, Russian River Valley, Estate Bottled

Sebastiani Vineyards, 1982, Sonoma Valley

BRONZE

Adler Fels, 1983, Sonoma County

Almaden Vineyards, 1982, San Benito County

R. Montali, 1982, Santa Maria Valley

Windsor Vineyards, 1983, Sonoma County

St. Francis, 1983, Sonoma Valley, Estate Bottled

Santa Ynez Valley, 1983, Santa Ynez Valley

GEWURZTRAMINER, (0.71% to 3.0% residual sugar) *1984 Rating*

($7.51 and Up—2 entries)

BRONZE

Chateau St. Jean, 1983, Sonoma County, Frank Johnson Vineyards

residual sugar) *1983 Rating*
($5.51 to $7.50—2 entries)
SILVER
Fetzer Vineyards, 1982, North Coast

GEWURZTRAMINER, (3.0% to 6.0% residual sugar) *1983 Rating*
($7.51 and Up—1 entry)
BRONZE
Smothers, 1982, Alexander Valley, Late Harvest

GEWURZTRAMINER, (3.0% to 6.0% residual sugar) *1984 Rating*
($7.51 and Up—2 entries)
GOLD
Dry Creek Vineyard, 1983, Sonoma County, Late Harvest

GEWURZTRAMINER, (6.1% to 12.0% residual sugar) *1981 Rating*
($7.01 and Up—6 entries)
GOLD
Chateau St. Jean, 1980, Alexander Valley, Belle Terre Vineyard, Selected Late Harvest
SILVER
Smothers, 1980, Alexander Valley, Late Harvest
BRONZE
Charles Le Franc, 1979, San Benito County, Pacines Vineyards, Late Harvest
Clos du Bois, 1979, Alexander Valley, Late Harvest

GEWURZTRAMINER, (6.1% to 12.0% residual sugar) *1982 Rating*
($7.51 and Up—3 entries)
SILVER
Navarro Vineyards, 1981, Mendocino, Late Harvest
Smothers, 1981, Alexander Valley, Late Harvest

GEWURZTRAMINER, (6.0% to 12.0% residual sugar) *1983 Rating*
($7.51 and Up—2 entries)
GOLD
Navarro Vineyards, 1982, Mendocino, Late Harvest

GEWURZTRAMINER, (6.0% to 12.0% residual sugar) *1984 Rating*
($7.51 and Up—2 entries)
SILVER
Villa Mt. Eden, 1983, Napa Valley, Late Harvest, Estate Bottled
BRONZE
Navarro Vineyards, 1983, Mendocino, Anderson Valley, Late Harvest

GEWURZTRAMINER, (12.1% and up residual sugar) *1981 Rating*
($7.01 and Up—1 entry)
SILVER
Grand Cru Vineyards, 1978, Alexander Valley, Garden Creek Ranch, "Induced Botrytis"

GEWURZTRAMINER, (12.1% and up residual sugar) *1982 Rating*
($7.51 and Up—2 entries)
GOLD
Chateau St. Jean, 1980, Alexander Valley, Late Harvest, Jimtown Ranch
BRONZE
Felton-Empire Vineyards, 1981, Santa Barbara, Tepusquet Vineyards, Select Late Harvest

GEWURZTRAMINER, (12.0% and up residual sugar) *1983 Rating*
($7.51 and Up—4 entries)
GOLD
Chateau St. Jean, 1981, Alexander Valley, Belle Terre Vineyards, Select Late Harvest
SILVER
Clos du Bois, 1981, Alexander Valley, Late Individual Bunch Selected
BRONZE
Chateau St. Jean, 1980, Alexander Valley, Robert Young Vineyard, Individual Bunch Selected, Late Harvest

GEWURZTRAMINER, (12.0% and up residual sugar) *1984 Rating*
($7.51 and Up—3 entries)
GOLD
ZD, 1983, Napa Valley, Select Late Harvest
Chateau St. Jean, 1982, Alexander Valley, Robert Young Vineyards, Special Select Late Harvest

GRENACHE ROSÉ, (0.0% to 0.70% residual sugar) *1983 Rating*
($3.51 to $5.00—1 entry)
BRONZE
Fortino Winery, 1982, California, Estate Bottled, Ruby

GRENACHE ROSÉ, (0.0% to 0.70% residual sugar) *1984 Rating*
($3.51 to $5.00—1 entry)
BRONZE
Fortino, 1981, Santa Clara County, Grenache-Ruby

GRENACHE ROSÉ, (0.71% to 3.0% residual sugar) *1983 Rating*
(Up to $3.50—3 entries)
GOLD
The Wine Cellars of Ernest & Julio Gallo, N.V., California, Rosé
SILVER
Sebastiani Vineyards, 1982, California, Vin Rosé

GRENACHE ROSÉ, (0.71% to 3.0% residual sugar) *1983 Rating*
($3.51 to $5.00—2 entries)
BRONZE
Windsor Vineyards, 1981, Mendocino County

GRENACHE ROSÉ, (0.71% to 3.0% residual sugar) *1984 Rating*
($3.51 to $5.00—1 entry)
SILVER
Windsor Vineyards, 1982, California

GREEN HUNGARIAN, *1983 Rating*
(Up to $4.00—2 entries)
GOLD
Delicato Vineyards, 1982, California

GREEN HUNGARIAN, *1983 Rating*
($4.01 to $5.50—3 entries)
BRONZE
Weibel Vineyards, N.V., California
Buena Vista Winery, 1980, Napa Valley

GREEN HUNGARIAN, *1984 Rating*
(Up to $4.00—1 entry)
GOLD
Delicato Vineyards, 1983, California

GREEN HUNGARIAN, *1984 Rating*
($4.01 to $5.50—4 entries)
GOLD
Parducci Wine Cellars, 1983, Mendocino County
SILVER
Buena Vista, 1980, Napa Valley
BRONZE
Weibel, N.V., California

GREY RIESLING, *1984 Rating*
($4.01 to $5.50—7 entries)
SILVER
Robert Pecota, 1983, Napa Valley
BRONZE
Bogle Vineyards, 1983, Clarksburg
Souverain, 1983, North Coast

GRIGNOLINO, *1983 Rating*
(Up to $5.00—1 entry)
GOLD
Heitz, N.V., Napa Valley

GRIGNOLINO, *1984 Rating*
(Up to $5.00—1 entry)
BRONZE
Heitz Cellars, 1979, Napa Valley

MALVASIA BIANCA, *1983 Rating*
($6.51 and Up—1 entry)
BRONZE
Stony Ridge, 1982, Livermore Valley, Estate Bottled

MALVASIA BIANCA, *1984 Rating*
($6.51 and Up—1 entry)
SILVER
Stony Ridge, 1982, Livermore Valley, Estate Bottled

MERLOT, *1981 Rating*
(Up to $6.00—9 entries)
SILVER
Lawrence Winery, 1980, California
BRONZE
Ridgewood, 1979, Santa Barbara County, James Flood Vineyard, Limited Bottling

MERLOT, *1981 Rating*
($6.01 to $10.00—15 entries)
GOLD
Clos du Val, 1978, Napa Valley
Rutherford Hill Winery, 1978, Napa Valley
SILVER
Carneros Creek Winery, 1978, Napa Valley, Turnbull/Fay Vineyards
BRONZE
Gundlach-Bundschu Winery, 1978, Sonoma Valley, Estate Bottled, Rhine Farm Vineyard
J. Carey Cellars, 1979, Santa Ynez Valley

MERLOT, *1981 Rating*
($10.01 and Up—4 entries)
GOLD
Stag's Leap Wine Cellars, 1978, Napa Valley, Stag's Leap Vineyards
SILVER
Veedercrest, 1978, Napa County, Mt. Veeder District
BRONZE
St. Francis Vineyards, 1979, Sonoma Valley, Estate Bottled

MERLOT, *1982 Rating*
(Up to $7.00—14 entries)
SILVER
Lawrence Winery, 1981, Southern Central Coast of California
BRONZE
Turner Winery, 1980, Lake County
Zaca Mesa Winery, 1979, Santa Ynez Valley
Farview Farm Vineyard, 1979, Templeton

MERLOT, *1982 Rating*
($7.01 to $11.00—12 entries)
GOLD
Gundlach-Bundschu Winery, 1979, Sonoma Valley, Rhinefarm Vineyards
Rutherford Hill Winery, 1979, Napa Valley
SILVER
Chappellet Vineyard, 1978, Napa Valley
Robert Keenan Winery, 1979, Napa Valley
BRONZE
Caparone Winery, 1980, Templeton, Unfined
Clos du Bois, 1979, Napa Valley
Calafia Cellars, 1979, Napa Valley

MERLOT, *1982 Rating*
($11.01 and Up—5 entries)
GOLD
Lawrence Winery, 1981, California, Bien Nacido Vineyard Selection
Stag's Leap Wine Cellars, 1979, Napa Valley, Stag's Leap Vineyard
SILVER
Duckhorn Vineyards, 1979, Napa Valley
BRONZE
Clos du Val, 1979, Napa Valley

MERLOT, *1983 Rating*
(Up to $7.00—11 entries)
SILVER
Santa Ynez Valley, 1980, Santa Maria Valley, Bien Nacido Vineyard
Charles Krug, 1978, Napa Valley
Louis M. Martini, 1980, North Coast
BRONZE
Crystal Valley Cellars, 1978, North Coast, Limited Reserve

MERLOT, *1983 Rating*
($7.01 to $11.00—15 entries)
GOLD
Mill Creek Vineyards, 1980, Sonoma County, Estate Bottled
Rutherford Hill, 1979, Napa Valley
Clos du Bois, 1980, Napa Valley
SILVER
Franciscan Vineyards, 1980, Napa Valley, Estate Bottled
Devlin Wine Cellars, 1981, California
Lakespring Winery, 1980, Napa Valley, Winery Lake Vineyard
BRONZE
Caparone, 1980, Templeton, Paso Robles
The Firestone Vineyard, 1979, Santa Ynez Valley, Ambassador's Vineyard

MERLOT, *1983 Rating*
($11.01 and Up—9 entries)
GOLD
Sterling Vineyards, 1980, Napa Valley
Gundlach-Bundschu, 1980, Special Selection, Estate Bottled
SILVER
Chateau Chevre, 1980, Napa Valley
BRONZE
Pine Ridge Winery, 1980, Napa Valley, Selected Districts

MERLOT, *1984 Rating*
(Up to $7.00—12 entries)
GOLD
El Paso De Robles, 1982, Paso Robles, Radike Vineyard
SILVER
Bel Arbres, 1981, Sonoma
BRONZE
Turner Winery, 1981, Lake County
Vincelli Cellars, 1982, Napa Valley, Limited Edition
Santa Ynez Valley, 1983, Santa Ynez Valley, "Merlot L'Enfant"
Louis M. Martini, 1981, North Coast
J. Patrick Doré, 1980, Napa Valley, Signature Selection

MERLOT, *1984 Rating*
($7.01 to $11.00—23 entries)
GOLD
Gundlach-Bundschu, 1981, Sonoma Valley, Rhinefarm Vineyards, Estate Bottled
Lakespring, 1981, Napa Valley
SILVER
Stephen Zellerbach Vineyard, 1980, Alexander Valley
Devlin Wine Cellars, 1982, San Luis Obispo County
Markham Vineyards, 1980, Napa Valley, Stag's Leap District
BRONZE
Mill Creek Vineyards, 1981, Sonoma County, Estate Bottled
Diablo Vista, 1981, Sonoma County, Dry Creek Valley
Rutherford Hill, 1980, Napa Valley
Chateau Chevre, 1981, Napa Valley

MERLOT, *1984 Rating*
($11.01 and Up—11 entries)
GOLD
****Inglenook Vineyards, 1980, Napa Valley, Limited Cask Reserve Selection, Estate Bottled
SILVER
Calafia Cellars, 1981, Napa Valley, Pickle Canyon Vineyard
Monterey Peninsula, 1981, Monterey County, Doctors Reserve
BRONZE
Stonegate Winery, 1980, Napa Valley, Spaulding Vineyard

MERLOT ROSÉ, *1983 Rating*
(Up to $4.50—1 entry)
BRONZE
The Firestone Vineyard, 1982, Santa Ynez Valley

MERLOT ROSÉ, (0.0% to 0.70% residual sugar) *1984 Rating*
(Up to $5.00—1 entry)
GOLD
The Firestone Vineyard, 1983, Santa Ynez Valley

MISSION, *1983 Rating*

($5.01 to $7.00—1 entry)
SILVER
Montevina Wines, N.V., Amador County, Estate Bottled

MUSCAT, (0.0% to 0.70% residual sugar)
1983 Rating
(Up to $5.50—1 entry)
BRONZE
ENZ Vineyards, 1982, San Benito County, Lime Kiln Valley, Estate Bottled, Orange Muscat

MUSCAT, (0.71% to 3.0% residual sugar)
1982 Rating
($5.01 to $7.00—4 entries)
SILVER
Pat Paulsen Vineyards, 1981, Sonoma County, (Muscat Canelli)
BRONZE
Concannon Vineyard, 1981, Livermore Valley, Estate Bottled, (Muscat Blanc)
Chateau St. Jean, 1981, Alexander Valley, Jimtown Ranch, (Muscat Canelli)

MUSCAT, (0.71% to 3.0% residual sugar)
1983 Rating
(Up to $5.50—1 entry)
BRONZE
Lost Hills Vineyards, 1981, California, "Pantelleria"

MUSCAT, (0.71% to 3.0% residual sugar)
1983 Rating
($5.51 to $7.00—2 entries)
GOLD
Sebastiani Vineyards, 1982, Sonoma Valley, Wildwood Vineyards
BRONZE
Pat Paulsen Vineyards, 1982, Sonoma County

MUSCAT, (0.71% to 3.0% residual sugar)
1984 Rating
($5.51 to $7.50—5 entries)
BRONZE
Monterey Peninsula, 1983, Monterey County, Arroyo Seco Vineyards, (Canelli)

MUSCAT, (3.0% to 6.0% residual sugar)
1982 Rating
(Up to $5.00—2 entries)
BRONZE
Beringer Vineyards, 1980, Madera, (Malvasia Amabile)

MUSCAT, (3.0% to 6.0% residual sugar)
1982 Rating
($5.01 to $7.00—7 entries)
SILVER
Souverain Winery, 1981, North Coast, (Muscat Canelli)
BRONZE
Ballard Canyon Winery, 1981, Santa Ynez, Estate Bottled, (Muscat Di Canelli)
Estrella River Winery, 1980, San Luis Obispo, Estate Bottled, (Muscat Canelli)
Parducci Wine Cellars, 1981, North Coast, (Muscat Canelli)

MUSCAT, (3.0% to 6.0% residual sugar)
1983 Rating
(Up to $5.50—1 entry)
GOLD
Sutter Home Winery, 1982, California, Muscat Amabile

MUSCAT, (3.0% to 6.0% residual sugar)
1983 Rating
($5.51 to $7.00—4 entries)
GOLD
Souverain, 1981, North Coast, Canelli
SILVER
Fetzer Vineyards, 1982, Lake County, Muscat Canelli
St. Francis, 1982, Sonoma Valley, Terra Pulchra Vineyard, Canelli

MUSCAT, (3.0% to 6.0% residual sugar)
1983 Rating
($7.01 and Up—5 entries)
SILVER
Estrella River Winery, 1981, San

Luis Obispo County, Estate Bottled
Inglenook Vineyards, 1982, Napa Valley, Estate Bottled, Muscat Blanc
BRONZE
Parducci Wine Cellars, 1981, Mendocino County, Canelli

MUSCAT, (3.0% to 6.0% residual sugar)
1984 Rating
($5.51 to $7.50—7 entries)
GOLD
Souverain, 1983, North Coast, (Canelli)
SILVER
Mastantuono, 1983, San Luis Obispo County, Paso Robles, (Canelli)
BRONZE
Parducci Wine Cellars, 1983, North Coast, (Canelli)

MUSCAT, (3.0% to 6.0% residual sugar)
1984 Rating
($7.51 and Up—5 entries)
GOLD
Inglenook Vineyards, 1983, Napa Valley, Estate Bottled, (Muscat Blanc)
SILVER
San Pasqual Vineyards, 1983, Paso Robles, (Canelli)
BRONZE
Robert Mondavi, 1982, Napa Valley, (Moscato d'Oro)

MUSCAT, (6.0% to 12.0% residual sugar)
1982 Rating
($5.01 to $7.00—3 entries)
SILVER
Papagni Vineyards, N.V., California, Estate Bottled, (Moscato d'Angelo)

MUSCAT, (6.0% to 12.0% residual sugar)
1982 Rating
($7.01 and Up—1 entry)
SILVER
Chateau St. Jean, 1980, Alexander Valley, Selected Late Harvest, Jimtown Ranch, (Muscat Canelli)

MUSCAT, (6.0% to 12.0% residual sugar)
1983 Rating
(Up to $5.50—2 entries)
BRONZE
Angelo Papagni, N.V., California, Moscato D'Angelo, Estate Bottled

MUSCAT, (6.0% to 12.0% residual sugar)
1983 Rating
($5.51 to $7.00—1 entry)
GOLD
Robert Pecota, 1982, Muscato Di Andrea

MUSCAT, (6.0% to 12.0% residual sugar)
1983 Rating
($7.01 and Up—2 entries)
BRONZE
San Pasqual Vineyards, 1982, San Pasqual Valley, Canelli

MUSCAT, (6.0% to 12.0% residual sugar)
1984 Rating
(Up to $5.50—2 entries)
SILVER
Round Hill Cellars, 1982, California, "Sweet", (Canelli)
Papagni Vineyards, N.V., California, Estate Bottled, (Moscato d'Angelo)

MUSCAT, (6.0% to 12.0% residual sugar)
1984 Rating
($5.51 to $7.50—1 entry)
BRONZE
Robert Pecota, 1983, California, (Muscato Di Andrea)

MUSCAT, (6.0% to 12.0% residual sugar)
1984 Rating
($7.51 and Up—1 entry)
BRONZE
V. Sattui, 1983, California

MUSCAT, (12.0% and up residual sugar)
1982 Rating
($7.01 and Up—1 entry)
BRONZE

Estrella River Winery, 1979, San Luis Obispo, Estate Bottled, Late Harvest, (Muscat Canelli)

MUSCAT, (12.0% and up residual sugar)
1984 Rating
($7.51 and Up—2 entries)
GOLD
Estrella River Winery, 1980, San Luis Obispo County, Estate Bottled, Late Harvest, (Canelli)
BRONZE
McLester, 1983, California, "Suite 13"

NEBBIOLO, *1984 Rating*
($5.01 to $7.50—1 entry)
BRONZE
Martin Brothers, 1982, California

PETITE SIRAH, *1979 Rating*
(Up to $4.00—12 entries)
GOLD
San Martin, 1975, Monterey County
SILVER
Barengo, N.V., Clarksburg
Round Hill, 1975, North Coast
BRONZE
Wente, 1974

PETITE SIRAH, *1979 Rating*
($4.01 to $6.50—21 entries)
GOLD
Caymus, 1975
Pope Valley, 1977, Spring Lane Vineyards
Sonoma Vineyards, 1975
SILVER
Pedrizzetti, 1977, Special Release, Shell Creek Vineyards
BRONZE
Fetzer, 1976, Mendocino, Special Reserve, (Syrah)
J. Lohr, 1976, Unfiltered
Parducci, 1975
Ridge, 1975, York Creek

PETITE SIRAH, *1979 Rating*
($6.51 and Up—8 entries)
GOLD
Stag's Leap Vineyard, 1975
SILVER
Concannon, 1974, Estate Bottled, (Limited Bottling)
Parducci, 1974, Cellarmaster's Selection
Sommelier, 1976, Solano
BRONZE
Burgess, 1976, Napa Valley
Hop Kiln, 1976, Russian River Valley

PETITE SIRAH, *1980 Rating*
(Up to $4.50—20 entries)
GOLD
Louis J. Foppiano, 1976, Russian River Valley
Fortino Winery, 1977, California
SILVER
Giumarra, N.V., California
Inglenook, 1975, Napa Valley, Estate Bottled
Wente, 1977, California
BRONZE
Bel Arbres, 1977, Mendocino

PETITE SIRAH, *1980 Rating*
($4.51 to $6.00—18 entries)
GOLD
Souverain, 1976, North Coast
SILVER
Round Hill Vineyards, 1977, Napa Valley
Sonoma Vineyards, 1975, Northern California
Stags' Leap Vineyards, 1976, Napa Valley
BRONZE
Rosenblum Cellars, 1978, Napa

PETITE SIRAH, *1980 Rating*
($6.51 and Up—8 entries)
SILVER
Burgess, 1977, Napa Valley
Fieldstone, 1977, Alexander Valley
BRONZE
Hop Kiln, 1977, Russian River Valley

PETITE SIRAH, *1981 Rating*

(Up to $5.00—21 entries)
GOLD
Wente Bros., 1979, Livermore Valley
Fetzer Vineyards, 1977, Mendocino
Vache, 1979, Temecula
SILVER
Emilio Guglielmo Winery, 1977, Santa Clara Valley, Mount Madonna
Inglenook, 1977, Napa Valley, Estate Bottled
BRONZE
Pedrizzetti Winery, 1977, California, Shell Creek Vineyards, Special Release
Bargetto, 1977, California
Giumarra Vineyards, 1979, California

PETITE SIRAH, *1981 Rating*
($5.01 to $7.00—17 entries)
GOLD
Ventana Vineyards, 1979, Monterey County, Estate Bottled
Caymus Vineyards, 1975, California
SILVER
Fieldstone Winery, 1978, Alexander Valley, Estate Bottled
Ranchita Oaks Winery, 1979, San Miguel
BRONZE
Cilurzo, 1979, Temecula
Sotoyome Winery, 1978, Sonoma County
Louis J. Foppiano Vineyards, 1978, Russian River Valley

PETITE SIRAH, *1981 Rating*
($7.01 and Up—9 entries)
GOLD
Veedercrest, 1974, Sonoma County, Cask YUG-77, Batch-2
Stag's Leap Wine Cellars, 1978, Napa Valley
SILVER
Mirassou Vineyards, 1977, California, Harvest Selection, Limited Bottling, Unfiltered
Fetzer Vineyards, 1978, Mendocino, Special Reserve
BRONZE
Geyser Peak Winery, 1976, California, Centennial Bottling, Unfiltered

PETITE SIRAH, *1982 Rating*
(Up to $5.00—13 entries)
GOLD
R. & J. Cook, 1979, Northern California
SILVER
Parducci Wine Cellars, 1978, Mendocino
Mirassou, 1978, Monterey County
BRONZE
Fortino Winery, 1978, Santa Clara County
Giumarra Vineyards, 1979, California
Emilio Guglielmo, 1977, Santa Clara Valley, Estate Bottled
Vache, 1980, Temecula

PETITE SIRAH, *1982 Rating*
($5.01 to $7.00—23 entries)
GOLD
Wente Bros., 1979, Livermore Valley, Estate Bottled
Fenestra, 1979, San Luis Obispo, Shell Creek Vineyard
Beringer Vineyards, 1977, Napa Valley, Spring Lane Vineyard
J. Lohr Winery, 1978, Northern California, Wilson, Spenker, Greenfield Vineyards
Topolos at Russian River Vineyards, 1979, Sonoma County, Hillside, Unfined
SILVER
Estrella River Winery, 1979, San Luis Obispo, Estate Bottled
Hultgren and Samperton, 1979, Sonoma County
Fetzer Vineyards, 1980, Mendocino
Paul Masson Vineyards, 1979, California
BRONZE
Pedrizzetti Winery, 1977, Shell

Creek
Trentadue Winery, 1977, Sonoma, Alexander Valley, Estate Bottled
Round Hill Cellars, 1978, Napa, Round Hill
Concannon Vineyard, 1978, California

PETITE SIRAH, *1982 Rating*
($7.01 and Up—13 entries)
GOLD
McDowell Valley Vineyards, 1979, Mendocino County, Estate Bottled
Hop Kiln, 1979, Russian River Valley
Parducci Wine Cellars, 1978, North Coast, Cellar Masters
SILVER
Cilurzo Vineyard & Winery, 1979, Temecula
Rosenblum Cellars, 1980, Napa, St. George & Rich Vineyard
BRONZE
Stag's Leap Winery, 1978, Napa

PETITE SIRAH, *1983 Rating*
(Up to $5.00—14 entries)
BRONZE
Inglenook Vineyards, 1979, Napa Valley, Estate Bottled, Centennial Vintage
Wente Brothers, 1980, Livermore Valley, Estate Bottled

PETITE SIRAH, *1983 Rating*
($5.01 to $7.00—23 entries)
SILVER
Beringer, 1978, Napa Valley
Louis J. Foppiano, 1980, Sonoma County
Sotoyome, 1979, Sonoma County
Guenoc, 1980, Lake and Napa Counties
BRONZE
Frick Winery, 1980, Monterey County
R & J Cook, 1981, Northern California, Clarksburg, Estate Bottled
Fortino Winery, 1978, California

PETITE SIRAH, *1983 Rating*
($7.01 and Up—11 entries)
GOLD
Fetzer Vineyards, 1980, Mendocino, Special Reserve
McDowell Valley Vineyards, 1980, McDowell Valley, Estate Bottled
SILVER
Callaway Vineyard and Winery, 1980, California, Estate Bottled
Hop Kiln Winery, 1980, Russian River Valley
BRONZE
Fieldstone Ranch & Vineyards, Estate Bottled

PETITE SIRAH, *1984 Rating*
(Up to $5.00—12 entries)
GOLD
Wente Brothers, 1980, Livermore Valley, Estate Bottled
SILVER
Parducci Wine Cellars, 1979, Mendocino County
BRONZE
Turner Winery, 1981, Lake County
Oak Ridge Vineyards, 1981, Northern California, (Gran Sirah)

PETITE SIRAH, *1984 Rating*
($5.01 to $7.00—20 entries)
GOLD
Rosenblum Cellars, 1982, Napa Valley, St. George Vineyard and Rich Vineyard
SILVER
Guenoc, 1981, Lake County
Roudon-Smith Vineyards, 1981, San Luis Obispo County, (Petite Syrah)
Round Hill Cellars, 1980, Napa Valley
BRONZE
Fawn's Glen, 1981, Mendocino
Cilurzo, 1980, Temecula
Winters, 1981, California, Naismith Vineyard
Piconi, 1982, Temecula
Gemello, 1977, California
Granite Springs, 1982, El Dorado,

Granite Hill Vineyards

PETITE SIRAH, *1984 Rating*
($7.01 and Up—8 entries)
GOLD
Hop Kiln, 1981, Russian River Valley
SILVER
Tobias Vineyards, 1981, Paso Robles, Jones Ranch
Stag's Leap Wine Cellars, 1980, Napa Valley
BRONZE
Parducci Wine Cellars, 1978, Mendocino County, Cellarmaster Selection
Preston Vineyards, 1981, Sonoma County, Dry Creek Valley, Estate Bottled

PETITE SIRAH ROSÉ, *1983 Rating*
($4.51 to $6.00—1 entry)
BRONZE
Fieldstone, 1982, Sonoma County, Alexander Valley, Redwood Ranch & Vineyards, Estate Bottled

PETITE SIRAH ROSÉ, (0.71% to 3.0% residual sugar) *1984 Rating*
($5.01 to $7.00—2 entries)
BRONZE
Fieldstone, 1983, Sonoma Valley, Carneros

PINOT BLANC, *1982 Rating*
(Up to $4.50)
GOLD
The Monterey Vineyard, 1979, Monterey County

PINOT BLANC, *1982 Rating*
($4.51 to $6.50—3 entries)
SILVER
Mirassou Vineyards, 1981, Monterey County, Labeled White Burgundy
BRONZE
Wente Bros., 1980, Monterey

PINOT BLANC, *1982 Rating*
($6.51 and Up—5 entries)
GOLD
Jekel Vineyards, 1980, Monterey, Estate Bottled
SILVER
Monterey Peninsula Winery, 1981, Monterey, Cobblestone Vineyards
BRONZE
Congress Springs Vineyards, 1980, Santa Cruz Mountains, St. Charles Vineyard

PINOT BLANC, *1983 Rating*
(Up to $6.00—2 entries)
BRONZE
Mirassou Vineyards, 1982, Monterey County, White Burgundy

PINOT BLANC, *1983 Rating*
($6.01 to $8.50—5 entries)
BRONZE
Jekel Vineyard, 1981, Monterey County, Home Vineyard, Estate Bottled

PINOT BLANC, *1983 Rating*
($8.51 and Up—3 entries)
GOLD
Chateau St. Jean, 1981, Sonoma Valley, St. Jean Vineyards

PINOT BLANC, *1984 Rating*
(Up to $6.00—1 entry)
GOLD
Mirassou, 1983, Monterey County, "White Burgundy"

PINOT BLANC, *1984 Rating*
($6.01 to $8.50—7 entries)
GOLD
J. Lohr, 1982, Monterey County, Greenfield Vineyards
SILVER
Fetzer Vineyards, 1982, Mendocino, Redwood Valley Vineyards
Jekel Vineyards, 1982, Arroyo Seco, Home Vineyard, Estate Bottled
BRONZE
Congress Springs, 1982, Santa Cruz Mountains

PINOT BLANC, *1984 Rating*
($8.51 and Up)
SILVER
Chateau St. Jean, 1982, Sonoma Valley, St. Jean Vineyards
Monterey Peninsula, 1982, Monterey County, Cobblestone Vineyards
Douglas Meador, 1982, Monterey

BLANC DE NOIR, (PINOT NOIR BLANC), *1984 Rating*
(Up to $4.50—4 entries)
BRONZE
Weibel, 1983, Mendocino County

BLANC DE NOIR, *1984 Rating*
($4.51 to $6.00—17 entries)
GOLD
Mark West Vineyards, 1983, Russian River Valley, Estate Bottled
Kenwood Vineyards, 1983, Sonoma Valley
Leeward Winery, 1983, Santa Maria Valley, "Coral"
BRONZE
HMR, 1983, Paso Robles, Estate Bottled
Fetzer Vineyards, 1983, Mendocino
Iron Horse Vineyard, 1982, Sonoma County, Green Valley

PINOT NOIR, *1979 Rating*
(Up to $3.50—6 entries)
BRONZE
J. Pedroncelli, 1975
Round Hill Cellars, 1976, North Coast

PINOT NOIR, *1979 Rating*
($3.51 to $6.50—49 entries)
SILVER
Rutherford Hill, 1976
Sonoma Vineyards, 1976
Souverain, 1976
Weibel, 1969, Estate Bottled
BRONZE
Beaulieu, 1973, Beau Velours
Beringer, 1975, Mendocino
Charles Krug, 1975
Fetzer, 1977, Mendocino
Heitz Cellars, 1973, Napa Valley
Robert Mondavi, 1976
J.W. Morris, 1977, Sonoma County
Joseph Phelps, 1975
Pope Valley, 1975, Yountville
San Martin, 1975
Simi, 1974

PINOT NOIR, *1979 Rating*
($6.51 and Up—12 entries)
GOLD
ZD, 1975, St. Clair Vineyards
SILVER
Burgess, 1976, Winery Lake Vineyards
Caymus, 1976, Estate Bottled
Hoffman Mountain Ranch, 1975, Estate Bottled
BRONZE
Joseph Phelps, 1974, Heinemann Mountain Vineyard
Zaca Mesa, 1975, Santa Ynez Valley

PINOT NOIR, *1980 Rating*
(Up to $5.00—24 entries)
GOLD
Clos du Bois, 1977, Sonoma County, Dry Creek
SILVER
Weibel, 1975, North Coast
BRONZE
Sebastiani Vineyards, N.V., Northern California
Cresta Blanca, 1975, California
Round Hill Vineyards, 1976, North Coast

PINOT NOIR, *1980 Rating*
($5.01 to $7.50—33 entries)
GOLD
Pendleton, 1978, Monterey
Beringer, 1975, Mendocino
SILVER
Inglenook, 1973, Napa Valley, Cask B-7
Gundlach-Bundschu, 1977, Sonoma County
Martin Ray, N.V, La Montana,

Cuvée 3
Sonoma, 1976, River East Vineyards
Trefethen, 1976, Napa Valley
Robert Mondavi, 1977, Napa Valley

BRONZE

Parducci, 1976, Cellar Master's, Mendocino
Caymus, 1976, Napa Valley
Rutherford Hill, 1976, Napa Valley

PINOT NOIR, ***1980 Rating***
($7.51 and Up—10 entries)

GOLD

Kenwood, 1977, Sonoma Valley, Jack London

SILVER

Beaulieu, 1976, Los Carneros, Napa Valley
Veedercrest, 1978, Sonoma County

BRONZE

HMR, 1976, Paso Robles, Estate Bottled

PINOT NOIR, ***1981 Rating***
(Up to $5.00—16 entries)

GOLD

Beaulieu Vineyard, 1978, Napa Valley, "Beau Velours," Estate Bottled
Perelli-Minetti Winery, 1974, California

SILVER

Tulocay Winery, 1977, Napa Valley

BRONZE

J. Pedroncelli Winery, 1977, Sonoma County
Weibel, 1975, North Coast
Stony Ridge Winery, 1979, California
Giumarra Vineyards, N.V., California

PINOT NOIR, ***1981 Rating***
($5.01 to $9.00—53 entries)

SILVER

Raymond Vineyard & Cellar, 1978, Napa Valley, Estate Bottled
Souverain Cellars, 1977, North Coast
Lawrence Winery, 1980, California
Alatera Vineyards, 1979, Napa Valley

BRONZE

Louis J. Foppiano Vineyards, 1977, Sonoma County
Villa Mt. Eden Winery, 1978, Napa Valley, Tres Ninos Vineyard, Estate Bottled
Sebastiani Vineyards, 1973, Northern California, "Proprietor's Reserve"
Felton-Empire Vineyards, 1979, California, Maritime Series
Sonoma Vineyards, 1977, Sonoma County, River East Vineyards, Estate Bottled
Kenwood Vineyards, 1978, Sonoma County

PINOT NOIR, ***1981 Rating***
($9.01 and Up—18 entries)

GOLD

Chateau Chevalier, 1978, Napa Valley, Stan's Patch
Santa Cruz Mountain Vineyard, 1977, Santa Cruz Mountains, Estate Bottled

SILVER

David Bruce Winery, 1978, Santa Cruz County
La Crema Vinera, 1979, Monterey County, Ventana Vineyards
DeLoach Vineyards, 1979, Sonoma County, Estate Bottled

BRONZE

Zaca Mesa, 1978, Santa Ynez Valley, Estate Grown, Special Selection
Veedercrest, 1978, Sonoma County

PINOT NOIR, ***1982 Rating***
(Up to $6.00—29 entries)

GOLD

Pendleton Winery, N.V., Monterey
J. Pedroncelli Winery, 1979, Sonoma

SILVER

Geyser Peak Winery, 1977, California
Fetzer Vineyards, 1978, Mendocino

Parducci Wine Cellars, 1979, Mendocino
Carmel Bay Winery, 1979, Monterey County, Sleepy Hollow Vineyards

BRONZE

Beringer Vineyards, 1978, Sonoma
Stony Ridge Winery, 1979, Monterey County, Tom Jones/Sleepy Hollow Vineyards
Sebastiani Vineyards, 1979, California, August Sebastiani Country, (1.5 liter size)
Whitehall Lane Winery, 1981, Fleur D'Helene
Christian Brothers Winery, N.V., Napa Valley

PINOT NOIR, *1982 Rating*
($6.01 to $10.00—46 entries)

GOLD

Lazy Creek Vineyards, 1979, Mendocino
Stony Ridge Winery, 1979, Monterey County, Vinco Vineyards

SILVER

Husch Vineyards, 1979, Anderson Valley, Mendocino, Estate Bottled
Tulocay, 1978, Napa Valley
Buena Vista Winery and Vineyards, 1979, Sonoma Valley, Carneros, Estate Bottled
Edmeades, 1980, Mendocino County, Anderson Valley
Louis M. Martini Winery, 1976, California, Special Selection
Frick Winery, 1979, Monterey County

BRONZE

Topolos at Russian River Vineyards, 1980, Sonoma County, Hillside, Unfined
Alatera Vineyards, 1979, Napa Valley
Mill Creek Vineyards, 1980, Sonoma County, Dry Creek Valley, Estate Bottled
Firestone Vineyard, 1978, Santa Ynez Valley

PINOT NOIR, *1982 Rating*
($10.01 and Up—18 entries)

GOLD

Zaca Mesa Winery, 1979, Santa Ynez Valley, Special Selection
Bacigalupi, 1979, Sonoma County

SILVER

Hacienda Wine Cellars, 1979, Sonoma Valley, Buena Vista Vineyard

BRONZE

Mount Eden Vineyards, 1979, Santa Cruz Mountains
Villa Mt. Eden, 1979, Napa Valley
ZD Wines, 1979, California
Robert Mondavi Winery, 1977, Napa Valley, Reserve
David Bruce Winery, 1979, Santa Cruz Mountain, Estate Bottled

PINOT NOIR, *1983 Rating*
(Up to $6.50—29 entries)

GOLD

Fetzer Vineyards, 1980, Mendocino

SILVER

Paul Masson Vineyards, 1980, Sonoma County
Parducci Wine Cellars, 1979, Mendocino County

BRONZE

Fritz Cellars, 1981, Sonoma County, Dry Creek Valley, Estate Bottled
Inglenook Vineyards, 1979, Napa Valley, Estate Bottled, Centennial Vintage
Sebastiani Vineyards, 1980, Sonoma Valley
Alexander Valley Vineyards, 1980, Alexander Valley, Estate Bottled
Zaca Mesa, 1981, Santa Barbara County

PINOT NOIR, *1983 Rating*
($6.51 to $10.50—48 entries)

GOLD

Tulocay, 1979, Napa Valley
DeLoach Vineyards, 1981, Sonoma County, Estate Bottled

Lazy Creek Vineyards, 1980, Mendocino
SILVER
Newlan Vineyards & Winery, 1980, Napa Valley, Estate Bottled
Felton-Empire Vineyards, 1980, California, Maritime
Donna Maria Vineyards, 1980, Sonoma County, Chalk Hill Region, Estate Bottled
Navarro Vineyards, 1980, Mendocino, Estate Bottled
BRONZE
Dehlinger Winery, 1980, Sonoma County
Davis Bynum, 1980, Sonoma County, Russian River Valley
Fenton Acres, 1980, Sonoma County, Joe Rochioli Vineyards
Louis M. Martini, 1979, Napa Valley, La Loma Vineyard, Vineyard Selection
Rutherford Hill, 1979, Napa Valley
Topolos at Russian River Vineyards, 1980, Sonoma County, Hillside, Unfined
Frick Winery, 1980, Monterey County
Husch Vineyards, 1980, Mendocino County, Anderson Valley, Estate Bottled

PINOT NOIR, *1983 Rating*
($10.50 and Up—22 entries)
GOLD
Santa Cruz Mountain Vineyard, 1979, Santa Cruz Mountains, Rider Ridge Vineyard
Acacia Winery, 1980, Napa Valley, Carneros District, Madonna Vineyard
Sunrise, 1980, Sonoma County, Glen Ellen Vineyard
SILVER
ZD Wines, 1980, Santa Barbara
Chateau Chevalier, 1981, Stanton's Pinot Patch
BRONZE
Acacia Winery, 1980, Napa Valley, Carneros District, Iund Vineyard
Carneros Creek, 1980, Napa Valley, Carneros District
ZD Wines, 1980, Napa Valley, Carneros

PINOT NOIR, *1984 Rating*
(Up to $6.50—36 entries)
GOLD
Zaca Mesa, 1981, Santa Barbara County
Caymus Vineyards, 1980, Napa Valley, Estate Bottled
Vincelli Cellars, 1979, Sonoma County
Villa Mt. Eden, 1980, Napa Valley, Estate Bottled
Donna Maria Vineyards, 1981, Chalk Hill, Estate Bottled
Santa Lucia Cellars, 1982, Paso Robles
SILVER
Almaden Vineyards, 1982, San Benito County
August Sebastiani Country, N.V., California
Cresta Blanca, 1980, Mendocino, Estate Bottled
Alexander Valley Vineyards, 1981, Alexander Valley, Estate Bottled
Weibel, 1976, Mendocino County
BRONZE
Breckenridge Cellars, N.V., California
Foppiano Wine Company, 1980, Russian River Valley
Edmeades Vineyards, 1981, Mendocino, Anderson Valley
Beringer Vineyards, 1979, Napa Valley
Fetzer Vineyards, 1981, Mendocino
Paul Masson Vineyards, 1982, Sonoma County
Sebastiani Vineyards, 1981, Sonoma County

PINOT NOIR, *1984 Rating*
($6.51 to $10.50—65 entries)
GOLD
Hacienda Del Rio, 1982, Sonoma County

Sanford Winery, 1981, Santa Maria Valley
SILVER
Buena Vista, 1980, Sonoma Valley, Carneros, Estate Bottled
DeLoach Vineyards, 1982, Russian River Valley, Estate Bottled
Topolos at Russian River, 1980, Sonoma County, Hillside
Austin Cellars, 1982, Santa Barbara County, Bien Nacido Vineyard
Calera Wine Company, 1981, Santa Barbara County
York Mountain, 1980, San Luis Obispo County
Sea Ridge, 1981, Sonoma County
BRONZE
Beaulieu Vineyards, 1979, Napa Valley, Los Carneros Region
Newland Vineyards, 1981, Napa Valley, Dry Creek Vineyard, Estate Bottled
Felton-Empire, 1981, California, "Tonneaux Americains"
Los Vineros, 1981, Santa Maria Valley
Saintsbury, 1982, Sonoma Valley
Sea Ridge, 1981, Sonoma County, Bohan Vineyards
Ventana Vineyards, 1982, Monterey County

PINOT NOIR, *1984 Rating*
($10.51 and Up—28 entries)
GOLD
Robert Stemmler, 1982, Sonoma County
Caymus Vineyards, 1981, Napa Valley, Special Selection, Estate Bottled
Winery Lake, 1981, Los Carneros
SILVER
Acacia, 1981, Napa Valley, Carneros District, Iund Vineyard
Chalone Vineyard, 1980, California, Estate Bottled
ZD, 1980, Santa Barbara
Iron Horse Vineyard, 1980, Sonoma County
BRONZE
Zaca Mesa, 1981, Santa Ynez Valley, American Reserve
Martin Ray, 1981, Napa Valley, Winery Lake Vineyard
Robert Mondavi, 1980, Napa Valley, Reserve
ZD, 1980, Napa Valley, Carneros
Buena Vista, 1981, Sonoma Valley, Carneros, Estate Bottled, Special Selection
Mayacamas, 1980, California
Hacienda, 1981, Sonoma Valley, Estate Bottled
Roudon-Smith Vineyards, 1981, Edna Valley

PINOT NOIR ROSÉ, *1983 Rating*
($4.51 to $6.50—1 entry)
BRONZE
Buena Vista Winery, 1982, Sonoma County, Carneros

PINOT NOIR ROSÉ, (0.71% to 3.0% residual sugar) *1984 Rating*
(Up to $5.00—4 entries)
GOLD
Buena Vista, 1983, Sonoma Valley, Carneros, Estate Bottled
SILVER
Souverain, 1983, North Coast

PINOT NOIR ROSÉ, (0.71% to 3.0% residual sugar) *1984 Rating*
($5.01 to $7.00—1 entry)
BRONZE
Buena Vista, 1983, Sonoma Valley, Carneros, "Pinot Jolie"

VINTAGE PORT, *1981 Rating*
($7.01 to $10.00—3 entries)
SILVER
Bargetto, 1965, California

VINTAGE PORT, *1981 Rating*
($10.01 and Up—8 entries)
GOLD
Quady Winery, 1978 Lot 1, Amador County
Berkeley Wine Cellars, 1978, Sonoma County, Kelley Creek,

Zinfandel Port
SILVER
Quady Winery, 1977, Amador County
Quady Winery, 1977, Paso Robles, Rancho Tierra Rejada
Quady Winery, 1978 Lot 2, Amador County
BRONZE
J.W. Morris Port Works, 1978, Sonoma County, Black Mt. Vineyard

PORT, (up to 6.0% residual sugar) *1984 Rating*
(Up to $5.00—2 entries)
GOLD
Barengo, N.V., California, Reserve, Tawny
SILVER
Breckenridge Cellars, N.V., California

PORT, (up to 6.0% residual sugar) *1984 Rating*
($5.01 to $8.00—1 entry)
BRONZE
Mount Palomar, 1978, California, Estate Bottled, Cabernet Sauvignon

PORT, (up to 6.0% residual sugar) *1984 Rating*
($8.01 and Up—4 entries)
GOLD
David Bruce Winery, 1980, California
V. Sattui, N.V., California, Madeira
BRONZE
Prager Winery and Portworks, 1981, Napa Valley, "Royal Consort"

TAWNY PORT, *1981 Rating*
(Up to $4.00—2 entries)
GOLD
Almaden Vineyards, N.V., California, "Tinta Tawny"
BRONZE
Louis M. Martini, N.V., California

TAWNY PORT, *1981 Rating*
($4.01 to $6.00—2 entries)
SILVER
Sebastiani Vineyards, N.V., Northern California, "Adagio"
BRONZE
Richert & Sons, N.V., California

VINTAGE TAWNY PORT, *1981 Rating*
($10.01 and Up—3 entries)
SILVER
Richert & Sons, 1967, "Silver Satin Port"
The Christian Bros., 1969, California, Mont La Salle

PORT (INCLUDES "RUBY"), *1981 Rating*
(Up to $4.00—7 entries)
SILVER
The Christian Bros., N.V., California, Ruby
BRONZE
Paul Masson Vineyards, N.V., California, "Rich Ruby Port", Ruby

PORT (INCLUDES "RUBY"), *1981 Rating*
($4.01 to $6.00—7 entries)
GOLD
J.W. Morris Port Works, N.V, "Founder's Port", Ruby
Shenandoah Vineyards, N.V., Amador County, Zinfandel
SILVER
Richard Carey Winery, N.V., California, Cabernet Sauvignon
BRONZE
Charles Le Franc, N.V., California, "Founder's Port"
Llords & Elwood, N.V., California, "Ancient Proverb"
Paul Masson Vineyards, N.V., California, "Rare Souzao Port"

PORT (INCLUDES "RUBY"), *1981 Rating*

($6.01 and Up—4 entries)
GOLD
Kosrof, N.V., California, Limited Bottling, Tinta Madeira
SILVER
Ficklin Vineyards, N.V., California, Estate Bottled, Tinta
BRONZE
Beringer Vineyards, N.V., Napa Valley, Cabernet Sauvignon

PORT, (6.0% to 12.0% residual sugar) *1984 Rating*
(Up to $5.00—10 entries)
GOLD
Almaden Vineyards, N.V., California, Solera
Ernest and Julio Gallo, N.V., California
SILVER
Livingston Cellars, N.V., California, Tawny
BRONZE
Trader Joe's, 1965, California, Tinta Madeira
Paul Masson Vineyards, N.V., California, Rich Ruby
Ernest and Julio Gallo, N.V., California, White

PORT, (6.0% to 12.0% residual sugar) *1984 Rating*
($5.01 to $8.00—8 entries)
GOLD
Shenandoah Vineyards, N.V. Lot 3, Amador County, Zinfandel
SILVER
Beringer Vineyards, 1979, Napa Valley, Cabernet Sauvignon
Paul Masson Vineyards, N.V., California, Rare Souzao
Ficklin Vineyards, N.V., California, Estate Bottled
BRONZE
Llords and Elwood, N.V., California, "Ancient Proverb"

PORT, (6.0% to 12.0% residual sugar) *1984 Rating*
($8.01 and Up—3 entries)
SILVER
Quady, 1981, Amador County, Clockspring Vineyard

RUBY CABERNET, *1983 Rating*
($5.01 to $7.00—1 entry)
BRONZE
Barengo, 1979, California, Estate Bottled

SAUVIGNON BLANC (FUMÉ BLANC), *1979 Rating*
(Up to $4.00—5 entries)
GOLD
Almaden, 1977, Monterey, Sauvignon Blanc
Inglenook, 1977, Estate Bottled, Fumé Blanc
SILVER
Ernest & Julio Gallo, N.V., Sauvignon Blanc
BRONZE
Christian Brothers, N.V., Sauvignon Blanc

SAUVIGNON BLANC (FUMÉ BLANC), *1979 Rating*
($4.01-$6.50—34 entries)
GOLD
Christian Brothers, Cuvée 806, Napa, Fumé Blanc
Robert Mondavi, 1977, Fumé Blanc
SILVER
Concannon, 1978, Estate Bottled, Sauvignon Blanc
Paul Masson, 1977, Estate Bottled, Monterey County, Fumé Blanc
Preston, 1977, Fumé Blanc
BRONZE
Beringer, 1977, Knight's Valley Estate, Fumé Blanc
Chateau St. Jean, 1978, Napa Valley, Fumé Blanc
Davis Bynum, 1977, Fumé Blanc
Spring Mountain, 1977, Sauvignon Blanc
Sterling, 1978, Sauvignon Blanc
Stonegate, 1977, Estate Bottled, Sauvignon Blanc
Wente, 1977, Estate Bottled,

Sauvignon Blanc

SAUVIGNON BLANC, *1979 Rating*
($6.51 and Up—5 entries)
GOLD
The Monterey Vineyard, 1975, Botrytised-Sweet, Sauvignon Blanc
SILVER
Montevina, 1978, Estate Bottled, Amador County, Sauvignon Blanc
BRONZE
Dry Creek, 1978, Fumé Blanc

SAUVIGNON BLANC, *1980 Rating*
(Up to $4.50—11 entries)
SILVER
San Martin, 1978, California, Limited Vintage
Parducci, 1979, Mendocino County
BRONZE
Bel Arbres, 1977, Napa

SAUVIGNON BLANC, *1980 Rating*
($4.51 to $7.00—35 entries)
GOLD
Franciscan, 1979, California
Santa Ynez Valley, 1979, California
SILVER
Louis J. Foppiano, 1978, Sonoma, Fumé
Chateau St. Jean, 1979, Sonoma County
BRONZE
Davis Bynum, 1978, Sonoma, Rochiolo-Harrison Reserve
Dry Creek, 1979, Sonoma County
Buena Vista, 1978, Mendocino
Fetzer, 1979, Mendocino

SAUVIGNON BLANC, *1980 Rating*
($7.01 and Up—4 entries)
GOLD
Robert Mondavi, 1978
BRONZE
Veedercrest, 1979, Sonoma County, Shiloh Vineyard

SAUVIGNON BLANC (FUMÉ BLANC), (0.0% to 0.70% residual sugar) *1981 Rating*
(Up to $5.00—7 entries)
GOLD
The Wine Cellars of Ernest & Julio Gallo, N.V., California
BRONZE
Mount Palomar Winery, 1979, California, "Dry"
Almaden Vineyards, 1978, Monterey

SAUVIGNON BLANC (FUMÉ BLANC), (0.0% to 0.70% residual sugar) *1981 Rating*
($5.01 to $8.00—70 entries)
GOLD
Dry Creek Vineyard, 1980, Sonoma, Fumé
Fenestra Winery, 1980, San Luis Obispo
The Christian Bros., N.V., Napa Valley, Cuvée 808, Napa Fumé
Kenwood Vineyards, 1980, Sonoma County
Pope Valley Winery, 1980, Lake County
Ventana Vineyards, 1980 Lot 2, Monterey County
Chateau St. Jean, 1980, Sonoma County, Fumé
J. Carey Cellars, 1980, Santa Maria Valley
SILVER
Stonegate, 1980, California
Montevina, 1980, Amador County, Estate Bottled
Joseph Phelps Vineyards, 1979, California
Chateau St. Jean, 1980, Napa Valley, Dry Fumé
Whitehall Lane Winery, 1980, Alexander Valley, Richard Feeney Vineyards
Geyser Peak Winery, 1980, Sonoma County, Estate Bottled, Fumé
BRONZE
Angelo Papagni Vineyards, 1979, Madera, Estate Bottled, Fumé
Fetzer Vineyards, 1980, Mendocino, Fumé-Dry
Louis J. Foppiano Vineyards, 1979,

Sonoma County, Sonoma Fumé
Stevenot Vineyards, 1980, Northern California
Cambiaso Winery and Vineyards, 1980, Sonoma County, Fumé
Mayacamas Vineyards, 1979, California
San Martin Winery, 1979, San Luis Obispo County, Select Vintage, Fumé
Charles Le Franc, 1979, Monterey County, Fumé
Santa Ynez Valley Winery, 1980, California

SAUVIGNON BLANC (FUMÉ BLANC), (0.0% to 0.70% residual sugar) *1981 Rating*
($8.01 and Up—5 entries)
GOLD
Chateau St. Jean, 1980, Sonoma Valley, St. Jean Vineyards II, Fumé
St. Clement Vineyards, 1980, Napa Valley
SILVER
Grgich Hills, 1979, Napa Valley, Fumé
Vose Vineyards, 1980, Sonoma County, Fumé
BRONZE
Robert Mondavi Winery, 1979, Napa Valley, Fumé

SAUVIGNON BLANC, (0.0% to 0.70% residual sugar) *1982 Rating*
(Up to $5.00—17 entries)
GOLD
Christian Brothers Winery, N.V., Napa Valley, Cuvée 809
Wine Warehouse, 1979, Sonoma County
SILVER
Filsinger Vineyards and Winery, 1980 Lot 1, Temecula, Estate Bottled, Fumé
Taylor California Cellars, N.V., California
The Wine Cellars of Ernest & Julio Gallo, N.V., California
BRONZE
Turner Winery, 1980, Lake County, Fumé
California Hillside Cellars, 1980, San Luis Obispo
Maddalena Vineyard, 1981, Temecula

SAUVIGNON BLANC, (0.0% to 0.70% residual sugar) *1982 Rating*
($5.01 to $8.00—72 entries)
GOLD
Napa Cellars, 1981, Napa Valley
Stephens, 1981, North Coast, Fumé
Husch Vineyards, 1980, Mendocino, Estate Bottled
Montevina Wines, 1981, Amador County, Estate Bottled
SILVER
Dry Creek Vineyard, 1981, Sonoma County, Fumé
Richard Carey, 1980, San Luis Obispo, Library Reserve, Fumé
Beringer Vineyards, 1981, Napa Valley, Fumé
Konocti Winery, 1981, Lake County, Isabel Downs Vineyard, Fumé
L. Foppiano Wine Co., 1980, Sonoma County, Fumé
BRONZE
San Martin Winery, 1980, San Luis Obispo, Fumé
Beringer Vineyards, 1980, Sonoma, Fumé
Kenwood Vineyards, 1981, Sonoma Valley
J. Pedroncelli Winery, 1981, Sonoma
Cambiaso Winery and Vineyards, 1980, Sonoma, Fumé
The Brander Vineyard, 1981, California
Monticello Cellars, 1981, Napa Valley
DeLoach Vineyards, 1981, Russian River Valley, Fumé

SAUVIGNON BLANC, (0.0% to 0.70% residual sugar) *1982 Rating*
($8.51 and Up—13 entries)

GOLD

H.N.W. Cellars, 1981, Napa Valley, H.N.W. Vineyards

Chateau St. Jean, 1981, Alexander Valley, Murphy Ranch, Fumé

SILVER

St. Clements, 1981, Napa Valley

BRONZE

Robert Mondavi Winery, 1980, Napa Valley, Fumé

Concannon Vineyard, 1981, Livermore Valley, Estate Bottled

Grand Cru Vineyards, 1981, Northern California

Wilson Daniels Cellars, 1980, Napa Valley

SAUVIGNON BLANC, (0.0% to 0.70% residual sugar) *1983 Rating*
(Up to $6.00—42 entries)

GOLD

Franciscan Vineyards, 1981, California, Fumé

Wente Bros., 1981, Livermore Valley, Estate Bottled

Trader Joe's, 1981, San Luis Obispo County, Fumé

SILVER

River Road Vineyards, 1981, North Coast, Fumé

Taylor California Cellars, N.V., California

Settler's Creek, N.V., California, Fumé

Bandiera, 1982, Mendocino

BRONZE

The Wine Cellars of Ernest & Julio Gallo, N.V, California

The Konocti Winery, 1981, Lake County

Angelo Papagni, 1981, Madera, Bonita Vineyard, Estate Bottled

Redwood Coast, 1982, California, Fumé

Stone Creek, 1981, North Coast

Delicato Vineyards, N.V., California, Mountain Style, Fumé

The Christian Bros., N.V., Napa Valley, Cuvée 811, Fumé

SAUVIGNON BLANC, (0.0% to 0.70% residual sugar) *1983 Rating*
($6.01 to $8.00—75 entries)

GOLD

Chateau De Leu, 1982, Solano County, Green Valley, Estate Bottled, Fumé

Husch Vineyards, 1981, Mendocino County, La Ribera Ranch, Estate Bottled

Beringer, 1982, Napa Valley, Fumé

Fenestra Winery, 1982, San Luis Obispo County

SILVER

Los Vineros, 1981, Santa Maria Valley

White Oak Vineyards, 1982, Dry Creek Valley

Cain Cellars, 1981, Napa Valley

Deer Park Winery, 1981, Napa Valley

BRONZE

Creston Manor Vineyards & Winery, 1982, San Luis Obispo County

Concannon Vineyard, 1982, California

Rutherford Hill, 1981, Napa Valley

Parducci Wine Cellars, 1982, Lake County, Fumé

William Wheeler, 1982, Sonoma County

Hart Winery, 1982, Temecula

SAUVIGNON BLANC, (0.0% to 0.70% residual sugar) *1983 Rating*
($8.01 and Up—38 entries)

GOLD

Chateau St. Jean, 1982, Sonoma County, La Petite Etoile Vineyards, Fumé

Kenwood Vineyards, 1982, Sonoma County

SILVER

Chateau St. Jean, 1982, Sonoma County, Fumé

Navarro Vineyards, 1982, Mendocino, Anderson Valley

Adler Fels, 1982, Sonoma County,

Salzgeber-Chan Vineyard
Dry Creek Vineyard, 1982, Sonoma County, Fumé
Girard, 1982, North Coast
Chateau St. Jean, 1982, Alexander Valley, Murphy Ranch, Fumé

BRONZE

DeLoach Vineyards, 1982, Sonoma County, Fumé
J. Carey Cellars, 1982, Central Coast
Santa Ynez Valley, 1981, Santa Ynez Valley, Reserve De Cave
Iron Horse, 1982, Alexander Valley
Souverain, 1981, North Coast, Estate Bottled
St. Clement, 1982, Napa Valley

SAUVIGNON BLANC (FUMÉ BLANC), (0.0% to 0.70% residual sugar)
1984 Rating
(Up to $6.00—49 entries)

GOLD

R.H. Phillips Vineyard, 1983, Yolo County, Dunnigan Hills
Bandiera, 1982, Mendocino
Orleans Hill, 1983, Yolo County, Reiff Vineyard

SILVER

Preston Vineyards, 1983, Sonoma County, Dry Creek Valley, Estate Bottled, Fumé
Robert James, 1982, North Coast
Chalk Hill, 1983, Sonoma County
Stratford, 1983, Napa Valley

BRONZE

Filsinger Vineyards, 1983, Temecula, Estate Bottled
Montevina, 1983, Shenandoah Valley, Estate Bottled, Fumé
Grand Cru Vineyards, 1983, California, Vin Maison
Karly, 1983, Amador County, Fumé
Santa Ynez Valley, 1982, California
Stone Creek, 1982, Sonoma, Special Selection, Fumé

SAUVIGNON BLANC (FUMÉ BLANC), (0.0% to 0.70% residual sugar)
1984 Rating
($6.01 to $8.50—115 entries)

GOLD

Obester, 1983, Mendocino County
Tijsseling Vineyards, 1982, Mendocino County
Glen Ellen, 1983, Sonoma County
Napa Cellars, 1983, Napa Valley
Hacienda, 1983, Sonoma County
Glen Ellen, 1983, Sonoma Valley, Fumé
Kenwood Vineyards, 1983, Sonoma County

SILVER

Clos du Val, 1983, Napa Valley
Shenandoah Vineyards, 1983, Amador County
Alderbrook, 1983, Sonoma County
Inglenook Vineyards, 1983, Napa Valley, Estate Bottled
Callaway Vineyards, 1982, California, (Dry)
Adler Fels, 1983, Sonoma County, Salzgeber Chan Vineyards, Fumé
Silverado Vineyards, 1982, Napa Valley
San Martin, 1982, San Luis Obispo County, Fumé
V. Sattui, 1983, Napa Valley
William Wheeler, 1983, Sonoma County
Smothers, 1983, Sonoma Valley, Remick Ridge Ranch
Rancho Sisquoc, 1982, Santa Maria Valley, Estate Bottled

BRONZE

Concannon Vineyard, 1983, California
Girard, 1982, North Coast
Foppiano, 1982, Sonoma County
HMR, 1982, Paso Robles, Rancho Tierra Rejada Vineyards
Groth Vineyards, 1982, Napa Valley
Pat Paulsen Vineyards, 1982, Alexander Valley, Estate Bottled
Husch Vineyards, 1983, Mendocino, La Ribera Ranch, Estate Bottled
Brander Vineyard, 1983, Santa Ynez Valley
Lakespring, 1982, California

Calafia Cellars, 1983, Napa Valley, Honig Vineyard
Whitehall Lane, 1982, San Luis Obispo County and Napa Valley

SAUVIGNON BLANC (FUMÉ BLANC), (0.0% to 0.70% residual sugar)
1984 Rating
($8.51 and Up)
GOLD
Chateau St. Jean, 1983, Sonoma County, "La Petite Etoile", Fumé
Mayacamas, 1982, California
St. Clement, 1983, Napa Valley
SILVER
Dry Creek Vineyard, 1983, Sonoma County, Fumé
Chateau St. Jean, 1983, Sonoma County, Fumé
Joseph Phelps Vineyards, 1982, Napa Valley
Robert Mondavi, 1982, Napa Valley, Fumé
Simi, 1982, Sonoma County
BRONZE
Monticello Cellars, 1982, Napa Valley
Zaca Mesa, 1982, Santa Barbara County, American Reserve
Douglas Meador, 1982, Monterey
Page Mill, 1983, San Luis Obispo County, French Camp Vineyard
Santa Ynez Valley, 1981, Santa Ynez Valley, Reserve de Cave
Charles F. Shaw Winery and Vineyard, 1983, Napa Valley, Fumé

SAUVIGNON BLANC (FUMÉ BLANC), (0.71% and up residual sugar)
1981 Rating
(Up to $5.00—3 entries)
SILVER
Papagni Vineyards, 1979, California, (1.5 liter size)
BRONZE
The Christian Bros., N.V., Napa Valley

SAUVIGNON BLANC (FUMÉ BLANC), (0.71% and up residual sugar)
1981 Rating
($5.01 to $8.00—6 entries)
SILVER
Beaulieu Vineyard, 1979, Napa Valley

SAUVIGNON BLANC (FUMÉ BLANC), (0.71% and up residual sugar)
1982 Rating
(Up to $5.00—4 entries)
BRONZE
Christian Brothers Winery, N.V., Napa Valley

SAUVIGNON BLANC (FUMÉ BLANC), (0.71% and up residual sugar)
1982 Rating
($5.01 to $8.50—5 entries)
GOLD
Parducci Wine Cellars, 1981, North Coast
SILVER
Konocti Winery, 1980, Lake County, Estate Bottled, Fumé
BRONZE
Callaway Vineyard & Winery, 1980, Temecula

SAUVIGNON BLANC (FUMÉ BLANC), (0.71% and up residual sugar)
1982 Rating
($8.51 and Up—2 entries)
BRONZE
J. Lohr Winery, 1981, Monterey, Greenfield Vineyards, Late Harvest

SAUVIGNON BLANC (FUMÉ BLANC), (0.71% to 3.0% residual sugar)
1983 Rating
(Up to $6.00—4 entries)
GOLD
Parducci Wine Cellars, 1982, Mendocino County

SAUVIGNON BLANC (FUMÉ BLANC), (0.71 to 3.0% residual sugar)
1984 Rating
(Up to $6.00—11 entries)
GOLD
Konocti, 1982, Lake County, Fumé

Almaden Vineyards, 1982, Monterey

SILVER

Parducci, 1983, North Coast

Ernest and Julio Gallo, N.V., California

BRONZE

Parducci Wine Cellars, 1983, California, "Summer Games Los Angeles 1984"

Granite Springs, 1983, El Dorado, "Sierra Gold"

SAUVIGNON BLANC (FUMÉ BLANC), (0.71 to 3.0% residual sugar) *1984 Rating*
($6.01 to $8.50—4 entries)

GOLD

Ballard Canyon, 1983, Santa Barbara County

SILVER

Boeger, 1983, El Dorado

BRONZE

R and J Cook, 1982, Clarksburg, Estate Bottled, "Reggeli Harmat"

Leeward Winery, 1983, Amador County, Shenandoah Valley

SAUVIGNON BLANC (FUMÉ BLANC), (6.0% to 12.0% residual sugar) *1984 Rating*
($8.51 and Up—4 entries)

GOLD

Chateau Chevre, 1982, Napa Valley, Late Harvest, Botrytis

Monticello Cellars, 1982, Napa Valley

BRONZE

J. Lohr, 1982, Monterey, Late Harvest

SAUVIGNON BLANC (FUMÉ BLANC), (12.0% and up residual sugar) *1984 Rating*
($8.51 and Up—2 entries)

GOLD

Austin Cellars, 1982, Santa Barbara County, Sierra Madre Vineyard, Botrytis

SEMILLON, *1983 Rating*
($7.01 and Up—1 entry)

BRONZE

Ahlgren Vineyard, 1981, Livermore Valley

SEMILLON, *1984 Rating*
(Up to $5.00—1 entry)

SILVER

Wente Brothers, 1981, Livermore Valley, Estate Bottled

SEMILLON, *1984 Rating*
($5.01 to $7.00—5 entries)

GOLD

Robert Pepi Winery, 1982, Napa Valley

Cranbrook Cellars, 1983, Napa Valley, Chevrier Blanc

SILVER

Alderbrook, 1983, Sonoma County

BRONZE

Stony Ridge, 1982, Livermore Valley, Estate Bottled, Chevrier

SEMILLON, *1984 Rating*
($7.01 and Up—1 entry)

SILVER

Congress Springs, 1983, Santa Cruz Mountains, Barrel Fermented

DRY SHERRY, *1980 Rating*
(Up to $3.00—11 entries)

GOLD

Assumption Abbey, Paildo, Pale Dry

SILVER

Almaden, Flor Fino, California, Solera

BRONZE

Concannon, Prelude

Paul Masson, Cocktail

Gallo, Old Decanter Blanc, Very Dry

Charles Krug, Pale Dry

Inglenook, Pale Dry

DRY SHERRY, *1980 Rating*
($3.01 to $4.50—13 entries)

SILVER
Weibel, Solera Flor, Dry Bin
BRONZE
Louis M. Martini, Dry
Setrakian, Solera, Dry

DRY SHERRY, *1980 Rating*
($4.51 and Up—3 entries)
GOLD
Angelo Papagni, Finest Hour
SILVER
Sebastiani Vineyards, Arenas

SHERRY, (0.0% to 0.70% residual sugar)
1983 Rating
(Up to $4.00—1 entry)
BRONZE
Paul Masson Vineyards, N.V., California, Cocktail

SHERRY, (0.0% to 0.70% residual sugar)
1983 Rating
($4.01 to $6.00—4 entries)
SILVER
Louis M. Martini, N.V., California, Dry Sherry
Cresta Blanca, N.V., California, "Triple Dry"
BRONZE
Barengo, N.V., California, Dry, Reserve

SHERRY, (0.0% to 0.70% residual sugar)
1983 Rating
($6.01 and Up—1 entry)
BRONZE
Chateau Julien, N.V., California, Dry

SHERRY, (0.0% to 0.70% residual sugar)
1984 Rating
($3.51 to $6.00—4 entries)
GOLD
Weibel, N.V., California, Solera Flor, "Dry Bin", (Cocktail)
Llords and Elwood, N.V., California, "Great Day", (Dry)
Louis M. Martini, N.V., California, (Dry)

SHERRY, (0.71% to 3.0% residual sugar)
1983 Rating
(Up to $4.00—4 entries)
SILVER
Almaden Vineyards, N.V., California, Solera Cocktail
Christian Bros., N.V., Cocktail Sherry

SHERRY, (0.71% to 3.0% residual sugar)
1983 Rating
($4.01 to $6.00—3 entries)
GOLD
Llords & Elwood, N.V., "Dry Wit"
BRONZE
Cresta Blanca, N.V., California, "Dry Watch"

SHERRY, (0.71% to 3.0% residual sugar)
1984 Rating
(Up to $3.50—3 entries)
SILVER
Christian Brothers, N.V., California, (Dry)

SHERRY, (0.71% to 3.0% residual sugar)
1984 Rating
($3.51 to $6.00—6 entries)
GOLD
Cresta Blanca, N.V., California, Dry Watch
SILVER
Christian Brothers, N.V., California, Cocktail
BRONZE
Paul Masson Vineyards, N.V., California, (Medium Dry)

SHERRY, (3.0% to 6.0% residual sugar)
1983 Rating
(Up to $4.00—2 entries)
GOLD
Almaden Vineyards, N.V., California, Solera Golden
BRONZE
Christian Brothers, N.V., Golden Sherry

SHERRY, (3.0% to 6.0% residual sugar)

1983 Rating
($4.01 to $6.00—2 entries)
GOLD
Weibel Vineyards, N.V., California, Solera Flor, (Medium)

SHERRY, (3.0% to 6.0% residual sugar)
1984 Rating
(Up to $3.50—4 entries)
BRONZE
Almaden Vineyards, N.V., California, Golden Solera

SHERRY, (3.0% to 6.0% residual sugar)
1984 Rating
($3.51 to $6.00—3 entries)
GOLD
****Llords and Elwood, N.V., California, "Dry Wit"
Weibel, N.V., California, Solera Flor, (Medium)

SHERRY, (3.0% to 6.0% residual sugar)
1984 Rating
($6.01 and Up—1 entry)
GOLD
Buena Vista, N.V., California, Golden Cream

SHERRY, (6.0% to 12.0% residual sugar)
1983 Rating
(Up to $4.00—5 entries)
GOLD
Christian Bros., N.V., Cream Sherry
Trader Joe's Winery, 1978, Santa Clara County, "Vinsanto"
BRONZE
The Wine Cellars of Ernest & Julio Gallo, N.V., California, Livingston Cream
San Antonio Winery, N.V., California, Solera Cream

SHERRY, (6.0% to 12.0% residual sugar)
1983 Rating
($4.01 to $6.00—7 entries)
BRONZE
Barengo, N.V., California, Cream, Reserve

SHERRY, (6.0% to 12.0% residual sugar)
1983 Rating
($6.01 and Up—3 entries)
SILVER
Paul Masson Vineyards, N.V., California, Rare Cream
BRONZE
Monterey Peninsula, N.V., California, Special Cream

SHERRY, (6.0% to 12.0% residual sugar)
1984 Rating
(Up to $3.50—6 entries)
GOLD
Livingston Cellars, N.V., California, Cream
BRONZE
Barengo, N.V., California, Reserve, Cream
San Antonio Winery, N.V., California, Solera Cream

SHERRY, (6.0% to 12.0% residual sugar)
1984 Rating
($3.51 to $6.00—8 entries)
GOLD
Louis M. Martini, N.V., California, Cream
SILVER
Indelicato, N.V., California, Private Stock, Cream
BRONZE
Llords and Elwood, N.V., California, "The Judge's Secret", Cream
Mount Palomar, N.V., California, Cream

SHERRY, (6.0% to 12.0% residual sugar)
1984 Rating
($6.01 and Up—2 entries)
GOLD
Paul Masson Vineyards, N.V., California, Rare Cream

CREAM SHERRY, *1979 Rating*
(Up to $2.00—3 entries)
BRONZE
Colony, Velvet Cream Sherry

CREAM SHERRY, *1979 Rating*
($2.01 to $3.50—16 entries)
SILVER
Almaden
Paul Masson, Golden Cream
BRONZE
Buena Vista, Golden Cream
Charles Krug

CREAM SHERRY, *1979 Rating*
($3.51 and Up—7 entries)
SILVER
Llords & Elwood, The Judge's Secret
Paul Masson, Cuvée 702-C, Rare Cream Sherry
BRONZE
Christian Brothers, Melroso
Sebastiani, Amore

SHERRY, (12% and up residual sugar) *1983 Rating*
($4.01 to $6.00—2 entries)
BRONZE
Weibel Vineyards, N.V., California, Solera Flor, Amber Cream

SHERRY, (12% and up residual sugar) *1983 Rating*
($6.01 and Up—1 entry)
GOLD
Christian Brothers, N.V., Cream Sherry, "Meloso"

SHERRY, (12.0% and up residual sugar) *1984 Rating*
($3.51 to $6.00—2 entries)
GOLD
Weibel, N.V., California, Solera Flor, Amber Cream

SHIRAZ, *1983 Rating*
(Up to $5.00—1 entry)
BRONZE
Mount Palomar Winery, N.V., Temecula, Estate Bottled

SYRAH, *1984 Rating*
($5.01 to $7.50—1 entry)
SILVER
Estrella River Winery, 1981, Paso Robles, Estate Bottled

SYRAH, *1984 Rating*
($7.51 and Up—3 entries)
SILVER
McDowell Valley Vineyards, 1981, McDowell Valley, Estate Bottled

SYLVANER, *1983 Rating*
(Up to $4.50—1 entry)
GOLD
Parducci Wine Cellars, 1982, Mendocino County

SYLVANER, *1984 Rating*
($5.01 to $6.50—1 entry)
BRONZE
Rancho Sisquoc, 1983, Santa Maria Valley, Estate Bottled, Franken Riesling

WHITE (JOHANNISBERG) RIESLING, (0.0% to 0.70% residual sugar) *1981 Rating*
(Up to $5.00—7 entries)
GOLD
Johnson's Alexander Valley Wines, 1979, Alexander Valley, Estate Bottled
Mt. Palomar Winery, 1979, California, "Dry"
SILVER
Vega Vineyards, 1979, Santa Ynez Valley

WHITE (JOHANNISBERG) RIESLING, (0.0% to 0.70% residual sugar) *1981 Rating*
($5.01 to $7.50—5 entries)
GOLD
Trefethen Vineyards, 1980, Napa Valley
SILVER
Gundlach-Bundschu Winery, 1980, Sonoma Valley, Estate Bottled

WHITE (JOHANNISBERG) RIESLING, (0.0% to 0.70% residual sugar) *1982 Rating*

(Up to $5.50—8 entries)
SILVER
San Antonio Winery, 1981, California
BRONZE
Vega Vineyards, 1981, Santa Ynez Valley, Labeled White Riesling

WHITE (JOHANNISBERG) RIESLING, (0.0% to 0.70% residual sugar) *1982 Rating*
($5.51 to $8.00—5 entries)
GOLD
Trefethen Vineyards, 1981, Napa Valley, Trefethen Vineyards
SILVER
Bargetto Winery, 1981, Sonoma

WHITE RIESLING, (0.0% to 0.70% residual sugar) *1983 Rating*
(Up to $5.50—6 entries)
BRONZE
Almaden Vineyards, 1981, San Benito County
Bandiera, 1982, Mendocino

WHITE RIESLING, (0.0% to 0.70% residual sugar) *1983 Rating*
($5.51 to $8.00—6 entries)
GOLD
Trefethen Vineyards, 1982, Napa Valley, Trefethen Vineyards, Estate Bottled
SILVER
Paul Masson Vineyards, 1982, Monterey County
BRONZE
Haywood, 1982, Sonoma Valley, Early Harvest

WHITE RIESLING, (0.0% to 0.705 residual sugar) *1984 Rating*
($5.51 to $8.00—4 entries)
BRONZE
V. Sattui, 1983, Napa Valley, "Dry"
Nevada City, 1983, El Dorado County

WHITE (JOHANNISBERG) RIESLING, (0.71% to 3.0% residual sugar) *1981 Rating*
(Up to $5.00—18 entries)
GOLD
Concannon Vineyard, 1979, "Livermore Riesling"
Vega Vineyards, 1980, Santa Ynez Valley
Boeger, 1980, El Dorado County
The Wine Cellars of Ernest & Julio Gallo, N.V., California
SILVER
Santa Ynez Valley Winery, 1980, California

WHITE (JOHANNISBERG) RIESLING, (0.71% to 3.0% residual sugar) *1981 Rating*
($5.01 to $7.00—64 entries)
GOLD
Obester Winery, 1980, Monterey County
Raymond Vineyard & Cellar, 1980, Napa Valley, Estate Bottled
Greenwood Ridge Vineyards, 1980, Mendocino County
Sycamore Creek Vineyards, 1980, Monterey County
Haywood, 1980, Sonoma Valley, Estate Grown
Robert Mondavi Winery, 1980, Napa Valley
SILVER
Mirassou Vineyards, 1980, Monterey County
Geyser Peak Winery, 1980, Sonoma County, Kiser Ranch, Estate Bottled, "Soft"
Bargetto, 1980, Santa Barbara County, Tepusquet Vineyards
Charles Krug Winery, 1979, Napa Valley
Beringer Vineyards, 1980, Monterey County, Estate Bottled
Chateau St. Jean, 1980, Alexander Valley, Robert Young Vineyards
Rutherford Vintners, 1980, Napa Valley
BRONZE
Zaca Mesa, 1980, Santa Ynez Valley, American Estate

Jekel Vineyard, 1979, Monterey County, Estate Bottled
Kenwood Vineyards, 1980, Sonoma Valley, Early Harvest
Concannon Vineyards, 1980, Livermore Valley, Estate Bottled
Souverain Cellars, 1979, North Coast
St. Francis Vineyards, 1979, Sonoma Valley, Estate Bottled
Chateau St. Jean, 1980, Sonoma County
Alexander Valley Vineyards, 1980, Alexander Valley, Estate Bottled

WHITE (JOHANNISBERG) RIESLING, (0.71% to 3.0% residual sugar) *1981 Rating*
($7.51 and Up—4 entries)
SILVER
Sonoma Vineyards, 1979, Sonoma County, Le Baron Vineyards, Estate Bottled
Veedercrest, 1979, Napa County, Mt. Veeder District, Miliken Hill Vineyard
BRONZE
Chateau St. Jean, 1980, Alexander Valley, Robert Young Vineyards, Early Harvest

WHITE (JOHANNISBERG) RIESLING, (0.71% to 3.0% residual sugar) *1982 Rating*
(Up to $5.50—20 entries)
GOLD
Veedercrest Vineyards, 1980, Napa, Steltzner and Winery Lake Vineyards
SILVER
Taylor California Cellars, N.V., California
BRONZE
Vega Vineyards, 1981, Santa Ynez Valley, Labeled Johannisberg Riesling

WHITE (JOHANNISBERG) RIESLING, (0.71% to 3.0% residual sugar) *1982 Rating*
($5.51 to $8.00—63 entries)
GOLD
J. Lohr Winery, 1981, Monterey, Greenfield Vineyards
Obester Winery, 1981, Monterey County, Ventana Vineyards
Sarah's Vineyard, 1981, Mendocino County, Victor Matheu Vineyards
Haywood Winery, 1981, Sonoma Valley, Estate Bottled, Early Harvest
Ventana Vineyards Winery, 1981, Monterey, Estate Bottled
SILVER
Ballard Canyon Winery, 1981, Santa Ynez, Estate Bottled
Conn Creek, 1979, Napa Valley
Chateau St. Jean, 1981, Sonoma County
Santa Ynez Valley Winery, 1981, California
Fieldstone Winery, 1981, Alexander Valley, Estate Bottled
Jekel Vineyards, 1981, Monterey, Home Vineyard, Estate Bottled
Chateau St. Jean, 1981, Alexander Valley, Robert Young Vineyards
Estrella River Winery, 1981, San Luis Obispo, Estate Bottled
Parsons Creek Winery, 1981, Mendocino, Anderson Valley, Philo Foothills Ranch
Zaca Mesa Winery, 1981, Santa Ynez Valley, American Estate
BRONZE
Jekel Vineyards, 1981, Monterey
Concannon Vineyard, 1981, Livermore Valley, Estate Bottled
Bargetto Winery, 1981, Santa Barbara County, Tepusquet Vineyard
Felton-Empire Vineyards, 1981, Santa Cruz Mountains
Fetzer Vineyards, 1981, Lake County
Smothers, 1981, California
Robert Mondavi Winery, 1980, Napa Valley
Raymond Vineyard & Cellar, 1981, Napa Valley, Estate Bottled

WHITE RIESLING, (0.71% to 3.0% residual sugar) *1983 Rating*
(Up to $5.50—21 entries)
GOLD
J. Pedroncelli, 1982, Sonoma County
Whitehall Lane Winery, 1982, Napa Valley
SILVER
The Konocti Winery, 1982, Lake County
Vega Vineyards, 1982, Santa Ynez Valley, Estate Bottled
BRONZE
Redwood Coast, 1982, California
Topolos at Russian River Vineyards, 1980, Mendocino
San Martin, N.V., California

WHITE RIESLING, (0.71% to 3.0% residual sugar) *1983 Rating*
($5.51 to $8.00—65 entries)
GOLD
Greenwood Ridge Vineyards, 1982, Mendocino
Obester Winery, 1982, Monterey County, Ventana Vineyard
Chateau St. Jean, 1982, Sonoma County
The Firestone Vineyard, 1982, Santa Ynez Valley
SILVER
Fieldstone, 1982, Sonoma County, Alexander Valley, Redwood Ranch & Vineyards, Estate Bottled
J. Lohr, 1982, Monterey, Greenfield Vineyards
Chateau St. Jean, 1982, Alexander Valley, Robert Young Vineyards
Felton-Empire Vineyards, 1981, California, Maritime
Jekel Vineyard, 1982, Arroyo Seco, Home Vineyard, Estate Bottled
BRONZE
Clos du Bois, 1982, Alexander Valley, Early Harvest
Hacienda Wine Cellars, 1982, Sonoma Valley, Estate Bottled
Zaca Mesa, 1982, Santa Barbara County
Santa Ynez Valley, 1982, Santa Ynez Valley
Watson Vineyards, 1982, Paso Robles, Estate Bottled
St. Francis, 1982, Sonoma Valley, Estate Bottled
Kendall-Jackson, 1982, Clear Lake
Llords & Elwood, 1981, "Castle Magic"

WHITE RIESLING, (0.71% to 3.0% residual sugar) *1983 Rating*
($8.01 and Up—2 entries)
GOLD
Hagafen Cellars, 1982, Napa Valley, Winery Lake Vineyard

WHITE RIESLING, (0.71% to 3.0% residual sugar) *1984 Rating*
(Up to $5.50—21 entries)
SILVER
North Coast Cellars, N.V., North Coast
Konocti, 1982, Lake County
Taylor California Cellars, N.V., California
BRONZE
The Monterey Vineyard, 1982, Monterey County
San Martin, 1983, California
Wente Brothers, 1982, Monterey
Rustridge Ranch, 1982, Napa Valley

WHITE RIESLING, (0.71% to 3.0% residual sugar) *1984 Rating*
($5.51 to $8.00—69 entries)
GOLD
Estrella River Winery, 1983, Paso Robles, Estate Bottled
Haywood, 1983, Sonoma Valley, Estate Bottled
Santa Ynez Valley, 1983, Santa Ynez Valley
Obester, 1983, Monterey County, Ventana Vineyard
SILVER
Zaca Mesa, 1983, Santa Barbara County
Kendall-Jackson, 1983, Lake

County
Ballard Canyon, 1983, Santa Ynez Valley, Estate Bottled

BRONZE

Jekel Vineyards, 1983, Monterey
Clos du Bois, 1983, Alexander Valley, Early Harvest
Fetzer Vineyards, 1983, Mendocino
Vega Vineyards, 1983, Santa Ynez Valley, Estate Bottled
V. Sattui, 1983, Napa Valley
Joseph Phelps Vineyards, 1983, Napa Valley
Stearns Wharf Vintners, 1983, Santa Ynez Valley, Vina De Santa Ynez
Windsor Vineyards, 1983, Sonoma County, River West Vineyard, Vineyard Selection
Chateau St. Jean, 1983, Sonoma County
The Firestone Vineyard, 1983, Santa Ynez Valley
Felton-Empire, 1982, California, Maritime Series
Chateau St. Jean, 1983, Alexander Valley, Robert Young Vineyards.

WHITE RIESLING, (0.71% to 3.0% residual sugar) *1984 Rating*
($8.01 and Up—2 entries)

BRONZE

Hagafen Cellars, 1983, Napa Valley, Winery Lake Vineyard
Felton-Empire, 1983, Santa Cruz Mountains, Hallcrest Vineyards

WHITE (JOHANNISBERG) RIESLING, (3.1% to 6.0% residual sugar) *1981 Rating*
(Up to $5.00—4 entries)

SILVER

Fetzer Vineyards, 1980, Mendocino
Weibel, 1980, Monterey

WHITE (JOHANNISBERG) RIESLING, (3.1% to 6.0% residual sugar) *1981 Rating*
($5.01 to $7.50—6 entries)

GOLD

The Firestone Vineyard, 1979, Santa Ynez Valley, The Ambassador's Vineyard

SILVER

Durney Vineyard, 1980, Monterey County, Carmel Valley
Konocti Cellars, 1980, Lake County
The Monterey Vineyard, 1979, Monterey County, "Soft"

WHITE (JOHANNISBERG) RIESLING, (3.1% to 6.0% residual sugar) *1981 Rating*
($7.51 and Up—5 entries)

SILVER

Long Vineyards, 1979, Napa Valley
Stony Ridge Winery, 1979 Lot 1, Monterey County, Selected Late Harvest

WHITE (JOHANNISBERG) RIESLING, (3.1% to 6.0% residual sugar) *1982 Rating*
(Up to $5.50—4 entries)

SILVER

Vache, 1981, Temecula, "Soft"

BRONZE

Lost Hills Vineyards, 1980, California

WHITE (JOHANNISBERG) RIESLING, (3.1% to 6.0% residual sugar) *1982 Rating*
($5.51 to $8.00—5 entries)

GOLD

Ballard Canyon Winery, 1981, Santa Ynez, Estate Bottled, "Reserve", Residual Sugar 3.9%

SILVER

Veedercrest Vineyards, 1980, Napa, Winery Lake Vineyards, Late Harvest, Sugar at Bottling, 4.7%
Beringer Vineyards, 1980, Sonoma, Residual Sugar 4.5%

BRONZE

Vega Vineyards, 1981, Santa Ynez Valley, Special Selection

WHITE (JOHANNISBERG) RIESLING, (3.1% to 6.0% residual sugar) *1982 Rating*
($8.01 and Up—3 entries)

SILVER

Smothers, 1980, Santa Cruz, Vine Hill Vineyard, Residual Sugar 4.2%

BRONZE

Charles Le Franc, 1977, San Benito,

Late Harvest, Residual Sugar 5.6%

WHITE RIESLING, (3.0% to 6.0% residual sugar) *1983 Rating*
($5.51 to $8.00—4 entries)
SILVER
Sarah's Vineyard, 1982, Mendocino County, Victor Matheu Vineyard
San Martin, 1982, California, Soft

WHITE RIESLING, (3.0% to 6.0% residual sugar) *1983 Rating*
($8.01 and Up—7 entries)
GOLD
Buena Vista Winery, 1981, Carneros, Special Selection
Ballard Canyon Winery, 1982, Santa Ynez Valley, Estate Bottled, Reserve
SILVER
Smothers, 1981, Santa Cruz, Vine Hill Vineyard
BRONZE
Robert Mondavi, 1981, Napa Valley, Special Selection
Grgich Hills Cellar, 1980, Napa Valley, Late Harvest
Shown & Sons Vineyards, 1981, Estate Bottled, Late Harvest

WHITE RIESLING, (3.0% to 6.0% residual sugar) *1984 Rating*
(Up to $5.50—2 entries)
GOLD
The Monterey Vineyard, 1983, Monterey County, Limited Vintage, "Soft White"

WHITE RIESLING, (3.0% to 6.0% residual sugar) *1984 Rating*
($5.51 to $8.00—2 entries)
GOLD
San Martin, 1983, California, Low Alcohol, "Soft"
BRONZE
Hidden Cellars, 1983, Mendocino County, Potter Valley

WHITE RIESLING, (3.0% to 6.0% residual sugar) *1984 Rating*
($8.01 and Up—4 entries)
GOLD
Robert Mondavi, 1983, Napa Valley, Special Selection
SILVER
Ballard Canyon, 1982, Santa Ynez Valley, Estate Bottled, Reserve

WHITE (JOHANNISBERG) RIESLING, (6.1% to 12.0% residual sugar) *1981 Rating*
($7.51 and Up—7 entries)
GOLD
Wente Bros., 1979, Monterey, Arroyo Seco
SILVER
Jekel Vineyard, 1979, Monterey County, Estate Bottled, Late Harvest
Joseph Phelps Vineyards, 1979, Napa Valley, Late Harvest
BRONZE
Chateau St. Jean, 1980, Alexander Valley, Robert Young Vineyards, Selected Late Harvest

WHITE (JOHANNISBERG) RIESLING, (6.1% to 12.0% residual sugar) *1982 Rating*
($5.51 to $8.00—4 entries)
GOLD
Bargetto Winery, 1981, Santa Barbara County, Late Harvest, Tepusquet
BRONZE
Christian Brothers Winery, 1980, Napa Valley, Late Harvest, Residual Sugar 7.3%
San Martin Winery, 1980, Santa Barbara, "Soft"

WHITE (JOHANNISBERG) RIESLING, (6.1% to 12.0% residual sugar) *1982 Rating*
($8.01 and Up—7 entries)
SILVER
Wente Bros., 1979, Monterey, Arroyo Seco Vineyards, Late Harvest
BRONZE

HMR Hoffman Vineyards, 1979, San Luis Obispo, Late Harvest, Residual Sugar 11.25%
Ventana Vineyards Winery, 1981, Monterey, Estate Bottled, Botrytised, Residual Sugar 9%
Charles Le Franc, 1979, San Benito, Selected Late Harvest, Advanced Botrytis, Residual Sugar 11.7%

WHITE RIESLING, (6.0% to 12.0% residual sugar) *1983 Rating*
(Up to $5.50—1 entry)
BRONZE
Rosenblum Cellars, 1982, Napa Valley, Late Harvest

WHITE RIESLING, (6.0% to 12.0% residual sugar) *1983 Rating*
($5.51 to $8.00—1 entry)
BRONZE
Vega Vineyards, 1982, Santa Ynez Valley, Estate Bottled, Special Selection

WHITE RIESLING, (6.0% to 12.0% residual sugar) *1983 Rating*
($8.01 and Up—11 entries)
SILVER
Bargetto Winery, 1982, Tepusquet Vineyard, Late Harvest
Navarro Vineyards, 1981, Mendocino, Late Harvest
Mirassou Vineyards, 1982, Monterey County, Late Harvest
J. Lohr, 1981, Monterey, Late Harvest
BRONZE
Nevada City, 1981, El Dorado County, Late Harvest

WHITE RIESLING, (6.0% to 12.0% residual sugar) *1984 Rating*
($5.51 to $8.00—2 entries)
SILVER
Vega Vineyards, 1982, Santa Ynez Valley, Special Selection, Estate Bottled
BRONZE
Chateau Robert Roumiguiere, 1982, Lake County, Late Harvest, "With Botrytis"

WHITE RIESLING, (6.0% to 12.0% residual sugar) *1984 Rating*
($8.01 and Up—12 entries)
GOLD
Chateau St. Jean, 1983, Alexander Valley, Late Harvest
Joseph Phelps Vineyards, 1982, Napa Valley, Late Harvest
Kenwood Vineyards, 1983, Sonoma Valley, Late Harvest
SILVER
J. Lohr, 1981, Monterey, Late Harvest
BRONZE
Smothers, 1983, Sonoma County, Late Harvest
Wente Brothers, 1979, Monterey, "Arroyo Seco Riesling"

WHITE (JOHANNISBERG) RIESLING, (12.0% and up residual sugar) *1981 Rating*
($5.01 to $7.50—1 entry)
GOLD
Felton-Empire Vineyards, 1979, Santa Barbara, Tepusquet Vineyard

WHITE (JOHANNISBERG) RIESLING, (12.0% and up residual sugar) *1981 Rating*
($7.51 and Up—4 entries)
GOLD
Stony Ridge Winery, 1979 Lot 2, Monterey County, Selected Late Harvest
SILVER
San Martin Winery, 1979, Santa Clara County, Late Harvest

WHITE (JOHANNISBERG) RIESLING, (12.1% and up residual sugar) *1982 Rating*
($8.01 and Up—7 entries)
GOLD
Milano, 1981, Mendocino, Philo, Anderson Valley Vineyards, Late

Harvest, Individually Bunch Selected, Residual Sugar 17.7%
Chateau St. Jean, 1980, Alexander Valley, Belle Terre Vineyards, Late Harvest, Individual Bunch Selected, Residual Sugar 17.9%
SILVER
Jekel Vineyards, 1980, Monterey, Late Harvest, Estate Bottled, Residual Sugar 16%
Felton-Empire Vineyards, 1981, Santa Barbara, Tepusquet Vineyards, Select Late Harvest, Residual Sugar 20%
BRONZE
Chateau St. Jean, 1981, Alexander Valley, Belle Terre Vineyards, Select Late Harvest, Residual Sugar 14.5%

WHITE RIESLING, (12.0% and up residual sugar) *1983 Rating*
($5.51 to $8.00—2 entries)
SILVER
San Martin, 1980, Santa Clara County, Late Harvest

WHITE RIESLING, (12.0% and up residual sugar) *1983 Rating*
($8.01 and Up—13 entries)
GOLD
Jekel Vineyard, 1981, Monterey County, Estate Bottled, Late Harvest
SILVER
Joseph Phelps Vineyards, 1980, Napa Valley, Selected Late Harvest
Chateau St. Jean, 1981, Alexander Valley, Robert Young Vineyards, Select Late Harvest
Chateau St. Jean, 1980, Alexander Valley, Robert Young Vineyards, Individual Dried Bunch Selected, Late Harvest
BRONZE
Milano, 1982, Mendocino County, Anderson Valley, Ordways Valley Foothills Vineyard, Late Harvest
Clos du Bois, 1981, Alexander Valley, Late Harvest, Individual Bunch Selected
Newlan Vineyards and Winery, 1981, Napa Valley, Late Harvest, Bunch Selected
Estrella River Winery, 1979, Santa Barbara County, Late Harvest

WHITE RIESLING, (12.0% and up residual sugar) *1984 Rating*
($8.01 and Up—14 entries)
GOLD
Susiné Cellars, 1982, El Dorado, Special Selection, Late Harvest
SILVER
Chateau St. Jean, 1982, Alexander Valley, Robert Young Vineyards, Special Select Late Harvest
BRONZE
Felton-Empire, 1983, Sonoma, Fort Ross, Select Late Harvest
Clos du Bois, 1983, Alexander Valley, Late Harvest, Individual Bunch Selected
Greenwood Ridge Vineyards, 1983, Mendocino, Select Late Harvest
Franciscan Vineyards, 1982, Napa Valley, Estate Bottled, Select Late Harvest

ZINFANDEL, *1980 Rating*
(Up to $3.50—25 entries)
GOLD
Parducci, 1978, Mendocino County
SILVER
Berkeley Wine Cellars, 1977, Kelley Creek, Sonoma County
Bel Arbres, 1975, Sonoma
BRONZE
Wente, 1978, California
Giumarra, N.V., California
River Oaks Vineyards, 1977, Alexander Valley
Fetzer, 1978, Lake County

ZINFANDEL, *1980 Rating*
($3.51 to $5.50—51 entries)
GOLD
Sierra Vista, 1978, El Dorado County
Preston, 1977, Sonoma County, Dry

Creek Valley
Zaca Mesa, 1977, Santa Ynez Valley
SILVER
Smothers, 1978, San Luis Obispo
Montevina, 1978, Amador County
Estrella River Winery, 1977, San Luis Obispo
Harbor, 1977, Shenandoah Valley, Deaver Vineyard
Ridge, 1977, San Luis III
Kenwood, 1977, Sonoma County
BRONZE
Lawrence, 1979, California
Parducci, 1979, Cellar Master's
Boeger, 1978, El Dorado County
Callaway, 1977, Temecula
Louis M. Martini, 1976, California
Napa Wine Cellar, 1978, Alexander Valley

ZINFANDEL, *1980 Rating*
($5.51 and Up—33 entries)
GOLD
Mastantuono, 1978, Templeton, San Luis Obispo, Dusi Vineyard, Unfined-Unfiltered
Gundlach-Bundschu, 1977, Sonoma Valley, Rhine Farm Vineyards
Clos du Val, 1977, Napa
Joseph Phelps, 1977, Alexander Valley, Black Mountain Vineyards
Milano, 1978, Mendocino County, Redwood Valley, Garzini Vineyards
SILVER
Chateau Montelena, 1976, North Coast
Cassayre-Forni, 1977, Sonoma County
Stony Ridge, 1977, Livermore Valley, Ruetz Vineyard
Sommelier, 1977 Lot 2, Lodi
HMR, 1976, Sauret Vineyards, San Luis Obispo
BRONZE
Lytton Springs, 1978, Sonoma County
Hop Kiln, 1978, Russian River
Cuvaison, 1976, Napa Valley
Davis Bynum, 1977, Sonoma
Ridge, 1978, Shenandoah, Esola Vineyard

ZINFANDEL, *1981 Rating*
(Up to $4.50—35 entries)
GOLD
Giumarra Vineyards, 1980, California
J. Pedroncelli Winery, 1978, Sonoma County
SILVER
Sierra Vista Winery, N.V., Sierra Foothills
Trader Joe's, 1979, Mendocino, Unfiltered & Unfined
BRONZE
Bel Arbres, 1978, Mendocino

ZINFANDEL, *1981 Rating*
($4.51 to $7.00—76 entries)
GOLD
Pope Valley Winery, 1978, Northern California
Ranchita Oaks Winery, 1979, California, San Miguel
Shenandoah Vineyards, 1978, Amador County
Rosenblum Cellars, 1979, Sonoma County, Batto Vineyards
Raymond Vineyard & Cellar, 1978, Napa Valley, Estate Bottled
McDowell Valley Vineyards, 1979, Mendocino County, Estate Bottled
Richard Carey Winery, 1978, Amador County, Library Reserve
SILVER
DeLoach Vineyards, 1979, Sonoma County, Estate Bottled
Fetzer Vineyards, 1979, Mendocino, Lolonis
Berkeley Wine Cellars, 1978, Sonoma County, Kelley Creek
J. Pedroncelli Winery, 1977, Sonoma Valley, Vintage Selection
Sierra Vista, 1979, El Dorado County
Rutherford Hill Winery, 1977, Napa Valley, Mead Ranch-Atlas Peak

San Martin Winery, 1978, Amador County
Montevina, 1979, Amador County, Estate Bottled
Carneros Creek Winery, 1979, Yolo County
Baldinelli Vineyard, 1979, Amador County

BRONZE

Ridge Vineyards, 1979, San Luis, Dusi Vineyard
David Bruce Winery, 1978, Santa Clara County
Fortino Winery, 1977, California
Fetzer Vineyards, 1978, Mendocino, Scharffenberger
Kenwood Vineyards, 1978, Sonoma Valley
Zaca Mesa, 1979, Santa Ynez Valley, Estate Grown
Preston Vineyards, 1978, Sonoma County, Dry Creek Valley, Estate Bottled
Devlin Wine Cellars, 1978, Livermore Valley
Mirassou Vineyards, 1978, Monterey County, Unfiltered

ZINFANDEL, *1981 Rating*
($7.01 and Up—32 entries)

GOLD

Lytton Springs, 1979, Sonoma County
Burgess Cellars, 1978, Napa Valley
Calera, 1979, California, Cienega District

SILVER

Napa Wine Cellars, 1979, Alexander Valley
Edmeades Vineyards, 1979, Mendocino
Milano Winery, 1979, Talmage-Mendocino County, Pacini Vineyard
Shenandoah Vineyards, 1979, Amador County, Special Reserve

BRONZE

Grgich Hills, 1978, Alexander Valley
Page Mill Winery, 1979, Napa Valley
Leeward Winery, 1979, Shenandoah Valley
Topolos, 1979, Sonoma County, Russian River Vineyards
Mastantuono, 1979, San Luis Obispo County, Templeton, Dusi Vineyard, Unfined & Unfiltered

ZINFANDEL, *1982 Rating*
(Up to $4.50—33 entries)

GOLD

Christian Brothers Winery, N.V., Napa Valley
Ridgewood Winery, 1979, Lytton Springs, Polk Vineyard (This wine was inadvertently judged in this category. It should have been judged in the $4.51 to $7.50 category.)

SILVER

Cresta Blanca Vineyards, 1978, Mendocino
Cribari & Sons Winery, N.V., California

BRONZE

Mirassou Vineyards, 1978, Monterey County, Unfiltered
Santino, 1980, Amador County, Shenandoah Valley
Fetzer Vineyards, 1980, Lake County
Trader Joe's Winery, 1979, Mendocino
Paul Masson Vineyards, 1980, California
Raymond Hill (Trader Joe's), 1979, Mendocino, Unfiltered, Unfined
J. Pedroncelli Winery, 1979, Sonoma
Story Vineyards, N.V., Amador County, Estate Bottled

ZINFANDEL, *1982 Rating*
($4.51 to $7.50—103 entries)

GOLD

Raymond Vineyard & Cellar, 1979, Napa Valley, Estate Bottled
Cordtz Brothers Cellars, 1980, Upper Alexander Valley

Montevina Wines, 1980, Amador County, Estate Bottled, Montino
Madrona Vineyards, 1980, El Dorado County, Estate Bottled
Ahern Winery, 1980, Amador County
Kenworthy Vineyards, 1979, Amador, Potter-Cowan Vineyards

SILVER
Richard Carey, 1979, Amador, Library Reserve
Parducci Wine Cellars, 1980, North Coast
Conn Creek, 1978, Napa Valley
Santino, 1979, Amador County, Shenandoah Valley
Dry Creek Vineyard, 1979, Sonoma County
San Martin Winery, 1979, Amador County

BRONZE
Shenandoah Vineyards, 1979, Amador County
McLester, 1980, San Luis Obispo County, Radike Vineyard
Obester Winery, 1979, Sonoma County, Forchini Vineyard
Round Hill Cellars, 1979, Napa, Round Hill
Hacienda Wine Cellars, 1979, Sonoma County
Bargetto Winery, 1980, San Luis Obispo County, Farview Farm Vineyard
Sutter Home Winery, 1979, Amador County
Santino, 1979, Amador County, Fiddletown
L. Foppiano Wine Co., 1978, Sonoma County

ZINFANDEL, *1982 Rating*
($7.51 and Up—39 entries)

GOLD
Sycamore Creek Vineyards, 1980, California, Estate Bottled
David Bruce Winery, 1979, El Dorado
Mastantuono, 1980, Paso Robles

SILVER
Montevina Wines, 1979, Amador County, Estate Bottled
Edmeades, 1980, Mendocino County, Pacini Vineyards
Sarah's Vineyard, 1980, San Benito County
Vose Vineyards, 1979, Napa Valley

BRONZE
Sarah's Vineyard, 1980, Sonoma County, Les Vignerous Vineyard
Burgess Cellars, 1979, Napa Valley, Estate Bottled
J.W. Morris Winery, 1980, Sonoma
Edmeades, 1980, Mendocino County, Ciapusci Vineyards

ZINFANDEL, *1983 Rating*
(Up to $5.00—50 entries)

GOLD
Estrella River Winery, N.V., San Luis Obispo County
August Sebastiani, N.V., California, Country

SILVER
Mirassou Vineyards, 1980, Monterey County
Round Hill Cellars, 1980, Napa Valley
Frey Vineyards, 1981, North Coast, Red Table
Winters Winery, 1980, Amador County, Bin "A"

BRONZE
Inglenook Vineyard, 1979, Napa Valley, Estate Bottled, Centennial Vintage
Sebastiani Vineyards, 1979, California
Stone Creek, 1980, North Coast
Cresta Blanca, 1979, Mendocino
Summerhill Vineyards, N.V., California

ZINFANDEL, *1983 Rating*
($5.01 to $7.50—104 entries)

GOLD
Haywood, 1981, Sonoma Valley, Chamizal Vineyard, Estate Bottled
David Bruce, 1980, Amador County
Rolling Hills Vineyards, 1981,

Amador County
Fitzpatrick, 1981, Shenandoah Valley, Jehling Vineyard

SILVER
Preston Vineyard & Winery, 1980, Sonoma County, Dry Creek Valley, Estate Bottled
Sierra Vista Winery, 1981, El Dorado County
Conn Creek, 1979, Napa Valley
Fritz Cellars, 1981, Sonoma County, Dry Creek Valley
Calera Wine Company, 1980, Cienega Valley, Reserve
Tulocay, 1980, Napa Valley
Callaway Vineyard & Winery, 1980, California, Estate Bottled
Whaler Vineyard, 1981, Mendocino County

BRONZE
Rutherford Hill, 1979, Napa Valley, Atlas Peak-Mead Ranch
Caché Cellars, 1980, Amador County, Baldinelli Vineyard
Winters Winery, 1980, Amador County, Bin "B"
Richard Carey, 1980, Amador County
Smothers, 1980, Sonoma County
Karly, 1981, Amador County
Shafer Vineyards, 1980, Napa Valley
Rosenblum Cellars, 1981, Sonoma Valley
Karly, 1980, Amador County, Vintage Reserve
McLester, 1980, San Luis Obispo County, Radike Vineyard
HMR, Ltd., 1979, San Luis Obispo

ZINFANDEL, *1983 Rating*
($7.51 and Up—44 entries)

GOLD
Chateau Montelena, 1980, Napa Valley, Estate Bottled
Topolos at Russian River Vineyards, 1980, Sonoma County, Sonoma Mountain, Dry, "Ultimo"
Mark West Vineyards, 1980, Sonoma County

SILVER
Napa Cellars, 1981, Alexander Valley
Monterey Peninsula, 1979, San Luis Obispo County, Willow Creek Wilpete Farms
Mastantuono, 1981, San Luis Obispo County, Templeton, Dante Dusi Vineyard

BRONZE
Edmeades Vineyard, 1980, Mendocino, DuPratt Vineyard
Sycamore Creek, 1981, California, Estate Bottled
Ravenswood, 1980, Sonoma County
Fetzer Vineyards, 1980, Mendocino, Scharffenberger

ZINFANDEL, *1984 Rating*
(Up to $5.00—54 entries)

GOLD
****Colony, N.V., California, "Classic"
Mendocino Estate, 1981, Mendocino, Estate Bottled
Stone Creek, 1980, Amador County, Shenandoah Valley
Paul Masson Vineyards, 1982, California

SILVER
Guenoc, 1981, Lake County
Round Hill Cellars, 1981, Napa Valley
Frey Vineyards, 1982, Mendocino
Vincelli Cellars, 1980, Amador County
Maddalena Vineyard, 1982, Sonoma County
Breckenridge Cellars, N.V., California
Delicato Vineyards, 1983, California

BRONZE
Sierra Vista, N.V., El Dorado County
Cranbrook Cellars, 1983, Napa Valley, "Premier"
Cresta Blanca, 1980, Mendocino, Estate Bottled

Cambiaso, 1981, Sonoma County, Chalk Hill
York Mountain, 1981, San Luis Obispo County, Farview Farm Vineyard
August Sebastiani Country, N.V., California
Parducci Wine Cellars, 1981, Mendocino County
Orleans Hill, 1982, Amador County, Clockspring Vineyard
Trader Joe's, 1980, Sonoma Valley, Private Reserve

ZINFANDEL, *1984 Rating*
($5.01 to $7.50—93 entries)
GOLD
Gemello, 1976, Sonoma County
Dry Creek Vineyard, 1980, Sonoma County
Story Vineyard, 1980, Amador County, Estate Bottled
Richardson, 1982, Napa Valley
SILVER
Karly, 1982, Amador County
Chanticleer Vineyards, 1979, Sonoma County
Santino, 1981, Amador County, Fiddletown, Eschen Vineyard, Special Selection
Roudon-Smith Vineyards, 1980, California, Chauvet Vineyard, 10th Anniversary Selection
BRONZE
Foppiano, 1980, Sonoma County
Rutherford Ranch, 1981, Napa Valley
Dehlinger, 1981, Sonoma County
Marietta Cellars, 1981, Northern Sonoma
McLester, 1981, San Luis Obispo County, Radike Vineyard
Calera, 1981, California, Templeton
Santino, 1981, Amador County, Shenandoah Valley
Smothers, 1981, North Coast
Stevenot, 1981, Calaveras County, Estate Bottled
Rutherford Hill, 1979, Napa Valley, Mead Ranch-Atlas Peak
V. Sattui, 1981, Napa Valley
Santa Barbara, 1981, Santa Ynez Valley
Rosenblum Cellars, 1982, Napa Valley, Morisoli Vineyard
Shown & Sons, 1981, Napa Valley, Shown Family Vineyards, Estate Bottled
Susiné Cellars, 1980, Amador County
Buehler Vineyards, 1981, Napa Valley, Estate Bottled
El Paso De Robles, 1982, Paso Robles, Benito Dusi Vineyard

ZINFANDEL, *1984 Rating*
($7.51 and Up—46 entries)
GOLD
Storybook Mountain, 1981, Napa Valley, Estate Reserve
Grgich Hills Cellars, 1981, Sonoma County
SILVER
Hidden Cellars, 1981, Mendocino County
Cuvaison Vineyard, 1981, Napa Valley
Topolos at Russian River, 1981, Sonoma County, Old Hill Ranch
Tobias Vineyards, 1981, Paso Robles, Dusi Ranch
BRONZE
Montevina, 1980, Amador County, Winemakers Choice, Estate Bottled
Edmeades Vineyards, 1981, Mendocino, Ciapusci Vineyards
Amador Foothill, 1980, Fiddletown, Eschen Vineyard, Special Selection
Mastantuono, 1981, San Luis Obispo County, Dante Dusi Vineyard, Templeton
Fetzer Vineyards, 1979, Mendocino, Scharffenberger
Ridge, 1981, Paso Robles
Grand Cru Vineyards, 1981, Sonoma Valley
Windsor Vineyards, 1980, Sonoma County, Alexander Valley
Congress Springs, 1981, Santa Cruz

Mountains
V. Sattui, 1980, Napa Valley, Reserve Stock

ZINFANDEL, LATE HARVEST, DRY, *1980 Rating*
($5.01 to $7.50—1 entry)
SILVER
Grand Cru, 1977, Sonoma, Late Picked

ZINFANDEL, LATE HARVEST, DRY, *1980 Rating*
($7.51 and Up—2 entries)
GOLD
Mount Veeder, 1978, Late Harvest, Napa County
BRONZE
Ridge, 1977, Late Harvest, Geyserville Trentadue Ranch

ZINFANDEL, LATE HARVEST, DRY, *1981 Rating*
($7.01 to $10.00—3 entries)
SILVER
Davis Bynum Winery, 1979 Lot 906-A, Sonoma, Late Harvest

ZINFANDEL, LATE HARVEST, DRY, *1981 Rating*
($10.01 and Up—2 entries)
BRONZE
Ridge Vineyards, 1978, Paso Robles, Dusi Ranch, Late Harvest
Grand Cru Vineyards, 1978, Sonoma Valley, Estate Bottled, Late Harvest

ZINFANDEL, LATE HARVEST, DRY, *1982 Rating*
($8.51 and Up—3 entries)
BRONZE
Monterey Peninsula Winery, 1977, Amador, Late Harvest
Monterey Peninsula Winery, 1978, Amador, Ferrerro Old Ranch, Late Harvest

ZINFANDEL, LATE HARVEST, (0.0% to 0.70% residual sugar) *1983 Rating*
($8.51 and Up—2 entries)
SILVER
Gundlach-Bundschu, 1980, Sonoma Valley, Barricia Vineyards, Late Harvest

ZINFANDEL, LATE HARVEST, (0.0% to 0.70% residual sugar) *1984 Rating*
($6.01 to $9.00—2 entries)
SILVER
Gundlach-Bundschu, 1980, Sonoma Valley, Barricia Vineyards, Late Harvest
BRONZE
Gemello, 1978, Amador County, Late Harvest Style

ZINFANDEL, LATE HARVEST, (0.71% to 3.0% residual sugar) *1983 Rating*
($6.01 to $8.50—3 entries)
SILVER
Santa Barbara, 1981, Santa Ynez Valley, Late Harvest
BRONZE
Montevina Wines, 1978, Amador County, Estate Bottled, Late Harvest
Milano, 1980, Mendocino County, Talmage, Pacini Vineyard, Late Harvest

ZINFANDEL, LATE HARVEST, (0.71% to 3.0% residual sugar) *1983 Rating*
($8.51 and Up—2 entries)
GOLD
Johnson's Alexander Valley, 1980, Alexander Valley, Sauers Vineyard, Late Harvest
BRONZE
Hop Kiln Winery, 1980, Russian River Valley, "Primitivo", Late Harvest

ZINFANDEL, LATE HARVEST, (0.71% to 3.0% residual sugar) *1984 Rating*
(Up to $6.00—1 entry)
BRONZE
Papagni Vineyards, 1978, Madera, Vallis Vineyard, Estate Bottled

ZINFANDEL, LATE HARVEST, (0.71% to 3.0% residual sugar) *1984 Rating*
($6.01 to $9.00—2 entries)
SILVER
Preston Vineyards, 1980, Sonoma County, Dry Creek Valley, Estate Bottled, Late Harvest
BRONZE
Vose Vineyards, 1982, Napa Valley, Estate Bottled, "Special Sweet Harvest"

ZINFANDEL, LATE HARVEST, SWEET, *1980 Rating*
($10.01 and Up—1 entry)
SILVER
Johnson's Alexander Valley, 1978 Lot 1, Alexander Valley, 5% Estate Bottled
(Fetzer Vineyards, Zinfandel, Lolonis, 1977 was awarded a gold medal but was judged in the wrong price category.)

ZINFANDEL, LATE HARVEST, SWEET, *1981 Rating*
($7.01 to $10.00—5 entries)
GOLD
Rutherford Ranch Brand, 1978, Napa Valley, Late Harvest
SILVER
Dry Creek Vineyards, 1979, Sonoma County, Late Harvest
BRONZE
Fetzer Vineyards, 1978, Mendocino, Ricetti, Late Harvest

ZINFANDEL, LATE HARVEST, SWEET, *1981 Rating*
($10.01 and Up—5 entries)
SILVER
Sutter Home Winery, 1977, Amador County, Deaver Vineyard, Late Harvest

ZINFANDEL, LATE HARVEST, SWEET, *1982 Rating*
($6.01 to $8.50—1 entry)
SILVER
Charles Le Franc, 1980, San Benito, Royale, (Sweet-3.1% Residual Sugar)

ZINFANDEL, LATE HARVEST, SWEET, *1982 Rating*
($8.51 and Up—3 entries)
BRONZE
Monterey Peninsula Winery, 1980, Amador, Sweet, Late Harvest

ZINFANDEL, LATE HARVEST, (3.0% to 6.0% residual sugar) *1983 Rating*
($6.01 to $8.50—1 entry)
SILVER
Trentadue, 1980, Alexander Valley, Estate Bottled, Late Harvest

ZINFANDEL, LATE HARVEST, (3.0% to 6.0% residual sugar) *1983 Rating*
($8.51 and Up—1 entry)
SILVER
Milano, 1980, Mendocino County, Talmage, Scharffenberger Vineyard, Late Harvest

ZINFANDEL ROSÉ, *1983 Rating*
(Up to $4.00—3 entries)
GOLD
J. Pedroncelli, 1981, Sonoma County
BRONZE
Casa de Fruta, 1982, Pacheco Pass Vineyards

ZINFANDEL ROSÉ, *1983 Rating*
($4.01 to $5.50—4 entries)
GOLD
The Konocti Winery, 1982, Lake County, "Alegre"

ZINFANDEL ROSÉ, *1984 Rating*
(Up to $4.00—8 entries)
GOLD
J. Pedroncelli, 1983, Sonoma County
SILVER
San Antonio Winery, 1983, California
Charles Krug, 1983, Napa Valley
BRONZE

Twin Hills Ranch, 1983, Paso Robles, Estate Bottled
Bandiera, 1983, Napa Valley

ZINFANDEL ROSÉ, *1984 Rating*
($4.01 to $5.50—3 entries)
BRONZE
Boeger, 1982, El Dorado County, 10th Anniversary

ZINFANDEL ROSÉ, *1984 Rating*
($5.51 and Up—1 entry)
BRONZE
Rapazzini, 1982, San Benito County

WHITE ZINFANDEL, (0.0% to 0.70% residual sugar) *1984 Rating*
(Up to $4.50—6 entries)
GOLD
Rosenblum Cellars, 1983, Napa Valley, Howell Mountain, Lamborn Vineyard
SILVER
Green and Red, 1983, Napa Valley, Estate Bottled
BRONZE
Pedrizzetti, 1983, California

WHITE ZINFANDEL, (0.0% to 0.70% residual sugar) *1984 Rating*
($4.51 to $6.00—6 entries)
SILVER
California Hillside Cellars, 1983, Russian River Valley
Davis Bynum, 1983, Sonoma County

WHITE ZINFANDEL, (0.71% to 3.0% residual sugar) *1983 Rating*
(Up to $4.50—12 entries)
GOLD
Bargetto Winery, 1982, Central Coast
SILVER
Estrella River Winery, N.V., San Luis Obispo County
BRONZE
Trader Joe's, 1982, Mendocino County
Weibel Vineyards, 1982, Mendocino County
Greenstone Winery, 1982, Amador County, Potter-Cowan Vineyard

WHITE ZINFANDEL, (0.71% to 3.0% residual sugar) *1983 Rating*
($4.51 to $6.00—13 entries)
SILVER
Bel Arbres, 1982, Mendocino
BRONZE
Sierra Vista Winery, 1982, El Dorado County
Stevenot, 1982, Amador County
Whaler Vineyard, 1982, Monterey County
Santino Wines, 1982, Amador County, Shenandoah Valley, White Harvest

WHITE ZINFANDEL, (0.71% to 3.0% residual sugar) *1984 Rating*
(Up to $4.50—23 entries)
GOLD
Trader Joe's, N.V., San Luis Obispo County, Black Label, "White Angel"
Cresta Blanca, 1983, Mendocino
SILVER
Bel Arbres, 1983, Mendocino
Sonoma Mission, 1983, California, "Caprice"
Franzia, N.V., California
Olson Vineyards, 1983, Mendocino County
Delicato Vineyards, 1983, California
BRONZE
Oak Ridge Vineyards, 1983, Lodi, "White Tail"
Bandiera, 1983, North Coast
Lost Hills Vineyards, N.V., California
St. Amant, 1983, Amador County

WHITE ZINFANDEL, (0.71% to 3.0% residual sugar) *1984 Rating*
($4.51 to $6.00—23 entries)
GOLD
Santino, 1983, Amador County, "White Harvest"

Santa Barbara, 1983, Santa Barbara County
SILVER
Boeger, 1983, El Dorado
Belli and Sauret, 1983, Paso Robles
Sutter Home, 1983, California
BRONZE
Amador Foothill, 1983, Amador County
Mission View, 1983, Paso Robles, Bacchanal Blanc
Buehler Vineyards, 1983, Napa Valley, Estate Bottled
Shenandoah Vineyards, 1983, Amador County

WHITE ZINFANDEL, (3.0% to 6.0% residual sugar) *1984 Rating*
($4.51 to $6.00—1 entry)
SILVER
Konocti, 1983, Lake County

SPARKLING WINE (CHAMPAGNE), *1981 Rating*
Note: Only Sparkling Wine "made in this bottle" (Methode Champenoise) was judged in 1981.

SPARKLING WINE (CHAMPAGNE), (0.0% to 0.50% residual sugar) *1981 Rating*
($9.51 to $13.50—6 entries)
GOLD
Korbel, N.V., California, "Blanc de Noir"
Mirassou Vineyards, 1978, Monterey County, "Au Naturel"
Korbel, N.V., California, Natural
SILVER
Beaulieu Vineyards, 1974, Napa Valley, "Champagne de Chardonnay"

SPARKLING WINE (CHAMPAGNE), (0.0% to 0.50% residual sugar) *1981 Rating*
($13.51 and Up—2 entries)
SILVER
Mirassou Vineyards, 1974, Monterey County, Late Disgorged

SPARKLING WINE, (0.0% to 0.50% residual sugar) *1982 Rating*
(Up to $5.00—3 entries)
BRONZE
Stanford, N.V., California, Governor's Cuvée, Dry

SPARKLING WINE, (0.0% to 0.50% residual sugar) *1982 Rating*
($5.01 to $9.50—11 entries)
SILVER
Weibel Vineyards, N.V., Bottled Fermented, Brut
Korbel, N.V., California, Brut
Stony Ridge Winery, N.V., California, Blanc de Noir
Trader Joe's Winery, N.V., Sonoma, Bottled Fermented, Brut
BRONZE
Weibel Vineyards, N.V., Bottled Fermented, Blanc de Pinot Noir

SPARKLING WINE, (0.0% to 0.50% residual sugar) *1982 Rating*
($9.51 and Up—13 entries)
GOLD
Schramsberg Vineyards, 1977, Napa Valley, Blanc de Noir
Mirassou Vineyards, 1978, Monterey County, Blanc de Noir
SILVER
Schramsberg Vineyards, 1979, Napa Valley, Cuvée de Pinot
Korbel, N.V., California, Natural, Extremely Dry
BRONZE
Mirassou Vineyards, 1979, Monterey County, Au Natural
Cresta Blanca Vineyards, N.V., California, Chardonnay

SPARKLING WINE, (0.0% to 0.50% residual sugar) *1983 Rating*
($6.01 to $10.00—5 entries)
GOLD
Geyser Peak, N.V., California, Brut Champagne
SILVER
Paul Masson Vineyards, 1981, California. Brut

Weibel Vineyards, N.V., California, Brut
BRONZE
Weibel Vineyards, N.V., California, Chardonnay, Brut

SPARKLING WINE, (0.0% to 0.50% residual sugar) *1983 Rating*
($10.01 and Up—8 entries)
GOLD
John Culbertson Winery, 1981, California, Natural
Piper Sonoma Cellars, 1980, Sonoma County, Blanc de Noir
BRONZE
Geyser Peak, N.V., California, Blanc de Noirs, Au Natural
F. Korbel & Bros., N.V., Blanc de Noirs, California

SPARKLING WINE, (0.0% to 0.50% residual sugar) *1984 Rating*
($6.01 to $10.00—6 entries)
SILVER
Paul Masson Vineyards, 1982, California, Brut
BRONZE
Weibel, N.V., California, Chardonnay, Brut
Geyser Peak, 1981, Sonoma County, Brut

SPARKLING WINE, (0.0% to 0.50% residual sugar) *1984 Rating*
($10.01 and Up—13 entries)
GOLD
Korbel and Bros., N.V., California, Blanc de Blancs
SILVER
Mirassou, 1980, Monterey County, Au Natural
John Culbertson, 1981, California, Natural
San Pasqual Vineyards, 1981, California, Blanc de Noir
BRONZE
Schramsberg, 1981, Napa Valley, Cuvée de Pinot
Korbel and Bros., N.V., California, Natural

Robert Hunter, 1980, Sonoma Valley, Brut de Noirs

SPARKLING WINE (CHAMPAGNE), (0.51% to 1.40% residual sugar) *1981 Rating*
(Up to $9.50—3 entries)
GOLD
Korbel, N.V., California, "Brut"
BRONZE
Shadow Creek, N.V., Sonoma County, "Brut", Cuvée N6

SPARKLING WINE (CHAMPAGNE), (0.51% to 1.40% residual sugar) *1981 Rating*
($9.51 to $13.50—7 entries)
GOLD
Domaine Chandon, N.V., Napa Valley, "Blanc de Noir"
Mirassou Vineyards, 1978, Monterey County, "Brut"
Domaine Chandon, N.V., Napa Valley, "Brut"
SILVER
Schramsberg Vineyards, 1978, Napa Valley, "Cuvée de Pinot"
BRONZE
Shadow Creek, 1978, Sonoma County, "Blanc de Blanc," Cuvée N1

SPARKLING WINE (CHAMPAGNE), (0.51% to 1.40% residual sugar) *1981 Rating*
($13.51 and Up—1 entry)
GOLD
Schramsberg Vineyards, 1975, Napa Valley, Reserve

SPARKLING WINE, (0.51% to 1.40% residual sugar) *1982 Rating*
(Up to $5.00—1 entry)
SILVER
Crystal Valley Cellars, N.V., California, Robins Glow, Blanc de Noir, Brut

SPARKLING WINE, (0.51% to 1.40% residual sugar) *1982 Rating*

($5.01 to $9.50—5 entries)
BRONZE
Papagni Vineyards, 1979, California, Estate Bottled, Brut

SPARKLING WINE, (0.51% to 1.40% residual sugar) *1982 Rating*
($9.51 and Up—3 entries)
BRONZE
Domaine Chandon, N.V., Napa Valley, Brut

SPARKLING WINE, (0.51% to 1.40% residual sugar) *1983 Rating*
(Up to $6.00—3 entries)
BRONZE
Crystal Valley Cellars, N.V., California, "Robin's Glow", Blanc de Noirs, Brut

SPARKLING WINE, (0.51% to 1.40% residual sugar) *1983 Rating*
($6.01 to $10.00—8 entries)
GOLD
Shadow Creek Champagne Cellars, N.V., Sonoma County, Brut Champagne
SILVER
F. Korbel & Bros., N.V, Brut, California Champagne
Windsor Vineyards, 1981, Sonoma County, Estate Bottled, Blanc de Noir

SPARKLING WINE, (0.51% to 1.40% residual sugar) *1983 Rating*
($10.01 and Up—17 entries)
GOLD
Wente Brothers, 1980, Monterey, Brut, 100th Anniversary
Shadow Creek Champagne Cellars, 1980, Sonoma County, Brut Champagne
Schramsberg, 1978, Napa Valley, Blanc de Noirs
Shadow Creek Champagne Cellars, 1980, Sonoma County, Blanc de Noir
Domaine Chandon, N.V., Napa Valley, Brut
SILVER
Sebastiani Vineyards, N.V., Sonoma County, Brut Champagne, Proprietor's Reserve
BRONZE
Shadow Creek Champagne Cellars, 1980, Sonoma County, Blanc de Blanc
Mirassou Vineyards, 1979, Monterey County, Brut
Schramsberg, 1980, Napa Valley, Blanc de Blancs
Domaine Chandon, N.V., Napa Valley, Blanc de Noir
Mirassou Vineyards, 1979, Monterey County, Blanc de Noir
John Culbertson Winery, 1981, California, Brut

SPARKLING WINE, (0.51% to 1.40% residual sugar) *1984 Rating*
(Up to $6.00—3 entries)
GOLD
Crystal Valley Cellars, N.V, California, "Robins Glow", Blanc de Noirs
SILVER
Taylor California Cellars, N.V, California, Brut
BRONZE
Guasti, N.V., California, Extra Dry

SPARKLING WINE, (0.51% to 1.40% residual sugar) *1984 Rating*
($6.01 to $10.00—6 entries)
SILVER
Korbel and Bros., N.V., California, Brut
BRONZE
Weibel, N.V., California, Blanc de Noir
Christian Brothers, N.V., California, Brut

SPARKLING WINE, (0.51% to 1.40% residual sugar) *1984 Rating*
($10.01 and Up—23 entries)
GOLD
Iron Horse Vineyard, 1981, Sonoma County, Green Valley, "Wedding

Cuvée", Blanc de Noir
Piper Sonoma Cellars, 1981, Sonoma County, Blanc de Noirs
Piper Sonoma Cellars, 1981, Sonoma County, Brut

SILVER
Shadow Creek Champagne Cellars, 1981, Sonoma County, Brut, Cuvée 1
Bargetto, 1981, Monterey County, Blanc de Noir

BRONZE
Mirassou, 1979, Monterey County, Blanc de Noirs
Windsor Vineyards, 1982, Sonoma County, Estate Bottled, Brut
Schramsberg, 1981, Napa Valley, Blanc de Blanc
Domaine Chandon, N.V., Napa Valley, Brut
Chateau St. Jean, 1980, Sonoma County, Blanc de Blanc
Shadow Creek Champagne Cellars, N.V., Sonoma County, Brut, Cuvée 1

SPARKLING WINE (CHAMPAGNE), (1.41% to 2.40% residual sugar) *1981 Rating*
(Up to $9.50—4 entries)

SILVER
Korbel, N.V., California, "Extra Dry"
Korbel, N.V., California, "Sec"

BRONZE
Korbel, N.V., California, "Rosé"

SPARKLING WINE, (1.41% to 2.40% residual sugar) *1982 Rating*
($5.01 to $9.50—4 entries)

GOLD
Weibel Vineyards, N.V., Bottle Fermented, Extra Dry

BRONZE
Christian Brothers Winery, N.V., California, Brut

SPARKLING WINE, (1.41% to 2.40% residual sugar) *1983 Rating*
(Up to $6.00—8 entries)

BRONZE
Chateau Napoleon, N.V., California, Brut

SPARKLING WINE, (1.41% to 2.40% residual sugar) *1983 Rating*
($6.01 to $10.00—7 entries)

SILVER
Weibel Vineyards, N.V., California, Extra Dry

BRONZE
F. Korbel & Bros., N.V., Sec, California, Champagne
Hanns Kornell Champagne Cellars, N.V., Demi Sec Champagne

SPARKLING WINE, (1.41% to 2.40% residual sugar) *1983 Rating*
($10.01 and Up—2 entries)

SILVER
F. Korbel & Bros., N.V., Rosé, California, Champagne

BRONZE
Hanns Kornell Champagne Cellars, N.V., Extra Dry Champagne

SPARKLING WINE, (1.41% to 2.40% residual sugar) *1984 Rating*
(Up to $6.00—7 entries)

GOLD
Stanford, N.V., California, Governor's Cuvée

SILVER
Crystal Valley Cellars, N.V., California, Extra Dry

BRONZE
Weibel, N.V., California, White Zinfandel

SPARKLING WINE, (1.41% to 2.40% residual sugar) *1984 Rating*
($6.01 to $10.00—9 entries)

SILVER
Korbel and Bros., N.V., California, Sec

BRONZE
Cresta Blanca, N.V., California, Brut

Korbel and Bros., N.V., California, Extra Dry
Paul Masson Vineyards, 1982, California, Extra Dry
Weibel, N.V., California, Brut
Stony Ridge, N.V., California, Cuvée 1, Blanc de Noirs

SPARKLING WINE (CHAMPAGNE), (2.41% and up residual sugar) *1981 Rating*
(Up to $9.50—4 entries)
SILVER
Hanns Kornell, N.V., California, "Rosé"
BRONZE
Hanns Kornell, N.V., California, "Demi-Sec"

SPARKLING WINE (CHAMPAGNE), (2.41% and up residual sugar) *1981 Rating*
($9.51 to $13.50—2 entries)
SILVER
Schramsberg Vineyards, 1978, Napa Valley, "Crement"

SPARKLING WINE, (2.41% and up residual sugar) *1982 Rating*
(Up to $5.00—15 entries)
SILVER
Crystal Valley Cellars, N.V., California, Spumante
BRONZE
Franzia Brothers Winery, N.V., California, Extra Dry

SPARKLING WINE, (2.41% and up residual sugar) *1982 Rating*
($5.01 to $9.50—10 entries)
SILVER
Papagni Vineyards, N.V., California, Estate Bottled, Spumante d'Angelo

SPARKLING WINE, (2.41% and up residual sugar) *1982 Rating*
($9.51 and Up—1 entry)
BRONZE
Schramsberg, 1979, Napa Valley, Cremant Demi Sec

SPARKLING WINE, (2.41% and up residual sugar) *1983 Rating*
(Up to $6.00—23 entries)
SILVER
Weibel Vineyards, N.V., California, Spumante
BRONZE
Crystal Valley Cellars, N.V., California, Spumante d'Francesca
Cribari & Sons, N.V., California, Extra Dry Champagne
Le Domaine, N.V., California, Extra Dry Champagne

SPARKLING WINE, (2.41% and up residual sugar) *1983 Rating*
($6.01 to $10.00—9 entries)
GOLD
Stony Ridge, N.V., California, Malvasia Bianca
SILVER
Cresta Blanca, N.V., California, Extra Dry Champagne
BRONZE
Angelo Papagni, 1980, California, Extra Dry Champagne, Estate Bottled
The Christian Brothers, N.V., California, Champagne Rosé
Almaden Vineyards, N.V., California, Extra Dry Champagne

SPARKLING WINE, (2.41% and up residual sugar) *1983 Rating*
($10.01 and Up—2 entries)
BRONZE
Schramsberg, 1980, Cremant, Demi Sec, Napa Valley

SPARKLING WINE, (2.41% and up residual sugar) *1984 Rating*
(Up to $6.00—23 entries)
GOLD
****Mogen David, N.V., California, Blackberry
Franzia, N.V., California, Almond
Lejon, N.V., California, Extra Dry
Weibel, N.V., California,

"Crackling Rosé"
SILVER
Weibel, N.V., California, Green Hungarian
Taylor California Cellars, N.V., California, Pink
Crystal Valley Cellars, N.V., California, Spumante D'Francesca
BRONZE
Weibel, N.V., California, Chenin Blanc
Jacques Bonet, N.V., California, Extra Dry
Franzia, N.V., California, Spumante

SPARKLING WINE, (2.41% and up residual sugar) *1984 Rating*
($6.01 to $10.00—6 entries)
SILVER
Stony Ridge, N.V., California, Malvasia Bianca
BRONZE
Christian Brothers, N.V., California, Extra Dry

GENERIC & PROPRIETARY RED, *1984 Rating*
(Up to $3.50—42 entries)
GOLD
Colony, N.V., California, "Classic Burgundy"
Sonoma Mission, N.V., California, Red Table Wine
Paul Masson Vineyards, N.V., California, "Burgundy"
SILVER
Windsor Vineyards, N.V., California, "Premium Burgundy"
Santa Lucia Cellars, 1981, Paso Robles, "Burgundy"
Round Hill Cellars, N.V., California, "Burgundy"
Delicato Vineyards, 1982, California, "Burgundy"
BRONZE
Mount Palomar, N.V., Temecula, Estate Bottled, "Burgundy"
Geyser Peak, 1980, Sonoma County, Sonoma Vintage Red
River Oaks Vineyards, N.V., Sonoma County, Premium Red Wine
Giumarra Vineyards, 1982, California, "Burgundy"
Sebastiani Vineyards, 1980, California, "Burgundy"
Shenandoah Vineyards, N.V., Amador County, Red Table Wine
Gundlach-Bundschu, N.V., Sonoma Valley, Sonoma Red Wine
Oak Ridge Vineyards, N.V., California, Barrel Red 50
La Mont, N.V., California, "Burgundy"
Christian Brothers, N.V., California, "Burgundy"
Franciscan Vineyards, N.V., North Coast, Cask 321, "Burgundy"
Seghesio, 1975, Northern Sonoma, Estate Bottled, "Marian's Reserve"

GENERIC & PROPRIETARY RED, *1984 Rating*
($3.51 to $5.50—38 entries)
GOLD
St. Carl, 1982, Santa Ynez Valley, Red Table Wine
SILVER
Guenoc, 1981, Lake County, Red Table Wine
BRONZE
Monterey Peninsula, N.V., California, Black Burgundy
Trefethen Vineyards, N.V., Napa Valley, Blend 283, Eshcol Red Wine

GENERIC & PROPRIETARY RED, *1984 Rating*
($5.51 and Up—10 entries)
GOLD
Hop Kiln, 1981, Russian River Valley, Marty Griffin's Big Red
SILVER
Joseph Phelps Vineyards, 1979, Napa Valley, Insignia
Opus One, 1980, Napa Valley
BRONZE
Whitehall Lane, 1982, Napa Valley, Fleur d'Helene

GENERIC & PROPRIETARY WHITE, (0.0% to 0.70% residual sugar) *1984 Rating*

(Up to $3.50—22 entries)

GOLD

Round Hill Cellars, N.V., California, "Chablis"

River Oaks Vineyards, 1982, Alexander Valley, Premium White Wine

SILVER

Riverside Farm, N.V., California, Premium Dry White

Ernest and Julio Gallo, N.V., California Reserve, "Chablis"

C K Mondavi, N.V., California, "Chablis"

BRONZE

Geyser Peak, 1982, Sonoma County, Sonoma Vintage White

Charles Krug, 1982, Napa Valley, "Chablis"

Sebastiani Vineyards, 1983, Sonoma County, "Chablis"

J. Pedroncelli, N.V., Sonoma County, Sonoma White Wine

Fetzer Vineyards, N.V., California, Premium White

GENERIC & PROPRIETARY WHITE, (0.0% to 0.70% residual sugar) *1984 Rating*

($3.51 to $5.50—31 entries)

GOLD

Chateau St. Jean, 1983, Sonoma County, Vin Blanc

SILVER

Shenandoah Vineyards, N.V., California, White Table Wine

Zaca Mesa, 1983, Santa Barbara County, "Toyon Blanc"

Kenwood Vineyards, 1983, California, Vintage White Wine

Navarro Vineyards, 1982, Mendocino, Vin Blanc

BRONZE

The Monterey Vineyard, 1983, Central Coast Counties, Dry White

Landmark Vineyard, 1983, California, Petite Blanc

Trefethen Vineyards, N.V., Napa Valley, Estate Bottled, Blend 184, Eshcol White Wine

Tepusquet Vineyards, 1982, Santa Maria Valley, Vin Blanc

Congress Springs, 1983, Santa Barbara County, Mont Blanc

Hop Kiln, 1983, Sonoma County, "A Thousand Flowers"

Fetzer Vineyards, 1983, Mendocino, Blanc de Blancs

Raymond Vineyard and Cellar, 1982, California, White Table Wine

GENERIC & PROPRIETARY WHITE, (0.71% to 3.0% residual sugar) *1984 Rating*

(Up to $3.50—21 entries)

GOLD

Paul Masson Vineyards, 1983, California, Emerald Dry

Colony, N.V., California, "Classic Chablis"

Gundlach-Bundschu, 1983, Sonoma Valley, Sonoma White Wine

SILVER

Christian Brothers, N.V., "California Sauterne"

Franzia, N.V., California, "Chablis Blanc"

Giumarra Vineyards, 1983, California, "Chablis"

BRONZE

La Mont, N.V., California, "Chablis"

Braren Pauli, N.V., Mendocino County, Vin Blanc

GENERIC & PROPRIETARY WHITE, (0.71% to 3.0% residual sugar) *1984 Rating*

($3.51 to $5.50—17 entries)

GOLD

Callaway Vineyards, 1983, California, Vin Blanc

SILVER

Buena Vista, 1982, Sonoma Valley, "Spiceling"

BRONZE

Parducci Wine Cellars, 1983, Mendocino County, Mendocino Riesling
Weibel, 1983, Mendocino County, Mendocino Riesling
Taft Street, N.V., California, White House Wine
Templeton, N.V., California, White Wine

GENERIC & PROPRIETARY WHITE, (3.0% to 6.0% residual sugar) *1984 Rating*
(Up to $3.50—3 entries)
GOLD
Colony, N.V., California, "Classic Rhine"
SILVER
Barengo, N.V., California, Vintners Reserve, "Haut Sauterne"
BRONZE
Paul Masson Vineyards, N.V., California, Rhine

GENERIC & PROPRIETARY WHITE, (6.0% to 12.0% residual sugar) *1984 Rating*
($5.51 and Up—1 entry)
SILVER
Chateau Marjon, 1982, Mendocino County, Talmadge Ranch, "Sauterne"

GENERIC & PROPRIETARY WHITE, (12.0% and up residual sugar) *1984 Rating*
($5.51 and Up-1 entry)
GOLD
Chateau St. Jean, 1982, Sonoma County, Select Late Harvest, Sauvignon D'Or

GENERIC & PROPRIETARY ROSÉ, (0.0% to 0.70% residual sugar) *1984 Rating*
(Up to $3.00—2 entries)
BRONZE
J. Pedroncelli, N.V., Sonoma County, Sonoma Rosé Wine

GENERIC & PROPRIETARY ROSÉ, (0.0% to 0.70% residual sugar) *1984 Rating*
($3.01 to $5.00—2 entries)
BRONZE
Robert Mondavi, 1983, California, Rosé

GENERIC & PROPRIETARY ROSÉ, (0.71% to 3.0% residual sugar) *1984 Rating*
(Up to $3.00—9 entries)
BRONZE
Weibel, N.V., California, Vin Rosé
Lost Hills Vineyards, N.V., California, Vin Rosé

GENERIC & PROPRIETARY ROSÉ, (0.71% to 3.0% residual sugar) *1984 Rating*
($3.01 to $5.00—9 entries)
SILVER
The Monterey Vineyard, 1983, Central Coast Counties, Rosé
BRONZE
August Sebastiani, N.V., California, Mountain Vin Rosé
Paul Masson Vineyards, N.V., California, Rosé

OTHER FORTIFIED WINES, (3.0% to 6.0% residual sugar) *1984 Rating*
(All Price Ranges—3 entries)
BRONZE
Richardson, 1982, Napa Valley, Zinfandel

OTHER FORTIFIED WINES, (6.0% to 12.0% residual sugar) *1984 Rating*
(Up to $8.00—3 entries)
BRONZE
Weibel, N.V., California, Black Muscat

OTHER FORTIFIED WINES, (6.0% to 12.0% residual sugar) *1984 Rating*
($8.01 and Up—2 entries)
SILVER
Woodbury Winery, 1980, Alexander Valley, Dessert Wine, Zinfandel
BRONZE

Woodbury Winery, 1980, Alexander Valley, Dessert Wine, Cabernet Sauvignon

OTHER FORTIFIED WINES, (12.0% and up residual sugar) *1984 Rating*
(Up to $5.00—1 entry)

SILVER

Breckenridge Cellars, N.V., California, Muscatel

OTHER FORTIFIED WINES, (12.0% and up residual sugar) *1984 Rating*
($8.01 and Up—2 entries)

GOLD

Quady, 1983, California, "Elysium", Black Muscat

SILVER

Quady, 1983, California, "Essensia", Orange Muscat

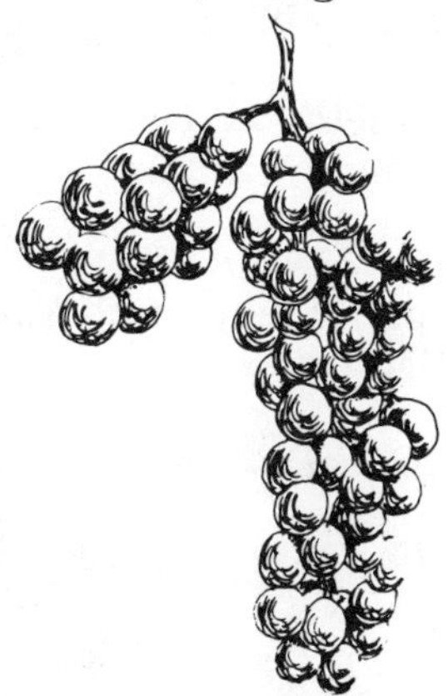

Zinfandel

PACIFIC NORTHWEST WINE COMPETITION 1977-1983

The Enological Society of the Pacific Northwest began evaluating the wines of Washington, Oregon and Idaho in 1977. This annual judging has grown to over 500 entries and 35 medals awarded in 1983. The following compilation of results of competitions for the seven years includes the judging of over 1,500 different wines and awarding of over 200 medals. All wines are tasted "blind" by a panel of experts representing all facets of the wine industry. The 1983 Panel of Judges were:

Merry Edwards, *Winemaker at Matanzas Creek, California*
Barbara Ensrud, *New York based wine journalist and author*
Frank Prial, *Internationally known wine journalist*
Mark Savage, *Master of Wine, Importer of Northwest wines to the U.K.*
Brooks Tish, *Northwest based wine journalist*
Dinsmoor Webb, *Chairman Emeritus of Viticulture & Enology, University of California at Davis*

NOTE: Where no medal is indicated, none was awarded in that category or variety.

CABERNET SAUVIGNON, *1977 Rating*

BRONZE

Associated Vintners, 1973

CABERNET SAUVIGNON, *1978 Rating*

GOLD

Chateau Ste. Michelle, 1975

SILVER

Associated Vintners, 1974

CABERNET SAUVIGNON, *1979 Rating*

SILVER

The Eyrie Vineyards, 1976

CABERNET SAUVIGNON, *1980 Rating*

SILVER

Hinzerling Vineyard, 1977

BRONZE

Cote des Colombe Vineyard, 1978
Hillcrest Vineyard, 1974

CABERNET SAUVIGNON, *1981 Rating*

SILVER

Chateau Ste. Michelle, 1977

BRONZE
Elk Cove Vineyards, 1978

CABERNET SAUVIGNON, *1982 Rating*
BRONZE
Chateau Ste. Michelle, 1976, Chateau Reserve
Neuharth Winery, Inc., 1979
Ste. Chapelle Vineyards, 1979, Mercer Ranch Reserve
Valley View Vineyard, 1978

CABERNET SAUVIGNON, *1983 Rating*
GRAND PRIZE
Quilceda Creek Vintners, 1979, Washington, Cabernet Sauvignon
SILVER
Chateau Ste. Michelle, 1978, Cold Creek Vineyards
Neuharth Winery, Inc., 1980
BRONZE
Hinzerling Vineyards, 1980, Sagemoor Vineyards
Preston Wine Cellars, 1978, Select Reserve

CHARDONNAY, *1977 Rating*
SILVER
The Eyrie Vineyards, 1975
BRONZE
Associated Vintners, 1975

CHARDONNAY, *1978 Rating*
SILVER
Knudsen Erath Winery, 1977
BRONZE
Ponzi Vineyards, 1976

CHARDONNAY, *1979 Rating*
GRAND PRIZE
Preston Wine Cellars, 1977
SILVER
Associated Vintners, 1977
BRONZE
Elk Cove Vineyards, N.V., Willamette Valley Appellation
Knudsen Erath Winery, 1977, Appellation Yamhill County

CHARDONNAY, *1980 Rating*
GRAND PRIZE
Preston Wine Cellars, 1977
GOLD
Ste. Chapelle Vineyards, 1978, Washington
Ste. Chapelle Vineyards, 1978, Idaho
BRONZE
Chateau Ste. Michelle, 1977
Columbia Cellars, 1979
E.B. Foote Winery, 1978, Yakima Valley
Knudsen Erath Winery, 1978, Yamhill County
Sokol Blosser Winery, 1978, Yamhill County
Tualatin Vineyards, 1978, Estate Bottled

CHARDONNAY, *1981 Rating*
GOLD
Ste. Chapelle Vineyards, 1979, Symms Family Vineyard
SILVER
Associated Vintners, 1979
E.B. Foote Winery, 1979, Yakima Valley, Unfiltered

CHARDONNAY, *1982 Rating*
SILVER
Valley View Vineyard, 1980
BRONZE
The Eyrie Vineyards, 1980, Willamette Valley
Ste. Chapelle Vineyards, 1980, Symms Family Vineyards, Reserve

CHARDONNAY, *1983 Rating*
GOLD
Chateau Benoit, 1982, Select Cluster
Ste. Chapelle Vineyards, 1981, Symms Family Vineyard
SILVER
Haviland Vintners, 1982, Columbia Valley, Dionysus Vineyard
BRONZE
Arbor Crest, 1982, Sagemoor Vineyard

The Associated Vintners, 1981
Knudsen Erath Winery, 1981, Willamette Valley
Tualatin Vineyards, Inc., 1980, Willamette Valley, Estate Bottled
Woodward Canyon Winery, 1981

CHENIN BLANC, *1977 Rating*
SILVER
Chateau Ste. Michelle, 1976

CHENIN BLANC, *1979 Rating*
SILVER
Preston Wine Cellars, 1978, Special Selection

CHENIN BLANC, *1982 Rating*
SILVER
Kiona Vineyards, 1981, Estate Bottled
Yakima River Winery, 1981
BRONZE
Tucker Cellars, 1981

CHENIN BLANC, *1983 Rating*
GOLD
The Hogue Cellars, 1982
SILVER
Chateau Ste. Michelle, 1982
Latah Creek, 1982
BRONZE
Yakima River Winery, 1982

GEWURZTRAMINER, *1977 Rating*
GOLD
Associated Vintners, 1975
BRONZE
Tualatin Vineyards, 1977

GEWURZTRAMINER, *1978 Rating*
SILVER
Associated Vintners, 1977
Preston Wine Cellars, 1977

GEWURZTRAMINER, *1979 Rating*
SILVER
Associated Vintners, 1977
BRONZE
Ste. Chapelle Vineyards, 1978

GEWURZTRAMINER, *1980 Rating*
GOLD
E.B. Foote Winery, 1978, (Dry)
SILVER
Associated Vintners, 1978, (Dry)
Sokol Blosser Winery, 1979, (Medium)
BRONZE
Bingen Wine Cellars, 1978, Mont Elize Vineyards, (Medium)

GEWURZTRAMINER, *1981 Rating*
BRONZE
Amity Vineyards, 1980
Bingen Wine Cellars, 1979, Mont Elise Vineyards

GEWURZTRAMINER, *1982 Rating*
SILVER
Hinman Vineyards, 1981, (Medium)
BRONZE
Hinzerling Vineyards, 1980, (Dry)
Tualatin Vineyards, 1980, Estate Bottled, (Dry)
Preston Wine Cellars, 1981, (Medium)

GEWURZTRAMINER, *1983 Rating*
BRONZE
Louis Vacelli Winery, 1982, Late Harvest
Henry Winery, 1982, Umpqua Valley, Henry Estates

MERLOT, *1979 Rating*
GOLD
Ste. Chapelle Vineyards, N.V.
SILVER
Elk Cove Vineyards, 1977
BRONZE
The Eyrie Vineyards, 1977

MERLOT, *1980 Rating*
BRONZE
Adelsheim Vineyard, 1978
The Eyrie Vineyards, 1978

MERLOT, *1981 Rating*

SILVER
Preston Wine Cellars, 1978

MERLOT, *1983 Rating*
GOLD
Knudsen Erath Winery, 1981

MULLER THURGAU, *1979 Rating*
BRONZE
Sokol Blosser Winery, 1978, Yamhill County, Muller Thurgau

MULLER THURGAU, *1983 Rating*
BRONZE
Mt. Baker Vineyards, Inc., 1980, Muller Thurgau

MUSCAT, *1978 Rating*
SILVER
Tualatin Vineyards, 1977, American Muscat of Alexandria

MUSCAT, *1979 Rating*
SILVER
Tualatin Vineyards, 1977, Muscat of Alexanderia

MUSCAT, *1981 Rating*
BRONZE
Tualatin Vineyards, 1980, Early Muscat, Estate Bottled

PINOT NOIR, *1977 Rating*
GOLD
Charles Coury Vineyards, 1974, Golden Cluster
Knudsen Erath Winery, 1975
SILVER
Amity Vineyards, 1976, Nouveau
The Eyrie Vineyards, 1974
BRONZE
Ponzi Vineyard, 1975

PINOT NOIR, *1978 Rating*
GOLD
Erath Vineyards, 1976
SILVER
The Eyrie Vineyards, 1975

PINOT NOIR, *1979 Rating*
BRONZE
Amity Vineyards, 1978, Nouveau

PINOT NOIR, *1980 Rating*
BRONZE
Knudsen Erath Winery, N.V., Yamhill County
Oak Knoll Winery, 1978

PINOT NOIR, *1981 Rating*
GOLD
Knudsen Erath Winery, 1979, Yamhill County, Vintage Select
SILVER
Knudsen Erath Winery, 1979, Yamhill County, Knudsen Vineyard, Estate Bottled
Sokol Blosser Winery, 1978, Yamhill County
BRONZE
Amity Vineyards, 1979
Tualatin Vineyards, 1978, Estate Bottled

PINOT NOIR, *1982 Rating*
SILVER
Alpine Vineyards, 1980, Benton County, Estate Bottled
Elk Cove Vineyards, 1979, Willamette Valley Appellation, Reserve
Ponzi Vineyards, 1979
BRONZE
The Eyrie Vineyards, 1979, Willamette Valley
Knudsen Erath Winery, 1979, Yamhill County, Maresh Vineyard
Salishan Vineyards, 1979

PINOT NOIR, *1983 Rating*
GOLD
Hidden Springs Winery, 1980
SILVER
Elk Cove Vineyards, 1980, Willamette Valley, Estate Bottled
The Eyrie Vineyards, 1980, Willamette Valley
BRONZE
Adelsheim Vineyard, 1981, Yamhill County

The Associated Vintners, 1979, Yakima Valley
Ponzi Vineyards, 1980, Reserve
Salishan Vineyards, 1980

PINOT NOIR BLANC, *1979 Rating*
SILVER
Reuter's Hill Vineyards, N.V.

PINOT NOIR BLANC, *1980 Rating*
GOLD
Chateau Ste. Michelle, 1974, Blanc de Noir Brut

PINOT NOIR BLANC, *1982 Rating*
BRONZE
Preston Wine Cellars, 1981
Shafer Vineyards Cellars, 1981, Willamette Valley

PINOT NOIR BLANC, *1983 Rating*
BRONZE
Tualatin Vineyards, Inc., 1982, Willamette Valley, Estate Bottled

JOHANNISBERG (WHITE) RIESLING, *1977 Rating*
SILVER
Ponzi Vineyards, 1976
Chateau Ste. Michelle, 1976
BRONZE
Hillcrest Vineyard, 1975
Knudsen Erath Winery, 1976, Erath White Riesling
Knudsen Erath Winery, 1976, Knudsen White Riesling
Ste. Chapelle Winery, 1976
Tualatin Vineyards, Inc., 1977

JOHANNISBERG (WHITE) RIESLING, *1978 Rating*
GOLD
Ponzi Vineyard, 1977
Preston Wine Cellars, 1977
Ste. Chapelle Vineyards, 1977
Tualatin Vineyards, Inc., 1978
SILVER
Knudsen Vineyards, 1977
Manfred Vierthaler Winery, 1976, Special Select Botrytis Cinerea
Ste. Chapelle Vineyards, 1977, The Idaho Vineyards, Special Harvest

JOHANNISBERG (WHITE) RIESLING, *1979 Rating*
GOLD
Chateau Ste. Michelle, Hand Selected Clusters, Botrytis Cinerea Affected Lot 1
Elk Cove Vineyards, 1978, Willamette Valley Appellation
SILVER
Hinzerling Vineyards, N.V., Selected Clusters
Oak Knoll Winery, 1978
Ponzi Vineyards, 1978, Willamette Valley
Sokol Blosser Winery, 1978, Yamhill County
BRONZE
Associated Vintners, 1977, Yakima Valley, Varietal
Hinzerling Vineyards, 1978, Yakima Valley
Preston Wine Cellars, 1978, Washington State
Reuter's Hill Vineyards, N.V.
Ste. Chapelle Vineyards, 1978
Tualatin Vineyard, Inc., 1977, Late Harvest

JOHANNISBERG (WHITE) RIESLING, *1980 Rating*
GOLD
Hillcrest Vineyards, 1978, Umpqua Valley, Medium
SILVER
Chateau Benoit Winery, 1979, Lane County Vineyards, Medium
Hinzerling Vineyard, 1979, Yakima Valley, Late Harvest
Ponzi Vineyards, 1979, Willamette Valley, Dry
Tualatin Vineyards, 1979, Estate Bottled, Medium
BRONZE
Elk Cove Vineyards, 1979, Willamette Valley Appellation, Dry
Preston Wine Cellars, 1979, Medium
Siskiyou Vineyards, N.V.

JOHANNISBERG (WHITE) RIESLING, *1981 Rating*

GOLD

Preston Wine Cellars, 1980, Select Harvest

SILVER

Alpine Vineyards, 1980, Benton County, Estate Bottled

Amity Vineyards, 1980, Dry

Hinzerling Vineyards, 1980, Yakima Valley

Ste. Chapelle Vineyards, 1980

Sokol Blosser Winery, 1980

BRONZE

Neuharth Winery, 1979

Paul Thomas Wines, 1980

Worden's Washington Winery, 1980

JOHANNISBERG (WHITE) RIESLING, *1982 Rating*

SILVER

Alpine Vineyards, 1981, Benton County, Estate Bottled, Off-Dry

Chateau Benoit, 1981, Dry

Facelli Vineyards, 1981, Sweet

Facelli Vineyards, 1981, Special Reserve

Paul Thomas Wines, 1981, Sagemoor Farms

Ste. Chapelle Vineyards, 1980, Special Harvest

BRONZE

Amity Vineyards, 1981

Associated Vintners, 1981, Dionysus Vineyards, Dry

Chehalem Mountain Winery, 1981, Mulhausen Vineyards, Estate Bottled

Elk Cove Vineyards, 1981, Bauer's Family's, Dundee Hill Vineyard, Willamette Valley, Sweet

Facelli Vineyards, 1981

Hinzerling Vineyards, 1980, Yakima Valley, Individual Selected Berries

Knudsen Erath Winery, 1981, Yamhill County, Vintage Select

Sokol Blosser Winery, 1981

Tualatin Vineyards, 1981

Tucker Cellars, 1981, Yakima Valley

Woodward Canyon Winery, 1981, Walla Walla County

Worden's Washington Winery, 1981

Yakima River Winery, Inc., 1981, Yakima Valley

JOHANNISBERG (WHITE) RIESLING, *1983 Rating*

GOLD

Alpine Vineyards, 1982, Willamette Valley, Estate Bottled

SILVER

Mt. Baker Vineyards, Inc., 1982, Okanogan Riesling

Quail Run Vintners, 1982, Yakima Valley

Woodward Canyon Winery, 1982

BRONZE

Amity Vineyards, 1982, Oregon, Dry

Elk Cove Vineyards, 1982, Willamette Valley, Selected Harvest

Glen Creek Winery, 1982

Hinman Vineyards, 1982

Kiona Vineyards Winery, 1982, Yakima Valley, Estate Bottled

Latah Creek Wine Cellars, 1982

Ste. Chapelle Vineyards, 1982, Symms Family Vineyards

Sokol Blosser Winery, 1982, Yamhill County, Select Harvest

ROSÉ, *1977 Rating*

SILVER

Tualatin Vineyards, Inc., 1977, Rosé of Pinot Noir

ROSÉ, *1978 Rating*

BRONZE

Calona Wines Limited, N.V., Sommet Rouge

Chateau Ste. Michelle, N.V., Grenache Rouge

Sokol Blosser Winery, 1978, Appellation American, Rosé of Pinot Noir

Ste. Chapelle Vineyards, 1978, Rosé of Cabernet Sauvignon

ROSÉ, *1983 Rating*
BRONZE
Kiona Vineyards Winery, 1982, Yakima Valley, Rosé of Merlot, Estate Bottled

SAUVIGNON BLANC (FUMÉ BLANC), *1977 Rating*
BRONZE
Preston Wine Cellars, 1976

SAUVIGNON BLANC (FUMÉ BLANC), *1978 Rating*
GOLD
Chateau Ste. Michelle, 1977
Preston Wine Cellars, 1977
The Eyrie Vineyards, 1978, Appellation American

SAUVIGNON BLANC (FUMÉ BLANC), *1979 Rating*
GOLD
Chateau Ste. Michelle, 1978
Sokol Blosser Winery, 1978
SILVER
The Eyrie Vineyards, 1978

SAUVIGNON BLANC (FUMÉ BLANC), *1980 Rating*
SILVER
Preston Wine Cellars, 1978

SAUVIGNON BLANC (FUMÉ BLANC), *1981 Rating*
SILVER
Worden's Washington Winery, 1980
BRONZE
Chateau Benoit, 1980
Preston Wine Cellars, 1980

SAUVIGNON BLANC (FUMÉ BLANC), *1982 Rating*
SILVER
Chateau Ste. Michelle, N.V.
BRONZE
Chateau Benoit, 1981

SAUVIGNON BLANC (FUMÉ BLANC), *1983 Rating*
GOLD
Arbor Crest, 1982, Bacchus Vineyard
Glen Creek Winery, 1982
SILVER
Sokol Blosser Winery, 1982
BRONZE
Preston Wine Cellars, 1981

SEMILLON, *1979 Rating*
BRONZE
Adelsheim Vineyards, 1978

SEMILLON, *1981 Rating*
SILVER
Chateau Ste. Michelle, 1980

SEMILLON, *1982 Rating*
BRONZE
Associated Vintners, 1980

SEMILLON, *1983 Rating*
GOLD
Adelsheim Vineyard, 1982

SYLVANER, *1982 Rating*
SILVER
Chehalem Mountain Winery, 1981, Mulhausen Vineyards

MISCELLANEOUS, *1977 Rating*
BRONZE
Ponzi Vineyard, 1976, Oregon, Harvest Wine

MISCELLANEOUS, *1978 Rating*
GOLD
Hinzerling Vineyard, 1977, Die Sonne
BRONZE
Sokol Blosser Winery, 1977, Bouquet Blanc

MISCELLANEOUS, *1979 Rating*
SILVER
Amity Vineyards, 1978, Solstice Blanc
BRONZE
Knudsen Erath Winery, N.V., Blanc des Blancs, White Table Wine

Tualatin Vineyards, Inc., 1975, Petite Sirah

MISCELLANEOUS, *1981 Rating*
GOLD
Associated Vintners, 1980, Valley White
SILVER
Bingen Wine Cellars, 1980, Mont Elise Vineyards, Gamay Beaujolais

MISCELLANEOUS, *1982 Rating*
SILVER
Neuharth Winery, Inc., 1982, Dungeness Red
BRONZE
Amity Vineyards, 1981, Solstice Blanc
Calona Wines, Inc., N.V., Haut Villages Dry White Wine

MISCELLANEOUS, *1983 Rating*
BRONZE
Quail Run Vintners, 1982, Yakima Valley, Aligote